T0393394

Lecture Notes in Electrical Engineering

Volume 641

The book series *Lecture Notes in Electrical Engineering* (LNEE) publishes the latest developments in Electrical Engineering - quickly, informally and in high quality. While original research reported in proceedings and monographs has traditionally formed the core of LNEE, we also encourage authors to submit books devoted to supporting student education and professional training in the various fields and applications areas of electrical engineering. The series cover classical and emerging topics concerning:

- Communication Engineering, Information Theory and Networks
- Electronics Engineering and Microelectronics
- Signal, Image and Speech Processing
- Wireless and Mobile Communication
- Circuits and Systems
- Energy Systems, Power Electronics and Electrical Machines
- Electro-optical Engineering
- Instrumentation Engineering
- Avionics Engineering
- Control Systems
- Internet-of-Things and Cybersecurity
- Biomedical Devices, MEMS and NEMS

For general information about this book series, comments or suggestions, please contact leontina. dicecco@springer.com.

To submit a proposal or request further information, please contact the Publishing Editor in your country:

China

Jasmine Dou, Associate Editor (jasmine.dou@springer.com)

India, Japan, Rest of Asia

Swati Meherishi, Executive Editor (Swati.Meherishi@springer.com)

Southeast Asia, Australia, New Zealand

Ramesh Nath Premnath, Editor (ramesh.premnath@springernature.com)

USA, Canada:

Michael Luby, Senior Editor (michael.luby@springer.com)

All other Countries:

Leontina Di Cecco, Senior Editor (leontina.dicecco@springer.com)

** Indexing: The books of this series are submitted to ISI Proceedings, EI-Compendex, SCOPUS, MetaPress, Web of Science and Springerlink **

More information about this series at http://www.springer.com/series/7818

Andrey A. Radionov · Alexander S. Karandaev
Editors

Advances in Automation

Proceedings of the International Russian
Automation Conference, RusAutoCon 2019,
September 8–14, 2019, Sochi, Russia

Set 2

 Springer

Editors
Andrey A. Radionov
Rectorate
South Ural State University
(National Research University)
Chelyabinsk, Russia

Alexander S. Karandaev
Nosov Magnitogorsk State
Technical University
Magnitogorsk, Russia

ISSN 1876-1100   ISSN 1876-1119   (electronic)
Lecture Notes in Electrical Engineering
ISBN 978-3-030-39224-6   ISBN 978-3-030-39225-3   (eBook)
https://doi.org/10.1007/978-3-030-39225-3

This Springer imprint is published by the registered company Springer Nature Switzerland AG
The registered company address is: Gewerbestrasse 11, 6330 Cham, Switzerland

Preface

International Russian Automation Conference (RusAutoCon) took place on September 8–14, 2019, in Sochi, Russian Federation. The conference was organized by South Ural State University (National Research University). The international program committee has selected 355 reports.

The conference was divided into 13 sections, including:

1. Process automation;
2. Modeling and simulation;
3. Control theory;
4. Machine learning, big data, Internet of things;
5. Flexible manufacturing systems;
6. Industrial robotics and mechatronic systems;
7. Computer vision;
8. Industrial automation systems cybersecurity;
9. Diagnostics and reliability of automatic control systems;
10. Communication engineering, information theory and networks;
11. Control systems;
12. Energy systems, power electronics, and electrical machines;
13. Instrumentation engineering;
14. Signal, image, and speech processing.

The International Program Committee has selected totally 125 papers for publishing in Lecture Notes in Electrical Engineering (Springer International Publishing AG).

The Organizing Committee would like to express our sincere appreciation to everybody who has contributed to the conference. Heartfelt thanks are due to authors, reviewers, participants, and to all the team of organizers for their support and enthusiasm which granted success to the conference.

Andrey A. Radionov
Conference Chair

Contents

Exergy Analysis of Single-Stage Heat Pump Efficiency Under Various Steam Condensation Conditions

S. V. Skubienko[(✉)], I. V. Yanchenko, and A. Yu. Babushkin

Platov South-Russian State Polytechnical University, 132 Prosveshcheniya,
Novocherkassk 346428, Russian Federation
skubienko@mail.ru

Abstract. The article describes the implementation of exergy method for the heat losses determination in one- and double-stage heat pump systems. The exergy method allows determining the energy loss in different parts of the heat pump. Exergy loss can be seen as an increase in energy as the exergy converts to the latter. Based on the above method a mathematical model was developed, and exergy losses under one- and double-stage condensation were analyzed by dividing the heat pump work into two temperature levels of heating from 10 °C to 35 °C, from 35 °C to 60 °C. The graphs show the temperature dependence t on the heat flow Q in the t-Q diagram were build. The results of exergy losses calculations obtained for single-stage and two-stage condensation were compared. According to the results of the calculations, the conclusions on energy efficiency for each system were drawn. It was found that using several stages of condensation, the transformation ratio of heat pump can improve significantly.

Keywords: Heat pump · Energy efficiency · Exergy method · Heat loses · Multi-stage condensation · Cycle optimization

1 Introduction

Heat recovery systems have become common all over the world. In many industries production comes with large energy losses which can recover and recycle for heating and cooling. For example, heat losses at thermal and nuclear power plants account for approximately 40 to 50%, mainly due to exhaustion steam after turbine must be condensing. Heat pumps can significantly improve the efficiency of power plant cycle [1]. For this reason, heat pumps are highly attractive anergy conversion devices for heat recovery. They are widely used in refrigerating, air-conditioning, space heating, hot water production, heat upgrading and waste heat recovery. The literature review has showed that the major part (approximately 70%) of heat pump applications are found in refrigeration, i.e. supermarket food cooling, household fridges/freezers, and cooling/air-conditioning/storage during transportation [2]. For the successful using of heat pump systems in the Russian industry, one of the most important tasks is to study the efficiency of their operation under various conditions of steam condensation, which the transformation coefficient value of the heat pump depends on. The article presents the

© Springer Nature Switzerland AG 2020
A. A. Radionov and A. S. Karandaev (Eds.): RusAutoCon 2019, LNEE 641, pp. 581–587, 2020.
https://doi.org/10.1007/978-3-030-39225-3_63

options for the heat pump calculation under one- and two-stage steam condensation, taking into account the exergy losses.

2 Historical Background

Thermodynamic laws derived by French scientist Sadi Carnot in 1824 are considered the prerequisite for creation a heat pump. However, heat pumps were developing very slowly, because the main refrigerant was ethyl oxide, which was highly explosive. It was Peter Ritter von Rittinger, who designed and successfully used the heat pump in salt production in 1857. Shortly before First World War in 1812, Swiss engineer Heinrich Zoelly patented heat pump, which used the earth's heat as an energy source [3]. The most active development in research of heat pump systems took place during the oil embargo in 1970s, when the government started subsidizing energy conservation.

Modern heat pumps are high-performance devices for heating and cooling of households and factories. Nowadays, about 12 million heat pumps operate in the world. The biggest number of heat pump systems are used in Germany and Japan. Recently, they have been used in industry, especially for waste heat recovery, heat upgrading, cooling and refrigeration in processes or for heating and cooling industrial buildings.

As presented by the International Energy Agency IEA, there is a wide demand for heating and cooling at different temperature levels in the world. In many countries, there is a high demand for low exergy heat from 60 °C to 100 °C, mainly in the pulp and paper, food and tobacco industries. For this reason, heat pumps are one of the most efficient way of energy conversion and energy saving [4, 5].

Any heat pump cycle is based on the reverse Carnot cycle (Fig. 1). In industry vapor-compression heat pumps received the widest application, with compressor being the main device for increasing the thermal energy of the cycle fluid. Supply of low-potential heat is carried out in the heat pump evaporator due to the refrigerant boiling under vacuum. When boiling, the working fluid vapor removes heat from the energy source and enters the compressor, where the compression process occurs with increasing their thermodynamic parameters. In the condenser, the working fluid vapor condenses, transferring heat energy to the consumer. Despite the consumption of additional electric energy needed for the compressor operation, heat pump is capable to release thermal energy by $2.5 \div 5.5$ times more, which is undisputable advantage [6, 7].

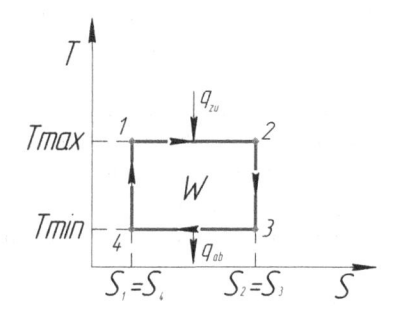

Fig. 1. Carnot cycle in T-s diagram.

The main disadvantages of vapor-compression heat pumps are limited temperature mode of operation and exergy loses.

For this reason, the article represents the exergy loses calculation for one stage heat pump system. Exergy analysis shows the thermodynamic efficiency and determines the energy losses for each element of heat pump.

3 Exergy Analysis of Heat Pump Operation

Energy is composed of two parts: exergy and anergy. Exergy is defined as follows: "Exergy is the part of energy, which can be completely transformed into any other form of energy [8, 9].

The inconvertible fraction of energy is called anergy. Simply put, exergy can be used as the technical works.

$$E_1 = -W_{t1b}^{rev}$$

The reversible heat exchange with the environment is understood as heat exchange at the ambient temperature level, with no driving temperature differences.

$$q_{1b}^{rev} + w_{1b}^{rev} = h_b + h_1 + g \cdot (z_b - z_1) + \frac{1}{2} \cdot (c_b^2 - c_1^2)$$

Neglecting kinetic and potential energy and T_b = const, we receive:

$$e_1 = h_1 - h_b - T_b \cdot (s_1 - s_b)$$

The full exergy of flowing liquid can be presented as mass rate multiplied by exergy:

$$E_1 = m \cdot e_1$$

Exergy loss occurs in all irreversible processes as well as in a real heat pump process. Exergy loss can be seen as an increase in anergy as the exergy converts to the latter. In reversible processes, however, no loss of exergy occurs and thus no increase in anergy [3].

4 Exergy Loss During Simple Condensation

Exergy loss through simple condensation can be derived from certain formulas. We take a number of assumptions necessary for calculating. These include the ambient temperature t_b, mass rate of flowing liquid m_w, temperature t_1 and t_2 minimum difference in temperatures Δt, and specific heat of water C_{pw}. In addition, it is assumed that the heat pump evaporator does not experience any overheating or condensate depression [8, 9].

Allowances: $t_b = 0$ °C; $m_w = 2$ kg/s; $t_1 = 10$ °C; $t_2 = 60$ °C; $\Delta t_{min} = 10$ K; $C_{pw} = 4,184$ kJ/(kg·K).

Heat balance in heat pump system is calculated by:

$$Q = m_w \cdot c_{pw} \cdot (t_2 - t_1)$$

Based on the above assumptions, the heat balance is:

$$Q = 2\frac{kg}{s} \cdot 4,184 \frac{kJ}{kg \cdot K} \cdot (60 - 10) = 418,4 \text{ kW}$$

In order to determine exergy losses, we make the energy balance.

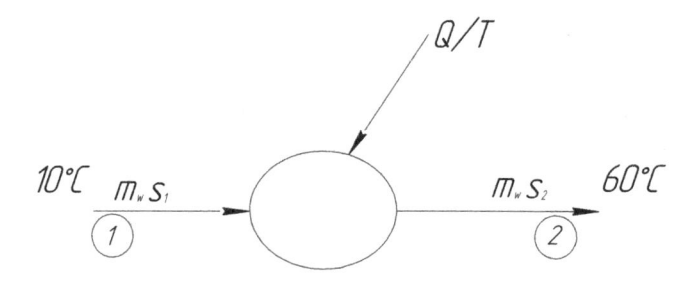

Fig. 2. Entropy balance.

The balance is important as the exergy loss depends on the ambient temperature in relation to the dissipated entropy difference.

$$E_v = T_b \cdot \Delta S_{diss}$$

Figure 2 shows the flows taken for the energy balance to be made. Flows m_{ws1} and Q/T flow into the system, and m_{ws2} out flows. This means that the energy balance is as follows:

$$S_{ab} = S_{zu} \cdot \Delta S_{diss}$$

$$\Delta S_{diss} = S_{ab} - S_{zu}$$

In addition

$$\Delta S_{diss} = m_{wS2} - \frac{Q}{T} - m_{wS1} = m_w \cdot (S_2 - S_1) - \frac{Q}{T}$$

For incompressible materials:

$$(S_2 - S_1) = c_{Pw} \cdot \ln \frac{T_2}{T_1}$$

As a consequence:

$$\Delta S_{diss} = m_w \cdot c_{Pw} \cdot \ln \frac{T_2}{T_1} - \frac{m_w \cdot c_{Pw} \cdot (t_2 - t_1)}{T} = m_w \cdot c_{Pw} \cdot (\ln \frac{T_2}{T_1} - \frac{t_2 - t_1}{T})$$

We substitute the corresponding values and obtain the losses resulting from condensation in the heat pump:

$$E_v = T_b \cdot m_w \cdot c_{Pw} \cdot \left(\ln \frac{T_2}{T_1} - \frac{t_2 - t_1}{t_2 + \Delta t_{min} + 273,15K} \right) = 273,15K \cdot 2 \frac{kg}{s}$$

$$\times 4,184 \frac{kJ}{kg \cdot K} \cdot (\ln \frac{331,15K}{283,15K} - \frac{(60 - 10)K}{60K + 5K + 273,15K}) = 33,72 \text{ kW}$$

Figure 3 shows the temperature dependence t on the heat flow Q in the t-Q diagram. It is clearly seen that the dependence is linear. According to the system operating conditions, the temperature difference is from 10 °C to 60 °C, while the maximum heat flow is Q = 418.4 kW. The resulting losses of exergy by water heating from 10 °C to 60 °C account for E_V = 33,72 kW.

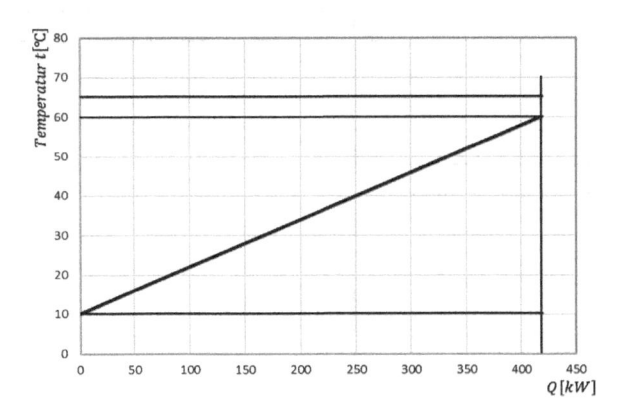

Fig. 3. t-Q-diagram for single-stage condensing.

The exergy losses calculated do not allow determining the system efficiency, for this reason we make similar calculation of exergy losses for two-stage condensation at identical temperatures.

The losses of exergy in two-stage condensation is calculated similarly, however it is important take into account the exergy loss for each step. For our purpose, we divide the heat pump work into two temperature levels of heating from 10 °C to 35 °C, from 35 °C to 60 °C (Fig. 4).

Exergy losses for the first condensation stage with temperature level from 10 °C to 35 °C are calculated by:

$$E_{vA} = T_b \cdot m_w \cdot c_{Pw} \cdot \left(\ln \frac{T_2}{T_1} - \frac{t_2 - t_1}{t_2 + \Delta t_{min} + 273,15K} \right) = 273,15K \cdot 2\frac{kg}{s}$$

$$\times \ 4,184 \frac{kJ}{kg \cdot K} \cdot (\ln \frac{35K}{10K} - \frac{(35-10)K}{35K + 5K + 273,15K}) = 10,92\,kW$$

For the second condensation stage, from 35 °C to 60 °C:

$$E_{vB} = T_b \cdot m_w \cdot c_{Pw} \cdot \left(\ln \frac{T_2}{T_1} - \frac{t_2 - t_1}{t_2 + \Delta t_{min} + 273,15K} \right) = 273,15K \cdot 2\frac{kg}{s}$$

$$\times \ 4,184 \frac{kJ}{kg \ \cdot K} \cdot (\ln \frac{60K}{35K} - \frac{(60-35)K}{60K + 5K + 273,15K}) = 9,31\ kW$$

Summing up the losses for two stages, we obtain:

$$E_v = E_{vA} + E_{vB} = 10,92 + 9,31 = 20,23\ kW$$

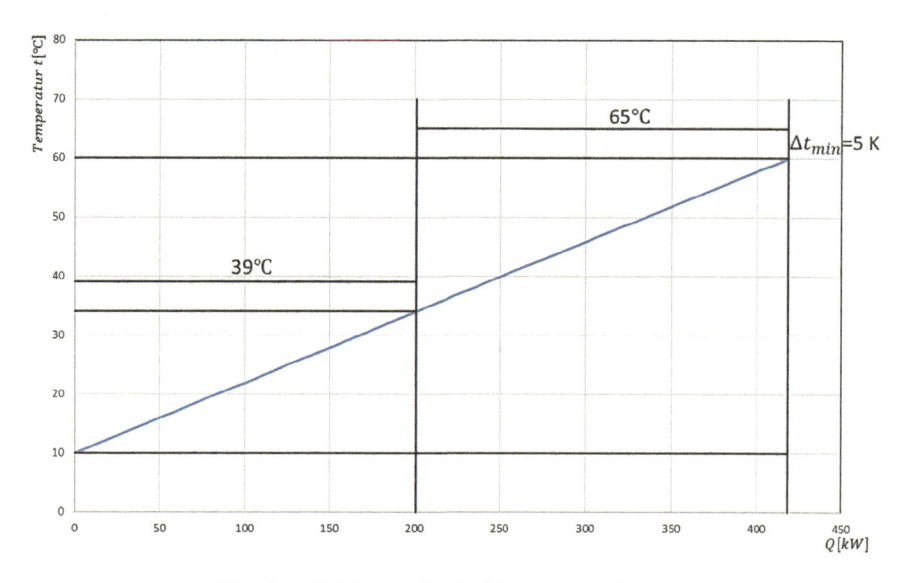

Fig. 4. t-Q-diagram for double-stage condensing.

5 Conclusion

Having compared the results of exergy losses calculations obtained for single-stage and two-stage condensation, we conclude that with heat pump operation in a two-stage scheme, the savings are by almost 40% compared to a single-stage operation. Thus, using several stages of condensation, the transformation ratio of heat pump can improve significantly.

Acknowledgements. The work has been carried out in the framework of implementation of the project № 13.9947.2017/5.2 «Increasing of thermal power plants efficiency by using the heat pump system», supported by the Grant from Russian Federation Ministry of Education and Science.

References

1. Skubienko, S.: Using an absorption heat pump in the regeneration system of turbine model K-300-240-2 manufactured by Kharkov turbo generator plant. In: 2nd International Conference on Industrial Engineering, Applications and Manufacturing (2016). https://doi.org/10.1109/icieam.2016.7911014
2. Cordin, A.: Multi-temperature heat pumps: a literature review. Int. J. Refrig. **69**, 437–465 (2016). https://doi.org/10.1016/j.ijrefrig.2016.05.014
3. Kuck, J.: Lecture thermodynamics 1. Hochschule für angewandte wissenschaften. Faculty of supply engineering. Unpublished transcript (2016)
4. IEA: Industrial heat pumps. IEA Heat Pump Cent. Newsl. **30**(1), 15 (2012)
5. Elbel, S., Hrnjak, P.: Ejector refrigeration: an overview of historical and present developments with an emphasis on air conditioning applications. In: International Refrigeration and Air Conditioning Conference, vol. 1, pp. 2351–2358 (2008)
6. Skubienko, S.V., Yanchenko, I.V.: Study of heat effectiveness of thermal power station equipped with turbine plant model K-300-240-2 KhTGP and absorption heat pump at variable-load operation modes. Izvestiya vuzov. Ser. Eng. Sci. **2**, 30–35 (2015). Severo-Kavkazskiy Region
7. Chen, X., Omer, S.: Recent developments in ejector refrigeration technologies. Renew. Sustain. Energy Rev. **19**, 629–651 (2013)
8. Kotas, T.: The Exergy Method of Thermal Plant Analysis. Paragon Publishing, London (2012)
9. Brodiansky, V.: The Exergy Method and its Applications. Energoatomizdat, Moscow (1988)

Comprehensive Comparison of the Most Effective Wind Turbines

E. Solomin[✉], X. Lingjie, H. Jia, and D. Danping

South Ural State University, 76 Lenina Avenue,
Chelyabinsk 454080, Russian Federation
e.solomin@bk.ru

Abstract. In this paper we analyzed the main features of vertical VAWT and horizontal HAWT axis wind turbines. Today modern HAWT use all the available progress in modern aviation engineering related to blades structure, pitch control, transmissions, and it is well known that their reliability is high. It may be not obvious but VAWTs promise even higher reliability. Great design impact in the beginning of XX Century caused such improvements as getting a generator to the ground, montage and operation simplification. And this includes the well-known advantages. The medium power of wind turbines in the World wind power industry is being increased during the last decades because of the technology quality increment, and the breaking costs taken from one square meter of swept area of energy as well as expenses on operation and technical servicing, are getting less year by year. The limitation of power of HAWT (up to 14 MW because of the limited strength features) today leads to the development of new technologies and in general, to the development of 20 MW turbines and bigger, which means the significant change of the design of wind turbines. The bird mortality from VAWT is closed to zero.

Keywords: Wind · Turbine · Comparison · Vertical · Axis · Power · Efficiency · VAWT · HAWT

1 Introduction

The World distribution between Horizontal Axis Wind Turbines (HAWT or propellers) and Vertical Axis Wind Turbines (VAWT) is 90:10, provided by over 100 companies on the market [1]. VAWT is being developed behind the VAWTs for several reasons: VAWT was developed later by HAWT (the Savonius rotor in 1929, the Darrieus rotor in 1931, the Musgrove rotor in 1975). Until recently, it was mistakenly thought that it is impossible to obtain a ratio of the maximum linear speed of the working parts (blades) to wind speed greater than 1:1 for the VAWT (for HAWTs this ratio is 5:1) [2]. The wrong conclusion was made that the maximum coefficient of wind usage or Power Coefficient Cp of VAWT is less than HAWT, and as a result - this type has not been developed for more than 40 years. These features are applicable only to low speed rotors of the Savonius type, which use Drag Torque difference or different blade resistance when traveling in and out of the wind flow. In 1960, Canadian and American scientists experimentally proved that the concept is incorrect for the Darrieus rotor,

© Springer Nature Switzerland AG 2020
A. A. Radionov and A. S. Karandaev (Eds.): RusAutoCon 2019, LNEE 641, pp. 588–595, 2020.
https://doi.org/10.1007/978-3-030-39225-3_64

which uses the power of lifting the blade or Lifting Force. The maximum ratio of the rotation speed (TSR) of the blades to the wind speed for these machines may exceed 6:1. The coefficient Cp is not less than of the modern HAWT [3].

To date, the wind turbines of the Darrieus type cover all countries [4, 5]. In the Russian Federation, the main researchers and manufacturers of vertical axial turbines are South Ural State University, Uralmet Scientific Research Institute, "Kuntsevo" Design Bureau, SRC-Vertical, CAGI, Hydroproject Holding, Institute of Electric Power. Several wind turbines, such as WPU-3(6), WPU-3(4), WPU-0.1(4), VL-2 M, VDD-16 and others, show positive promising results.

Why do scientists and designers prefer vertical wind turbines for development more and more often?

The literature widely mentions the independence on the direction of the wind as the main feature or advantage. Therefore, the rotor structure could greatly be simplified [6]. For developing countries, this is more attractive, since there are restrictions on the usage of high towers. As their main argument, the adepts of VAWT show that its design does not require any yawing systems and pitch assemblies. Experience in the design, development and operation of VAWT has shown that this is not just one or just main advantage of Axis vertical turbines.

These types of wind turbines are cardinally different, they use different technical solutions, and many parameters cannot be applied from HAWT to VAWT and vice versa [7].

Some distinctive parameters and features of both types are presented below and discussed from different points of view. Basically it's about comparing the classic HAWT and VAWT.

The capacity in the World wind energy of wind power plants is constantly growing. This success appeared due to the following reasons: due to the growth of the size, the cost of energy extracted from 1 m^2 of the covered zone, as a consequence, decreases, while the maintenance and servicing costs as well as the wind farm area decrease. As the result in general the efficiency of the entire construction is growing as well [8].

2 Power

The increase of HAWT power is obviously limited [9]. There is a predictable for today upper limit of 10 MW, since the blades are affected by bending and centrifugal forces. They are also alternating in direction and magnitude being a part of the whole construction. This fact reduces the reliability of HAWT and reduces the observed time of work and imposes restrictions on the dimensions of the blades. Therefore, the increase in power will lead to the variations of the design of wind turbines [10]. At the moment, the most popular and resource-intensive solution is the use of a vertical axis structure, which technical and economic boundaries far exceed the HAWT features, in accordance with theoretical developments.

3 Dependence of Efficiency on Wind Direction

The maximum achievable efficiency of HAWT occurs when the rotor axis and wind direction are collinear. This requirement for orientation to wind currents requires additional control systems and mechanisms to orient the continuous monitoring of the state and direction of the wind, to search for the maximum potential of the directional wind, to turn the wind wheel in the given direction and to maintain it in the most effective position [11]. Orientation or yaw management system makes the wind turbine more complex and reduces its reliability (practice shows about 13% failures in the orientation control system). These upgrades reduce the reliability of the wind turbine and increase the cost of depreciation. The efficiency of the VAWT is independent of the direction of the wind [12, 13]. That is, there is no need for an orientation system and additional mechanisms that create additional value. A slight equalization of the torque on the blade is caused by uneven wind flows in height.

4 Coefficient of Wind Power Use

The starting torque of HAWT is considered doesn't equal to zero, which means that it may start with no outside power source. However, practice showed that the propeller may starts by itself when it is directed enough on the wind. The HAWT cannot start on the side wind. But it means that some outside power source is required for yawing system activation and turning the gondola on the wind.

It was considered for a long time that the starting torque of VAWT is either negative or equal to zero [14]. It basically means that the VAWT cannot start by itself. However, we at South Ural University had broken this opinion when developed Darrieus type rotor which starts by itself on 1.5 m/s wind speed due to multi-tier rotor design and some other important features. The self-starting torque of VAWT is based on the lift forces and positive even on small wind speeds. One more feature of these turbines is that they start from even from small wind gust. However, the more complex the wind turbine, the less the reliability of the whole structure [15], which means that the self-starting feature is provided by special aerodynamic profiles of the blades, and rotation frequency is stabilized by aerodynamic governors which reduce the power of Wind Turbine on the strong winds. All these extra improvements increase the number of potential points for failure.

5 Rationality of Power Structure of Wind Turbine

The HAWT structure considers that the inertial loading on the blade is being applied along the whole blade length, which reflects the most disadvantageous approach. Hub and support bearing are rather compact and relatively small due to the high angular speed. Inertial loading on the VAWT blade is directed in perpendicular to the blade length or across the blade, in general along traverse. Hub and bearing module of VAWT has relatively big dimensions. Therefore, the VAWT has not as rational structure as HAWT has. Due to the masses of all components is heavier than HAWT.

However, for big megawatt class turbines it is necessary to count the alternating loading on the blades. At first, aerodynamic load on the HAWT blade is different in the top and bottom positions because the wind speed along the length of the blades is also different. The blade operates in different torques and transfers the pulsing vibrations to the hub which may cause damage and other failures. Second, the influence of gravitation increases and distributes strength forces along the whole length of HAWT blade which is moving up or down, especially being parallel to the ground. The pulsing gravitational and aerodynamic loads reduce the blade vibro-stability, and also may damage the hub and support-transmission system. Coriolis forces also increase when the turbine is yawing on the wind.

6 Design of Blade

All cross-sections of HAWT rotor are operating in different status of energy as they have different speeds of rotation and each blade has its angle of attack different from others. This difference would be much less if the cross-sections could be rolled relatively each other. On one hand the inertial loading leads to the necessity of narrowing the profile from butt to the end. Therefore, the HAWT blade is initially more complex than straight VAWT blade which is symmetrical relatively the chord plane. On another hand the assembling of fiber plastic blade segments of VAWT is a complex task because the flange joints arrangement is not a simple task taking into account further side loading.

7 Pitch Control

Pitch control or turning the HAWT blades from their normal or operating position is used not only for braking of the wind rotor as the alternative to the general frictional brake, but mainly the mean of searching of optimal angle of blade setting, to keep the wind rotor angular speed on the level of optimal number of revolutions in accordance with the required tip speed ratio, without causing of any damage [16]. The system of blade pitch control makes the design of the Wind Turbine a lot more complex as it requires the systems of continuous tracing the rotation speed and activating the devices with the synchronized drives for each blade, managing the automatic control of blade pitch angle. Such systems for HAWT rotors are required to avoid the unsafe operation duty.

The turn of the HAWT blade would be efficient not just for braking only, but in some cases supporting the optimal angle of attack in all positions of the blade on the whole rotation circle. The Wind Turbines designed using this approach, are not distributed widely because the blade becomes heavy and while spinning round along the circle, it should make several swinging, to manage the angle of attack and orient on the wind. In addition the devices and mechanisms are extremely complex and the production is very slow and expensive. Finally the Wind Turbine becomes also too complex and expensive.

Instead the VAWT efficiency is constant without any blade turn, and similar to the efficiency of HAWT, but in the most efficient duty. It's a main advantage of VAWT in general.

However, in different conditions VAWT can be considered as more efficient than HAWT and vice versa.

8 Swept Area

The swept area of HAWT is the circle made by rotating ends of the blades. For VAWT this area is equal to the area of rectangle with one side is the length of blade and the other side is the diameter of Wind Turbine.

Therefore, the swept area of VAWT is more flexible than of HAWT and it can be adjusted or changed by not only by the blade length size, but also by the whole diameter of turbine, which makes the tactic features of these turbines much wider.

Energy extracted from the blade length unit of HAWT, is different along the whole length from butt to the end, because the angular velocities are different, from 0 in the butt to maximum value in the end.

The value of energy extracted from the VAWT blade is not really changing along the blade length. The small difference in practice appears due to the difference of wind flow and inconstancy of wind speed along the blade length or actually height.

There are several ways of energy losses based on the VAWT blade concept: not optimal angle of attack in different blade positions, reduction of the integral rotating torque when the blade is moving directly against the flow, and reduction of torque coming from blade which is going in the aerodynamic shadow of the mast or tower.

In general it can be considered that the efficiency of extraction energy of the wind flow with the help of blades of both types HAWT and VAWT, will be about the same. It's necessary to note that for the smaller Wind Turbines up to 5 kW, the angle of attack should be set up optimal depending on the wind conditions in local region.

Big VAWTs can be equipped with the pitch angle control which adjusts the angle of attack depending on the duty of operation. When starting it should be bigger, when the angle velocity increases, the angle of attack is decreasing. Such system would make the efficiency of big turbine higher.

9 Rotation Speed Features

The most distributed turbines among the HAWTs, are the 3-bladed high-speed turbines with 5–7 modules of tip speed ratio (TSR). They provide the highest coefficient of energy usage; therefore, they are the most efficient. The high degree of high-speed requires the number of extra special devices, mechanisms, controls and aggregates which control the angular velocity of wind rotor in accurate limits and avoid the damage of rotor and transmission, which makes the turbine more complex and as said above, more unsafe.

The constancy of high rotation speed simplified the transmission connections between wind wheel and alternator [12, 17, 18], and the quality of energy becomes higher without extra converting and transforming circuits.

At the same time the operating rotation speed constancy is limited by the inertial blade strength, which means the limitation of the range of wind speed (usually within 20–25 m/s). In addition to that the operation of turbine in optimal duty on the fixed wind speed reduces the efficiency [19].

VAWTs do not have such problems as their design provides initially low speed operation [20]. In all known experiments including the well-known with the search of maximum efficiency of wind usage [20], the tip speed ratio didn't increase 2.5–2.8 modules. The significance of this fact is more understandable if the designers will take into account that all VAWT characteristics including the power coefficient, are almost the same as of HAWT [14, 21, 22]. The VAWT with straight blades can be high-speed, i.e. to use the high tip speed ratio up to 7 modules, even higher than HAWT. The only limitation is the blade strength on transversal inertial loading in combination with vibration loading. The tendency of development of maximally light and high durable materials makes the development of high-speed VAWTs of Darrieus and similar types like multi-tier and helicoid, more and more attractive.

10   Alternator and Gearbox (Multiplier) Location

The big advantage of VAWT is the down position of alternator and optional multiplier which could be located on the base of turbine, which can make the requirements for repairs and servicing simpler because of no limitation on mass and dimensions of the said aggregates. In this case the requirements for operation become also more light because of no vibration and no jerk. When the equipment is located on the ground, the montage and transmission of energy conditions improve greatly. The energy production becomes simpler. However, the transmission of rotating torque should be more powerful. In this case the construction may be multi-module.

It is considered that the angle transmission would reduce the reliability of HAWT turbine [23]. Thus, the equipment is located in yawing gondola. The complexity of montage, operation and repairs increase a lot. It's necessary to note that the rotating torque transmission on the foundation level can be transmitted by long transmission shaft.

However, the advantage of the bottom-based equipment compensates the shaft complex design, even if the shaft will be high-speed post-gearbox. When the shaft is low speed off-gearbox, it would not make the design more complex [24–26].

11   Conclusion

It is obvious to anybody that to cover the needs of even one region in wind power is impossible by one size or one type of Wind Turbines, in analog to Automotive Industry, where each class, type and sort of automobile finds its customer. Wind Power Industry will become competitive when several different directions will be approached, including different ranges in power, types and classes. This will allow creating the really big Wind Turbine market with reliable and efficient products.

Acknowledgements. The work was supported by Act 211 Government of the Russian Federation, contract № 02.A03.21.0011.

References

1. WWEA Half-year report (2016) World wind energy association (2018). http://www.wwindea.org. Accessed 20 Mar 2018
2. Solomin, E.: Iterative approach in design and development of vertical axis wind turbines. In: Applied Mechanics and Materials: Energy Systems, Materials and Resigning in Mechanical Engineering, vol. 792, pp. 582–589 (2015). https://doi.org/10.4028/792.582
3. Kirpichnikova, I.M.: Simulation of a generator for a wind–power unit. Russ. Electr. Eng. **84** (10), 46–49 (2013). https://doi.org/10.3103/s1068371213100076
4. Spinato, F., Tavner, P.J., Bussel, G.J.W., et al.: Reliability of wind turbine sub-assemblies. Power Gener. **3**(4), 387–401 (2009)
5. Bussel, G.J.W., Zaaijer, M.B.: Reliability, availability and maintenance aspects of large-scale offshore wind farms, a concepts study. In: IMarEst, MAREC Conference, Newcastle Upon Tyne (2001)
6. Shires, A.: Design optimization of an offshore vertical-axis wind turbine. Proc. ICE – Energy **166**, 7–18 (2013)
7. Keller, A.V., Korobatov, D.V., Solomin, E.V., et al.: Development of algorithms of rapid charging for batteries of hybrid and electric drives of city freight and passenger automobile transportation vehicles. In: Proceedings XI International Siberian Conference on Control and Communications (2015)
8. Martyanov, A.S., Solomin, E.V., Korobatov, D.V.: Development of control algorithms in Matlab/Simulink. Procedia Eng. J. **129**, 922–926 (2015). https://doi.org/10.1016/j.proeng.2015.12.135
9. Chavan, D.S., Kulhari, P., Kadaganchi, N.: Prediction of power yield from wind turbines for hilly sites. In: IEEE 2nd International Future Energy Electronics Conference, pp. 1–5 (2015)
10. Chavan, D.S., Karandikar, P.B.: Assessment of flicker due to vertical wind shear in a wind turbine mounted on a hill with linear approach. In: 4th International Conference on Artificial Intelligence with Applications in Engineering and Technology, pp. 259–263 (2014)
11. Muhaimeen, B., Mehler, R.W.: Wind shear detection for small and improvised airfields. In: IEEE Aerospace Conference, pp. 1–8 (2012)
12. Martyanov, A.S., Korobatov, D.V., Solomin, E.V.: Research of IGBT–transistor in pulse switch. Procedia Eng. J. Chelyabinsk (2016). https://doi.org/10.1109/icieam.2016.7911470
13. Korobatov, D.V., Sirotkin, E.A., Troickiy, A.O., et al.: Wind turbine power plant control. In: X International IEEE Scientific and Technical Conference Dynamics of Systems, Mechanisms and Machines. Omsk State Technical University, Omsk, 5–17 November 2016, pp. 1–5 (2016). https://doi.org/10.1109/dynamics.2016.7819031
14. Solomin, E.V., Sirotkin, E.A., Martyanov, A.S.: Adaptive control over the permanent characteristics of a wind turbine. Procedia Eng. J. **129**, 640–646 (2015). https://doi.org/10.1016/j.proeng.2015.12.084
15. Chavan, D.S., Karandikar, P.B.: Linear model of flicker due to vertical wind shear for a turbine mounted on a green building. In: 4th International Conference on Artificial Intelligence with Applications in Engineering and Technology, pp. 253–258 (2014)

16. Volovich, G.I., Solomin, E.V., Topolskaya, I.G., et al.: Modeling and calculation of adaptive devices of automation, control and protection for intellectual electric grid in SCILAB freeware computer mathematic package. Bull. South Ural State Univ. Sect. Math. Model. Program. South Ural State Univ. Chelyabinsk **8**(4), 76–82 (2015). https://doi.org/10.14529/mmp150406

17. Topolskiy, D., Topolskiy, N., Solomin, E.: Modeling of data exchange process in automation system on the base of universal SCADA–system. In: IOP Conference Series: Materials Science and Engineering, Tomsk Polytechical University, Tomsk, vol. 124(1), 1–4 December 2015. https://doi.org/10.1088/1757–899x/124/1/012104

18. Sirotkin, E.A., Martyanov, A.S., Solomin, E.V.: Emergency braking system for the wind turbine. Procedia Eng. J. Chelyabinsk (2016). https://doi.org/10.1109/icieam.2016.7911451

19. Kozlov, S.V., Sirotkin, E.A., Solomin, E.V.: Wind turbine rotor magnetic levitation. In: Procedia Engineering Journal of 2nd International Conference on Industrial Engineering, Applications and Manufacturing, Chelyabinsk, 19–20 May 2016. https://doi.org/10.1109/icieam.2016.7911477

20. Kirpichnikova, I.M., Korobatov, D.V., Martyanov, A.S., et al.: Diagnosis and restoration of Li–Ion batteries. J. Phys: Conf. Ser. **803**(1), 012070 (2017). https://doi.org/10.1088/1742-6596/803/1/012070

21. Solomin, E.V., Topolskiy, D.V., Topolskiy, N.D.: Integration of adaptive digital combined current and voltage transformer into digital substation ethernet grid. In: Control and Communications. International Siberian Conference, Omsk, pp. 1–4, 21–23 May 2015. https://doi.org/10.1109/sibcon.2015.7147242

22. Solomin, E.V., Topolsky, D.V., Topolsky, N.D.: Arrangement of data exchange between adaptive digital current and voltage transformer and SCADA–system under IEC 61850 standard. Procedia Eng. J. **129**, 207–212 (2015). https://doi.org/10.1016/j.proeng.2015.12.034

23. Chavan, D.S., Karandikar, P.B., Pande, A.K.: Assessment of flicker owing to turbulence in a wind turbine placed on a hill using wind tunnel. In: International Conference on Circuits, Power and Computing Technologies, pp. 560–566 (2014)

24. Van, T.L., Nguyen, T.H., Lee, D.C.: Flicker mitigation in DFIG wind turbine systems. In: Proceedings of the 2011 14th European Conference on Power Electronics and Applications, pp. 1–10 (2011)

25. Chavan, D.S., Karandikar, P.B., Pande, A.K.: Computation of flicker as a result of turbulence in a wind turbine sited on a green building using wind tunnel. In: International Conference on Circuits, Power and Computing Technologies, pp. 554–559 (2014)

26. Walling, R.A., Clark, K., Miller, N.W.: Advanced controls for mitigation of flicker using doubly-fed asynchronous wind turbine-generators. In: 18th International Conference and Exhibition on Electricity Distribution, pp. 1–5 (2005)

The Development of Cost Assessment Models for Overhead Power Transmission Lines

V. V. Cherepanov, N. S. Bakshaeva[✉], and I. A. Suvorova

Vyatka State University, 41 Preobrazhenskaya Street,
Kirov 610020, Russian Federation
bakshaeva@vyatsu.ru

Abstract. Discounted cost assessment models for overhead power transmission lines using wires comprised of a steel core and twisted aluminum strands, conforming to GOST 839-80 (AS-type wires) and self-supporting insulated wires complying with GOST 31946-2012 (SSIW-type wires) rated at voltages of 10, 20 and 35 kV have been developed. The models are based on discounted cost linearization method and constitute simplified mathematical models generalized to the whole range of cross section areas available for a wire of a given specific type, which can be consequently used in the constructed power transmission lines (hereinafter referred to as cross section group). Coefficients of linearization are not dependable on rated power, which facilitates significantly the designing of power grids. The method of linearisation of discounted costs of power grid components enables the reduction of time and money required for the selection of the optimum solution. The suggested analytical representation of the function of grid component costs is perfectly simple and suitable for all the items in a cross section group. The coefficients of the costs function are constant for any given cross section group and can be determined in advance, which makes the calculation of discounted costs substantially easier and faster.

Keywords: Assessment models · Discounted costs linearization method · Overhead power transmission lines

1 Problem Statement

In the present-day Russia, a large number of new urban districts and residential locations comprising cottage-style houses appear. Numerous power transmission lines are being constructed in the rural areas. Power grids are undergoing an upgrade as 6- and 10-kV overhead power transmission lines existing in the majority of Russian towns and cities prove unable to carry the increased load and are often worn out. As a result, the issue of voltage choice arises.

The requirements as to the quality and timelines of designing are becoming increasingly stringent because the complexity of objects designed and the importance of functions fulfilled by them are growing. It is impossible to meet these requirements by simply involving more designers as the possibility of carrying out designing works in a concurrent manner is limited, and the number of engineering personnel of design organizations in the country cannot be significantly increased. The solution to the

A. A. Radionov and A. S. Karandaev (Eds.): RusAutoCon 2019, LNEE 641, pp. 596–605, 2020.
https://doi.org/10.1007/978-3-030-39225-3_65

problem lies in the automation of designing processes, i.e. in a wide use of computer equipment [1].

Software, once developed, can be used repeatedly in different situations coming up in the process of designing of various objects. Engineers using CAD software must be familiar with the methods and algorithms utilized in it, which will help avoid mistakes in the setting of tasks, the selection of input data and the interpretation of results, and will also help obtain the latter at a minimum expenditure of overall and data processing time [2–4].

The designing of power grids implies the processing of large volumes of various data. The labour-intensity of designing increases dramatically if the optimum modes and parameters are determined through multivariate calculations of electric character-istics and a feasibility study [5–8].

When CAD software is used, technical decisions are made in the designer-PC interaction mode, on the basis of mathematical modelling of objects designed. In this case, a large number of problems are sorted out: the accuracy is enhanced, calculation errors are eliminated, the optimum choice is ensured and the preparation of design documents is sped up [4, 9, 10].

The right voltage choice calls for making calculations on a number of values. The making of full-scale calculations is a labor-intensive and expensive process. Design-related calculations generally account for 10% of the total cost of the object designed and are time-consuming [1].

In comparative economic assessments of engineering solutions total discounted costs are presently used.

Consequently, to reduce the expenditure of time and money a new simplified procedure enabling the assessment of discounted costs of electrical components rated at various power distribution grid voltages, making it possible to determine the optimum voltage for a given power grid setup without carrying out detailed calculations and with the help of computers must be developed.

2 The Development of Assessment Models for the Determination of Overhead Power Line Costs

The choice of a design solution requires the application of assessment methods. These methods help make the selection of the optimum voltage for a power distribution network a faster process. The use of a simplified procedure, algorithm and software reduces the time required for designing. Calculation methods adapted for CAD soft-ware must be applied [11].

The advantage of assessment models consists in the fact that they factor in indi-vidual features of systems designed. In addition to that, requirements as to reliability are also taken into account in the process of calculations made with regard to every single component compared. Assessment models make it possible to carry out a full-scale feasibility study of every pre-selected alternative. Once the study is accomplished, the choice of the optimum solution from the variety of pre-selected ones does not take much effort [9, 10, 12, 13].

In comparative economic assessments of engineering solutions, total discounted costs measured in thousand rubles are used [14, 15]. They are given by:

$$3 = K_t + \sum_{t=1}^{T_p} \frac{И_{m.r.t} + И_{los.t}}{(1+E)^t} \tag{1}$$

where T_p is the accounting period taken as equal to the life cycle of the object designed or to other time interval within the life cycle of the object; t is discounting interval; K_t is capital investments with reference to the considered power supply layout at a discounting interval of t (rub.); $И_{m.r.t}$ is the costs of object maintenance and repair (rub.); $\Delta И_{los.t}$ is the costs of compensation of electric power losses with regard to the considered power supply layout at a discounting interval of t; and E is the discount rate, which represents the minimum required rate of capital profit and can take per-unit values in a range of 0.1–0.15 [16].

Expressed in terms of parameters of power grid components and their operational modes, (1) takes the form:

$$3 = K_t + \sum_{t=1}^{T_p} \frac{\alpha_{maint.} \cdot K_t + \dfrac{C_0 \cdot S^2 \cdot \tau \cdot \rho \cdot l}{U^2 \cdot F}}{(1+E)^t}, \tag{2}$$

where α_{maint} is the maintenance and repair cost rate (%), C_0 is the cost of electric power (rub/kW $\cdot$ h), S is rated load power, ρ and F are specific resistivity and conductor cross section area respectively, U is nominal voltage (kV) and l is line length (km).

At a certain cross section area, F, the discounted costs of a power transmission line falling into the cross section group considered represent a quadratic function of rated power:

$$3_{OL} = A + BS^2, \tag{3}$$

Where

$$A = K_t + \sum_{t=1}^{T_p} \frac{\alpha_{maint} \cdot K_t}{(1+E)^t}, \tag{4}$$

$$B = \sum_{t=1}^{T_p} \frac{\dfrac{C_0 \cdot \tau \cdot \rho \cdot l}{U^2 \cdot F}}{(1+E)^t}, \tag{5}$$

Using (4) and (5), the discounted costs of overhead power transmission lines rated at 10, 20 and 35 kV have been calculated. The calculations have been made for power lines 1 km long. The costs of electric power correspond to the data provided by Kirovenergosbyt OAO for 2017. The discount rate, E, equals 0.12. The accounting period, T_p, is 5 years. The results of calculations are given in Fig. 1.

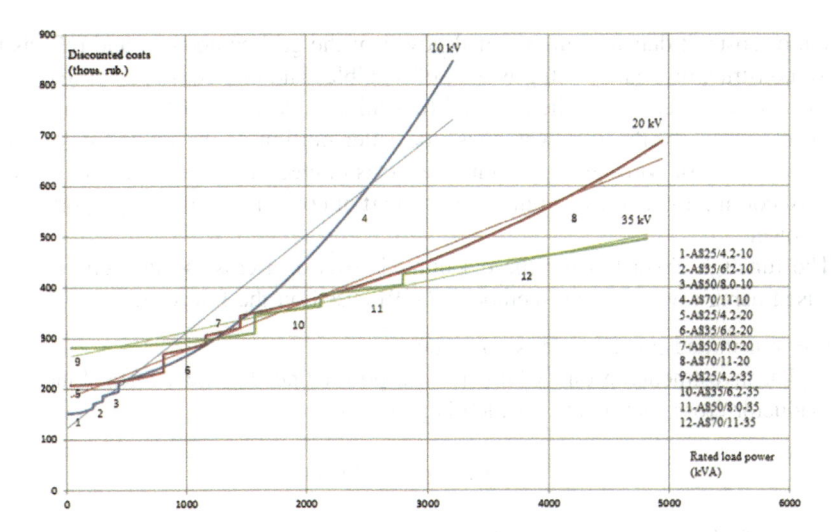

Fig. 1. The discounted costs of 1 km of an overhead power line as a function of transmitted power and voltage approximated by straight lines.

In accordance with the original expression, i.e. (2), the discounted costs of power grid components are piecewise smooth non-linear functions of rated power. The presence of kinks in the function curves makes them unsuitable for use in a feasibility study. Owing to this, the mathematical model of discounted costs of a specific cross section group based on the economic interval function is rarely applied in feasibility studies.

Such functions can be approximated by means of linear dependences on rated power at a relatively low mean error.

$$3 = b + cS, \tag{6}$$

Using expressions for the calculation of discounted costs, allowances for various power grid components enabling the determination of coefficients b and c for linear functions are worked out.

The first summand of (6) characterises the part of discounted costs dictated by the presence of a power grid component. This portion of costs is not dependable on the transmission capacity of the component and, consequently, on its rated power. It represents the construction, mounting and commissioning works relevant to every power grid component used. The costs of such works are determined solely by the fact of component incorporation and are not influenced by its transmission capacity related to its position in the cross section group.

As regards overhead power transmission lines, this portion of discounted costs reflects the expenses incurred in the preparation of power line routes and in land transfer, and indicates the costs of line supports, insuator strings and their mounting.

The second summand appearing in (6) represents the costs associated with the ability of the power grid component to transmit a specific rated power. As for this portion of costs, it depends on the rated power of the grid component and reflects the part of construction and mountng works, of tangibles and of electric power losses, the cost of which is dictated by the required transmission capacity of the component.

In respect of overhead power lines, the latter portion of discounted costs partly reflects the price of wires and represents the costs of electric power losses occurring in them. According to linearisation method, the part of costs in question is proportional to rated power.

The function characterising the discounted costs of a cross section group can be linearised using a well-known method [17–19] based on the following allowances:

1. The transmitted power scale is continuous.
2. Capital investments in one kilometre of a power line, K_0, are a linear function of conductor cross section area, which is given by

$$K_0 = \lambda L + \gamma L F ,\tag{7}$$

The coefficients λ and γ are constant for every single cross section group and can be approximated by means of least squares method and data on capital investments at various standard cross section areas, F.

Once the allowances are factored in, the calculation of the economically feasible cross section area of conductors based on the minimum discounted costs condition can be made using

$$\frac{d3}{dF} = \lambda l + \gamma l F + \sum_{t=1}^{T_p} \frac{\alpha_{ma\,int}\left(\lambda l + \gamma l F\right) + \dfrac{C_0 \cdot S^2 \cdot \tau \cdot \rho \cdot l}{U^2 \cdot F}}{(1+E)^t} = 0 ,\tag{8}$$

Expression (8) implies

$$F_{EF} = \sqrt{\frac{3 \cdot \rho \cdot \sum\limits_{t=t_0}^{T_p} \tau \cdot \frac{S^2}{3U^2} \cdot C_o (1+E)^{-t}}{\gamma\left(1 + \sum\limits_{t=t_0}^{T_p} \alpha_{maint}(1+E)^{-t}\right)}} = \sqrt{\frac{\rho \cdot \sum\limits_{t=t_0}^{T_p} \tau \cdot \frac{1}{U^2} \cdot C_o (1+E)^{-t}}{a\left(1 + \sum\limits_{t=t_0}^{T_p} \alpha_{maint}(1+E)^{-t}\right)}} \cdot S,\tag{9}$$

$$F_{EF} = a \cdot S,$$

$$a = \sqrt{\frac{\rho \cdot \sum\limits_{t=t_0}^{T_p} \tau \cdot \frac{1}{U^2} \cdot C_o (1+E)^{-t}}{\gamma\left(1 + \sum\limits_{t=t_0}^{T_p} \alpha_{maint}(1+E)^{-t}\right)}} = \frac{1}{U} \cdot \sqrt{\frac{\rho \cdot \sum\limits_{t=t_0}^{T_p} \tau \cdot C_o (1+E)^{-t}}{\gamma\left(1 + \sum\limits_{t=t_0}^{T_p} \alpha_{maint}(1+E)^{-t}\right)}},\tag{10}$$

After the substitution of (7) and (10) into (2), the discounted costs are given by

$$
3 = \lambda L + \gamma L \cdot a \cdot S + \sum_{t=t_0}^{T_p} \frac{\alpha_{ma\,int}\left(\lambda L + \gamma L \cdot a \cdot S\right) + \dfrac{3 \cdot C_0 \cdot S^2 \cdot \tau \cdot \rho \cdot L}{3U^2\left(a \cdot S\right)}}{(1+E)^t} =
$$

$$
= \lambda L + \gamma L \cdot a \cdot S + \sum_{t=t_0}^{T_p} \frac{\alpha_{ma\,int}\left(\lambda L + \gamma L \cdot a \cdot S\right) + \dfrac{C_0 \cdot S \cdot \tau \cdot \rho \cdot L}{U^2 a}}{(1+E)^t} = \tag{11}
$$

$$
= \lambda L + \sum_{t=t_0}^{T_p} \frac{\alpha_{ma\,int} \cdot \lambda L}{(1+E)^t} + \gamma L \cdot a \cdot S + \sum_{t=t_0}^{T_p} \frac{\gamma L \cdot a \cdot S + \dfrac{C_0 \cdot \tau \cdot \rho \cdot L}{U^2 a}}{(1+E)^t}
$$

$$
b = \lambda L + \sum_{t=t_0}^{T_p} \frac{\alpha_{maint} \cdot \lambda \cdot L}{(1+E)^t}, \tag{12}
$$

$$
c = \gamma La + \sum_{t-t_0}^{T_p} \frac{\left(\gamma La + \frac{C_0 \cdot \tau \cdot \rho \cdot L}{a \cdot U^2}\right)}{(1+E)^t} \cdot S, \tag{13}
$$

In (12), the costs of mounting works required to lay 1 km of an overhead line are not taken into account. What follows is a formula, in which the mounting works are factored in:

$$
b = \lambda L + \sum_{t=t_0}^{T_p} \frac{\alpha_{maint} \cdot \lambda \cdot L}{(1+E)^t} + b_0 L, \tag{14}
$$

where b_0 is the costs of mounting works required to lay 1 km of an overhead line (thous. rub.) and I is overhead line length (m).

It is necessary to evaluate how well the obtained assessment models approximate the input data. To do so, the R^2 goodness-of-fit test is used.

The R2 test can only take on values in the range from zero to one. Generally, the closer to one its value is, the better the parametric model approximates the input data. The results of calculations are presented in Table 1.

Table 1. Equations of approximating curves.

U, kV	Mathematical model of discounted costs, thous. rub.	R^2, per unit
$U = 10$ kV	$3 = 122.182 + 0.189 \cdot S$	$R^2 = 0.993$
$U = 20$ kV	$3 = 137.093 + 0.079 \cdot S$	$R^2 = 0.991$
$U = 35$ kV	$3 = 196.534 + 0.034 \cdot S$	$R^2 = 0.978$

Thus, an analysis of the obtained R^2 values enables us to conclude that the mathematical models of discounted costs ensure maximum approximation to the input data.

The mathematical model of discounted costs of power grid components, which has been developed on the basis of discounted costs linerisation method, is a simplified mathematical model generalised to all the items in a cross section group. Its simplification is achieved due to the fact that deviations of individual parameters of separate power grid components from the values determined by the law of their variation generalised to the entire cross section group are not factored in. However, the generalised law retains and reflects the main tendency of discounted costs variation within every single cross section group, which consists in the existence of a cost portion not dependable on rated power and of a part of costs increasing with rated power enhancement [20].

In a similar way, dependencies for SSIW wires have been determined (Fig. 2).

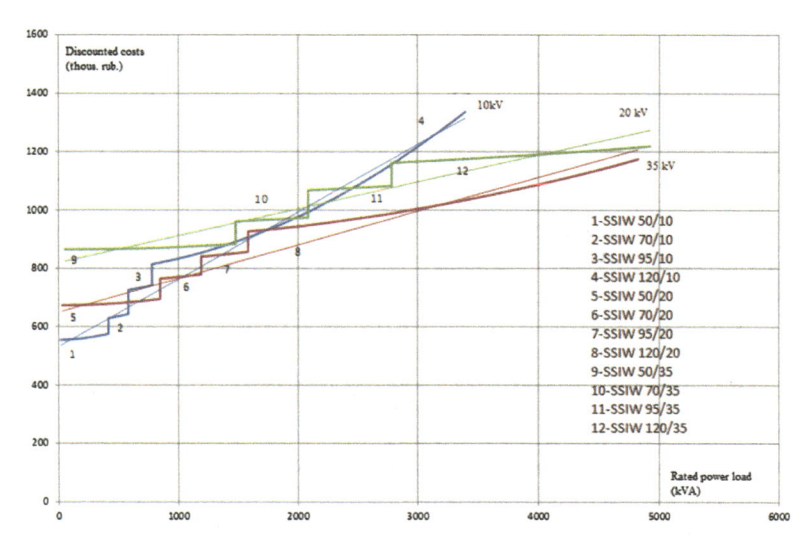

Fig. 2. The discounted costs of 1 km of an overhead power line as a function of transmitted power and voltage approximated by straight lines.

As is the case with overhead power lines using AS-type wires, the presence of kinks in the function curves makes their differentiation impossible. To simplify calculations, a linear function is used as an approximating curve. The calculations are made according to (6–13). The calculation results are given in Table 2.

Table 2. The mathematical model of ssiw-type wire costs.

U, kV	Mathematical model of discounted costs, thous. rub.	R^2, per unit
$U = 10$ kV	$3 = 0.249 \cdot S + 150.038$	$R^2 = 0.967$
$U = 20$ kV	$3 = 0.095 \cdot S + 181.286$	$R^2 = 0.924$
$U = 35$ kV	$3 = 0.049 \cdot S + 262.959$	$R^2 = 0.918$

Thus, an analysis of the obtained R^2 values enables us to conclude that the mathematical models of discounted costs ensure maximum approximation to the input data.

3   The Determination of Discounted Costs Calculation Error Using a Simplified Method with Regard to Overhead Power Lines

The simplification of the mathematical model of power transmission line costs and its generalisation to the entire cross section group leads to the appearance of errors in the calculation results.

The allowances considered in the process of development of a mathematical model intended for the solution of a specific problem, regardless of its kind, should be assessed on the basis of the value of the error attributable to the model, which is present in the results of the problem solution. It should not exceed the errors caused by other factors such as the solution method applied and the inaccuracy of input data.

The error observed in the results of feasibility studies of conventionally applied design methods is mostly acribable to the following factors:

- In case of determination of capital investments in power supply system components, a 5–10% error is inevitable as the actual prices will always be more or less different from the ones accepted for the needs of designing [15]
- As an analysis of design experience [17] has shown, the mean error in rated loads worked out through the application of existing methods comes out at 16%. A rated load error that high leads to an inaccuracy in costs determination of 5–7%.

On average, the inevitable error in the results of a feasibility study carried out in the course of designing works amounts to 5–10% [15].

The relative error in costs worked out by means of a linearised function can be found using the formula:

$$\Delta 3 = \frac{3 - 3'}{3} \cdot 100\%, \tag{15}$$

where 3 is discounted costs calculated in accordance with the original relation given by (1) (thous. rub.) and $3'$ is discounted costs determined by means of models for the assessment of costs of power supply system components (thous. rub.).

The results of the calculation are presented in Tables 3 and 4.

Table 3. Error in the results of discounted costs calculation made by means of a simplified method intended for as-type wires.

Error type	Voltage, kV		
	10	20	35
δ_{mean}, %	5.74	4.23	1.84
δ_{max}, %	9.12	8.87	1.84

Table 4. Error in the results of discounted costs calculation made by means of a simplified method intended for ssiw-type wires.

Error type	Voltage, kV		
	10	20	35
δ_{mean}, %	2.17	0.56	2.98
δ_{max}, %	13.98	7.33	11.87

The mean relative error for every voltage value is within the limits of permissible accuracy of determination, which allows us to conclude that the results of calculations of overhead power line parameters obtained by using the procedure discussed here are acceptable.

4   Conclusion

1. The discounted costs of overhead power lines are piecewise smooth non-linear functions of rated power, which are difficult to use in calculation automation.
2. Mathematical models of discounted costs of overhead power transmission lines have been put forward. The models have been developed on the basis of discounted costs linearisation method and constitute simplified mathematical models generalised to all the items in single specific cross section groups.
3. The suggested analytical representation of the function of power grid component costs is perfectly simple and suitable for all the items in any given cross section group. The costs function coefficients b and c are constant for every single cross section group.
4. The use of the models suggested enables the calculation of discounted costs of any single item in any specific cross section group without precise determination of the item parameters. In other words, there is no need to work out the standard cross section area for a power transmission line. The parameters of power grid components are determined accurately only for the optimum solution chosen.
5. A mathematical model of discounted costs of overhead power lines based on linearisation method has been put forward. The discounted costs are determined taking into consideration the technical limitations, i.e. the permissible loss of voltage and power.
6. The suggested algorithm of discounted costs calculation based on linearisation method can be put into practice as computer software.

References

1. Lukyanov, M.M., Konoshenko, A.V.: Design of Electrical Installations. Kniga, Chelyabinsk (2008)
2. Andreev, L.N., Bortyakov, D.E., Meshcheryakov, S.V.: CAD System. SPbGTU, St. Petersburg (2002)

3. Seifi, H., Sepasian, M.S.: Electric Power System Planning: Issues, Algorithms and Solutions. Springer, Moscow (2011)
4. Kossov, V.V., Livshits, V.N., Shakhnazarov, A.G.: Methodological Recommendations for Assessing the Effectiveness of Investment Projects. Economics, Moscow (2000)
5. Bebko, V.G.: Selection of the optimal scheme of electrical networks. Application of mathematical methods and computer engineering in power engineering, vol. 2. Academy of Sciences of the Moldavian SSR, Kishinev (1968)
6. Blok, V.M.: The choice of optimal cable sections considering economic indicators. Power plant **9–10**, 8–12 (1945)
7. Belyaev, B.I., Tavtadze, M.N.: Error Theory and Least Squares Adjustment. Nedra, Moscow (1992)
8. Idelchik, V.I.: Calculations and Optimization of Modes of Electric Networks and Systems. Energoatomizdat, Moscow (1988)
9. Venikov, V.A.: Similarity and Modeling Theory. High School, Moscow (1976)
10. Venikov, V.A., Arzamastsev, D.A.: On the construction of economic and mathematical models of electrical systems. Energy Transp., 76–85 (1970)
11. Fyodorov, A.A., Nikulchenko, A.G., Sadchikov, S.V.: To the question of optimization of construction of a network of industrial power supply. Proc. MPEI **446**, 10–14 (1980)
12. Geraskin, O.T.: Methods of one-dimensional optimization of electric power systems. Energy Transp., 52–60 (1980)
13. Apenko, V.I., Khodak, I.Y.: Development of CAD of power supply based on information model. Ind. Power Eng. **6**, 35 (1983)
14. Fomina, V.N.: Energy Economics. GUU, Moscow (2005)
15. Gitelson, S.I.: Economic Solutions in the Design of Power Supply of Industrial Enterprises. Energy, Moscow (1974)
16. Zuev, E.N.: Determination of economic current density based on the criterion of minimum discounted costs. Bull. MPEI **3**, 59–61 (2000)
17. Gamazin, S.I., Cherepanov, V.V.: Application of Mathematical Programming Methods in the Design of Power Supply Systems. GUU, Gorky (1980)
18. Cherepanov, V.V., Suvorova, I.A.: Solving problems of design of distribution electric networks using the method of linearization of discounted costs. News of High Schools. Electromechanics **3**, 75–76 (2014)
19. Suvorova, I.A.: Development of a program for calculating discounted costs for the design of urban, rural and industrial networks. Mod. Probl. Sci. Educ. **1**, 146–152 (2013)
20. Suvorova, I.A.: Using the minimum discounted cost criterion to select the supply voltage of the power supply system element. Electrician **9**, 26–28 (2012)

Energy Efficiency of Electric Vacuum Systems: Induction Motor – Water Ring Pump with an Ejector

A. R. Denisova$^{(\boxtimes)}$, A. I. Rudakov, and N. V. Rozhentcova

Kazan State Power Engineering University, 51 Krasnoselskaya Street,
Kazan 420066, Russian Federation
denisova_ar@mail.ru

Abstract. The article addressed and resolved issues of improving the energy efficiency of vacuum units: motor - liquid ring pump with an upstream ejector. The purpose of the study was to determine the relationship between the required power N and the magnitude of the residual pressure of the vacuum water-ring pump BBH-0,25 p_{sat}, the pre-actuated stationary BBH-0.25E and the pulsed BBH-0.25EP ejector. The way the system modeling asynchronous motor - liquid ring pump - ejector ballasts (BP-VVNE), which is based on a joint consideration of the electrical, hydraulic and gas dynamic subsystems. Simulation System AD-VVNE carried out on the basis of the equivalent circuit, which is not done by the authors of known works devoted to the modeling of water ring pumps with ejectors. At the same time the actual asynchronous car with electromagnetic interactions between windings is replaced with rather simple electric circuit that allows to simplify calculation of characteristics significantly.

Keywords: Energy efficiency · System · Induction motor · Liquid ring pump · Ejector ballasts

1 Introduction

Energy efficiency is considered to the rational use of energy resources, in which a power reduction is achieved while maintaining the load power level [1, 2].

The efficiency of electromechanical systems consisting of an electric motors used as an actuators and executors, for example, pumps of various types, fans, compressors, etc., is usually considered separately. But it is the collateral consideration of the interrelated electromechanical, hydraulic and gas-dynamic processes occurring in these devices, mainly defines as operating capacity of system, so its reliability and quality [3–6].

In the process of energy conversation, a part it is lost in the form of heat. The amount of the lost energy is defined by power factors. Use of energy efficient electric motors and actuators, allows to significantly reduce energy consumption and the maintenance of carbon dioxide in the environment [7–9].

Increase in energy efficiency of electromechanical system an induction motor - the water ring pump with an ejector can be carried out at a design stage by modeling of the

A. A. Radionov and A. S. Karandaev (Eds.): RusAutoCon 2019, LNEE 641, pp. 606–612, 2020.
https://doi.org/10.1007/978-3-030-39225-3_66

system being designed. In this case it is advisable to consider the general electrome-chanical system in the form of separate subsystems [2, 4, 6, 10]:

- The electric subsystem modeling processes in the electric motor
- The mechanical subsystem modeling processes of the electromechanical converter of energy and mechanical losses
- The hydrogasdynamic subsystem modeling processes in the preincluded ejector and a liquid ring of the water ring pump [11]

2 Calculation of Characteristics of the Induction Motor

For calculation of characteristics of an induction motor and a research of various modes of its work equivalent circuits are used. At the same time the actual asynchronous car with electromagnetic interactions between windings is replaced with rather simple electric circuit that allows to simplify calculation of characteristics significantly.

In work [8] as model of an electric subsystem it is used the T-shaped equivalent circuit of an induction motor which is used for calculation of characteristics of an induction motor and a research of various modes of its operation. In this work the hydraulic subsystem modeling processes in a flowing part of an impeller pump of the engine is considered and researches of various modes of its work are used equivalent circuits.

Parameters have to be known when calculating characteristics of an induction motor with use of its equivalent circuit.

The T-shaped scheme completely reflects the physical processes occurring in the engine, but has a nodal point between resistances that complicates calculation of currents at various values of sliding. Therefore, as model of an electric subsystem the

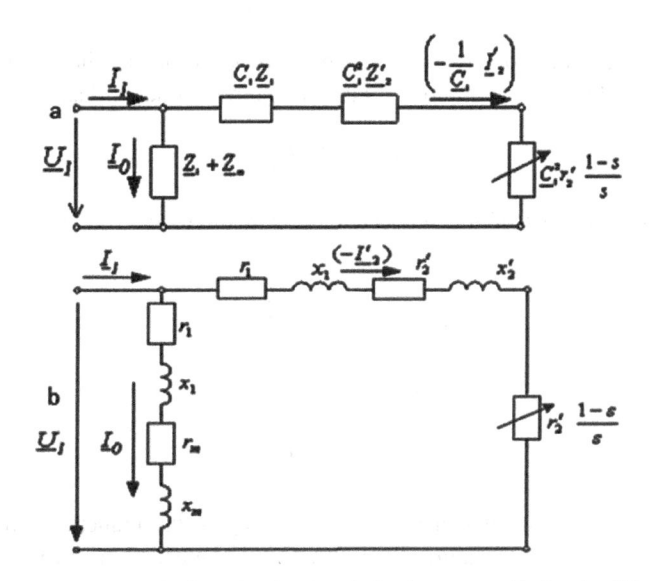

Fig. 1. The Gamma-type equivalent circuit of an induction motor (a), its simplified version (b).

Gamma-type equivalent circuit of an induction motor (Fig. 1) is used. In the provided scheme the magnetizing branch is connected immediately to the input, where the voltage U_1 is applied.

In this case, the equivalent circuit is characterized by three parameters, the active resistance of the rotor and stator, as well as the active resistance of the magnetizing circuit and three parameters, the reactance of the rotor and stator, as well as the reactance of the magnetizing circuit.

The electrical subsystem is characterized by processes in the windings of the electric motor.

As the rotor winding during the work of an induction motor is short-circuited, then:

$$E_2^I = I_2^I\left(\frac{r_2^I}{s} + jx_2^I\right), \tag{1}$$

The quantity r_2^I/s, can be presented how

$$\frac{r_2^I}{s} = r_2^I + r_2^I\frac{1-s}{s}, \tag{2}$$

Then (1) and (2) are transformed into:

$$E_1^I - I_1^I\left(r_1^I + jx_1^I\right)_1 = I_1^I r_1^I \frac{1-s}{s}$$

or

$$E_1^I - I_1^I z_1^I = I_1^I r_1^I \frac{1-s}{s}, \tag{3}$$

where z_1^I – the complex of the total reduced impedance of the phase winding of the rotor. For the reduced asynchronous motor, the current equation has the form

$$I_1 = I_0\left(-I_0^I\right), \tag{4}$$

The basic equations of the reduced induction motor can be represented in the following form:

$$E_1^I - I_1^I z_1^I = I_1^I r_1^I \frac{1-s}{s};$$

$$U_1 = \left(-E_1\right) + I_1; \tag{5}$$

$$I_1 = I_0\left(-I_0^I\right).$$

We will take into account that $\cos\varphi_1$ at loading of the induction motor is much higher than $\cos\varphi_0$ of its idling. It follows that unloaded asynchronous motors significantly degrade the power factor in the network, therefore it is necessary to aim for their full load.

From the simplified Gamma-type equivalent circuit the given current of a rotor winding is defined:

$$I_2^I = \frac{U_1}{\sqrt{\left(r_1 + \frac{r_2^I}{s}\right)^2 + \left(x_1 + x_2^I\right)^2}}, \tag{6}$$

Further the Eq. (6) is used in deriving the magnitude of the electromagnetic moment of an induction motor, which is proportional to the squared voltage, decreases with increasing frequency.

Maximum (critical) moment

$$M_{max} = \pm \frac{pm_1 U_1^2 2}{2c_1 \omega \left[\pm r_1 + \sqrt{r_1^2 + \left(x_1 + c_1 x_2^I\right)^2}\right]}, \tag{7}$$

$$s_{\kappa p} = \pm \frac{c_1 r_1^I}{\sqrt{r_1^2 + \left(x_1 + c_1 x_1^I\right)^2}}, \tag{8}$$

$$\frac{M}{M_{max}} = \frac{2}{\frac{s_{\kappa p}}{s} + \frac{s}{s_{kp}}}.$$

The ideal electromechanical converter of energy described by the equation is considered by an analog of a mechanical subsystem:

$$P_{el} = P_{mech}, \tag{9}$$

where $P_{el} = m\left(I_2^I\right)^2 R_2^I \frac{1-s}{s}$ – power produced by an asynchronous motor; m – number of phases; s – sliding; $P_{mech} = M\omega_{pom}$ – mechanical power; M – torque; ω_{pom} – angular velocity.

In comparison with a centrifugal pump, the hydro-gas-dynamic system is more complex, since it contains hydraulic processes in the liquid ring, which to some extent corresponds to the processes in the centrifugal pump, and the associated gas-dynamic processes in the induction ejector [3].

This complicates the mathematical model, since a new link appears - the jet apparatus (the pre-actuated ejector). If we take the work of the liquid ring as the work of a centrifugal pump, then we have:

$$I_{zhk} = P_{zhk}, \tag{10}$$

I_{zhk} – current, analog an expense in a liquid ring;

$$I_{nom} = Q_{nom}, \tag{11}$$

I_{nom} – current, analog of hydraulic losses in a liquid ring.

3   Determination of Parameters of the Gas Dynamic Subsystem

Determination of the parameters of the gas-dynamic subsystem (the pre-actuated ejector of the water-ring vacuum pump BBH-0.25, BBH-0.25E and BBH-0.25EP) was carried out experimentally with the subsequent approximation of the obtained dependences [2].

Figure 2 shows a three-dimensional graph for feeding the BBII-0.25 (Q) pump against the pressure (P) at a various water temperature t.

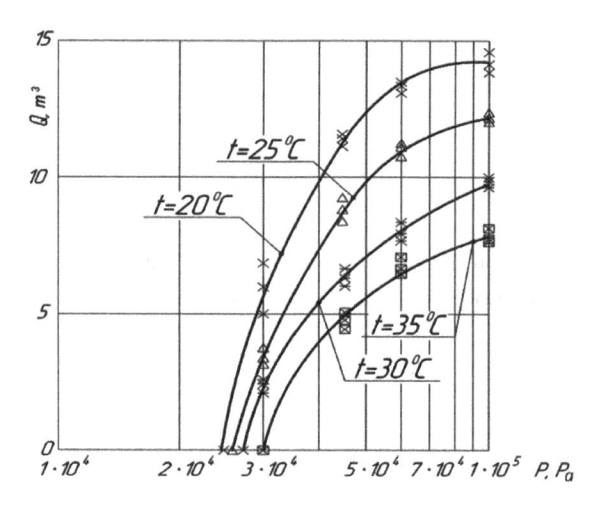

Fig. 2. Dependence of the supply *(Q)* of the water-ring vacuum pump BBH-0.25 pump against the pressure *(P)* at various water temperatures *t*.

The equation for a certain temperature has the form:

$$Q = 3,1242 + 0,289p + 0,010p^2, \tag{12}$$

Taking into account the temperature change, Eq. (2) has the form:

$$Q = 2,6543 + 0,289p + 0,51t + 0,322pt - 0,19p^2 \cdot 0,05t^2. \tag{13}$$

The purpose of the study was to determine the relationship between the required power N and the magnitude of the residual pressure of the vacuum water-ring pump BBH-0,25 p_{sat}, the pre-actuated stationary BBH-0.25E and the pulsed BBH-0.25EP ejector.

The analysis of these dependences shows a slight increase in the required power with a significant decrease in the residual pressure in the system.

Note that physical quantities, their dimensions and symbols of the induction electric motor are adopted according to international standards (SI system) [1] (Fig. 3).

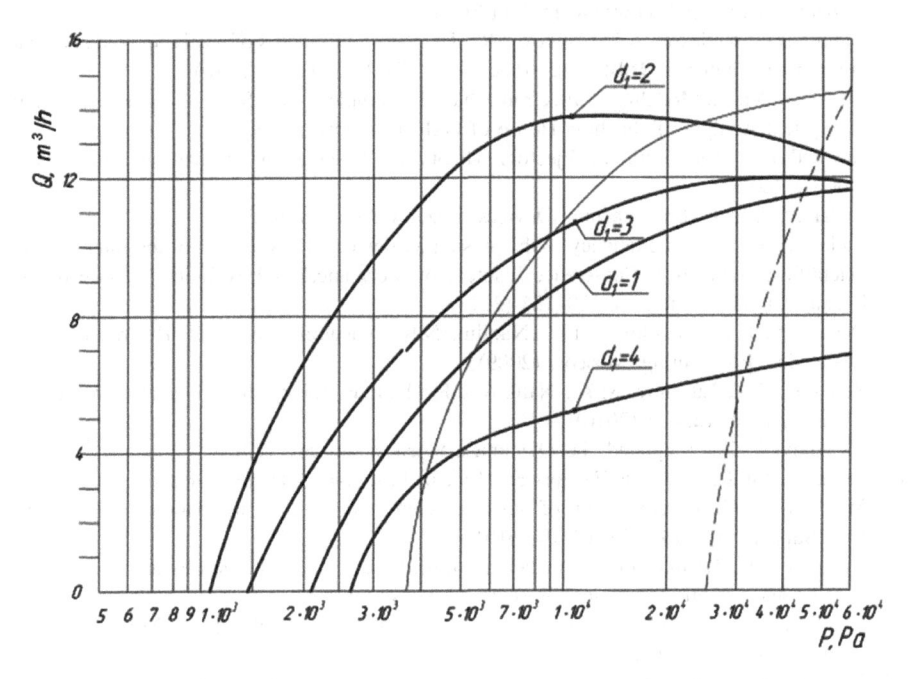

Fig. 3. Dependence of the supply of a water-ring vacuum pump with a pre-actuated stationary ejector: $d_1 = 1$; 2; 3 and 4 mm.

4   Conclusion

The energy efficiency of electromechanical systems consisting of an electric motor (actuator) and executors (in our case it is a water-ring pump with an ejector) must be considered together [12–14]. It is the collateral consideration of the interrelated electromechanical, hydraulic and gas-dynamic processes occurring in these devices, mainly defines as operating capacity of system, so its reliability and quality [15].

Increase in energy efficiency of electromechanical system an induction motor - the water ring pump with an ejector can be carried out at a design stage by modeling of the system being designed [16, 17]. In this case it is advisable to consider the general electromechanical system in the form of separate subsystems.

References

1. GOST: Power efficiency. Indicators. General provisions (2000)
2. Rudakov, A.I., Rozhentcova, N.V., Denisova, A.R.: Modern technical means of improving the energy efficiency of water ring machines. Ind. Power **5**, 27–30 (2014)
3. Arsenyev, V.M., Meleychuk, S.S., Levchenko, D.A.: Aggregation by the ejector of the water ring vacuum pump. Compressor and power engineering. Bags **13**(17), 64–67 (2009)
4. Gamazin, S.I., Zhukov, V.A., Pupin, V.M., et al.: Dynamic stability of electromechanical complexes with synchronous and asynchronous motors at the enterprises with a continuous production cycle. Ind. Power **4**, 15–20 (2011)
5. Ivanov, B.L., Rudakov, A.I., Nafikov, I.R.: The theory of process of receiving cold and heat in a tube the Wound – Hilsha. Sci. Pract. Conf. **77**(2), 174–176 (2010)
6. Kovalyov, Yu.Z.: Modeling of Electromechanical Complexes and Systems from Positions of the System Analysis. Publishing House of OMGTU, Omsk (2006)
7. Alexandrov, V.Yu.: Optimum Ejectors. Theory and Calculation. Mechanical Engineering, Moscow (2012)
8. Rudakov, A.I.: Jet Low-Vacuum Devices. Kazan GAU, Kazan (2008)
9. Rudakov, A.I.: Power efficiency of the pulsing jet devices. In: Materials of the International Scientific and Practical Conference Institute of Mechanical and Technical Service of the Kazan State University, pp. 129–131 (2013)
10. Demikhov, K.E., Panfilov, Yu.V., Nikulin, N.K.: Vacuum Equipment: Reference Book. Mechanical Engineering, Moscow (2009)
11. Rudakov, A.I., Lushnov, M.A., Nafikov, I.R.: Devices for creation of a vacuum. Mech. Electr. Agricul. **12**, 4–5 (2010)
12. Abramovich, G.N.: Applied Gas Dynamics. Science, Moscow (2011)
13. Arkadov, Yu.K.: New Gas Ejectors and Ejector Processes. Fizmatlit, Moscow (2001)
14. Martynov, A.V.: Determination of power efficiency of devices, installations and systems. Heat Supply News **10**(122), 17–19 (2010)
15. Girsanov, I.V.: Lectures on the mathematical theory of extreme tasks. Research Center SIC Regular and Chaotic Dynamics, Moscow – Izhevsk (2003)
16. Kondrashov, B.M.: Essentially new jet energy technologies. Small Energy, Moscow, 11–14 October 2005, p. 7 (2005)
17. Nafikov, I.R., Ivanov, B.L., Rudakov, A.I.: Results of tests of the pulsing ejector as a part of the water ring vacuum pump. In: Materials of the XI Scientific-Practical Conference Effective Tools of Modern Science, pp. 16–19. Publishing House Education and Science, Prague (2015)

Cooling System Oil-Immersed Transformers with the Use of a Circulating Sulfur Hexafluoride

M. G. Bashirov, A. S. Khismatullin, and E. V. Sirotina[✉]

Salavat Branch of Ufa State Petroleum Technical University, 22 Gubkina Street,
Salavat 453265, Russian Federation
katrina.sirotina18@yandex.ru

Abstract. The question of this article is the increase of efficiency of power transformers cooling systems with significant long and short-term overloads. The data presented the causes of failures of transformers of different voltage classes and different periods of operation. It is revealed that one of the main causes of transformers failure is the ineffectiveness of the oil cooling system during the summer operation period and with short-term but significant overloads. To improve the efficiency of the cooling system of power transformers, the authors propose to implement mixing cooling oil circulating sulfur hexafluoride and subsequent cooling of the sulfur hexafluoride thermoelectric refrigerator. Laboratory experiments were carried out which confirm the effectiveness of the method. The design of the proposed laboratory setup is described in detail. Stamps of the thermocouple and the brand of the compressor and an analog-to-digital are also selected. Due to the fact that the device does not require regular replacement of consumable parts, and also does not require periodic maintenance, the system is maintenance-free.

Keywords: Bubbly liquid · Cooling · Sulfur hexafluoride · Transformers

1 Introduction

In large-capacity transformers, a large amount of heat is generated, for the removal of which special oil-air coolers are used. These air coolers are blown by air with a fan and equipped with pumps for forced circulation of oil. Forced oil circulation allows obtaining more uniform temperature along the tank height and increases cooling efficiency of the windings and magnetic core of the transformer.

The term of natural wear of power transformers TM3, TM, TAM, operating in nominal mode, is approximately 30 years.

Since the acquisition of new transformers is highly cost-effective, enterprises take additional measures and extend their service life. This increases the risk of emergencies. As it can be seen in [1], a significant number of failures of power transformers are due to unsatisfactory operation (more than 50% of all failures).

The main operational reasons leading to damage (failure) of transformers include overheating of the active part due to the ineffectiveness of the oil cooling system during

A. A. Radionov and A. S. Karandaev (Eds.): RusAutoCon 2019, LNEE 641, pp. 613–621, 2020.
https://doi.org/10.1007/978-3-030-39225-3_67

the summer operation period and with short-term but significant overloads. The deterioration of the quality of the oil itself is also a significant factor.

Overheating of the insulation of windings and magnetic core reduces the resource of the transformer, and the operation of the gas relay often leads to unreasonable tripping of the transformer during short-time overloads. Therefore, the problem of improving the efficiency of cooling oil transformers is topical.

The aim of this work is to investigate the effectiveness of the method of a transformer cooling based on bubbling transformer oil with SF6 gas.

The principle of SF6 gas bubbling of transformer oil is the basis of the utility model "The installation for cooling the oil transformer" [2]. Its essence is in the fact that the heat removal from the active part of the transformer is carried out by transformer oil with circulating SF6 gas. Gas bubbles, floating up in transformer oil, carry with them particles of oil, which are sequentially separated from SF6 gas in a tank-expander and filters of coarse and fine cleaning. Further, SF6 gas is cooled in the refrigerator and goes into a special container, from where it is fed by the compressor to the mineral oil tank through the tubes with valves-distributors evenly located in the lower part of the tank.

A temperature sensor is installed directly in the area of the transformer winding, whose output is connected to the input of the control unit.

When the active part of the transformer (winding and magnetic core) is heated above the set value, the control unit gives the activation signals to the control inputs of the compressor and the refrigerator. When the temperature of the active part of the transformer drops below the set value, the control unit applies switch-off signals to the control inputs of the compressor and the refrigerator [2]. In the absence of overloads, and accordingly, overheating of the active part, the transformer operates in the normal mode, i.e. cooling of the active part is carried out by circulating transformer oil. The use of the refrigerator makes it possible to increase the efficiency of cooling of the SF6 gas, and, therefore, the transformer oil and the active part of the transformer. The refrigerator is a thermoelectric module, which operation is based on the phenomenon of thermoelectric emission. The thermoelectric module contains series-connected p- and n-type semiconductors that form p-n junctions between the ceramic plates. In turn, each of these transitions has thermal contact with one of the two radiators. As a result of the passage of an electrical current of a certain polarity, a temperature difference forms between the radiators of the module: one radiator acts as a refrigerator, the other radiator heats up and serves to remove heat.

The main advantage of the thermoelectric module is that it makes it possible to obtain a significant temperature difference in a few tens of degrees, i.e. to ensure effective cooling of SF6 gas, and this, in turn, provides more efficient cooling of the active part of the transformer [1–6].

2   The Theoretical Part

Let's consider the heat flux in a rectangular parallelepiped (Fig. 3). There is the oil at a temperature of $T_0 = 15\ °C$, inside the tank at the initial moment, which gradually over time acquires a temperature of $T_H = 50\ °C$, corresponding to the temperature of the

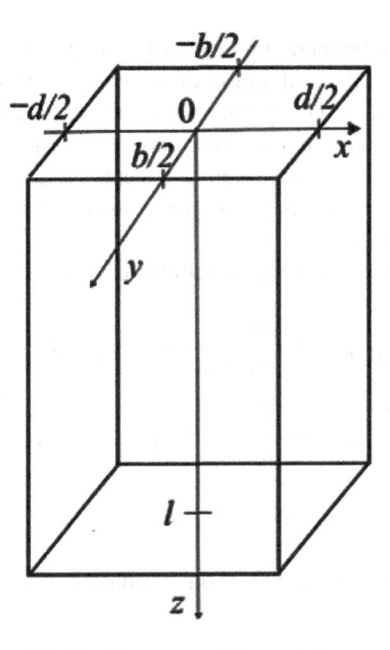

Fig. 1. Geometry of the problem.

heater. Let's consider a rectangular parallelepiped bounded in x, y, and z, respectively. The temperature is found by solving the heat equation [7] (Fig. 1).

$$\frac{\partial T}{\partial t} = a \left(\frac{\partial^2 T}{\partial x^2} + \frac{\partial^2 T}{\partial y^2} + \frac{\partial^2 T}{\partial z^2} \right) \tag{1}$$

$$0 < x < \frac{d}{2}, \ 0 < y < \frac{b}{2}, \ 0 < z < l, \ t > 0$$

with the following initial condition:

$$T|_{t=0} = T_0 \tag{2}$$

where $a = \lambda / c\rho$ – is the thermal diffusivity, λ – conductivity and T_0 – temperature of the environment. The heat exchange with the environment on the surface S is described by Newton's law:

$$-\lambda \frac{\partial T}{\partial x}\bigg|_S = \alpha \left(T|_S - T_0 \right)$$

where S – surface of the wall, α – heat transfer coefficient of the medium (oil – metal – air).

In the experiment, the temperature reaches the programmed value above the heater within 10 min. Therefore, if the duration of the experiment is more than 30 min, we can assume that T_H = const. The ambient temperature is also constant, because it is assumed that the average daily temperature variation during the year is approximately 10 °C. And the time of the experiments, except for experiments without pop-up bubbles, is less than two hours. Therefore, it is considered that the ambient temperature does not change T_0 = const.

Let us denote, then the boundary conditions can be written as

$$\frac{\partial T}{\partial x}\big|_{x=0} = 0, \ \frac{\partial T}{\partial x}\big|_{x=\frac{d}{2}} + h\left(T\big|_{x=\frac{d}{2}} - T_0\right) = 0;$$

$$\frac{\partial T}{\partial y}\big|_{y=0} = 0, \ \frac{\partial T}{\partial y}\big|_{y=\frac{b}{2}} + h\left(T\big|_{y=\frac{b}{2}} - T_0\right) = 0;$$

$$T\big|_{z=l} = T_0, \ T\big|_{z=0} = T_H$$

Solving the problem using the method of convolution, we finally obtain for the calculations:

$$T = 16\left[\sum_{n=0}^{\infty}\sum_{m=0}^{\infty}\frac{\sin(\chi_n d/2)\sin(\mu_n b/2)\cos(\chi_n x)\cos(\mu_n y)}{(\chi_n d + \sin(\chi_n d))(\mu_n b + \sin(\mu_n b))}\times\left\{\frac{sh\left(\sqrt{\chi_n^2 + \mu_n^2}\cdot(l-z)\right)}{sh\left(\sqrt{\chi_n^2 + \mu_n^2}\cdot l\right)} + \right.\right.$$

$$\left.\left. +2\sum_{k=1}^{\infty}\frac{\pi k\sin\left(\frac{\pi k}{l}z\right)\exp\left(-a\left(\chi_n^2 + \mu_n^2 + (\pi k/l)^2\right)t\right)}{\left((\chi_n^2 + \mu_n^2)l^2 + \pi^2 k^2\right)}\right\}\right]\cdot(T_H - T_0) + T_0$$

where are the equations to determine χ_n and μ_n are recorded as

$$-\chi_n\sin(\chi_n\frac{d}{2}) + h\cos(\chi_n\frac{d}{2}) = 0, \ h\cos(\mu_n\frac{b}{2}) - \mu_m\sin(\mu_n\frac{b}{2}) = 0,$$

where μ_n and χ_n – constants transcendental equations.

3 Experimental Part

The installation for cooling the oil transformer" [8] includes series-connected tubes filled with SF6 gas, a compressor, pipes with valves evenly spaced on them, a dilator tank, and fine and coarse filters. A refrigerator is turned on between the output of the fine filter and the entrance to the tank with SF6 gas. A temperature sensor is installed in the zone of the winding of the transformer, the output of which is connected to the input of the control unit, and the outputs of the control unit are connected to the control inputs of the compressor and the refrigerator.

In Fig. 2 there is a block diagram of an installation for cooling an oil transformer.

According to GOST (All-Union State Standard) 12.1.007-76 SF6 gas belongs to the 4th class of danger on the degree of exposure to the body, which low hazard substances belong to.

During the experimental studies air was used instead of SF6 gas since it is forbidden to use SF6 gas in laboratory premises in safety rooms. Its use is dangerous for people [9].

A laboratory installation was developed to investigate the proposed method for cooling oil transformers (Fig. 2).

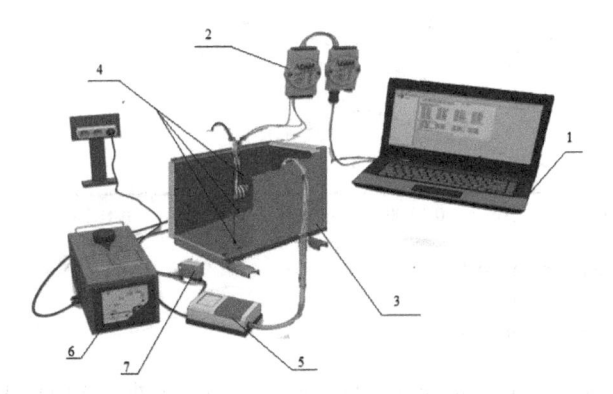

Fig. 2. Scheme of the experimental setup: 1 – computer; 2 – analog-to-digital Converter; 3 – the oil tank of the transformer; 4 – thermocouple; 5 – pump; 6 – autotransformer; 7 – relay.

In a series of experiments, the transformer oil was heated by a 450 W heating element at an initial oil temperature of 20 °C.

Thermocouples TXK-0292 (L-type) are installed to study the topography of the thermal field in the transformer tank at certain distances, the signals from which are fed to the ADAM 4018 B analog-digital converter (manufactured by Advantech), the signals from which enter the computer 1 for subsequent processing and analysis. The signal of the upper thermocouple (1) by means of a special program controls the oil compressor 5 (manufactured by Abac Pole Position 241) supplying gas to the ceramic distributors located at the bottom of the tank. The compressor automatically switches on when the oil temperature in the upper layers reaches 55 °C and switches off when the temperature drops to 30 °C.

In case that the transformer oil is cooled by the air bubbles, the air flow rate is 1.3 l/min. The oil was heated to a temperature of 70 °C and the time for cooling the oil was recorded without bubbling with sulfur hexafluoride. The oil temperature has dropped to a room temperature of 25 °C in 100 min.

During the cooling every 5 min, the temperature was fixed. A graph of the natural cooling of the oil was constructed based on the data obtained (Fig. 4).

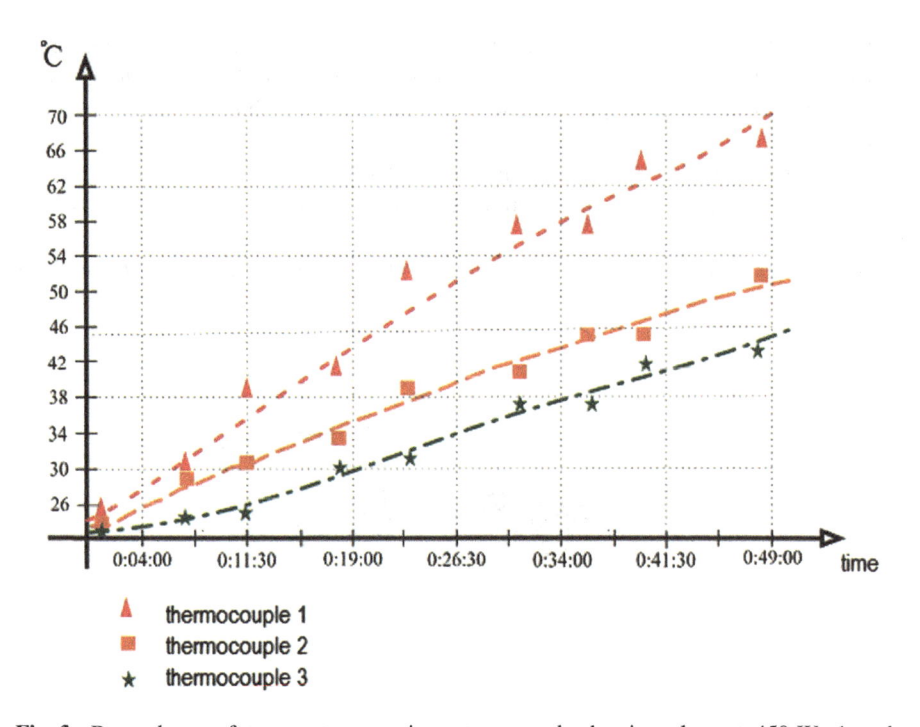

Fig. 3. Dependence of temperature on time at power the heating element 450 W: 1 – the temperature near the heater, 2 – at a distance of 0.11 m from the heater, 3 – 0,21 m from the heater.

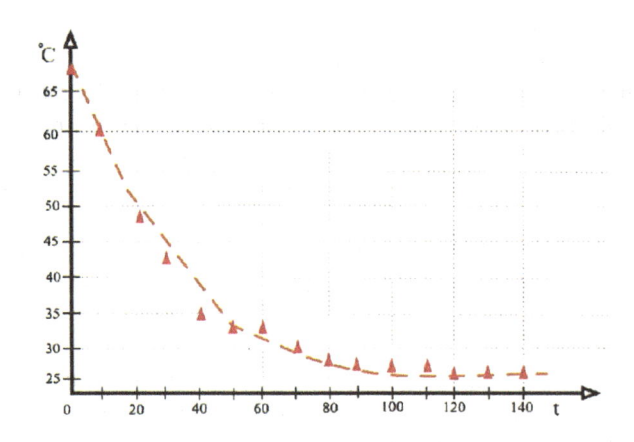

Fig. 4. Graph of the natural oil cooling.

In the second experiment, a compressor is used to bubble the oil with air bubbles. After heating the oil to 70 °C, the compressor was turned on (Fig. 5).

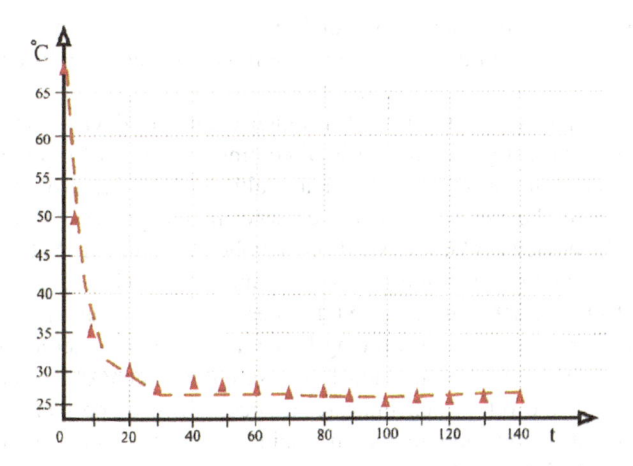

Fig. 5. Graph of cooling oil when the bubbling.

This experiment showed that cooling oil with bubbles is more effective than without them. Comparing graphs Figs. 4 and 5, it can be seen that a transformer oil with a volume of 0.018 m^3 with a temperature of 70 °C is reduced to 25 °C in 30 min with bubbling, while without bubbling it cools down in 100 min.

The heat transfer parameter was determined experimentally, the value of which on the oil surface was $h = 0.02 \pm 0.003$ M^{-1}, where $h = \alpha/\lambda$ (here λ is the effective coefficient of thermal conductivity of the medium in the tank, α is the heat transfer coefficient through the medium transformer oil/the metal case of the transformer/air). The data are listed in Table 1.

The obtained experimental data testify to the operability of the installation. The coefficient of molecular thermal diffusivity is determined, which is $a \sim 10 - 4$ m^2/s, where $a = \frac{\lambda}{c \cdot \rho}$ (here c is the specific heat of transformer oil, ρ is density of transformer oil). These coefficients are determined using the Teplo.exe program.

Table 1. Results of experiment.

The location of the thermocouple	Thermocouple 1	Thermocouple 2	Thermocouple 3
The thermal diffusivity	$2.5 \cdot 10^{-4}$ m^2/s		
The heat transfer coefficient	0.02 m^{-1}	0.005 m^{-1}	0.0045 m^{-1}

As it can be seen from the table the heat transfer coefficient reduces with depth due to a decrease in the temperature gradient.

Based on the results of the experiments, it can be said that the obtained data confirm the cooling efficiency by using bubbles. The disadvantage of air bubbling is that air inherently has the property of penetrating into the molecules of transformer oil, thereby worsening its quality. When SF6 bubbling is used, this phenomenon is not observed.

SF6 gas is much more efficient as a dielectric than air. Due to the physical properties of the SF6 gas, the oil transformer may be more compact than its counterparts in air bubbling.

The system requires the presence of a digital automatic device that regulates the consumption of electrical gas on the basis of continuously input from the temperature and pressure measuring instruments, and a multilateral analysis of the gas in the working volume of the device is also necessary. In the process of solving it was decided to use the device HYDROCAL 1008 (supplied by company MTE). This device serves for online monitoring of the state of oil transformers [10].

The HYDROCAL 1008 is designed for permanent installation on an oil-filled transformer and serves as an early warning function for the pre-emergency state of the transformers. The instruments continuously measure the content of the most important gases and water in the oil, which indicate the occurrence of problems in the transformer. Instruments allow you to record readings, set individual alarm thresholds, connect external devices and have various communication interfaces.

The device has programmable control outputs. Due to the fact that the device does not require regular replacement of consumable parts, and also does not require periodic maintenance, the system is maintenance-free.

The HYDROCAL 1008 is a permanently installed multilateral gas analysis system with transformer monitoring functions. It allows individual measurements of moisture and key gases such as hydrogen (H_2), carbon monoxide (CO), carbon dioxide (CO_2), methane (CH_4), acetylene (C_2H_2), ethylene (C_2H_4) and ethane (C_2H_6) dissolved in transformer oil [11].

4 Conclusion

The results of the analysis of oil transformer failures make it possible to conclude that it is necessary to increase the efficiency of their cooling system, since the main reason for the failure of the transformers is overheating. Experimental studies show that bubbling sulfur hexafluoride in the oil allows more efficient cooling of the oil, and accordingly the active part of the oil transformers. This method of cooling power oil transformers is a modern, more efficient solution for use in enterprises and also it is a new direction in the development of transformer engineering. The installation is effective especially in a hot period of time when the oil cooling system cannot cope with the heat removal.

References

1. Krivokoneva, O.O., Kudoyarov, R.I., Mamleyev, E.Y., et al.: Extension of life of oil-immersed transformers with long term operation. Herald SUSU Series Energy **17**(3), 60–66 (2017)
2. Nigmatuli, R.I., Filippov, A.I., Khismatullin, A.S.: Transcillatory heat transfer in a liquid with gas bubbles. Thermophys. Aeromech. **19**(4), 589–606 (2012)
3. Khismatullin, A.S., Vakhitov, A.H., Feoktistov, A.A.: Cooling System of transformer oil on the basis of transcillatory transfer have been heat, energy security and energy efficiency **4**, 43–46 (2016)

4. Khismatullin, A.S., Khismatullin, A.G., Kamalov, A.R.: Study of heat transfer in industrial power transformers, gas-insulated cooling. Ecol. Syst. Devices **2**, 29–33 (2017)
5. Bashirov, M.G., Gribovsky, G.N., Gallyamov, R.W., et al.: Recommendations for improving the reliability of power supply of the industrial site line production Department of main gas pipelines. Electr. Complexes Syst. **2**(31), 23–26 (2016)
6. Bashirov, M.G., Khismatullin, A.S., Pereverzev, A.I.: Cooling oil of the transformer. Patent for useful mode 167206, 08 Dec 2016
7. Khismatullin, A.S., Gareev, I.M.: Calculation of three-dimensional temperature field in oil transformers, gas cooled. Fundam. Study **10**(3), 534–537 (2015)
8. Khismatullin, A.S., Sourakov, M.R., Shintemirov, A.: Improving the cooling oil of power transformers by bubbling bubbles of sulfur hexafluoride. Eng. Phys. **6**, 27–31 (2017)
9. Maltsev, I.G., Pavlov, M.I.: Effects of sulfur hexafluoride on the environment and human health, condition and prospects of development of the electricity and heat technologies. In: Materials of International Scientific-Technical Conference Dedicated to the 175th Anniversary from the Birthday N. N. Benardos, pp. 137–140 (2017)
10. Bashirov, M.G., Khismatullin, A.S.: Electrotechnological installations and electrotechnical systems of oil and gas complex. Chronicles of the united fund of electronic resources of science and education, vol. 12(79), p. 113 (2015)
11. Mandrosov, V.V.: Instrumentation for online condition monitoring of oil-filled transformers. Autom. IT Energy **10**(63), 18–20 (2014)

Assessment Method of Technical Condition of Small Refrigerating Machine Using Programmable Controller

M. A. Lemeshko, S. R. Urunov[✉], and A. V. Kozhemyachenko

Don State Technical University in Shakhty,
147 Shevchenko Street, Shakhty 346500, Russian Federation
salavat4you@gmail.com

Abstract. The article provides the prerequisites for creating a method for assessing the technical condition of a small refrigeration machine using a programmable controller, as well as a description of a programmable controller algorithm that allows monitoring the current performance of the small refrigeration machine in more detail, based on the existing method for detecting changes in the specific energy consumption of a small refrigeration machine by measuring the performance of a small refrigeration machine before its operation and after it is regulated Foot period of its operation. An important point in this approach to assessing the technical condition of a small refrigeration machine is the procedure for creating identical measurement conditions before and after the period of operation. For example, it is necessary to obtain data for comparison at the same ambient temperature. The article draws conclusions and recommendations on creating a programmable controller with a given algorithm for assessing the technical condition of a small refrigeration machine.

Keywords: Small refrigerating machine · Controller · Heat and power performance · Operation · Algorithm

1 Introduction

During operation, small compression-type refrigeration equipment, including household refrigerators, after a certain period of operation, they can manifest deviations from thermal power regulated performance specifications (nameplate values).

These abnormalities may be due to changes in operating conditions of the refrigerating machine, or malfunctions or temporary changes in the subsystems of the refrigerator [1].

One of the urgent tasks to improve small refrigerating machine, which include household refrigerating appliances is a challenge to ensure timely detection of deviations heat power indicators of the refrigerator operated by the regulated values. For example, the detection and display of increase in the share (average daily), low power consumption of the chiller, the detection of the temperature level changes in the cooling chambers, or establishing allowable norms fluctuations of the temperature level.

A. A. Radionov and A. S. Karandaev (Eds.): RusAutoCon 2019, LNEE 641, pp. 622–629, 2020.
https://doi.org/10.1007/978-3-030-39225-3_68

When upgrading refrigerator solve the problem to reduce the operating costs for a cold [2], and then also considers the stability characteristics of small thermal power refrigerator.

Most household refrigerators did not have the means to control a change in their energy efficiency during operation. In some models of refrigerators indicated the actual temperature in the chamber or the cooling chamber, but not compared with its desired value of temperature [3]. The deviation from the normalized values of temperatures can determine the user visually, but such control and the comparison is not carried out in real conditions of small refrigerating machine. At the same time, the probability of such rejections due to incorrect operation of the natural environment and temporal variations of the refrigeration unit components and the cooling chamber [4]. The number of household refrigerating appliances used in households, hotels and restaurants, on public and fast food very much.

The total power consumption of these devices such as a city with a population of 200 thousand people is about 100 thousand kWh/day, or about 3 million kWh per month.

Possible operating loss [5] may thus be up to 10% of the power consumption, i.e. 0.3 million KWh per month. For the conditions of the cost of electricity - 3 rubles/1 kWh losses in monetary terms could reach 0.9 million rubles per month.

For a city with a population of 1 million people loss can be about 4.5 mln rubles per month. This is a substantial loss to be excluded or minimized. To achieve such a goal can be achieved by monitoring the technical condition [6] during the operation of the refrigerator and in a timely manner to exclude the causes of increased power consumption [7].

2   Materials and Methods

One of the methods to detect specific changes (daily, annual) consumption small refrigerating machine (in comparison with the passport data) is a method of measuring the performance of the refrigerator before the start of its operation and, after a specified period of its operation [8]. An important point in this approach to the assessment of the technical state of refrigerator is the procedure for creating identical conditions of measurement before and after the period of operation. For example, to obtain data necessary for comparison with the same ambient temperature [9].

Existing methods of determining the technical condition of the small refrigerating machine to the technical operation of the period characterized by the use of bulky measuring instruments, test duration, the use of scale pressure gauge, manual measurement processes, which leads to a relatively high measurement error. They typically do not account for the influence of ambient temperature [10].

Considering the above, it is currently of particular interest is the solution of the autonomy of the process of determining the technical condition of the small refrigerating machine to increase the reliability of evaluation of its technical condition while reducing the time spent on diagnosis.

Assessment method of technical condition of a small refrigerating machine includes the steps of placing a temperature sensor in the test chamber of the refrigerator and ambient temperature sensor, connect your device against the clock operation of the compressor and use interface for data collection and processing, the measurement process, the calculation process and an indication of the technical condition of the small refrigerating machine. Essence the method explained in the charts of the cooling process shown in Fig. 1.

As shown in the graphs (Fig. 1) temperature-time operation of the compressor at different ambient temperatures can be described by a linear function, and the cooling rate (line angle) is the same at different ambient temperatures. Thus, by measuring the cooling rate at various ambient temperatures can be judged on the technical condition of the refrigerator. The cooling rate depends on the cooling capacity of the cooling unit, capacity refrigerated chamber, the volume of the cooled product and the technical condition of all subsystems of the refrigerator.

In the diagnosis or in determining the technical condition of the refrigerating machine studied chamber is not loaded products. Thus, for every new volume of the chamber (the reference) and test speed cooling refrigerator characterizes technical state of all its subsystems collectively, and from the deviation of the reference rate of cooling of the refrigerator cooling rate in the test is evaluated refrigerator its technical condition. Modifications of this method of determining the technical condition of the small refrigerating machine allow testing of various subsystems: the refrigeration unit, hermetic motor-compressor, filter drier and other elements of the subsystem.

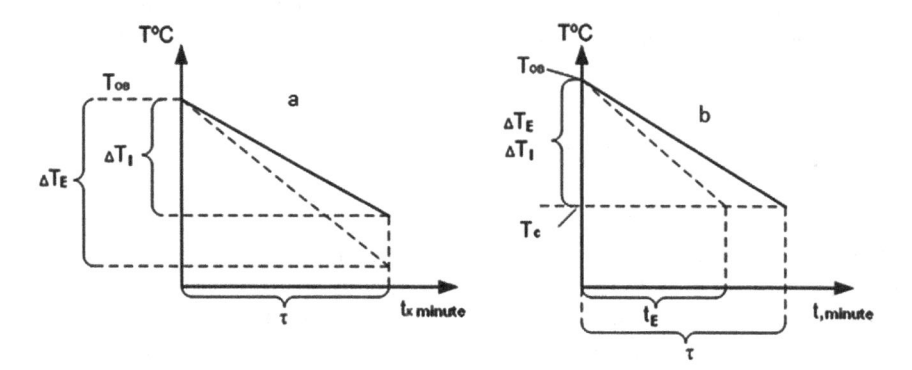

Fig. 1. (a) Depending on the temperature of the reference and the test of the refrigerator in the cold chamber of the compressor running time t_k for a fixed period of time τ; (b) the graph of temperature change of the reference and the test of the refrigerator to the set temperature cooling T_c.

In each case, the conclusion of the technical condition of the test subsystem is its integral evaluation: for the respective actual speed and the reference cooling.

The cooling rate is determined by the expression:

$$V_c = \frac{\Delta T}{\tau},\tag{1}$$

where ΔT – range of temperature change in the cooling compartment for a set time τ of the refrigeration unit. For example:

$$\Delta T = T_{OB} - 0 \,°C, \tag{2}$$

Where T_{OB} – ambient temperature. For reference refrigeration unit, the cooling rate is equal to:

$$V_E = \frac{\Delta T_E}{\tau_E}, \tag{3}$$

To test the refrigerator, the cooling rate is equal to:

$$V_I = \frac{\Delta T_I}{\tau}, \tag{4}$$

The difference between the cooling rates:

$$\Delta V_c = V_E - V_I = \frac{\Delta T_E}{\tau_E} - \frac{\Delta T_I}{\tau_I}, \tag{5}$$

ΔV_c – characterizes the degree of conformity of test reference refrigeration unit.

The magnitude of this deviation is determined by the technical condition of the small refrigeration machine. You can also use the corresponding coefficients of the test of the refrigerator:

$$K_{V_C} = \frac{\Delta V_C}{V_E} \cdot 100\% \tag{6}$$

where K_{V_C} – the compliance rate of test reference refrigeration unit on the cooling rate.

Technical condition of refrigeration unit may also be determined (using the concept of a cooling rate) time of the compressor for lowering the temperature from the initial temperature value (for example, ambient temperature) to the target (for example up to 0 °C). This is measured and compared to the work of the compressor for the reference τ_E of refrigeration unit and τ_I of the test refrigeration unit.

The convergence or divergence of the measurement results is possible to determine the absolute units $\Delta\tau_C = \tau_E - \tau_I$ or relative. In this case it is advisable to introduce the concept of the coefficient of compliance test reference cooler on the cooling rate:

$$K = \tau_E - \frac{\tau_I}{\tau_E} \, m \tag{7}$$

The technical condition of the small refrigeration machine can also be determined (using the concept of cooling rate) obtained by comparing the temperature of the reference and test a refrigerator for the same period of time of the compressor ($\tau_E = \tau_I$).

That is to measure:

$$\Delta T = \Delta T_E - \Delta T_I \tag{8}$$

If while

$$\begin{cases} \Delta T_E = T_{OB} - T_1 \\ \Delta T_I = T_{OB} - T_2 \end{cases}, \Rightarrow \Delta T_{OB} = T_1 - T_2 \tag{9}$$

Then, the technical condition of the refrigerator, you can determine the coefficient of deviation of the actual state of the refrigeration unit by the reference refrigeration unit:

$$K_{\Delta T} = \frac{\Delta T}{T_1} \cdot 100\% \tag{10}$$

This value characterizes the compliance or deviation of the technical condition of the reference and test a refrigerator at $\tau_E = \tau_I$. For the described embodiment of determination of cooling rates measured ΔT_E and ΔT_I.

Determination of technical condition for the above two methods (for $\Delta \tau_{OB}$ and ΔT_{OB}) is a variant of the method of evaluation of the technical state of the refrigerator on the cooling rate.

The proposed method can simplify the measurement process and eliminate the presence of the operator when removing the characteristics of the chiller.

In the method to eliminate the human factor on the result of determination of technical condition of the object. In addition, the proposed method simplified the process of determining the technical condition of the refrigerator due to the lack of appropriate control valves and monitors the performance of devices.

According to the degree of divergence of the measured and reference indicators can be defined evaluate the technical condition of the refrigerating machine as a whole.

The proposed method can be applied in the study of refrigerating machines such as the study of various refrigerants as well as for diagnosing the operational location of the refrigerator at operation using his portable PC.

Another option is the evaluation of the technical state of control in specific energy consumption of small refrigerating machine.

3 Results

Below is a list algorithm stage of a method of evaluating changes in small refrigerating machine for energy efficiency during the operation [11]. This algorithm can be implemented under the control of a programmable microcontroller, incorporated into the structure of a refrigerator control unit. The process of determining the deviation of the compression in the refrigerator comprises the following steps:

- Standard test conditions are created before operating. It creates the conditions in which the values are determined by the specific model passport refrigerator plant supplier
- The measurement indicator of the average daily energy intensity for a limited period of time using the appliance in stable environments. For example, within 1–3 days at a stable ambient temperature is measured by the average daily power consumption, which is taken as the base value for comparison with the index over a long period of time of operation
- Measurement results are saved in the first microcontroller memory cell as the reference value of specific energy consumption. In the future, apparently, it is useful when upgrading small refrigerating machines in the refrigerator control unit to enable this record as an electronic passport specific refrigerator models (for comparison with this indicator after a certain period of operation)
- Activated timer operation with a time limit of the refrigerator. It sets the operating period at the end of which, another measurement of the refrigerator will be performed. This period can vary from one month to several years, and established a specialist in diagnosing refrigerating machine, taking into account the probability of failure or the probability of occurrence of deviations
- Timer signal is generated on the expiry of the deadline of operation, after which user must perform the measurement of the test indicator
- Create a standard test conditions identical test conditions refrigerating machine before its operation
- Performs the first control measurement indicator of the average daily energy intensity for a limited period of time using the appliance in stable environments
- Compares the reference values of the specific energy usage with the value of this parameter after the first period of operation of the test
- The results of the comparison are displayed. If there are significant deviations possible inclusion of an alarm. In the absence of abnormalities include the current time timer of the refrigerator until the second control measurement of test parameter
- Next, the algorithm of action cycle is repeated many times

Thus, regulated by the operating period, such as a year, it may be accomplished more, for example 12, the average daily consumption of control measurements. Or over 10 years of operation to perform 10 measurements. Identifying deviations and timely maintenance of the refrigerating machine will provide increased resorce his work [12]. The microcontroller during the regulated period of operation will enable to organize and compile information on the trends of the test index, and in the event of significant deviations will produce information about the need for maintenance or repair of the refrigerator.

The method makes it possible not only to continuously measure and record the value of the temperature in the chambers, measure and record the specific energy consumption (daily, monthly, yearly average), identify deviations, using a sensor temperature of the ambient air and the sensor door opening chambers of the refrigerator to perform a detailed analysis of the effect of operating conditions on energy consumption of the refrigerator to perform long-term studies with a view to modernizing the small refrigerating machine [13].

4 Discussion and Conclusion

Considered method of assessing the technical condition of the small refrigerating machine using a programmable controller allows to monitor the current performance of the refrigerator works in more detail. The method makes it possible not only to continuously measure and record the value of the temperature in the chambers, measure and record the specific energy consumption (daily, monthly, yearly average), identify deviations nose using ambient air temperature sensor and the sensor opening the doors of the refrigerator chambers to carry out a detailed analysis of the effect of operating conditions on the refrigerator power consumption to perform long-term studies with a view to modernizing the small refrigerating machines.

Using the programmable controller will also evaluate the technical condition of the compression refrigerator for stability factor of working time. The theoretical justification for applying the method of evaluation of the technical condition of the small refrigerating machine using a programmable controller is given in [14].

Analysis of factors affecting the thermal power of the working characteristics of the refrigerator revealed the main reasons for their change for a certain period of operation.

Establishing the state of the refrigerator before failure condition subsystems or integrated assessment of changes in the refrigerator subsystems allows time to service a small refrigerating machine, and repair or replace any element or sub-block if necessary. This increases the operational reliability of the refrigeration machine.

Experimental studies on the specialized stand were studied questions clogged filter driers small refrigerating machines, as well as the impact of fouling on heat and power performance of the refrigeration unit [15]. As conducted experimental studies have revealed, during continuous operation of the compression refrigeration unit, there is clogging the filter drier, which leads to an increase in hydraulic resistance to the refrigerant flow and an increase in specific energy consumption by 5–15%. This also found to influence the rate of ambient temperature.

References

1. Rozhentsev, A.: Refrigerating machine operating characteristics under various mixed refrigerant mass charges. Int. J. Refrig. **31**(7), 1145–1155 (2008). https://doi.org/10.1016/j.ijrefrig.2008.03.001
2. Jacobson, V.B.: Small Refrigerating Machines. Food industry, Moscow (1977)
3. Lawton, R.: How refrigerated containers work. In: Reference Module in Food Science (2016). https://doi.org/10.1016/B978-0-08-100596-5.03159-0
4. James, S.J.: Food refrigeration and thermal processing at Langford: 32 years of research. Food Bioprod. Process. **77**(4), 261–280 (1999). https://doi.org/10.1205/096030899532556
5. Kozhemyachenko, A.V., Fomin, Yu.G., Lemeshko, M.A., et al.: Theoretical determination of diagnostic parameters of technical condition of the chokes compression refrigerators. Proc. High. Educ. Inst. Technol. Text. Ind. **2**(362), 173–178
6. Mansouri, R., Bourouis, M., Bellagi, A.: Steady state investigations of a commercial diffusion-absorption refrigerator: experimental study and numerical simulations. Appl. Therm. Eng. **129**, 725–734 (2017). https://doi.org/10.1016/j.applthermaleng

7. Carrilho, D.G., Silva, P.D., Pires, L.C., et al.: Quantification of the thermal resistance variation in evaporators surface due to ice formation. Energy procedia **142**, 4151–4156 (2017). https://doi.org/10.1016/j.egypro.2017.12.339
8. Yuan, W., Yang, B., Yang, Y., et al.: Development and experimental study of the characteristics of a prototype miniature vapor compression refrigerator. Appl. Energy **143**, 47–57 (2015). https://doi.org/10.1016/j.apenergy.2015.01.001
9. Alekperov, I.D.: Development of rational cooling systems for a sealed unit of a small refrigerating machine. Moscow (2001)
10. Dowing, R.O.: Refrigerants: service pointers. Refrig. Serv. Contract. **39**(10), 40–41 (1971)
11. Lemeshko, M.A., Kozhemyachenko, A.V., Urunov, S.R.: Algorithm of monitoring of the technical condition of compression refrigerator. In: Innovations in Technologies of Cultivation of Agricultural Crops Materials of the International Scientific and Practical Conference, pp. 360–364 (2015)
12. Sullivan, G.P., Pugh, R., Melendez, A.P.: Operations and maintenance. Best practices. A guide to achieving operational efficiency (2016). https://www.pnnl.gov/main/publications/external/technical_reports/PNNL-13890.pdf. Accessed 12 Oct 2017
13. Petrosov, S.P., Lemeshko, M.A., Kozhemyachenko, A.V.: Monitoring of the energy performance of domestic refrigerators in operation period. Tech. Technol. Serv. Probl. **4**(30), 20–25 (2014)
14. Lemeshko, M.A.: The method for determining the technical condition of the compression refrigerator according to mode of operation of the compressor. In: Innovations in Technologies of Cultivation of Agricultural Crops, pp. 339–344 (2015)
15. Kozhemyachenko, A.V., Petrosov, S.P., Lemeshko, M.A., et al.: Theoretical principles of the technical state of household refrigerating appliances during their technical operation. Proc. High. Educ. Inst. North-Caucasian Region. Series: Engineering **3**(172), 107–109 (2013)

Research and Mathematical Modeling of the Thermal and Power Performance of Resistance Furnaces at Metallurgical Enterprises

R. V. Klyuev[(✉)], I. I. Bosikov, and A. D. Alborov

North Caucasian Institute of Mining and Metallurgy, 44 Nikolaeva Street,
Vladikavkaz 362021, Russian Federation
kluev-roman@rambler.ru

Abstract. Metallurgical industry, which is characterized by high power consumption, is the most energy-intensive industry for the Russian Federation. Primary role in the optimization of electricity consumption by a powerful technological equipment is given to development and introduction of a complex of energy-saving measures, obtained during the study of energy consumption parameters on the basis of power inspection (power audit) in large industrial enterprises. This allows to track the entire process of electricity production and to explore the effect of various technical and economic factors on it. The importance of the factors in the whole cycle can be estimated in the mathematical modeling of power consumption by means of active and passive experiments, thermal and energy analysis, the use of consumer-regulator (CR), etc. The article presents the results obtained during the energy audit at the enterprise of ferrous metallurgy for the production of hard alloys. The comprehensive methodology for the investigation of thermal and power characteristics for consumers of non-ferrous metallurgy for the production of hard alloys, which allows to obtain statistical estimates of power consumption in static and dynamic work conditions of the enterprise have been developed. Minimized energy cost component in production costs by improving the quantitative objective function: the use of consumer-regulator (CR) of electricity in the peak areas of the day; reduction of specific energy consumption in stationary and dynamic operating conditions of the process equipment. The mathematical model for calculating and forecasting of specific consumption of electricity in the range of technological equipment protection was obtained.

Keywords: Temperature · Power · Electricity · Resistance furnace · Mathematical model

1 Introduction

The most effective way for optimization of the electricity consumption in enterprises is the implementation of energy saving measures that can be obtained in the course of power audit [1–16]. A comprehensive power audit of main technological processes of

© Springer Nature Switzerland AG 2020
A. A. Radionov and A. S. Karandaev (Eds.): RusAutoCon 2019, LNEE 641, pp. 630–636, 2020.
https://doi.org/10.1007/978-3-030-39225-3_69

production was conducted on the example of large industrial enterprise of non-ferrous metallurgy "Pobedit" [17–19].

According to the results of cluster analysis of expert assessments were identified important factors that have the greatest impact on energy consumption during the manufacture of hard metal. Ranked analysis of factors conducted on the basis of H-distribution correlates well with peer review. A comprehensive study of energy characteristics of these consumers can be seen as the first stage of the energy audit, when experimentally studied the work of the process equipment in a stationary mode.

Experimental investigation of thermal power characteristics of the production equipment at the operating plant of JSC "Pobedit" was performed for the first time [17, 18]. During the experimental studies the following equipment was used: comprehensive control device PKK-57, power consumption analyzer AR5 and energy tester PKE. The developed methodology provides measurement of electrical quantities: current, voltage, power, energy consumption, computer processing of measured values.

It should be noted, that in the process of production of hard alloys the major technological (non-electrical) value is the temperature of working medium in a rotating chamber furnaces (RCF) and tube furnaces (TF) in the modes of heating and cooling, so when measuring power consumption was simultaneously measured temperature and the individual heat zones of the furnace. During the process of energy survey (power audit) the company determined the following consumers as the most effective CR that have a significant impact on reducing full loads in the peak of a day and specific energy consumption: rotating chamber resistance furnaces (type RCF), tube resistance furnace (type TF), plating baths, special electrothermal equipment (furnaces of sintering, carbidization, slotted, calcining and welding machines).

Based on the results of experimental studies formed a representative sample of the following parameters: current $\{I\}$, voltage $\{U\}$, active power $\{P\}$ energy $\{W\}$, fluid temperature $\{T\}$, over a time period t ($t = 0.5$–24 h). Research results of thermal and energy characteristics of RCF furnaces are listed below.

2 Research of Thermal and Power Characteristics of Resistance Furnaces

The following diagrams of changes in heat energy characteristics in time were set up on the basis of data obtained by the results of power audits of RCF furnaces: power $P = f(t)$, electric energy consumption $W = f(t)$ (Fig. 1) and the temperature of heating zones $T = f(t)$ (Fig. 2).

From the relationship $T_i = f(t)$ (Fig. 2) it follows, that the minimum temperature of 560–574 °C corresponds to the first (T_1) and fourth (T_4) thermal zones. For heating zones 2 and 3 the temperature is in the range 580–595 °C. In the dynamic regime of heating of RCF values of power $P = f(t)$, power consumption $W = f(t)$ and the

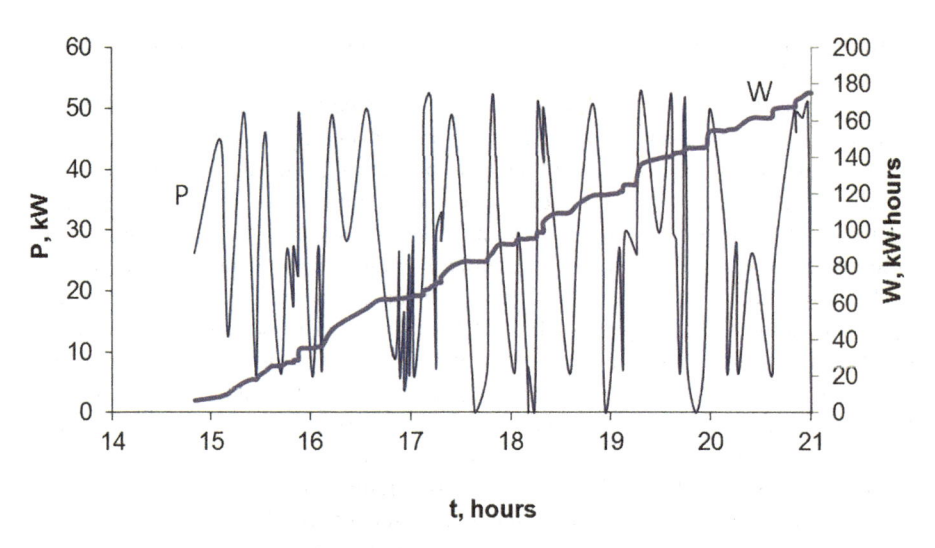

Fig. 1. Diagram of power and electricity stationary (static) furnace operation.

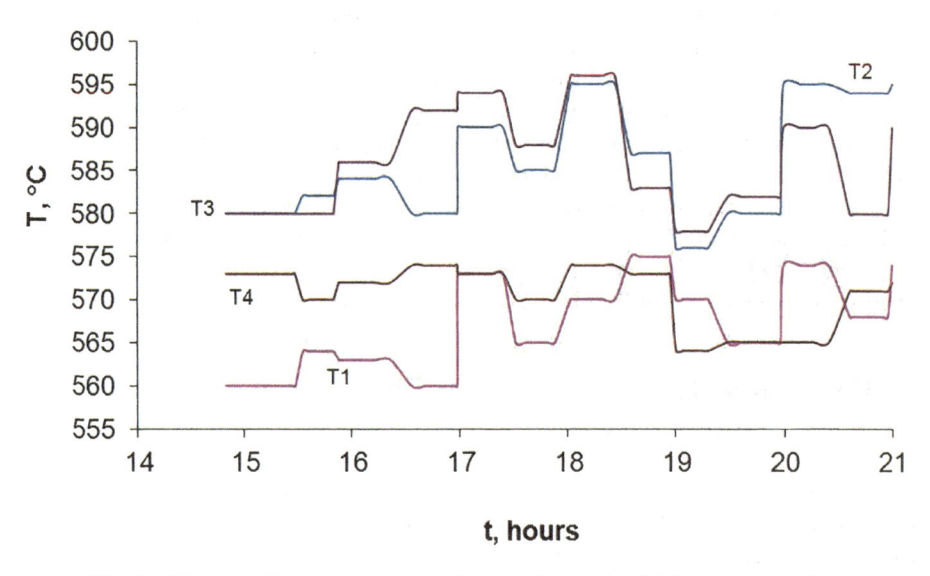

Fig. 2. Diagram of temperature zones in a stationary (static) furnace operation.

temperature $T_i = f(t)$ (with built for it by the least squares method (LS) equations of the trend, the coefficients of determination (R^2) are in the range: $R^2 = 0,95 - 0,98$) on 4 zones presented on the diagram (Fig. 3).

For the average power of the furnace was as follows: $P = 25.6$ kW, unbiased assessment standard: SD (data) = 16.4 kW.

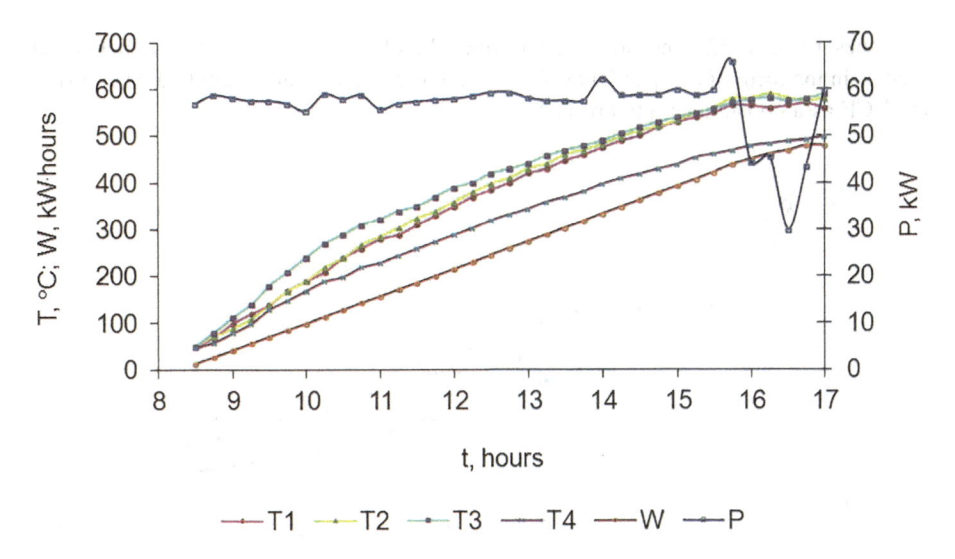

Fig. 3. Diagram of power change $P = f(t)$, the electric power $W = f(t)$ and temperatures of individual zones of $T_i = f(t)$ in RCF furnace during its heating period.

3 Construction of Mathematical Models of Thermal Characteristics Furnaces

Mathematical models in the form of the equations of trend line changes depending on the temperature of furnace in time $T_i = f(t)$ for individual zones during the heating of the RCF and the coefficients of determinism: $T_1 = 63.35t - 435.94$ ($R^2 = 0.9664$); $T_2 = 65.23t - 451.96$ ($R^2 = 0.9657$); $T_3 = 60.5t - 373.64$ ($R^2 = 0.9461$); $T_4 = 53.83t - 372.14$ ($R^2 = 0.9803$); $T_{average} = 60.73t - 408.42$ ($R^2 = 0.9661$). Mathematical model of energy consumption depending on the average temperature zones: $W = 0.914T_{average} - 75.06$ ($R^2 = 0.972$).

Period of furnace heating temperature from 500 to 5700 is about 8 h, with $P_{average} = 25.6$ kW, electricity consumption during this time is 480 kWh. Load on phases is uneven. The current in the phase C ($I_{average} = 120,8$ A) exceeds two times the currents in the phases A and B ($I_{average} = 120,8$ A). The minimum rate of increase of temperature corresponds to zones 1 and 4 (T1, T4).

Mathematical models in the form of trend equations $y = f(x)$ for the dependence of furnace temperature in time $T_i = f(t)$ during the cooling of RCF and the coefficients of determination: $T_1 = 1214,6e^{-0,1191t}$ ($R^2 = 0,9706$); $T_2 = 1221,8e^{-0,1098t}$ ($R = 0,9921$); $T_3 = 1275,6e^{-0,1153t}$ ($R^2 = 0,9857$); $T_4 = 1108,4e^{-0,1009t}$ ($R^2 = 0,9882$).

Obtained dependence of the temperature in 4 zones $T_i = f(t)$ (Fig. 4) and built trend equations using the least square method, the coefficients of determination are in the range $R^2 = 097 - 0,99$.

Over a period of 6 h after shutdown, the lower temperature in the zones was: for zones 1 and 4: from 525 to 248 °C; for zones 2 and 3: from 550 to 271 °C. Over a period of 3 h (time of peak load in the power system) residual temperature zones was:

for zones 1 and 4: 320 and 365 °C; for zones 2 and 3 365 and 352 °C. Mathematical models in the form of a dependency $T = a \cdot t + b$ и $T = a \cdot e^{-t}$ can be applied when using the RCF as a consumer regulator [20].

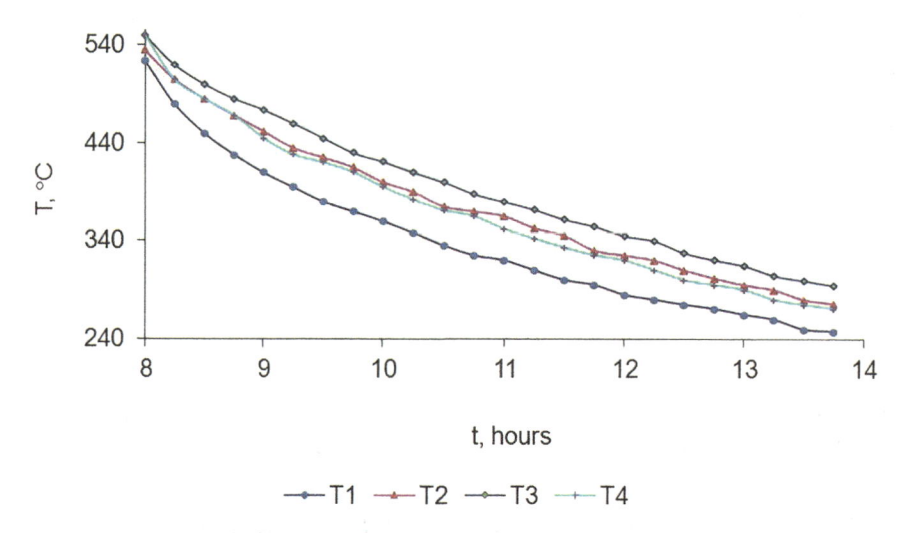

Fig. 4. Diagram of individual zones of RCF temperature during its cooling.

4 Mathematical Model for Calculation of Specific Electricity Consumption of Furnaces on the Basis of the Full Factorial Experiment (FFE)

For the conduction of FFE as technological factors that have the most significant effect on the specific energy consumption, significant expert assessment of the following factors were taken: x_1 - mass of ammonium paramolybdate loaded into RCF furnace, γ, kg; x_2 - amount of hydrogen, V, m³ [18]. The algorithm for obtaining the mathematical model based on the FFE, type $N = 2^2$ is given above. Table 1 shows the planning matrix of FFE and results of the experiment [21–24].

Table 1. Planning matrix of FFE type $N = 2^2$ and results of the experiment.

№	Coded scale			Natural scale		Specific energy consumption $Wspec$, kWt·hour/kg			The average specific energy consumption $Wspec$, average, kWt·hour/kg
	x_0	x_1	x_2	γ, kg	V, m³	1	2	3	
1	1	−1	−1	110	8	0.91	0.95	0.86	0.91
2	1	1	−1	130	8	0.78	0.81	0.74	0.78
3	1	−1	1	110	10	1	1.05	0.98	1.01
4	1	1	1	130	10	0.84	0.87	0.83	0.85
Σ		0	0						3.55

After conducting a FFE a mathematical model based on $W_{spec} = f(\gamma, V)$ has the form:

$$W_{spec} = 1,375-7,25 \cdot 10^{-3}\gamma + 4,25 \cdot 10^{-2}V, \text{ кWt·hour/kg.}$$

The analysis of the mathematical model (1) shows that the optimization of specific consumption of electricity $W_{spec} \to$ min is possible, within the limits of technological protection, due to the higher utilisation of furnaces with paramolybdate ammonium and reduce the amount of consumed hydrogen, which is a restorative environment in the production of the dioxide of molybdenum and tungsten.

5   Conclusion

To improve the efficiency of electricity consumption in the enterprise of non-ferrous metallurgy a comprehensive program for calculation and forecasting of electricity consumption has been developed and implemented. This program includes: methods of statistical processing and analysis of experimental data, allowing to estimate a static power consumption mode; the method of constructing mathematical models of specific energy consumption based on full factorial experiment (FFE).

Type $N = 2^2$, allowing to estimate the dynamic power consumption mode in online and, subject to the requirements of technological protection of processes of production, optimize and predict energy consumption for individual units and the enterprise as a whole; the use CR electricity, allowing to adjust the hourly electricity consumption in semi-peak and peak zones and the acceptable range of normal technological process. A comprehensive program is included in the database of the automated system of commercial electricity metering (ASCEM) and in the dispatching service in the on-line mode, that provides the greatest efficiency of CROSS in adjusting the daily load curve.

Acknowledgements. The results were reflected in the Grant of the President of the Russian Federation for support of young scientists: MK-1324.2007.8 on "Research and development of mathematical models of power quality in the non-ferrous metallurgy enterprises".

References

1. Fresner, J., Morea, F., Krenn, C., et al.: Energy efficiency in small and medium enterprises: lessons learned from 280 energy audits across Europe. J. Clean. Prod. **142**(4), 1650–1660 (2017)
2. Tso, G.K.F., Liu, F., Ke, L.: The influence factor analysis of comprehensive energy consumption in manufacturing enterprises. Procedia Comput. Sci. **17**, 752–758 (2013)
3. Cagno, E., Trianni, A.: Evaluating the barriers to specific industrial energy efficiency measures: an exploratory study in small and medium-sized enterprises. J. Clean. Prod. **82**, 70–83 (2014)
4. Anisimova, T.: Analysis of the reasons of the low interest of Russian enterprises in applying the energy management system. Procedia Econ. Financ. **23**, 111–117 (2015)

5. Kluczek, A., Olszewski, P.: Energy audits in industrial processes. J. Clean. Prod. **142**(4), 3437–3453 (2017)
6. Henriques, J., Catarino, J.: Motivating towards energy efficiency in small and medium enterprises. J. Clean. Prod. **139**, 42–50 (2016)
7. Andersson, E., Arfwidsson, O., Bergstrand, V., et al.: A study of the comparability of energy audit program evaluations. J. Clean. Prod. **142**(4), 2133–2139 (2017)
8. Makridou, G., Andriosopoulos, K., Doumpos, M., et al.: Measuring the efficiency of energy-intensive industries across European countries. Energy Policy **88**, 573–583 (2016)
9. Bosikov, I.I., Klyuev, R.V.: System analysis methods for natural and industrial system of mining and metallurgical complex. IPC IP Copanova A. Ju. Publ., Vladikavkaz (2015)
10. Muhamad, N.A., Phing, L.Y., Arief, Y.Z.: Pre-study on potential electrical energy savings in Malaysia dairy manufacturing industry of medium enterprises. Energy Procedia **68**, 205–218 (2015)
11. Oseni, M.O., Pollitt, M.G.: A firm-level analysis of outage loss differentials and self-generation: evidence from African business enterprises. Energy Econ. **52**(B), 277–286 (2015)
12. Kostka, G., Moslener, U., Andreas, J.: Barriers to increasing energy efficiency: evidence from small-and medium-sized enterprises in China. J. Clean. Prod. **57**, 59–68 (2013)
13. Li, M.J., Tao, W.Q.: Review of methodologies and polices for evaluation of energy efficiency in high energy-consuming industry. Appl. Energy **187**, 203–215 (2017)
14. Santana, P.H., Bajay, S.V.: New approaches for improving energy efficiency in the Brazilian industry. Energy Rep. **2**, 62–66 (2016)
15. Tantisattayakul, T., Soontharothai, J., Limphitakphong, N., et al.: Assessment of energy efficiency measures in the petrochemical industry in Thailand. J. Clean. Prod. **137**, 931–941 (2016)
16. Ozkara, Y., Atak, M.: Regional total-factor energy efficiency and electricity saving potential of manufacturing industry in Turkey. Energy **93**(1), 495–510 (2015)
17. Youn, R.B., Klyuev, R.V., Bosikov, I.I., et al.: The petroleum potential estimation of the North Caucasus and Kazakhstan territories with the help of the structural-geodynamic prerequisites. Sust. Dev. Mt. T. **9**, 172–178 (2017)
18. Klyuev, R.V.: Analysis of energy consumption in the non-ferrous metallurgy enterprises. Math. Univ. Electromech. **2**, 65–67 (2012)
19. Vasiliev, I.E., Klyuev, R.V.: Methodological bases of an energy audit on the mining and metallurgical combine. Mt. Inf. Anal. Bull. MGGU Sep. Release **8**, 131–134 (2009)
20. Klyuev, R.V., Bosikov, I.I., Youn, R.B.: Analysis of the functioning of the natural-industrial system of mining and metallurgical complex with the complexity of the geological structure of the deposit. Sustain. Dev. Mt. T. **8**, 222–230 (2016)
21. Boorman, A.P.: Managing Electricity Flows and Increase the Efficiency of Electric Power Systems. MPEI Publishing House, Moscow (2012)
22. Varnavskiy, B.P., Kolesnikov, A.I.: Energy Audit of Industrial and Municipal Enterprises. ASEM, Moscow (1999)
23. Tange, E., Shao, Y., Fan, X., et al.: Application of energy efficiency optimization technology in steel industry. J. Iron. Steel Res. Int. **21**(1), 82–86 (2014)
24. Tanaka, K.: Review of policies and measures for energy efficiency in industry sector. Energy Policy **39**(10), 6532–6550 (2011)

Comparative and Optimizing Calculations of Energy Efficiency Indicators for Operation of CHP Plants Using the Normative Characteristics and Mathematical Models

N. V. Tatarinova and D. M. Suvorov[(✉)]

Vyatka State University, 36 Moskovskaya Street,
Kirov 610020, Russian Federation
dmilar@mail.ru

Abstract. The article deals with the problem of the correct calculation of energy efficiency indicators for the operation of CHP plants in the combined production of heat and electric energy. The authors review traditional methods, which are more often used in Russia and in the countries with developed markets of electricity and heat. They develop a new method for calculating energy efficiency indicators for the operation of cogeneration steam turbines using mathematical models based on the actual power and flow characteristics of turbine stages and compartments. The problem solutions are based on generalized experimental data and theoretical positions of technical thermodynamics, the theory of heat and mass transfer, gas dynamics, the theory of turbomachines and mathematical modeling of thermal power plants, numerical and physical experiments and proven techniques of technical and economic research in the energy sector under market conditions. The authors compare the proposed method for calculating variable operating conditions of steam turbines using formulated principles and calculation results based on regulatory characteristics, and the most effective methods are identified. The proposed approach can be implemented and used to optimize the operation of CHP plants in the Russian wholesale electricity and capacity market.

Keywords: CHP plant · Cogeneration steam turbine · Computational model · Characteristics of compartments · Mode diagram · Mathematical modeling · Optimization

1 Introduction

Increasing the efficiency of producing energy has always been a priority in research in the energy sector. At the same time, the task of providing reliable energy supply to consumers while maintaining acceptable technical and economic indicators is becoming urgent for many operating CHP plants due to the fact that a significant part of their main equipment has fulfilled the design life or close to it [1, 2]. Due to the variety of types of cogeneration steam turbines and their operating modes, there are many opportunities for optimizing the thermal schemes and operating modes of turbines and

© Springer Nature Switzerland AG 2020
A. A. Radionov and A. S. Karandaev (Eds.): RusAutoCon 2019, LNEE 641, pp. 637–646, 2020.
https://doi.org/10.1007/978-3-030-39225-3_70

CHP plants in general, improving the operating methods and design of the main and auxiliary turbine equipment. It allows to obtain fuel economy practically without additional capital expenditures [3, 4]. Therefore, the development of new low-cost ways to increase the cost-effectiveness of existing turbines that have been in operation for a long time is an urgent problem and one of the ways to increase the competitiveness of CHP plants in the heat and electricity markets [5]. The rational distribution of heat and electrical loads between cogeneration turbine units, which determines the most economical in-house modes of CHP plants, requires a complex solution of two main tasks. According to available computing means, it is necessary to find the most accurate expression of turbine power characteristics and to choose (or develop) and apply the appropriate method for distributing loads between turbines [6, 7].

The regulation of modern TPP, in particular with cogeneration turbines, is a very laborious task, which requires calculating hundreds of regimes for their optimal distribution. Leading scientific organizations have been developing a number of mathematical methods for distributing loads between turbines of CHP plants (methods of dynamic programming, methods for finding the extremum of functions of several variables, etc.) [8, 9]. And if the development of techniques is given great attention, then the methods of representing the turbine power characteristics are not developed well enough [10, 11].

2 The Aim of the Study

The aim of the study is to identify the most effective methods for determining indicators of variable operating modes of cogeneration turbine plants for solving research problems, optimizing and improving the energy efficiency of CHP plants.

Following tasks were set to achieve it:

1. To describe advantages and disadvantages of two different approaches for determining the energy indicators of CHP plants using mathematical modeling methods based on actual energy characteristics of steam turbines, on the one hand, and regulatory characteristics, including mode diagrams, on the other hand;
2. To carry out a detailed study of the energy efficiency of variable operating modes of cogeneration turbine plants using the example of the T-50-130 turbine;
3. To compare the effectiveness of both methods with respect to short-term optimization studies based on the formulated principles and numerical calculations.

3 Methodological Framework

The underlying normative energy characteristics have a number of significant drawbacks; therefore, their use to solve optimization problems in some cases is illegal, for the following reasons:

(a) The construction of the mode diagram was carried out by the manufacturer on the basis of calculations of heat balances of variable regimes of the turbine. However, turbines which are used today were made 30–50 years ago on standard projects and, therefore, they cannot fully comply with modern operating conditions.

(b) Currently, to determine the technical and economic performance of cogeneration turbine plants, we continue to widely apply normative characteristics obtained from the average results of turbine tests under the design thermal scheme. These characteristics are linear (polynomials of the first degree) or piecewise-linear functions (linear equations with kinks). Their basic part refers to the nominal parameters, and all deviations from them are described by additional amendments, which, as a rule, are also linear in nature. The final result of using normative characteristics is to obtain values of the heat consumption for the turbine unit and the specific heat consumption for the generation of electricity for the given heat load, electric power, steam pressure in the chamber of controlled heating selection (the temperature of delivery water), parameters (pressure and temperature) of the steam in front of the turbine, pressure in the condenser. But it is clearly not enough, since full-scale tests confirm the complex and essentially non-linear influence of independent variables on the technical and economic performance and characteristics of turbine stages and compartments (especially in the low-pressure section).

(c) Characteristics which separate electrical load into components (condensation and heating) complicate calculations by introducing additional variables [12, 13].

(d) The temperature of the return delivery water, which in actual operating conditions does not remain constant, but has a lesser effect on the heat efficiency of the turbine unit, is considered as a given parameter. Its numerical value is taken in the form of a dependence built in accordance with the standard thermal schedule of heat networks.

At the same time, modern tasks of optimizing the operation of the CHP plants has a high computational complexity, the reasons for which are: (1) the complex form of the function, (2) the large number of controllable (variable) parameters, (3) the large number of constraints imposed on values of controlled parameters [14, 15].

Thus, there is a discrepancy between the methods used and the means of solving the problem, on the one hand, and the initial data, on the other. Therefore, the main tasks of standardization are not fulfilled: ensuring the use in the energy sector of technically sound normative values of fuel, heat and electric energy consumption for the economy, an objective analysis of the operation of the TPP equipment, and determination of ways to reduce the unreasonable fuel consumption. Taking into account the simplified approach in obtaining normative characteristics, the question arises: is their application suitable for carrying out computational studies of the variable operating modes of the cogeneration steam turbine and for the solution of optimization problems.

The authors propose to use nonlinear mathematical models (MM), which are described in detail in [16, 17], taking into account the turbine operation parameters, the actual state of turbine equipment and actual characteristics of their stages and compartments [18, 19]. It became possible due to the improvement of the computational methods of mathematical modeling, namely the method of solving systems of nonlinear equations of large dimension (with the number of independent variables greater than 50), the modernization of existing algorithms and numerical methods applied to the calculation of all possible modes of operation of steam turbine installations [20, 21].

4 Computational Studies of the Variable Operating Modes of Cogeneration Steam Turbines Based on Mathematical Models and Normative Characteristics

The choice of the criterion of the efficiency of the transition to variable operating modes of the turbine installation is justified by the following. One of the most important indicators of the economy of a steam turbine plant is the specific heat consumption for generating electrical energy. For steam turbines, this indicator is defined as [18]:

$$q = (Q_o - Q_t)/N_e, \tag{1}$$

where Q_o is the heat consumption to the turbine; Q_t is the heat supply from controlled taps (heat load); N_e is the electric power.

While studying variable regimes, first of all, it is necessary to assess the change in the economy of a turbine installation with a change in its operating mode and besides, the use of the absolute value of q is not convenient. It is more productive to use a comparative indicator, namely, the specific change in heat consumption with a change in electricity generation

$$q_{add} = (\Delta Q_o - \Delta Q_t)/\Delta N_e, \tag{2}$$

where $\Delta Q_o = Q_o - Q_{oo}$; $\Delta Q_t = Q_t - Q_{to}$; $\Delta N_e = N_{eo} - N_{eo}$; Q_{oo}, Q_{to}, N_{eo} and Q_o, Q_t, N_e relate to some initial and new operating modes of the turbine.

Value q_{add} characterizes the energy efficiency of changes in the operating mode of the turbine, which is very important for its correct operation under conditions of variable heat and electric load schedules.

When carrying out a calculation study, the following conditions were met:

- The identity of the heat load specified for normative characteristics and mathematical models (depending on the set level of steam pressure in the controlled tap) and its constancy when changing the electric power
- The identity of the electric power values given for normative characteristics and mathematical models at each given level of steam pressure in the controlled tap
- The identity of the absolute values of the heat flow rate to the turbine for normative characteristics and mathematical models in the initial modes in determining the dependences q_{add} from ΔN_e

In given conditions $\Delta Q_t = 0$, and $q_{add} = \Delta Q_o/\Delta N_e$, that is q_{add} is the specific heat consumption for obtaining additional electric power.

The results of the calculations showed that both in the case of a double-stage (Fig. 1) and single-stage delivery water heating the dependence of q_{add} on the increase of electric power ΔN_e and the pressure in the controlled tap (respectively upper tap P_2 and lower tap P_1), obtained by normative characteristics (q_{add_nch}) and mathematical models (q_{add_mm}), are essentially different, both quantitatively and qualitatively. The level q_{add_mm} in comparison with q_{add_nch} was, as shown in Table 1, much larger and significantly dependent on the power increase ΔN_e. The latter circumstance is very important, as it predicts the possibility of optimizing the operating conditions of a group of turbines. The results of previous calculations and experimental studies confirmed this possibility even with respect to the same type of turbines. As for the values of q_{add_nch}, the increase in electric power has practically no effect on them for actual regimes, which is a consequence of the use of linearized energy characteristics in normative characteristics.

The difference in the absolute heat flow rates to the turbine in identical operating modes, obtained from normative characteristics and mathematical models, (Fig. 2), as well as q_{add_mm}, essentially depends on the power increase, increasing with ΔN_e with the intensity determined by the operation mode (single- or double-stage delivery water heating) and steam pressure in the controlled taps (P_1 or P_2).

As an example, Fig. 3 shows the results of the calculated comparison of the specific heat consumption for electricity generation, obtained by mathematical models (q_{mm}) and normative characteristics (q_{nch}) for the operation modes of the T-50-130 turbine with a two-stage heating of the delivery water, depending on the electrical power N_e at various specified values of the heat load Q_t and the steam pressure in the upper tap P_2.

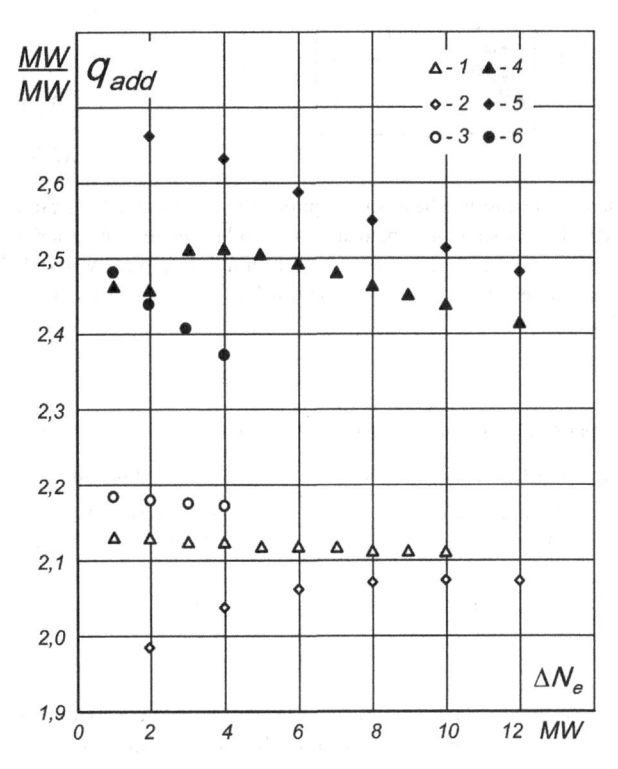

Fig. 1. Dependence of the specific heat consumption on the generation of additional electric power (q_{add}) on the increase in electric power (ΔN_e) the T-50-130 turbine and the steam pressure in the upper tap (P_2) with a two-stage heating of delivery water (1, 2, 3 - calculation by normative characteristics; 4, 5, 6 - mathematical model calculation; 1, 4 – $P_2 = 0,06$ MPa, 2, 5 – $P_2 = 0,08$ MPa, 3, 6 – $P_2 = 0,12$ MPa).

As can be seen, the difference between q_{nch} and q_{mm} can be very significant, and it doesn't have a systematic nature, it is determined by parameters of the mode.

The comparison was made for the following groups of modes. The simplest and most widely used method for obtaining peak electrical power is to open the sliding grid in the low-pressure section.

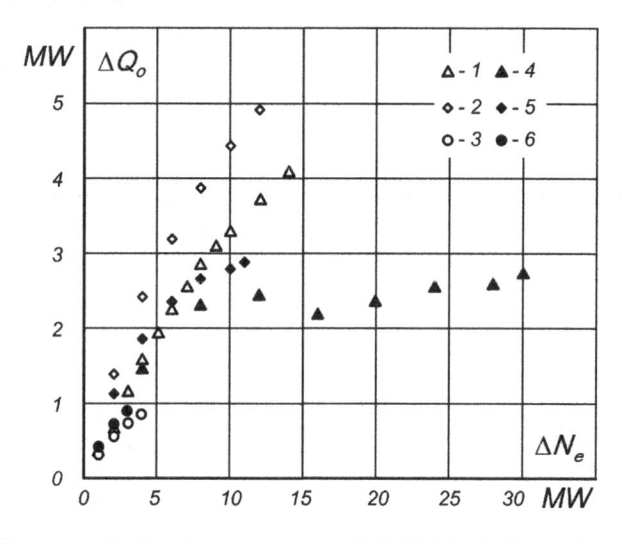

Fig. 2. The difference in absolute heat consumption (ΔQ_o) in identical operating modes of the T-50-130 turbine, calculated using mathematical model and normative characteristics depending on the increase in electric power (1, 2, 3 - two-stage heating of delivery water; 4, 5, 6 - one-stage heating of delivery water; 1, 4 - P_1 (P_2) = 0,06 MPa, 2, 5 - P_1 (P_2) = 0,08 MPa, 3, 6 - P_1 (P_2) = 0,12 MPa).

Table 1. Comparison of normative characteristics and mathematical models in value q_{add}.

P_2 (P_1), MPa	Two-stage heating of delivery water	One-stage heating of delivery water
	q_{add_mm}/q_{add_nch}	q_{add_mm}/q_{add_nch}
0,6	1,14 ÷ 1,18	1,16 ÷ 1,04
0,8	1,34 ÷ 1,20	1,23 ÷ 1,11
1,2	1,14 ÷ 1,09	1,16 ÷ 1,12

Calculations showed that the value q_{add} varies in a very wide range and with a decrease in the flow of delivery water through the increase in pressure in the upper tap, it also decreases and may become lower than that for condensation turbines.

From this we can conclude that when the delivery water flow is less than a certain level, the operation of the cogeneration steam turbines with the sliding grid in conditions of electric power deficit becomes inappropriate, since the reduction in electricity

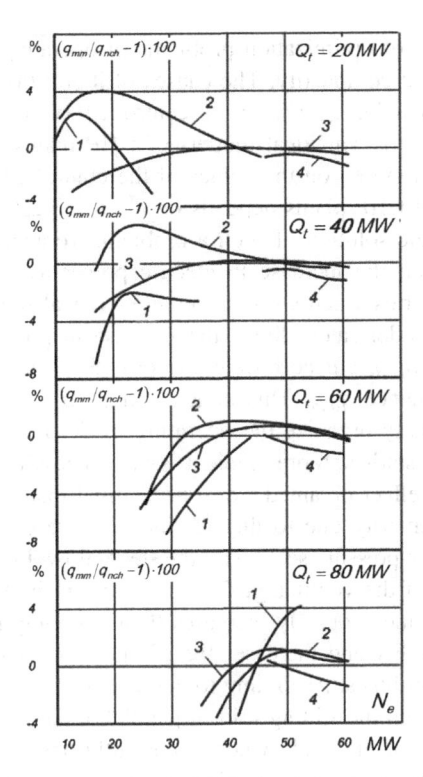

Fig. 3. Comparison of specific heat consumption for generation of electricity obtained by mathematical model (q_{mm}) and normative characteristics (q_{nch}) for turbine operation modes T-50-130 with a two-stage heating of delivery water, depending on the electrical power N_e for different specified values of heat load Q_t and steam pressure in the upper tap P_2.

generation in this case will be compensated for by the less economical condensing capacity. When calculating by the mode diagram, the values q_{add} were found to be almost 1.5 times lower, and the influence of the mode parameters on them is much weaker.

This is due to the fact that the mode diagram is built on the basis of the linearized characteristics of the turbine compartments, which, for sufficiently large deviations of the operating mode from the nominal one, significantly differ from the actual ones. As can be seen, the use of normative characteristics gives significant errors in comparison with the detailed calculation when using mathematical models.

5   Optimization Research on the Mathematical Model

Optimization of work of the combined heat and power plant is an important technical and economic task aimed to increase the efficiency of natural resources use, which serve as fuel for the CHP plants, and to improve the economic efficiency of the operation in the electricity and heat market. Criteria for the optimality of the CHP plants depend on the conditions of its operation.

Modern formulation of optimization problems of operating modes of CHP plants has a high computational complexity. The causes of it are: (a) a complex form of the objective function; (b) a large number of controlled (variable) parameters, which depends on the applied mathematical models of CHP plants aggregates; (c) a large number of constraints imposed on the values of the controlled parameters. Choice of mathematical models of CHP plants depends on the task [22].

As an example of the solution of such a problem, we analyzed the change in the value q_{add} depending on the increase in electric power as a result of the different opening of the sliding grids when working on an electrical schedule.

According to the model calculations, the increase in power under all conditions leads to a decrease in q_{add}, but according to normative characteristics this increase practically doesn't influence q_{add}. This conclusion is of great practical importance, since it allows to correctly organize the operating mode of the same type of turbine units. As shown by the studies, more efficient is the successive opening of the sliding grid (Fig. 4) [23]. The effect obtained from the optimal distribution of the peak load between turbines is primarily due to the nonlinear nature of the dependence of the internal power of the low-pressure section on the steam flow. In this regard, it should be noted that a reduction of the ventilation flow of steam in the low-pressure section of cogeneration steam turbines not only increases their economy in the basic mode, but also expands the possibility and increases the efficiency of optimizing variable operating conditions of the turbines. According to the regulatory characteristics, it turns out that it is necessary to open the sliding grid in parallel at all turbo-units, i.e. optimal is the uniform distribution of the load, which is currently observed in most cases when operating the same turbines and causing significant economic damage.

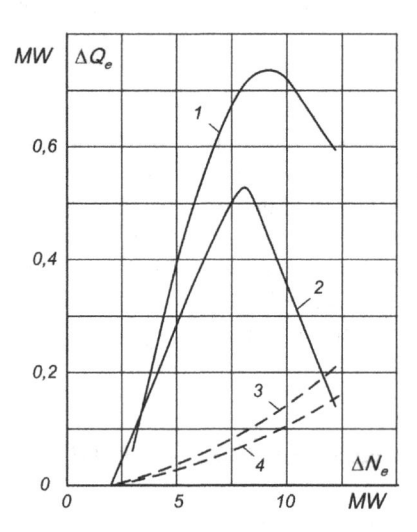

Fig. 4. Efficiency of optimizing peak load distribution between T-50-130 turbines (1, 3 - between three turbines; 2, 4 - between two turbines) using normative characteristics (3, 4) and mathematical models (1, 2) (ΔQ_e - the total economy in absolute heat consumption for several turbines; ΔN_e - the total increase in electric power for several turbines).

6 Conclusion

1. The authors have reviewed and described the main disadvantages in the application of normative characteristics in the calculation of variable operating modes of cogeneration steam turbines.
2. The authors have analyzed the results of determining the values q_{add} on the example of a T-50-130 turbine with the use of normative characteristics and the developed mathematical model in operating modes for heat and electrical graphs different than nominal. The results testify to the inexpediency of using the normative characteristics for an adequate assessment of the performance of the cogeneration steam turbines under the conditions described above, and the increase of q_{add} in approximately 1.5 times when calculated by normative characteristics.
3. The authors have developed recommendations aimed to optimize the distribution of loads between turbine units, and proposals for rational operating modes based on the comparison of calculation results with the use of normative characteristics and mathematical models. The authors underlined incorrectness of the application of normative characteristics for the solution of optimization problems.

References

1. Tatarinova, N.V., Efros, E.I., Sushchikh, V.M.: Calculation results on mathematical models of variable operating modes of steam-turbine-type CHP plants under actual operating conditions. Prospects Sci. **3**(54), 95–100 (2014)
2. Efros, E.I., Tatarinova, N.V.: Efficiency of obtaining additional condensing power in steam turbines. Electr. Stations **10**, 26–32 (2006)
3. Shempelev, A.G., Iglin, P.V., Sushchikh, V.M.: Estimation of the influence of operational factors on the oxygen content of the turbine condensate at the outlet from the condenser of steam turbine. Probl. Regional Energ. **2**(34), 81–89 (2017)
4. Tveit, T.M., Fogelholm, C.J.: Multi-period steam turbine network optimization. Part II: development of a multi-period MINLP model of a utility system. Appl. Therm. Eng. **26**(14–15), 1730–1736 (2006). https://doi.org/10.1016/j.applthermaleng
5. Trukhny, A.D., Lomakin, B.V.: Heating Steam Turbines and Turbine Units. MPEI, Moscow (2006)
6. Chuchueva, A., Inkina, N.E.: Optimization of CHP plants operation in the wholesale electricity and capacity market of Russia. Sci. Educ. Bauman MSTU Electron. J. **8**, 195–238 (2015)
7. Tatarinova, N.V., Suvorov, D.M.: Development of adequate computational mathematical models of cogeneration steam turbines for solving problems of optimization of operating modes of CHP plants. In: 2nd International Conference Industrial Engineering Application Manufacturing, pp. 1–6 (2016). https://doi.org/10.1109/icieam.2016.7911578
8. Shchinnikov, P.A., Safronov, A.V.: Enhancing the calculation accuracy of performance characteristics of power-generating units by correcting general measurands based on matching energy balances. Therm. Eng. **61**(12), 898–904 (2014)

9. Tveit, T.M., Fogelholm, C.J.: Multi-period steam turbine network optimization. Part I: simulation based regression models and an evolutionary algorithm for finding D-optimal designs. Appl. Therm. Eng. **26**(10), 993–1000 (2006). https://doi.org/10.1016/j.applther maleng.2005.10.025

10. Simoyu, L.L., Efros, E.I., Gutorov, V.F., et al.: Heating Steam Turbines: Increase in Economy and Reliability. Energotech, St. Petersburg (2001)

11. Ledukhovsky, G.V., Zhukov, V.P., Barochkin, E.V., et al.: Algorithms for striking material and energy balances in calculating the technical-and-economic indicators of thermal power plant equipment based on the ill-posed problem regularization method. Therm. Eng. **62**(8), 607–614 (2015)

12. Efros, E.I., Kalinin, B.B., Tatarinova, N.V.: Comparative computational studies of variable operating modes of a heating turbine installation using the proposed mathematical model and typical regulatory characteristics. Adv. Sci. **1**(4), 133–142 (2014)

13. Smirnov, D.K., Galashov, N.N.: A software package for visual modeling of heat power plants. News Tomsk. Polytech. Univ. **320**(4), 36–40 (2012)

14. Vob, A., Hlawenka, A., Haug, M., et al.: Making the most of available assets. How intelligent add-on technology helps to upgrade boiler perfomance. VGB PowerTech, p. 6 (2009)

15. Likhvar, N.V.: Flexible mathematical models of power plants for optimization of CHP plants modes. Perfection of turbine units using mathematical and physical modeling methods: SC. Works IPMash NAS of Ukraine, pp. 413–419 (2003)

16. Berdyshev, V.I., Letun, V.M., Volkova, T.V., et al.: Mathematical modeling: optimization of operational modes of thermal power plants. Bull. Ural. Branch RAS **1**, 25–34 (2013)

17. Tatarinova, N.V., Suvorov, D.M., Shempelev, A.G.: Approaches to building computational mathematical models based on the flow and power characteristics of cogeneration steam turbine stages and compartments. In: 3rd International Conference Industrial Engineering Application Manufacturing, pp. 1–6 (2017). https://doi.org/10.1109/icieam.2017.8076463

18. Tatarinova, N.V.: Mathematical modeling of heating turbines for solving problems of increasing the energy efficiency of the CHP plants. Dissertation, Ural Federal University (2014)

19. Efros, E.I.: Efficiency and reliability of high-power heating turbines and ways to increase them. Dissertation, All-Russian Thermotechnical Institute (1998)

20. Simoyu, L.L.: Gas-dynamic calculations of steam turbine flow paths: methods, computer programs, and application practice. Therm. Eng. **58**(6), 464–470 (2011)

21. Simoyu, L.L., Indurskii, M.S., Efros, E.I.: Calculation of the performance of the LP section in a cogeneration steam turbine under variable operating conditions. Therm. Eng. **47**(2), 105–110 (2000)

22. Tatarinova, N.V., Suvorov, D.M.: Mathematical modeling of the effect of operational factors on the level of steam humidity in the low-pressure section of steam turbines. Saf. Reliab. Power Ind. **4**(31), 53–56 (2015)

23. Tatarinova, N.V., Sushchikh, V.M.: Study of flow and power characteristics of the last compartments of cogeneration steam turbines. Probl. Regional Energ. **3**(44), 79–90 (2019)

Description of Complex Hierarchical Systems by Matrix-Predicate Method

V. S. Polyakov$^{(\boxtimes)}$ and S. V. Polyakov

Volgograd State Technical University, 28 Lenina avenue,
Volgograd 400005, Russian Federation
vladstrix@mail.ru

Abstract. Multi-unit, multi-link, and multi-channel complex systems which have a hierarchical structure and represent a multitude of interacting objects, subsystems, components, etc., are necessary for management solutions and decision-making. In this work the possibility of describing complex systems of a hierarchical structure using a matrix-predicate method is considered. This will facilitate the interaction of structure elements and help to avoid possible iso-morphism when performing logical and theoretical operations on the units of complex systems. It is offered to present a technological process performing object as a finite number (r) of interacting components K. Components are easy to create in the matrix-predicate form as they are small, well-accessible and easily corrected with the use of matrices. The interaction of components is determined by r-relation which is assigned in the form of a compatibility table. In this work the possibility of describing complex hierarchical production system using a matrix which elements represent the true value of pentad predicate. The possibility of the description by square matrix in the matrix-predicate form will be provided by an example of a two-level hierarchical structure.

Keywords: Graph · Incidentor · Matrix · Predicate · Component · Matrix-predicate form · Structure · Hierarchy

1 Introduction

The systemized analysis of the modern complex systems (CS), which are multivariate, multi-level multi-criteria systems, divides them into four classes of complex technical systems [1–5].

1. Multi-unit multi-criteria systems.
2. Hierarchical systems.
3. Polyhierarchical systems of 'diamond-shaped' structure.
4. Polyhierarchical CS of conceptual hierarchical multi-agent structure.

   Hierarchical systems can be classified according to their complexity:

- Operation complexity hierarchy (by layers)
- Mathematical models hierarchy (by strata)
- Organizational hierarchy (by tiers)

A. A. Radionov and A. S. Karandaev (Eds.): RusAutoCon 2019, LNEE 641, pp. 647–656, 2020.
https://doi.org/10.1007/978-3-030-39225-3_71

When studying CS with hierarchical structure different degree of intersection or matching of considered hierarchies is possible [6–9]. For CS with hierarchical structure the solution for design objectives and the setting of efficient functioning is typical.

In this work the possibility of describing complex systems of hierarchical structure using matrix-predicate method is considered. Thus, it will facilitate the interaction of structure elements and avoid possible isomorphism when performing logical and theoretical operations on the units of CS.

For this purpose, it is proposed to use a mathematical apparatus of the representation of the technological process (TP) in the form of matrix-predicate, which was created by the authors [10–14].

A CS, performing TP, is represented as a finite number (r) of interacting components K^{ρ}. Components are easy to create in the matrix-predicate form as they are small, well-accessible and easily corrected with the use of matrices [16–20]. The interaction of components is determined by r-relation which it is most convenient to assign in the form of a compatibility table [14].

2 Complex Hierarchical Production Systems

Now we will consider a complex hierarchical production system (CHPS). While analyzing of such a system interacting modules are easily defined at every hierarchical level. These modules are blocks, units, components, etc. each of which is described by a suitable graph $G^{\rho}(Y; X; H)$, and they make up a set:

$$G = \{ G^{\rho} \} \tag{1}$$

Where $\rho = 1, 2, ..., r$.

Then, we make an assumption: the work of the components of the low level is carried out only given a certain state of the components of the highest level. A basic two-level CHPS is to be examined, and for its description it is set as follows:

- A graph of the high level $G(Y; X; P)$, presented in the picture (Fig. 1)
- The set of graphs $K = \{A, B, C, D\}$ of the low level, presented in the pictures (Figs. 2, 3, 4 and 5)

For describing a G component of the high level, a quadratic predicate $G(Y; X; P_B t_i)$ is used instead of the triadic one $G(Y; X; P_B)$.

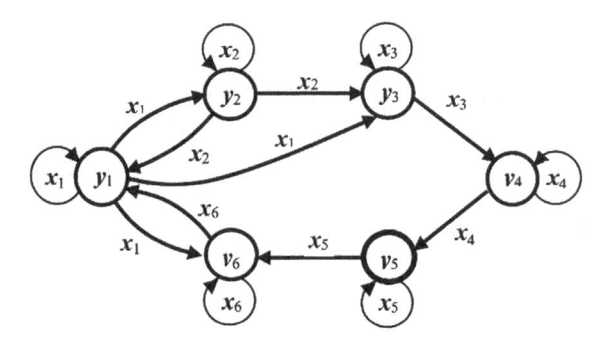

Fig. 1. A graph of the structure of the high-level component $G(Y; X; P)$

A matrix-predicate representation of a high-level component is shown in the formula (2).

$$
P = \begin{Vmatrix}
y_1x_1y_1t_1 & y_1x_{12}y_2{}_- & y_1x_{13}y_3{}_- & 0 & 0 & y_1x_{16}y_6{}_- \\
y_2x_{21}y_1{}_- & y_2x_2y_2t_2 & y_2x_{23}y_3{}_- & 0 & 0 & 0 \\
0 & 0 & y_3x_3y_3t_3 & y_3x_{34}y_4{}_- & 0 & 0 \\
0 & 0 & 0 & y_4x_4y_4t_4 & y_5x_{45}y_5{}_- & 0 \\
0 & 0 & 0 & 0 & y_5x_5y_5t_5 & y_5x_{56}y_6{}_- \\
y_6x_{61}y_1{}_- & 0 & 0 & 0 & 0 & y_6x_6y_6t_6
\end{Vmatrix} \qquad (2)
$$

The low level is set by a multitude of one-component structures consisting of four components $K = \{A, B, C, D\}$, the structure of which is described by graphs in the pictures (Figs. 2, 3, 4, and 5), and their matrix-predicate representation is also given in the formulas (3–6).

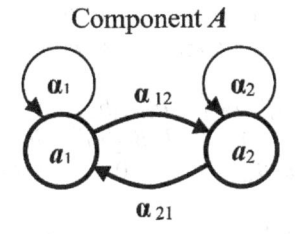

Fig. 2. The graph $G_A(A;\ \alpha;\ P_A)$ of the structure of the low-level component A.

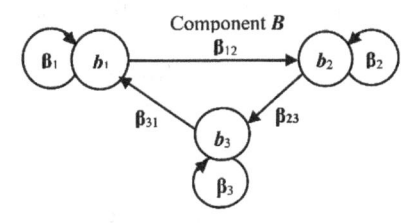

Fig. 3. The graph $G_B(B;\ \beta;\ P_B)$ of the structure of the low-level component B

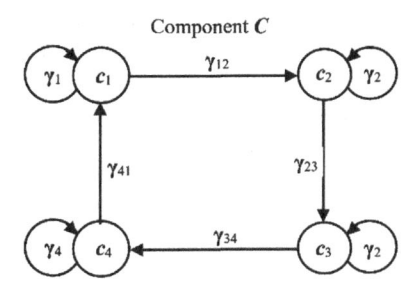

Fig. 4. The graph $G_C(C;\ \gamma;\ P_C)$ of the structure of the low-level component C.

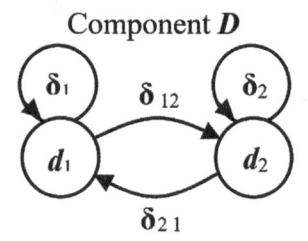

Fig. 5. The graph $G_D(C; \delta; P_D)$ of the structure of the low-level component D.

3 The Matrix-Predicate Representation of the Components

The matrix-predicate representation of the components is depicted in the following formulas (3–6).

$$P_A = \left\| \begin{matrix} a_1\alpha_{11}a_1t_1^A & a_1\alpha_{12}a_{2-} \\ a_2\alpha_{21}a_{1-} & a_2\alpha_{22}a_2t_2^A \end{matrix} \right\| \tag{3}$$

$$P_B = \left\| \begin{matrix} b_1\beta_{11}b_1t_1^B & b_1\beta_{12}b_{2-} & 0 \\ 0 & b_2\beta_{22}b_2t_2^B & b_2\beta_{23}b_{3-} \\ b_3\beta_{31}b_{1-} & 0 & b_3\beta_{33}b_3t_3^B \end{matrix} \right\| \tag{4}$$

$$P_C = \left\| \begin{matrix} c_1\gamma_{11}c_1t_1^C & c_1\gamma_{12}c_{2-} & c_1\gamma_{13}c_{3-} & 0 \\ 0 & c_2\gamma_{22}c_2t_2^C & c_2\gamma_{23}c_{3-} & 0 \\ 0 & 0 & c_3\gamma_{33}c_3t_3^C & c_3\gamma_{34}c_{4-} \\ c_4\gamma_{41}c_{1-} & 0 & 0 & c_4\gamma_{44}c_4t_4^C \end{matrix} \right\| \tag{5}$$

$$D = \left\| \begin{matrix} d_1\delta_{11}d_1t_1^D & d_1\delta_{12}d_{2-} \\ d_2\delta_{21}d_{1-} & d_2\delta_{22}d_2t_2^D \end{matrix} \right\| \tag{6}$$

4 The Structure of the Two-Level CHPS

The functioning of CHPS is as follows: some of the high-level operators cause the low-level components to function. The structure of the two-level CHPS, which is shown in the picture (Fig. 6), and its matrix-predicate description (3) characterizes the functioning of the TP.

The structure of a non-zero element of any matrix component is the true value of the quadratic predicate P and has the form:

$$y_i\, x_{ij}\, y_j\, t_i,$$

where y_i; $y_j \in Y$ – a set of the graph vertices of the component; $x_{ij} \in X$ – a set of the graph ribs of the component; t_i – the time of the functioning of the current operator of the action of the component.

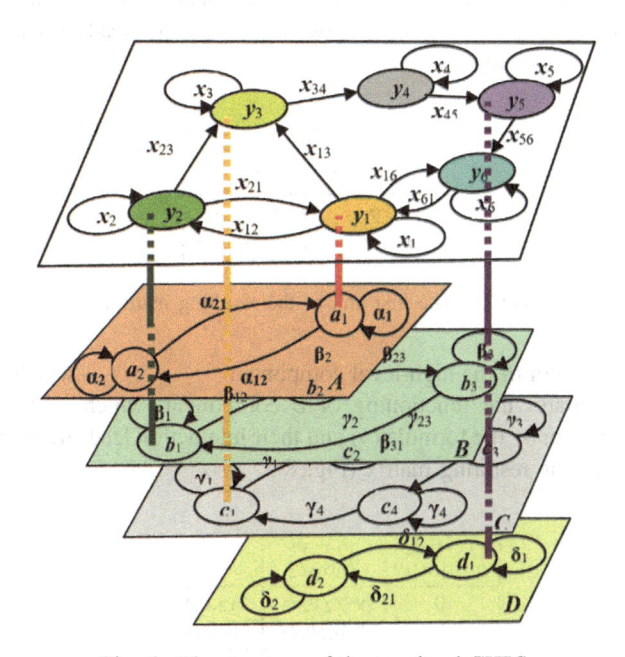

Fig. 6. The structure of the two-level CHPS.

5 The Matrix-Predicate Representation of the Two-Level CHPS

The matrix-predicate representation of the two-level CHPS is shown in the formula (7). Dashed lines indicate six operators of the action of the top-level component, four of which interact with four low-level components.

$$M^{S_{1-4}} = \begin{Vmatrix} y_1x_1y_1t_1A & y_1x_{12}y_2 \text{ }_{--} & 0 & 0 & 0 & y_1x_{16}y_6 \text{ }_{--} \\ y_2x_1y_1 \text{ }_{--} & y_2x_2y_2t_2B & y_2x_{23}y_3 \text{ }_{--} & 0 & 0 & 0 \\ 0 & 0 & y_3x_3y_3t_3C & y_3x_{34}y_4 \text{ }_{--} & 0 & 0 \\ 0 & 0 & 0 & y_4x_4y_4t_4 \text{ }_- & y_4x_{45}y_5 & 0 \\ 0 & 0 & 0 & 0 & y_5x_5y_5t_5D & y_5x_{56}y_6 \text{ }_{--} \\ 0 & 0 & 0 & 0 & 0 & y_6x_6y_6t_6 \text{ }_- \end{Vmatrix}$$

$$(7)$$

The description of the two-level CHPS structure (S_{1-4}) is a matrix (7), which elements are defined by pentadic predicate $y_i\, x_j\, y_i\, t_i\, A$, where $y_i\, x_j\, y_i$ – are the true value of the triadic predicate which defines the i-condition of the high-level of the CS, t^i – the functioning time of the i-condition of the top-level of the CS, A – the fifth place of the predicate which defines the functioning of the component of the low-level of the CS.

The action operator of the high-level component – $y_1 x_1 y_1 t_1$ – which acts during t_1 - time period, initiates the functioning of A component, which is described in the matrix-predicate form by the formula (3) and their interaction leads to the emergence of the first unit of the resulting matrix (Fig. 7).

$$
\begin{array}{|c|c|}
\hline
\begin{array}{c} y_1 x_1 y_1 t_1 \\ a_1 \alpha_{11} a_1 t_1{}^A \end{array} & \begin{array}{c} y_1 x_1 y_1 \\ a_1 \alpha_{12} a_2 \_ \end{array} \\
\hline
\begin{array}{c} y_1 x_1 y_1 \_ \\ a_2 \alpha_{21} a_1 \_ \end{array} & \begin{array}{c} y_1 x_1 y_1 t_1 \\ a_2 \alpha_{22} a_2 t_2{}^A \end{array} \\
\hline
\end{array}
$$

Fig. 7. The first unit of the resulting matrix.

The action operator of the high-level component – $y_2 x_2 y_2 t_2$ – which acts during t_2 - time period, initiates the functioning of B component, which is described in the matrix-predicate form by the formula (4) and their interaction leads to the emergence of the second unit of the resulting matrix (Fig. 8).

$$
\begin{array}{|c|c|c|}
\hline
\begin{array}{c} y_2 x_2 y_2 t_2 \\ b_1 \beta_{11} b_1 t_1{}^B \end{array} & \begin{array}{c} y_2 x_2 y_2 \\ b_3 \beta_{31} b_1 \_ \end{array} & \begin{array}{c} 0 \\ 0 \end{array} \\
\hline
\begin{array}{c} 0 \\ 0 \end{array} & \begin{array}{c} y_2 x_2 y_2 t_2 \\ b_2 \beta_{22} b_2 t_2{}^B \end{array} & \begin{array}{c} y_2 x_2 y_2 \_ \\ b_3 \beta_{31} b_1 \_ \end{array} \\
\hline
\begin{array}{c} y_2 x_2 y_2 \_ \\ b_3 \beta_{31} b_1 \_ \end{array} & \begin{array}{c} 0 \\ 0 \end{array} & \begin{array}{c} y_2 x_2 y_2 2 \\ b_3 \beta_{33} b_3 t_3{}^B \end{array} \\
\hline
\end{array}
$$

Fig. 8. The second unit of the resulting matrix.

The action operator of the high-level component – $y_3 x_3 y_3 t_3$ – which acts during t_2 - time period, initiates the functioning of C component, which is described in the matrix-predicate form by the formula (5) and their interaction leads to the emergence of the third unit of the resulting matrix (Fig. 9).

$$
\begin{array}{|c|c|c|c|}
\hline
\begin{array}{c} y_3 x_3 y_3 t_3 \\ c_1 \gamma_{11} c_1 t_1{}^C \end{array} & \begin{array}{c} y_3 x_3 y_3 \_ \\ c_1 \gamma_{12} c_2 \_ \end{array} & \begin{array}{c} y_3 x_3 y_3 \_ \\ c_1 \gamma_{13} c_3 \_ \end{array} & \begin{array}{c} 0 \\ 0 \end{array} \\
\hline
\begin{array}{c} 0 \\ 0 \end{array} & \begin{array}{c} y_2 x_2 y_2 t_3 \\ c_2 \gamma_{22} c_2 t_2{}^C \end{array} & \begin{array}{c} y_3 x_3 y_3 \_ \\ c_2 \gamma_{23} c_3 \_ \end{array} & \begin{array}{c} 0 \\ 0 \end{array} \\
\hline
\begin{array}{c} 0 \\ 0 \end{array} & \begin{array}{c} 0 \\ 0 \end{array} & \begin{array}{c} y_2 x_2 y_2 t_3 \\ c_3 \gamma_{33} c_3 t_3{}^C \end{array} & \begin{array}{c} y_3 x_3 y_3 \_ \\ c_3 \gamma_{34} c_3 \_ \end{array} \\
\hline
\begin{array}{c} y_3 x_3 y_3 \_ \\ c_4 \gamma_{41} c_1 \_ \end{array} & \begin{array}{c} 0 \end{array} & \begin{array}{c} 0 \end{array} & \begin{array}{c} y_3 x_3 y_3 t_3 \\ c_4 \gamma_{44} c_4 t_4{}^C \end{array} \\
\hline
\end{array}
$$

Fig. 9. The third unit of the resulting matrix.

The action operator of the high-level component $- y_5 x_5 y_5 t_5 -$ which acts during t_5-time period, initiates the functioning of D component, which is described in the matrix-predicate form by the formula (6) and their interaction leads to the emergence of the fifth unit of the resulting matrix (Fig. 10).

$$
\begin{array}{|c|c|}
\hline
y_5 x_5 y_5 t_5 & y_5 x_5 y_5 t_5 \\
d_1\delta_{11}d_1t_1{}^D & d_1\delta_{12}d_2 \\
\hline
y_5 x_5 y_5 t_5 & y_5 x_5 y_5 t_5 \\
d_2\delta_{21}d_1 & d_2\delta_{22}d_2t_2{}^A \\
\hline
\end{array}
$$

Fig. 10. The fifth unit of the resulting matrix.

The action operators of the high-level component $- y_4 x_4 y_4 t_4$ and $y_6 x_6 y_6 t_6 -$ which acts during t_5 and $t_6 -$ time period respectively, do not initiate the functioning of the low-level components and in the resulting matrix they will be represented only by the operators of the high-level component.

A full matrix-predicate representation of the two-level CHPS is carried out in two parts. During the first stage we create a 'diagonal' matrix from the received units (Figs. 7, 8, 9 and 10) and two components of the high-level component $- y_4 x_4 y_4 t_4$ and $y_6 x_6 y_6 t_6$ (Fig. 11).

Every generalized action operator of the matrix unit.

Fig. 11. A 'diagonal' matrix of the two-level CHPS.

Every generalized action operator of the matrix unit element $y_2\,x_2\,y_2\,t_2/b_1\beta_{11}b_1t_1^B$ (Fig. 11) consists of true values of two quadratic predicates и means the following:

- The first one characterizes the action operator of the top-level component, $y_2\,x_2\,y_2\,t_2$ in this case
- The second one characterizes the action operator of the low-level component, $b_1\beta_{11}b_1t_1^B$ of component B in this case
- Characterizes the joint cooperation of the components

A transition from one matrix unit (Fig. 11) to another is as follows (will be considered on the example of the transition between units two and three).

The completion of the second unit is characterized by $y_2\,x_2\,y_{2\_}/b_3\beta_{31}b_{1\_}$ element (Fig. 12).

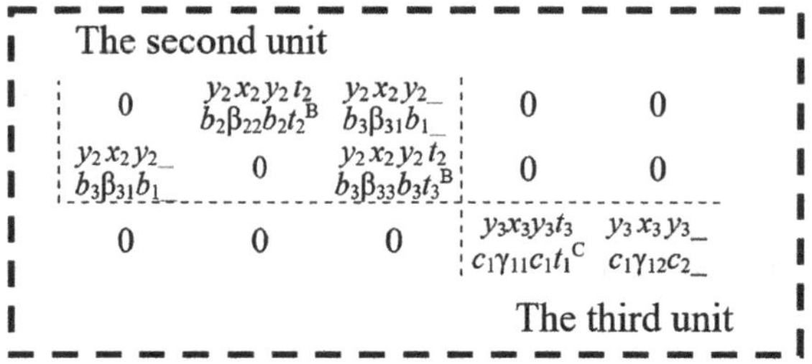

Fig. 12. A fragment of the second and third units of the 'diagonal' matrix of the two-level CHPS.

Now we will replace this element with «0» and transfer it to the intersection of units two and three. We will replace its $y_2x_2y_2$ element with the condition of the transition between $y_2x_2y_2t_2B$ and $y_3x_3y_3t_3C$ operators of the structure of the two-level CHPS (7) - y2 x23y3. Then we will have (Fig. 13):

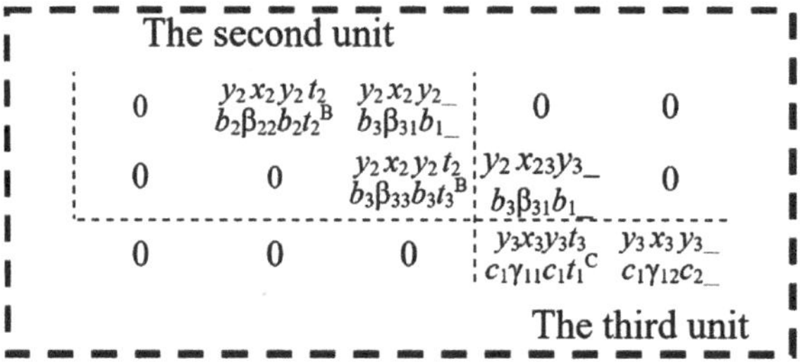

Fig. 13. The transition between the second and third units of the two-level CHPS (fragment).

After conducting this operation with all units, we will have the description of the functioning of the considered CHPS which is shown in the picture (Fig. 14).

$$
M = \begin{bmatrix}
\frac{y_1x_1y_1t_1}{a_1a_{11}a_1t_1}{}^A & \frac{y_1x_1y_1}{a_1a_{12}a_2} & 0 & 0 & 0 & 0 & 0 & 0 & 0 & 0 & 0 & 0 & 0 \\
0 & \frac{y_1x_1y_1t_1}{a_2a_{22}a_2t_2}{}^A\,\frac{y_1x_{12}y_2}{a_2a_{21}a_1} & \frac{y_1x_{12}y_2}{a_2a_{21}a_1} & 0 & 0 & 0 & 0 & 0 & 0 & 0 & 0 & 0 & \frac{y_1x_{16}y_6}{a_2a_{21}a_1} \\
0 & 0 & \frac{y_2x_2y_2t_2}{b_1\beta_{11}b_1t_1}{}^B\,\frac{y_2x_2y_2}{b_3\beta_{31}b_1} & \frac{y_2x_2y_2}{b_3\beta_{31}b_1} & 0 & 0 & 0 & 0 & 0 & 0 & 0 & 0 & 0 \\
0 & 0 & 0 & \frac{y_2x_2y_2t_2}{b_2\beta_{22}b_2t_2}{}^B\,\frac{y_2x_2y_2}{b_3\beta_{31}b_1} & \frac{y_2x_2y_2}{b_3\beta_{31}b_1} & 0 & 0 & 0 & 0 & 0 & 0 & 0 & 0 \\
\frac{y_2x_1y_1}{b_3\beta_{31}b_1} & 0 & 0 & 0 & \frac{y_2x_2y_2t_2}{b_3\beta_{33}b_3t_3}{}^B\,\frac{y_2x_{23}y_3}{b_3\beta_{31}b_1} & \frac{y_2x_{23}y_3}{b_3\beta_{31}b_1} & 0 & 0 & 0 & 0 & 0 & 0 & 0 \\
0 & 0 & 0 & 0 & 0 & \frac{y_3x_3y_3t_3}{c_1\gamma_{11}c_1t_1}{}^C\,\frac{y_3x_3y_3}{c_1\gamma_{12}c_2} & \frac{y_3x_3y_3}{c_1\gamma_{12}c_2} & 0 & 0 & 0 & 0 & 0 & 0 \\
0 & 0 & 0 & 0 & 0 & 0 & \frac{y_3x_3y_3t_3}{c_2\gamma_{22}c_2t_2}{}^C\,\frac{y_3x_3y_3}{c_2\gamma_{23}c_3} & \frac{y_3x_3y_3}{c_2\gamma_{23}c_3} & 0 & 0 & 0 & 0 & 0 \\
0 & 0 & 0 & 0 & 0 & 0 & 0 & \frac{y_3x_3y_3t_3}{c_3\gamma_{33}c_3t_3}{}^C\,\frac{y_3x_3y_3}{c_3\gamma_{34}c_4} & \frac{y_3x_3y_3}{c_3\gamma_{34}c_4} & 0 & 0 & 0 & 0 \\
0 & 0 & 0 & 0 & 0 & 0 & 0 & 0 & \frac{y_3x_3y_3t_3}{c_4\gamma_{44}c_4t_4}{}^C\,\frac{y_3x_{34}y_4}{c_4\gamma_{41}c_1} & \frac{y_3x_{34}y_4}{c_4\gamma_{41}c_1} & 0 & 0 & 0 \\
0 & 0 & 0 & 0 & 0 & 0 & 0 & 0 & 0 & \frac{y_5x_4y_4}{d_1\delta_{11}d_1t_1} & \frac{y_5x_{45}y_5}{d_1\delta_{12}d_2} & 0 & 0 \\
0 & 0 & 0 & 0 & 0 & 0 & 0 & 0 & 0 & 0 & \frac{y_5x_5y_5}{d_1\delta_{11}d_1t_1}{}^D\,\frac{y_5x_5y_5}{d_1\delta_{12}d_2} & \frac{y_5x_5y_5}{d_1\delta_{12}d_2} & 0 \\
0 & 0 & 0 & 0 & 0 & 0 & 0 & 0 & 0 & 0 & 0 & \frac{y_5x_5y_5}{d_2\delta_{22}d_2t_2}{}^A\,\frac{y_5x_{56}y_6}{d_2\delta_{21}d_1} & \frac{y_5x_{56}y_6}{d_2\delta_{21}d_1} \\
0 & 0 & 0 & 0 & 0 & 0 & 0 & 0 & 0 & 0 & 0 & 0 & \frac{y_6x_6y_6}{d_6}
\end{bmatrix}
$$

Fig. 14. The description of the functioning of the two-level CHPS in the matrix-predicate form.

6 Conclusion

The developed matrix-predicate methods of information processing provide a possibility of describing CHPS with the matrix which elements are the true value of the pentadic predicate. It will enable to describe any multi-level hierarchical systems by matrices, perform logical and theoretical operations and facilitate working on these systems on the computer.

References

1. Sumin, V.I., Smolentseva, T.E., Vasilchenko, D.A., et al.: The Method of Complex System Partition into Hierarchical Structures. Elibrary, Moscow (2015)
2. Volkova, V.N., Denisov, A.A.: The Theory of Systems and Systematical Analysis. Urait, Moscow (2010)
3. Dubov, V.M., Capustyanskaya, T.I.: The Issue of Complex Systems. Elmor, St. Petersburg (2006)
4. Cron, G.: The study of complex systems in parts. Science, Moscow (1972)
5. Cron, G.: Tenser Analysis of Networks. Soviet radio, Moscow (1978)
6. Sukhov, Y.I., Danilov, A.M.: The development of hierarchical structures of complex systems. vol. 2, pp. 46–48. New University Engineering (2014)
7. Vanin, A.V., Voronov, E.M., Carpunin, A.A.: The managing optimization of two-level hierarchical system of stabilizing aircraft guidance. Nat. Sci. **6**, 19–42 (2012)
8. Voronov, E.M.: The optimization of multi-unit and multi-criteria systems management based on sustainable and effective game solutions. MSTU, Moscow (2001)
9. Petrov, J.K.: A hierarchical asynchronous model of graph representation in parallel and distributed systems. Youth scientific and technical Bulletin, pp. 30–35 (2014)

10. Polyakov, S.V.: The representation of the description of an object conducting a technological process by Berge's graph. The issues of mechanization and technology of the construction operation. Questions of mechanization and technology of building production, pp. 24–31 (1978)
11. Polyakov, S.V.: The methodology of formal describing of navigational lock functioning. In: Proceedings of the Leningrad Institute of Water Transport Organization and Management of the Transport Process in Water Transport, pp. 33–39 (1975)
12. Polyakov, S.V.: About locking process formalization. In: Proceedings of the Gorky Institute of Water Transport Engineers, vol. 201, pp. 18–24 (1984)
13. Polyakov, S.V., Slastinin, S.B., Polyakov, V.S.: Exception of isomorphism when operating graphs describing technological process. Control diagnosis, vol. 1, pp. 46–48 (2006)
14. Polyakov, S.V., Polyakov, V.S.: Modeling of the parallel processes by the units of interacting components. Control diagnosis, vol. 8, pp. 70–72 (2008)
15. Malynin, L.I., Malynina, N.L.: Graph Isomorphism in Theories and Algorithms. URSS, Moscow (2009)
16. Berge, C.: The Theory of Graphs and Its Applications. Wiley, New York (1962)
17. Mclikhov, A.N.: Directed graphs and state machines. Science, Moscow (1971)
18. Harary, F.: Graph Theory. Addison Wesley, Boston (1994)
19. Zykov, A.A.: The Theory of Finite Graphs. Science, Moscow (1968)
20. Carelin, V.P.: Graph theory models and methods in decision support systems. Bulletin of Taganrog Institute of Management and Economy, vol. 2, pp. 69–73 (2014)
21. Gapanyuk, J.E., Revunkov, G.I., Fedorenko, J.S.: The predicate description of metagraphic data model. Information, measuring and control systems, vol. 12, pp. 122–131 (2016)

Modeling and Automation of the Hydro-Transport System of Water-Coal Fuel at Negative Ambient Temperatures

K. V. Osintsev, O. G. Brylina[(✉)], and Yu. S. Prikhodko

South Ural State University, 76 Lenina Avenue,
Chelyabinsk 454080, Russian Federation
teolge@mail.ru

Abstract. The article considers the option of creating an installation for hydrotracking water-coal fuel to study the effect of thermophysical and rheological properties of the pumped material on heat and mass transfer processes during cooling and heating. The studied scheme is divided into three lines, through which water-coal fuel with different characteristics is pumped. In the simulation, it is proposed to use the refrigerant at the first experimental site, and low-boiling liquid in the heat pump at the second experimental site. The process hydrotransport automation possibilities are also shown. A scheme for controlling the technological process and controlling the main parameters is proposed. It includes elements of automated electric drive and control based on the multi-zone regulator. In addition, the implementation of the controller based on the neural network software allows achieving an increase in efficiency by reducing unnecessary heat losses. The article may be interesting to specialists in the field of power electronics and information electronics, electric drives and process automation.

Keywords: Water-coal fuel · Hydrotransport · Control system · Multi-zone regulator · Neural network

1 Introduction

Hydrotransport systems are used for pumping liquids with various thermophysical characteristics: density, viscosity, temperature. The prospects of hydraulic pipeline transport of various materials are determined not only by high efficiency, but also by indicators such as continuity and uniformity of the transport process, high productivity, environmental friendliness and the possibility of full automation of the entire process. In the coal mining regions of developing countries remote from railways and with an undeveloped automobile network in mountainous areas, high technology is being developed for pumping coal-water fuel [1] and flaring of water-coal suspensions [2] in power steam generators of industrial enterprises and thermal power plants. Scientists are approaching the decision of the choice of fuel, first of all, from the economic point of view [3]. For example, in China, the leader in the development of water-coal technologies, the largest at the moment in Asia Maoming thermal power station is built and operates on the basis of suspension preparation plant [4]. Previously, similar

© Springer Nature Switzerland AG 2020
A. A. Radionov and A. S. Karandaev (Eds.): RusAutoCon 2019, LNEE 641, pp. 657–666, 2020.
https://doi.org/10.1007/978-3-030-39225-3_72

projects were successfully implemented in the USSR, for example, the Min-Kushskaya CHPP in the Kyrgyz ASSR [5, 6], the coal lines Belovo-Novosibirsk CHPP-5 [7] and the Jubilee hydroelectric power plant - the West Siberian Metallurgical Combine [8]. During the construction of coal pipelines, a number of problems arise, one of which is transportation in winter, since the water-coal suspension freezes at zero degrees. Underground pipe laying in the first place cannot always guarantee the non-freezing in different climatic conditions of the Russian Federation. And secondly it leads to difficulties in maintenance of the pipeline. In addition, the equipment used at many sites commissioned decades ago has already become obsolete, and all works have ceased at a number of facilities, in particular, the Belovo-Novosibirsk coal pipelines has been stopped since 1993 [9]. At the same time, domestic and foreign experts have proved the technical feasibility and economic advisability of the main transport of coal through pipelines [10–13]. Therefore, issues related to the creation of reliable and energy-efficient automated water transport systems for water-coal fuel become urgent [14].

2 Research Relevance

The work urgency is due to the need to increase the competitiveness of the Russian Federation in the market of alternative fuels.

Since suspensions can be prepared not only from high-quality raw materials for metallurgical production, but also from coal enrichment wastes for transportation over short distances and direct disposal in flaring. Thus, the costs of industrial enterprises own needs are reduced. For example, concentrating mills and factories. For comparison, the share of coal-fired power generation in the US is 52%, in Germany - 54%, China - 72%, Poland - 94%. At the same time, the share of natural gas in the total world consumption of primary energy resources has increased significantly in recent years, primarily due to its use in thermal power plants. The main difficulty lies in the high cost of initial investment in the construction of coal pipelines. However, according to experts, the cost of building a coal pipeline should be offset by the profitability (compared to other modes of transport) of its subsequent exploitation.

3 Research Objectives Setting

To study the water-coal fuel properties during its transportation in conditions of negative ambient air temperatures, it is proposed to use combined modeling: physical and computer. Following tasks are solved during the work: measurement of water-coal mixture parameters for determination of fluidity and determination of electric energy consumption for pumping fuel at various characteristics, and also selection of the optimal (from the point of view of providing specified characteristics) automatic control system for the hydrotransport complex.

4 Scientific Novelty. Study Features

The experiments are conducted in two sections: the first section is for cooling water-coal fuel with freon, the second one is for heating in pairs of low-boiling liquid in a heat pump. For the first time it is proposed to use a thermal transformer as a unitized unit.

5 Experimental Setup Description

Consider the option of creating an installation for hydrotracking water-coal fuel to study the effect of thermophysical and rheological properties of the pumped material on heat and mass transfer processes during cooling and heating, Fig. 1.

According to the scheme in Fig. 1, it is necessary to consider the mixture motion along three lines, with the water-coal mixture having different characteristics on each of them. For example, on the first line a cavitator [12] is installed, the operation principle of which is based on the shock waves action. The second line moves a mixture of water and coal coarse fractions (dry grinding in a hammer mill) [13]. On the third line there is a wet coal-grinding mill with reduced energy consumption [14]. That is, on all lines, there are three main ways of preparing a water-coal mixture, which greatly facilitates the task of data verification with existing energy objects and systems.

Control of the experimental setup is reduced to one unit. The use of neural network algorithms in the control unit allows for the consideration of implicit dependencies during the operation of the transport system, such as the adhesion of coal particles to welds and shut-off valves, depending on the particle size, fuel rate, pressure in the pipeline and the temperature difference between the slurry and the ambient air. Reducing the chances of such a situation reduces the emergency potential of the system. There are many empirical formulas linking the temperature and viscosity of liquid petroleum products. Among them: the formula of Poiseuille, Bingham, Slott, Darcy, Le Chatelier, Vogel-Fulcher-Tamman, Raman, Walter, Bachinsky. Different formulas have different limits of applicability, which is related to the structure of the oil product, the behavior of real liquids in certain temperature ranges and interpolation errors. Poiseuille formula can be used to estimate flow effect on viscosity.

$$V = \frac{\pi R^4}{8\eta} \frac{\Delta \rho}{l}$$

where η is the dynamic viscosity, V is the volume per unit time, $\Delta \rho$ is the required pressure difference, l is the tube length, R is the pipe radius. Vogel-Fulcher-Tammann equation can be used as a basic viscosity dependence on temperature, describing in the form of a continuous function discrete experimental data.

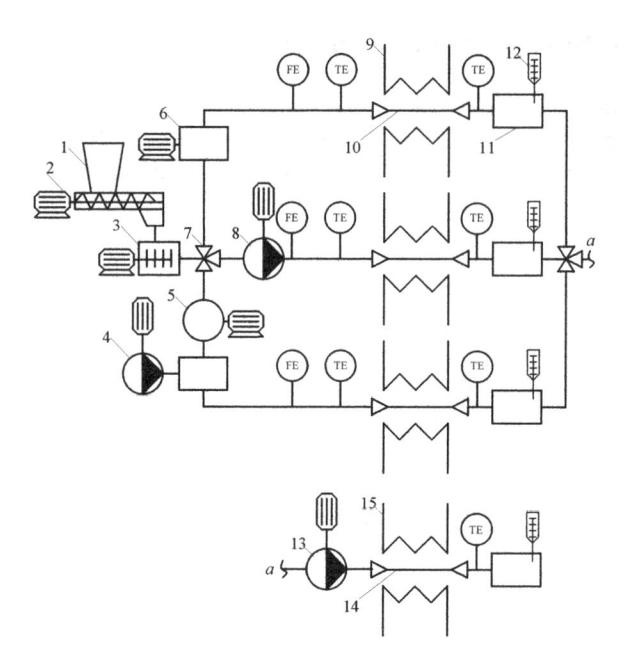

Fig. 1. The scheme of the experimental stand: 1 – bunker and feeder of coal or water-coal mixture; 2 – electric drive of the feeder; 3 – water-coal fuel mixer; 4 – pump-doser of additives; 5 – wet coal grinding mill with reduced energy consumption; 6 – cavitator; 7 – a three-way valve; 8 – pulp pump; 9 – cooler; 10 – the first experimental site; 11 – intermediate tank of the finished product; 12 – viscometer; 13 – second lift pump; 14 – the second experimental site; 15 – heat pump.

6 Example of the Control System

From the electric drive point of view in the hydrotransport system, as a rule, normal and emergency modes of operation are singled out [15–17]. Normal operating mode usually involves two (or more) serially connected pumps with all open sluices of the straight pressure line. Their joint work provides the specified characteristics of the moving mixture. In this case, the speed of each pump is synchronously regulated, for example, by a frequency-controlled converter. Emergency mode of operation provides for two series-connected pump with the open valves of the bypass (reserve) line. In this case, two options are possible: each pump control performs its frequency-controlled converter, providing the specified characteristics; if one frequency-controlled converter fails, the pump is controlled by the remaining working converter. It accelerates the first pump to the nominal constant speed, and then switches to the second pump and adjusts the frequency of its rotation in the specified range. To regulate the pipeline hydro-transport operation in working condition is also possible on the basis of throttle control at a constant shaft rotation speed. This is the simplest, but less economical way. In order to increase reliability indicators, in particular those related to solving diagnostics and reservation issues, it is possible to propose a control system for electric drives of the hydro-transport system of water-coal fuel on the basis of a multi-zone integrating regulator [18–21]. The functional scheme of a control system of a group of electric drives of the water-coal fuel hydro-transport system is shown in the Fig. 2. The functional scheme includes a multi-zone regulator (MR) with frequency-pulse-width modulation, a group of electric motors M1-Mn and pumps P1-Pn, connected to them, with return gates RG1 - RGn, output trunk T and pressure sensor PS, located in a dictating point of the trunk T. The start of M1 - Mn is carried out from thyristor control stations TS1 - TSn, realizing a mode of "soft" start-up of electric motors for the score of phase regulation of power at a stage of M1 - Mn dispersal up to a moment of their output on the natural characteristic [22–26].

The structure of an integrating MR (Fig. 3) includes adders $\Sigma 1$, $\Sigma 2$, an integrator I with a constant of time T_I and a group of relay elements $RE1 – REn$. It is considered that $n \geq 3$ is the odd number, and for obtaining the required number "k" of modulation zones $n = k - 1$ is necessary.

The relay elements have a non-inverting closed hysteresis loop, symmetric about the zero level, and thresholds of switching satisfying the expression $|\pm b1| < |\pm b2| < \ldots < |\pm bn|$. There is the index at «b» defined by a sequence number of RE. The signal at the output of all relay elements varies discretely within the bounds of $\pm A/n$. We shall be limited henceforward to the case where $n = 3$. Here and later it is considered, that the transfer coefficient of MR is equal to 1, and the change of a level of an input signal coincides with the beginning of the next cycle of sweep conversion.

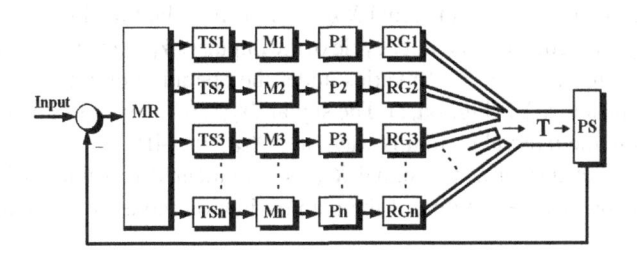

Fig. 2. The functional scheme of a control system of a group of electric drives of the hydro-transport system of water-coal fuel: MR – multi-zone regulator; TS1 - TSn – thyristor control stations; M1-Mn – electric motors; P1-Pn – pumps; RG1 – RGn – gates; T – trunk; PS – pressure sensor.

If the input signal X_{IN} is equal to zero and MR is turned on, then relay elements are installed arbitrarily. For example, the output signals of relay elements are equal to $+A/3$ (Fig. 4c–e). The $RE1$ and $RE2$ are switched sequentially in position $-A/3$ (Fig. 4c, d, time moments t_{01}, t_{02}), under the action of the sweep signal $Y_1(t)$ from the output I (Fig. 4b). After that sweep conversion direction is changing and the sweep signal $Y_1(t)$ begins to increase in a positive direction. Subject to the expression $Y_I(t) = b1$ being true, then MR is enabled in sustained auto oscillation mode, wherein the maximum value of the sweep signal $Y_I(t)$ is limited of $RE1$'s ambiguity zone. The output signals of $RE2$ and $RE3$ are disposed in fixed and inverse on the sign conditions $Y_{RE2}(t)$, $Y_{RE3}(t)$ (Fig. 4d, e). The MR's output signal is generated by the changing of $RE1$'s condition (Fig. 4c) within

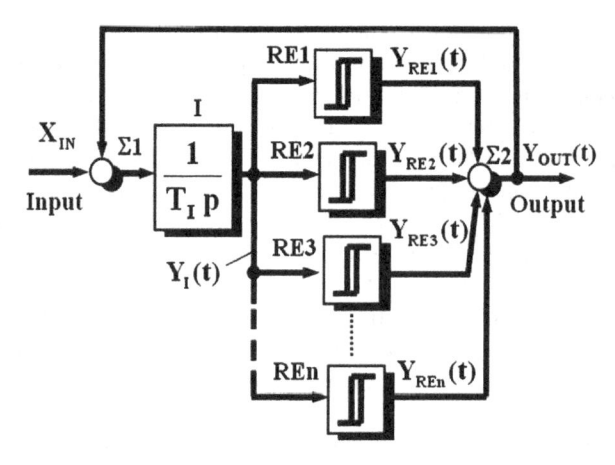

Fig. 3. The structure scheme of an integrating MR: X_{IN} – input signal; $\Sigma 1$, $\Sigma 2$ – adders; I – integrator; T_I – constant of time of the integrator; $Y_I(t)$ – output signal of the integrator; RE1 - REn – relay elements; $Y_{RE1}(t)$ - $Y_{REn}(t)$ – output signal of relay elements; Y_{OUT} – output signal of MR.

the first modulation zone. That is in the range $\pm A/3$ (Fig. 4f). In the absence of signal X_{IN} (Fig. 4a, time moment $t < t_0$), the average value Y_0 of the output pulses $Y_{OUT}(t)$ is equal to zero. The availability of input signal $X_{IN} < A/3$ (Fig. 4a, time interval $t_0 < t < t_0^*$) entails a change of frequency and duty cycle of the impulses $Y_{OUT}(t)$, since in an interval t_1 (Fig. 4c) the signal $Y_I(t)$ (Fig. 4b) varies under the operation of a difference of signals submitted on the adder $\Sigma 1$ (Fig. 4a, f). The signal $dY_I(t)/dt$ depends upon the amount of these impacts in the interval t_2. In this way $Y_0 \equiv X_{IN}$ (Fig. 4f).

Assume that at instant t_0^* the signal X_{IN} is incremented to value $A/3 < X_{IN} < A$ (Fig. 4a). This breaks the stability of the auto oscillation mode in the modulation zone number one.

MR reorientation of the *RE2* and *RE3* states begins, it comes to an end at instant t_{03}, when the *RE3* is changed over to the state $-A/3$ (Fig. 4e). The output signal $Y_{OUT}(t)$ is equal to the value $-A$ (Fig. 4f), and the MR passes into the modulation zone number two. At the time moments t_1, t_2 (Fig. 4b, c) the formation rate of the sweep function $Y_I(t)$ (Fig. 4b) is also defined by the sum or the difference of signals affecting the adder $\Sigma 1$. In this case the signal Y_0 contains a constant value $-A/3$ of the modulation zone number one and the average value of pulse stream $Y_{OUT}(t)$ of the second modulation zone of MR. In the control system (Fig. 2), with help of an input signal, MR is transferred into a state, in which all groups of pumps «Mi - Pi» are turned on. With the growth of pressure in the trunk T under the operation of a negative feedback signal from an output of PS, the relay elements REi (Figs. 3 and 4) are oriented in such a manner that a part from in a parallel way working electric drives are switched off, and other pass into a mode of multi-zone regulation. For example (Fig. 5), groups «RE2 - TS2 – M2 – P2» and «RE3 – TS3 - M3 – P3» are constantly in the included state, and the cascade «RE1 - TS1 - M1 - P1» is in a mode of periodic activations, supporting nominal pressure N_{NOM} in the trunk T with an allowable error is equal to $\pm \Delta P$.

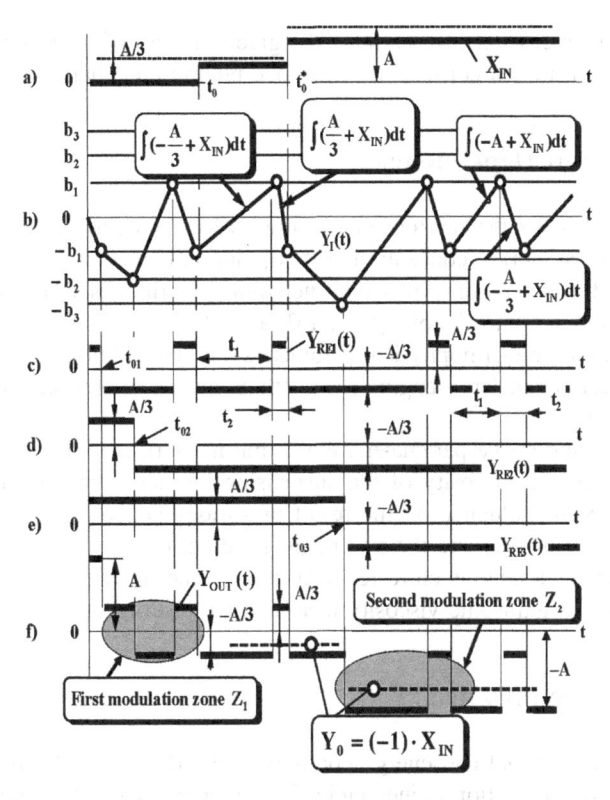

Fig. 4. The time diagrams of a mode of digital-pulse control of water pumps group: X_{IN} – input signal; $Y_I(t)$ – output signal of the integrator; $Y_{RE1}(t)$ - $Y_{REn}(t)$ – output signal of relay elements; Y_{OUT} – output signal of MR; Y_O – output signal of MR.

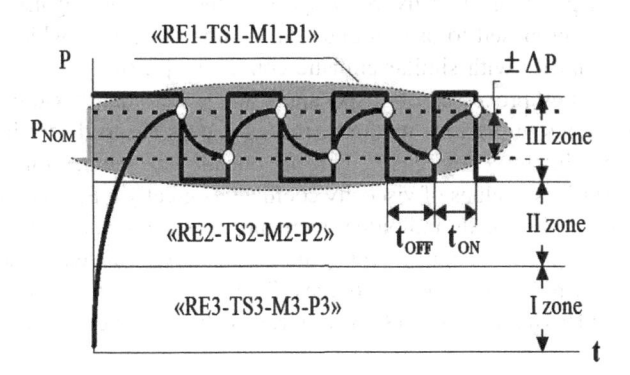

Fig. 5. The time diagrams of a mode of digital-pulse control of a group of water pumps: t_{ON} and t_{OFF} are the durations of the on state and off states of the electric drive, respectively.

For the uniform allocation of a number of start - brake control modes between a group of electric drives, one of the "relay" algorithms can be used, when through a defined slice of time or on expiration of the given number of inclusions the digital - pulse mode is transmitted consistently from one pump to another.

7 Basic Design Dependencies

Control parameters for the mixture preparation is its mass and the amount of water added to the coal at a given concentration [5, 6]; during transportation - the pressure in the pipeline and the speed of movement, including the critical speed of the mixture is governed by the recommendations [5, 6, 14], as well as the viscosity of the mixture, which depends on the thermophysical parameters and the addition of surfactants [27]. Among electrical parameters - engine speed and pump impeller turn, head and feed at each point in time.

By changing the above parameters, the neural network controller can achieve the optimum value of the viscosity of the suspension by calculating it through various formulas. Degree of influence on the result of various factors, such as temperature, pressure, the share of coal and the size of its grinding, is evaluated during the calculations. The obtained value is corrected according to the actual viscosity index, taken with the help of an ultrasonic viscosity measurement instrument [28].

8 Practical Significance

Diversification of the fuel and energy complex of the Russian Federation under conditions of import substitution is inevitable. The strategic plan for the development of electricity and heat generating capacities of the country should include the possibility of switching to alternative energy sources, which, of course, includes water-coal fuel. It is necessary to develop energy technology complexes for hydrotransport and utilization of water-coal suspensions. For hydrotransport in the case of negative ambient air temperatures, it is proposed to use ground heat pumps as heaters, which have proven themselves in countries with similar climatic conditions [5, 6].

Study of the flow processes of suspensions and water-coal mixtures, obtained in laboratory conditions by various methods, makes it possible to determine the dependence of the media viscosity coefficient on outside air temperature during their movement in the pipes. Values of viscosity coefficients greatly influence hydrotransport methods and the design of its individual elements. Research results can be tested on industrial hydrotransport systems, which in the end will increase their reliability through the use of renewable energy sources. Such solutions are especially important for the regions of Russia, China and CIS countries with difficult climatic conditions and terrain.

The automated electric drive of a hydrotransport system typically includes a group of parallel operating channels, including a backup channel. Many issues related to the complexity of control and coordination, backup and diagnosis of individual components of the drive can be solved using the methods of multi-zone regulation [18–21].

To save thermal energy for heating the suspension at low temperatures, it is possible to use ready-made developments in the field of short-term local weather forecasting using neural networks. This model allows you to predict future weather conditions at a certain point in the terrain more accurately than conventional sensors, reducing the necessary overheating from 5 °C to 2 °C [29, 30].

9 Conclusion

1. In the course of physical modeling it is necessary to measure the parameters of water-coal suspensions in order to determine fluidity and determine the cost of electricity for pumping fuel at various characteristics.
2. To solve the issues of reliability and energy efficiency of the water-coal fuel hydro-transport system, it is necessary to identify its optimal structure and management methods.
3. The use of neural network technologies allows minimizing the thermal and economic costs of the transport system.

Acknowledgements. The work was supported by Act 211 Government of the Russian Federation, contract № 02.A03.21.0011.

References

1. Staron, A., Kowalski, Z., Staron, P.: Properties of coal-water suspensions. Przemysl Chemiczny **93**(5), 748–751 (2014)
2. Redkina, N.I., Khodakov, G.S., Gorlov, E.G.: Rheology of coal-water slurries. Solid Fuel Chem. **43**(6), 341–350 (2009)
3. Mohapatra, S., Kumar, S., Chakravarty, A.: Transportation performance of highly concentrated coal-water slurries prepared from Indian coals. In: Applied Mechanics and Materials, pp. 592–594 (2014)
4. Wang, H., Guo, S., Yang, L., et al.: Surface morphology and porosity evolution of CWS spheres from a bench-scale fluidized bed. Energy Fuels **29**(5), 3428–3437 (2015)
5. Khidiyatov, A.M., Osintsev, V.V., Dzhundubaev, A.K.: Modeling of heat transfer in conditions of pollution of furnaces of pulverized coal boilers, transferred to burning of coal-water suspensions. Influence of mineral part of energy fuels on working conditions of steam generators. In: 4-th AUSTC, pp. 63–68 (1986)
6. Dzhundubaev, A.K., Kozmin, G.V., Bulanakov, Yu.K.: Experimental-industrial installation for the preparation and combustion of coal-water slurries of Kavak brown coal. Pipeline hydro transport of solid materials. AUSTC Hydro Transport, pp. 65–66 (1986)
7. Feldman, V.G.: Automation of reception, storage and burning of water-coal suspension at Novosibirsk CHP-5. Pipeline hydro transport of solid materials. AUSTC Hydro Transport, pp. 170–171 (1986)
8. Delyagin, G.N., Ivanov, V.M., Kantorovich, B.V.: Ways of efficient burning of watered fuels in the form of dispersed fuel systems and the prospects for the creation of a fuel and energy complex. Issues Hydraul. Coal Min. **8**, 157–167 (1986)
9. Kijo-Kleczkowska, A.: Research on destruction mechanism of drops and evolution of coal-water suspension in combustion process. Arch. Min. Sci. **55**(4), 923–946 (2010)
10. Aidarova, S., Bekturganova, N., Kerimkulova, M., et al.: The influence of surfactants to the stability of coal water suspension. Periodica Polytech. Chem. Eng. **58**, 21–26 (2014)

666   K. V. Osintsev et al.

11. Kirkby, W.A.: Combustion of coal-water suspension. Combust. Flame **8**(4), 112–118 (1964)
12. Glushkov, D.O., Strizhak, P.A.: Ignition of composite liquid fuel droplets based on coal and oil processing waste by heated air flow. J. Clean. Prod. **165**, 0087–0095 (2017)
13. Valiullin, T.R., Vershinina, K.Yu., Lyrshchikov, S.Yu., et al.: Ignition of fuel based on filter cake. Coke Chem. **60**(3), 127–132 (2017)
14. Zhou, M., Pan, B., Yang, D., et al.: Rheological behavior investigation of concentrated coal-water suspension. J. Dispersion Sci. Technol. **31**(6), 838–843 (2010)
15. Braslavskii, I.Ya., Plotnikov, Y.V., Ishmatov, Z.S. et al.: Evaluation of the technical and economic efficiency of implementation of frequency-controlled electric drives with capacitive storage in traversing gears. Russ. Electr. Eng. **5**(9), 559–563 (2014)
16. Lei, Zh., Jing, Zh.: Implement of increment-model PID control of PLC in constant-pressure water system. In: 8th International Conference on the Electronic Measurement and Instruments, vol. 4, pp. 336–339 (2007)
17. Xue, Z., Shi, L.: Modeling and experimental investigation of a variable speed drive water source heat pump. Tsinghua Sci. Technol. **15**(4), 34–440 (2010)
18. Brylina, O.G.: Multi-zone pulse regulator. In: 2nd International Conference on Industrial Engineering, Applications and Manufacturing, pp. 1–4 (2016)
19. Tsytovich, L.I., Brylina, O.G.: Features of modes of a multizone integrating controller with an even number of relay elements. Russ. Electr. Eng. **87**(12), 672–676 (2016)
20. Brylina, O.G., Saprunova, N.M.: The dynamic characteristics of a multizone regulator with frequency–width–pulse modulation at harmonic modulation of a relay element's switching thresholds. Russ. Electr. Eng. **86**, 689–693 (2015)
21. Dudkin, M.M., Brylina, O.G., Ponosov, D.A.: Adaptive thyristor voltage converter for a smooth start of induction motors supplied by stand-alone power plants. In: International Conference on Industrial Engineering, Applications and Manufacturing, pp. 1–6 (2017)
22. Battiston, A., Miliani, E.H., Pierfederici, S.: Soft-switched quasi-Z-source inverter topology for variable speed electric drives. In: 17th European Conference on Power Electronics and Applications, pp. 1–12 (2015)
23. Fulin, W., Huajie, X., Jing, C., et al.: A new hydro-electro-mechanical transmission system with soft start-up/stepless speed regulation. In: International Conference on Mechanic Automation and Control Engineering, pp. 2376–2379 (2010)
24. Thanyaphirak, V., Kinnares, V., Kunakorn, A.: Soft starting control of single-phase induction motor using PWM AC chopper control technique. In: International Conference on the Electrical Machines and Systems, pp. 1996–1999 (2013)
25. Jiang, Z., Huang, X., Lin, N.: Simulation study of heavy motor soft starter based on discrete variable frequency. In: 4th International Conference on the Computer Science and Education, pp. 560–563 (2009)
26. Tsytovich, L.I., Brylina, O.G.: Pulse-width and pulse-frequency-width sweeping converters for potential separation of DC circuits. Autom. Control. Comput. Sci. **49**(5), 293–302 (2015)
27. Drezin, E.L.: On the mechanism of asymmetric aluminum particle combustion. Combust. Flame **115**, 809–850 (1999)
28. Urazmetov, S., Kraev, V.P., Verevkin, A.P., et al.: Method for determining the viscosity of nonlinear viscous liquids and the device for its implementation. Fed. Intellect. Prop. Serv. Off. Bull. **16**, 24–32 (2013)
29. Fevralev, A.A., Prikhodko, Y.: Short-term local weather forecast in case of solving a problem of increasing efficiency of heating system. Bull. South Ural. State Univ. Constr. Eng. Arch. **16**(2), 48–51 (2016)
30. Fevralev, A.A., Prikhodko, Y., Babaylova, D.M.: Neural network usage for solving the problem of short-term local forecast of outdoor temperature. Bull. South Ural. State Univ. Constr. Eng. Arch. **17**(3), 48–53 (2016)

Modelling and Controlling the Temperature Status of the Turbine T-125/150 CCGT 450 Flow Part at the CCGT Operation in the GTU Based CHP Mode with Steam Turbine in the Motoring Drive Mode

E. K. Arakelyan[1]([✉]), K. A. Andryushin[1], and F. F. Paschenko[2]

[1] Moscow Power Engineering Institute, 14 Krasnokazarmennaya Street, Moscow 111250, Russian Federation
edik_arakelyan@inbox.ru
[2] Institute of Control Sciences. V.A. Trapeznikova RAS, Moscow, Russian Federation

Abstract. The technical possibility of transferring the T-125/150 steam turbine of 450 CCGT unit into motoring drive mode is considered as an alternative to stopping the steam turbine when transferring the CCGT unit to the GTU based CHP mode. To simulate the temperature state of the flow part of a steam turbine in steam-free and motoring drive modes, it is suggested to take into account the design features of a steam turbine. To illustrate the possibility of controlling the temperature of steam turbine stages, the results of model calculations show the temperature distribution graphs after their stabilization along the flow-through part of T-125 turbine for two modes of turbine operation - nominal and motoring drive modes. The possibility of adjusting the steam temperature at the inlet to the steam turbine and at the outlet from the LP cylinder is shown, whereby the duration of the starting operations of the steam turbine and the loading of the CCGT unit 450 to the nominal mode in this case is reduced by 20 min in comparison with the steam turbine shutdown mode.

Keywords: Combined cycle gas turbine · GTU based CHP mode · Shutdown · Motoring drive mode · Modelling · Temperature status · Design features · Control · Reduction · Duration · Load

1 Introduction

Given the high efficiency of the combined cycle gas turbine unit (CCGT unit), especially the heating type, they were designed for operation mainly in the basic mode with a minimum number of starts and load changes. However, the limited possibilities of load regulation in power systems due to the absence of special maneuvering power plants led to the fact that in reality the modes of operation of the CCGT unit differ significantly from the basic ones, and the low electricity tariffs at night led to the need for their deep unloading. At the same time, the work of power plants in the electricity and capacity market led to an increase in the demands of the power system to

A. A. Radionov and A. S. Karandaev (Eds.): RusAutoCon 2019, LNEE 641, pp. 667–674, 2020.
https://doi.org/10.1007/978-3-030-39225-3_73

maneuverability and reliability, especially during the hours of load failure. At the same time, the restricted limits on the regulation of the electric power of the CCGT unit, especially when operating in the heating mode, have led to the need for ways to expand the control range of CCGT unit [1–4]. One of the possible ways to solve this problem when the CCGT unit is operating in the heating mode is to transfer it to the GTU-CHP mode.

When CCGT unit working in GTU based CHP mode, all the steam generated in the heat recovery steam generators is discharged into the district heating water heaters in addition to the steam turbine, which, during the CCGT unit operation, stops in this mode. The need for such a mode arises in situations where the CCGT unit is required to be uploaded by electric power according to the power system schedule of electric loads, and the thermal load remains at the same level or increases (for example, during the passage of the winter time schedule load failures). When switching to GTU based CHP mode, the electric power of CCGT unit is reduced by the power of the steam turbine at the time of transfer of CCGT unit to this mode, and the generation of thermal power increases to the total heat volume of the high and low pressure steam streams at the entrance to the steam turbine for the deduction of thermal losses in steam pipelines, control units and network heaters. The operational disadvantage of this mode is the need to stop the steam turbine with subsequent start-up, which is associated with fuel losses during the start-up period, delayed loading of the CCGT unit as a whole due to the considerable duration of the start-up operations of the steam turbine.

2   General Information About the Motoring Drive Mode

In the presented report, instead of stopping the steam turbine, it is proposed to transfer it to motoring drive mode (MD). The motoring drive mode means the turbine is operated in the forced idle rotation mode due to the consumption of a certain power by a synchronous generator that remains in the network and a small amount of steam is supplied to the flow part of the turbine, moving at a low speed, performs only the cooling function of the turbine stages and guide blades [5].

Experimental studies and operation experience gained have shown that the transfer of turbine units in MD for periods of nighttime failures in the electric load schedule has a number of operational advantages in comparison with the stopping and starting conditions that significantly improve the maneuverability and reliability of equipment when operating in the load control mode of the power system [6–9].

In addition, the longevity of the steam-driven elements of the steam turbine increases due to the minimum amplitude of the thermal stresses during the whole cycle of load variation and the cyclic temperature stresses are reduced, which removes the limitations on the number of turbine unit shifts to standby mode and the reduction of its life-time parameters [10].

When operating in MD, elimination of the temperature changes, occurring at the turbine hot start, which considerable (140–200 °C) and sharp (at 18–20 °C/min) downstream of the control stage, in intermediate and last stages of HP cylinder, IP cylinder and LP cylinder. A higher level of steam temperatures downstream of the control stage and less dampening of the HP cylinder stages in the area of the first stages during turbine start-up is explained by the absence of stages of turn-over, ramp-up, idling, synchronization and switching of the generator into a network with low steam consumption that results in substantially faster turbine loading process.

One of such advantages is the ability to control the temperature of the flow part of the turbine by varying the parameters and flow rates of the cooling steam, which creates favorable conditions for the subsequent start-up of the steam turbine from an arbitrary temperature state.

When the CCGT unit is operating in the condensing mode [11], the motoring drive mode is proposed as a method of power back-up of the steam turbine when the CCGT unit is stopped as a whole during the passage of the power consumption failures, with emergency and short (for 8–10 h) time of gas turbines' shutdown. The heat balance diagram (Fig. 1) and the algorithm for transferring the steam turbine to MD are justified there, a technique for calculating the temperature state of the turbine blades in a steamless (without feeding the cooling steam) and in MD modes with steam supply is proposed, and the parameters and necessary costs of the cooling steam streams are determined.

3 Modeling the Temperature State of the Flow Parts of a Steam Turbine

On the basis of the mathematical model presented in [12] for estimating the steam temperature in the steam turbine stages, the following calculated dependence was obtained in [13, 14].

$$T_{i+1} = T_{0i} + \left(\sqrt{\frac{\beta_{STi+1} p_{i+1}}{c_{pi+1} GR} + \left(\frac{T_{0i}}{2}\right)^2} - \frac{T_{0i}}{2} \right) \cdot \left[1 - \exp\left(-\frac{c_{pi+1} G}{c_m m_{i+1}} \tau \right) \right], \quad (1)$$

where β_{ST} is the coefficient, which depends on the geometric parameters of the stage and is constant for a given stage; G is a mass of steam in the volume of the stage; m is the mass of the metal of the stage; $T_{0,i}, T_{i+1}$ - the temperature of the steam at the exit from the i-th stage and at the inlet to the stage $(i + 1)$; c_p, c_m - heat capacity of steam and metal.

When CCGT unit operates in the heating mode with a steam turbine in MD, the cooling circuit of the turbine flowpart is the same as when CCGT unit operates in the condensing mode, but has the following specifics:

- As gas turbines and heat recovery steam generators remain in operation, the steam for cooling the flow part of the steam turbine is extracted from the relevant steam supply lines (in addition to the steam turbine) leading to the network heaters
- When modeling the temperature state of steam in turbine stages in MD, it is necessary to take into account the design features of the steam turbine T-125/150. It consists in the fact that the HP cylinder of the steam turbine is made as double-flow, double-hull with a loop circuit of steam flow and during the flow of steam from the 8th stage to the 9th, heat exchange takes place between the steam and the external environment through the external HP cylinder casing, which is associated with the

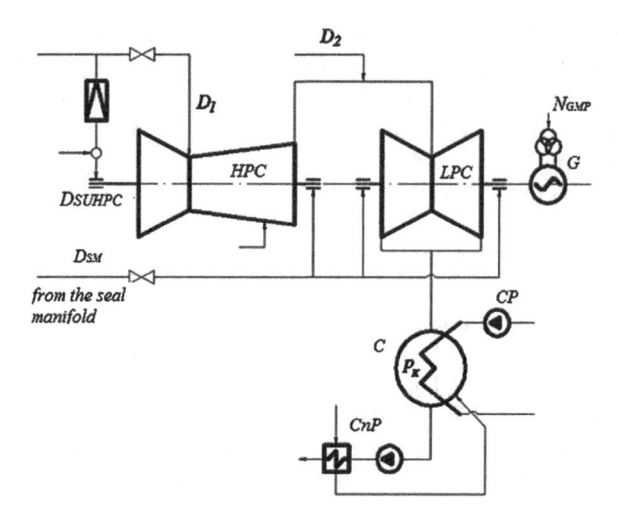

Fig. 1. Simplified scheme of steam supply to the steam turbine in MD in condensing mode of CCGT unit: HPC – high pressure cylinder; LPC – low pressure cylinder; G – generator; CP – circulating pump; CnP – condensate pump; C – condenser.

thermal losses. Although the absolute value of these losses is small, but at low steam flow rates, they will also affect the steam temperature at the inlet of the 9th stage and neglecting this component of the heat loss can lead to an unreasonable increase in the steam temperature at the inlet of the 9th stage [11].

Taking into account this peculiarity, the following equation is proposed for calculating the steam temperature at the inlet of the 9th stage:

$$t_{9in} = t_8 + \frac{\sum_{j=1}^{j=8} Q_{nj} - Q_{os}}{c_p G}, \tag{2}$$

where Q_{nj} – loss of heat through the surface of the inner body along the width of the j-th stage, determined by the heat transfer equation:

$$Q_{nj} = k_{Hj} F_{Hj} \left(t_{hj} - t_{os} \right), \tag{3}$$

where k_H is the heat transfer factor; F_H – the surface of the inner case of the body along the width of the j-th stage t_{hj}, t_{os} – temperature of steam at the exit from the j-th stage and outside the surface; Q_{os} – heat loss through the outer surface of the HP cylinder body:

$$Q_{os} = k_{os} F_{os} (t_h - t_{oa}), \tag{4}$$

where *kos* is the average value of heat transfer factor from the steam to the outside air; *Fos*– the outer surface of the body in a width of 1–8 steps; t_h, t_{oa} – average temperature of steam and outside air.

The use of calculation Eqs. (2–4) is associated with a significant amount of computation. For their reduction, it is proposed to calculate heat losses through the internal and external surfaces (taking into account the linear dependence of the steam temperature in the stages) based on the average steam temperature, i.e.

$$\sum_{j=1}^{j=8} Q_{nj} = k_H F_H \left(t_{at1-8} - \frac{t_8 + t_{9in}}{2} \right),$$ (5)

$$Q_{os} = k_{os} F_{os} \left(\frac{t_8 + t_{9in}}{2} - t_{oa} \right),$$ (6)

where t_{at1-8} – average temperature of steam in stages 1–8.

4 Turbine Steam Temperature Control

To illustrate the possibility of controlling the temperature of the steam turbine stages, the graphs of the temperature distribution after their stabilization along the flow part of T-125 turbine are shown below for the two modes of turbine operation: in a steam-free mode with steam supplied only to the turbine seals and in MD mode with the supply of cooling steam also to the flow part of the turbine at a condenser pressure of 0.004–0.005 MPa in the amount that ensure a deviation of the metal temperature of the turbine stages by no more than 35 °C in relation to the nominal operation mode temperature values (Fig. 2).

Control of steam temperature in the flow part of the turbine is considered in two aspects:

1. Ensuring reliable operation of the steam turbine during the backup period at stabilized temperatures of the turbine stages. For this mode, the temperature of the last stage of LP cylinder is selected as the adjustable one.
2. Creation of the required temperature state necessary for the rapid start-up of the steam turbine after operation in the MD, when it is necessary to load the CCGT unit by electric load.

The need to control the temperature of the last stage of LP cylinder is determined by the following factors:

- An excessive increase in temperature may lead to its unacceptable increase and to a decrease in the life of the turbine unit
- A decrease in temperature can lead to the formation of condensate, which at high speed of rotation of the turbine can cause mechanical damage to the blades

The nominal temperature of the metal body of the last stage of LP cylinder of the turbine is equal to 55 °C. For start-ups, the temperature deviations should not exceed

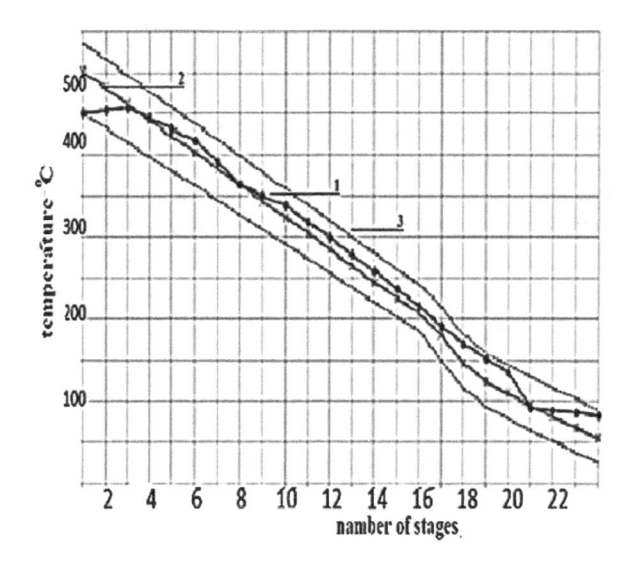

Fig. 2. Temperature distribution over the flow part of the turbine T-150-7 in the motor mode: 1 – Motor mode; 2 – Nominal mode; 3 – allowable variation.

35 °C. That is, when working in MD, it is necessary to maintain a temperature value of not more than 90 °C.

For the second mode, the steam temperature at the inlet of the steam turbine is the determining one. As can be seen from these graphs, the control of the steam temperature upstream of the first stage can be done in the temperature range from 310 to 450 °C.

5 Duration of Startup Operations

The start-up duration of the steam turbine after a stop for 6–10 h with operating gas turbines and heat recovery steam generators is at least 30 min. After the ramp-up, synchronization and switching of the generator into the network, the loading of the steam turbine to a power of 55–60 MW is performed in accordance with the start-up of the CCGT unit with parallel start-up of the gas turbines, after which the gas turbines and steam turbines are parallel loaded to the maximum load at a given thermal load of the steam turbine (Fig. 3a).

The duration of the start-up operations of the steam turbine and the loading of the CCGT unit to its rated power, as seen in Fig. 3a is 30.5 min.

The peculiarity of starting a steam turbine when leaving the MD mode is that the thermal state of the flow part of the turbine is such that steam of high and low pressures can be supplied, which accelerates the loading of the steam turbine and CCGT unit as a whole (Fig. 3b).

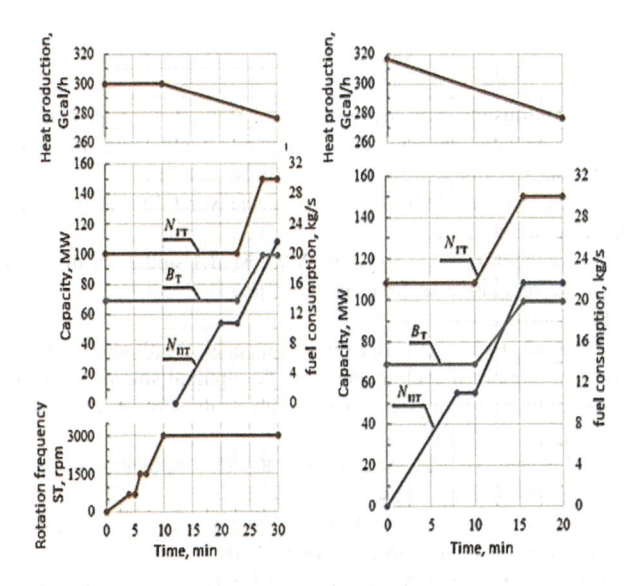

Fig. 3. Simplified curves: a—start-up of steam turbine and loading of CCGT unit in the GTU based CHP mode with shutdown of steam turbine; ∂—CCGT unit loading in mode GTU based CHP with transition of steam turbine into MD.

The total duration of the CCGT unit reaching the designed design load for electricity and heat in this version is estimated as 15.5 min and thus the duration of the start and load of CCGT unit is reduced by 20 min.

6 Conclusion

Transfer of the T-150-7 CCGT unit-450 steam turbine into the motoring drive mode when the CCGT unit is operating in the GTU based CHP mode allows:

- To provide long-term operation of the steam turbine in the hot backup mode with the possibility of temperature control in the steamer and at the outlet of the LP cylinder at the levels necessary for accelerated loading of the steam turbine when entering the normal operation mode
- The loading duration of the CCGT unit from the GTU based CHP mode with the transfer of the steam turbine to the motoring drive mode in comparison with the similar mode with the stop of the steam turbine is reduced by 20 min.

Acknowledgement. The reported study was funded by RFBR according to the research project № 18-08-01090.

References

1. Arakelyan, E.K., Andriushin, A.V., Burtsev, S.Yu.: Methods for expanding the control range of steam and gas installations and their comparative efficiency in terms of economy, maneuverability and reliability. Bull. MPEI **6**, 20–30 (2017)
2. Radin, Yu.A.: Research and improvement of maneuverability of combined-cycle plants. Dissertation, Moscow (2013)
3. Davydov, N.I., Zorchenko, N.I., Davydov, A.V.: Model studies of the possibility of PSU participation in the regulation of frequency and power flows in the UES of Russia. Heat Power Eng. **10**, 11–16 (2009)
4. Arakelyan, E.K., Andriushin, A.V., Burtsev, S.Yu., et al.: Methodology for consideration of specific features of combined-cycle plants with the optimal sharing of the thermal and the electric loads at combined heat power plants with equipment of a complex configuration. Thermal Eng. **62**(5), 335–340 (2015)
5. Madoyan, A.A., Levchenko, B.L., Arakelyan, E.K.: The application of the motor regime in thermal power stations. Energy, Moscow (1980)
6. Kapelovich, D.B., Kobzarenko, L.N., Baltjan, Y.N.: Research and technology scheme of motor turbine mode to-160-130. Electr. Stations **7**, 16–21 (1987)
7. Starshinov, V.A., Margaryan, L.V., Carasov, A.L.: Power characteristics of the turbine generators of thermal power plants in synchronous compensator mode. Electr. Stations **2**, 22–24 (1982)
8. Bjoorgar, A.B.: Economic problems in connection with production reactive power. Report and power 109 (1962)
9. Kulichikhin, V.V., Gutorov, V.F.: The use of a motor mode of the turbine units at thermal power plants. SPO ORGRES, Moscow (1977)
10. Truhnij, A.D., Kobzarenko, L.N., Madoyan, A.A.: Low-cycle reliability of turbine rotors to-200-130 with different ways of their output in night/reserve. Heat Power Eng. **10**, 33–38 (1982)
11. Arakelyan, E.K., Andriushin, A.V., Andryushin, K.A.: Increased reliability, manoeuvrability and durability of steam turbines through the implementation of the generator driving mode. WIT Trans. Ecol. Environ. **205**, 95–105 (2016)
12. Arakelyan, E.K., Starshinov, V.A.: Improving the economy and maneuverability of thermal power plant equipment. Publishing house MPEI, Moscow (1993)
13. Arakelyan, E.K., Sakharov, V.K.: Study of the temperature state of the HPC stages of steam turbine T-125/150 CCGT-450 at work in low-steam mode. New Russian Electric Power Ind. **1**, 5–17 (2013)
14. Arakelyan, E.K., Andriushin, K.A., Bezdelgin, I.Yu.: Investigation of the temperature state of the flow part of a steam turbine T-125/150 when operating it in a steam-free and motoring drive mode. Electr. Stations **6**, 21–26 (2015)

Analysis of the Secure Data Transmission System Parameters

M. O. Tanygin$^{(\boxtimes)}$, M. A. Efremov, and Ya. A. Hyder

Southwest State University, 94 Pyatdesyat Let Oktyabrya Street,
Kursk 305040, Russian Federation
`tanygin@yandex.ru`

Abstract. In this paper we describe the method which allows the separation of legal software data from the data that is sent by extraneous software sources in order to increase the reliability of legal software data. This is done through the use of a buffer to store legal software data by using a set of mathematical equations. The combination of reversible and irreversible transformations is the basis of secure data messages formation algorithm. Having finished the reception, the receiver starts the analysis of the data written in the buffer and makes chains of words. We also describe the mathematical models which allow us to get the numerical values of the presented method. The article show the size of the buffer influences the security of data transmission system. The best correlations between the parameters of formation of secure data transmission algorithm are determined. At the same time the article reveals the problem of collisions during the transmission of secure messages. By collision in this work we understand situation, when hardware selects two or more different chains from buffer. It is shown that by variation of parameters of secure data analysis algorithm we can reduce the possibility of authentication mistakes by a factor a 10.

Keywords: Analysis · Secure · Data · Transmit · Hash · Buffer

1 Introduction

In modern hardware, the hardware can't work without special software which provides compatibility with the operating system and supports the user interface [1–3]. It is necessary to control the software commands in order to run the correct firmware. Alternatively, data which is sent by extraneous software can be received and processed by hardware. It may cause errors or failures in the hardware processor or firmware [4–10]. Data is the backbone of today's communication. To ensure that data is secured and does not go to unintended destination. Therefore, it is important to protect the data transferred between software and hardware.

A. A. Radionov and A. S. Karandaev (Eds.): RusAutoCon 2019, LNEE 641, pp. 675–683, 2020.
https://doi.org/10.1007/978-3-030-39225-3_74

2 Method of Data Control

The software sends data combined in command pools. Each word of each pool is analyzed by the hardware. In case of a mistake, it is necessary to recurring poll transmission. The method under discussion allows us to the separation of legal software data from the data that sent by extraneous software sources [11–17].

Let $\{A|B|...|N\}$ – be concatenation result of words A, B, N. Our task is to transmit word Si from the pool of software words to hardware. The number of this word is i. These two words must be transformed as in relation to secret word S_{sec}. We suppose that secret word S_{sec} is known only by legal software and hardware.

Legal software executes the following set of procedures:

1. Forms word $i' = F_A(S_{sec}, i)a$;
2. Forms word $S_i' = F_B(S_i, i')$;
3. Forms word $i'' = F_C(S_i', i')$;
4. Sends word $\{S_i'|i''\}$ to hardware.

Hardware executes the following set of procedures:

1. Receives word $\{S' | i''\}$ from software;
2. Defines number $i_{rec} = F_C^{-1}(S_i', i'')$;
3. Defines number $i'_{rec} = F_A(S_{sec}, i_{rec})$;
4. Defines received word $S_i = F_B^{-1}(S_i', i'_{rec})$.

In these formulas: $F_A(A, B)$ – irreversible transformation; $F_B(A, B)$ and $F_C(A, B)$ – reversible transformation of word B in relation to key A; $F_B^{-1}(A, B)$ and $F_C^{-1}(A, B)$ – the reverse transformation of $F_B(A, B)$ and $F_C(A, B)$ [18].

3 Buffering and Verification

All received words are buffered in hardware. Let i_{rec} be received word tier. If current received software's word is sent by legal software then $i_{rec} = i$. In common case (if hardware receives some words sent by extraneous software) number of received words is $j \geq m \geq i$, where m – the current maximum number of tiers in hardware buffer. Condition of recording to bufferUse (1). In another case, the word will be ignored.

$$i_{rec} = F_C^{-1}(S_i', i'') \leq m + 1, \tag{1}$$

After pool transmission to hardware to select one software words from each of M hardware buffer tiers. Each software word should contain information about all previous words of pool for correct implementation of this procedure [19]. The simplest way to solve this problem is to use cryptography hashing. Each legal software word S_i

consists of information part S_i^{inf} and hash code S_i^{hash}, formed by all information parts of previous words $S_1^{\text{inf}} - S_{i-1}^{\text{inf}}$, Use (2):

$$S_i = \left\{ S_i^{\text{inf}} \middle| F_{hash}\left(S_{i-1}^{\text{inf}}, \ldots, S_1^{\text{inf}}\right)\right\}, \tag{2}$$

where F_{hash} – cryptography hashing function.

The algorithm of software word forming is shown schematically at Fig. 1.

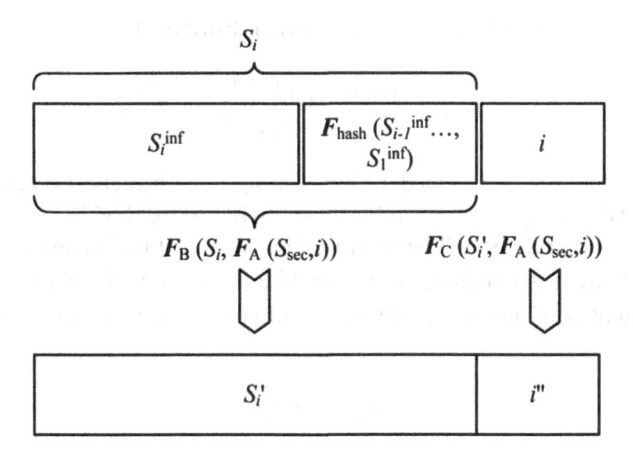

Fig. 1. Software word forming.

Let: $S_{j,r}$ be software word, which is received by hardware under number j and recorded on tier r, $S_{j,r}^{\text{inf}}$ be its information part; $S_{j,r}^{\text{hash}}$ – hash code.

Hardware has to determinate the chain of M words $S_{j(1),1}^{\text{inf}}, S_{j(2),2}^{\text{inf}}, \ldots, S_{j(M),M}^{\text{inf}}$ which satisfiers the following conditions (3).

$$j(1) < j(2) < \ldots j(M); \; S_{j(r),r}^{hash} = F_{hash}\left(S_{j(r-1),r-1}^{\text{inf}}, \ldots, S_{j(1),1}^{\text{inf}}\right), \tag{3}$$

where $r = 2 \ldots M$. Algorithm of words selection from buffer is based on conditions (3).

Described data control algorithm is adaptive. It means that transmitting conditions (intensity of extraneous software command words (ESCW), hardware errors intensity) define transmitting parameters: buffer size, numbers of buffer tiers, words length. Intensity of ESCW is informative and easy detected transmitting characteristic. Increasing of ESCW numbers increase erroneous chains forming probability in that time.

4 Buffer Overflow

Let's discuss the correlation between buffer size, numbers of buffer tiers and "buffer overflow" error probability. Receiving ESCW and legal command words (LCW) can be represented as probable Poisson process [20]. Assume now ESCW receiving probability P_{ESCW} is K times greater than LCW receiving probability P_{LCW}:

$$P_{ESCW} = K \times P_{LCW}, \tag{4}$$

Number of received ESCW n_{ESCW} is Poisson distributed:

$$p(n_{ESCW}) = \frac{(K \times n_{LCW})^{n_{ESCW}} \times e^{-K \times n_{LCW}}}{n_{ESCW}!}, \tag{5}$$

where n_{LCW} – number of received LCW; $K \times n_{LCW}$ – theoretic average number of received ESCW; $p(n_{ESCW})$ – probability of receiving n_{ESCW} ESCW.

Random formed ESCW will be record to buffer's tier if tier's number will be equal to $F_C^{-1}(S_i', i'')$ (as it determinate by transmitting algorithm). Probability of that p_{rec} depends on number of bits in i_{rec} or length L of command word field, which contains i (Fig. 1).

$$p_{rec} = 2^{-L}, \tag{6}$$

ESCW number n_R, recorded to a buffer tiers is binomial disturbed. Condition of buffer overflow is $n_R \geq N$ – maximal tier size. Now get the buffer's tier overflow probability (7):

$$p_{owfl} = \sum_{j=N}^{\infty} \sum_{i=j}^{\infty} \left\{ \left[C_i^j \cdot (p_{rec})^j \cdot (1 - p_{rec})^{i-j} \right] \cdot \frac{K \cdot n_{LCW}^i \cdot e^{-K \cdot n_{LCW}}}{i!} \right\}, \tag{7}$$

At Fig. 2 shown relationships between p_{owfl} and n_{LCW} (logarithmic scale) with different L, K and N.

Analysis shows that function graph form of p_{owfl} is depend on product $2^{-L} \cdot K$. Dependency diagrams between p_{owfl} and n_{LCW}, N and $2^{-L} \cdot K$ are shown at Fig. 3, where $1 - 2^{-L} \cdot K = 0{,}25$; $2 - 2^{-L} \cdot K = 0{,}5$.

In actual practice, we can preset values of parameters L, N M, but cannot determinate ESCW's activity K. We can assume that parameter K takes some value of an estimate defined value buffer overflowing frequency. Of course, values of parameters L, N M have to provide desired value of overflowing probability. If observed frequency of overflowing will be different than estimated, our hypotheses about parameter K value would be wrong, and we should change values of parameters L, N M.

It means, if for some values of the parameter we have detected a multiple overflow of the buffer in width (the overflow frequency exceeded the theoretically calculated example), this indicates a high intensity of ESCW issuance. Therefore, to return the

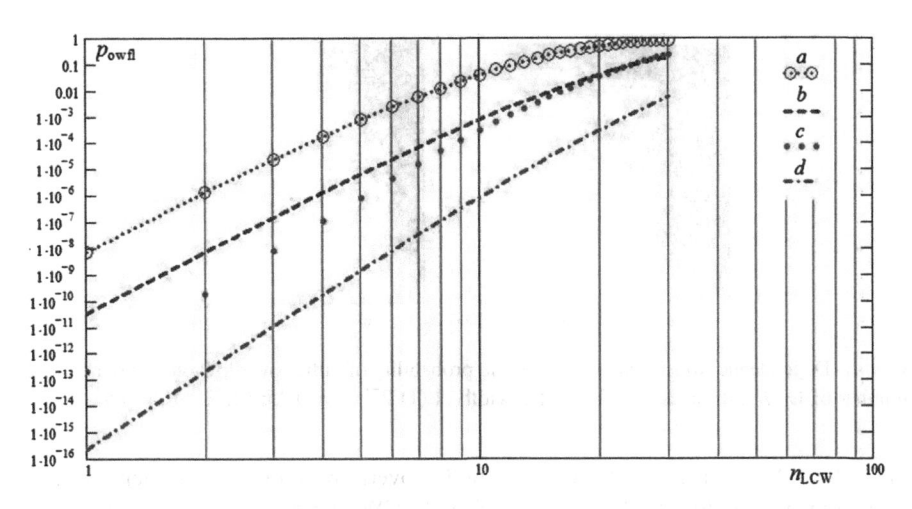

Fig. 2. Relations between buffer's tier overflow and LCW number (logarithmic scale) (a) L = 3, K = 3, N = 8; (b) L = 4, K = 3, N = 8; (c) L = 3, K = 2, N = 10; (d) L = 4, K = 2, N = 10.

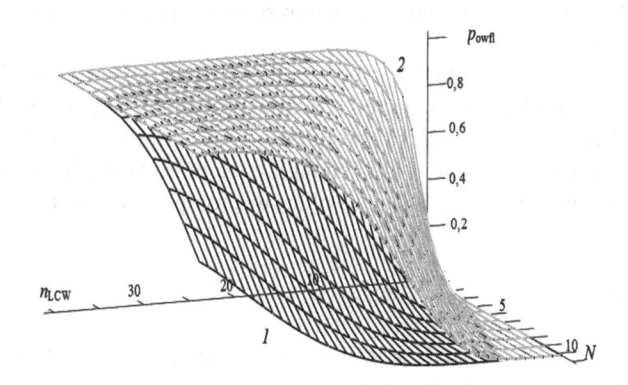

Fig. 3. Dependency diagrams between buffer's tier overflow and LCW number and maximal tier size.

overflow frequency to an acceptable range, it is necessary either to increase the length of the synchronization field L, or to change the parameters N and M.

Conversely, if overflows do not occur, we can increase the pool length (if it is necessary to issue long series of LCW), reduce its width (to increase the processing speed of the contents of the buffer) and reduce the length of the synchronization field (to reduce information redundancy).

The graph of the dependence of p_{owfl} on n_{LCW} and N for fixed values of the product $2^{-L} \times K$ is given in Fig. 4, where the probability of buffer overflow in width is displayed in shades of gray (10 intervals from 0 to 1, white corresponds to the interval from 0.9 to 1, black – from 0 to 0.1).

It is seen that for fixed buffer overflow probabilities, a linear relationship is observed between the buffer width N and the number of LCW received (or the depth of

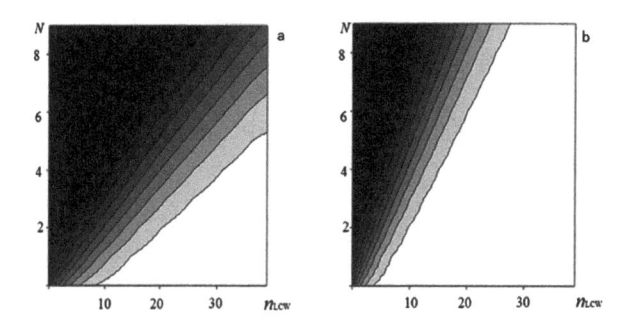

Fig. 4. Dependence (in shades of gray) of the probability of buffer overflow on one tier from the number of LCW estimates and the buffer width N (1) $2^{-L} \cdot K = 0.25$; (2) $2^{-L} \cdot K = 0.5$.

the buffer M). It means that any fixed buffer overflow frequency, the length of the synchronization field, and the intensity of the ESCW issuance corresponds to a certain ratio between the width and depth of the buffer. In other words: $N \approx \alpha \times M$, where $\alpha = f(2^{-L} \times K, v)$ is a function of the parameter $2^{-L} \times K$ and the observed buffer overflow frequency (which, with a large number of transmission cycles, is almost equal to the overflow probability).

Using the relationships found, it is possible to vary the length of the CW pool at a fixed buffer width, reaching the required overflow probability of any tier. Thus, we can vary the ratio between the length of the buffer and its width, achieving the required level of buffer overflow frequency. It is enough to store in the device memory a set of coefficients α for different L, K and v, in order to promptly change the N/M ratio.

5   Collisions

By collision in this work we understand situation, when hardware selects two or more different chains from buffer. Only one of these chains contains words of legal software.

The simplest collision is the selection of two chains shown on Fig. 5:

- First (legal): $S_{j(1),1}^{\text{inf}}, \ldots, S_{j(e-1),e-1}^{\text{inf}}, S_{j(e),e}^{\text{inf}}, S_{j(e+1),e+1}^{\text{inf}}, \ldots, S_{j(M),M}^{\text{inf}}$
- Second (illegal): $S_{j(1),1}^{\text{inf}}, \ldots, S_{j(e-1),e-1}^{\text{inf}}, S_{p,e}^{\text{inf}}, S_{j(e+1),e+1}^{\text{inf}}, \ldots, S_{j(M),M}^{\text{inf}}$

where: $S_{j(i),i}^{\text{inf}}$, $i = 1 \ldots M$ – information part of legal software word number i, received by hardware under number $j(i)$ and recorded to tier number i; e – number of tier where collision was detected; $S_{p,e}^{\text{inf}}$ – information part of extraneous software word, received by hardware under number p and recorded to tier number e.

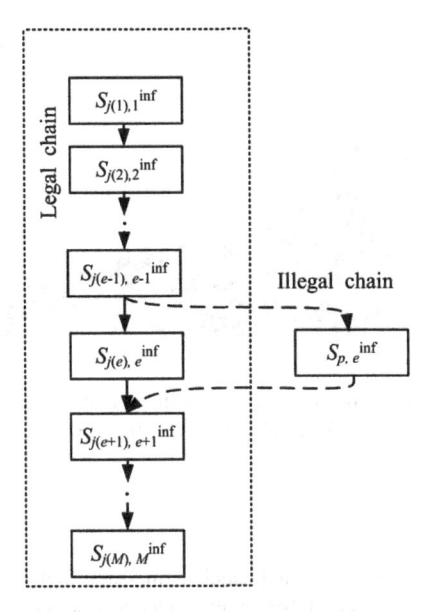

Fig. 5. Collision example.

Conditions of this collision:

$$S_{j(e),e}^{\text{inf}} \neq S_{p,e}^{\text{inf}};$$

$$F_{hash}\left(S_{j(k),k}^{\text{inf}}, \cdots S_{j(e+1),e+1}^{\text{inf}}, S_{j(e),e}^{\text{inf}}, S_{j(e-1),e-1}^{\text{inf}} \cdots S_{j(1),1}^{\text{inf}}\right)$$

$$= F_{hash}\left(S_{j(k),k}^{\text{inf}}, \cdots S_{j(e+1),e+1}^{\text{inf}}, S_{p,e}^{\text{inf}}, S_{j(e-1),e-1}^{\text{inf}} \cdots S_{j(1),1}^{\text{inf}}\right); \qquad (8)$$

$$j(e-1) < p < j(e+1);$$

where $k = e + 1 \ldots M$.

To prevent the collision formation of this type the Secure Data protocol, it is possible to provide the final integrity control generated by algorithms independent of the algorithm of integrity and authenticity control that would allow to reduce the risk from this kind of collision to a minimum.

At Fig. 6 shown a graph of the function of the probability p_{chain} of the collision on the buffer length M and the width of the buffer N.

It can be seen that the length of the hash sequence field has a very strong effect on the probability of a ESCW chain building: increasing it by only 1 bit reduces the probability by 2 orders of magnitude. At the same time, the rate of issuance of the ESCW has a significant effect on the probability value.

It is seen that for $N < M$ the probability p_{chain} is practically zero, for $N \approx M$, the probability increases, and then, as N increases, the probability value grows insignificantly. This is explained by the fact that for small N, the probability of buffer overflow in width is large, at which chain construction does not occur at all. This explains the presence of local maxima in the probability curves in the region $N \approx M$ (Fig. 6), which are so significant that they exceed the probability values in the region of small values of the length of the buffer M.

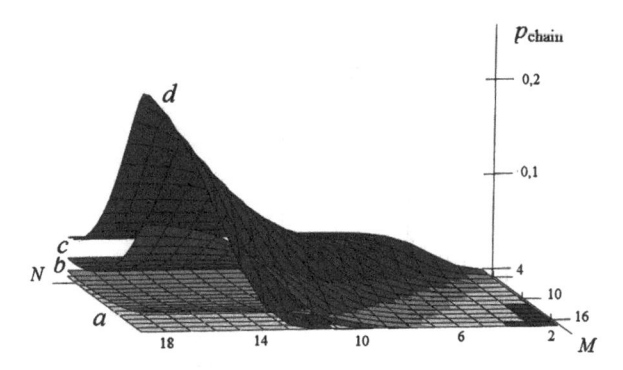

Fig. 6. Dependence of the probability p_{chain} of the collision on the buffer length M and the width N(a) $K = 2$, $L = 3$; (b) $K = 2$, $L = 3$; (c) $K = 4$, $L = 3$; (d) $K = 5$, $L = 3$.

References

1. McGrew, D., Viega, J.: Flexible and efficient message authentication in hardware and software (2005). http://www.cryptobarn.com/gcm/gcm-paper.pdf. Accessed 26 Nov 2017
2. Bellare, M., Kilian, J., Rogaway, P.: The security of the cipherblock xhaining message authentication code. J. Comput. Syst. Sci. **61**(3), 362–399 (2000)
3. Black, J., Rogaway, P., Shrimpton, T.: CBC MACs for Arbitrary-Length Messages. Advances in Cryptology. Santa Barbara (2000)
4. Black, J.: Authenticated encryption. In: Encyclopedia of Cryptography and Security. Springer, Boston (2005)
5. Iwata, T., Kurosawa, K.: One-key CBC MAC. In: Fast Software Encryption, pp. 137–161 (2003)
6. Jonsson, J.: On the security of CTR CBC-MAC. In: Proceedings of Selected Areas in Cryptography, pp. 76–93 (2002)
7. McGrew, D., Viega, J.: The security and performance of the Galois. In: Proceedings, Indocrypt, pp. 343–355 (2004)
8. National Institute of Standards and Technology: Recommendation for block ciphermodes of operation, Gaithersburg (2001)
9. National Institute of Standards and Technology: Recommendation for block ciphermodes of operation, Gaithersburg (2004)
10. National Institute of Standards and Technology: Recommendation for block ciphermodes of operation, Gaithersburg (2007)
11. Rogaway, P., Wagner, A.: A critique of CCM. Cryptology ePrint archive: report 070 (2003). http://eprint.iacr.org. Accessed 12 Oct 2017
12. Stallings, W.: The advanced encryption standard. Cryptologia **26**, 165–188 (2002)
13. Stallings, W.: NIST block cipher modes of operation for confidentiality. Cryptologia **34**(2), 163–175 (2010)
14. Diffie, W., Hellman, M.: Privacy and authentication: an introduction tocryptography. In: Proceedings of the IEEE, pp. 397–427 (1979)
15. Knudson, L.: Block chaining modes of operation. In: NIST First Modes of Operation Workshop (2000). http://csrc.nist.gov/groups/ST/toolkit/BCM/workshops.html. Accessed 23 Dec 2017

16. Lipmaa, H., Rogaway, P., Wagner, D.: CTR mode encryption. In: NIST First Modes of Operation Workshop (2000). http://csrc.nist.gov/groups/ST/toolkit/BCM/workshops.html. Accessed 23 May 2009
17. Voydock, V., Kent, S.: Security mechanisms in high-level network protocols. Comput. Surv. **15**, 135–171 (1983)
18. Tanygin, M.O.: Search and elimination of collisions in information exchange through open communication channels. In: Collection of Articles of the Xth International Scientific and Technical Conference. Privolzhsky House of Knowledge, Penza, pp. 62–64 (2010)
19. Tanygin, M.O.: Method of control of data transmitted between software and hardware. In: Computer Science and Engineering: Materials of the IV International Conference of Young Scientists, pp. 344–345 (2010)
20. Tipikin, A.P., Tanygin, M.O.: Methods of authentication of information protection systems and controlling software. Telecommun. Radio Eng. **66**(5), 453–463 (2007)

Geometric Modeling and CAD System to Solve Tectonics-Related Tasks Using Core Pole

T. S. Guriev, A. V. Kalinichenko$^{(\boxtimes)}$, and M. M. Tsabolova

North Caucasian Institute of Mining and Metallurgy, 44 Nikolayeva street,
Vladikavkaz 362021, Russian Federation
kalinichenkoalla@mail.ru

Abstract. Proper and rational borehole utilization in various aspects of prospecting and outlining of mineral deposits is undoubtedly a vital task. Geometric modeling and a widespread use of a CAD system could facilitate the solution of various mining, geological, and engineering problems. The geometric modeling of boreholes improves the process of prospecting and outlining of mineral deposits and supports their assignment to automation system of project planning. This research addresses one of the aspects of subsoil geometry associated with widespread use of computers to solve various mining-and-geological and engineering problems. A core pole provides various data about the geological structure of a blanket deposit gap. It reveals mechanical and structural properties of mineral deposits and helps to evaluate mineral reserves. To obtain this data a number of exploration holes is bored on a studied gap of a mineral deposit. The geometric modeling of boreholes is based on their approximation to the classic geometric images. From the geometric point of view, a borehole is a cylinder. Therefore, some positional and metric problems could be solved on a core pole using descriptive geometry. A developed algorithm and its mathematical formulation contribute to designing of a program for automatic problem solving and measuring of strike and dip using a core pole.

Keywords: Geometric modeling · Mining and geological tasks · Borehole · Core pole

1 Introduction

From a geometric point of view, mineral deposits are a combination of various geometric images of many possible space forms. They could take forms of isolated irregular geometric bodies (e.g. lenses and stockworks), stretched irregular geometric bodies (veins) or stretched regular bodies confined by planes (blanket deposits) [1–3].

Diverse space forms of mineral deposits are most pronounced in the setting of mountainous terrain, which can be explained by genesis of accumulation and its tectonics. Surface irregularities have a great impact on the geometric structure of a subsoil mineral deposit, which is most noticeable in the setting of mountainous terrain. From a geometric point of view, blanket deposits have a simpler form that allows for a relatively accurate estimation of their metric characteristics and a higher quality mining.

© Springer Nature Switzerland AG 2020
A. A. Radionov and A. S. Karandaev (Eds.): RusAutoCon 2019, LNEE 641, pp. 684–692, 2020.
https://doi.org/10.1007/978-3-030-39225-3_75

2   Recent Research and Publications Analysis

Research suggests that improved methods of geological analysis and project planning are key to proper and rational borehole utilization. Computer-aided design systems are now more and more often used in this field [4–10].

So for example, in the study of [7] a point is made that building of 3D geologic visual and computational models is of great importance for mining and geological projects. A 3D model can illustrate spatial properties of boreholes and a visual model transformed into computational provides for engineering analysis. A method of building a 3D geologic model of solid mine surface powered by AutoCAD is analyzed in [9]. The article reviews optimization issues based on a geologic model and technology mining. The author of [10] suggests a new visualization method based on spatial database and graphical analysis.

A core pole provides various data about geological structure of a blanket deposit gap. Core poles reveal mechanical and structural properties of mineral deposits and help evaluate mineral reserves [11, 12].

Moreover, boring of several exploration holes on a studied gap makes it possible to estimate the deposit shape, size and volume and, as a result, its reserves. This task is especially relevant for mineral deposits found in the setting of mountainous terrain.

3   Objective Statement

Computer-aided design systems are now frequently used in geology. To improve the results of mineral deposits prospecting and outlining the authors introduce computer programs addressing respective goals, which help to obtain information about deposit parameters before completion of prospecting and exploration operations and cut down material expenses for field observations and reduce time of data collection.

The goal was to solve the following problems:

- Research and analysis of geometric modeling of exploratory bores and methods for approximating the surfaces of mineral deposits with regular geometric shapes
- Development of geometric algorithms for solving exploration and outlining problems of mineral deposits by means of geometric modeling based on approximation techniques by regular geometric images (geometric algorithms are considered to be a sequence of geometric operations for solving a problem expressed in symbols)
- Investigation of the possibility of using the obtained geometric algorithms for the automated solution of exploration and outlining problems of mineral deposits
- Development of a mathematical description of the presented graphic algorithms for solving the tasks
- Development of algorithms for the automated solution of the main tasks of exploration and outlining of mineral deposits based on the completed mathematical description and software development

4   Main Part

Geometric modeling of boreholes is based on their approximation to the classic geometric images. From the geometric point of view, a borehole is a cylinder. So, some positional and metric problems could be solved using a core pole using descriptive geometry [13–15].

Suppose boreholes of 120–140 mm diameter are bored in solid rocks. Let us create a geometric model of a core pole on a complex diagram. A core pole in Fig. 1 has pronounced contact surface of materials shown as parallels prescribed by projections of ellipses.

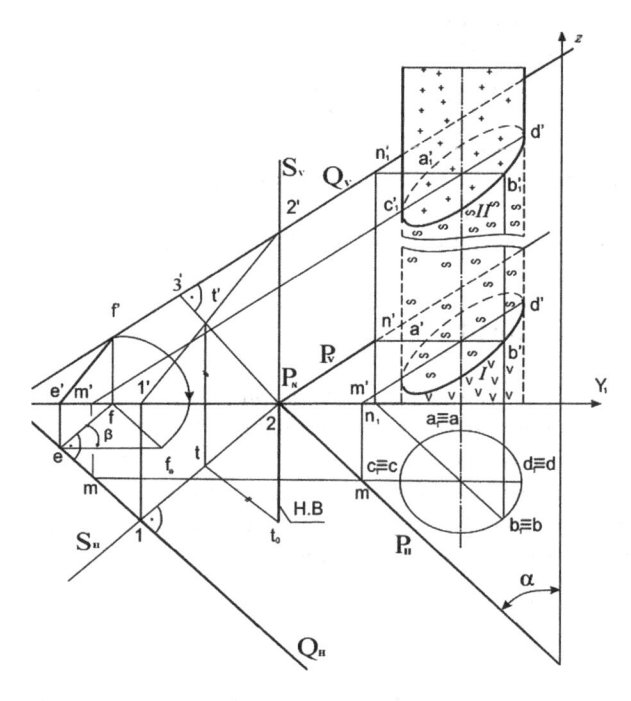

Fig. 1. Strike and dip measurements for blanket deposit using a core pole. Contact surfaces – parallels.

Suppose a mineral deposit is found between contact surfaces. With the help of descriptive geometry, we define deposit parameters.

Assume planes P and Q are contact surfaces of geological materials. Let us create diagonal lines $AB||H$, $A_1B_1||H$ and frontals $CD||V$, $C_1D_1||V$ in these planes. Then we form traces of contact planes: P_V and P_H – traces of contact plane P, Q_V and Q_H – traces of contact plane Q. We obtain angle α — an angle of stretch. We define angle β, an angle of dip, with the line of largest inclination. By dropping a perpendicular from plane P to plane Q we obtain t_02 – reservoir capacity.

While we know the coordinates of points A, B, C, D of contact surface and core pole diameter D we are able to provide an analytical description to the geometrical algorithm for solution of this problem.

We need to find angle of dip β and angle of stretch α.

To solve a problem represented by Fig. 1 we follow these steps:

1. Find coordinates of n_1', pierced by frontal trace Q_V of contact plane Q:

$$\left(\frac{x_b - x_a}{y_b - y_a} \cdot \left(x_a \frac{y_b - y_a}{x_b - x_a} - y_a \right); z_{a_1'} \right) \tag{1}$$

2. Define an equation of straight-line Q_V, prescribing frontal trace of contact plane Q:

$$z - z_{a_1'} = \frac{z_{d_1'} - z_{c_1'}}{x_{d_1'} - x_{c_1'}} \left(x - \frac{x_b - x_a}{y_b - y_a} \cdot \left(x_a \frac{y_b - y_a}{x_b - x_a} - y_a \right) \right) \tag{2}$$

3. Find vanishing point coordinates of Q plane trace on axis OX. Coordinates of point in question:

$$\left(x_a - y_a \frac{x_b - x_a}{y_b - y_a} - z_{a_1'} \frac{x_{d_1'} - x_{c_1'}}{z_{d_1'} - z_{c_1'}}; 0 \right) \tag{3}$$

4. Define a straight-line equation of horizontal trace of Q_H hanging wall plane:

$$y = \frac{y_b - y_a}{x_b - x_a} \left(x - x_a + y_a \frac{x_b - x_a}{y_b - y_a} + z_{a_1'} \frac{x_{d_1'} - x_{c_1'}}{z_{d_1'} - z_{c_1'}} \right) \tag{4}$$

5. Angle of dip β is found from triangle $\Delta eff0$:

$$tg\beta = \frac{ff'}{ef} \Rightarrow \beta = arctg \frac{ff'}{tf} \tag{5}$$

6. We express a straight-line equation of horizontal trace of hanging wall plane P_H:

$$y = \frac{y_b - y_a}{x_b - x_a} \left(x - x_a + y_a \frac{x_b - x_a}{y_b - y_a} + z_{a_1'} \frac{x_{d'} - x_{c'}}{z_{d'} - z_{c'}} \right) \tag{6}$$

7. We find stretch angle α:

$$\alpha = arctg \left(-\frac{x_b - x_a}{y_b - y_a} \right) \tag{7}$$

8. Calculate reservoir capacity, equal to length of the segment $P_x t_0$:

$$m = |y_{px} - y_{t0}|, \tag{8}$$

where y_{px} and y_{t0} – ordinates of points P_x and t_0.

The developed algorithm and its mathematical formulation help to design a program for automatic solving of this problem [16–20].

Figure 2 is a flow diagram of automatic strike and dip measurement using a core pole when contact surfaces of materials are parallel.

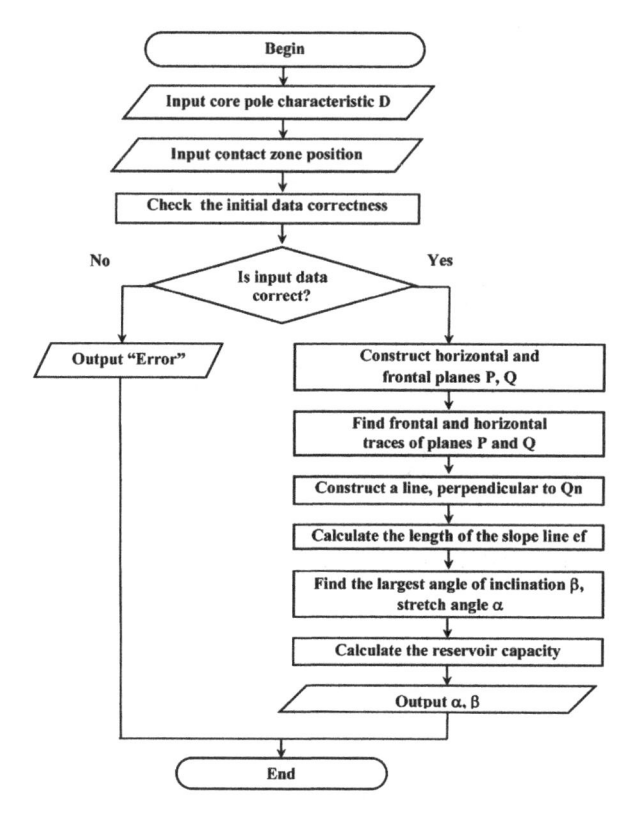

Fig. 2. Algorithm for strike and dip measurement by contact surface on a core pole. Contact surfaces – parallel.

When contact surfaces are non-parallel solution algorithm is the same. Suppose a middle section of a core pole between contact surfaces is where a mineral deposit crosses a borehole.

In Sect. 1 we establish a horizontal DE and frontal DF to find traces of Q plane. Largest angle of inclination on β_1 plane defines a dip and direction of horizontal line and horizontal trace demonstrate its strike α_1.

We find traces of P plane in Fig. 3 with horizontal AB and frontal BC. Piercing of planes P and Q characterizes pinching-out of a deposit constrained by these planes. Angle α_2 defines the strike, and β_2 is angle of dip.

Geometric algorithm for solving of this problem:

1. A, B, C $\in$ P; 1-2 $\perp$ PH; 1-2 — line of largest inclination P; β - angle of dip of plane P;
2. *D, E, F $\in$ Q; 3-4 $\perp$ Q_H; 3-4* — line of largest inclination Q; β_1 — angle of dip of plane *Q;*
3. $P \cap Q = 6\text{-}7$; 6-7 — line of deposit pinching-out; β_2 dip angle of pinching-out line; $o_1 h \perp 6\text{-}7$; $o_1 h = L$ — cross-cut between borehole axis and pinching-out line.

Figure 4 is a geometrical algorithm and a mathematical formulation for problem solution when contact surfaces of materials are non-parallel.

As a result of the research, software modules for solving the main problems of exploration and outlining of mineral deposits are presented in Fig. 5.

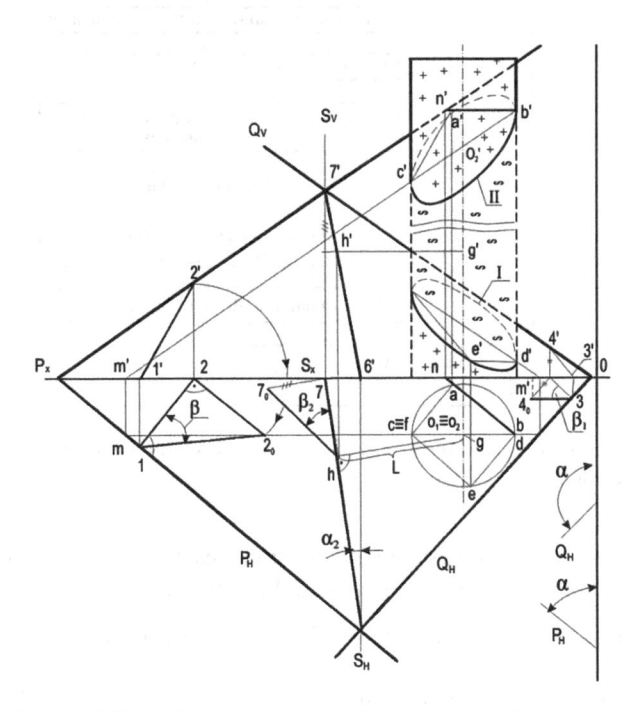

Fig. 3. Strike and dip measurement for cross beds of a core pole.

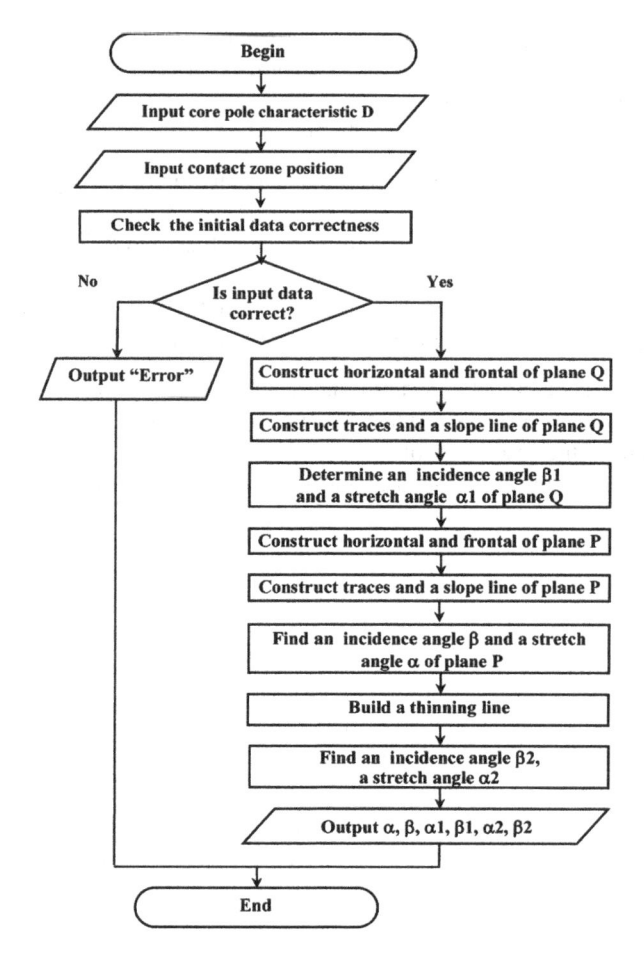

Fig. 4. Algorithm for strike and dip measurement by contact surface on a core pole. Contact surfaces - non-parallel.

The developed automated system for solving the main problems of exploration and outlining of mineral deposits consists of the following subsystems:

- The input data input subsystem, responsible for input data and verification of their correctness
- The subsystem for creating graphic images allows you to define (build) a front and a horizontal projection of the original data
- A subsystem of calculations, in which the solution of such problems as determination of the intersection points of the bores with contours of the mineral deposit, determination of the elements of bedding along the pole core, determination of the boundaries and parameters of the deposit output to the earth's surface are realized

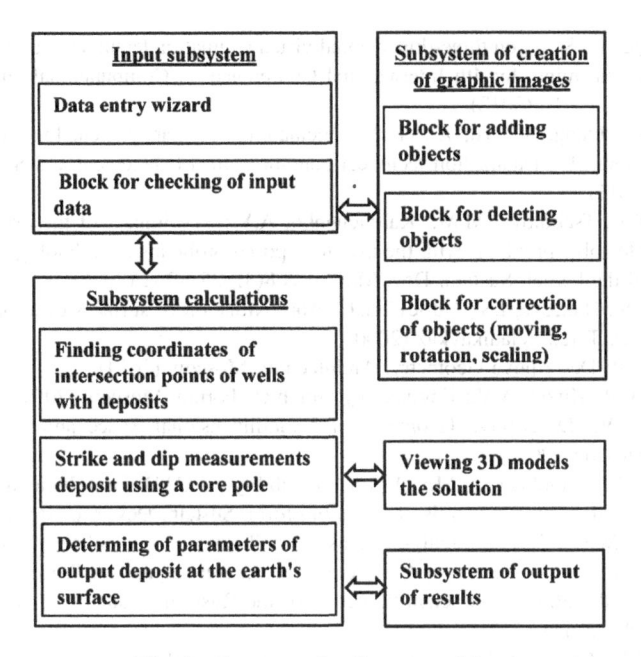

Fig. 5. Structure of software modules.

5 Conclusion

To conclude, geometric modeling helps solve the problems related to deposits of minerals by methods of descriptive geometry. This makes the application of CAD system possible and reduces the costs of field observations.

References

1. Bukrinskij, V.A.: The Geometry of the Subsoil. Mining book, Moscow (2012)
2. Lomonosov, G.G.: Mountain Qualimetry. Moscow State Mining University, Moscow (2007)
3. Gordon, V.O., Semencov-Ogievskij, M.A.: Course Descriptive Geometry. Nauka, Moscow (2008)
4. Bouma, W., Fudos, I., Hoffmann, C.M., et al.: Geometric Constraint Solver. Comput. Aided Des. **21**, 6 (1995)
5. Lin, V.C., Gossard, D.C., Light, R.A.: Variational Geometry in Computer-Aided Design. ACM, New York (1981)
6. Liu, J, Qin, J.: A visualized drilling geological design method based on spatial database. In: Intelligent human-machine systems and cybernetic, pp. 34–37 (2012)
7. Wang, H., Univ, H., Xu, W.: 3-D Geological visual model and numerical model based on the secondary development technology of CAD. In: Software Engineering, Second World Congress, vol. 1, pp. 175–178 (2010)
8. Cuiping, L., Zhenming, S., Zhongxue, L., et al.: Design and implementation of an integrated system of mining GIS and AutoCAD. In: IEEE 3rd International Conference on Communication Software and Networks, pp. 442–445 (2011)

9. Bai, R., Qu, Y.: Study on three-dimensional visual simulation technology and its application in surface coal mine. In: 4th International Conference on Computational and Information Sciences, pp. 14–16 (2012)

10. Wang, W., Huang, S., Wu, X., et al.: Calculation and management for mining loss and dilution under 3D visualization technical condition. In: Management and Service Science, pp. 1–8 (2011)

11. Guriev, T.S., Tsabolova, M.M., Kalinichenko, A.V.: The impact of the use of geometric modeling to solve problems with the use of ill-posed problems of technological analysis on the core of the kernel. Sustain. Dev. Mt. Areas **8**(3), 255–262 (2016)

12. Guriev, T.S., Hazeeva, I.S., Guriev, G.T.: Approximation of surfaces of a random form to regular ones. Terek, Vladikavkaz (2000)

13. Koroev, YuI: Descriptive Geometry. Architecture, Moscow (2014)

14. Kulikov, V.P., Kuzin, A.V.: Engineering Graphics. Forum, Moscow (2009)

15. Filippov, P.V.: Descriptive Geometry of Multidimensional Space and its Applications. Lenand, Moscow (2016)

16. Guriev, T.S., Tsabolova, M.M.: On one possibility of CAD for solving the problem of delineating a deposit of minerals of a random form. Sustain. Dev. Mt. Areas **4**, 5–7 (2012)

17. Tsabolova, M.M.: Tectonic analysis of useful collisions and postulate of the nucleus and application of their research for CAD covering measurements. In: Proceedings of young scientists of the all-Russian scientific center of the Russian Academy of Sciences, vol. 2, pp. 346–349 (2011)

18. Averin, V.N.: Computer Graphics. Academy, Moscow (2012)

19. Norenkov, I.P.: Fundamentals of Computer Aided Design. MSTU Publishing, Moscow (2009)

20. Kalinichenko, A.V.: Development of applications for CAD-systems AutoCAD with use of technologies ActiveX (COM-automation). In: Collection of States of the International Scientifically-Practical Conference, pp. 20–25 (2015)

Computer-Aided Design and Construction Development of the Main Elements of Aviation Engines

D. A. Akhmedzyanov and A. E. Kishalov[(✉)]

Ufa State Aviation Technical University, 12 Karla Marksa street,
Ufa 450077, Russian Federation
kishalov@ufanet.ru

Abstract. In the article it is described developed database of aviation materials and expert system of decision-making of computer-aided design of aviation engines and ground power plants on their basis. The database of aviation materials contains information on the properties and characteristics of a large number of materials used in the design of aircraft engines. The system allows automated selection of materials for parts and assembly units. In this case, the thermogasdynamic calculation of the engine as a whole, a detailed aerodynamic calculation of the assembly units and its strength calculation are performed. In the article it is shown description of mathematical model scheme of expert system. There are simulation results of low-pressure compressor (with different laws of profiling flow passage and different number of stages) for turbojet bypass engine with afterburner of the fourth generation of military high-maneuverable aircraft; there are shown also simulation results of high-pressure compressor for turbofan engine with a high bypass ratio for a civil aircraft, and also results of modelling of rotor blade of fan. Results of simulation were compared with construction of series-produced engines. Using the developed system, it is possible to optimize the weight dimension characteristics of the main components of aircraft engines in the early stages of design.

Keywords: Aviation internal-combustion engines · Optimization of the cycle parameters · Expert system · Database of aviation materials · Simulation of compressor

1 Introduction

The development of modern aviation internal-combustion engines, both gas turbine and piston is a challenging and complex task. The engine development begins with the performance of thermogasdynamic and hydraulic calculations, optimization of the cycle parameters, thereafter it is followed the structure design and the performance of various strength calculations. As a result, it is generated a constructive look of the engine, its main characteristics, overall dimensions and weight. There are often such situations when the aerodynamic engine quality and its individual components are perfect, but individual parts can't withstand the applied loads, thermal stresses or the product can't withstand the intended resource. Then you have to change the design of

© Springer Nature Switzerland AG 2020
A. A. Radionov and A. S. Karandaev (Eds.): RusAutoCon 2019, LNEE 641, pp. 693–702, 2020.
https://doi.org/10.1007/978-3-030-39225-3_76

the part (to weight it), change the material, apply various types of preparation and surface treatment, various coatings [1, 2]. When the design has been changed, the hydraulic and thermogasdynamic calculations and the calculations specifying strength must be carried out again. Thus, the process of aircraft engines designing is represented by a set of cyclically repeated consecutive calculations, the number of iterations of which is limited only by the function of the target [3]. The use of various bundled software at the early stages of design makes it possible to speed up the design process and significantly reduce its cost [4]. The purpose of this study is to develop methods and tools for the automated design and development of the main units and parts design of aircraft engines by the way of example of gas turbine engines.

2   Database and Expert System

For the automated selection of the elements material and the design of the aircraft engine construction on the basis of the DVIGw simulation system, an expert system (ES) "AM" [5] has been developed, which contains separate structural elements (SE) for modeling and carrying out of strength calculations of the main air-gas channel elements of the engine (Fig. 1).

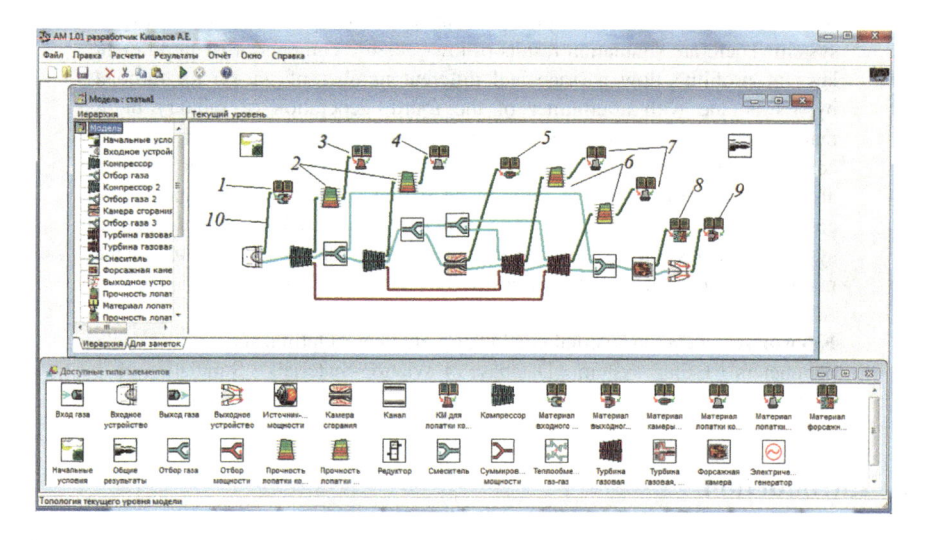

Fig. 1. Topological model turbofan bypass engine with afterburner in ES AM, where 1 – SE for material selection input unit; 2 – SE for the strength analysis of the compressor's rotor blade; 3 – SE for selection of composite material for compressor rotor blade; 4 – SE for selection of compressor material; 5 – SE for selection of the material combustion chamber; 6 – SE for strength analysis of turbine rotor blade; 7 – SE for selection the material of turbine rotor blade; 8 – SE for selection of the material of afterburner; 9 – SE for selection of materials jet nozzle; 10 – information flow between the SE.

Table 1. Structure of the developed materials database.

Material	Chemical labelling of material	Working temperature	Mechanical properties				Physical properties
			Test temperature, °C	Elastic modulus, MPa	Yield strength, MPa	Ultimate strength, MPa	Density, kg/m^3
VG102	X15N30VMT	800	20	205000	720	1040	8290
			600	166000	560	900	
			700	160000	500	700	
			800	152000	400	540	
EP202	XN67VMTY	750	20	210000	630	720	7800
			750	185000	460	560	
EI961	11X12N2VMF 13X11N2V2MF	600	20	200000	850	950	7820
			300	175000	720	830	
			500	145000	570	650	
			600	10900	500	530	

To select the material of the air-flow channel main units and to perform the calculations, the expert system needs the properties of various aviation materials that are contained in the database (DB). The database contains following information as the main material data: the name and chemical labelling of the material, the working temperature (if the temperatures acting on the part exceed the working temperature, the material is assigned negative points), the elasticity modulus, the yield stress and the ultimate strength of the material, depending on working temperature. An example of the structure of developed materials database is given in the Table 1 [6]. For some engine elements, for example rotor blades (RB) of compressors and turbines, it is provided the design mode of isotropic unidirectional composite material (CM).

In [7], the operation of the ES is described in detail, and in Fig. 2 is a graphic description of its mathematical model. SE for material selection receives information of flow parameters and geometry at the input and output from the node, turns to the DB of materials, obtains material characteristics at the temperatures acting on the part. Each material is assigned points, according to the results of the work, the ES displays a list of materials that have received the maximum number of points.

For compressor and turbine units, the ES forms the structural face of the unit, distributes the work and efficiency among stages, calculates the velocity distribution, and profiles the blades for height [8, 9]. In SE for strength analysis, calculation of rotor blades for static strength is performed for each of the materials under consideration.

The ES determines the optimal type of connection of the rotor blades with the disk and calculates the strength of connection [10]. The results of the calculation are transmitted to the SE for selecting of the blade material.

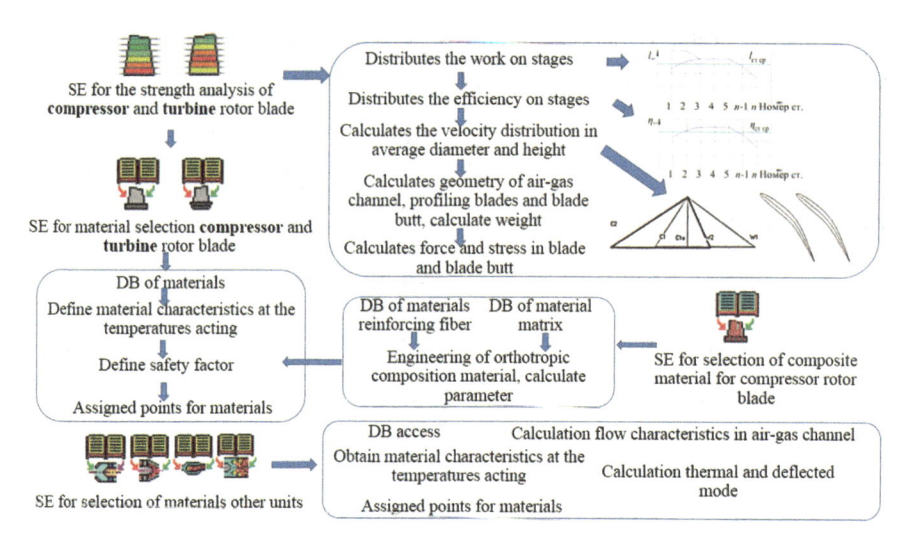

Fig. 2. Graphical description of the mathematical model of ES.

3   Simulation Results

With the help of ES, it is possible to optimize the overall mass characteristics of the compressor and turbine units [11]. In Figs. 3, 4 and 5 and Table 2 results of design modeling of a low-pressure compressor (LPC) of the current generation turbojet bypass

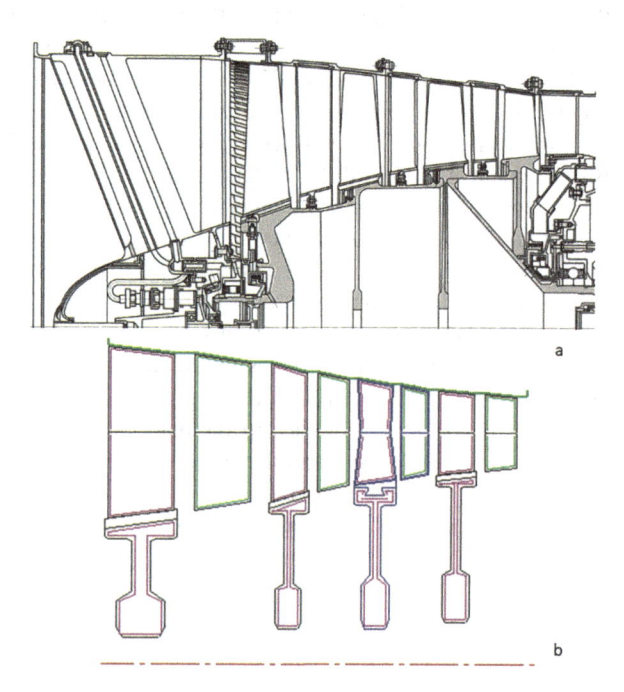

Fig. 3. Diagram of the flow passage of the LPC (a) the simulated engine; (b) with the law of profiling with a constant average diameter (number of compressor stages – 4, design stage – 3, recommended number of stages – 4, recommended number of supersonic stages – 1).

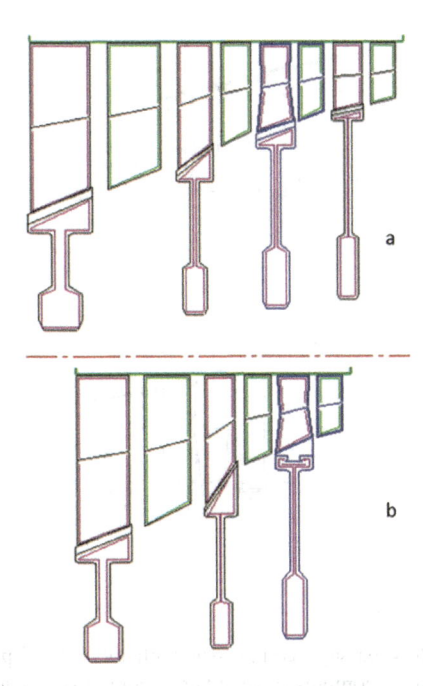

Fig. 4. Diagram of the flow passage compressor with the law of profiling with a constant external diameter (a) four-stage compressor (number of compressor stages – 4, design stage – 3, recommended number of stages – 3, recommended number of supersonic stages – 4); (b) three-stage compressor (number of compressor stages – 3, design stage – 3, recommended number of stages – 3, recommended number of supersonic stages – 3).

engine with afterburner of the fourth generation for a military high-maneuverable aircraft with various laws of flow passage profiling and a different numbers of stages are shown [12, 13]. In Fig. 3a is a diagram of the simulated LPC, in Fig. 3b it is given the construction of the flow passage proposed by the ES.

In Fig. 4a it is given the scheme of the LPC with the law of profiling with a constant outer diameter. As with the given law of profiling, the peripheral velocities at the last stages increase (in comparison with the law of the flow passage profiling with a constant average diameter), and, therefore, their work also increases, the expert system recommended the construction of a compressor with three supersonic stages (Fig. 4b). But, as the design of supersonic stages is more complicated and heavier, the length and total weight of the compressor increased by 20%. Also, with such a flow passage scheme, there are difficulties in locating of the aggregate and compressor devices.

When modeling the compressor under the law of profiling with a constant internal diameter (Fig. 5a), the expert system recommended the construction of a compressor with five stages (because the peripheral velocities decrease with this law of profiling at the last stages of the compressor), Fig. 5b. Such a scheme of a compressor with four stages is a little longer and one and half as heavy again than the prototype. The five-stage compressor scheme has a 10% smaller length and 16% more weight than the prototype (load-bearing element and compressor design become more complex).

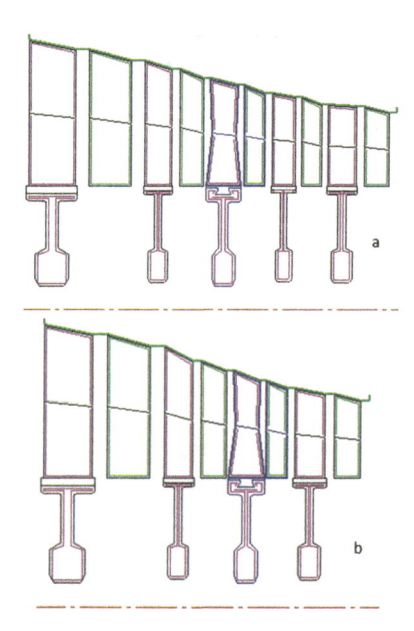

Fig. 5. Diagram of the flow passage compressor with the law of profiling with a constant internal diameter (a) four-stage compressor (number of compressor stages – 4, design stage – 3, recommended number of stages – 5, recommended number of supersonic stages – 1); (b) five-step compressor (number of compressor stages – 5, design stage – 3, recommended number of stages – 5, recommended number of supersonic stages – 1).

But the great advantage of such a scheme is the ease of locating of the motor units and the compressor mechanization. Due to the reduction of the chord of the blades and their number increase in one blade row, the ES proposed to switch to the method of connecting of blades with dovetail discs in the annular groove (for the design stage).

In Fig. 6a is given scheme of the flow passage of the 13-stage high-pressure compressor (HPC) of the fourth-generation turbofan engine with a high by-pass ratio for a civil aircraft [2, 13, 14]. The flow passage of the compressor is carried out according to the law with variable diameter, in the ES when the compressor simulating it is divided into two parts: I–IV stages of the compressor are modeled according to the law with a constant average diameter (Fig. 6b), the V–XIII stages – with constant external diameter (Fig. 6c).

In Fig. 7a it is shown a 3D model of a fan blade of fourth-generation turbofan engine with a high bypass ratio for a civil aircraft modeling in the NX CAD system [2, 14]. In Fig. 7b is a diagram of the flowing passage of the fan.

As the engine has a high bypass ratio, the fan blade has a sufficiently large elongation; the work of the blade (and therefore pressure ratio) is different in height. In modeling, the blade is artificially divided into two parts: the upper part (the air after which passes into the outer contour) and the bottom part (the air after which passes

Table 2. Results of design modeling of LPC.

Law of flow passage compressor profiling	Constant average diameter				Constant external diameter							Constant internal diameter								
Stage	I	II	III	IV	I	II	III	IV	I	II	III	I	II	III	IV	I	II	III	IV	V
Length stage, m	0,1397	0,1145	0,1135	0,0967	0,1645	0,1023	0,0982	0,0452	0,1537	0,1858	0,1473	0,1723	0,1158	0,1168	0,0677	0,1469	0,0813	0,0846	0,0497	0,0624
Length compressor, m	0,4612 (prototype: 0,4280)				0,4102				0,4868			0,4726				0,4249				
Recommended number of supersonic stages	1 (prototype: 1)				4				3			1				1				
Weight of blade row, kg	3,93	3,11	2,07	1,44	17,72	7,80	5,05	1,59	15,85	23,51	9,39	18,86	11,00	8,70	3,63	14,38	5,71	4,29	2,41	2,34
Weight of stage, kg	32,26	24,61	23,83	19,36	72,21	42,76	37,90	14,39	65,04	118,2	73,58	68,86	39,53	35,03	15,05	53,27	21,63	20,60	9,89	11,50
Weight of compressor, kg	100,06				167,26				256,85			158,47				116,89				

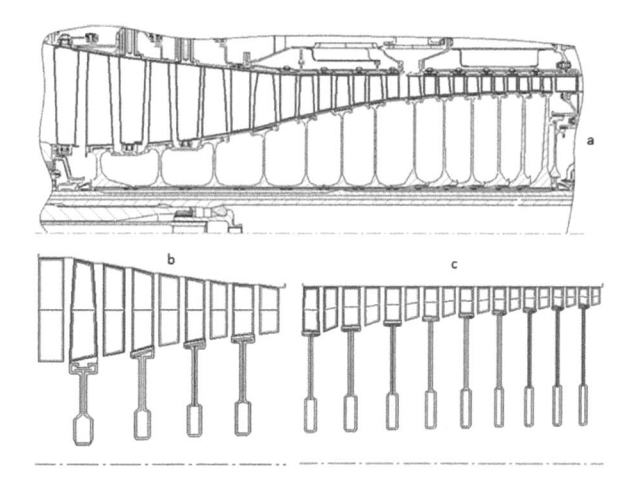

Fig. 6. Diagram of the flowing passage of the HPC (a) the simulated engine; (b) the I–IV stage with the law of profiling with a constant average diameter; (c) V–XIII stage with the law of profiling with a constant outer diameter.

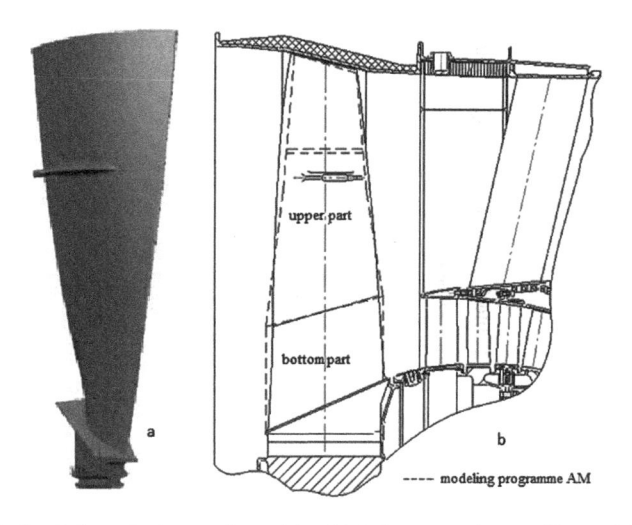

Fig. 7. Fan blade of the turbofan engine with a high bypass ratio: a - 3D model in the CAD-system NX, b - flowing passage of the blade with the scheme of model.

through the supporting stages into the inner contour), Fig. 7b. As can be seen from the simulation results (when the model is applied to the design scheme), the simulation results are in good agreement with the prototype sizes (taking into account that these results are suitable for models used in the early design stages). The blade weight calculated by ES is 0.2% less than the weight of the prototype blade, determined by its 3D model.

4 Conclusion

The developed database and expert system are designed for the thermogasdynamic calculation of the engine, automated design of its main units, the preliminary strength analysis, the selection of the most probable materials, coatings and other types of surface preparation of the main parts and assembly units of the aircraft engine flowing passage.

Taking into account the high speed of calculations, when modeling new engines, it is possible to analyze a large number of possible designs in a short span of time and to choose the most optimal one. The simulation results of aviation GTE units of various schemes show a fairly good qualitative, and in some parameters also a quantitative coincidence with the geometry of the real design.

As a result of the developed ES and DB application, even at the early stages of the design of aviation GTEs, it becomes possible to optimize the design scheme and the overall mass characteristics of the main GTE units.

References

1. Lozinskiy, L.P., Vetrov, A.N., Doroshko, S.M., et al.: Design and Strength of Aviation Gas Turbine Engine. Air Transport, Moscow (1992)
2. Inozemtsev, A.A.: Aviation Engine PS-90A. Fizmatlit, Moscow (2007)
3. Krivosheev, I.A., Sapognikov, A.Yu., Karpov, A.V.: Automatic creation of aviation GTE sketch. In: Proceedings of the Institute of Higher Education, Aeronautical Engineering, vol. 1, pp. 75–76 (2003)
4. Krivosheev, I.A., Sapozhnikov, A.Y., Karpov, A.V.: Organization of data base for systematic computer-aided design of aviation gas turbine engine. In: Proceedings of the Institute of Higher Education, Aeronautical Engineering, vol. 1, pp. 69–72 (2004)
5. Kishalov, A.E., Markina, K.V.: Expert system for automated design nodes and selection of materials the main components of jet engines. Official registration certificate of the computer program, no. 2016663846 (2016)
6. Akhmedzyanov, D.A., Kishalov, A.E., Markina, K.V.: Computer-aided design of aviation gas turbine engines, and materials selection of basic details. Bull. Samara State Aerosp. Univ. **14**(1), 101–111 (2015)
7. Akhmedzyanov, D.A., Kishalov, A.E., Markina, K.V.: Computer-aided engineering design of main aviation GTE units. In: 2nd International Conference on Industrial Engineering, Applications and Manufacturing, Chelyabinsk, 20 May 2016
8. Rzhavin, Y.A.: Axial and Centrifugal Compressors of Aircraft Engines. MAI, Moscow (1995)
9. Holschevnikov, K.V.: Theory and Design of Aviation Bladed Machines. Mashinostroenie, Moscow (1986)
10. Hronin, D.V.: Design and Projecting Aviation GTE. Mashinostroenie, Moscow (1989)
11. Krivosheev, I.A., Rozkov, K.E.: Development of simulation technique and computer-aided design of compressors. Bull. Samara State Aerosp. Univ. **5–2**(47), 150–158 (2014)

12. Kishalov, A.E., Markina, K.V.: Computer-aided design of structural elements of modern turbofan compressors. Procedia Eng. **206**, 367–372 (2017). https://doi.org/10.1016/j.proeng. 2017.10.487. International Conference on Industrial Engineering
13. Electronic encyclopedia: Power units: aviation, rocket, industrial 1944–2000. AKS-Konversalt, Moscow (2000)
14. Akhmedzyanov, D.A., Kishalov, A.E., Markina, K.V.: Automated material selection of the PS-90A fan blade. In: 30th Congress of the International Council of the Aeronautical Sciences, DCC, Daejeon, 25–30 September 2016

Complex Engineering System Acceptability Domain: Worst-Case Analysis via Fuzzy Forecast Technique

P. A. Zinovev$^{(\boxtimes)}$ and I. I. Ismagilov

Kazan National Research Technical University, 68 Karla Marksa Street,
Kazan 420015, Russian Federation
pazinoviev@gmail.com

Abstract. This paper is devoted to very important problem of contemporary engineering. The problem is how to ensure and to maintain complex technical systems possibilities to be reliable and survivable during whole system's life cycle. In this way heuristic approach based on group-expertise fuzzy forecasting procedure may be perfectly fruitful. Worst-case investigation of complex system's evolution behavior through exploring of its acceptability domain fuzzy bounds drift is now presented. For employment of expert's skills in order to realize its application via fuzzy forecasting toolkit we use composite aggregative criteria for quantitative as well as for qualitative estimation of complex engineering system main performability characteristics. These forecasting experiments as well as fuzzy shaping of acceptability domain configuration through group-expertise technique allow system's creators to be almost entirely convinced that within well determined frame of acceptability domain nearly all design specifications, main constructive restrictions and other technical requirements either quantitative or qualitative will be satisfied. In conclusion few perspective directions of complex system acceptability analysis and further enhancement of proposed fuzzy forecasting technique based on worst-case group-expertise procedure are observed.

Keywords: Acceptability reserves · Design technical requirements · Domain fuzzy bounds · Group-expertise approach · Heuristic forecasting technique · System performability evaluation · System life cycle · Worst-case prognoses

1 Introduction

Nowadays contemporary engineering society comes upon serious and very complicated challenge. The problem is how to design, to construct and to develop various type of complex engineering systems (CES) so reliable and survivable, as it possible [1–6]. During system's life cycle the basic trend of CES's evolution may be either positive (progressive/evolvable) or negative (in other words, regressive and degradable). Hence system's dependability assurance during an overall exploitation period may be considered as one of the important constituents of CES's worst-case design and development working process. Functional performability assurance of CES (in other words, its acceptability) is very essential, indefeasible and troublesome part of the general

A. A. Radionov and A. S. Karandaev (Eds.): RusAutoCon 2019, LNEE 641, pp. 703–714, 2020.
https://doi.org/10.1007/978-3-030-39225-3_77

problem of complex system's engineering [7–10]. Originally, CES's acceptability may be determined by threshold values of performance quality measures destined to appreciate how successfully the new system being developed will carry out all prescribed functions during its life cycle. And successive realization of all initially specified CES's design functions indicates that all primary system mission goals will be eventually achieved.

Because many notable performance metrics (NPM) of CES have not only quantitative nature but also qualitative one, thus it will be reasonable to introduce the concept of acceptability domain, within frame of which the entire set of performance quantity/quality measures (PQ^2M) have to meet to all design technical requirements and restrictions during total system's life cycle (SLC). And then it will be expedient to use multifunctional complex aggregative criteria for quantitative as well as for qualitative estimation of system's substantial characteristics and investigation of acceptability domain configuration by its borderlines shaping.

Several useful examples and fruitful results of practical employment of acceptability domain concept and its fuzzy bounds shaping were recently established in [11]. These results were concerned to mostly rational choice of real CES evolutionary project. Nevertheless, in the bounds of approach implementation an ultimate decision about most convenient project approval has to be adopted by CES's design authorities or top crew of decision takers. To promote this idea in CES's design and development process an approach is proposed hereafter which may be employed for the system worst-case acceptability evaluation.

2 The Problem of System's Acceptability Evaluation

2.1 Performability Reserves of Evolvable/Degradable CES

Earlier in [12] for estimation of CES performability it was proposed to use exclusively quantitative (i.e. numerical) NPM measures, based on design technical requirements (DTR), specified for various primary design characteristics of CES has to be developed. In accordance with these requirements, performance quality of CES is defined by set of the system's destination indices (SDI) $Q = \{q_1 \dots q_J\}$, which have to be located within correspondent technical bounds or other limits and restrictions predetermined by very skilled design experts. Usually these specifications are represented in the form of design constraints (DC) in Design Technical Brief (DTB).

Such solely quantitative evaluation of CES performability and working efficiency may be carried out on the base of various numerical metrics, characterized system progressive evolution or degradation during certain period of its SLC. In [13] it was proposed to define the global criterion that CES will strongly operate within precisely restricted performability domain, represented in the form of the next condition:

$$\min_j \ \ z_j(X) \geq 0, \quad j = \overline{1, J}, \tag{1}$$

where z_j – certain normalized performability quantitative reserve, corresponded to j-th NPM in aggregated estimation of the system operational capability; X – vector of the system's internal parameters, directly influencing on NPM value; and J – quantitative NPM number.

Unfortunately, our attempts to solve the problem (1) for real CES were encountered with many hardly overcoming problems related not only with awful volume of computation, but also with existence in PQ^2M not merely quantitative but qualitative specifications, which are very substantial and haven't any numerical measurements.

2.2 Qualitative Evaluation of CES Acceptability

Solution of this problem it is reasonable to realize via fuzzy forecasting of the transformation and variation of CES acceptability domain configuration in the course of SLC. This approach seems to be fruitful for prediction of configuration shape for CES's acceptability domain with primarily shaping of corresponding fuzzy visual images of it. But first of all we are interested in shaping and investigation of System Acceptability Domain (SAD), which represents an enclosed region in a normalized space of system parameters $Z(X)$, defined as acceptability reserves. Within the borderlines of this region all design specifications, restrictions, conditions and requirements either quantitative or qualitative ones have to be entirely satisfied. And moreover, all prescribed system functions and performance quantity/quality measures must be in feasible ranges and may be masterfully realized in practice.

To really implement this approach and to organize rational exploration of SAD's configuration it seems to be expedient and fruitful to apply slightly modified fuzzy forecasting approach to CES's evolution process investigation [14–16]. This approach is founded on complex aggregative criteria for CES's level of development estimation, which includes some partial criteria for system characteristics evaluation. This multicriterion vector comprises several metrics for assessment of system stability in physical and intellectual sense.

2.3 SAD's Frame Borderline Drift Fuzzy Simulation

After ranging of alternatives and intermediate decision obtaining it will be necessary to determine basic development process trend of CES. It is reasonable to realize such process via fuzzy classification by specifying values of membership function to main development type of certain trend directions: progressive, regressive or neutral. First of all, it is necessary to define dominated type of these trends. It may be done in accordance with some individual expert's evaluation approach previously established in [17].

Within the framework of here proposed forecasting technique required trend definition was inherently based on mutual group-expertise procedure as a convolution of fuzzy estimation values obtained from the set of collective expert judgments. Time series data mining seems to be also very useful toolkit in this case. But from the point of view of worst-case analysis first of all we are interested in investigation of the pessimistic trends in SAD's fuzzy borderlines drift during their position's simulation.

3   Worst-Case Analysis of System's Acceptability

3.1   Fuzzy Forecasting via Group-Expertise Procedure

Very often it is not possible to obtain exact values of NPM metrics due to essential complexity of many real contemporary engineering systems. In other words, most received CES investigation results demonstrate considerable presence of uncertainty. Consequently, principles and techniques of decision making in CES fuzzy environment seem to be very fruitful and perspective. It's necessary to take into account, that predicted PQ^2M of investigated systems sometimes are established in the form of time series data. To explore such unusual objects fuzzy forecasting methods are successfully used, including fuzzy trend analysis techniques and expert groups prognosis.

Moreover, in prediction process seems to be helpful to use so-called combined fuzzy forecast technique. Combined fuzzy forecasting is based on employment of the multiple set of fuzzy estimation values, which are results of aggregative consolidation not only fuzzy prognosis from expert groups but also variety of fuzzy particular prognosis made by individual experienced experts, who are not members of any groups.

Hereinafter in our fuzzy forecasting technique we employ heuristic algorithm for membership function (MF) realization and embodiment [17]. In this regard MF for fuzzy variable values, approximately equal to K, is represented by

$$\mu_K(u) = e^{-\alpha(K-u)^2}, u \in U,$$

where U – universum media for u (in general case the set of all real numbers); α – some predetermined MF fuzzy parameter.

Value of α may be defined as

$$\alpha = \frac{-4\ln 0.5}{(\beta(K))^2},$$

in this case $\beta(K)$ – is the distance between argument values for $\mu_K(u)$, where MF meaning is equal to 0.5.

Then expression for fuzzy variable MF may be defined as

$$\mu_K(u) = e^{4\ln 0.5((K-u)/\beta(K))^2}, u \in U$$

The substance of the approach in our case is in sequential narrowing of original collective-predicted interval sizes to appropriated fuzzy estimation values performed by subgroups of experts.

For possible values of system's level of development specifications on forecasting period duration original interval size is determined on the base of group-expertise procedure by definition of minimum and maximum limits for left and right borderline of corresponding horizon prognostic interval. This interval is divided on s overlapping subintervals with equal length. Then investigation will be done both by individual experts also by expert's group to define normalized priorities of these subintervals based on Saaty's hierarchy analyses method [18]. According to these subinterval priorities expert's division on subgroups may be implemented and then some convenient procedure of evaluation of initial prognostic interval's values may be realized. Further step of this approach implements certain iteration procedure to find appropriated fuzzy variable values to each prognostic interval.

Iteration procedure of prognostic interval narrowing to fuzzy variable values may be described as follows. Prognostic interval on running iteration is divided on s equal length and half-overlapping subintervals, and then determination of its priorities may be executed. With the account of these priorities further prognostic interval narrowing is carried out and its comparison with confidence interval (α-cut) of fuzzy variable on pre-assigned level α is performed. As an expected average meaning of it medium value of prognostic interval is accepted. In the case of prognostic interval inclusion into confidence interval – iteration process has to be stopped, in other case prognostic interval narrowing procedure must be continued.

Thus, analyzed initial prognostic interval consists of three half-overlapping subintervals:

$$[\underline{m}, \underline{m} + 2\delta), [\underline{m} + \delta, \overline{m} - \delta), [\overline{m} - 2\delta, \overline{m}],$$

where $\underline{m}, \overline{m}$ – left and right borderline bounds of initial prognostic interval length respectively,

$$\delta = \frac{\overline{m} - \underline{m}}{4}.$$

Each iteration based on group-expertise procedure reduces initial length of prognostic interval exactly half-and-half. Proceeding from this assumption, prognostic interval length dependence from total iteration number may be represented as follows:

$$\Delta_i = \Delta_0 2^{-i}, \ i = 1, 2, \ldots, n,$$

where Δ_0, Δ_i – prognostic interval's length after i-th iteration.

In the same manner in general case for segmentation of initial prognostic interval length on k equal intervals and on ($s = 2k - 1$) half-overlapping subintervals respectively, aforementioned expression may be represented as follows

$$\Delta_i = \Delta_0 k^{-i}, \ i = 1, 2, \ldots, n.$$

Let's determine an estimation values for maximum limit of iteration number for narrowing of prognostic interval length. It may be done from conditions for prognostic values to be embedded in confidence intervals for three fuzzy variables, located on left/right bounds and medium point of initial prognostic interval. These conditions may be formulated as follows:

$$\Delta_0 k^{-i_1} \leq \beta_L(\underline{m}), \quad \Delta_0 k^{-i_2} \leq \beta_C(\tfrac{\overline{m} + \underline{m}}{2}), \quad \Delta_0 k^{-i_3} \leq \beta_R(\overline{m}), \quad i = 1, 2, \ldots, n, \quad \text{where}$$

$\beta_C(\frac{\overline{m} + \underline{m}}{2})$ – medium value of forecasting interval, $\beta_L(\underline{m})$ and $\beta_R(\overline{m})$ confidence intervals for correspondent fuzzy variables.

After non-complicated transforming of these expressions, we derive next equations for iterations number evaluation:

$$i_1 = \log_k \Delta_0 - \log_k \beta_L(\underline{m}), i_2 = \log_k \Delta_0 - \log_k \beta(\frac{\overline{m} + \underline{m}}{2}), i_3 = \log_k \Delta_0 - \log_k \beta_R(\overline{m}).$$

Worst-case estimation value of iteration number is defined as maximum value from i_1, i_2, i_3. So, preliminary evaluation of total iteration volume may be executed under operations of prognostic interval division on k equal subintervals.

3.2 CES's Development Trend Fuzzy Identification

Most reasonable for estimation of CES's elaboration trend direction will be division of initial prognostic interval on three equal intervals: (a) evolvable or progressive, (b) neutral or stable and (c) degradable or regressive.

In this regard it is necessary to appreciate complex system environment and development situation. Consequently, perspective of five alternatives for SAD's borderline drift is produced.

Let's consider certain expressions for alternatives priorities calculations by using Saaty's approach in application to concrete case of prognostic interval division on three equal subintervals. Let's we have next system of equations for expert's alternative priorities estimation:

$$AW = \Lambda W, w_1 + w_2 + w_3 + \ldots + w_n = 1,$$

where A – alternatives pairwise comparison matrix (PCM), $W = (w_1, w_2, w_3, \ldots, w_n)^T$ – vector of expert's priorities, Λ – maximum eigenvalues vector of PCM.

As the result of prognostic interval division on three equal half-overlapping subintervals next expressions for alternatives priorities estimation may be easily received:

$$w_1 = \frac{t_1 t_6 + t_2 t_7}{t_1(t_3 + t_5 + t_6 + t_8) + t_2(t_4 + t_5 + t_7 + t_9)}$$

$$w_2 = \frac{t_1 t_8 + t_2 t_9}{t_1(t_3 + t_5 + t_6 + t_8) + t_2(t_4 + t_5 + t_7 + t_9)}$$

$$w_3 = \frac{t_1 t_3 + t_2 t_4}{t_1(t_3 + t_5 + t_6 + t_8) + t_2(t_4 + t_5 + t_7 + t_9)}$$

$$w_4 = \frac{t_5 t_1}{t_1(t_3 + t_5 + t_6 + t_8) + t_2(t_4 + t_5 + t_7 + t_9)}$$

$$w_5 = \frac{t_2 t_5}{t_1(t_3 + t_5 + t_6 + t_8) + t_2(t_4 + t_5 + t_7 + t_9)}$$

Values of all aforementioned coefficients t_k, $k = \overline{1,9}$ may be received from expert's evaluation of PCM elements a_{ij}. Formulas for its calculation were given earlier in [19].

3.3 Main Stages of the Worst-Case Analysis Technique

Let's consider basic substantial features of the proposed technique for SAD shaping and exploration. Main stages of this group-expert fuzzy forecast procedure may be described as follows.

1. Specification of original prognostic tasks and formulas for calculation or fuzzy estimation of normalized elements of the system's PQ^2M acceptability reserves vector $Z(X)$.
2. Selection and organization fellowship of M skilled experts. Specification of the expert's competence coefficients (weights) vector E as described in [20, 21].
3. Specification of fuzzy set based on individual expert judgments: $A = \{(\underline{a_i}, \overline{a_i}), i = \overline{1,M}\}$, where $\underline{a_i}, \overline{a_i}$ – left and right bounds of i-the expert prognostic interval respectively. Obtainment of initial individual expert's estimation values for prognostic interval bounds:

$$(\underline{m},\ \overline{m}), \underline{m} = \min\{\underline{a_i},\ i = \overline{1,M}\}, \overline{m} = \max\{\overline{a_i},\ i = \overline{1,M}\}.$$

4. Specification of fuzzy variable value $\tilde{P}$ "approximately equal" to $(\underline{m}+\overline{m})/2$ and its confidence interval β length finding on the level of 0,5 MF.
5. Checking of prognostic interval bounds on its embedding into confidence interval: $(\overline{m} - \underline{m}) \leq \beta$. If this condition is satisfied, then go to stage 23, otherwise – to stage 6.
6. Preliminary estimation of the number of prognostic interval length half-narrowing iteration. If such estimation number is larger than predetermined limit it's necessary to divide initial prognostic interval into three or five equal subintervals. In this case the number of prognostic alternatives to be analyzed may be five or more. For example, for three subintervals we have next alternatives:

$$A_1 = [\underline{a_1}, \overline{a_1}) = [\underline{m}, \underline{m}+2\delta), A_2 = [\underline{a_2}, \overline{a_2}) = [\underline{m}+\delta, \underline{m}+3\delta),$$
$$A_3 = [\underline{a_3}, \overline{a_3}) = [\underline{m}+2\delta, \overline{m}-2\delta), A_4 = [\underline{a_4}, \overline{a_4}) = [\underline{m}+3\delta, \overline{m}-\delta),$$
$$A_5 = [\underline{a_5}, \overline{a_5}] = [\overline{m}-2\delta, \overline{m}],$$

where $\delta = (\overline{m} - \underline{m})/6$.

In other case it's necessary to divide initial prognostic interval into two equal subintervals with three prognostic alternatives to be analyzed.

7. Obtainment of expert's prognoses set made by individual judgment manner for elements of PCM arranged in accordance with alternatives preference degree growth. It will be expert's priorities set.
8. Calculation of individual expertise alternative's priorities vector: $L_i = (l_{i1}, l_{i2}, \ldots, l_{is}), i = \overline{1,M}$, for $s = \{3,5\}$.
9. Obtainment of local individual interval's prognoses: $(\underline{n_i},\ \overline{n_i})$, $\underline{n_i} = \sum_{j=1}^{s} l_{ij}\underline{a_j}$,

$$\overline{n_i} = \sum_{j=1}^{s} l_{ij}\overline{a_j}, i = \overline{1,M}.$$

10. Calculation of overlapping coefficients for local individual interval's prognoses K_{ij}, $i = \overline{1,M}, j = \overline{1,3}$ with next intervals bounds:
for $s = 3$: $[\underline{a_1}, \overline{a_1}), [\underline{a_2}, \overline{a_2}), [\underline{a_3}, \overline{a_3}]$;
for $s = 5$: $[\underline{b_1}, \overline{b_1}), [\underline{b_2}, \overline{b_2}), [\underline{b_3}, \overline{b_3}]$;
where $\underline{b_1} = \underline{a_1}, \overline{b_1} = \overline{a_1}, \underline{b_2} = \underline{a_3}, \overline{b_2} = \overline{a_3}, \underline{b_3} = \underline{a_5}, \overline{b_3} = \overline{a_5}$.

11. Specification of expert groups in accordance with its members "attraction" to basic intervals as follows:

$$E_1 = \{e_i \big| \max_j \{K_{ij}\} = K_{i1}\}, E_2 = \{e_i \big| \max_j \{K_{ij}\} = K_{i2}\},$$

$$E_3 = \{e_i \big| \max_j \{K_{ij}\} = K_{i3}\}, i = \overline{1,M}, E_1 \cup E_2 \cup E_3 = E_0.$$

12. Checking of condition $E_i \neq \phi, i = \overline{1,3}$. Definition of weight coefficients v_i for each expert's group and then execution stages from 13 to 22 for all non-empty expert groups.

13. Construction of expert group alternatives combined PCM by calculation of its elements as geometric mean value of respective components of individual PCM.

14. Calculation of expert group alternative's priorities vector: $P = (p_1, p_2, \ldots, p_s)$ $s = \{3,5\}$.

15. Calculation of expert group's combined interval prognostic bounds as $(\underline{m}, \overline{m})$:

$$\underline{m} = \sum_{j=1}^{s} p_j \underline{a_j}, \overline{m} = \sum_{j=1}^{s} p_j \overline{a_j}, s = \{3,5\}.$$

16. Specification of fuzzy variable value $\tilde{P}$ "approximately equal" to $(\underline{m} + \overline{m})/2$ and its confidence interval length β finding on the level of $0,5$ MF.

17. Checking of expert group's interval prognostic bounds on its embedding into confidence interval: $(\overline{m} - \underline{m}) \leq \beta$. If this condition is satisfied, then go to stage 23, otherwise – to next stage 18.

18. Preliminary estimation of the remaining number of interval half-narrowing iteration. Rational choice of overlapping subintervals set $s \in \{3,5\}$ to perform initial prognostic interval division and respective alternatives definition.

19. Obtainment of expert's individual prognoses for elements of PCM in accordance with alternatives preference growth.

20. Obtainment of group-expertize aggregative PCM by calculation of its elements as geometric mean values of respective components of individual PCM.

21. Calculation of expert group alternative's priorities vector: $P = (p_1, p_2, \ldots, p_s)$ $s = \{3,5\}$.

22. Obtainment of expert group's aggregative interval prognostic bounds as $(\underline{m}, \overline{m})$:

$$\underline{m} = \sum_{j=1}^{s} p_j \underline{a_j}, \overline{m} = \sum_{j=1}^{s} p_j \overline{a_j}, s = \{3,5\}.$$

Then return to stage 16.

23. Formulation of expert prognosis estimation in the form of fuzzy variable value $\tilde{P}_{zp}$ and its MF. If we came to this stage from stage 5, then go to 25, otherwise - to stage 24.
24. Aggregation of fuzzy prognoses from expert groups with account of group's weight coefficients and obtainment of final fuzzy forecast prognosis $\tilde{P}_l = \oplus v_i \tilde{P}_i$.
25. Specification of MF shape for aggregative group-expertize prognosis and calculation of its confidence interval.

Obtainment of expert group's prognoses in qualitative form for system's acceptability domain shaping may be performed subsequently in accordance of seven or ninth gradation scale. In this case primary values transforming to normalized values of secondary scale may be carried out by using Harrington scale and Harrington desirability function (HDF) [22, 23].

4 SAD: Fuzzy Forecasting and Shaping

4.1 SAD's Shaping with Qualitative Reserves Estimation

For successful SAD's shaping first of all it's necessary to know in detail all features of the previously determined trend of system's development. This trend may be either progressive (evolvable) or regressive (degradable). Furthermore, system's development trend may be neutral (or stable). Then qualitative worst-case forecasting process may be successfully introduced by toolkit of established above group-expertise procedure with the employment of certain specific modification.

Obtainment of PCM-elements values to determine worst-case bounds of SAD is carried out with applying of group-expert judgment method. Thus, we receive fuzzy estimations of qualitative criteria properties, described by its MF.

Appraisal of correspondence of alternative variants of CES development with earlier predicted trend is performed in the space of development indexes (physical and intelligence). Left bounds of forecasted normalized values of acceptability reserves will correspond to required worst-case SAD's shape.

For normalizing CES's qualitative variable values let's use next estimation for qualitative acceptability reserves:

$$z_j = \exp(-\exp(-q_{ij})) , i = \overline{1,N}, j = \overline{1,K},$$

where q_{ij} – intermediate estimations of PQ2M, defined via corresponded preliminary values in initial scale y_{ij}; K – total criteria number (qualitative and quantitative).

Conversion from original estimation scale to normalized values is carried out by applying a worst-borderline values of intervals in Harrington scale. Transformation from numerical to normalized values is performed by using generalized HDF. Then values z_j for quantitative criteria group maybe re-specified in correspondence with Table 1.

Table 1. Correspondence between values of original estimation scale and their normalized values.

Value y_{ij}	1	2	3	4	5	6	7	8	9
Value z_j	0,1	0,2	0,28	0,37	0,5	0,63	0,71	0,8	0,9

To determine/estimate qualitative criteria values q_{ij} let's use next relationships, formulated for those acceptability criteria, that need to be maximized:

$$
q_{ij} = \begin{cases}
0.35 \frac{(y_{ij}-y_j^*)}{(y_{1j}^*-y_j^*)} - 0.83, & y_{ij} \in [y_j^*, y_{1j}^*], \\[2mm]
0.24 \frac{(y_{ij}-y_{1j}^*)}{(y_{2j}^*-y_{1j}^*)} - 0.48, & y_{ij} \in (y_{1j}^*, y_{2j}^*], \\[2mm]
0.246 \frac{(y_{ij}-y_{2j}^*)}{(y_{3j}^*-y_{2j}^*)} - 0.24, & y_{ij} \in (y_{2j}^*, y_{3j}^*], \\[2mm]
0.37 \frac{(y_{ij}-y_{3j}^*)}{(y_{4j}^*-y_{3j}^*)} + 0.06, & y_{ij} \in (y_{3j}^*, y_{4j}^*], \\[2mm]
0.394 \frac{(y_{ij}-y_{4j}^*)}{(y_{5j}^*-y_{4j}^*)} + 0.376, & y_{ij} \in (y_{4j}^*, y_{5j}^*], \\[2mm]
0.3 \frac{(y_{ij}-y_{5j}^*)}{(y_{6j}^*-y_{5j}^*)} + 0.77, & y_{ij} \in (y_{5j}^*, y_{6j}^*], \\[2mm]
0.43 \frac{(y_{ij}-y_{6j}^*)}{(y_{7j}^*-y_{61j}^*)} + 1.07, & y_{ij} \in (y_{6j}^*, y_{7j}^*], \\[2mm]
0.75 \frac{(y_{ij}-y_{7j}^*)}{(y_j^{**}-y_{7j}^*)} + 1.5, & y_{ij} \in (y_{7j}^*, y_j^{**}],
\end{cases}
$$

where y_j^*, y_j^{**} – the worst-case and the best-case group-expertise values for j-th partial criterion respectively; y_{kj}^*, $k = \overline{1,7}$ – right borderlines of intervals values of j-th criterion, corresponded to linguistic variable values "very low", "low", "average", "above average", "good", "very good", "high". It should be noted, that optimistic value "very high" corresponds to half-open subinterval $(y_{7j}, y_j^{**}]$.

It may be reasonable to proceed from quantitative scale to initial verbal scale comprised nine grades and hereinafter to employ relationships for qualitative criteria to calculate values of quantitative reserves z_j. Evidently, that by using proposed approach of normalized criteria evaluation all elements in pairwise comparison matrix on PCM will in interval from 1 to 9. An essential difference from Saaty's scale is that, in our case elements of pairwise comparison matrix may have arbitrary values from aforementioned diapason.

Finally, as the general condition that CES acceptability domain will be adequate in worst-case sense let's demand:

$$
\min_j \; z_j(X) \geq 0.25, \quad j = \overline{1,K}. \tag{2}
$$

For well-controlled evolutionary process all intermediate states along desirable development trajectory must satisfy these requirements. But for most CES with regressive evolutionary trend their SAD configuration will be volume-reduced.

4.2 Application of Acceptability Domain Fuzzy Shaping

Several non-trivial modeling experiments concerning to application of the aforementioned technique to fuzzy shaping of acceptability domain for real CES were carried out. As a result of our investigation some patterns of SAD visual fuzzy images in Z-space for real CES were obtained.

Better part of the experimentally received patterns of SAD-shapes corresponded to worst-case system development project may be considered as an initial variant for rational choosing of evolution way for investigated CES. Although some analyzed patterns belong to stable but nearly regressive forecasting trend, they are still quite acceptable for realization, because they completely satisfy to acceptability condition (2) and as consequence their SAD shapes are not empty. So, they would be considered as basic feasible solutions for the rational choice of CES future development.

As usual system acceptability domain represents polytope in normalized multi-criteria Z-space of CES. The next problem will be how to choose most acceptable and satisfactory configuration for this domain from worst-case point of view. Evidently, that consistence of CES development processes may be determined comparatively by its convergence to predicted desirable trajectory of system elaboration as a most reasonable template. Furthermore, we should take into account also all possible alternatives from progressive or stable trends as a perspective and completely fruitful patterns.

If all criteria in normalized Z-space z_j, $j = \overline{1, K}$ of CES are arranged in the order of decreasing of its importance in a clockwise sense beginning from vertical ordinate axis, then each non-even-numerical direction in normalized Z-space will be more preferable than following even-numerical direction. Due to such suggestion in result of SAD-shape configuration analysis we shall receive final conclusion what shape is much more preferable. Indeed, if SAD-shape of certain project may have essential preferences in all non-even directions while SAD-shapes of other projects may have some advantages along subsidiary even-numerical axes then ultimate decision about concrete system development way will be evident.

5 Conclusion

Let's suppose that the set of degradable alternatives will be empty. Then it will be necessary to make final decision on the base of SAD visual analysis. And after all, ultimate decision has to be done with regard of expert group's chief individual opinion about worst alternatives which may belong to "stable" development trend.

Main issues for further research and development will be:

- Essential strengthening of expert group judgments and prognoses validity via estimation of expert's competence
- Application of fuzzy neural networks collective's opinions technique to support forecasting approach described above and its practical realization
- Possible ways for improving of existing fuzzy shape of SAD after final evaluation of CES's acceptability reserves.

References

1. Longbottom, R.: Computer System Reliability. Wiley, Chichester (1980)
2. Gulyaev, V.A., Dodonov, A.G., Pelekhov, S.P.: Organization of survivable computing structures. Naukova Dumka, Kiev (1982)
3. Norenkov, I.P., Zinovev, P.A.: Multilevel optimization of a large-scale engineering systems. Electron. Model. **6**, 49–53 (1987)
4. Cherkesov, G.N.: Methods and modelling techniques for complex systems survivability estimation. Znanie, Moscow (1987)
5. Dodonov, A.G., Kuznetsova, M.G., Gorbachik, E.S.: An introduction to theory of computing systems dependability. Naukova Dumka, Kiev (1990)
6. Zinovev, P.A., Moiseev, V.S., Ginatullin, I.A., et al.: Topical problems of corporative control in aircraft manufacturing. Russ. Aeronaut. **50**(2), 204–209 (2007)
7. Meyer, J.F.: On evaluating the performability of degradable computing system. IEEE Trans. Comput. **29**(8), 720–731 (1980)
8. Haverkort, B.R.: Performability Modelling: Techniques and Tools. Wiley, Chichester (2001)
9. Nicola, V.F., Shahabuddin, P., Nakayama, M.: Techniques for fast simulation of models of highly dependable systems. IEEE Trans. Reliab. **50**(3), 246–264 (2001)
10. Nicol, D.M., Sanders, W.H., Trivedy, K.S.: Model-based evaluation: from dependability to security. IEEE Trans. Dependable Secure Comput. **1**(1), 48–65 (2004)
11. Ismagilov, I.I., Khasanova, S.F., Zinovev, P.A.: Complex engineering systems: rational choice of evolutionary projects. Revista Publicando **5**(16), 409–420 (2018)
12. Zinovev, P.A.: Modelling of the corporative IT-system's performability domain fuzzy shape. Dyn. Syst. Mech. Mach. **4**(1), 15–18 (2016)
13. Zinovev, P.A.: The simulation of fuzzy border drift of corporative IT-system performability area. In: Proceedings of 2-th International Conference on Industrial Engineering, pp. 452–457 (2016)
14. Sugeno, M., Tanaka, K.: Succesive identification of fuzzy model and its applications to prediction of a complex systems. Fuzzy Sets Syst. **42**, 315–334 (1991)
15. Fodor, J., Roubens, M.: Fuzzy preference relations and multicriteria decision support. Kluwer Academic Publishers, Dordrecht (1994)
16. Baldwin, J.F., Martin, T.P., Rossiter, J.M.: Time series modelling and prediction using fuzzy trend information. In: Proceedings of Fifth International Conference on Soft Computing Information Intelligent System, pp. 499–502 (1998)
17. Ismagilov, I.I.: Decision making based on quantitative and qualitative criteria of alternatives description. Trans. Inform. Res. **6**, 21–28 (2003)
18. Saaty, T.L.: The Analytic Hierarchy Process: Planning, Priority Setting, Resource Allocation. McGraw-Hill, New York (1980)
19. Ismagilov, I.I., Bichurin, R.V.: Fuzzy forecasts: classification and method of their development based on procedure of a group expertise. Fundam. Res. Tech. Sci. **11**, 1240–1247 (2014)
20. Bagaturia, G., Tabatadze, M.: Estimation of expert's competence for the tasks of forecasting. J. Bus. **1**, 5–8 (2012)
21. Mirkin, B.G.: The problem of group choice. Science, Moscow (1974)
22. Ismagilov, I.I., Zinkin, V.A.: Fuzzy forecasting of the complex systems quantitative characteristics. Trans. Inform. Res. **11**, 49–56 (2007)
23. Ismagilov, I.I., Zinovev, P.A., Zinkin, V.A.: Estimation of development level of the complex engineering systems: technique and its applications to corporative IT-systems. J. Inf. Technol. **6**, 64–71 (2008)

AR Guides Implementation for Industrial Production and Manufacturing

A. Ivaschenko[1(✉)], V. Avsievich[1], and P. Sitnikov[2]

[1] Samara State Technical University,
244 Molodogvardeyskaya Street, Samara 443001, Russian Federation
anton.ivashenko@gmail.com
[2] ITMO University, 49 Kronverksky Avenue,
St. Petersburg 197101, Russian Federation

Abstract. This paper introduces an original approach for the implementation of Augmented Reality automated systems for industrial production and manufacturing. Based on analysis of the up-to-date problems of AR devices practical use like limited usability of AR goggles, low performance and quality of image recognition in real time there is proposed a new approach of accented visualization. The paper addresses some problems of machine vision application for production quality control, which helps solving the challenges for AR guides and improving the efficiency of their implementation in practice. In particular, augmented reality allows users to undergo interactive training and receive active assistance during the execution of technological operations without the need to refer to a set of documentation in paper or electronic form. Within the developed framework there is presented a new solution architecture that combines the capabilities of neural networks, Ontologies and Big Data processing facilities. The proposed approach is recommended to improve the quality of recognizing production operations and manufactured products using machine vision technologies and neural networks in industrial applications.

Keywords: Augmented reality · Factory 4.0 · Automation · Accented visualization · Maintenance · Decision making support

1 Introduction

Augmented Reality (AR) implementation is highly perspective in industrial production and manufacturing. Possible functionality of such a solution includes visualization of product design, equipment information for maintenance and execution management, production operations tracking and coordination, inventory control and others. AR is a powerful technology of Factory 4.0 paradigm that describes a number of current trends in automation and data exchange in manufacturing.

Modern user interfaces aim at providing maximum usability and performance. Augmented Reality (AR) occupies a special niche of computer-human interaction introducing new features of virtual content mixture with real objects in real time. The principles of AR interface design and utilization differ from UI recommendations. Virtual and real objects and control elements can be distributed in space, cover each other and overlap, which makes it important to study and consider new requirements of AR user interfaces.

© Springer Nature Switzerland AG 2020
A. A. Radionov and A. S. Karandaev (Eds.): RusAutoCon 2019, LNEE 641, pp. 715–723, 2020.
https://doi.org/10.1007/978-3-030-39225-3_78

One of the possible approaches to study the features of AR user interfaces and peculiarities of their practical use is based on accented visualization [1, 2] that consists in tracking of the user's focus and context. Accented visualization allows adapting the UI of AR guides and extending the possibilities of their implementation in industrial production and manufacturing according to the modern principles of Industry 4.0. Based on the experience of AR solutions development for production enterprises in this paper there are proposed some technical recommendations of AR application in practice.

2   State of the Art

Augmented Reality remains a new and challenging area of research nowadays. Its practical implementation makes it possible to introduce new possibilities to improve user interfaces [3, 4]. In addition to well-known possibility to enrich real scenes with virtual elements, AR brings new ideas to computer human interaction.

AR is usually implemented by head mounted devices like goggles that give a new opportunity to track and capture head movement under the context of a current situation. This makes it beneficial to use AR technology as a platform for adaptive user interfaces [5, 6]. Under this context AR has the same problems as traditional user interfaces, such as user overload with redundant information and increased complexity of actions required to fulfil ordinary processes.

The concept of Industry 4.0 is intensively and massively described in modern literature [7, 8]. Modern goals of the digital economy development distinguish the concept of "Industry 4.0" as one of the key technological trends aimed at improving the quality and competitiveness of products. This concept is based on the development of cyber physical systems capable of tracking real physical processes, complementing them with specially generated virtual objects and providing contextual and decentralized decision-making support. These functions require the implementation of innovative user interfaces suitable for processing, analysis, virtualization and presentation of image data based on augmented reality and the Internet of things.

Through the Internet of things [9], cyber physical systems interact with each other and people in real time, both within the organization and with external services. Using multiple devices with various functions in the cloud can improve the quality of monitoring and decision support. Modern protocols and architectures of wireless networks allow implementing various topologies at a technical level.

The practical implementation of Industry 4.0 is related to the possibility of processing Big Data [10, 11], which is currently a separate area of research. The combination of the Internet of things as the main source of data and big data analysis technologies as a powerful tool for processing information has been successfully used in modern manufacturing enterprises. However, analyzing big data is challenging in industrial applications because of the need to process unstructured volumes of data in real time. The results of this analysis should be presented to decision makers in the form of illustrative information graphics that give an impact on the relevant indicators and their critical changes.

3 Industrial Challenges for AR Guides

Modern trends in the organization of production in mechanical engineering are often presented in the form of the fourth industrial revolution, associated with the active and mass introduction of cyber physical systems in production. However, expectations from the application of specific solutions in this area in practice should be determined not so much by their innovativeness as by the possibility of a real increase in the functioning efficiency and quality of the organization of production systems.

The main criterion for the effectiveness of the use of intelligent technologies in mechanical engineering is to improve the quality of products. This is especially true in small-scale, pilot or tool production, when the range of executions varies depending on incoming orders. Ensuring high quality products with its diversity can increase the competitiveness of the enterprise, however, it requires high coordination of production processes, the supply of components and materials, information support for the product and the use of intelligent technology to control technological operations at all stages of production.

The growing capabilities of technologies for processing large volumes of poorly structured information in real time, as well as intelligent machine vision and pattern recognition technologies based on the implementation of artificial neural networks, allow reliable image recognition. Moreover, the practical application of these technologies in engineering is difficult due to the lack of a theoretical solution to the problem of using neural networks to control the quality of production processes and products in the case of a wide variety of designs and complexity of production processes.

For example, in the production problem domain [12], modern enterprises must provide a variety of items and the complexity of the technology. At the same time, quality control must be carried out not only during testing, but also during technological operations. For this, machine vision technologies can be applied, for which it is necessary to solve the problems of their practical implementation at a machine-building enterprise.

The implementation of augmented reality also allows users to undergo interactive training and receive active assistance during the execution of technological operations without the need to refer to a set of documentation in paper or electronic form. The main requirement for AR guides is to avoid overloading secondary information and to adapt a holistic visual representation of complexly structured information in accordance with the user's context.

Within this framework, there were developed new methods and means of analysis and quality control of products using modern intelligent machine vision technologies that were implemented within a specialized software platform.

4 AR Application for Quality Control

Existing technologies of machine vision and pattern recognition using artificial neural networks currently do not allow to comprehensively solving the problem of quality control of production processes and manufactured products. In particular, there is no single mechanism for preparing data for training a neural network, the accuracy of training and the quality of interpretation of the results are rather low, and there are no methods for the practical implementation of such solutions.

Within the developed framework there was developed new solution architecture (see Fig. 1) that combines the capabilities of neural networks, ontologies and big data processing facilities. Combining this technology stack is a promising step in ensuring the accuracy of pattern recognition and the practical utility of the resulting hardware-software complex.

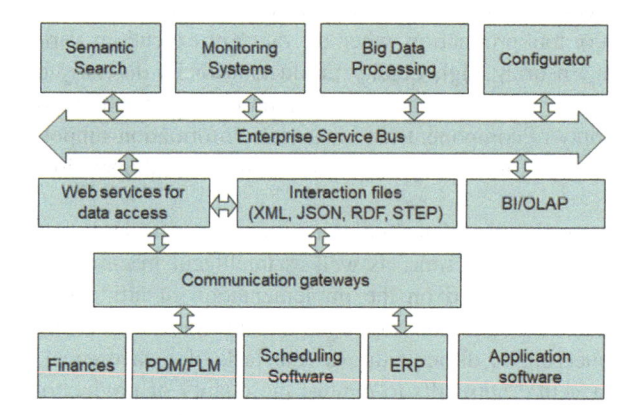

Fig. 1. Solution architecture.

Practical use case is illustrated by Figs. 2, 3 and 4. A number of cameras are used to track operations in accordance with the technological process and determine the objects that will be used in a real scene (parts and assemblies).

Fig. 2. Tablet implementation.

Intelligent software provides recognition of images of objects and their comparison with the corresponding description in the knowledge base. Video panels or augmented reality glasses are used to present relevant contextual information to the operator.

The general solution is used to identify gaps and failures of the operator in real time, predict possible errors in the work and suggest the best procedures based on comparing the sequence of actions with the experience of highly qualified operators included in the knowledge base. As a part of the proposed solution, there was developed an original algorithm of production items' identification. Industrial production Ontology was introduced to describe critical characteristics or features of each objects form and shape.

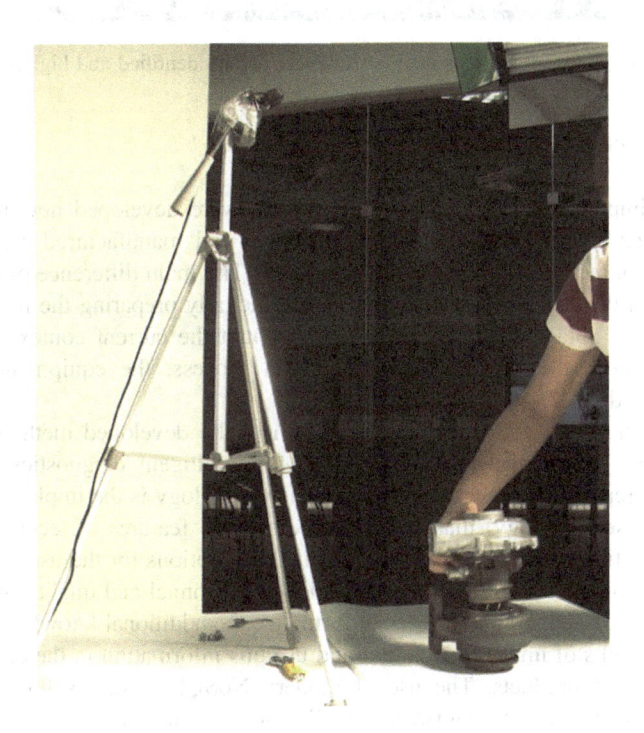

Fig. 3. Practical use example.

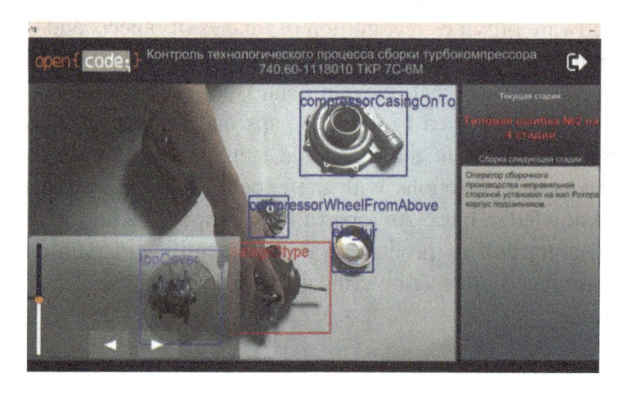

Fig. 4. Software implementation. Relevant object identified and highlighted.

5   Implementation Results

During the implementation of the projects, there were developed new methods and algorithms for recognizing production operations and manufactured products using machine vision technologies and neural networks. The main difference of this result is an increase in the quality of recognition by additionally preparing the initial data and interpreting the obtained results taking into account the current context, the technological operation, the stage of the production process, the equipment, tools and equipment used.

As a result of the software implementation of the developed methods and algorithms, a new knowledge-based technology for intelligent diagnostics and quality control was created. A distinctive feature of the technology is the implementation of a knowledge base (ontology) that describes the specific features of technological processes (in addition to technological documentation), options for the use of equipment, tools and equipment, individual characteristics of personnel and quality requirements.

The proposed approach allows us to formalize the additional knowledge needed to process the results of image recognition and use this information in the quality control of processes and products. The use of modern NoSQL solutions for data storage, parallel data processing mechanisms, as well as logical inference tools will allow real-time processing of poorly structured data.

The proposed technology was implemented by SEC "Open code" in a specialized hardware-software complex of machine vision (see Figs. 5 and 6), which allows application of intelligent systems for diagnostics and quality control of production operations and manufactured products within the workstation of assembly-line operators.

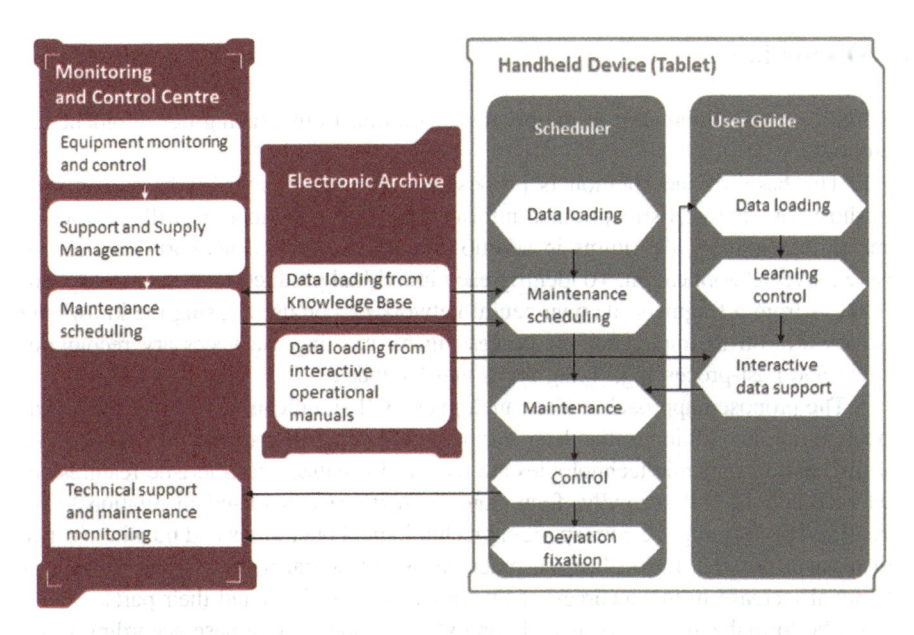

Fig. 5. Implementation solution.

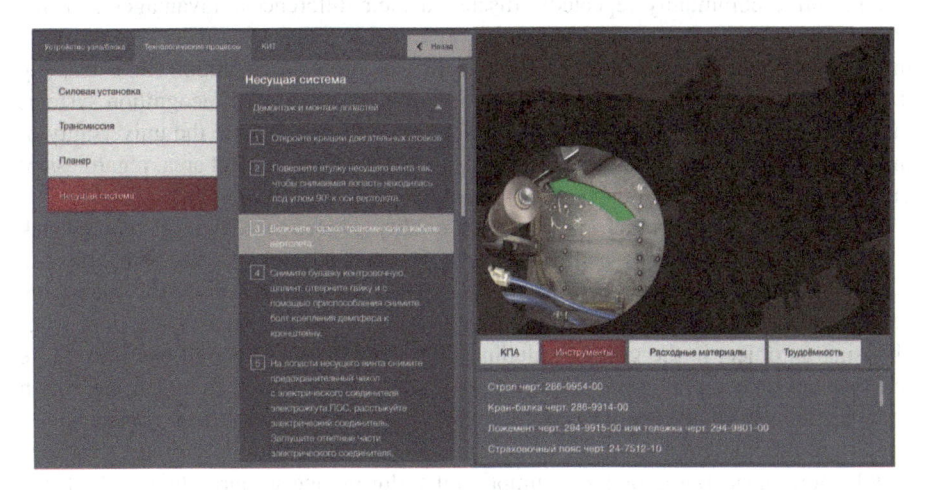

Fig. 6. Software implementation of AR guide.

The hardware-software complex will become a platform for the implementation of specific applied solutions for quality control in the production of modern technique and other engineering industries.

A series of experiments were carried out for a number of various production parts. Out of a set of 74 various objects there were successfully identified up to 66 items, that gives 89% of algorithm efficiency. One of the critical benefits of the proposed approach is its possibility to function in real time.

6 Discussion

The fundamental differences of the proposed solution from existing developments are as follows:

1. The basis of the solution is proposed to use the knowledge base - an open repository of characteristic product information. This information will be used in semantic recognition algorithms in addition to artificial neural networks, which will ensure universal application. To identify each individual product, part or assembly unit, as well as from a fragment, its own neural network can be used. Using the knowledge base allows you to implement pre-processing by creating the necessary recognition context and post-processing, fixing focus and feedback.

2. The proposed approach, unlike analogues, will significantly improve the quality of recognition in addition to the deep learning capabilities of artificial neural networks. For this, firstly, semantic technologies are used to formalize characteristic features and their recognition, and secondly, focus and context are considered in relation to the scenario of performing the corresponding technological operations and transitions. This will make it possible to localize the practical use of neural networks and provide an additional increase in the accuracy of identification of objects and their parts.

3. The formalization of focus and context in the knowledge base according to the concept of accented visualization forms know-how, successfully probated in practice. The scientific community repeatedly discussed their differences, advantages and disadvantages, which allows us to judge the presence of unique features. This development defines the main difference between the proposed solution and existing developments, providing a key to solving the problem that image recognition systems face in practice. As a result, the proposed solution allows to ensure the universality of the use of interactive electronic technical manuals in a given subject area, regardless of the number of types of products, their parts and versions.

4. The proposed approach improves the quality of recognition based on feedback from the user, which uses hardware and software for tracking user attention and augmented reality glasses. According to the results of the analysis of user attention, objects are prioritized, the association of focus and context increases the priority of choosing an object in the group (including in the conditions of partial overlap). More relevant are objects that have already been viewed by the user, or regarding which certain actions have been performed. Irrelevant results are excluded from consideration, which reduces the computational complexity of the algorithms. Improving relevance may be non-linear. Based on the assumption that during one scenario the user finds the last one, the last selected object will have the highest priority. Interactive visualization tools are actively used to generate prompts to users, control the execution of technological operations and transitions, and obtain feedback.

7   Conclusion

Augmented Reality technology allows you to develop interactive and context-sensitive user interfaces that provide real-time computer vision and object recognition capabilities. These functions make AR a powerful tool for implementing Industry 4.0 in practice, which can significantly improve the ability to control the quality of production processes and products.

References

1. Ivaschenko, A., Milutkin, M., Sitnikov, P.: Accented visualization in maintenance AR guides. In: Proceedings of SCIFI-IT Conference, Bruges, pp. 42–45 (2017)
2. Ivaschenko, A., Sitnikov, P., Milutkin, M., et al.: AR optimization for interactive user guides. In: Arai, K., Kapoor, S., Bhatia, R. (eds.) Advances in Intelligent Systems and Computing, vol. 869, pp. 948–956. Springer, Cham (2019)
3. Krevelen, R.: Augmented reality: technologies, applications, and limitations. Vrije universiteit, Amsterdam (2007)
4. Navab, N.: Developing killer apps for industrial augmented reality. IEEE Comput. Graph. Appl. **14**(7), 15–17 (2004). IEEE Computer Society
5. Julier, S., Livingston, M.A., Swan, I.I., et al.: Adaptive user interfaces in augmented reality. Swan and Bailot, New Jersey (2004)
6. Singh, M.P.: Augmented reality interfaces. Natural web interfaces. IEEE Internet Comput. **17**, 66–70 (2013)
7. Kagermann, H., Wahlster, W., Helbig, J.: Recommendations for implementing the strategic initiative Industry 4.0. Final Report of the Industry 4.0 Working Group, p. 82. Forschungsunion, Berlin (2013)
8. Lasi, H., Kemper, H.G., Fettke, P.: Industry 4.0. Bus. Inf. Syst. Eng. **4**(6), 239–242 (2014)
9. Gubbi, J., Buyya, R., Marusic, S., et al.: Internet of things: a vision, architectural elements, and future directions. Future Gener. Comput. Syst. **29**(7), 1645–1660 (2013)
10. Baesens, B.: Analytics in a Big Data World: The Essential Guide to Data Science and Its Applications. Wiley, New Jersey (2014)
11. Bessis, N., Dobre, C.: Big Data and Internet of Things: A Roadmap for Smart Environments. Studies in Computational Intelligence. Springer, Cham (2014)
12. Ivaschenko, A., Khorina, A., Sitnikov, P.: Accented visualization by augmented reality for smart manufacturing applications. In: IEEE Industrial Cyber-Physical Systems, ITMO University, Saint Petersburg, pp. 519–522 (2018)

The Modelling and Optimization of Machine Management System with Computer Numerical Control

I. M. Yakimov, A. P. Kirpichnikov, V. V. Mokshin$^{(\boxtimes)}$,
and Z. T. Yahina

Kazan National Research Technical University, 31 Karla Marksa Street,
Kazan 420111, Russian Federation
vladimir.mokshin@gmail.com

Abstract. The process of modeling and optimization of machine management system with computer numerical control and its findings are given. The production process is presented as IDEF3 diagram in BPwin system, structural and simulation modeling is shown in Arena system. Mathematical model has been constructed in accordance with simulation modeling findings, by which optimization has been carried out and formulae for calculation of optimized factor values on objective factor values have been obtained. In the present study, we consider a complex system whose behavior is characterized by set of various time-dependent factors. Some of these factors can characterize the external influences on the system, whereas other factors contain information generated by system. We demonstrate that time dependence of these factors can be reproduced by the nonlinear regression model. This allows us to predict a possible behavior of the system and to identify the so-called significant factors that have a significant impact on the behavior of the system. To demonstrate validity of the method, we apply it to analyze the data characterizing a manufacturing company.

Keywords: Tools · Computer numerical control · IDEF3 · BPwin · Arena · Simulation modeling · Mathematical model · Optimization

1 Introduction

In this research the results of simulation modeling and optimization of machine management system with computer numerical control (CNC) according to statistic data of the part process production on the helicopter plant have been given. The part process is made on CNC machine. It is thought that the optimization in the plant shop might allow to cut down the expenses on equipment and the lead time as well as increase the machine loading ratio that can lead to profit rise and productiveness of enterprise.

Structurally the aim of the research is obtaining the formulae for value estimations of optimizing factors: tool optimum quantity with CNC and center of shared usage of equipment. According to the values of objective factors: desired level of producing parts and planned production time. In the furtherance of this goal the method was given consisting of the following stages [1–5]:

A. A. Radionov and A. S. Karandaev (Eds.): RusAutoCon 2019, LNEE 641, pp. 724–731, 2020.
https://doi.org/10.1007/978-3-030-39225-3_79

1. The preliminary analysis of manufacturing process and the indicator selection of production efficiency and influencing factors on then;
2. The task assignment;
3. The development of structural and simulation model of production;
4. The simulation experiments planning;
5. Simulating modeling fulfillment according to strategic plan;
6. Correlation analysis;
7. The mathematical scheme development;
8. The production optimization due to mathematical scheme;
9. The formula construction for value calculations of the optimized factors according to objective factor values.

2 The Preliminary Analysis of Manufacturing Process and the Indicator Selection of Production Efficiency and Influencing Factors on then

The production process is developed in the form of IDEF3-diagram. The diagram is constructed in BPwin system [6, 7] (Fig. 1).

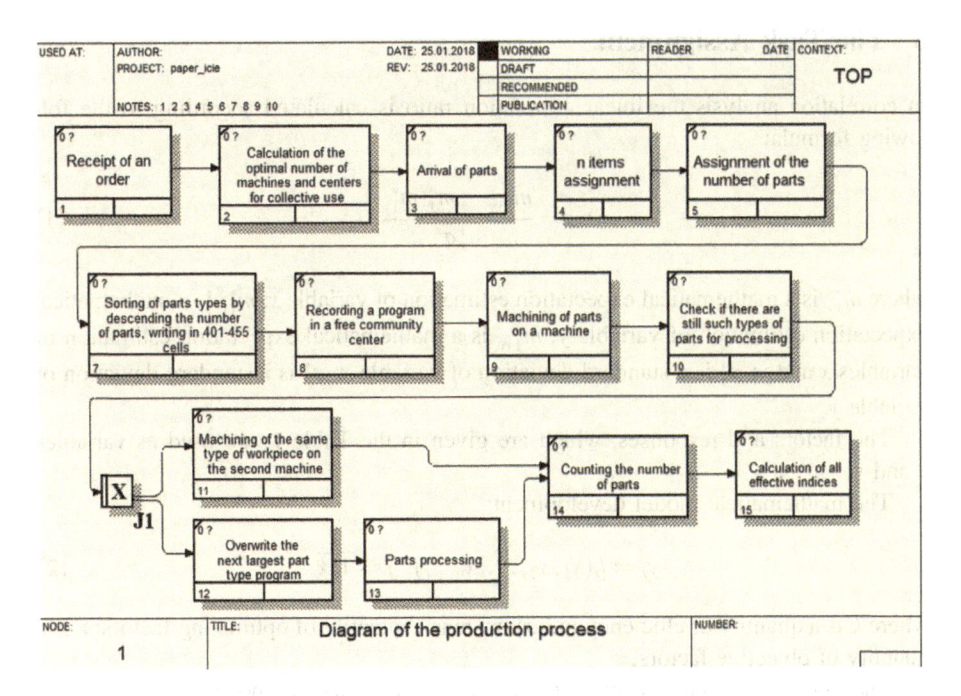

Fig. 1. IDEF3-diagram of the production process.

The Table 1 is the list of variables selected for research.

Table 1. List of variables selected for research.

№	Values	Naming
1	x_1	The quantity of tools with CNC
2	x_2	The quantity of center of shared usage of equipment
3	x_3	The quantity of sets in batch
4	x_4	The planned production time of a set per minute
5	y_1	The cost of equipment according to a part batch in rubles
6	y_2	The production time of a batch per minute
7	y_3	The production time deviation on a batch per minute
8	y_4	The tool use ratio with CNC
9	y_5	The center of shared usage of equipment ratio
10	y_6	The production time of a set per minute
11	y_7	The standard time deviation of producing making single set in min
12	y_8	The probability of order making on planned time

In the Table 1 the values x_1 and x_2 are optimizable, x_1 and x_2 are objective. Variable y_i, $i = \overline{1, k}$ are the efficiency indicators (responses).

3 The Task Assignment

In correlation analysis the linear correlation ratio is calculated according to the following formula:

$$r_{xy} = \frac{m_{1xy}^* - m_{1x}^* m_{1y}^*}{\sigma_x^* \sigma_y^*}, \tag{1}$$

where m_{1x}^* is a mathematical expectation estimation of variable x; m_{1y}^* is a mathematical expectation estimation of variable y; m_{1xy}^* is a mathematical expectation estimation of variables x and y; σ_x^* is a standard deviation of variable x; σ_y^* is a standard deviation of variable y.

The factors and responses, which are given in the Table 1, are used as variables x and y.

The mathematical model development:

$$y_j = f_j(x_1, x_2, \ldots, x_{m+r}); \ j = \overline{1, k}, \tag{2}$$

where k is a quantity of efficiency indicator; m is a quantity of optimizing factors; r is a quantity of objective factors.

The optimization task is set according to the mathematical model (2).

The target maximizing (minimizing) function is selected from the list of efficiency indicator (responses). The rest responses and optimizing factors are restricted.

$$f_q(x_1, x_2, \ldots, x_m) \rightarrow \max(\min);$$
$$a_j \leq f_j(x_1, x_2, \ldots, x_m) \leq b_j;$$
$$c_i \leq x_i \leq d_i; \quad i = \overline{1, m}; \qquad ; \qquad (3)$$
$$x_g = const; \quad g = \overline{1, r}.$$

where k is a quantity of efficiency indicators; m is a quantity of optimizing factors; r is a quantity of objective factors.

The formula construction for value calculations of the optimizing factors according to objective factor values:

$$x_j = f_j(x_{m+1}, x_{m+2}, \ldots, x_{m+r}); \quad j = \overline{1, m} \qquad (4)$$

where x_j is an optimizing factor j; x_{m+1} is an objective factor i; m is a quantity of optimizing factors; r is a quantity of objective factors.

4 Development of Structural and Simulation Model of Production

The structural model of production process in system of structure and simulation modeling (SSSM) Arena [7–9] was developed. The part blanks are delivered to warehouse – Sklad. Having read in fed information about a kind of a batch on center of shared usage of equipment, it transfers required part blank processing software to tools with CNC by which the part blank processing is carried out. The performance statistic figures are registered after completing the operation.

The programming simulation model is created on SM Siman language and carried out according to structural model.

5 The Simulation Experiments Planning

It is considered acceptable that the dependence of efficiency indicators on four influencing factors on them are non-linear and not above the second level. To construct mathematical model the optimum design – D is used consisting of 49 variants: the central part of four-dimensional cube, 16 corners and 32 middle ribs [10].

The quantity of simulation model implementations can be calculated by formula, obtained on base of main limiting theorem of the theory of probability [10]:

$$n = \frac{\sigma^2}{\varepsilon^2} t_\beta^2 \qquad (5)$$

By formula (5) the required implementation quantity can be calculated for average time of one-part batch processing with probability $\beta = 0,95$ for which $t_\beta = 1,96$.

$$\varepsilon = 0{,}05 \cdot m_1^* = 0.05 \cdot 2497 = 124{,}85 \text{ and } \sigma = \sigma^* - 2012$$

$$n = \frac{\sigma^2}{\varepsilon^2} t_\beta^2 = \frac{2012.67^2 \cdot 1,96^2}{124{,}85^2} = 998{,}334 \approx 1000$$

are accepted according to the results of modeling in main variant of strategic plan.

The quantity (1000) of model implementations are accepted in every variant of strategic plan.

6 Simulating Modeling (SM) Fulfillment According to Strategic Plan

The table is created by the results of SM, in which independent variable value – factors and values of dependent variables – efficiency indicator, obtained in simulating modeling process for all variants of strategic plan. This table is applied for further simulating modeling construction.

7 Correlation Analysis

The value calculation of linear correlation ratio among 8 efficient indicators and 4 factors allowed to define that all factors significantly have influence on responses. Essential mutual factor influence on efficient indicators of production process can be seen as well. That is why the decision was taken to include in equations the regression factors in the first and second levels of pair-wise factor products.

8 The Mathematical Scheme Development

As all variables, used in the research, are analog quantity, then in this case it is better to use regression analysis based on the least squire method [10], which requires deviations quire sum of experimental values from calculated ones to be smallest. The regression equations, joining the efficient indicators with production influencing factors, were obtained with help of multiple regression procedure of Statistica 10.0 program [11]. As example an obtained regression equation for calculating y, which is a cost of equipment of producing single part batch, is provided.

$$\begin{aligned} y_1 = &- 7348572{,}054 - 26483{,}756x_1 - 500220{,}757x_2 + 1228859{,}468x_3 + 24884{,}296x_4 \\ &+ 751{,}361x_1x_2 - 15821{,}548x_1x_3 + 5 \cdot 10^{-11}x_1x_4 - 2586{,}623x_2x_3 - 2{,}9 \cdot 10^{-10}x_2x_4 + 817{,}121x_1^2 \\ &- 132007{,}187x_2^2 - 23214{,}904x_3^2 - 16{,}371x_4^2 \end{aligned}$$

$$(6)$$

The rest seven regression equations of mathematical model consist of total all of the piece variables and diverge considerably from each other by ratio values at variables.

9   The Production Optimization Due to Mathematical Scheme

The problem of optimization is solved under limitation on efficient indicators and optimizing factors.

$$y_1 = f_1(x_1, x_2) \rightarrow \min;$$

$$300 \leq f_2(x_1, x_2) \leq 5000; \quad 100 \leq f_3(x_1, x_2) \leq 500;$$

$$0,7 \leq f_4(x_1, x_2) \leq 1; \quad 0,7 \leq f_5(x_1, x_2) \leq 1;$$

$$250 \leq f_6(x_1, x_2) \leq 1300; \quad 30 \leq f_7(x_1, x_2) \leq 70; \tag{7}$$

$$0,8 \leq f_8(x_1, x_2) \leq 1;$$

$$31 \leq x_1 \leq 65; \quad 1 \leq x_2 \leq 3.$$

Objective factors don't vary at optimization, their values are accepted according to strategic plan variations.

The Newton method is selected for optimization, and appropriate optimization process is applied which is available in application program package Excel [11]. Optimized factor values have been obtained as well as value of efficient indicators y_1 in all 49 strategic plan variants.

10   The Formula Construction for Value Calculations of the Optimized Factors According to Objective Factor Values

In accordance with optimization findings, the regression equations for value calculations of optimized factors by objective factor values. The formulae for optimum value of optimized factors:

Optimal quantity of tools with CNC:

$$\begin{aligned} x_{1opt} = &- 8690,26 + 43,2x_3 + 5,62x_4 - 2,21x_3^2 - 0,01x_4^2 - 91,51ln(x_3) \\ &+ 1479,43ln(x_4) - 36,721/x_3 + 396403,411/x_4 \end{aligned} \tag{8}$$

Optimal quantity of center of shared usage of equipment:

$$\begin{aligned} x_{2opt} = &- 1598,99 + 5,91x_3 + 0,72x_4 - 0,30x_3^2 - 0,001x_4^2 \\ &- 12,61ln(x_3) + 227,66ln(x_4) - 5,02/x_3 + 95989,721/x_4 \end{aligned} \tag{9}$$

We consider the result obtaining on the following example. There is an order with the following parameters: the quantity of parts sets in a single batch $x_3 = 7$; single set producing time $x_4 = 780 \min = 13$ h. Optimized factor value is calculated by formulae (8) and (9): $x_{1opt} = 52$ tools; $x_{2opt} = 2$ center of shared usage of equipment, and equipment cost equipment according to a part batch in rubles is calculated by formula (7) $y_1 = 421312{,}1$ rubles. The following values of the rest efficient indicators of production process have been obtained: $y_2 = 3376$ min. $\approx 56{,}3$ h; $y_3 = 194$ min $\approx 3{,}2$ h; $y_4 = 0{,}85$; $y_5 = 0{,}86$; $y_6 = 487$ min ≈ 8 h; $y_7 = 33$ min; $y_8 = 0{,}88$.

The obtained results match up properly to the limitation range.

11   Conclusion

Based on the research findings the following conclusions can be drawn.

1. Proposed method for researching the production of part process on tools with CNC can be used successfully for other production processes as well.
2. Obtained formulae for calculating the optimized factor values can be applied for defining the main parameters of production in part process on the tools with CNC provided that objective factor values are within limits the ranges, accepted at modeling.
3. The production mathematical model of part process on the tools with CNC allows us to define 8 efficient indicators of the production.
4. The research findings have been used for optimization of particular production process on helicopter plant.

The introduced method for the analysis of complex systems can also be used in [13].

References

1. Yakimov, I.M., Kirpichnikov, A.P.: Simulation modeling in production and installation of ventilation and plumbing equipment. Bull. Kazan Technol. Univ. 16(20), 295–302 (2013)
2. Yakimov, I., Kirpichnikov, A., Mokshin, V., et al.: The comparison of structured modeling and simulation modeling of queueing systems. Commun. Comput. Inf. Sci. 800 (2017). https://doi.org/10.1007/978-3-319-68069-9_21
3. Mokshin, A.V., Mokshin, V.V., Sharnin, L.M.: Adaptive genetic algorithms used to analyze behavior of complex system. Commun. Nonlinear Sci. Numer. Simul. 71, 174–186 (2019). https://doi.org/10.1016/j.cnsns.2018.11.014
4. Tutubalin, P.I., Mokshin, V.V.: The evaluation of the cryptographic strength of asymmetric encryption algorithms. In: Second Russia and Pacific Conference on Computer Technology and Applications, Vladivostok, 25–29 September 2017, pp. 180–183 (2017). https://doi.org/10.1109/rpc.2017.8168094
5. Mokshin, V.V., Saifudinov, I.R., Sharnin, L.S., et al.: Parallel genetic algorithm of feature selection for complex system analysis. IOP Conf. Ser. J. Phys. Conf. Ser. 1096, 012–089 (2018). https://doi.org/10.1088/1742-6596/1096/1/012089
6. Yakimov, I.M., Kirpichnikov, A.P., Mokshin, V.V., et al.: Integrated approach to complex system modeling with BPwin-Arena system. Bull. Kazan Technol. Univ. 17(6), 287–292 (2014)

7. Yakimov, I.M., Trusfus, M.V., Mokshin, V.V., et al.: AnyLogic, ExtendSim and Simulink overview comparison of structural and simulation modelling systems. In: Proceedings of the 3rd Russian-Pacific Conference on Computer Technology and Applications (2018). https://doi.org/10.1109/rpc.2018.8482152
8. Yakimov, I.M.: Computer Modeling. Teaching Medium. Kazan Technical University Press, Kazan (2008)
9. Neylor, T.: Machine Simulation Tests with Economical System Models. Mir, Moscow (1975)
10. Borovikov, V.P., Borovikov, I.P.: Statistic Analysis and Data Proceeding in Windows. Statistika, Moscow (1998)
11. John, W.: Formulae in Excel 2013. Wiley, Moscow (2014)
12. Saifudinov, I.R., Mokshin, V.V., Tutubalin, P.I., et al.: Visible structures highlighting model analysis aimed at object image detection problem. In: CEUR Workshop Proceedings, vol. 2210, pp. 139–148 (2018)

Measuring Systems Application Analysis in Engineering

M. I. Kovalev[✉]

Novocherkassk Electric Locomotive Works, 7A Mashinostroiteley street,
Novocherkassk 346413, Russian Federation
kovalev@gmail.com

Abstract. This work is devoted to the analysis of measurement systems as a tool to improve the quality of measurements. Measurement System Analysis (MSA) in the sector of railway engineering is performed using the following methods: variance analysis; ranges; average and range method. A number of experiments for measuring systems are carried out with coordinate measuring machines ACCURA MASS 16/24/14 and a digital caliper. The results obtained are of universal character; and the methodology described can be applied to other measuring systems. This method is needed to quantify the measurement systems. MSA is the constituent of statistical management of processes of manufacture intended for a research of measuring processes for the purpose of an assessment of their acceptability. MSA allows drawing conclusions about the suitability of measuring processes. This gives you the opportunity to make corrective actions to improve the measuring system and, as a consequence, to guarantee the quality of the measured parameters [1–3].

Keywords: Measuring piston · Measurement process · Coordinate measurement machine · Average and range method · Repeatability · Reproducibility · Analysis of variance (ANOVA) · Range method

1 ISO Standards are Based on Facts

Improving the quality and competitiveness of products and satisfaction of customers are an important and urgent task for the industrial enterprises in the conditions of contemporary market economy. MSA is widespread in the quality management systems (QMS) of the railway industry and is one of the basic techniques required when introducing ISO 9001-2015 standard requirements: "Quality Management System. Requirements" par. 8.2 "Monitoring and measurement"; Russian National Standard ISO/TS 16949-2009 "Quality Management Systems. Particular requirements for the application" par. 7.6.1 "Analysis of measuring systems" and the IRIS "International Railway Industry Standard" par. 8.2 "Measurement and analysis of the data". IRIS is the international standard for the railway industry based on the universal standard for the ISO 9001 quality management system. The importance of MSA application in the industry is forced by the necessity to increase customer satisfaction by improving the product quality.

© Springer Nature Switzerland AG 2020
A. A. Radionov and A. S. Karandaev (Eds.): RusAutoCon 2019, LNEE 641, pp. 732–743, 2020.
https://doi.org/10.1007/978-3-030-39225-3_80

One of the ISO 9000 methods is the fact-based decision-making. To make a decision, first you must gather facts (accurate information). Monitoring, measurement and analysis are used for this purpose. However, information derived from these processes is not always objective and accurate for various reasons, and the use of inaccurate information leads to wrong decisions. Thus, the normal functioning of the QMS, within the framework of ISO/TS 16949, is possible only provided data repeatability. Modern measurement processes are based on complex measuring technology. Traditional approach to assess its validity and creditability is inapplicable to it. New measuring equipment purchasing and its introduction to manufacturing must be accompanied by specific measurement processes surveys. That is when MSA is of use. MSA is a collection of experiments and statistical methods applied for measurement results validity assessment. [4].

Measurement System Analysis is a method designed to prove validity and applicability of measuring systems using quantification of their characteristics. Measurement System (MS) is a set of instruments or tools, standards, operations, methods, fixtures, software, personnel, environment and assumptions used to quantify a measurement unit or fix assessment to the feature characteristic being measured; it is also the complete process used to obtain measurements [5].

The analyzes of the measuring systems carried out in this article uses coordinate measuring machines (CMMs) of ACCURA MASS 16/24/14 and a digital caliper. This CMM has robust design due to the granite table and rigid structure of portal made of a heat-stable composite material. All axes are equipped with 4-sided air bearings; and the X and Y-axis guides are completely closed from the drive side.

Caliper is a universal tool designed for high-precision measurements of external and internal dimensions, as well as of hole depths. Caliper is one of the most common tools of measurement due to its simple design, ease and speed of use [6].

The present study is conducted at an industrial enterprise of railway locomotive production. The enterprise is a manufacturing complex with significant technical and scientific potential, qualified personnel, extensive experience in effective cooperation of industry with science. It also provides a unique opportunity to create a completely new locomotive from design development to high-quality high-end certified mass production.

Measuring systems analysis begins with its purpose and measurement process understanding. All sources of chaotic information and unacceptable errors should be eliminated. The study follows Deming concepts:

- To determine the major error sources and to eliminate them
- To allow one or more factors to change
- To measure several times
- To analyze actions results

A measuring system can be influenced by various sources of variability. Thus, the results of repeated measurements of the same part can be different from each other due to common and special causes of deviations. The effect of different deviation sources in the measuring system should be evaluated during short and long time periods.

Measurement System Validity as well as the process usability is the result of long-term assessment of measurement system variability.

The measuring system is valid for use when:

- The measurement process is statistically controlled i.e., it is stable and unchangeable, or only common causes of variability exist
- The system is aimed at the target (no offset or deflection)
- Process variability (repeatability and reproducibility) is acceptable and is within the expected range

2 Variability in Interaction with the Environment

One purpose of measurement system studying is to obtain information on measurements variability types and its magnitude caused by the measuring system interaction with the environment.

The Method of Average and Range is an approach that provides an estimate of repeatability and reproducibility for the measurement system. However, the analysis does not include variability induced by the interaction of an operator and a manufactured part.

In the experiment under consideration, the inner diameter of a locomotive brake system component called threaded sleeve is chosen as a measured variable. The component has the tolerance range of 38h14, and it is used in the pneumatic installation of the compressor and the main brake tanks in a locomotive. Measurements of ten threaded sleeves samples are carried out three times. Measured samples are selected at random. Measurements are carried out with a digital caliper of 0.01 division value. The required number of measurements in one cycle is $n = 3$. Control checklists are made on recommended calculations. Data processing is carried out in MS Excel spreadsheet program.

Checklists are used to analyze the measurement system stability. Several cycles of experiments are carried out. Each cycle of the experiment is to measure the selected sample parameter a certain number of times by the operator. Results-based average chart (Fig. 1) and range chart (Fig. 2) are plotted [7].

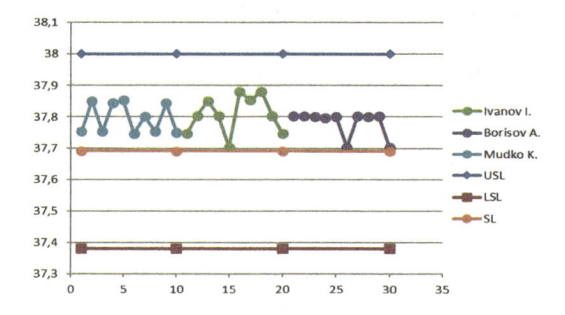

Fig. 1. X-R – Average chart.

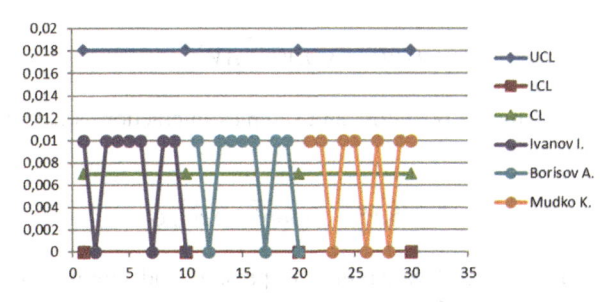

Fig. 2. R – Range chart.

Average chart indicates that the measurement results for all the operators are located above the center line, but they do not exceed upper tolerance limit. The tendency of overcoming upper tolerance limits is observed due to the variability of the process. Range chart shows measurement process stability, no trends are observed. Range chart is periodic. The measurement process variability components are calculated within the level of significance of $\alpha = 0.99$:

Repeatability is variability of measurement results obtained with the same measuring instrument used by the same controller several times to measure the same dimensional characteristics of the same component. It usually includes dispersion of an instrument [8, 9].

Repeatability is calculated using the following formula:

$$EV = K_\alpha S_e,$$

where K_α – is a coefficient used to calculate confidence interval for the true value of the sample measuring parameter, when the level of significance is α; Se – is an estimate of standard deviation (SD) for measuring process convergence.

Reproducibility is variability of the measurement results average; results being obtained by different controllers using the same measuring instrument to measure the same dimensional characteristics of the same component.

$$AV = K_\alpha S_o,$$

where S_o – is the estimate of standard deviation of the measurement process reproducibility.

The variability of the samples is calculated using the following formula:

$$PV = K_\alpha S_p,$$

where S_p – is assessment of standard deviation of the sample measurement process variability.

Gage Repeatability and Reproducibility, GRR, is the combined estimate of measurement system repeatability and reproducibility.

$$GRR = \sqrt{EV^2 + AV^2}$$

Variability caused by the operators and samples interaction is calculated using the following formula:

$$INT = 5.15 \cdot \sqrt{(Sop^2 + Se^2)/Q},$$

where Sop^2 – is an estimate of variability of operators and samples interactions.

The total variability of Measurement System is calculated using the following formula:

$$TV = \sqrt{GRR^2 + PV^2}$$

3 Components of Variability

To analyze the measurement process fully and thoroughly, relative values of the variability components are calculated as follows:

$$\%EV_{TV} = EV/TV \cdot 100 \tag{1}$$

$$\%AV_{TV} = AV/TV \cdot 100 \tag{2}$$

$$\%PV_{TV} = PV/TV \cdot 100 \tag{3}$$

$$\%INT_{TV} = INT/TV \cdot 100 \tag{4}$$

$$\%GRR_{TV} = GRR/TV \cdot 100 \tag{5}$$

Ratio of process components relative variability to tolerance range is calculated using formulas (1)–(5) when total variability TV is substituted with tolerance range. The calculation results are shown in Table 1. Based on the relative repeatability and reproducibility value of $\%GRR_{SL} = 11.4\%$, we can make conclusions on the measurement process usability/reliability to evaluate tolerance compliance. The conclusion is that the measuring process is acceptable to assess tolerance compliance [4].

Acceptability assessment of the measurement system is carried out in the second experiment using analysis of variance. Variance analysis (ANOVA) is a standard statistic technique used to analyze measurement errors and reasons for data variability when a measurement system is studied [10, 11]. In variance analysis, the variance comprises the following four categories: parts, controllers, interaction between parts and controllers, and equipment error at repetitive measurements.

Table 1. Average and range method calculation results.

Variability component	Standard deviation (SD)	Variability component estimation (5.15·SD)	Variability regarding tolerance field	Total variability proportion
Repeatability (repeatability, variability of measurement system)	0.00413	0.0213	3.4	15.4
Reproducibility (variability depending on controller)	0.01307	0.0673	10.9	48.6
Repeatability and reproducibility	0.01371	0.0706	11.4	50.9
Sample parameter variability	0.0269	0.1386	22.3	100.0

Advantages of ANOVA as compared to Average and Range Method are the following:

- Its capability to assess any kind of experimental equipment
- Variances estimates and calculations are more accurate
- Experimental data are more informative (e.g. provides information on parts and controllers interaction effect)

Disadvantages of ANOVA are:

- Complexity of numerical calculations
- Users require certain statistical knowledge to interpret the results
- The ANOVA method is recommended for application when computer is available due to the complexity calculations required

During the experiment, operators successively perform measurements of all samples using calipers, and record them. The following formulas are used to calculate standard deviation (SD) constituents' values:

Measurement equipment variance estimate:

$$S_e^2 = \frac{1}{NM(Q-1)} \sum_{i=1}^{N} \sum_{j=1}^{M} \sum_{k=1}^{Q} \left(X_{ijk} - \bar{X}_{ij*}\right)^2, \tag{6}$$

where M, N, Q – are the quantity of operators, samples, and measurements, respectively; $\bar{X}_{ij*}$ – is the average value of the measurement results by each operator.

Operators' variance/reproducibility estimate:

$$S_o^2 = \frac{NQ}{M-1} \sum_{j=1}^{M} \left(\bar{X}_{*j*} - \bar{\bar{X}}_{***}\right)^2, \tag{7}$$

where $\bar{X}_{*j*}$ – is the average value of the measurement by each operator; $\bar{\bar{X}}_{***}$ – is the average for all measurements.

Samples variance estimate:

$$S_p^2 = \frac{MQ}{N-1} \sum_{j=1}^{N} \left(\bar{X}_{i**} - \bar{\bar{X}}_{***} \right)^2, \tag{8}$$

where $\bar{X}_{i**}$ – is the average value of the measurement of each sample by all operators.

Operators and samples interaction variance estimate:

$$S_{op}^2 = \frac{Q}{(N-1)(M-1)} \sum_{j=1}^{N} \sum_{j=1}^{M} (\bar{X}_{ij*} - \bar{\bar{X}}_{i**} - \bar{\bar{X}}_{*j*} + \bar{\bar{X}}_{***})^2 \tag{9}$$

The results of calculation are shown in Table 2. It should be noted that the variation constituent cannot be measured by the method of Average and Range. Since samples variance equals the standard deviation of samples, the variance calculated with formulas (5)–(9) can be easily converted into the corresponding standard deviations to be used as analogue to the Average and Range Method.

The significance of operator and sample interaction effect for the measurement results variability can be measured with the following formula:

$$F = \frac{S_{op}^2}{S_e^2}$$

The critical value $F\alpha$ (k_1, k_2) distribution is calculated according to the Table of F-distribution (Fischer-Snedecor distribution) at the level of significance of $\alpha = 0.05$, where k_1 – is the number of degrees of freedom for larger variance, k_2 – is the number of degrees of freedom of the less variance. The number of degrees of freedom k_1 and k_2 are equal to (N-1) (M-1) and NM (Q-1), respectively, where M is the number of operators, N is number of samples and Q is number of retries. The results of calculation are listed in Table 2. As $F > F\alpha$ (k_1, k_2), the operator and samples interaction variability effect are recognized as important and is considered in further calculations [12].

The formula for the calculation of reproducibility, if operator and samples interaction variability effect is recognized as significant, is as follows:

$$AV = 5.15 \cdot \sqrt{(Sop^2 + Se^2)/Q}$$

The formula for the calculation samples variability, if operator and samples interaction variability effect is recognized as important, is the following:

$$PV = 5.15 \cdot \sqrt{\frac{\left(Sp^2 - S_{op}^2 \right)}{MQ}}$$

Variability conditioned by the operator and samples interaction variability effect is calculated according to the formula:

$$INT = 5.15 \cdot \sqrt{\frac{\left(S_{op}^2 - S_e^2\right)}{Q}}$$

The formula to calculate repeatability and reproducibility, when ANOVA method is used to analyze variability, and if operator and samples interaction variability effect is recognized as important, is the following:

$$GRR = \sqrt{EV^2 + AV^2 + INT^2}$$

Considering the obtained value of relative repeatability and reproducibility of %GRR$_{SL}$ = 46.10% (see Table 2) we draw the following conclusion on validity of the measurement process to estimate meeting tolerance: the measurement process needs improvement. To reduce defects, to prevent faulty production and to lower variability of the process, the following actions are developed: to train lathe operators to adjust size setting to the middle of tolerance range; and to instruct them to align the lathe to the middle of the tolerance range when manufacturing the parts [13–15].

Table 2. ANOVA analysis results.

Variability constituents	SD score	Variability constituent score (5.15·SD)	Variability in tolerance range	Total variability share
Repeatability (repeatability and variability of MS)	0.0265	0.1363	22.0	47.6721
Reproducibility (variability of the operator)	0.0693	0.0000	0.0	0
Sample and operator interaction	0.0885	0.2513	40.52	87.9054
Repeatability and reproducibility		0.2858	46.10	100
Sample parameter variability	0.0852	0.0000	0.0	0
Total variability of measurement process (MS)		0.2858	46.0998	100

Automatic check on the gross error is done to protect against errors. The check results show probability of gross error in measurement results. Very often gross errors are caused by the operators' errors in measurements. The most common operators' errors are the following: wrong reading of the measuring device scale; incorrect entry of observations result (misprint or lapsus or slip of the pen); incorrect entry of individual measurement values when a set of tools is used; repetitive mis-manipulations with equipment; etc. The check during the survey described proved no gross errors.

In the third experiment, each operator measures samples one time using CMM ACCURA MASS 16/24/14. Each operator measures five samples of a detail called a "capsule", which is a part of the locomotive's engine. The parameter under consideration is the internal diameter. Calculation of repeatability and reproducibility is carried out with the Method of Range. The experiment is a modified measurement system study, which provides quick approximation of measurement variability. This method provides only the overall assessment of the measurement system. It does not show repeatability and reproducibility as constituents of the variability.

This method is recommended for use in case of strict time restrictions for testing. Operators successively perform measurements of all sample, the samples are taken at random. All operators measure each sample once. The results of measurements and of preliminary calculations are presented in Table 3.

Repeatability and reproducibility SD estimation of the measurement process is determined according to the following formula:

$$S_m = \frac{\bar{R}}{D_2},$$

where $\bar{R}$ – is the Average Range of the measurement results, D_2 – is constant to calculate the standard deviation using the Range.

Sample variability SD estimation of the measurement process is calculated with using the following formula:

$$S_p = \frac{R_p}{D_2},$$

where R_p – is the samples parameters values range.

Repeatability and reproducibility is calculated with using the following formula:

$$GRR = K_\alpha S_m$$

The calculation results are summarized in Table 3.

Table 3. Calculation of variability.

D_2 constant of SD	1.19
S_m repeatability and reproducibility of the range method	0.0001849
D_2 constant of SD	2.48
S_p score of sample variability	0.00010081
K	5.15
Repeatability and reproducibility of R&R	0.0009521
%GRR$_{SL}$	3.2831063

On the basis of the relative repeatability and reproducibility $\%GRR_{SL} = 3.28\%$, we can conclude that measuring system is valid and applicable.

4 Displacement of Measurement Process

Above product "capsule" has been selected for determining the displacement measuring system as the test sample. The measurements were made using CMM in manual mode. Offset - the difference between the reference value and the average measurements. Thus, to obtain the bias estimation requires knowledge of the reference value, and the need to be analyzed was stable.

The reference value - is the value of the object or group that serves as a comparison sample approved. The average value of several measurements made high-level equipment (such as a reference laboratory) can be selected as the reference value.

To determine the reference value of 10 was performed sequentially diameter measurements. The results are recorded in the checklist.

Next, 10 measurements were performed consecutively investigated parameter five samples automatically. Calculated arithmetic means of all the measurements, which is equal to - X = 239.96463 mm. Determine the bias of each reading:

$$B = X - X_{onop}$$

The relative displacement is determined as a percentage of tolerance:

$$\%B = 100B/T$$

The shift is not a criterion for the admissibility of the MS, it should be taken into account when carrying out further measurements using the test measurement process.

Under the linearity of displacement measuring system is understood ramp displacement within the operating range of the measurement process. The linearity of the displacement estimated direct inclination value that best approximates the dependence of the average offset values for various samples of their estimated true (reference) values.

To offset this measured at several points within the operating range, and studied the dependence of the displacement from the reference (the estimated true value) of the sample value [7].

Verify linearity MS displacement performed by the above product "capsule" and comprised the following stages of the experiment and data processing:

1. Operator measured each sample 10 times in random order.
2. The calculation of the average value for each sample.
3. Implemented calculation of the absolute value of each sample and the average displacement of the displacement for each part.
4. The calculation of the correlation coefficient between the reference values of the respective offsets.
5. The schedule of linear displacement measurement process (Fig. 3).

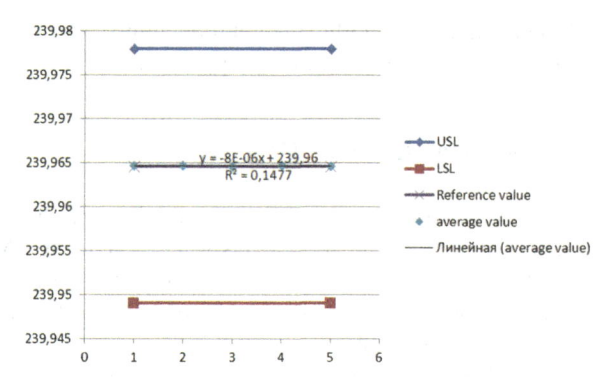

Fig. 3. Graph linear displacement of measurement process.

Evaluation of Communication (quality approach) was performed on the correlation coefficient, which amounted to $R^2 = 0.971123245$. For values of (0.75, 0.9) - a linear relationship between the average values (change the offset within the operating range can be considered linear), and at (0.9, 1.0) - a linear relationship between the values of the strong (the change in bias linearly within the working range).

The calculation results are recorded in Table 4. The correlation coefficient is equal to 0.971123245, therefore, a linear relationship between the estimated true values of the measured parameters and the corresponding strong shift of the measuring process.

Table 4. Results of calculations.

$X_i * B_i$	0.05999118
X_i^2	287915.12
B_i^2	0.00000012
$B*$	0.00005
R^2	0.97112325
a	0.00000021
b	0.0000000000023
Linearity (absolute)	0.00000001
Linearity (relative)	0.00002084
Bias (absolute)	0.00005
Offset (relative)	0.44827586

The experiments allow to determine the degree of operators' influence on the total variability of the measurement process. They also help to draw conclusions on the need for corrective and preventive actions for the measuring process, and to determine the manufacturing process variability, which is an integral part of the measurement process. The results described above prove the need of MSA application fully and comprehensively to confirm the applicability and validity of all measuring systems at an enterprise. MSA allows to prove and demonstrate validity and applicability of measurement processes at manufacturing. The system enables planning and implementation of corrective and preventive actions to improve the measurement system and, consequently, to guarantee the quality of the measurements.

References

1. Borisenko, E.V., Buyanova, I.V.: Carrying out of measurement system analysis on OJSC «BSW – management company of holding «BMC». Cast. Metall. **2**(87), 54–57 (2017)
2. Richter, E.V.: Analysis of measuring systems. Actual Prob. Aviat. Cosmonautics **2**(12), 199–201 (2016)
3. Zimina, E.V., Kaynova, V.N.: Analysis of the quality measurement systems in automotive industry. Bull. Nizhny Novgorod State Univ. Eng. Econ. **5**(72), 7–16 (2017)
4. Russian National Standard 51814.5-2005: Quality management systems in the automotive industry. Analysis of the measurement and control processes. Standartinform, Moscow (2005)
5. National State Standard 166-89. Calipers. Specifications. Standartinform, Moscow
6. Mesarovic, M.D., Takahara, Y.: General Systems Theory: Mathematical Foundations, vol. 113. Academic Press, New York (1975)
7. Chrysler, Ford, General Motors: Measurement systems analysis. Reference manual, AIAG (2010)
8. Russian National Standard 50779.42-99: Statistical methods. Shewhart control chart. Standartinform, Moscow (1999)
9. Russian National Standard 2450-2016: Management systems. Measurement management. Analysis of measuring systems. Standartinform, Moscow (2016)
10. Chrysler, Ford, General Motors: Analysis of measuring systems. MSA Reference manual. Priority LLC, N. Novgorod (2012)
11. Kastorskaya, L.V.: Analysis of Measuring Systems. Priority LLC, N. Novgorod (2006)
12. Kasyanov, S.V., Yamalieva, R.A., Fattakhova, G.R.: Practical issues of assessment of acceptability of measuring processes. Prod. Qual. control **5**, 33–37 (2017)
13. Khabibrakhmanova, A.R., Vyacheslavova, O.F., Parfenieva, I.E.: Analysis of measuring systems as a tool to improve the quality of measurements. In: Collection of Scientific Papers on the Materials of the VIII International Scientific Conference, vol. 3, pp. 48–55 (2018)
14. Isaev, S.V.: Implementation of methods of statistical process control and analysis of measuring systems. Qual. Manage. Meth. **9**, 39–41 (2006)
15. Kovalev, M.I., Pototskaya, E.A.: Application of analysis of measuring systems and statistical quality management methods for monitoring the production processes of a machinery. Notes of the Tula state university. Technical sciences, 12-2, pp. 58–65 (2017)

Test Bench with Controlled Impact Action for Analyzing Rotor-Bearing Assemblies

A. V. Gorin$^{(\boxtimes)}$, A. V. Sytin, and A. Y. Rodichev

Orel State University, 95 Komsomolskaya street,
Orel 302026, Russian Federation
`gorin57@mail.ru`

Abstract. The article outlines the background and the process of creating a test bench for analyzing external impact actions applied to the rotor-bearing assemblies during their operation. Control systems for the test bench and the impact device are presented. The latter designed on the basis of a hydraulic cylinder with a hydraulic directional valve. The impact device can operate either in manual or programmable modes. Precise variation in velocity and magnitude of the impact (according to a predetermined law of variation) allows us to cover a wide range of possible external impacts applied to the rotor assembly. Automatic control of the test bench modules as well as the collection and processing of the experimental data is accomplished by a computer program developed in the LabView visual programming environment. The drivers for matching the frequency converter and the controller as well as all the software for the data measuring system operation, including the experiment itself and data processing was written in graphic programming language "G".

Keywords: Impact action · Test bench · Control system · Rotor-bearing assembly · Sleeve bearing

1 Introduction

Finding measures to ensure the vibrational reliability of rotor assemblies still remains an important issue as of this day. Specified amplitude-frequency responses and the operating capability of such systems are primarily related to the bearing assemblies [1, 2]. The use of elastically compliant sleeve bearings allows to provide a reliable operation of the rotor assembly in a wide range of frequencies and loads by means of the adaptive bearing surface, required rigidity and high shock-absorbing capability [3, 4]. The preferred direction of the rotor assemblies' development is improving efficiency by the increase in running frequencies while minimizing the weight and size. However, to ensure steady operation of the rotor at high speeds, it is necessary to take into account a large number of parameters in the rotor-bearing system, the determination of which allows to optimize the design to a large extent [5, 6]. Moreover, in addition to the internal oscillations of the rotor in the bearings caused by the imbalance, the effect of the rotary actuator, etc. there are situations of external impact actions applied to the rotor assembly itself. These situations, despite the probabilistic nature, can lead to a significant change in the rotor trajectory, which at considerable speeds can lead to a direct

© Springer Nature Switzerland AG 2020
A. A. Radionov and A. S. Karandaev (Eds.): RusAutoCon 2019, LNEE 641, pp. 744–752, 2020.
https://doi.org/10.1007/978-3-030-39225-3_81

contact with the bearing surface thus causing deformations and/or destruction of the rotor-bearing assembly and followed by the failure of the rotor assembly [7]. Therefore, experimental studies with modeling an external impact are important. At the same time, it is necessary to control the impact force to save time and obtain a complete picture in the object under study.

2   Development of a Test Bench with Controlled Impact Action Applied to the Rotor-Bearing Assembly

The loads applied to the rotor can be divided into independent and dependent on the motion of the rotor [8]. The first group includes constant in magnitude and direction loads (i.e. gravity, aerohydrodynamic forces, etc.), as well as various external forces such as forces from maneuvering, oscillation of the foundation, etc. The second group consists of forces from imbalance, non-conservative aerohydrodynamic forces, reactions of bearings and seals, electromagnetic forces, etc. [9]. The dynamic behavior of the rotor under transient operating modes can be complex. The "rotor-sleeve bearings" system reacts to various kinematic and dynamic perturbations [10].

The rotating shaft is a source of oscillations and at the same time an element subjected to oscillations itself [11]. In order to study the effect of a controlled impact action applied to the rotor-bearing assembly, a special test bench was used as a basis (Fig. 1). The body of the test bench is a rigid welded structure consisting of a thick-walled metal tube, two vertical frames reinforced with angle bars and a lower base. On each of the pipe ends there are threaded holes for attaching the bearing housings and additional equipment. The lower base has grooves for securing the body to the base. The test bench includes a rotor assembly with two hydrodynamic bearings, one of which is a hybrid journal bearing with radial lubricant supply, and as the second bearing can serve either a hybrid journal bearing with a radial lubricant supply or a

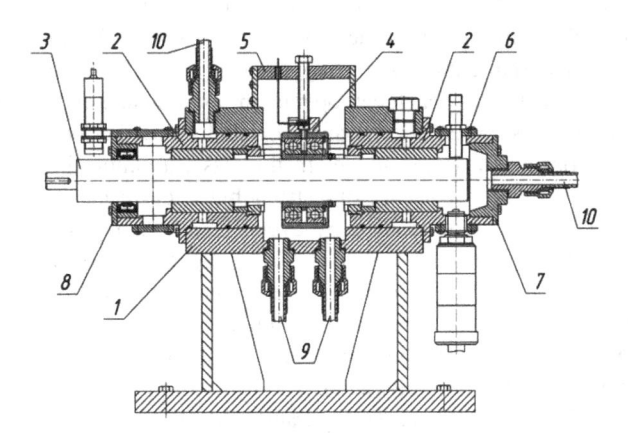

Fig. 1. Test bench: 1 – test bench housing; 2 – support bearing assembly; 3 – rotor (shaft); 4 – steady loading unit; 5 – cover; 6 – measuring system mounting elements; 7 – rear cover; 8 – front cover; 9 – orifices for outflowing lubricant; 10 – orifices for lubricant supply.

smooth cylindrical bearing with axial throttling of lubricant. In addition, there is enough room on the surface of the module to place primary data measuring system transducers and fasteners.

In order to simulate single and periodic external impacts the "SD 500" test bench shown in Fig. 2 was designed and manufactured by The Scientific Laboratory of "Mechatronics, Mechanics and Robotics" Department. The dynamic forces applied to the sleeve bearing housing 7 were modeled with a hydraulic cylinder 4. Hydraulic schematic in simplified form is shown in Fig. 1. It is based on a hydraulic drive.

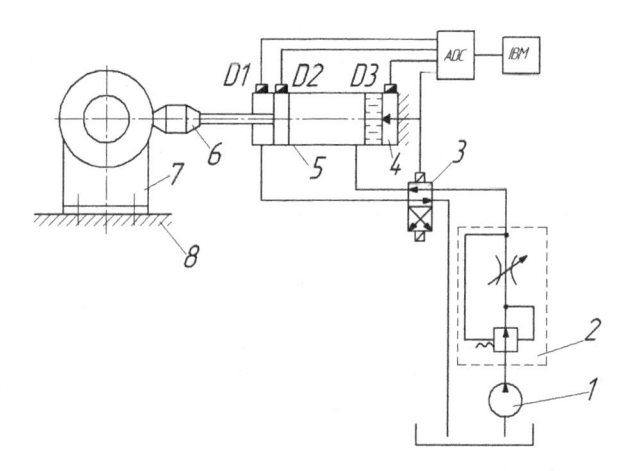

Fig. 2. General view of the controllable hydrodynamic impact device.

The hydraulic drive allows to change the speed of the piston rod of the hydraulic cylinder 4 quite easily, steplessly and in a wide range, and consequently to obtain the required energy of the dynamic action. The change in the speed of the hydraulic cylinder piston rod is carried out by the flow control valve 2. Another way to change the energy of the external action is to vary the weight of the tip 6 of the hydraulic cylinder piston rod [12]. Test bench can operate in both manual and automatic modes. Single external action is easier to perform in manual mode. Periodic external actions are to be carried out in automatic mode. In this case, a simple impact can cause a structural failure due to the occurrence of a strong but brief overstresses in the material [13]. A complex impact, accompanied by cyclic or alternating stresses can lead to accumulation of fatigue micro deformations. The operation of the hydraulic schematic shown in Fig. 1 is as follows. The required energy and the periodicity of the dynamic action are determined from the calculation formulae.

Based on the magnitude of the dynamic action, the speed of movement and the weight of the tip of the piston rod are determined. The flow control valve 2 sets the required displacement speed of the piston rod. The hydraulic directional valve 3 enters the position of the working stroke and the piston rod of the hydraulic cylinder 4 starts its movement. Then, the tip 6 and the sleeve bearing housing 7 collide. The amount of energy of the external dynamic action is controlled by the readings of the hydropneumatic damper 4 using transducer D3. The speed of the piston rod of the hydraulic cylinder is controlled by the speed transducer D1, and the position of the piston rod is controlled by the transducer D2. The signals received from the transducers and the control system of the hydraulic circuit are sent to the analog-digital converter board (ADC). The results of measurements and further control are carried out by a computer.

Mechanical shock is characterized by a rapid release of energy, resulting in local elastic or plastic deformations, stress waves and other effects, which in some cases may lead to a disruption in the functioning of the rotor-bearing assembly and of the entire unit [14, 15]. Therefore, for automatic control of the test bench as well as for collecting and processing the obtained experimental data, a data measuring system (DMS) (Fig. 3) was developed based on the hardware and software developed by National Instruments [16].

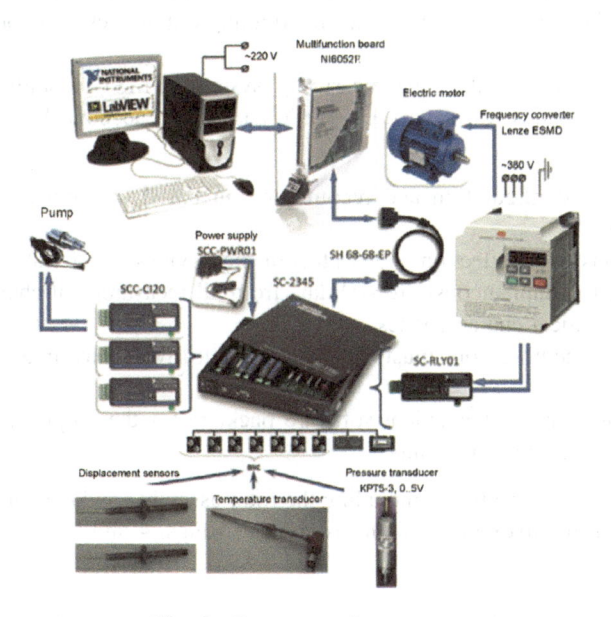

Fig. 3. Data measuring system.

The basis of DMS is the multifunction board NI6052E, which supports multi-channel digital and analog I/O and has counters-timers.

The functional purpose of the board is to provide input-output signals, digitizing and execution of commands as well as a control of the test bench power modules. Signals from different transducers are sent to one- or two-channel analog SCC matching modules.

The control of the electric motor and the pump, as well as the implementation of emergency power down, is carried out through the relay modules SC-RLY01. For a more compact arrangement and subsequent processing of signals, the SCC and SC-RLY01 modules are mounted in a single SC-2345 block module, where the unified signals from the transducers are converted into a range perceived by the analog-digital converter (0..5 V).

In our particular case, the displacement sensors and power modules of the test bench are connected to the SC-2345 via BNC, 9-Pin D-sub and Strain Relief interface connectors. The SC-2345 module requires an external 5 V SCC-PWR01 power supply. The NI6052E board and the SC-2345 module are connected by a single cable SH 68-68-EP. The data from the NI6052E multifunction board with advanced synchronization and timing capabilities is directed to the controller for further processing or writing to the hard drive.

The experimental procedure includes a minimum of three tests for the same operating parameters. Each test consists of accelerating the rotor to the running frequency, working for a while on steady-state operation, impact action and stopping.

The features of the test bench allow to track the rotation frequency, the trajectory of the center of the rotor shaft in the sleeve bearing clearance, the time-frequency characteristics of the rotor in acceleration and running-out modes, and other important parameters [17].

The development of the data measuring system (DMS) to study rotor system dynamic characteristics was carried out while taking into account the following basic requirements:

- Simplicity of configuration and reconfiguration based on unification of primary transducers
- Simultaneous data collection and control of the devices under test
- The ability to simultaneously record data from all measurement channels
- High metrological characteristics
- The displacement sensors must be noncontact to avoid any interference on the rotating rotor
- DMS should have a convenient software package for the output, processing and analysis of measurement results.

The picture of the test bench including the rotor assembly and the controlled impact device based on the hydraulic cylinder is shown in Figs. 4 and 5.

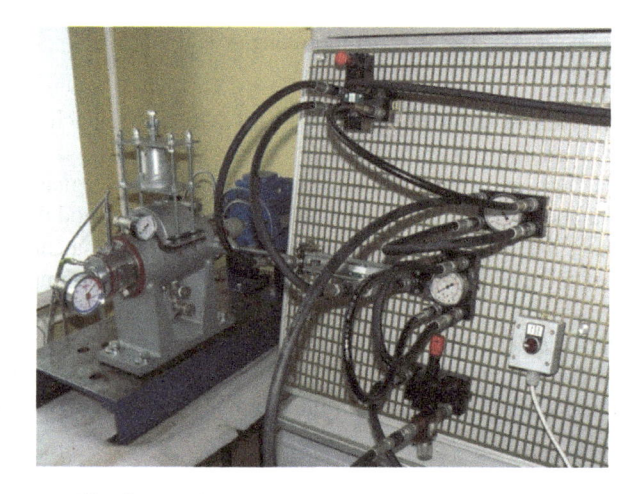

Fig. 4. Test bench with controllable impact system.

Fig. 5. Test bench with controllable impact system.

Eddy current probes AP2100A-C-05.05.1. are used as the primary transducers. The eddy current probe consists of a contactless probe (AE051.00.07 primary transducer), an extension cable (ARKZ950/3 connector) and an electronic module (D210A-C.05.05 matching unit). The electronic module generates an excitation signal of the probe and carries out the selection of the informative parameter. The output signal is an electrical signal directly proportional to the distance from the end of the probe to the monitored object. The eddy current probe is often called eddy current sensor system. Two such sensors are installed in the bearing assembly (Fig. 6). With their help, the position of the rotor is tracked in two mutually perpendicular planes.

The tool for automatic control of the test bench modules as well as the collection and processing of experimental data is software developed in the LabView visual programming environment. The drivers for matching the frequency converter and the controller as well as all the software for the data measuring system operation, including the experiment itself and data processing was written in graphic programming language "G".

The software for conducting experimental studies consists of two modules: (1) setting up and running the experiment; (2) processing of experimental data. The module for setting up the data measuring system and for carrying out the experiment in real-time mode, integrated with the data retention module and automated experimentation, serves to visualize signals from all primary transducers and their primary detuning.

The software for conducting full-scale experiments includes an interface that displays settings for the output of the results, a report generation panel and sweeps of oscillations from displacement sensors.

Fig. 6. Layout diagram of eddy current displacement sensors.

As a result of conducting a controlled experiment, the trajectories of motion of the rotor in the ascent and steady motion mode are obtained (Fig. 7).

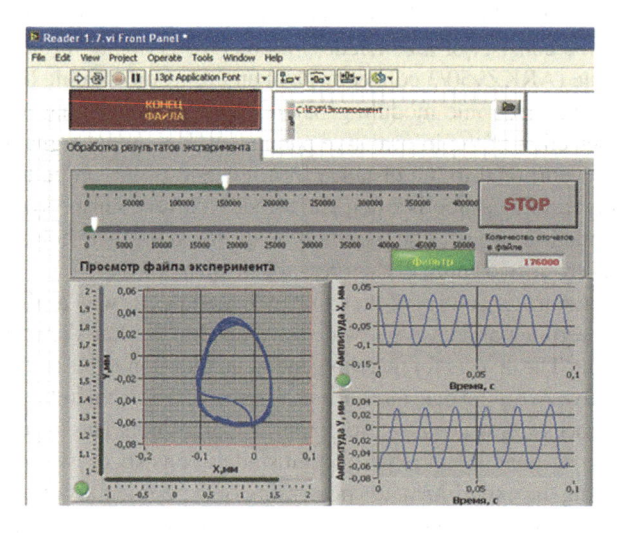

Fig. 7. Motion trajectory of the rotor without external action.

Under the controlled impact action, the trajectory of the rotor in the sleeve bearings changes sharply (Fig. 8) and becomes chaotic.

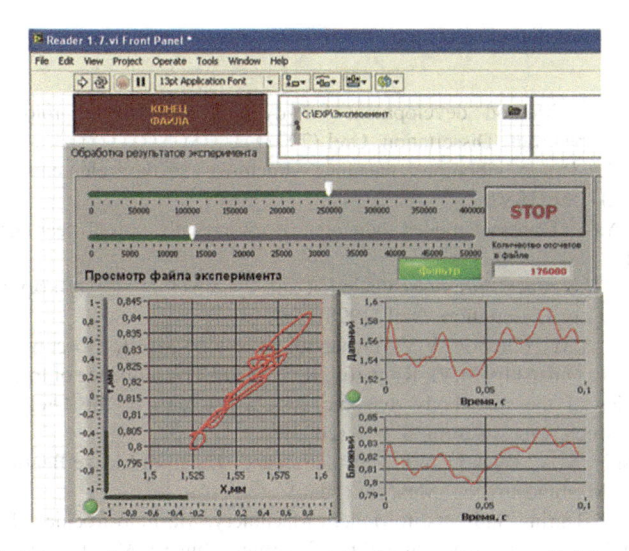

Fig. 8. Trajectory of the motion of the rotor under external controlled impact action.

Depending on the impact force, the geometric parameters of the rotor-bearing assembly and the kinematic parameters of motion, there are two possible scenarios:

- Loss of stability of motion, leading to an increase in the amplitude of oscillations, followed by a collision of the rotor on the inner surface of the bearing
- Vibration damping by elastic forces of the lubricating layer of the sleeve bearing, gradual decrease of the amplitude of oscillations and return of the rotor to the orbit of steady motion.

In the framework of this study, a separate software module has been developed which is intended for subsequent processing of stored measured data. As a filter for processing the signals from the sensors, the Fourier analysis routines are built into the software.

3   Conclusion

The test bench with controlled impact action presented in the article allows to study the influence of external force factors on the trajectory and stability of the rotor movement in sleeve bearings in manual and automatic modes, with the ability to record and then analyze motion trajectories. Precise variation in velocity and magnitude of the impact (according to a predetermined law of variation) allows us to cover a wide range of possible external impacts applied to the rotor assembly.

Acknowledgements. This work was supported by the Ministry of Education and Science of the Russian Federation under the project No. 9.2952.2017/4.6.

References

1. Solomin, O.V.: Method development for dynamic analysis of rotor systems with hydrodynamic bearings. Dissertation, Orel (2007)
2. Poznyak, E.L.: Rotors vibrations machines, structures and their elements vibrations. Vib. Mach. **3**, 130–189 (1980)
3. Pugachev, A.O.: Transient modes dynamics of rotors with journal bearings. Dissertation, Orel (2004)
4. Savin, L.A.: Theoretical basis of calculation and dynamics of sleeve bearings with liquid-vapor lubrication. Dissertation, Orel (1998)
5. Aleksandrov, A.M., Filippov, V.V.: Dynamics of Rotors. MPEI, Moscow (1995)
6. Antipov, V.A., Duletskiy, M.V., Komarov, R.N., et al.: The influence of force factors on the characteristics of the rotor-body dynamic system. Bulletin of the Orel State Technical University. Mech. Eng. Instrum. **3**, 6–9 (2003)
7. Voskresenskiy, V.A., Dyakov, V.I.: Calculation and Design of Sleeve Bearings with Liquid Lubrication. Mashinostroenie, Moscow (1980)
8. Gaevik, D.T.: Bearing Supports in Modern Machinery. Mashinostroenie, Moscow (1985)
9. Kostyuk, A.G.: Dynamics and Durability of Turbomachinery. MPEI, Moscow (2000)
10. Barkov, A.V., Barkova, N.A., Azovtsev, A.Yu.: Vibration monitoring and diagnostics of rotary machines. SPb STU, St. Petersburg (2000)
11. Gorin, A.V., Eshutkin, D.N., Gorina, M.A.: Hydraulic impact devices utilization for ground wells development. State University-UNPK, Orel (2015)
12. Kotylev, U.E., Eshutkin, D.N.: Applied Hydraulic Impact Devices Theory. Mashinostroenie, Moscow (2007)
13. Ushakov, L.S., Kotylev, U.E., Kravchenko, V.A.: Hydraulic Impact Devices. Mashinostroenie, Moscow (2000)
14. Bashta, T.M.: Hydraulics, Hydraulic Machines and Hydraulic Drives. Mashinostroenie, Moscow (1992)
15. Gorin, A.V., Eshutkin, D.N., Zhuravleva, A.V.: Simulation of the drive of a static-dynamic machine for trenchless construction of pipelines. Fundam. Appl. Probl. Eng. Technol. **3** (287), 20–26 (2011)
16. Gorin, A.V., Eshutkin, D.N., Gorina, M.A.: Volumetric hydraulic drive of combined machines for ground wells development. State University-UNPK, Orel (2015)
17. Gorin, A.V.: Pressure and shock mechanisms with volumetric hydraulic drive parameters of the machines for ground wells development. Dissertation, Orel (2012)

Method of Lathe Tool Condition Monitoring Based on the Phasechronometric Approach

D. D. Boldasov$^{(\boxtimes)}$, A. S. Komshin, and A. B. Syritskii

Bauman Moscow State Technical University, Vtoraya Baumanskaya street,
Moscow 105005, Russia
boldasovd@gmail.com

Abstract. This paper presents the results of a correlation study of cutting tool deterioration and the measurement results of a phase chronometric system. The significance of this work is because a tool failure can be a reason for defects and even for the failures of machine components. Phase-chronometric approach has been implemented and showed good results in such complex technical objects as turbines and hydraulic units and is considered as a possible alternative or complement to existing methods of the tool condition diagnostics. We provide a brief description of the phase-chronometric method, its advantages and theoretical basis, as well as the main components and operating principle of the phase-chronometric system. The paper describes how to obtain experimental measurement data, its mathematical processing and the data that supports the possibility of studying the cutting process by the phase-chronometric method, as well as the obtained experimental results correlated with the lathe tool deterioration in the determined cutting process conditions.

Keywords: Phasechronometric method · Monitoring · Machining · Tool wear

1 Introduction

Special attention to the cutting tool diagnostics is due to the fact that a missed tool failure is the main reason for defects, as well as the reason for the failures of the gears running on subsequent operations, and, sometimes, for the failures of machine components. Assessment of the actual condition of the tool cutting edge allows preventing these events.

Efficient operation of machines with computer numerical control (CNC) is impossible without the tool condition diagnosing, as the tool condition is the weakest link in the technological system. The cutting tool as an element of the metal processing system is characterized by wear rate. It is significantly higher than the wear rate of other machine parts and components. The tool failure, in contrast to other damages, inevitably leads to a failure of the whole technological system [1], and this, at best, leads to an increase of the primary (machine) and auxiliary processing time for one workpiece, and, at worst, - to the increase of defect proportion.

The problem of primary time reduction is always relevant, because it leads to an increase in productivity. The most rational way of solving this problem is intensification

© Springer Nature Switzerland AG 2020
A. A. Radionov and A. S. Karandaev (Eds.): RusAutoCon 2019, LNEE 641, pp. 753–762, 2020.
https://doi.org/10.1007/978-3-030-39225-3_82

of cutting modes. However, both in domestic and foreign practice, the cutting modes are often set 20–30% below standards [2] in order to improve reliability.

On the other hand, automatic control and diagnostics of the cutting tool condition allow you to do the following:

1. Increase the reliability of metal processing (determine the correctness of its workflow, automatically restore the machine functionality in case of failures);
2. Reduce the tool consumption;
3. Improve processing quality and reduce defects;
4. Secure the machine mechanisms and components from damage and early loss of accuracy;
5. Increase the processing modes;
6. Implement a technology that does not require humans.

Despite the long and numerous works in the field of cutting tool condition monitoring, the problem of creating and large-scale implementing such monitoring systems in the manufacturing process remains relevant [3–7]. A great number of research and industrial teams are engaged in this problem all over the world, there is a large number of proposed methods to estimate the cutting tool wear parameters indirectly [7–10]. Hereinafter, we will call such methods traditional. However, each proposed approach has its own features, difficulties for industrial implementation, and cannot be considered a unique solution of the problem described above.

In this situation, a relevant problem is to find new methods of the tool monitoring that lack the aforementioned disadvantages. According to the authors of this article, the phasechronometric method (PCM) to retrieve information on cyclic machinery and mechanisms operation can be viewed as one of such new methods. This approach has been developed at the department of "Metrology and interchangeability" of Bauman Moscow State Technical University. It has been implemented and showed good results in such complex technical objects as turbines, hydraulic units, clockworks [11–13]. The method is characterized by low cost of primary converters and high-precision measurement of time intervals [14, 15]. The latter is caused by the fact that today the highest precision is ensured in the field of time and frequency measurement.

2 Phasechronometric Approach

The phasechronometric approach can be attributed to the phase methods of control and diagnostics (the information parameter is phase); it is completely different from the amplitude methods (the information parameter is measured amplitude). Due to these differences, the PCM has the major advantages over the available analogues.

Firstly, the established modern level of chronometry allows obtaining measurement data in industrial environment with errors of no more than 10^{-7} s. This provides a large margin of accuracy compared to the amplitude methods, and also gives an opportunity to reduce the cost of the system's measuring channel components due to reduction of the accuracy by one or two orders of magnitude without losing competitiveness. For reference, the accuracy of the amplitude analog method for measuring current, voltage, pressure, temperature and other physical values is limited to their noise, while the error is more than 0.1% depending on the noise level of the measured object, the sensors used, filters and other components of the control system. In addition, the sensor readings are significantly affected by the temperature dependence of the structural material parameters of converters.

Secondly, when using the PCM, the measured physical values are the time intervals corresponding to the phases of the operating cycle. Thus, the method is strictly connected with the operating cycle of the technical system, i.e. with the most sustainable process, since all technical measures are aimed at providing it during the whole life cycle from development to operation. This defines the characteristics that accompany the device at the testing steps, during manufacturing and the entire operation period, whereas their quantitative changes define the state of a particular serial copy. Phase-chronometry generates the time series that reflects the influence of the device operation dynamics on the time interval, the variations of which are determined by the operating cycle instability.

This method, unlike the analog ones, is much less affected by amplitude noise, it does not directly measure the values of current, voltage and other noisy parameters. The operating principle of the phasechronometric system is shown in Fig. 1.

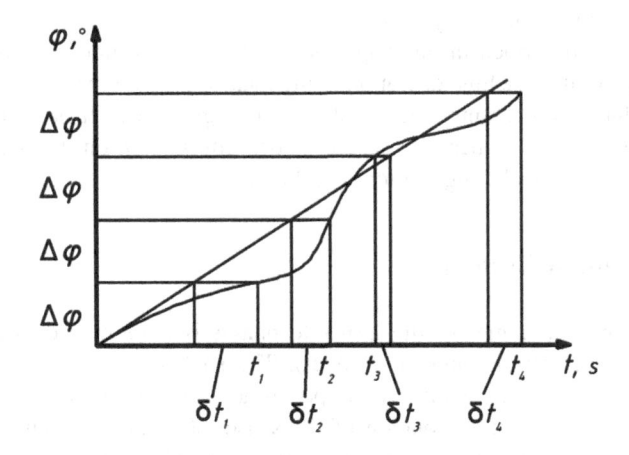

Fig. 1. Operating principle of the phase chronometric system.

In this method, measurements are done for the time intervals $\delta t_1 \ldots \delta t_i$, which correspond to the rotation (displacement) of the machine running gear undergoing a cycle (the machine spindle in this case) at a certain angle $\Delta \phi$ (operating cycle phase).

To implement phasechronometric analysis of cyclic systems, we have to use the measuring equipment that ensures the registration of the time point when the given level of sensor signal is reached – it should correspond to the boundary of a certain phase in the operating cycle. Thus, the measuring procedure is based on a uniform quantification by level and sampling by time.

At the same time, it is possible to get highly accurate readings, due to preliminary precision set of coordinates of contrasting markers in statics, e.g. at a fixed rotor or limb.

For the registration of the given angle of rotation of rotor, we can use the contrasting elements on its surface. These may be embossed labels, specially applied labels made of magnetic material, optical labels and other assisting elements rigidly fixed to the running gear of the controlled cyclic machine. The angular rotation velocity of the aforementioned labels coincides with the angular velocity of the running gear; the moment of running gear rotation at an angle corresponding to the label is recorded by the recording converter. Such labels may be positioned on the rotor circumference, typically at equal angular intervals $\Delta\varphi$.

In contrast to the vibroacoustic method, the accuracy of which is limited by the amplitude noise of the measured vibration, the phasechronometric method accuracy is determined by the technical means of time measurement; it can be no worse than 10^{-7}–10^{-8} s in conditions of an operating object. Here, the amplitude noise of the sensor that records the labels or continuity defects on the rotor surface, which are bound to the operating cycle phase, does not directly affect the accuracy of the phasechronometric system. This is because the sensor is not used for measuring the signal amplitude from the labels or continuity defects – it is used for recording the moment when the rotor reaches certain angular positions (operating cycle phases) corresponding to the given labels or continuity defects. In this case, the amplitude noise, which prevents the analog systems from operating with high precision, does not have a critical influence on the accuracy of the chronometric system.

The approach described in this paper is based on the assumption that we get the information about the machine operation while measuring the time intervals of spindle turn phases during processing. The rotation chronogram, which is a graphic representation of time series values $\delta t_1 \ldots \delta t_i$, provides information on the operation of all machine components including the tool condition.

3   Measurement System

Virtual partition of a single spindle turn into phases is done by means of a circular raster of the angle sensor (angular encoder). The shaft of the primary converter is connected to the machine spindle via a special adapter that eliminates slippage and thereby provides synchronous rotation of the sensor shaft and machine spindle. The sensor case is attached to the machine body by means of a mounting.

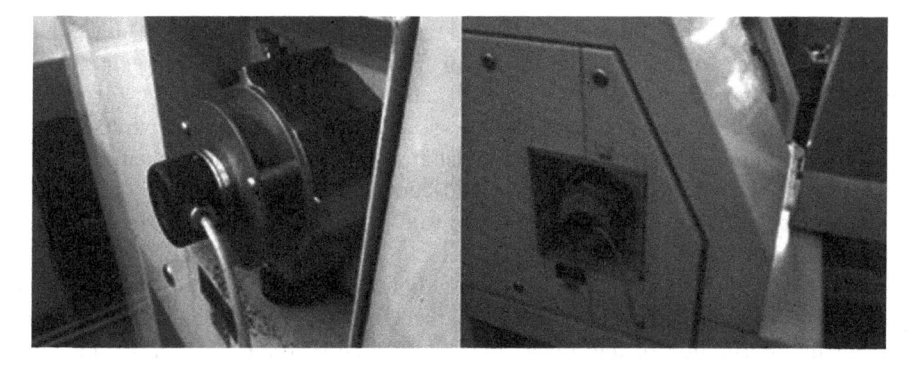

Fig. 2. Installation of the angle sensor on the lathe machine.

The installation of the monitoring system primary converter (Fig. 2) has been developed in order to conduct experimental studies to validate the system performance. Using the angle sensor as a primary converter in such a setting is possible only if bars are not fed into the machine work zone from the headstock during the machine operation. In other cases, it is necessary to use the angle sensors with hollow shafts or other solutions described above.

An analog sinusoidal signal comes from the sensor to the input of the measurement information processing unit (Fig. 3).

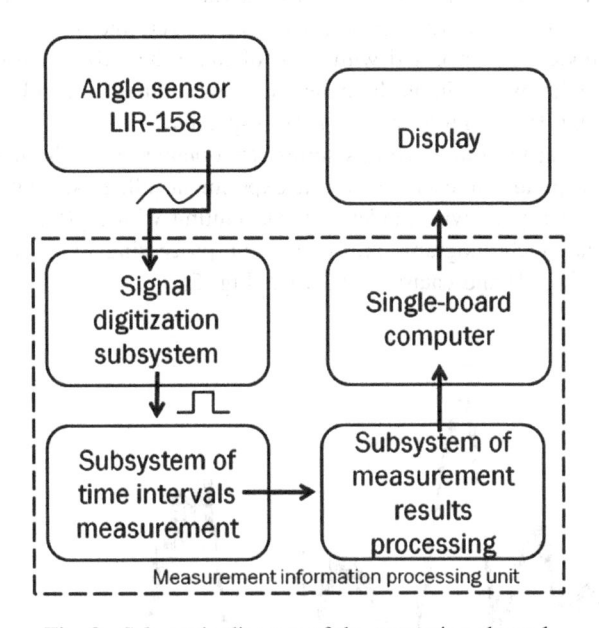

Fig. 3. Schematic diagram of the measuring channel.

The period of signal for this type of sensor is not constant and is correlated with the non-uniformity of the machine spindle rotation. The signal is digitalized by analog comparators, after that the duration of each impulse is measured; for that reason, the time intervals measurement subsystem is provided with a crystal oscillator with frequency of 50 MHz. The subsystem of measurement data processing and single-board computer provide formation and recording of the received time series and their processing in the form of constructing a rotation chronogram (a graph of phase time change depending on the measurement number) or a frequency spectrum of this function. The measurement system arranged this way is characterized by a relative error of no worse than $2 \cdot 10^{-2}\%$.

The main task of the time interval measurements processing is to extract information about the cutting tool. This requires finding a parameter that varies over time along with deterioration.

4   Experimental Results

In order to confirm the hypothesis described above and to test the monitoring system prototype, we studied the processing of a workpiece made of stainless steel by a lathe tool with removable cutting plates of triangular shape (the material is an alloy T15K6). The experiment was designed in such a way that we could obtain the measurement data about the spindle rotation unevenness (rotation speed 315 rpm) of the machine UT16P at the idle mode and at three conditionally different processing modes. The conditional difference was in using a removable plate with a varying degree of cutting edge wear: without visible signs of wear (new plate), with wear of 100 microns on the back surface (a plate with moderate wear), and with wear of more than 400 microns on the back surface (catastrophic wear). In all the cases, the longitudinal feed and the depth of cut remained constant: 0.217 mm/r and 0.2 mm, respectively.

After processing the time intervals series, we obtained four chronograms of rotation, which corresponded to each type of the experiment (idle mode, cutting with a new plate, cutting with a plate with moderate wear, cutting with a plate with catastrophic wear). The rotation chronograms are shown for processing by a cutting tool with moderate wear (Fig. 4) and catastrophic wear (Fig. 5).

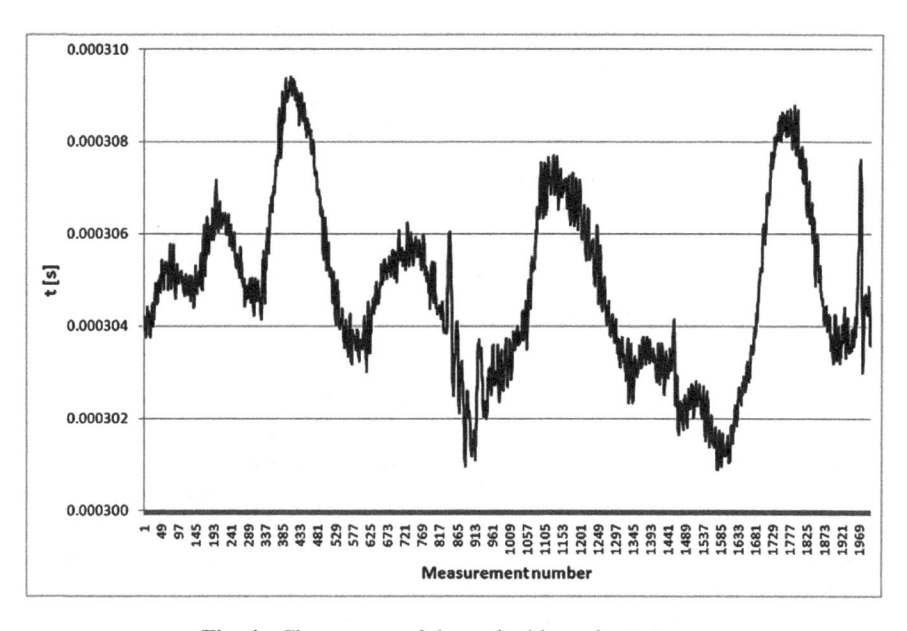

Fig. 4. Chronogram of the tool with moderate wear.

As it has already been mentioned above, the rotation of the machine spindle is unstable, also within one turn. It can clearly be seen on the provided chronograms, which, for further mathematical processing, are presented as signals with a complex shape; they can be represented as a sum of simple harmonic oscillations.

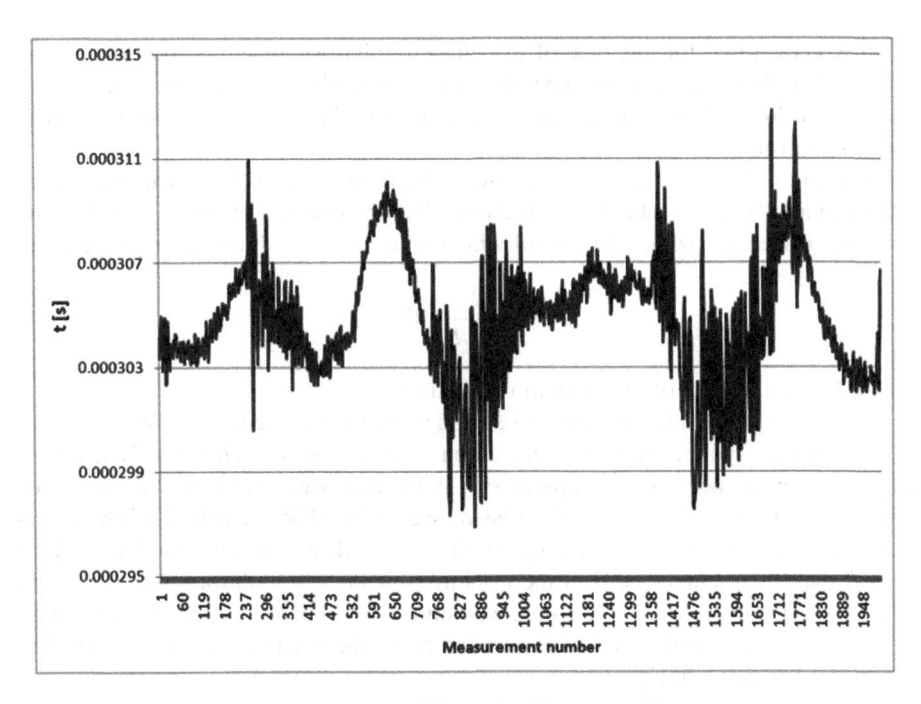

Fig. 5. Chronogram of the tool with catastrophic wear.

In Fig. 6, you can see that high-frequency components of time interval oscillations become more evident upon the increase of the cutting tool wear. It should also be noted that the experiment was conducted so that apart from the wear of the cutting plate, the system "machine-device-tool-piece" was not changed, whereas the effect of changes from external factors was negligible, because there were no more than 5 min between the moments of chronograms recording.

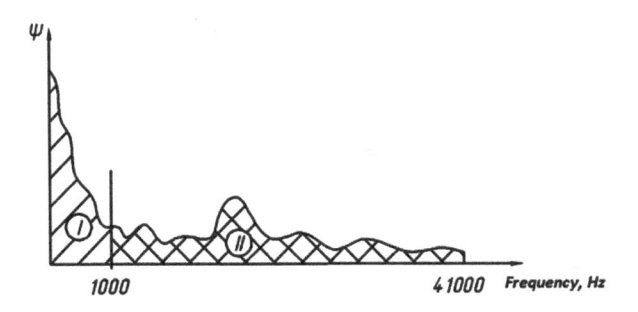

Fig. 6. Principle of power ratio estimation.

We have chosen the parameter that formalizes the above-mentioned results to be the change of the relative proportion of high-frequency oscillations in the range of the rotation chronogram (Fig. 5).

It is known that for this type of machines, oscillations of up to 1000 Hz include information about non-uniform operation of the machine electric motor, defects of gearing and lubrication. Whereas the oscillations associated with the wear of the lathe tool are high-frequency.

Therefore, dividing the spectrum power from the range of high-frequency oscillations (zone II) by the total power of the oscillation spectrum (zone I + zone II), you can get a percentage ψ, the increase of which is correlated with the cutting edge wear:

$$\psi = \frac{II}{I+II} \cdot 100\% \tag{1}$$

where I is total power of the oscillation spectrum, W;

II is spectrum power from the range of high-frequency oscillations, W.

The experimental research was focused on confirmation of the hypothesis described above. The experiments involved processing of the cylindrical surface of the workpiece with a diameter of 70 mm, which had been made of steel 45 on 16K20 machine, and recording chronometric data during cutting. The depth of cut a = 1 mm, feed f = 0.2 mm, the rotation speed of the machine spindle n = 1000 rpm. At regular and successive time intervals of processing, we estimated the changes of relative proportion of high-frequency oscillations in the spectrum of the rotation chronogram ψ. The results are presented in Fig. 7:

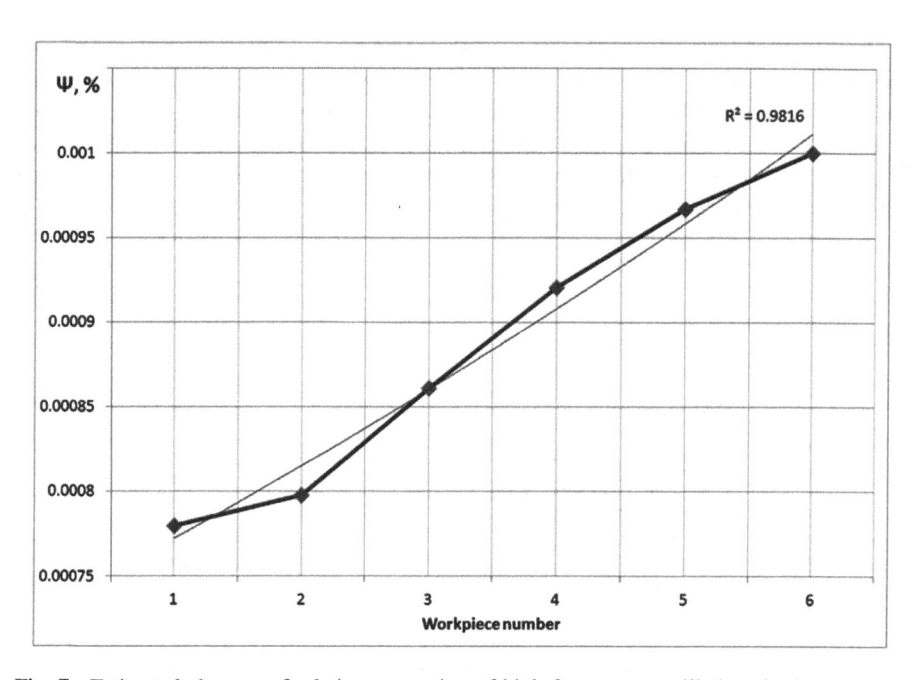

Fig. 7. Estimated changes of relative proportion of high-frequency oscillations in the spectrum of the rotation chronogram.

In Fig. 7, the trend line increases in line with the tool wear, so this method of the obtained measurement data processing is the most promising currently.

After the analysis of the experimental curve shown in Fig. 7, we concluded that the power of the high-frequency components of the machine spindle rotation oscillation increases relatively to the power of the low-frequency components – in line with the development of the cutting tool wear. This conclusion can be taken as a basis for choosing a criterion of the cutting tool blunting.

5 Conclusion

The results given in article prove a phasechronometric method possibility of use in case of cutting tool wear. Such approach is effective especially in case of large-size and unique details machining where tool failure can lead to considerable financial losses. Also, application of PCM in automated machines is very actual.

Application of this method in metal working is new perspective area of researches and further we plan to develop a technique of forecasting of a condition of the tool in processing using PCM measurement information.

Acknowledgements. This work supported by Research Program supported by the Department of Science and Education № 9.4968.2017/BCh, Russian Federation.

References

1. Brzhozovskii, B.M., et al.: Reliability of modified cutting tools. Russ. Eng. Res. **34**(12), 769–772 (2015)
2. Altintas, Y.: Manufacturing automation. University of British Columbia, Vancouver (2012)
3. Ghasempoor, A., Moore, T.N., Jeswiet, J.: Online wear estimation using neural networks. Proc. Inst. Mech. Eng. **212**, 105–112 (1998)
4. Silva, R.G., Reuben, R.L., Baker, R.J., et al.: Tool wear monitoring of turning operations by neural network and expert system classification of a feature set generated from multiple sensors. Mech. Syst. Sig. Process. **12**, 319–332 (1998)
5. Kozochkin, M.P., Sabirov, F.S.: Attractors in cutting and their future use in diagnostics. Measur. Tech. **52**(2), 166–171 (2009)
6. Scheffer, C., Heyns, P.S.: An industrial tool wear monitoring system for interrupted turning. Mech. Syst. Sig. Process. **18**, 1219–1242 (2004)
7. Castejon, M., et al.: Online tool wear monitoring using geometric descriptors from digital images. Int. J. Mach. Tools Manuf **47**, 1847–1853 (2007)
8. Kurada, S., Bradley, C.: Machine vision system for tool wear assessment. Tribol. Int. **30**, 295–304 (1997)
9. Sick, B.: Review online and indirect tool wear monitoring in turning with artificial neural networks: a review of more than a decade of research. Mech. Syst. Sig. Process. **16**, 487–546 (2002)
10. Sharma, V.S., Sharma, S.K., Sharma, A.K.: Cutting tool wear estimation for turning. J. Intell. Manuf. **19**, 99–108 (2007)
11. Kiselev, M.I., Pronyakin, V.I.: A Phase method of investigating cyclic machines and mechanisms based on a chronometric approach. Measur. Tech. **44**(9), 898–902 (2001)

12. Kiselev, M.I., et al.: Multifactorial mathematical models of the functioning of gas-turbine aviation engines in a phase-chronometric representation. Measur. Tech. **54**(9), 1081–1090 (2011)
13. Boldasov, D.D., Potapov, K.G., Syritskii, A.B.: Phase–chronometric diagnostics of metal-cutting lathes. Russ. Eng. Res. **36**(8), 668–672 (2016)
14. Syritskii, A.B.: Measurement of the wear of a cutting tool by phase chronometer method in the course of working. Measur. Tech. **59**(6), 595–599 (2016)
15. Syritskii, A.B.: Phasechronometric measuring technologies application for turning tool wear monitoring. In: MATEC Web of Conferences, vol. 129, p 01049 (2017)

Depersonalization of Personal Data in Information Systems

D. V. Primenko, A. G. Spevakov$^{(\boxtimes)}$, and S. V. Spevakova

Southwest State University, 94 Pyatdesyat let Oktyabrya street,
Kursk 305040, Russian Federation
aspev@yandex.ru

Abstract. The analysis of the methods of anonymization of personal data processed by automated information systems was carried out, their shortcomings were revealed. To eliminate the shortcomings, a method of anonymization of personal data has been developed, based on the method of introducing identifiers using hashing of critical data and the key of the institution. This method has the following advantages: the data becomes impersonal, what allows reducing the cost of personal data information systems protection; the impossibility of determining a specific subject in personal data information systems based on known unique attributes of the subject; the operator, when the subject refers to his personal data, gets access only to one record of the personal data information system. The use of this method is considered using the famous characters of Alice and Bob. The conducted experiments have confirmed the proposed software operability and allow recommending it for use in practice for solving the problems of diagnosis and automatic classification on the features. The prospects for further research may include the creation of parallel methods for calculation of set of proposed indicators, the optimization of their software implementations, as well as a experimental study of proposed indicators on more complex practical problems of different nature and dimensionality.

Keywords: Depersonalization · Personal data · Hash identifier · Hash algorithm

1 Introduction

In modern automated systems a large volume of personal data (PD) of various security classes is processed. Provision of information security requires significant material costs. For cost minimization the opportunity of personal data depersonalization in information systems is provided [1–3].

It usually takes a long time to convert confidential personal data into anonymous non-confidential sequence. It also has low resistance to attacks and has limits in the process of working with large amount of personal data. At the same time anonymous data are used for archival storage without a possibility of confirming whether they are true and their analytical purposes. The most interesting things are methods of depersonalization which allow dividing all data into two groups: confidential and anonymous with the possibility of checking them. The use of such methods allows working with

© Springer Nature Switzerland AG 2020
A. A. Radionov and A. S. Karandaev (Eds.): RusAutoCon 2019, LNEE 641, pp. 763–770, 2020.
https://doi.org/10.1007/978-3-030-39225-3_83

personal data in an anonymous form. It reduces the costs on creating an information security system at the automation facility, that's why the development of a method, an algorithm and a software module of personal data's depersonalization is an urgent scientific and technical task.

In accordance with an order of Roskomnadzor from the 5th of September 2013 № 996 «About approval of requirements and methods of personal data depersonalization» next methods of PD depersonalization are signed out: method of identifiers implementation, method of change of composition or semantic, method of decomposition, mixing method [4].

1. Method of identifiers implementation is a replacement of personal data values with creation of a table (guide) of conformity of identifiers with the initial data [5]. The disadvantages of this method are:

   - In the request and in the response to this request the type of representation of PD attributes that were replaced with identifiers is changed
   - In the records the relations between attributes of depersonalized data and PD attributes corresponding to them are saved
   - It is applicable to a small amount of PD attributes and the small volume of a PD array.

2. Method of change of composition or semantics is the change of composition or semantics of personal data by replacement with statistic processing, transformation, compilation or replacement of some information [6]. This method has the next disadvantages:

   - Application of this method is uneffective for PD depersonalization, because during PD attributes extracting it is necessary to consider the possibility of depersonalization with the usage of these attributes
   - During basic replacement of values of separate attributes only change of PD composition can happen, but not depersonalization
   - In record relations between attributes of depersonalized data and the attributes of personal data corresponding to them are partially saved
   - Applicable when processing tasks do not require personalization of depersonalized data, if it is needed this process can be used on small data arrays.

3. Method of decomposition is division of an array of personal data into several sub-arrays with subsequent separate storage of sub-arrays [7]. The basic disadvantages are:

   - It saves relations between attributes of depersonalized data and PD attributes corresponding to them in records of each storage
   - Is applicable on large arrays of PD
   - Resistance to attacks depends on the complexity of setup of relations between tables.

4. Mixing method is a reshuffle of separate values or groups of values of personal data attributed in an array of persona data [8–10]. This method has these disadvantages:

- This method does not save relations between attributes of depersonalized data and personal data attributes corresponding to them in records
- Resistance to attacks increases with growth of the size of the array of personal data
- In applicable to large arrays of personal data with frequent changes in data.

To address these shortcomings was developed a method of personal data depersonalization based on the method of identifiers implementation with the usage of hashing of critical data. As initial data a table of personal data $D_N(d_1, d_2, \ldots, dM)$ is used, where M is the number of attributes, N is the table row count, A_1 and A_2 are data arrays related to key and non-key data respectively [11].

In this, at the first step by expert way critical data and data clearly identifying the personal data subject is defined. Corresponding attributes are defined as ley ones.

At the second step the initial array D according to chosen key attributes is split into two non-intersecting sub-arrays A_1 and A_2. It is worth noting that into each of sub-arrays an additional attribute d_0 is added, by which value later the comparison of depersonalized data with the personal data subject is conducted. As the result the number of key values is equal to K patients $K(0 < K < M)$. In this, in A_2 is stored depersonalized data that is not interesting for the intruder, so it does not require protection and is stored in the clear.

At the third step for the set of key data of each row $a_{i1}, a_{i2}, \ldots, a_{i3} \in A1$, where $i = 1, 2, \ldots, N$ the value of an attribute procedure $d_0 - a_{i0} = F(a_{i1}, a_{i2}, \ldots, a_{i3})$ is calculated where F is an unique function unknown for the user. The hash function works by the principle of cryptographic sponge [12, 13]. The structure of the sponge is shown at the Fig. 1. It provides for to basic steps:

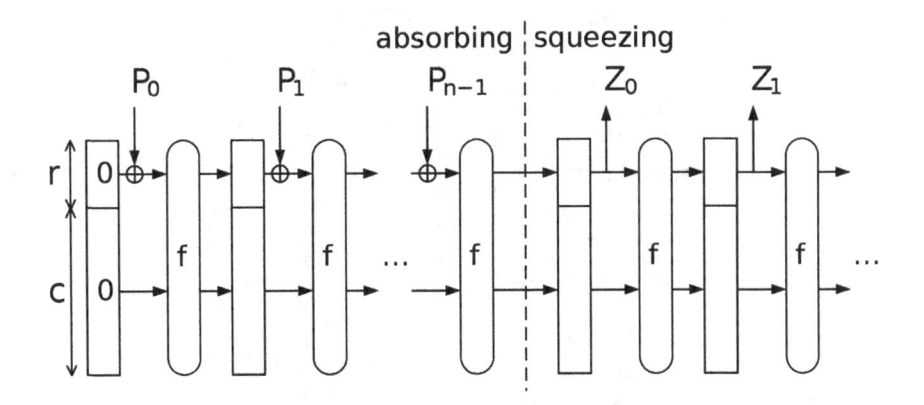

Fig. 1. The structure of cryptographic sponge.

Absorbing. The initial message P is subject to multi-round reshuffles f, accumulation and processing of all blocks of the message from which the hash will be developed is conducted.

Squeezing. The output of the received value of Z as the shuffle result, the development of the hash value and the output of the results until the necessary length of the hash is reached [14].

In the absorbing phase first is set the initial state from the zero vector with the size up to 1600 bits [15]. Next is conducted the operation xor of a fragment of the initial message p_0 with the fragment of the initial state with the size of r, the remaining part of the state with capacity of c remains the same [16].

The result is processed by the f function which is a multiround non-key pseudo-random reshuffle and repeats till the initial message blocks exhaust [17]. Next comes the squeezing phase at which it is possible to extract a hash of a random length. The flow chart of the hashing algorithm is shown at the Fig. 2.

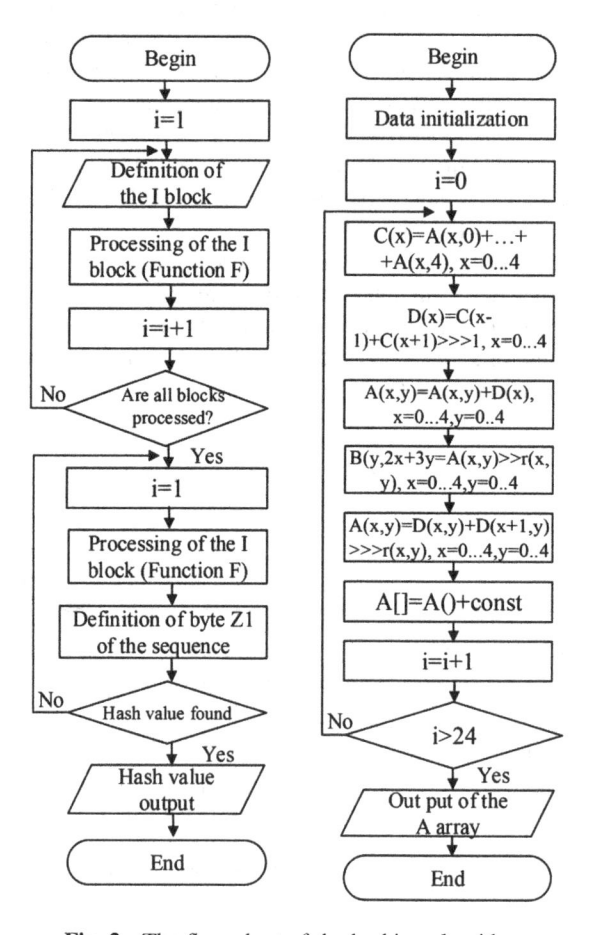

Fig. 2. The flow chart of the hashing algorithm.

The function $F(\ )$ in this algorithm executes 24 rounds, one round includes the work of five functions Theta, Chi. Pi, Rho, Iota, consistently processing the inner state at each round [18].

The function Theta is represented by the next expressions (1):

$$C[x] = A[x, 0] \oplus A[x, 1] \oplus A[x, 2] \oplus A[x, 3] \oplus A[x, 4], x = 0 \ldots 4;$$
$$D[x] = C[x - 1] \oplus (C[x + 1] >>> 1), x = 0 \ldots 4; \tag{1}$$
$$A[x, y] = A[x, y] \oplus D[x], x = 0 \ldots 4, y = 0 \ldots 4,$$

The function Chi is represented by the next expression (2):

$$A[x, y] = B[x, y] \oplus (\sim B[x + 1, y] \, \& \, B[x + 2, y], x = 0 \ldots 4, y = 0 \ldots 4, \tag{2}$$

The functions Pi, Rho are represented by the next expression (3):

$$B[y, 2x + 3y] = A[x, y] >>> r(x, y), x = 0 \ldots 4, y = 0 \ldots 4, \tag{3}$$

The function Lota is represented by the next expression (4):

$$A[0, 0] = A[0, 0] \, xor \, RC, \tag{4}$$

Where B is a temporary array having the same structure as the state array; C and D are the temporary arrays each containing 5 64-bit words; r-array defining the number of bits of spinage for each word of the state; inversion of the value $\sim B[x + 1, y]$.

Step1: at the beginning of the algorithm data initialization is conducted. The size of the state is 1600 bits. Next to the variable i the value 0 is assigned.

Step2: after this the processing of the array with functions $C[x], D[x], A[x, y],$ $B[y, 2x + 3y], A[x, y]$ begins, and besides these operations is conducted the summation of the xor-round constant RC with the word $A[0, 0]$.

Step 3: after data processed with subfunctions goes the check for the rounds count. If the condition $i > 24$ is true then the output of the A array is conducted. If not then we increment by 1 and make the operations until this condition is true [19].

As an example, let's review a database of patients of some treatment institution (Table 1).

Table 1. Patient database.

Last name	First Name	Patronymic	Sex	Date of birth	Medical insurance	Diagnosis
Ivanov	Ivan	Ivanovich	M	12.12.1992	12345678910	Pneumonia
Petrov	Denis	Yurievich	M	11.11.1990	46548677684	Pyelonephritis

For example, for the patient Ivanov the critical personal data is: first name, last name, patronymic, date of birth. For the hash identifier preparation we will use this data: {Ivanov, Ivan, Ivanovich, 12.12.1995} + {bPeShVkYp3s6v9y$B&E)H@McQfTjWn-Zq}, where the second addend is the private key of the treatment institution. After the calculation we get the hash identifier: 1628b3db5c13865aea5856a630a7 36653059fc7e-2d7c49f897b636428c62a26b.

In the depersonalized database the hash identifier and the depersonalized personal data are stored (Table 2).

Table 2. Depersonalized database.

Hash identifier	Medical insurance	Diagnosis
1628b3db5c13865aea5856a630a736653059fc7e2d7c49f897b636428c62a26b	12345678910	Pneumonia
4d949d630cfaafe3dd151a2e06d7345a44a61889a8c097622abfd6ca0f515a7f	46548677684	Pyelonephritis

In the secure database the hash identifier and the critical personal data are stored (Table 3).

Table 3. Secure database.

Hash identifier	Last name	First name	Patronymic	Date of birth
1628b3db5c13865aea5856a630a736653059fc7e2d7c49f897b636428c62a26b	Ivanov	Ivan	Ivanovich	12.12.1992
4d949d630cfaafe3dd151a2e06d7345a44a61889a8c097622abfd6ca0f515a7f	Petrov	Denis	Yurievich	11.11.1990

The practical usage of this method assumes that personal data is stored in two information systems. One is protected and the other is depersonalized. In the protected information system of personal data (ISPD) all personal data and hash identifier formed of critical personal data identifying the PD subject is contained. In depersonalized ISPD there is no PD allowing to identify the PD subject by the intruder, and for identification the hash identifier is introduced which is calculated from critical personal data.

Let's consider the application of this method sing the famous characters Alice and Bob [20].

Alice came to see Doctor Bob. To identify Alice, she shows Bob the critical PD from her initial documents (passport and medical insurance). Bob using the calculator for hash identifier inserts this data and the key of the hospital and forms the hash identifier that allows getting the access to Alice's patient file. After diagnosing and prescribing treatment Bob inserts data into the information system and sings it with his electronic signature.

A curious staff member Eva wanted to know Alice's diagnosis but can't find her card in the information system because she does not know the hash identifier as well as Alice's critical data.

Mallorie found out Alice's critical PD and got the access to the calculator for hash identifier, but she does not know the hospital's key for calculating Alice's identifier.

This method has the next advantages:

- Data becomes depersonalized which reduces costs of ISPD protection
- It is impossible to define the presence of a certain subject in ISPD by known unique attributes
- Operator during subject's application by his PD gets access only to one record of ISPD
- The context analysis is impossible

The distinctive feature of this method is that after the process of depersonalization, the possibility of processing the personal data of the subject remains with strictly confidentially. It prevents the leakage of reliable personal data and penalty on a data operator. Nowadays there are a huge number of examples of such leakage and their further publication on the Internet. The process of depersonalization allows to eliminate the connection between the data and the subject.

Usage of this method allows providing protection of personal data from unsanctioned access, also from information compromising during its leakage though technical channels. It also provides guaranteed access to personal data at legitimate address.

References

1. Tanygin, M.O.: Search and elimination of collisions in information exchange through open communication channels. In: Problems of Informatics in Education, Management, Economics and Technology: a Collection of Articles of the Xth International Scientific and Technical Conference, pp. 62–64 (2010)
2. Tanygin, M.O.: Method of control of data transmitted between software and hardware. In: Computer Science and Engineering: Materials of the IV International Conference of Young Scientists, pp. 344–345 (2010)
3. Tipikin, A.P., Tanygin, M.O.: Methods of authentication of information protection systems and controlling software. Telecommun. Radio Eng. **5**, 453–463 (2007)
4. Spevakov, A.G.: Basics of information security: training manual in two parts. Part 1. Southwest State University, Kursk (2013)
5. Spevakov, A.G.: Basics of information security: training manual in two parts. Part 2. Southwest State University, Kursk (2013)
6. Primenko, D.V.: A method of personal data depersonalization. Papers of the All-Russian Scientific Conference Intellectual and Information Systems, p. 150 (2016)
7. Spevakova, S.V., Primenko, D.V.: A method of personal data depersonalization in automated systems. In: Compilation: Optoelectronic Devices in Pattern Recognition Systems, Image Processing and Symbol Information, Recognition. Compilation of Materials of the XIII International Scientific-Technical Conference, pp. 330–333 (2017)
8. Spevakov, A.G.: Depersonalization of personal data during processing of information in automated. Telecommunications **10**, 16–20 (2016)
9. Evseeva, A.A., Klutskii, I.V., Spevakov, A.G.: Comparative analysis of Russian and Chinese legislation in the sphere of personal data processing and protection. News of the Southwest State University. Series: Management, Computer Equipment, Informatics. Medical Device Engineering, pp. 233–237 (2013)

10. Popykin, A.V.: Information security of Russia in the system of multipolar cooperation. In: Actual Problems of International Relations in Conditions of a Multipolar World Formation. Compilation of Scientific Articles of the III International Scientific Conference, pp. 129–131 (2015)

11. Nozdrina, A.A., Spevakov, A.G., Primenko, D.V.: Method of personal data depersonalization. RU patent 2016126867, 4 July 2016 (2017)

12. Altuknova, V.A., Anfilova, E.B., Tezik, K.A.: Hashing algorithm GOST R 34.11-2012 and SHA-3. In: Compilation: Scientific and Education Space: Development Perspective. Compilation of Materials of the II International Scientific Conference, pp. 259–260 (2016)

13. Avezova, Y.E.: Modern approaches to hash-function formation using the example of the finalists of the competition SHA-3. In: Works of the International Symposium on Reliability and Quality, vol. 1, pp. 164–168 (2015)

14. Babenko, L.K., Shapovalova, E.R.: Analysis of the hashing function Keccak. Inf. Oppos. Terror. Threat. **23**, 240–244 (2014)

15. Dobritsa, V.P., Spevakov, A.G., Gubarev, A.A.: Algorithm of exclusive transformation of data. News Kursk State Tech. Univ. **1**(30), 49–54 (2010)

16. Potapenko, A.M., Marukhlenko, A.L., Konarev, D.I., et al.: Personalized system of information search with the function of topic definition and meaning values analysis. In: Compilation: Infocommunications and Information Security: State, Problems and Solutions. Materials of the II International Scientific Conference, pp. 181–187 (2015)

17. Klyucahryov, P.G.: Cryptographic hash-functions based on generalized cellular automatons. Sci. Educ. Sci. Ed. MSTU **1**, 161–172 (2013)

18. Bilduk, D.M., Salomatin, S.B.: Cryptographic low-grade codes in information systems with code distribution of channels. DSPA: Questions of application of digital signal processing **6** (1), 85–88 (2016)

19. Birlikkyzy, G.: Search using hash-functions. Sci. Perspect. **5**, 96–97 (2014)

20. Barakat, M., Eder, C., Habke, T.: An introduction to cryptography. https://www.mathematik.uni-kl.de/~ederc/download/Cryptography.pdf. Accessed 20 Sept 2018

Complex Evaluation of Information Security of an Object with the Application of a Mathematical Model for Calculation of Risk Indicators

A. L. Marukhlenko, A. V. Plugatarev$^{(\boxtimes)}$, and D. O. Bobyntsev

Southwest State University, 90 Pyatdesyat let Oktyabrya street,
Kursk 305040, Russian Federation
aplugatarev@bk.ru

Abstract. In modern information systems designing, great attention is paid to information security issues, and therefore minimizing the risk of unauthorized access and failure of individual elements of the computer network, which characterizes the possible danger of an unfavorable outcome, a combination of the likelihood and consequences of an event, both from the attacker and the work of the information system elements. In this regard, the problem of building an appropriate mathematical model for calculating risk is relevant. The need to evaluate indicators is dictated by the statistics of cases of unauthorized access, failure of system elements and the consequences of reducing the level of information security. Thus, the problem of a calculation option search, which takes into account a complex of factors that are not tied to a specific subject area and allows considering the frequency of occurrence of abnormal situations and the relative probabilities of adverse events is considered. A version of the risk calculation indicators of the information system is considered, shortcomings in the design of the computing environment and the allocation of information environment vulnerabilities are highlighted, a mathematical model is developed for a comprehensive assessment of the object's information security. The main idea is to analyze vulnerabilities, the degree of performance indicators variability in relation to the variation of environmental input parameters.

Keywords: Information security · Risk analysis · Unauthorized access · Comprehensive evaluation

1 Introduction

The rapid development of science and technology, large industrial organizations and private companies have led to a more rational use of information resources and significantly increased the level of automation in most stages of activity. With the increasing number and complexity of information systems, there is a need to ensure comprehensive information security. Nowadays, it is impossible to imagine a private or public organization which does not have services using the Internet. In recent years, the increasing frequency of attacks on the infrastructure force companies to ensure the security of the network perimeter stronger [1, 2].

© Springer Nature Switzerland AG 2020
A. A. Radionov and A. S. Karandaev (Eds.): RusAutoCon 2019, LNEE 641, pp. 771–778, 2020.
https://doi.org/10.1007/978-3-030-39225-3_84

Today the problem of providing comprehensive information security includes consideration of all stages of the computing system: the data processing mechanism, methods of information flows transferring, ways to improve the performance and resistance of the system to failures of its elements, as well as minimizing potential leaks and unauthorized access. Analysis of existing methods of calculation and evaluation of the risks of unauthorized access to elements of corporate systems has shown the lack of affordable, within small organizations, scalable solutions that allow to execute the operational generation, processing and analysis of the results while maintaining the necessary level of flexibility in setting up the system as a whole and the ability to manage information security support tools [3]. Important attention should be paid to minimizing the risk of unauthorized access both inside and outside the system and characterizing the possible danger of an unfavorable outcome, the combination of the likelihood and consequences of an event both from the attacker and as the result of work of the elements of the information system itself, as well as characterizing the situation, having an uncertainty of outcome with the obligatory presence of adverse consequences. All this is possible using a mathematical model that allows calculating quantitative risk indicators and their subsequent evaluation [4, 5].

2 The Task of Comprehensive Analysis in Information Systems

To determine the indicators of individual and collective risks, on the basis of which a conclusion about the compliance of an object with information security requirements is made, it is necessary to conduct a set of calculations by analogy with other areas of science and industry [6]. Requirements of the need to conduct a risk evaluation are dictated by negative statistics of intrusion into the work of information systems with a number of consequences both within the individual and in the organization. Questions of methodological support for the development of requirements, as well as the quality of work, do not have a universal solution due to the lack of a clear declaration. Typical errors include an incorrect definition or an incomplete list of potential sources of threats, equipment malfunctions and instability of interaction channels, incorrect behavior of staff and users, flaws at the level of software used [7, 8].

An integral part of conducting an information security audit is penetration testing (security analysis). The information security specialist is faced with the task of checking the organization's infrastructure for vulnerabilities in the system for further compromise. There are several approaches to conducting security analysis. In this context, a "box" is a penetration testing system. Therefore, based on the initial information about the infrastructure, an information security specialist is faced with a particular method of conducting a security analysis, which affects further actions in the analysis. In the case of a black box, the researcher knows a small amount of information about the system, it can be a web service domain name or a small range of network addresses. In the case of a gray box, the researcher is provided with information about user accounts, limited information about the system as a whole. In the case of a white box, the researcher is provided with complete information about the system - the source codes of the software, the topology of the organization's networks, a complete list of system users, the protective mechanisms used, and technological specifications. Black box security analysis is an effective way to test your organization's infrastructure for penetration when an attack comes from the Internet. This method allows you to fully emit the actions of an external intruder.

Information gathering is the initial stage of an attack. The information security specialist collects comprehensive information about the organization to determine further attack vectors. An example of the information collected: a list of employee e-mail addresses, information from employee social networks, identification of external services, determination of the technologies used and their versions, port scanning, etc. Depending on this stage, further actions are formed to penetrate the system, as well as to prepare the necessary tools for carrying out an attack, therefore it is important for a specialist to collect as much information as possible.

Preparing tools for an attack - the second stage of an attack. After collecting information about the object of study and identifying vulnerable components of the system, it is necessary to prepare the appropriate tools for the attack. At this stage, malware is being developed to exploit the vulnerability found, a delivery system is being formed, etc. Malware delivery is the third stage of the attack. After preparing the necessary tools for exploiting the vulnerability found, it is necessary to deliver it for further penetration into the infrastructure. One of the most effective delivery methods is phishing. Phishing is the process of sending pre-formed (tampered) messages to e-mail addresses in order to launch malicious software on the recipient's side. Since employees of a non-technical nature often work with emails, this method of delivery is the most common. An infected web resource may also be a delivery method. When collecting information, the attacker analyzes the behavior of the organization's employees, thereby can find weaknesses in the web resources that the personnel visit, resulting in placing malicious software on another company's resource for further delivery. Exploitation of vulnerability (penetration) - the fourth stage of the attack. After the successful delivery of malware, the vulnerability is exploited directly, as a result of which information about the research object is re-collected, but from the internal part of the network. Fixation in the system - the fifth stage of the attack. Using the information received at the last stage and access to the machine in the infrastructure of the organization, it is necessary to gain a foothold in the system. At this stage, backdoor and trojans are installed, privileges are increased, as well as actions are invisible and the presence in the system is hidden. Setting feedback - the next stage of the attack. After fixing in the system, the attacker needs to set up feedback for sending remote commands. Execution of commands - the finish stage of the attack. As soon as the attacker has configured the reverse communication channel for direct control of the machine, it becomes possible to spread further throughout the infrastructure. Thus, at this stage, the attacker gained full access to the compromised machine and collects information from the internal network to search for new points of further penetration.

Existing software tools for analysis and evaluation of information security, such as MethodWare, Buddy System, RiskWatch, COBRA, CRAMM, and others, are highly expensive and do not have the flexibility to change calculations according to the expansion of possible sources of threats.

This raises the task of developing a mathematical model that meets the require-ments of flexibility in setting up and completing the calculation steps based on the set of parameters of the considered information system and the potential actions of the attacker [9]. Thus, the problem of a calculation option search, which takes into account a complex of factors that are not tied to a specific subject area and allows taking into account the frequency of occurrence of abnormal situations and the relative probabilities of adverse events is considered.

3 Method Aimed at Carrying Out a Comprehensive Evaluation of Information Security of an Object

The designed mathematical model takes into account many scenarios of adverse factors development, the main idea of the assessment is to analyze vulnerabilities, the degree of changeability of performance indicators in relation to variation of the input parameters of the information system (probability distribution, areas of variation of these or those values). Usually the conclusions of the study of the project input parameters sensitivity reflect the degree of reliability obtained in the analysis of project results [10, 11]. In case of divergence between the calculation results and safety requirements, a list of the most dangerous factors is formed, for which it is necessary to clarify the parameters, inaccuracy or incomplete assignment of which is most significant in obtaining the result, or to take a number of measures leading to a decrease in risk indicators. The system approach is associated with the construction of a model aimed at calculating risk indicators that are not adapted for a particular area. As a result of the evaluation of such a calculation model, a description of the functioning of the real subsystem is formed [12].

Conditional probability of a successful attack of h-terminals i-th threat in condition of the scenario realization is calculated by the formula:

$$P_{ji}(h) = \frac{N_i!}{h! \cdot (N_i - h)!} R_{ji}^h (1 - R_{ji})^{N_i - h}, \tag{1}$$

where R_{ji} is a conditional probability of a successful attack of a certain threat, N_i is a total number of attacks.

In this case, if we suppose that the events of occurrence of unauthorized access by different scenarios are disjoint (if there if a domino effect, then it will be reviewed as a separate scenario) frequency of realization of scenarios with hack is calculated by the formula [13, 14]:

$$F_1(n) = \sum_{j=1}^{J} v_j \cdot h_j \cdot P_j(n), \tag{2}$$

where h_j is specific duration of an attack; $F_1(n)$ frequency of realization of scenarios with successful attacks; v_j is frequency of scenario realization, $P_j(n)$ is probability of getting unauthorized access.

Also, it is important to consider that events of attack occurrence by different scenarios throughout a year are independent for each type of threat, n, a frequency of realization of scenarios with hack is calculated as follows:

$$F_2(n) = \sum_{\substack{k_1, k_2, \ldots, k_J \\ k_l \in \overrightarrow{0 \ldots N} \\ k_1 + k_2 + \ldots + k_J = n}} \prod_{j=1}^{J} v_j \cdot h_j \cdot P_j(k_l), \tag{3}$$

where $P_j(k_l)$ is probability of hack of k terminals.

Thus, the indicator of collective risk is calculated according to the ratio:

$$R_{num} = \sum_{n=0}^{N} n \cdot \sum_{j=1}^{J} v_j \cdot h_j \cdot P_j(n) = \sum_{n=0}^{N} \sum_{j=1}^{J} n \cdot v_j \cdot h_j \cdot P_j(n) = \sum_{j=1}^{J} \sum_{n=0}^{N} n \cdot v_j \cdot h_j \cdot P_j(n)$$

$$= \sum_{j=1}^{J} v_j \cdot h_j \cdot \sum_{n=0}^{N} n \cdot P_j(n) = \sum_{j=1}^{J} v_j \cdot h_j \cdot \sum_{n=0}^{N} n \cdot \sum_{\substack{k_1, k_2, \ldots, k_M \\ k_l \in \overrightarrow{0 \ldots N_i} \\ k_1 + k_2 + \ldots + k_M = n}} \prod_{i=1}^{M} P_{ji}(k_l)$$

$$= \sum_{j=1}^{J} v_j \cdot h_j \cdot \sum_{k_1=0}^{N} \cdots \sum_{k_M=0}^{N} \left(k_1 \cdot \prod_{i=1}^{M} P_{ji}(k_i) + k_2 \cdot \prod_{i=1}^{M} P_{ji}(k_i) + \ldots + k_M \cdot \prod_{i=1}^{M} P_{ji}(k_i) \right)$$

$$= \sum_{j=1}^{J} v_j \cdot h_j \cdot \left(\sum_{k_1=0}^{N} k_1 \cdot P_{j1}(k_1) \cdot \sum_{k_2} \sum_{k_M=0}^{N} \prod_{i=1}^{M} P_{ji}(k_i) + \ldots + \sum_{k_M=0}^{N} k_M \cdot P_{jM}(k_M) \cdot \sum_{k_1} \sum_{k_{M-1}=0}^{N} \prod_{i=1}^{M} P_{ji}(k_i) \right)$$

$$= \sum_{j=1}^{J} v_j \cdot h_j \cdot \left(\sum_{k_1=0}^{N} k_1 \cdot P_{j1}(k_1) + \ldots + \sum_{k_M=0}^{N} k_M \cdot P_{jM}(k_M) \right)$$

$$= \sum_{n=0}^{N_1} n \cdot \sum_{j=1}^{J} v_j \cdot h_j \cdot P_{j1}(n) + \ldots + \sum_{n=0}^{N_M} n \cdot \sum_{j=1}^{J} v_j \cdot h_j \cdot P_{jM}(n)$$

$$\tag{4}$$

4 Numeric Simulation

Individual risk is a quotient of the collective risk and the number of those who risk only for groups of users, but not for the object as a whole and is calculated by the formula:

$$R_{indi} = \sum_{j=1}^{J} v_j \cdot h_j \cdot R_{ji} = \frac{\sum_{n=0}^{N_i} n \cdot \sum_{j=1}^{J} v_j \cdot h_j \cdot P_{ji}(n)}{N_i} = \frac{R_{coli}}{N_i}, \tag{5}$$

where R_{ji} is a conditional probability of a separate technology unit failure; N_i is a number of recipients of a risk; R_{coli} is a collective risk.

In terms of modeling, using the example of computer network of department of information security of SWSU, an option of a visual representation of calculation results was received, which lies in forming of correlation of an adverse event frequency with the number of damaged or working in off-design mode due to development of negative factors of the elements.

As a result of this assessment, a set of functioning outcomes is formed with reference to the real subsystem. The modeling results allowed to obtain the dependence of the failure rate of individual elements of the system on their number, which showed a decrease in the frequency of simultaneous adverse effects with an increase in the number of system element failures, which is important to consider during evaluation of potential vulnerabilities and stability of work of the system in general [15, 16] (Fig. 1).

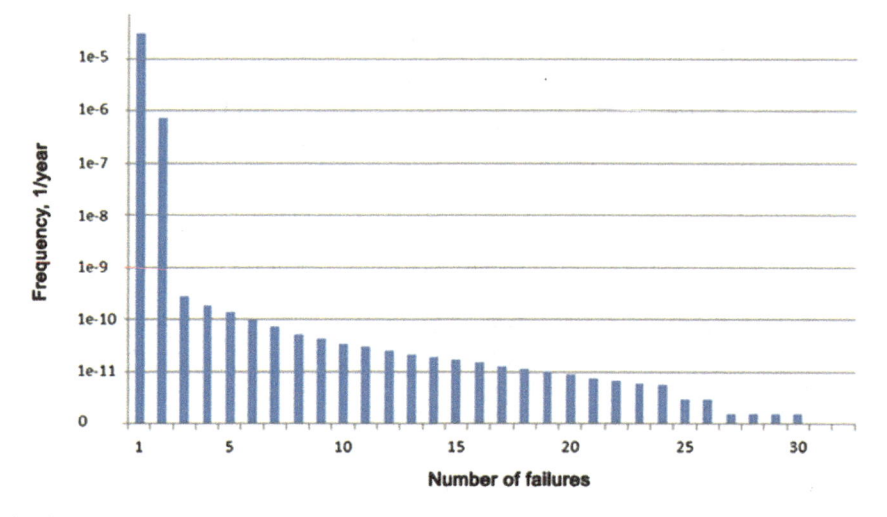

Fig. 1. Dependence of adverse event frequency on the abnormally working elements number.

The main idea is to analyze vulnerabilities, the degree of variability of performance indicators in relation to the variation of environmental input parameters. As a result of this assessment, a set of functioning outcomes is formed with reference to the real subsystem.

Analysis of the received correlation has shown that with increase of the number of simultaneous failures comes the sharp decline of the level of frequency of their occurrences, the chart is non-linear, the maximum indicator about 10^{-5} can be considered as permissible as in the scale of information environment accounting and processing of confidential data are not recorded.

The need to evaluate indicators is dictated by the statistics of cases of unauthorized access, failure of system elements and the consequences of reducing the level of information security.

5 Conclusion

Thus, the obtained results allow executing the calculation of risk indicators in the development of information security policy and evaluation of vulnerability of the information system as a whole. Analysis of the received dependency of frequency of failure of separate system elements from their number, which has shown the decline of the level of frequency of simultaneous failures' occurrences at increase of the number of failures of the system elements, what is important to consider during the evaluation of potential vulnerabilities and stability of the system's work as a whole.

References

1. Marukhlenko, A.L., Mirzakhanov, P.S.: Program complex for modeling of the process of transferring and processing of network flows of data. News of the Southwest State University. Series: management, computer equipment, informatics. Med. Instrum. **2–3**, 175–180 (2012)
2. Tanygin, M.O., Marukhlenko, A.L., Marukhlenko, L.O., et al.: Technology and software implementation of a software module for localizing potentially dangerous objects on a graphic substrate using neural networks. In: Infocommunications and Space Technologies: Status, Problems and Solutions. The Collection of Scientific Articles Based on the Materials of the II All-Russian Scientific and Practical Conference, pp. 23–28 (2018)
3. Efremov, M.A., Kalutskii, I.V., Tanygin, M.O., et al.: Personal data security, social networks and commercials in the Internet. News of the Southwest State University. Series: management, computer equipment, informatics. Med. Instrum. **7**(1), 27–33 (2017)
4. Tanygin, M.O., Marukhlenko, A.L., Marukhlenko, L.O., et al.: Analysis of potential vulnerabilities and modern methods of protecting multi-user resources. In: Infocommunications and Space Technologies: State, Problems and Solutions. The Collection of Scientific Articles Based on the Materials of the II All-Russian Scientific and Practical Conference, pp. 136–140 (2018)
5. Lepina, N.V., Tanygin, M.O., Kalutskii, I.V.: About features of providing of information security of information computing systems of universities compilation: Infocommunications and information security: state, problems and solutions. In: Materials of the II International Scientific Conference, pp. 246–249 (2015)
6. Dobritsa, V.P., Lipunov, A.A., Savenkova, E.S.: System of access control and management in complex of information security systems. News of SWSU. Series: management, computer equipment, informatics. Med. Instrum. **2**(2), 87–90 (2012)
7. Agapov, A.A., Khlobystova, I.O., Marukhlenko, S.L., et al.: Hardware and software system "toxi + meteo" for assessing the consequences of possible accidents taking into account data on current weather conditions. Labor safety in industry, pp. 22–25 (2011)
8. Degtyarev, S.V., Marukhlenko, A.L., Marukhlenko, S.L.: Program model for automation of calculation of the risk of technological accidents. Inf.-Measur. Manag. Syst. **8**(11), 35–39 (2010)
9. Potapenko, A.M., Marukhlenko, A.L., Konarev, D.I., et al.: Personalized system of information search with the function of topic definition and meaning values analysis. In: Compilation: Infocommunications and Information Security: State, Problems and Solutions. Materials of the II International Scientific Conference, pp. 181–187 (2015)

10. Harper, A., Regalado, D., Linn, R., et al.: Gray Hat Hacking: The Ethical Hacker's Handbook. McGraw-Hill Education, New York (2018)
11. Lisanov, M.V.: Errors of rationing of quantitative criteria of a tolerable risk. Methods Conform. Assess **9**, 41–43 (2009)
12. Tanygin, M.O.: A proposal for forming of threat model for telecommunication systems. In: Compilation: Optoelectronic Equipment and Devices in the Systems of Image Recognition, Image Processing and Symbol Information, pp. 341–343 (2017)
13. Marukhlenko, A.L., Marukhlenko, S.L.: Mathematical model of a system approach for evaluation of risk of technological accidents. News of SWSU. Series: management, computer equipment, informatics. Med. Instrum. **2**, 60–64 (2013)
14. Marukhlenko, S.L., Degtaryov, S.V.: System analysis in solving the tasks of risk analysis. News of SWSU. Series: management, computer equipment, informatics. Med. Instrum. **2**, 33–37 (2012)
15. Khalin, Yu.A., Marukhlenko, A.L., Marukhlenko, L.O.: Development of secure corporate systems based on client-server technology. SWSU, Kursk (2018)
16. Marukhlenko, L.O., Marukhlenko, A.L., Kerimbaeva, K.M., et al.: Variant of ensuring information security by increasing the fault tolerance of the hardware firewall. In: Infocommunications and Space Technologies: State, Problems and Solutions. The Collection of Scientific Articles Based on the Materials of the II All-Russian Scientific and Practical Conference, pp. 10–14 (2018)

Application of Simulink and SimEvents Tools in Modeling Marketing Activities in Tourism

A. N. Kazak[1]($\boxtimes$), D. V. Gorobets[1], and D. V. Samokhvalov[2]

[1] Crimean Federal University, 4 Vernadskogo Avenue, Simferopol 295007, Russian Federation
`kazak_a@mail.ru`
[2] Saint Petersburg Electrotechnical University, 5 Professora Popova Street, St. Petersburg 197376, Russian Federation

Abstract. The article shows how to simulate the activities of a travel agency in the SimEvents environment and how to calculate the system load. This model allows you to optimally load the system by regrouping services. The considered example of building the architecture of the system model, taking into account the logistics of moving the groupings of objects, can be generalized to the tasks of planning the optimal workload of work centers and other production management tasks. In our case, the agency operates mainly through the Internet. The object of the study is the process of servicing several types of clients of a travel agency. The distribution of various types of tourist destinations is under study. A simulation of a deterministic system, including a servicing device (server) with a discrete time for servicing applications and a request generation unit, which also has a constant arrival time, was carried out. The intelligence of the model was achieved by creating a program code that ensured full automation of the modelling process. The simulation results indicate the quality of the constructed model.

Keywords: Service system · Simulation modeling · System efficiency indicators · Optimization · Queue · Management

1 Introduction

A large number of original products are created in the tourism market through the efforts of many companies. Note that each of the companies uses its own technologies, tools and methods of marketing policy for the promotion and sale of tourism products. Therefore, travel companies should focus on the use of the latest tools and technologies in the complex marketing activities in the tourism market. We made an attempt to build a model of a tourist service system in which the object of research is the input flow and distribution of customers using tourism services in accordance with requests.

A. A. Radionov and A. S. Karandaev (Eds.): RusAutoCon 2019, LNEE 641, pp. 779–786, 2020.
https://doi.org/10.1007/978-3-030-39225-3_85

2 Methodology and Data

There were applied methods of synthesis, economic modeling, and simulate the activities of a travel agency in the Simulink and SimEvents environment [1–3].

3 Results

The block diagram of the model is shown in Fig. 1. It identifies the functional areas, individual blocks and shows the relationships between them.

The model uses different types of customer flow generators, each of which operates according to a separate law of statistical distribution. This allows you to study the specifics of various groups and target audiences of clients served in a tourist destination.

The model allows you to organize incoming various orders in accordance with the selected priorities. As a result, the processed order flows can be sorted with a high degree of efficiency in a minimum amount of time [4]. The order of incoming applications is taken into account in two blocks of the model [5]:

- FIFO queue
- Priority Queue

This approach in this aspect provides the opportunity to take into account the features of customer service in the tourism business.

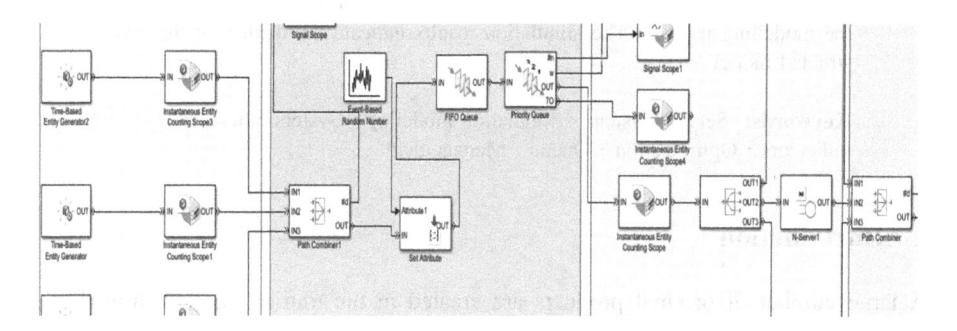

Fig. 1. Block diagram of the model.

A separate scenario considered in the model describes the state of a sudden peak load, which is characterized by such a sharp change in the flow of application traffic that it becomes impossible to quickly respond to them organizationally. In our model experiment, the orders with a low average transaction price are discarded in the priority queue block, and are placed in priority, on the decomposer block, where the excess stream of orders normalized by the dump is serviced [6].

The simulation results are shown in Fig. 2 "Peaks and declines in the incoming customer flow associated with the system's ability to process orders" and Fig. 3 "The number of customers" lost "by the processing system due to peak situations".

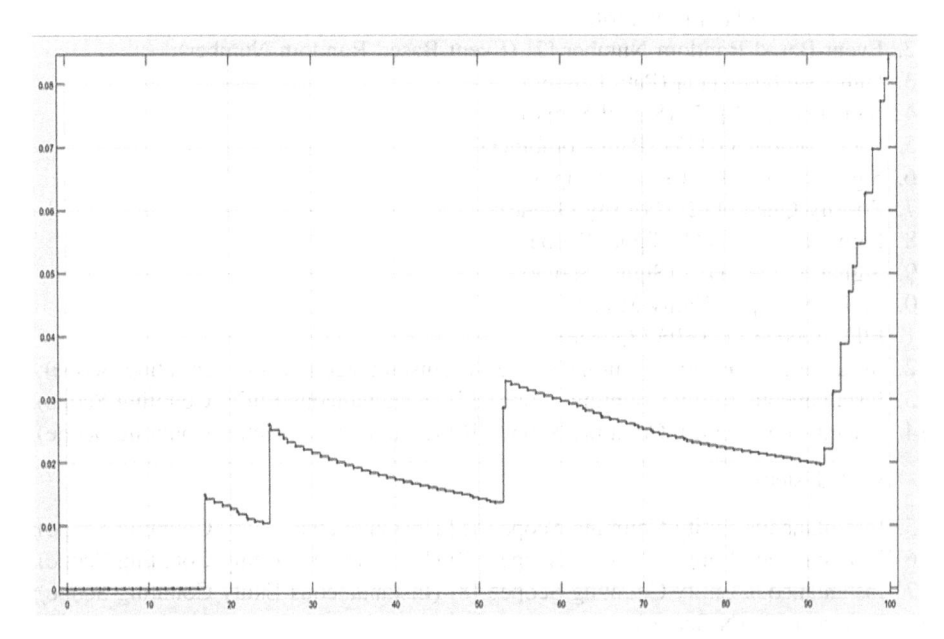

Fig. 2. Peaks and declines in the incoming customer flow associated with the system's ability to process requests.

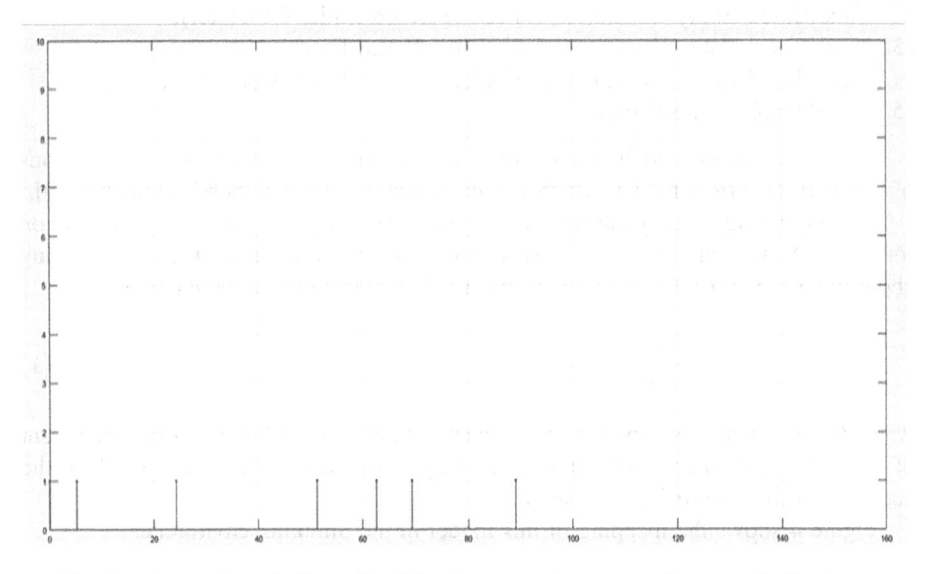

Fig. 3. The number of clients "lost" by the processing system due to peak situations.

Design description of this model contains, in particular, the section "Block execution order".

Block Execution Order:

1. Set Attribute [13] (SetAttribute);
2. Event-Based Random Number [2] (Event Based Random Number);
3. Path Combiner [11] (Path Combiner);
4. Signal Scope2 [15] (Signal Scope);
5. Path Combiner1 [12] (Path Combiner);
6. Signal Scope [14] (Signal Scope);
7. Priority Queue [12] (Priority Queue);
8. Signal Scope1 [14] (Signal Scope);
9. Signal Scope3 [16] (Signal Scope);
10. Entity Sink [2] (Entity Sink);
11. FIFO Queue [4] (FIFO Queue);
12. Instantaneous Entity Counting Scope [4] (Instantaneous Entity Counting Scope);
13. Instantaneous Entity Counting Scope1 [5] (Instantaneous Entity Counting Scope);
14. Instantaneous Entity Counting Scope2 [6] (Instantaneous Entity Counting Scope);

Root System

15. Instantaneous Entity Counting Scope3 [6] (Instantaneous Entity Counting Scope);
16. Instantaneous Entity Counting Scope4 [7] (Instantaneous Entity Counting Scope);
17. Instantaneous Entity Counting Scope5 [8] (Instantaneous Entity Counting Scope);
18. N-Server [9] (N Server);
19. N-Server1 [10] (N Server);
20. N-Server2 [11] (N Server);
21. Replicate [13] (Replicate);
22. Time-Based Entity Generator [17] (Time Based Entity Generator);
23. Time-Based Entity Generator1 [18] (Time Based Entity Generator);
24. Time-Based Entity Generator2 [18] (Time Based Entity Generator);
25. SESubgraph0 (SESubgraph).

We also consider modeling marketing activities in tourism based on modifications of the Nerlove-Arrow model, which is a continuation of the authors' research [7–9].

It is interesting to study the effect of advertising on the choice of a resort destination (or hotel). N (t) is the number of people who receive information through advertising about this resort (or a separate hotel), where N- is a solution to the equation:

$$\frac{dN(t)}{dt} = [a_1(t) + a_2(t)N(t)](N_0 - N(t))_1, \tag{1}$$

where $0 < a_1$ - is the coefficient of the intensity of advertising, $0 < a_2$ - is the coefficient of the intensity of communication of the country's population with each other, N_0 is the audience of the advertising campaign.

Figure 4 shows the mapping of this model in the Simulink environment.

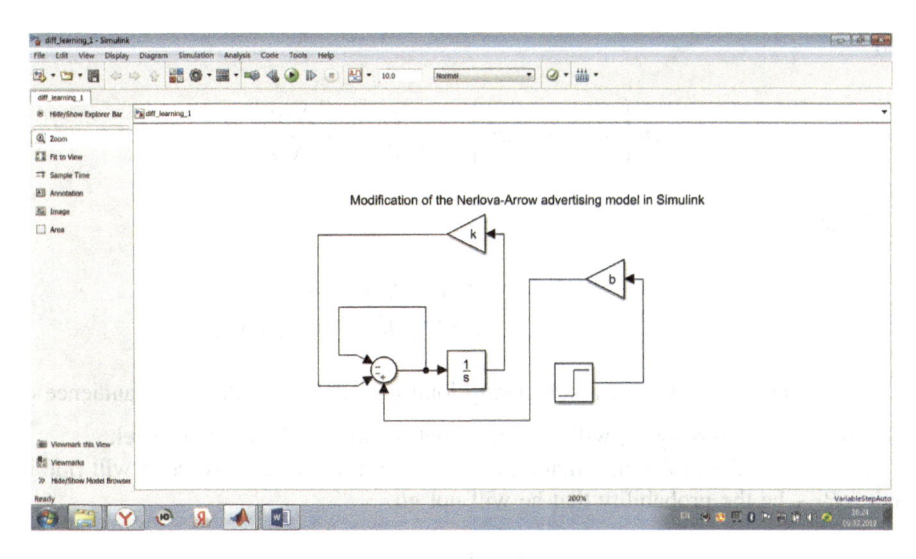

Fig. 4. The modification of the Nerlove-Arrow model in the Simulink environment.

If a $a_1(t) \gg a_2(t)$, then Eq. (1) takes the form:

$$\frac{dN(t)}{dt} = a_1(t)(N_0 - N(t)) \tag{2}$$

The solution of Eq. (2) has the form:

$$\int_{N(0)}^{N(t)} \frac{dX}{(N_0 - X)} = \int_0^t d\theta a_1(\theta) = -\ln\frac{N_0 - N(t)}{N_0 - N(0)},$$

or

$$N(t) = N_0 - (N_0 - N(0))I_0^{\int_0^t d\theta a_1(\theta)} \tag{3}$$

If a $a_2(t) \gg a_1(t)$, then (1) has the form:

$$\frac{dN(t)}{dt} = a_2(t)N(t)(N_0 - N(t)), \tag{4}$$

The solution of Eq. (4) has the form:

$$\int_{N(0)}^{N(t)} \frac{dx}{x(N_0 - x)} = \int_0^t a_2(\theta)d\theta = \frac{1}{N_0}\int_{N(0)}^{N(t)} \left(\frac{1}{x} + \frac{1}{N_0 - x}\right)dx = \frac{1}{N_0}\ln\frac{N(t)(N_0 - N(0))}{N(0)(N_0 - N(t))}$$

and

$$exp\left\{N_0 \int_0^t a_2(\theta)d\theta\right\} = \frac{N(t)(N_0 - N(0))}{N(0)(N_0 - N(t))} \tag{5}$$

What gives:

$$N(t) = \frac{N(0)N_0 exp\{N_0 \int_0^t a_2(\theta)d\theta\}}{(N_0 - N(0) + exp\{N_0 \int_0^t a_2(\theta)d\theta\}N(0))} \tag{6}$$

Since $\lim_{t\to\infty} N(t) = N_0$, for a sufficiently long period of time the entire audience of the advertising company N_0 will have information about this resort or hotel.

Let P_1 – be the probability that a person who knows about this resort will ride it, and, a P_2 – be the probability that he will not go.

$$P_1 + P_2 = 1, \tag{7}$$

where P_1 – is a solution of the equation:

$$\frac{dP_1}{dt} = -\gamma_1 S_1 P_1 - \gamma_2 S_2 P_2 = -\gamma_1 S_1 P_1 - \gamma_2 S_2(1 - P_1), \tag{8}$$

Here S_2 – is the size of the standard of living in the country, S_1 – is the average cost of rest, $\gamma_1, \gamma_2 > 0$.

So, Eq. (8) takes the form:

$$\frac{dP_1}{dt} = (\gamma_2 S_2 - \gamma_1 S_1)P_1 - \gamma_2 S_2$$

or

$$\frac{dP_1}{dt} = (\gamma_2 S_2 - \gamma_1 S_1)\left(P_1 - \frac{\gamma_2 S_2}{\gamma_2 S_2 - \gamma_1 S_1}\right) \tag{9}$$

The solution of Eq. (9) has the form:

$$P_1 = \frac{\gamma_2 S_2}{\gamma_2 S_2 - \gamma_1 S_1} + l^{(\gamma_2 S_2 - \gamma_1 S_1)t}\left(P_1(0) - \frac{\gamma_2 S_2}{\gamma_2 S_2 - \gamma_1 S_1}\right) \tag{10}$$

For formula (10) to be true, it is necessary that $\gamma_2 S_2 > \gamma_1 S_1$, in the opposite case (if $\gamma_2 S_2 < \gamma_1 S_1$):

$$\lim_{t\to\infty} P_1(t) = \frac{\gamma_2 S_2}{\gamma_2 S_2 - \gamma_1 S_1} < 0$$

Based on (6) and (10), we can estimate the number of tourists T, who went to this resort as a result of advertising:

$$T(t) = P_1(t)N(t) = \left[\frac{\gamma_2 S_2}{\gamma_2 S_2 - \gamma_1 S_1} + l^{(\gamma_2 S_2 - \gamma_1 S_1)t} \left(P_1(0) - \frac{\gamma_2 S_2}{\gamma_2 S_2 - \gamma_1 S_1} \right) \right]$$

$$\frac{N(0)N_0 exp\{N_0 \int_0^t a_2(\theta)d\theta\}}{(N_0 - N(0) + N(0)exp\{N_0 \int_0^t a_2(\theta)d\theta\}} \tag{11}$$

Thus, the following conclusions may be drawn.

4 Conclusion

A simulation of a deterministic system, including a servicing device (server) with a discrete time for servicing applications and a request generation unit, which also has a constant arrival time, was carried out. The intelligence of the model was achieved by creating a program code that ensured full automation of the modeling process. The simulation results indicate the quality of the constructed model.

The article shows how to simulate the activities of a travel agency in the SimEvents environment and how to calculate the system load. This model allows you to optimally load the system by regrouping services.

The considered example of building the architecture of the system model, taking into account the logistics of moving the groupings of objects, can be generalized to the tasks of planning the optimal workload of work centers and other production management tasks. In our case, the agency operates mainly through the Internet. The object of the study is the process of servicing several types of clients of a travel agency. The distribution of various types of tourist destinations is under study.

References

1. Mathworks Matlab Documentation: SimEvents getting started guide (2015). http://in.mathworks.com/help/simevents/getting-started-with-simevents.html. Accessed 24 Sept 2017
2. Cassandras, C.G., Lafortune, S.: Introduction to Discrete Event Systems. Kluwer Academic Publishers, Boston (1999)
3. Bose, S.K.: An Introduction to Queueing Systems. Plenum Publishers, New York (2002)
4. Boucherie, R.J., Dijk, N.M.: Queueing Networks - A Fundamental Approach. International Series in Operations Research and Management Science. Springer, Cham (2011)
5. Manitz, M.: Analysis of assembly/disassembly queueing networks with blocking after service and general service times. Ann. Oper. Res. **226**, 417–441 (2015)
6. Zeigler, B.P., Praehofer, H., Kim, T.G.: Theory of Modeling and Simulation: Integrating Discrete Event and Continuous Complex Dynamic Systems. Academic Press, San Diego (2000)

7. Kazak, A.: Investigation of properties of the dynamic model of tourism development. In: Proceedings of 20th IEEE International Conference on Soft Computing and Measurements, pp. 827–829 (2017)
8. Kazak, A.: Qualitative analysis of the mathematical model of tourism development, proposed by Casagrandi and Rinaldi. In: Proceedings of 20th IEEE International Conference on Soft Computing and Measurements, pp. 823–826 (2017)
9. Kazak, A., Lukyanova, Ye., Chetyrbok, P.: One of the regions of the southern federal district touristy flows dynamics modeling. In: IEEE II International Conference on Control in Technical Systems, pp. 103–105 (2017)

Cascade Windows in Intellectual Agents of Multichannel Images Classification

I. A. Malyutina[1]($\boxtimes$), S. A. Filist[1], and A. R. Dabagov[2]

[1] Southwest State University, 94 Pyatdesyat let Octyabrya street,
Kursk 305040, Russian Federation
imalutina2019@yandex.ru
[2] Medical Technologies Ltd., 31 Ibragimova street,
Moscow 105318, Russian Federation

Abstract. The paper suggest using not slid cascade windows generating "weak" qualifier for the classification of hardly structured images. At each level of processing of a parental cascade window it is divided into several window descendants, for analysis of which the intellectual agents are used. The agents use two approaches to form descriptors, creating two groups of qualifiers based on these descriptors in a cascade window and the subsequent aggregation of the decision made by these qualifiers. It is shown that the use of cascade windows in the analysis and classification of multichannel images or their segments as integration is the most efficient approach. In this case, is possible in the narrow spatial range of a cascade window. In those channels, in which the image has good spatial resolution, the descriptors are formed based on the analysis of border contours of segments of the corresponding raster. For the analysis and classification of the allocated contours rasters are used in the channels with the good permission in the field of spatial frequencies or in the field of an electromagnetic range. The use of cascade windows in each channel allows forming a set of qualifiers on one channel in the form of trees of decisions with the subsequent aggregation of decisions both in the channel, and between channels. As a result, the network structure of qualifiers (cellular qualifiers) is formed. Its parameters are defined by means of training on the basis of expert estimates or hybrid methods.

Keywords: Multichannel images · Cellular processors · Informative signs · Qualifiers · Walsh's transformation · Cascade windows · The sliding window

1 Introduction

Nowadays a presentation of information in the form of the image is very often used in the systems of decision-making on management of difficult objects. In the result of the analysis of the image the vectors of informative signs, which are used as input information for decision-making modules, are formed. The majority of methods of allocation of an object on the image is constructed on the basis of pixel-by-pixel classification [1–3]. Classification process, in this case, is presented by two stages: at the first stage objects of the set class (interest objects), and on the second one their properties are allocated which can indicate the belonging of an object to a subclass

A. A. Radionov and A. S. Karandaev (Eds.): RusAutoCon 2019, LNEE 641, pp. 787–796, 2020.
https://doi.org/10.1007/978-3-030-39225-3_86

inside the allocated class. For example, after the allocation of a uniform element of blood against the background of plasma it is necessary to carry the element to an erythrocyte, or to a leukocyte. In the same way the leukocyte can be segmented with the subsequent allocation of the classes of the received segments [4–6].

The essence of classification of the first stage means the binarization of the image, which is its division into an object and background. For this purpose, scanning of a raster of the image with decision-making on reference of each pixel of a raster to any of two classes is carried out. The second stage can repeat the first, use other methods of segmentation or in general be absent. At this stage prior information on the studied object is very often used, in particular about its anatomic or morphological properties [7, 8]. However, the information provided by the image is not very often clear, and the object of interest masks hindrances (with the objects on the same image which are not of interest).

One of decisions on improvement of quality of recognition of images is an increase the number of rasters which submit the image of an interest object. It can be reached by the selection of range of electromagnetic radiation by means of use of multizone and hyper zone pictures or use for recognition of a series of panchromatic pictures.

For example, the radiographic images, received by means of digital plane-parallel x-ray receivers, which allow to receive a series of pictures with considerably reduced dose loading [9].

Both in that, and in the other case two problems apper: a problem of aggregation of decisions on rasters and a problem of an increase in efficiency of decision-making as an increase in number of the analyzed rasters can lead to an inadmissible increase in time of decision-making, for example, at control of robots, unmanned aerial vehicles, in the systems of monitoring of a condition of the patient at expeditious treatment.

Using multiraster or multichannel images hardens the process of pixel-by-pixel classification. It is caused by difficulties of aggregation of decisions in various channels as channels can distinguish both the physical, and geometrical principles of formation of the image, which results in "not compatibility" of "vicinities" of pixels on which belonging to a required segment the decision is made.

In this case, it is expedient to determine segment coordinates in general or at least rectangular area, in which there is a required segment, and at integration of images use these coordinates, it is much simpler to them to find compliance on different images than to each separate pixel. At the same time such approach opens ample opportunities for use of prior information on the studied image and its segments.

2 Materials and Methods

2.1 Cascade Windows

The concept of a cascade window is entered for the purpose of creation a theoretical base of not pixel qualifiers of the image. It is supposed that at the release of such a qualifier not the required image will be submitted, for example, in a binary form, but only the coordinates of a required class of images, for example, in the form of coordinates of the described rectangle.

We believe that after the allocation of a segment it has to be classified. At the same time the qualifiers can be constructed by methods of pixel-by-pixel classification [4, 6, 10] or by methods of the analysis of borders of a segment [4, 6, 11].

The formation procedure of a cascade window consists that the raster image decomposed on a quantity of levels. Each level possesses a great number of descendants - images. The image of the top level is the "maternal" image for a set of images descendants of the lower level, as well as the image of the lower level is "maternal" for images of the following (lower in relation to this image) level.

Each image in the received hierarchy is coded by the corresponding number which describes its status in this hierarchy, which is its relation to images of the previous (higher) level. Each level of decomposition forms the "weak" qualifier. The aggregation of "weak" qualifiers into "strong" can be carried out across (in one level) and down, forming "trees" of "weak" qualifiers. For example, as a way of decomposition the rule "2" can be used^{2n}. It means that at the first level we have 4 images descendants, on the second-16, on the third-64, etc.

For the creation of qualifiers at the levels ("weak" qualifiers) any paradigm can be used [9, 12]. If it is necessary to define whether there is a segment of a certain class on the initial image, it is enough to pass on "branch" of "weak" qualifiers which "voted" for this class. If the image is multichannel the integration is carried out at the horizontal levels with strengthening the corresponding "weak" qualifiers.

2.2 Cellular Processors

In the approach basis trying to find the solution of classification of a cascade task window the development of the method of intellectual agents for segmentation of hardly structured images stated in work [13–15] is put. The replacement of threshold decisive rules by the decisive rules constructed on methodology of a busting [16, 17] was further development of this approach. The advantage of this modification of the intellectual agent was that, unlike Viola-Jones's method, it was succeeded to receive acceptable speed due to use of spectral transformation of Walsh with the subsequent selection of spectral components for a specific objective here and also to receive multidimensional structures of intellectual operators, we will call them cellular processors, at the expense of multialternative ways of decomposition of a raster.

The idea of an increase in speed of processing of multichannel images consists in the use of the system of the processing of images constructed on the basis of a great number of calculators distributed in virtual hyperspace - the cellular processors which are carrying out rather simple algorithms in parallel.

The qualifier of multichannel images or their segments consists of a matrix of weak qualifiers (a matrix of cellular processors) for which several ways of construction can be offered.

The concept of a cascade window is put to the basis of the offered method.

2.3 Multidimensional Structures of Intellectual Agents

In Fig. 1 the structure of the cascade qualifier of a cascade window constructed on the basis of the analysis of signs (for example, Haar's veyvlet) in cascade multiscale windows is presented. Such a structure is an elementary cell on the basis of which strong qualifiers are under construction. This cell is established in each of cascade windows. The aggregation of decisions is carried out by means of neural networks or fuzzy logic of decision-making, or on hybrid algorithms [5, 18].

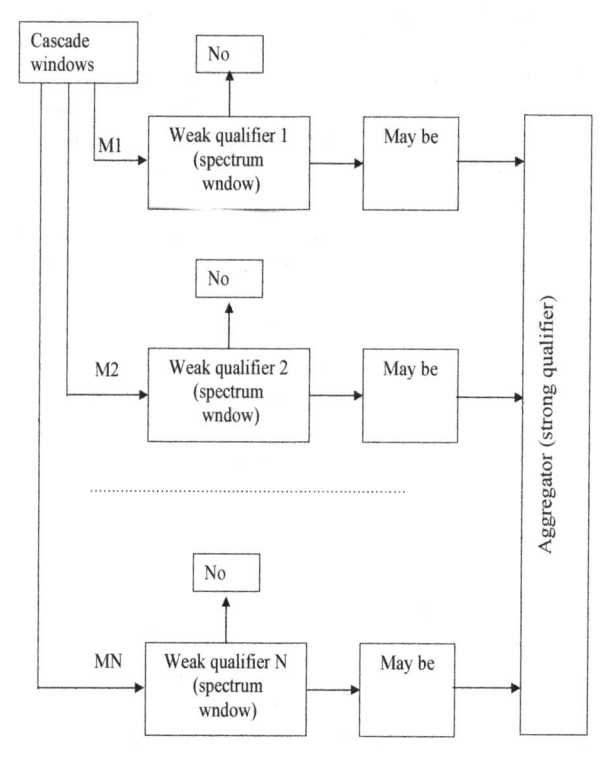

Fig. 1. Structure of the cascade qualifier for one channel constructed on the basis of the analysis of signs in the sliding multiscale windows.

2.4   Coding of the Image at the Hierarchical Levels

At each hierarchical level we have 2^{2n} cascade windows where the n is a level number. Each cascade window is coded, so it belongs to a certain class. The subsequent decomposition of cascade windows, in case of its need, is defined by the code of a cascade window.

The essence of classification of a cascade window consists in the assignment to it a certain status. If the status of a cascade window is low, then it is not classified and gets to "a gray zone", that is does not participate in formation of krone (branches). If the status of a cascade window is high, then it does not make descendants and forms "leaves". Thus, cascade windows only with determined by the status: not too "old" and not too "young" can be parents.

For the coding of cascade windows at the levels spectral methods can be used.

In Fig. 2 the structure of the cellular processor constructed on the basis of the spectral analysis of the image is presented, the image is received in a cascade window or its segment, the structure is expedient for using for images with the good permission in spectral area when the optimum size of a window is a priori known.

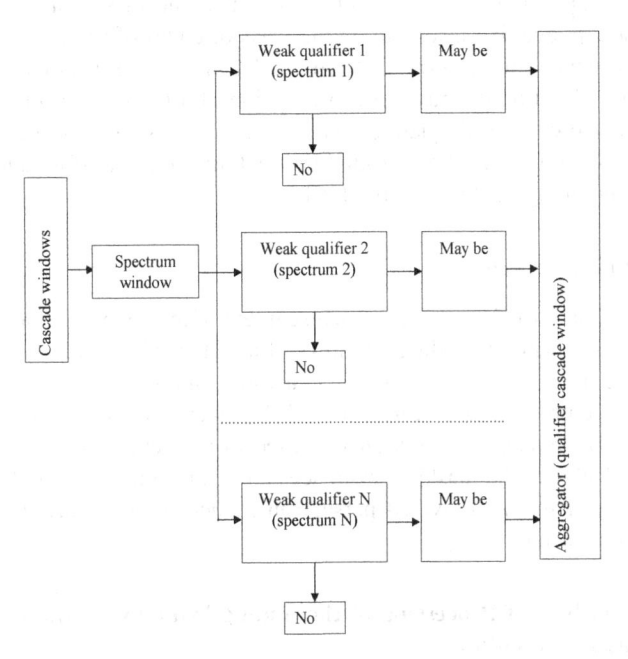

Fig. 2. Structure of the cellular processor constructed on the basis of the analysis of a range of Walsh of a window.

In this case for each channel we receive so many weak qualifiers how many spectral segments (a spatial range) are allocated on the two-dimensional spectral plane.

This structure can assume a stage of the prospecting analysis at which the spatial frequency segments relevant for the studied class of images are allocated or, otherwise, the frequency plane randomly breaks into rectangular segments the number of which defines the number of weak qualifiers.

For multichannel rasters similar cascade windows and cellular processors similar to Figs. 1 and 2 are under construction.

Each raster gives the certain areas belonging to the required class on the initial image. If areas of rasters coincide with identical codes, then it is enough to construct linear model and to choose threshold function for making decision on belonging of area of the image to a required class.

3   Results of a Research

The cascade model of classification of pictures carries out a purposeful allocation of segments of interest in the picture. The segment of interest is determined by a cascade window at an appropriate level in such a way that the classification of a cascade window of the top level is carried out on the basis of codes of cascade windows of the lower level for which this window is "maternal". Reference of a cascade window to a segment is carried out on the basis of classification of a fragment of the picture which has got to a window. "Weak" qualifiers in cascade windows are constructed on Viola-Jones's ideas with use of the two-dimensional discrete transformation of Walsh (TDDTW) of counting of this window [19].

3.1   Choice of a Segment

According to Viola-Jones's technique, each segment of interest is allocated by scanning of the initial image with the sliding windows of a certain M1 × M2 size [18, 20]. The same technique is used when processing a cascade window.

For automatic processing of a fragment of the image which has got - in the sliding window, multi-stage adaptive filtration was used in frequency area.

Both TDDTW, and the masking sequences are realized by the neural network built by the principle of the multilayered perseptron trained on an algorithm of the return distribution of a mistake [21].

3.2   The Procedure of Processing of the Sliding Windows of the Image of a Cascade Window

Each cascade window of a hierarchical level has the identifier of coordinates and the identifier of a class (the identifier of the status). For definition of the identifier of a class of a cascade window the qualifiers constructed on the basis of neural network technologies [10, 19] were used.

The counting of a fragment of cascade $x_1, \ldots x_{M1 \times M2}$ window with hindrances moves on entrances of neural network. In the first layer TDDTW of an entrance signal is carried out as a result of which corresponds to each neuron of the second layer components of a range of Walsh. In the second layer the filtration of a signal in spectral area is carried out for the purpose of detuning from hindrances. At the exit of a layer the response is formed without hindrances.

On the following layer of network there is directly a filtration in a spectral area directed to allocation of that spectral component which correlates in the greatest way with the studied class. The choice of synoptic scales provides revenues to the following layer of signals only of those neurons that enter the bandwidth of the considered filter. By the control of the third layer filter parameters were flexibly changing that allowed to carry out the filtration of spectral coefficients on classes [16, 22].

In the fourth layer of neural network pixel accessory decides on coordinates of i, j to the segment belonging to the required class. The exit of a neuron of this layer is binary.

If the neural network is ready for allocation of a segment of interest in a cascade window descendant, then it works as the filter which passes the set fragments of the image and blocks passing of fragments of other class.

3.3 Experimental Check

For an experimental check of efficiency of classification by means of cellular processors the analysis of a series of pictures of roentgenograms of a thorax of the patient with pneumonia was carried out. The roentgenograms were received with a speed of 3 frames per second, all the video stream included 30 frames. In Fig. 3 this ten second interval with the corresponding fragments of roentgenograms is shown.

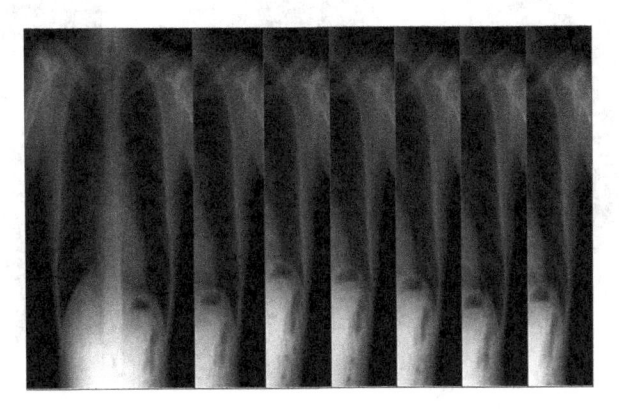

Fig. 3. Fragments of the roentgenograms received on the ten second interval.

3.4 Walsh's Range

If the status of a cascade window does not allow it to have descendants, then it is segmented with use of cellular processors.

For segmentation of cascade windows, the analysis of a two-dimensional range of Walsh in the sliding window was used. The range of the sliding window is processed by the sequence of the filters constructed on the basis of various paradigms of processing of images. As a result of this analysis, the decision was made on accessory of pixel, in the neighborhood of which the sliding window is investigated, to a required class of segments.

In Fig. 4a one of pictures of the roentgenograms of a thorax in which the circle designated an image fragment with manifestation of pneumonia, is shown. The borders of a cascade window are designated by a white rectangle.

In Fig. 4b the fragment of the image of a cascade window, which got to the sliding window, and in Fig. 4c – its range of Walsh is shown.

If necessary, the number of cellular processors can be increased due to the variation of scale of the sliding window for the purpose of creation of qualifiers of hierarchical structure. The decisions received by the cellular processor in the sliding window can be considered as "weak" qualifiers. Changing the size of the sliding window in borders of a cascade window we receive a set of "weak" qualifiers by the aggregation of which we receive "strong" qualifiers.

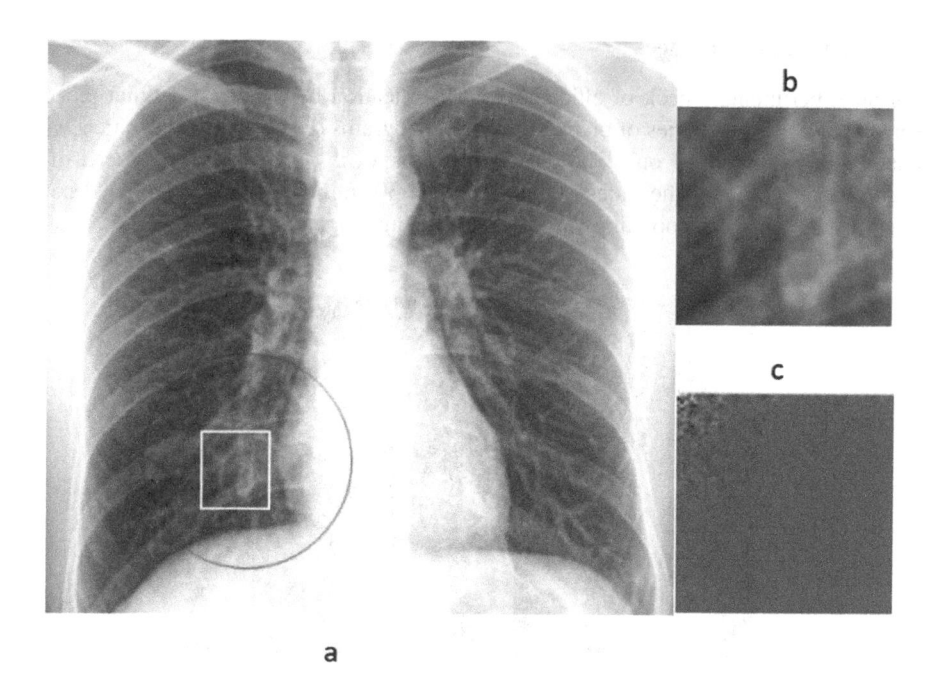

Fig. 4. (a) roentgenogram of a thorax with pneumonia, borders of the sliding window are designated by a white rectangle; (b) the corresponding contents of the sliding window with the manifestations of pneumonia and (c) a two-dimensional range of Walsh of the contents of this window (c).

As a result of the aggregation of "weak" qualifiers and the integration of decisions of strong qualifiers on pictures, we receive the image at the release of the qualifier of a cascade window which is presented in Fig. 5.

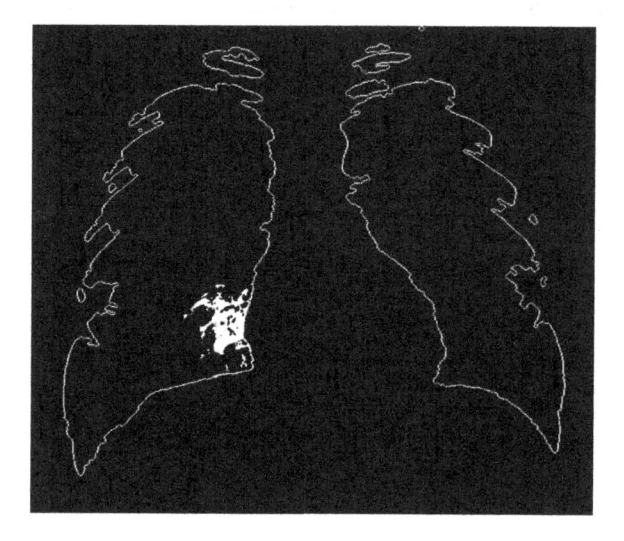

Fig. 5. Of the image at the release of the strong qualifier.

4 Conclusion

In the proposed solution of the problem of classification of multichannel images it is carried out by means of the analysis of images in cascade windows. For the classification of a cascade window or determination of its status the intellectual agents constructed on the basis of the sliding windows - cellular processors are used. The current pixel in the sliding window with its subsequent reference to two alternative classes, by means of cellular processors as which neural networks were used is analyzed. Each level of decomposition will give the number of cellular processors equal to number of channels. The number of cages in NxM matrix, where N-number of channels, M-number of levels of decomposition (the number of scales of the sliding windows). Each "cage" is an independent agent (processor) who makes the decision on pixel belonging to a segment. The solution of cellular processors is exposed to threshold processing (an analog of function of activation in neurons of neural networks) and arrives on the aggregator of decisions. Therefore, we have 2D - decision-making model. If we use two methods of the organization of the cellular processor, then we have a 3D model, etc., however an increase in computing complexity of a task does not lead to a significant increase in time of its performance, and leads only to an increase in the number of cellular processors. The structure and parameters of cellular processors can be adapted to specific features of images in channels.

The planimetric analysis of the panchromatic image allows to receive one more layer in the aggregator of solutions of cellular processor which hierarchical level is defined by the concrete subject domain which is a source of pictures.

References

1. Dyudin, M.V., Zhilin, V.V., Kudryavtsev, P.S.: The method of selection of the contour of the image of the lungs on a chest X-ray. J. Proc. Southwest State Univ. Ser. Comput. Sci. Comput. Eng. Manag. **4**, 107 (2014)
2. Dyudin, M.V., Filist, S.A., Kudryavtsev, P.S.: The method of isolation and classification of the contours of the lungs on the images of chest X-ray. High Technol. **15**(12), 25–30 (2014)
3. Kudryavtsev, P.S., Kuzmin, A.A., Savinov, D.Y., et al.: Simulation of morphological formations on chest radiographs in intelligent diagnostic systems for medical purposes. Casp. J. Manag. High Technol. **3**(39), 109–120 (2017)
4. Tomakova, R.A., Filist, S.A., Yemelyanov, S.G.: Theoretical bases and methods of processing and analysis of microscopic images of biomaterials. SWSU, Kursk (2011)
5. Tomakova, R.A., Filist, S.A.: The method of processing and analyzing complexly structured images based on the built-in functions of the MATLAB environment. J. Proc. Chita State Univ. **1**(80), 3–9 (2012)
6. Filist, S.A., Tomakova, R.A., Yemelyanov, S.G.: Intellectual technologies of segmentation and classification of biomedical images. SWSU, Kursk (2012)
7. Filist, S.A., Tomakova, R.A., Degtyarev, S.V., et al.: Hybrid intelligent models for segmentation of chest radiograph images. Med. Equip. **5**(305), 41–45 (2017)
8. Tomakova, R.A., Filist, S.A., Pykhtin, A.I.: Development and research of methods and algorithms for intelligent systems for complex structured images classification. J. Eng. Appl. Sci. **12**, 6039–6041 (2017)

9. Wilkie, J.R., Giger, M.L., Chinander, M.R.: Comparison of radiographic texture analysis from computed radiography and bone densitometry systems. Med. Phys. **31**, 882–891 (2004)
10. Filist, S.A., Tomakova, R.A., Shatalova, O.V., et al.: Classification method of complex structured images based on self-organizing neural network structures. Radiopromyshlennost, Moscow (2016)
11. Filist, S.A., Tomakov, R.A., Rudenko, V.V.: Fuzzy network model of the morphological operator for the formation of segment boundaries. Belgorod State Univ. Sci. Bull. Econ. Inf. Technol. **1**(96), 188–195 (2011)
12. Kudryavtsev, P.S., Kuzmin, A.A., Filist, S.A.: X-ray image correction method based on taking into account global information about their structure. In: The 12th International Scientific Conference Physics and Radio Electronics in Medicine and Ecology, vol. 1, pp. 160–164 (2016)
13. Belobrov, A.P., Borisovsky, S.A., Tomakova, R.A.: Neural network models of morphological operators for segmentation of images of biomedical signals. Izvestiya SFedU. Eng. Sci. **109**(8), 28–32 (2010)
14. Dyudin, M.V., Kudryavtsev, P.S., Podmasteryev, K.V., et al.: Mathematical models for intelligent classification systems of chest radiographs. J. Proc. Southwest State Univ. Ser. Comput. Sci. Comput. Eng. Manag. **2**(19), 94–107 (2016)
15. Belykh, V.S., Efremov, M.A., Filist, S.A.: Development and a research of a method and algorithms for the intellectual systems of classification of hardly structured images. J. Proc. Southwest State Univ. Ser. Comput. Sci. Comput. Eng. Manag. **2**(19), 94–107 (2016)
16. Dyudin, M.V., Zuev, I.V., Filist, S.A.: Automatic classifiers of highly structured images based on multi-method multi-criteria selection technologies. Questions of radio electronics. Series Systems and means of information display and control of special equipment **1**, 130–140 (2015)
17. Filist, S.A., Kudryavtsev, P.S., Kuzmin, A.A.: Development of boosting methodology for the classification of chest x-rays. Biomed. Electron. **9**, 10–15 (2016)
18. Malyutina, I.A., Kuzmin, A.A., Shatalova, O.V.: Methods and algorithms for the analysis of chest radiographs using local windows in pathology detection tasks. Casp. J. Manag. High Technol. **3**(39), 131–138 (2017)
19. Filist, S.A., Kabus, A., Kuzmin, A.A.: Formation of attribute space for classification problems of complex-structured images based on spectral windows and neural network structures. J. Proc. Southwest State Univ. Ser. Comput. Sci. Comput. Eng. Manag. **4**(67), 56–68 (2016)
20. Tomakova, R.A., Filist, S.A., Pykhtin, A.I.: Automatic fluorography segmentation method based on histogram of brightness submission in sliding window. Int. J. Pharm. Technol. **1**, 28220–28228 (2017)
21. Tomakova, R.A., Filist, S.A., Durakov, I.V.: Software for automatic classification of chest radiographs based on hybrid classifiers. Hum. Ecol. **6**, 59–64 (2018)
22. Dyudin, M.V., Tomakova, R.A., Tomakov, M.V.: Neural network decision-making models for the diagnosis of lung diseases based on the analysis of chest x-ray fluorograms. Biomed. Electron. **9**, 12–16 (2014)

Developing a Technical Diagnostic Systems for Internal Combustion Engines

L. A. Galiullin[✉] and R. A. Valiev

Kazan Federal University, 18 Kremlyovskaya street,
Kazan 420008, Russian Federation
newcastle2017@yandex.ru

Abstract. This article describes the methods of developing a technical diagnostic system for internal combustion engines. The automotive industry plays a leading role in the economy of any state. The history of the global automotive industry development is closely linked with the development of many branches of engineering. Thus, by the beginning of the 20th century, the automobile industry began to consume half of the steel and iron produced, three-quarters of rubber and leather, a third part of nickel and aluminum, and a seventh part of wood and copper. Autobuilding came in first place in terms of production among other branches of engineering, it began to seriously impact the national economies. By the beginning of World War I, the global number of cars reached about 2 million. Of these, 1.3 million were in the USA, 245 thousand in England, 100 thousand in France, 57 thousand in Austria-Hungary, 12 thousand in Italy, 10 thousand in the Russian Empire.

Keywords: Engine · Diagnostic · Model · Information system

1 Introduction

By the mid-20s, the number of cars exceeded 30 million cars, of which 20 million in the United States, one million in England and France, half a million in Germany, and 100–150 thousand each in several capitalist countries. The share of trucks and buses averaged 15% (up to 25% in Europe). Each car had five residents in the USA, forty to sixty in other countries. In the USA, about 150 workers and employees were required for every thousand cars produced annually, in Europe: 500–600 people. Millions of man-years of labor and a significant amount of materials and fuel were spent on the production and maintenance of automobiles.

According to the data of the International Union of Automotive Engineers, by the beginning of the XXI century, the number of passenger cars in the world exceeded 520 million units. Estimates show that the average global passenger car with a mass of 1,390 kg has a mileage of about 190 thousand km and a service life of 12 years. At the same time, over the full term of its operation, such a vehicle consumes more than 14 thousand liters of fuel, more than 200 L of oil, more than 2 tons of solid waste, about 200 kg of liquid waste and emits more than 50 tons of total emissions into the atmosphere. In this case, the main contribution is made by gasoline internal combustion engines, which are in personal use.

A. A. Radionov and A. S. Karandaev (Eds.): RusAutoCon 2019, LNEE 641, pp. 797–805, 2020.
https://doi.org/10.1007/978-3-030-39225-3_87

At present, environmental pollution is becoming increasingly important, where the share of road transport accounts for more than half of the harmful effects.

The composition of the exhaust gases (exhaust gases) of engines for 99.0–99.9% consists of the products of complete combustion of fuel (carbon dioxide and water vapor), unused oxygen and nitrogen of the air. But it is the remaining part of the exhaust gases (no more than 1% of the total exhaust gas flow) that determines the ecological level of the engines, i.e. degree of harmful effects on the environment.

Some components of the exhaust gas that have a harmful effect on the environment: nitrogen oxides NO_x, carbon monoxide CO, hydrocarbons C_nH_m, aldehydes RCHO, sulfur compounds SO_2 and SO_3, soot C, lead oxides PbO etc.

All of these substances (except soot) when they reach a certain concentration in the air can be fatal. Although, of course, the degree of harmfulness of these substances is different and, accordingly, their permissible concentration in the air is different.

Carbon dioxide CO_2 emissions from exhaust gases contribute to the development of the greenhouse effect - the difference between the average temperature of the planet's surface and its radiation temperature in space. Decrease in concentration CO2 in exhaust gases may be reduced by reducing fuel consumption.

Along with environmental safety, the reduction of the level of fuel and energy consumption becomes an equally important problem. The share of road transport accounts for not only the bulk of harmful emissions, but also more than half of the resources consumed by the main modes of transport.

Another equally important problem is traffic safety and operational safety for human life, due to the use of highly flammable materials, a high degree of compression of the working mixture.

The development of modern engine-building occurs along the path of improving the economic, environmental and operational performance of engines. This is due, primarily, with the use of electronic control systems - fuel injection and ignition control systems, which can significantly reduce the toxicity and energy consumption of vehicles. In addition, for example, according to AUDI, one of the priority areas for development is the improvement of technical diagnostics tools.

2 Methods and Tools for Diagnosing Internal Combustion Engines

The limited ability to continuously monitor the technical condition of the engine and the vehicle as a whole directly during operation leads to the fact that developing defects and faults are detected only when they manifest themselves to a significant degree [1]. Indeed, the addresses of the majority of vehicle owners at the service station occur already after the detection of malfunctions related mainly to such signs as increased fuel consumption, extraneous sounds during engine operation, reduced power (acceleration dynamics), inability to start the engine, instability work on a particular mode, increasing smoke or the smell of exhaust gases.

Along with the inability to continuously monitor the technical condition of the engine, the situation is aggravated by the fact that most motorists are unavailable for maintenance and repair at dealerships or large service stations equipped with modern, expensive diagnostic tools, due to the predominance of small garage-type enterprises. Equipping the latter with technical means leaves much to be desired, as well as the qualifications of the service personnel, who often do not have special knowledge in the field of the design and operation of the internal combustion engine. This is due, first of all, to the high complexity of the internal combustion engine as a system, and hence its mathematical models [2, 3].

At the same time, the efficiency of internal combustion engines, their fuel efficiency, reliability and durability, power and environmental performance largely depend on the quality of the workflow in the engine cylinder, which is influenced not only by the piston group and timing, but also by the fuel equipment ignition and electronic control systems.

The main malfunctions of these systems in varying degrees affect the working process of the engine. Worn parts of the cylinder-piston group and the gas-distributing mechanism, loss of compression in the cylinders lead to deterioration of the processes of fuel formation and combustion of the fuel-air mixture, which, in turn, slows down the pre-flame reactions, the ignition delay period increases, and a significant portion of the working mixture burns out in tact extensions. As a result, reduced efficiency, environmental performance.

Failure of the fuel system also leads to an increase in fuel consumption and the content of harmful substances in the exhaust gas - a decrease in pressure due to a malfunction of the high-pressure fuel pump, improper operation of the pressure regulator, high pollution of the spray nozzles.

The state of the sensors and mechanisms of the electronic control system has a significant impact on the operation of the engine.

Thus, the preservation of the normal performance of automotive engines in operation is largely determined by timely and high-quality maintenance using modern tools and diagnostic methods.

Diagnostic methods are classified depending on the nature and physical essence of the recognizable features and the measured parameters of the technical state of the objects.

Analysis of the literature and patent search showed that currently there are a number of methods and tools for diagnosing an internal combustion engine, which are based on various aspects and patterns of operation of the internal combustion engine and its systems. In this regard, it is possible to identify the following fixed assets and methods based on:

- Measuring and analyzing vibroacoustic oscillations [4–6]
- Measuring and analyzing composition of the exhaust gases [7]
- Measuring and analyzing the composition of the oil [8, 9]
- Measuring and analyzing the pressure of crankcase gases
- Measuring and analyzing optical radiation of the ignition system [10, 11]
- Measuring and analyzing mechanical stresses in bolts and studs of the engine [12]
- Measuring and analyzing irregularity of rotational speed of the engine crankshaft [13–16]
- Measuring and analyzing the starter current when the engine is started [17]
- Misfire measurement and analysis [18]

- Measuring and analyzing the ion current of the ignition system [19, 20]
- Measuring and analyzing power and technical and economic characteristics [21–24]

Engine testers and integrated diagnostics systems should be singled out as a separate group. The latter are widely used in modern ICEs equipped with an electronic control system, the functional part of which is the monitoring of the status of sensors and self-diagnostics. When a malfunction is detected, the system memorizes the characteristic error code, and on the dashboard a light comes on, informing the driver of the occurrence of a malfunction. The same class of devices includes scanners - devices that interact with the control system through a special protocol that allows you to receive and monitor on the monitor the main parameters of the engine and control system, such as engine speed, fuel injection time, air consumption, advance angle ignition, condition of actuators, etc. The world leader in the development and production of scanners is the German company BOSCH.

The advantage of this approach is that the scanner allows you to look at the operation of the engine control system with the eyes of the control unit itself. After all, part of the displayed parameters are primary information from sensors, on the basis of which the unit produces control actions. Also, the advantages of such devices include the ability to graphically display the parameters of the engine in the form of their graphs of changes over time, as well as compactness.

The disadvantages of devices that implement the capabilities of in-line diagnostics include a limited number of monitored parameters—no more than what the developers laid into the control system, a low degree of universality. In the part of the criterion for determining malfunctions (self-diagnostics), the possibility of detecting faults at the high/ low value of the signal from the sensor is used, and the deviations from the signal from the norm should be maintained for a long time. Short-term deviations of the signals are not recorded.

Engine tester – class diagnostic devices are console devices equipped with their own sensors. Engine tester can measure a wide range of engine operating parameters regardless of the control system - rotational speed, ignition timing, uneven rotation, battery voltage, primary and secondary ignition system, etc. The world leaders in this field are Sun Electric and BOSCH concerns. Engine tester developed by them allows diagnosing up to 27 brands and 158 car systems, including Russian ones. The disadvantages of such systems are their high cost (up to several tens of thousands of dollars) and significant size.

Engine testers, like scanners, do not have the diagnostic capabilities of the engine itself.

The modern level of development of information technology and computer technology has determined the possibility of combining diagnostic devices of different classes into a single complex. Such systems can be equipped with a digital oscilloscope, for direct control of signals in electrical circuits, built-in expert systems for monitoring the deviation of parameters from the set, functions of test diagnostics.

Diagnostic tools with broader and more versatile capabilities include devices based on methods for measuring power and technical and economic characteristics. These characteristics include indicator diagrams and external speed characteristics.

Indicator diagram (ID) - a graphical representation of a set of thermodynamic processes that make up the working cycle of an internal combustion engine, in coordinates "pressure-volume", "pressure-temperature". There are theoretical and actual indicator diagrams. Theoretical ID is a mathematical model obtained from the calculated parameters of the working fluid at the end points of the processes. The actual ID is obtained from experimental studies of the real engine. Based on a comparison of the theoretical and actual ID, we can conclude about the nature of the flow of work processes in the engine cylinders, and, consequently, the conclusion about the state of the engine itself.

Obtaining a valid ID is associated with the removal of the engine from the car and installing it in a special stand on which the main indicators of the engine are measured. The obvious disadvantage of the method is an increase in the time and labor costs for diagnostics. Therefore, this approach is carried out mainly in the design and development of the engine. It is also worth noting the complexity of mathematical models of the internal combustion engine, as well as any theoretical studies using preliminary assumptions.

External speed characteristics (ESC) - the dependence of the main engine parameters (effective power, power loss, effective torque, fuel and air consumption, ignition timing) on the crankshaft rotational speed when the body controlling the fuel supply is stationary and unchanged.

ESC describes the engine as a dynamic system. It reflects such characteristics as elasticity, adaptability, efficiency, efficiency, efficiency.

The conclusion about the state of the engine is made on the basis of a comparison of the obtained characteristics with the characteristics inherent in a serviceable engine. In this case, the human operator (diagnostician) acts as an expert.

Diagnostic systems based on the measurement of ESC, were embodied in the load stands. At the same time the car is installed on special rolling rollers and fixed. The load generated by the electric motor-brake is transmitted through the drive wheels and the transmission to the engine. By changing the position of the body that controls the fuel supply (usually the throttle valve), the engine is accelerated from the minimum steady speed to the maximum. At the same time, the angular velocity of rotation of the motor shaft and the driving wheels is measured, on the basis of which the effective power and torque on the engine shaft and on the wheels, the effective power loss on the transmission is determined.

Such systems carry out engine diagnostics in a wide range of engine rotational speeds, allow you to simulate various load conditions (level road, climb uphill). This allows you to identify faults that do not manifest themselves or are weakly manifested in static modes of operation.

The disadvantages of such systems include inaccurate modeling of road conditions (for example, tire adhesion coefficients, air resistance) and incompleteness of the obtained characteristics, as well as the need to use a loading mechanism, which increases the cost of loading stands, and, consequently, the cost of diagnostics.

Test (break-in) stands are used mainly in the design and study of engines. In this case, the engine is fixed in a special frame, and the load is created by an electric motor. The effective power is determined based on the measurement of the electric currents of the electric motor. At the same time, it is possible to measure fuel and air consumption with the help of special sensors. When using a special cylinder head, it is possible to measure the pressure inside the cylinders and calculate indicator diagrams. Test benches allow determining various parameters of ICE operation in a wide range of rotational frequencies at various engine operating modes with high accuracy. But the use of such devices for diagnosis is difficult, due to the need to dismantle the engine from the car.

Systems based on the brake-free method described in [25] allow to avoid such drawbacks. The method is based on measuring the acceleration of free acceleration, when the load is the moment of resistance of the flywheel and the moving parts of the engine itself, reduced to the engine shaft, or the reduced mass of the vehicle during acceleration on one of the gears. This method does not require the use of expensive load stands, dismantling the engine or installing special sensors, which significantly reduces the time and labor costs for diagnostics.

The systems that implement the above method [26] are limited mainly to the determination of effective moments and power at the motor shaft. Moreover, their values are determined for some characteristic rotational speeds - for the nominal frequency and for the frequency at which the maximum torque is reached. This is primarily due to the need to calculate the angular acceleration of the motor shaft during its acceleration, that is, the need to numerically differentiate the values of the angular velocity of rotation, which leads to increased noise.

Also a difficult task is to determine the rotational speed, despite the fact that most of the processes in the engine are cyclical. The most widely used method, based on the calculation of the time of rotation of the engine at equal angular values, when measuring the primary or secondary voltage of the ignition system. The error of this method of determining the rotational speed is associated with a change in the ignition advance angle.

Thus, the development of a diagnostic system based on a brake-free method is a promising direction for monitoring and timely detection of deviations of parameters leading to a deterioration of the economic, environmental and effective performance of the engine.

Improving the accuracy of calculations can significantly increase the applicability of this method. In addition, the equipment of modern engines with electronic injection and ignition control systems makes it possible to complement the ESC set with functions for changing fuel consumption, air flow and ignition timing reflecting the operation of the control system itself.

3 Conclusion

The most complete picture of the technical condition of an internal combustion engine can be obtained by using comprehensive assessments that comprehensively reflect its state and complement each other. Therefore, for a reliable assessment of the technical condition of the engines, technical diagnosis systems are currently used.

The development of such a system is preceded by a statistical analysis of the most common failures of the engine, its components and systems; the study of physical processes leading to the occurrence of defects and their accompanying; formation of the object of diagnosis; development of a diagnostic method and algorithm taking into account functional or statistical relationships between the engine design, mode and diagnostic parameters, as well as taking into account the technical and economic possibilities of its practical use.

The block diagram of the engine diagnostics system can be presented in the form of the diagram shown in Fig. 1.

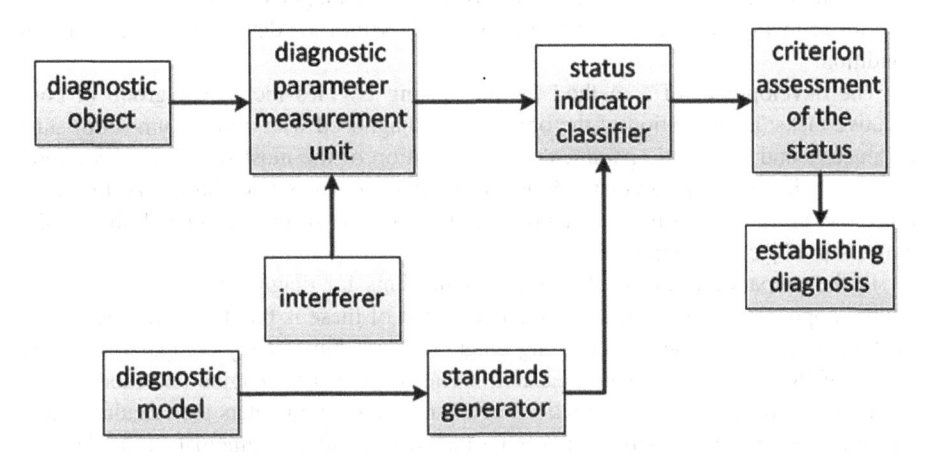

Fig. 1. Block diagram of the engine diagnosis system.

The layer determines the degree to which the numeric value corresponds to a fuzzy label. At this level, the base range and the number of fuzzy labels on it are defined, as well as the type of the membership function. The number of fuzzy labels depends on the accuracy of the control. The more of them, the more precise the control, but at the same time the time of filling the knowledge base increases. The source of technical condition signals (state informant) in the diagnostic system is the object of diagnosis. The measurement unit supplies the initial information about the state of the object contained in the measured signal. It includes the primary transducers of diagnostic signals from the object of diagnosis to their electrical equivalents, as well as amplifiers and devices for converting primary signals into unified signals for further processing. The signal carrying diagnostic information is more or less interfered with, which must be considered when developing a diagnostic algorithm.

Due to the fact that an internal combustion engine has many technical states, in fact, during diagnosis it is necessary to break this set into a finite number of recognizable classes of states, combining states with the same physical nature in each class. For example, it is possible to refer to one class the states of the engine characterized by defects of knots of the valvate mechanism, bearing knots of a cranked shaft and so on. This class of states is characterized not only by a single sign of the state, but also by a single method of diagnosis. Based on the statistical analysis of failures of the object being diagnosed, standards for each class of technical states are formed (the values of diagnostic parameters averaged for this class).

To form a system of diagnostic parameters and standards, diagnostic models of the object can be used, in some cases facilitating the process of searching for informative components in the signal under study.

The final element of the engine diagnostic system is the decision subsystem, which, based on the values of the diagnostic parameters, evaluates the technical condition of the engine and its elements (diagnosis) using various criteria.

The main tasks in creating a system of technical diagnostics are the choice of a diagnostic method and the construction of an algorithm for determining the technical condition.

The development of a method for diagnosing engines includes a group of consecutive tasks: a description of the object to be diagnosed with a minimum set of state parameters and diagnostic parameters; identification of the most sensitive to common defects of diagnostic parameters; division of technical states into classes. At the same time, the tasks of measuring diagnostic parameters, ensuring the operability of the diagnostic object are being solved.

In the preparation and application of algorithms for diagnosing engines, the following approaches can be distinguished. The first of these is that the measured values of the parameters of the engine being diagnosed are immediately compared with the values of the same parameters measured earlier on the same engine and mode. As a result of the comparison, the deviations of the measured parameters are calculated, and all further diagnostic operations involving the corresponding mathematical models are carried out with the indicated deviations.

The second approach is that, using the parameter values measured on the engine being diagnosed, the mathematical model calculates the values of other parameters that are not directly measured, which are compared with the values obtained earlier using the same or similar models for the same engine model in the same mode, and based on their analysis, diagnostic recommendations and decisions are made.

References

1. Galiullin, L.A., Valiev, R.A.: Modeling of internal combustion engines test conditions based on neural network. Int. J. Pharm. Technol. **8**(3), 14902–14910 (2016)
2. Galiullin, L.A., Valiev, R.A.: An automated diagnostic system for ICE. J. Adv. Res. Dyn. Control Syst. **10**, 1767–1772 (2018)
3. Galiullin, L.A., Valiev, R.A.: Method for neuro-fuzzy inference system learning for ICE tests. J. Adv. Res. Dyn. Control Syst. **10**, 1773–1779 (2018)
4. Galiullin, L.A., Valiev, R.A.: Modeling of internal combustion engines by adaptive network-based fuzzy inference system. J. Adv. Res. Dyn. Control Syst. **10**, 1759–1766 (2018)
5. Galiullin, L.A., Valiev, R.A.: Optimization of the parameters of an internal combustion engine using a neural network. J. Adv. Res. Dyn. Control Syst. **10**, 1754–1758 (2018)
6. Galiullin, L.A., Valiev, R.A., Mingaleeva, L.B.: Development of a neuro-fuzzy diagnostic system mathematical model for internal combustion engines. HELIX **8**, 2535–2540 (2018)
7. Galiullin, L.A., Valiev, R.A.: Mathematical modelling of diesel engine testing and diagnostic regimes. Turk. Online J. Des. Art Commun. **7**, 1864–1871 (2017)
8. Galiullin, L.A., Valiev, R.A.: Diagnosis system of internal combustion engine development. Rev. Publicando **4**(13), 128–137 (2017)

9. Galiullin, L.A., Valiev, R.A.: Diagnostics technological process modeling for internal combustion engines. In: International Conference on Industrial Engineering, Applications and Manufacturing, pp. 5648–5651 (2017). https://doi.org/10.1109/ICIEAM.2017.8076124

10. Galiullin, L.A., Valiev, R.A., Mingaleeva, L.B.: Method of internal combustion engines testing on the basis of the graphic language. J. Fundam. Appl. Sci. **9**, 1524–1533 (2017)

11. Galiullin, L.A.: Development of automated test system for diesel engines based on fuzzy logic. In: IEEE 2nd International Conference on Industrial Engineering, Applications and Manufacturing, pp. 1322–1325 (2017). https://doi.org/10.1109/ICIEAM.2016.7911582

12. Galiullin, L.A., Valiev, R.A.: Automation of diesel engine test procedure. In: IEEE 2nd International Conference on Industrial Engineering, Applications and Manufacturing, pp. 1325–1328 (2016). https://doi.org/10.1109/ICIEAM.2016.7910938

13. Galiullin, L.A.: Automated test system of internal combustion engines. In: International Scientific and Technical Conference Innovative Mechanical Engineering Technologies, Equipment and Materials-IOP Conference Series-Materials Science and Engineering, vol. 86, pp. 012–018 (2015)

14. Galiullin, L.A., Valiev, R.A.: Automated system of engine tests on the basis of Bosch controllers. Int. J. Appl. Eng. Res. **10**(24), 44737–44742 (2015)

15. Valiyev, R.A., Galiullin, L.A., Iliukhin, A.N.: Methods of integration and execution of the code of modern programming languages. Int. J. Soft Comput. **10**(5), 344–347 (2015)

16. Valiyev, R.A., Galiullin, L.A., Iliukhin, A.N.: Approaches to organization of the software development. Int. J. Soft Comput. **10**(5), 336–339 (2015)

17. Valiev, R.A., Galiullin, L.A., Dmitrieva, I.S., et al.: Method for complex web applications design. Int. J. Appl. Eng. Res. **10**(6), 15123–15130 (2015)

18. Valiyev, R.A., Galiullin, L.A., Iliukhin, A.N.: Design of the modern domain specific programming languages. Int. J. Soft Comput. **10**(5), 340–343 (2015)

19. Biktimirov, R.L., Valiev, R.A., Galiullin, L.A., et al.: Automated test system of diesel engines based on fuzzy neural network. Res. J. Appl. Sci. **9**(12), 1059–1063 (2014)

20. Zubkov, E.V., Galiullin, L.A.: Hybrid neural network for the adjustment of fuzzy systems when simulating tests of internal combustion engines. Russ. Eng. Res. **31**(5), 439–443 (2011)

21. Valiev, R.A., Khairullin, AKh, Shibakov, V.G.: Automated design systems for manufacturing processes. Russ. Eng. Res. **35**(9), 662–665 (2015)

22. Guihang, L., Jian, W., Qiang, W., et al.: Application for diesel engine in fault diagnose based on fuzzy neural network and information fusion. In: IEEE 3rd International Conference on Communication Software and Networks, pp. 102–105 (2011). https://doi.org/10.1109/ICCSN.2011.6014398

23. Yu, Y., Yang, J.: The development of fault diagnosis system for diesel engine based on fuzzy logic. In: Proceedings of 8th International Conference on Fuzzy Systems and Knowledge Discovery, pp. 472–475 (2011). https://doi.org/10.1109/FSKD.2011.6019556

24. Wei, D.: Design of web based expert system of electronic control engine fault diagnosis. In: Proceedings of International Conference on Business Management and Electronic Information, pp. 482–485 (2011). https://doi.org/10.1109/ICBMEI.2011.5916978

25. Shah, M., Gaikwad, V., Lokhande, S., et al.: Fault identification for I.C. engines using artificial neural network. In: Proceedings of International Conference on Process Automation, Control and Computing, pp. 764–769 (2011). https://doi.org/10.1109/PACC.2011.5978891

26. Li, X., Yu, F., Jin, H., et al.: Simulation platform design for diesel engine fault. In: Proceedings of the International Conference on Electrical and Control Engineering, pp. 4963–4967 (2011). https://doi.org/10.1109/ICECENG.2011.6057562

Automation of the ICE Testing Process

L. A. Galiullin[✉], R. A. Valiev, and D. I. Valieva

Kazan Federal University, 18 Kremlyovskaya Street,
Kazan 420008, Russian Federation
newcastle2017@yandex.ru

Abstract. The paper considers and automated system for testing diesel engines based on a neural network. A neuro-fuzzy system has been designed to form fuzzy rules for diesel engine control during its testing expansion of production of cars, tractors and their increasing role in meeting the needs of modern society leads to a continuous improvement of power units of cars - diesel engines. Declared capacity, economy, toxicity and other evaluation parameters of the diesel engine, as well as its reliability and durability, are established by testing in stand and operating conditions. Currently, all newly created, upgraded and serial engines of cars and tractors subject to different types of tests, the essence, volume and content of which is determined by their purpose and stipulated by GOST. Tests constitute the final stage of the complex process of creating and improving diesel engines. All kinds of new, modernized and serial engines are subjected to various types of tests in this connection.

Keywords: Engine · Automation · Test · Information system

1 Introduction

At present, there are a number of methods and means of testing an internal combustion engine, based on various aspects and patterns of its operation [1]. The most common are motor testers and integrated diagnostic systems. Devices that realize the possibility of self-diagnostics are aimed at detecting malfunctions of the electronic unit and sensors of the engine management system [2]. Motor-testers have a set of their own sensors, are able to measure various parameters of the engine systems and implement test modes of diagnostics. At the same time, these devices have weak capabilities for assessing the general state of ICE and are intended primarily to locate and locate faults or fault locations already after they occur.

The estimation of the general condition of the engine is made on the effective parameters of its operation, which include the effective torque and power on the motor shaft, fuel and air consumption, ignition timing, and harmful emissions in the exhaust gases [3]. The work of systems implementing this approach is based on braking and non-brazing methods.

Brake methods involve the use of special loading stands with running drums. This method was not widely used due to the high cost of equipment [4]. Non-brake methods are simpler and do not require the use of special braking devices. In this case, the angular acceleration is measured when the engine is accelerated without an external

A. A. Radionov and A. S. Karandaev (Eds.): RusAutoCon 2019, LNEE 641, pp. 806–815, 2020.
https://doi.org/10.1007/978-3-030-39225-3_88

load from a minimum stable speed to a maximum due to the sudden opening of the throttle (fuel pump rail - in diesel engines) [5]. This method allows to carry out diagnostics in real operating conditions, and equipping modern ICE with electronic control systems to increase the number of monitored parameters.

Disadvantages of systems implementing the best-mode method are low accuracy due to the need for numerical differentiation of the angular velocity variation, incompleteness and narrow range of the rotational frequencies of the obtained characteristics [6].

Thus, the task is to develop a test system for a modern internal combustion engine, taking into account its features and in real operating conditions, according to its high-speed characteristics, for the timely detection of deviations in parameters leading to deterioration in the economic, environmental and efficient performance of the engine [7].

This is achieved by solving the following tasks:

- Determination of the structure of the ICE test system for real operating conditions. Selection of the test mode
- Determination of the minimum required composition of sensors and actuators of the engine management system, whose signals contain the necessary information for determining the characteristics of ICE [8]
- Development of algorithms and information processing software that guarantee high accuracy in calculating the effective performance of the engine [9].

The tests allow to evaluate the quality of the diesel engine and compare its performance with the performance of other engines. In the process of testing determine the traction-dynamic, economic, environmental and other parameters of the engine and establish the compliance of these indicators with standards and technical conditions. During the tests, the peculiarities of this diesel are revealed, and comparing the results of tests of various types of engines, it is possible to evaluate the efficiency of design features, the quality of manufacture or their technical condition [10].

At present, testing of diesel engines is a complex and time-consuming technological process, which differs little from an experimental study. Therefore, automated testing systems (ATS) for engines are created.

In modern society, the problems of saving fuel and energy resources become acute. Road transport consumes more than 30% of produced petroleum products, and fuel costs make up about 20% of the cost of any product.

The pollution of the environment is also an acute problem, and motor transport accounts for more than half of the harmful effects.

The development of modern engine building is on the way to improving the economic, environmental and operational performance of engines. This is primarily due to the use of electronic control systems - fuel injection and ignition control, which can significantly reduce the toxicity and energy consumption of vehicles.

In the process of car operation, wear and aging of its components and assemblies inevitably results in deterioration of its economic, ecological and effective indicators. In this regard, in order to maintain the engine in good condition and to timely detect the deviation of parameters, leading to a deterioration in the economic, environmental and

efficient performance of its work, the leading place belongs to the system of maintenance and repair, its scientific validity and excellence. At the same time, technical
diagnostics is of great importance [11].

At present, there are a number of methods and means for diagnosing an internal
combustion engine (ICE), based on various aspects and patterns of its operation. The
most common are motor testers and systems built-in diagnostics.

The use of mathematical and table models for controlling the diesel engine during
the tests is not effective [12], since the process of creating such models is time-
consuming and time consuming, and such models have a high complexity of describing
the properties of the engine because of the complexity of real processes.

In general, the mathematical model of a diesel engine cannot describe the process
of operation of a particular engine because of the complexity of the processes occurring
in it [13]. In such cases, the solution to this problem can be reduced to isolating in the
object significant input and output characteristics and conducting a series of experiments to obtain data on the functioning of the object in special cases. The task of
controlling the engine in the process of its testing is the task of forecasting, as well as
non-linear and control.

To control the ATS of a diesel engine, it is thus more appropriate to use methods of
artificial intelligence. Among the methods of artificial intelligence, hybrid neural-fuzzy
modeling has been identified, which makes it possible to create a universal approxi-
mator of any "input-output" dependence [14].

Devices that realize the possibility of self-diagnostics are aimed at detecting malfunctions of the electronic unit and sensors of the engine management system.

Motor-testers have a set of their own sensors, are able to measure various
parameters of the engine systems and implement test modes of diagnostics.

At the same time, these devices have weak capabilities for assessing the general
state of the ICE and are intended primarily to search for and localize faults or failure
sites already after they occur.

Assessment of the general condition of the engine is necessary in the organization
of preventive works. At the same time, it is envisaged to monitor the technical condition of the engine with a specified periodicity. If during monitoring the actual value
of one of the parameters goes beyond the permissible limits, then only in this case
restoration work is carried out for localization and elimination of the malfunction. This
approach reduces the downtime of the vehicle and increases the reliability of its
operation [15].

The estimation of the general condition of the engine is made on the effective
parameters of its operation, which include the effective torque and power on the motor
shaft, fuel and air consumption, ignition timing, and harmful emissions in the exhaust
gases. The work of systems implementing this approach is based on braking and non-
brazing methods [16].

Brake methods involve the use of special loading stands with running drums. This
method was not widely used due to the high cost of equipment.

Non-brake methods are simpler and do not require the use of special braking
devices [17]. At the same time, the angular acceleration is measured when the engine is
accelerated without an external load from the minimum stable speed to the maximum
due to the sudden opening of the throttle.

This method allows to carry out diagnostics in real operating conditions, and equipping modern ICE with electronic control systems to increase the number of monitored parameters.

The disadvantages of systems implementing the best-mode method are the low accuracy associated with the need for the numerical differentiation of the angular velocity change function, the incompleteness and narrow range of the rotational frequencies of the received characteristics [18].

Thus, the task of developing a diagnostic system for a modern automotive internal combustion engine, taking into account its features, is topical and requires modern approaches when solving it [19].

2 Methods

The properties of the internal combustion engine, as a dynamic system, are described by a set of external speed characteristics (ESC) - the dependencies of the change in the main operating parameters on the rotational speed (angular velocity). The composition of the ESC includes the following characteristics (as a function of the rotational speed) [20]: fuel consumption; air consumption; effective power and torque developed on the motor shaft.

Distinguish complete and partial ESC. A full characteristic is obtained by ensuring the maximum filling of the engine cylinders with air (maximum fuel delivery) at a constant load on the shaft. Partial characteristics - respectively, with incomplete fuel supply. Important here is the fact that the values of the partial characteristics lie within the region bounded by the values of the total characteristic [21].

It should also be noted that in order to obtain adequate model, a necessary condition is the constancy of the position of the fuel supply control body. The magnitude of the fuel supply and the driver's control action is the degree of opening of the fuel pump rail in percent. This value is more convenient for perception and interpretation than the angle of rotation of the crankshaft in degrees [22].

The selection of the test mode is reduced to providing such an operating mode of the engine, in which its properties are presented more fully. This mode corresponds to the mode of full fuel supply [23]. This is due primarily to the maximum wide frequency range of the engine and the maximum work of inertial forces and frictional forces.

Thus, as a test mode, the engine operating mode is selected with full fuel delivery. At the same time, it is necessary to develop a decision-making mechanism on the sufficient provision of such a regime.

High degree of equipping of modern electronic control systems of the ICE work process by primary converters determines the saturation of information flows between the electronic control unit (ECU), sensors and actuators. ECU based on signals coming from the primary transducers (sensors), determines the mode of operation of the engine (idling, power mode) and forms control actions on the actuators.

Based on the composition of the ESC, it is possible to determine the list of sensor signals and actuators of the engine management system that contain the necessary information for the indirect evaluation of each parameter.

When studying ICE and constructing its mathematical model, as a rule, the problem arises of obtaining the law of the functioning of an object as a whole or some of its parts [24]. Most often, the model cannot be built on the basis of known regularities and the form of the object's functioning law is unknown. In such cases, the solution to this problem can be reduced to isolating in the object significant input and output characteristics and conducting a series of experiments to obtain data on the functioning of the object in special cases.

The implemented structure of ATS ICE can be divided into static and dynamic parts.

The static part of the ATS developed ICE takes a direct part in the process of testing ICE and does not change.

The dynamic part of the ATS ICE is used for setting the ICE test modes and for obtaining fuzzy knowledge base rules using the ANFIS hybrid neural networks.

The construction of the base of the rules of fuzzy inference in this scheme is connected with certain difficulties of conceptual nature. the task of controlling ICE in the process of its testing is the task of forecasting, as well as non-linear and control.

To solve this problem, it is proposed to use hybrid neural networks to configure fuzzy systems. The merits of models built on the basis of neural networks include the possibility of obtaining new information in the form of a prediction [25]. For example, the forecast of the control vector of the test of an unknown ICE model.

Fuzzy logic can be used to control the engine during testing. The determination of the control effect is carried out with the help of fuzzy rules.

To obtain fuzzy rules, experts can be attracted, or it is possible to automatically create fuzzy rules using an unclear neural network such as ANFIS (adaptive network for fuzzy inference system). The basic idea behind the model of fuzzy neural networks is to use the existing data sample to determine the parameters of the membership functions of input and output variables to fuzzy sets.

The diesel test program includes a description of the engine operating modes. The modes of operation of the engine are determined by the type of tests [26]. The task of ATS is to maintain the current mode of the engine in accordance with the test program by developing a control action.

The input parameters are presented in the test program in the form of their changes in time, for example Fig. 1.

Let the input parameters for the ATS in the test program are given: the speed of the crankshaft – n (min^{-1}), torque – M_H (Nm), hourly fuel consumption – G_T (kg/hr). However, it is possible to specify more input parameters.

The test program of the diesel engine is formed in accordance with GOST 18509 88 "Diesel tractors and combine harvesters. Methods of bench tests". This standard assumes the measurement of the parameters of the engine in a stationary mode. The running time of the engine with the specified parameters depends on the type of tests and is specified in the test program.

All graphic images representing changes in the input parameters of the diesel engine in time are reduced to a tabular form (Table 1).

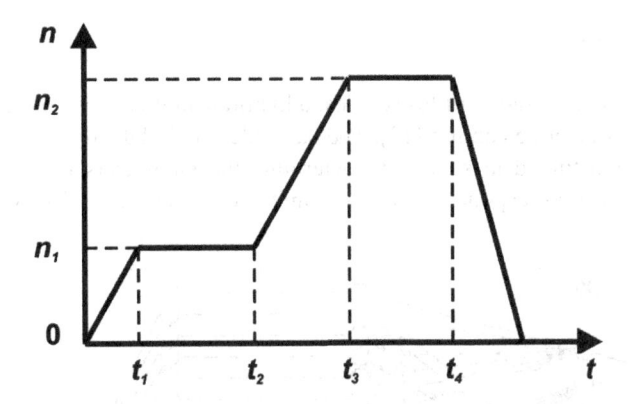

Fig. 1. An image of tests on turns of a cranked shaft n.

Table 1. Input vector.

№	Test time, min	n, rpm	MH, Nm	GT, kg/hr
Engine start				
1	0–15	700	72	18
2	15–30	900	91	19
3	30–45	1100	108	22
4	45–60	1300	120	33
5	60–75	1500	124	37
6	75–90	1700	122	40
7	90–105	1900	117	47
8	105–120	2100	113	51
9	120–135	2300	108	52
10	135–150	2500	111	56
11	165–180	700	72	18
Engine shutdown				

Control action, allowing to maintain the specified test mode of the diesel, is the displacement of the rail of the fuel pump of high pressure (HP pump), h. By changing the value of h at each particular moment of time, the ATS maintains the engine operating mode specified by the input vector.

In work [26], the definition of the control effect on the diesel engine is made with the help of fuzzy rules for the management of the ATS knowledge base, which have the form.

Fuzzy rules can be set by experts, can be obtained by direct measurement or with the help of a self-learning neural network. In the latter case there is a complete automation of the filling process of the ATS database of the diesel engine.

3 Conclusion

For the automatic generation of fuzzy rules, a hybrid neural network must be designed, which is a multi-layer perceptron [13]. The basic idea behind the hybrid neural network is to use the existing data sample to determine the parameters of the membership functions that best correspond to the fuzzy inference system (Fig. 2).

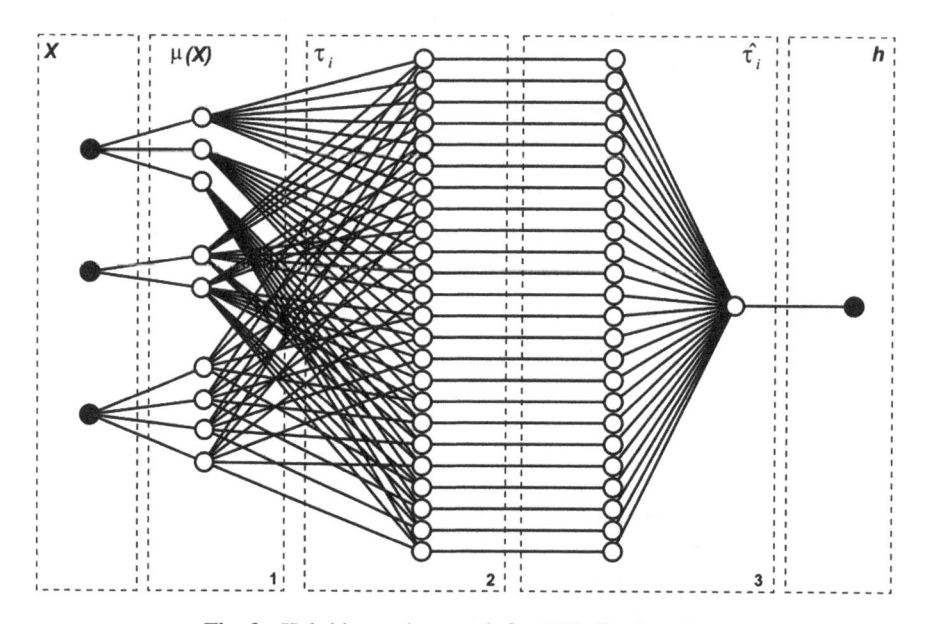

Fig. 2. Hybrid neural network for ATS diesel engine.

The layer determines the degree to which the numeric value corresponds to a fuzzy label. At this level, the base range and the number of fuzzy labels on it are defined, as well as the type of the membership function. The number of fuzzy labels depends on the accuracy of the control. The more of them, the more precise the control, but at the same time the time of filling the knowledge base increases.

To obtain the required control accuracy, 3 fuzzy labels for the parameter n, 2 fuzzy labels for the parameter M_H and 4 fuzzy labels for the parameter G_T.

Layer 2 determines the degree to which the values of the input parameters correspond to the conditions of the rules:

The layer's neurons combine the fuzzy sets of input parameters to form the right-hand side of fuzzy rules.

Layer 3 generates a clear output value.

The signal at its output is the sum of the products of the weights $w_c^{(k)}$ and the normalized degrees of activity of the rules $\hat{\tau}_k$. Weights of bonds $w_c^{(k)}$ correspond to the conclusion $c^{(k)}$ in fuzzy rules. Prior to training, they have zero initial values, which reflects the fact that there are no exceptions before the training of the network.

This layer determines the degree to which the values of the input signals match the conditions of the rules. The outputs of this layer represent the normalized degrees of activity of the rules.

The total number of fuzzy rules, which should be obtained with 3 fuzzy labels of the parameter n, 2 fuzzy parameter labels M_H and 4 fuzzy parameter labels G_T, equals $3 \cdot 2 \cdot 4 = 24$. In the layer 2, 24 neurons are determined, respectively (Fig. 2).

For a given input vector (Table 1), the hybrid network after learning by the method of back propagation of the error received a control vector.

According to the obtained values n, M_H and G_T built external speed characteristics (ESC) (Fig. 3).

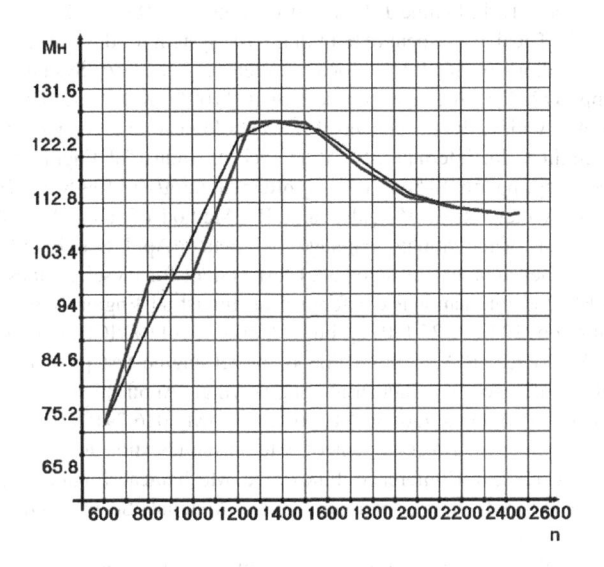

Fig. 3. ESC of the moment of loading.

The errors in the calculated indicators are due to the appropriate selection of the training sample, which was used in the training of the hybrid network.

The maximum relative error of control of the same diesel engine in terms of engine speed is 0.4% at 800–1500 rpm, which is due to the nonlinearity of the engine speed in this section.

Neuro-fuzzy ATSs allows to automate the technological process of testing diesel engines, which is one of the main tasks of increasing the technological level of production and the quality of the produced engines.

The implementation of the diesel control model based on the hybrid neural-fuzzy network in the ATS will increase the productivity of the technologist-testers and ensure the required accuracy of diesel control during testing.

References

1. Galiullin, L.A., Valiev, R.A., Mingaleeva, L.B.: Development of a neuro-fuzzy diagnostic system mathematical model for internal combustion engines. HELIX **8**, 2535–2540 (2018)
2. Galiullin, L.A., Valiev, R.A.: An automated diagnostic system for ICE. J. Adv. Res. Dyn. Control Syst. **10**, 1767–1772 (2018)
3. Galiullin, L.A., Valiev, R.A.: Method for neuro-fuzzy inference system learning for ICE tests. J. Adv. Res. Dyn. Control Syst. **10**, 1773–1779 (2018)
4. Galiullin, L.A., Valiev, R.A.: Optimization of the parameters of an internal combustion engine using a neural network. J. Adv. Res. Dyn. Control Syst. **10**, 1754–1758 (2018)
5. Galiullin, L.A., Valiev, R.A.: Mathematical modelling of diesel engine testing and diagnostic regimes. Turk. Online J. Des. Art Commun. **7**, 1864–1871 (2017)
6. Yu, Y., Yang, J.: The development of fault diagnosis system for diesel engine based on fuzzy logic. In: Proceedings of 8th International Conference on Fuzzy Systems and Knowledge Discovery, pp. 472–475 (2011). https://doi.org/10.1109/FSKD.2011.6019556
7. Galiullin, L.A., Valiev, R.A.: Diagnostics technological process modeling for internal combustion engines. In: International Conference on Industrial Engineering, Applications and Manufacturing, pp. 5648–5651 (2017). https://doi.org/10.1109/ICIEAM.2017.8076124
8. Galiullin, L.A., Valiev, R.A., Mingaleeva, L.B.: Method of internal combustion engines testing on the basis of the graphic language. J. Fundam. Appl. Sci. **9**, 1524–1533 (2017)
9. Galiullin, L.A.: Development of automated test system for diesel engines based on fuzzy logic. In: IEEE 2nd International Conference on Industrial Engineering, Applications and Manufacturing, pp. 1322–1325 (2017). https://doi.org/10.1109/ICIEAM.2016.7911582
10. Galiullin, L.A., Valiev, R.A.: Automation of diesel engine test procedure. In: IEEE 2nd International Conference on Industrial Engineering, Applications and Manufacturing, pp. 1325–1328 (2016). https://doi.org/10.1109/ICIEAM.2016.7910938
11. Galiullin, L.A.: Automated test system of internal combustion engines. In: International Scientific and Technical Conference Innovative Mechanical Engineering Technologies, Equipment and Materials. IOP Conference Series - Materials Science and Engineering, vol. 86, pp. 012–018 (2015)
12. Galiullin, L.A., Valiev, R.A.: Automated system of engine tests on the basis of Bosch controllers. Int. J. Appl. Eng. Res. **10**(24), 44737–44742 (2015)
13. Valiyev, R.A., Galiullin, L.A., Iliukhin, A.N.: Methods of integration and execution of the code of modern programming languages. Int. J. Soft Comput. **10**(5), 344–347 (2015)
14. Valiyev, R.A., Galiullin, L.A., Iliukhin, A.N.: Approaches to organization of the software development. Int. J. Soft Comput. **10**(5), 336–339 (2015)
15. Valiev, R.A., Galiullin, L.A., Dmitrieva, I.S., et al.: Method for complex web applications design. Int. J. Appl. Eng. Res. **10**(6), 15123–15130 (2015)
16. Valiyev, R.A., Galiullin, L.A., Iliukhin, A.N.: Design of the modern domain specific programming languages. Int. J. Soft Comput. **10**(5), 340–343 (2015)
17. Galiullin, L.A., Valiev, R.A.: Modeling of internal combustion engines test conditions based on neural network. Int. J. Pharm. Technol. **8**(3), 14902–14910 (2016)
18. Biktimirov, R.L., Valiev, R.A., Galiullin, L.A., et al.: Automated test system of diesel engines based on fuzzy neural network. Res. J. Appl. Sci. **9**(12), 1059–1063 (2014)
19. Zubkov, E.V., Galiullin, L.A.: Hybrid neural network for the adjustment of fuzzy systems when simulating tests of internal combustion engines. Russ. Eng. Res. **31**(5), 439–443 (2011)
20. Valiev, R.A., Khairullin, AKh, Shibakov, V.G.: Automated design systems for manufacturing processes. Russ. Eng. Res. **35**(9), 662–665 (2015)

21. Galiullin, L.A., Valiev, R.A.: Diagnosis system of internal combustion engine development. Rev. publicando **4**(13), 128–137 (2017)
22. Guihang, L., Jian, W., Qiang, W., et al.: Application for diesel engine in fault diagnose based on fuzzy neural network and information fusion. In: IEEE 3rd International Conference on Communication Software and Networks, pp. 102–105 (2011). https://doi.org/10.1109/ICCSN.2011.6014398
23. Wei, D.: Design of web based expert system of electronic control engine fault diagnosis. In: Proceedings of International Conference on Business Management and Electronic Information, pp. 482–485 (2011). https://doi.org/10.1109/ICBMEI.2011.5916978
24. Shah, M., Gaikwad, V., Lokhande, S., et al.: Fault identification for I.C. engines using artificial neural network. In: Proceedings of International Conference on Process Automation, Control and Computing, pp. 764–769 (2011). https://doi.org/10.1109/PACC.2011.5978891
25. Galiullin, L.A., Valiev, R.A.: Modeling of internal combustion engines by adaptive network-based fuzzy inference system. J. Adv. Res. Dyn. Control Syst. **10**, 1759–1766 (2018)
26. Li, X., Yu, F., Jin, H., et al.: Simulation platform design for diesel engine fault. In: Proceedings of the International Conference on Electrical and Control Engineering, pp. 4963–4967 (2011). https://doi.org/10.1109/ICECENG.2011.6057562

Noise-Robust Method to Determine Speech Prosodic Characteristics to Assess Human Psycho-Emotional State in Free Motor Activity

A. K. Alimuradov[(⊠)], A. Yu. Tychkov, and P. P. Churakov

Penza State University, 40 Krasnaya Street, Penza 440026, Russian Federation
alansapfir@yandex.ru

Abstract. The paper proposes noise-robust method to determine prosodic characteristics of speech. The nature of the method consists in decomposing a speech signal into informative noise and signal frequency components by means of improved complete ensemble empirical mode decomposition with adaptive noise, and selection of the component containing the pitch with the subsequent determination of prosodic characteristics. Decomposition used in the method does not need any a priori information (presetting) on the analyzed noisy speech signals. The study was conducted using a verified base of pure and noisy speech signals recorded in 220 males and females, aged 18 to 79 years, with signs of psycho-emotional disorders. The results show that the algorithm provides good noise robustness of various intensities (signal-to-noise ratio is from 0 dB to 30 dB) and can be successfully tested in actual conditions of an "aggressive" noise environment for assessment of a human psycho-emotional state in free motor activity.

Keywords: Speech signal · Noise-robust processing · Noise robustness · Decomposition · Prosodic characteristics · Psycho-emotional state

1 Introduction

The use of speech as an interface of man-machine interaction has gained a wide practical popularity in various information systems (voice control, voice identification, voice assistant, etc.). In the last decade, assessment of a human psycho-emotional state by speech has been actively developed. This is especially important in the areas of human activity that are fraught with the risk of emergency situations: pilots of civil and military aviation, airport dispatchers, dispatchers at hazardous production facilities (chemical and nuclear industry), employees of special services (police, military), etc.

There are commercial versions of such systems in the speech technology market [1, 2], allowing to assess a human emotional state in an automated mode with a certain accuracy.

Speech is a complex acoustic non-stationary signal generated by the articulation department of the speech apparatus for the purpose of communication, through certain language constructs. Speech is very sensitive to impaired functioning of the nervous system and, depending on the psychological situation, "encodes" the emotional state of a person into certain relevant informative parameters.

A. A. Radionov and A. S. Karandaev (Eds.): RusAutoCon 2019, LNEE 641, pp. 816–826, 2020.
https://doi.org/10.1007/978-3-030-39225-3_89

In the conditions of human free motor activity and current "aggressive" noise environment, all speech signals are to some degree noisy. Depending on the intensity and type, noise can significantly distort the results of the assessment of a human psycho-emotional state. To improve noise robustness of psycho-emotional state assessment systems, it is urgent to create novel speech processing methods that are adaptive to modern noise environments.

Currently, the activity in the field of noise-robust speech processing is quite high. A large number of algorithms, methods, techniques and tools based on classical approaches have been developed: enhancing the clarity (adjustment) of speech, well known as spectrum subtraction methods, Wiener filters, Cepstral Mean Subtraction (CMS), Cepstral Variance Normalization (CVN), etc. [3–6]. Studies of existing filtering approaches have revealed that the residual noise problem has not been completely solved [7]. In conditions of aggressive noise environment with signal-to-noise ratio values of 10 dB, 5 dB and 0 dB, the known methods do not provide the required noise suppression. The reason is the use of non-adaptive methods for processing noisy speech signals, which need a priori information on noise (presetting).

Based on the analysis [7–9], a noise-robust speech signal processing method has been developed that is applicable in systems for assessing a psycho-emotional state in noisy environments [10].

In particular, the algorithm is designed to determine prosodic characteristics and is based on the adaptive technology for processing non-stationary data, namely, empirical mode decomposition (EMD) [11]. The EMD technology is used in various areas of research, including an assessment of a human psycho-emotional state [12, 13].

The article is the development of previously published works of the authors [9, 10], and is structured as follows. The second section briefly presents information on prosodic speech characteristics and the EMD technology. The third and fourth sections briefly describe and investigate the developed method. In the fifth section, the presented research results and conclusions are made.

2 Materials and Methods

2.1 Prosodic Characteristics

As noted earlier, the type and severity of a psycho-emotional state is encoded into relevant informative speech parameters, which can be divided into three conditional groups [14]: amplitude-frequency, spectral-temporal, and cepstral. Each group of informative parameters is relevant to specific features of speech signals. The amplitude-frequency parameters relevant to low and high psycho-emotional arousal include prosodic characteristics of speech [10].

Prosodic characteristics describe the melody of speech, its temporal and timbral features, rhythm, verbal tones and intonations (i.e., the background tone at the level of phrases). Speech signals can be divided into sections consisting of voiced speech, unvoiced speech, and pauses [15]. The first and second sections are formed as a result of periodic and non-periodic oscillations of the vocal cords, respectively. Pauses are formed during the period of tranquility of the speech apparatus in accordance with certain linguistic constructions. Periodic oscillations are called the pitch (P). The frequency of cord oscillations is the pitch frequency (PF), which is an important prosodic characteristic of speech.

In psycho-emotional arousal, oscillations of the vocal cords are characterized by irregularities resulting from incomplete closure at the beginning and at the end of voiced sections of speech. At extremely high and low arousal, the change in the PF can reach 30%–40% of the nominal value corresponding to a neutral psycho-emotional state.

The prosodic characteristics used in the algorithm, and most fully reflecting information about a human psycho-emotional state, include:

- The PF mean value (measured in Hz), which is the average value of the frequencies calculated for each P period in all voiced sections of the recorded speech:

$$f_{0,mean} = \frac{1}{P} \sum_{p=1}^{P} f_{0,p},$$
(1)

where f_0, mean is the PF mean value; p is the PF value at a certain period; $p = 1, 2, ..., P$ is the P period number.

- Maximum $\max(f_{0,p})$ and minimum $\min(f_{0,p})$ values of the PF (measured in Hz) in all voiced sections of the recorded speech

- The standard PF deviation (measured in Hz), which is the deviation (error) between the current and average PF value:

$$SD_{f0} = \frac{1}{P-1} \sum_{p=1}^{P} \left(f_{0,p} - f_{0,mean} \right)^2$$
(2)

- A range of background frequencies, representing a band from $\min(f_{0,p})$ to maximum $\max(f_{0,p})$ PF values on a logarithmic scale:

$$PFR = 12 \times \frac{\log\left(\frac{\max(f_{0,p})}{\min(f_{0,p})}\right)}{\log 2}$$
(3)

- A mean absolute value of jitter, which is the average value of the difference in absolute value between the current and previous PF values:

$$MAJ = \frac{1}{P-1} \sum_{p=P-1}^{1} |f_{0,p+1} - f_{0,p}|$$
(4)

- A jitter, which is the change value (vibration) of the P frequency modulation:

$$J = \frac{MAJ}{f_{0,mean}}$$
(5)

- The relative average perturbation of the PF (smoothed over three P periods), representing the ratio of the difference in the PF mean value during the three P periods to the PF mean value for all voiced sections of the recorded speech:

$$RAP = \frac{\frac{1}{P-2}\sum_{p=2}^{P-1} \left| \left(f_{0,p+1} + f_{0,p} + f_{0,p-1}/3 \right) - f_{0,p} \right|}{f_{0,mean}} \times 100 \qquad (6)$$

- Pitch perturbation quotient (smoothed over five P periods), representing the ratio of the difference in the PF mean value during the five P periods to the PF mean value for all voiced sections of the recorded speech:

$$PPQ = \frac{\frac{1}{P-4}\sum_{p=3}^{P-2} \left| \left(\sum_{k=p-2}^{p+2} f_{0,k}/5 \right) - f_{0,p} \right|}{f_{0,mean}} \times 100 \qquad (7)$$

2.2   An Adaptive EMD Technology

The results of detailed studies of decomposition technologies have revealed the promise of using the improved complete ensemble empirical mode decomposition with adaptive noise (CEEMDAN) [16], based on the classical empirical mode decomposition (EMD). EMD is a unique adaptive technology in the field of time-frequency analysis, which does not need any a priori information about the analyzed non-stationary signals. The principle of the EMD is to decompose a signal into a sum of functions with a limited band, called intrinsic mode functions (IMF), or modes. During decomposition, the signal model is not specified in advance, and an IMF is calculated during the elimination, taking into account local features (such as extremes and signal zeros), and the internal structure of each particular signal. Thus, an IMF does not have a rigorous analytical description, but should satisfy two conditions that guarantee a certain symmetry and narrowband basic functions:

- The total number of extremes equals the total number of zeros with an accuracy of one
- The average value of the upper envelope interpolating local maxima, and the lower envelope interpolating local minima, should be approximately zero

In addition to adaptability, the EMD technology has other important properties:

- locality, that is the ability to take into account local features of the signal
- Orthogonality, that is ensuring signal recovery with a certain accuracy
- Completeness, as a guarantee of a finite number of basic functions at a finite signal duration

An analytical expression of the EMD is as follows:

$$x(n) = \sum_{i=1}^{I} IMF_i(n) + r_I(n), \qquad (8)$$

where $x(n)$ is the original signal; $IMF_i(n)$ is the IMF; $r_I(n)$ is a residue; $i = 1, 2, ..., I$ is the IMF number; n is discrete timing ($0 < n \leq N$, N is a number of discrete samples in the signal).

A distinctive feature of the improved CEEMDAN is the addition of controlled noise to the original signal to create new extremes:

$$x_j(n) = x(n) + w_j(n), \tag{9}$$

where $x_j(n)$ are the noise copies of the speech signal; $w_j(n)$ is the implementation of white noise,

$$x_j(n) = \sum_{i=1}^{I} IMF_{ji}(n) + r_{jI}(n) \tag{10}$$

$$IMF_i(n) = \sum_{j=1}^{J} \frac{IMF_{ji}(n)}{J} \tag{11}$$

$$r_I(n) = \sum_{j=1}^{J} \frac{r_{jI}(n)}{J} \tag{12}$$

where $j = 1, 2, ..., J$ is the amount of white noise implementations.

The disadvantages of the existing types of decompositions are eliminated by the addition of controlled noise: mode mixing; lack of decomposition completeness (all received noise copies are decomposed independently without communication with each other); residual noise in the IMF; "parasitic" IMFs in the early stages of decomposition. A detailed analysis of the advantages and disadvantages of the EMD technologies for speech signal processing are considered in the works of the authors [8–10].

3   Description of the Method

Figure 1 presents a simplified block diagram for the noise-robust method to determine the prosodic characteristics of speech signals for assessing a human psycho-emotional state. The nature of the algorithm is to decompose the signal into informative noise and signal frequency components, and the selection of the P containing component followed by the determination of prosodic characteristics.

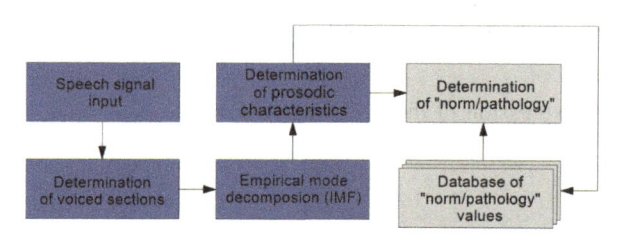

Fig. 1. Simplified block diagram for the noise-robust method to determine the prosodic characteristics of speech signals.

Let us consider briefly some of the processing steps of the method. A detailed mathematical description of the method and its functional operation is presented in [10].

Decomposition by the improved CEEMDAN method is the basis of the algorithm, which allows filtering a noisy speech signal in an automated mode. The EMD is a unique adaptive technology in the field of time-frequency analysis that does not need any a priori information (selection of the basic function, etc.) on the analyzed non-stationary signals.

Figure 2 presents the result of the decomposition of a voiced section of a speech signal "a" with duration of 100 ms using the improved CEEMDAN. Figure 2 shows that the voiced speech section is decomposed into eight IMFs. The first five IMFs are informative signals, and the last three are compensating trend ones.

The first IMF is of high frequency, and the subsequent modes are of low frequency (descending). If we assume that the noise component of the signal will have the highest frequency, then it will be extracted into the first IMF, which can be eliminated in subsequent processing steps. An idea of noise robustness is based on this principle of the improved CEEMDAN technology.

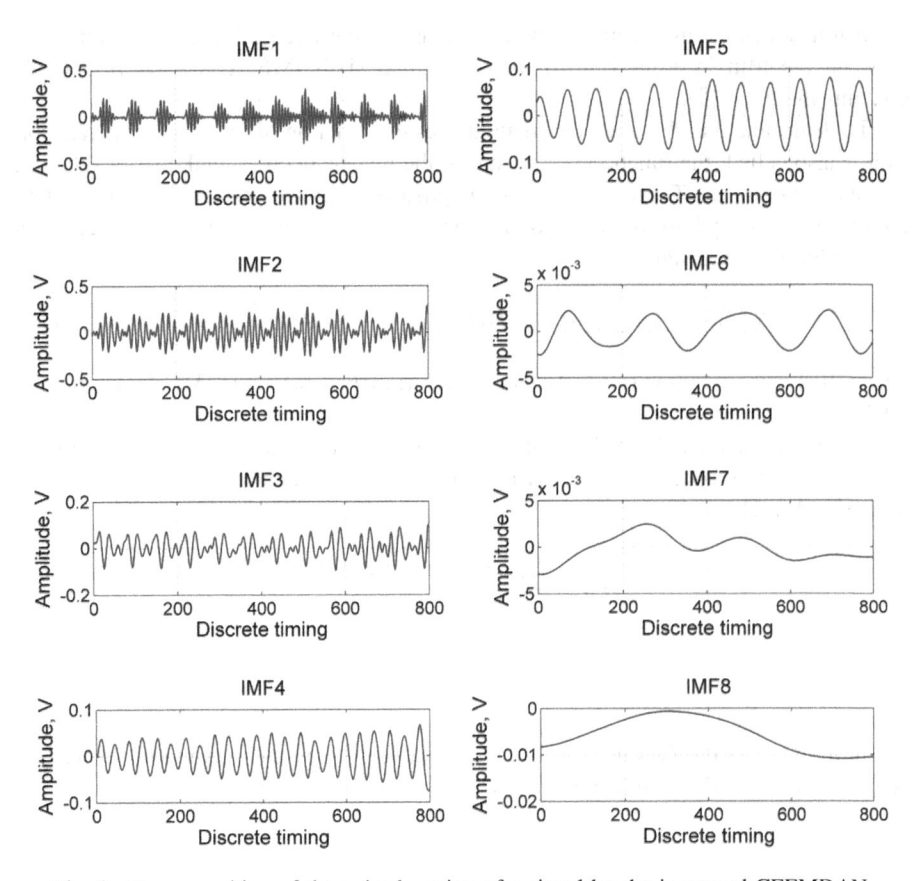

Fig. 2. Decomposition of the voiced section of a signal by the improved CEEMDAN.

Determination of the informative signal mode containing the pitch consists in sequent modulo calculating the difference between the values of the energy logarithms of the current and subsequent modes [10] (Fig. 4). The energy base 10-logarithm operation is used to compress the values of the IMF energy in a wide dynamic range (Fig. 3).

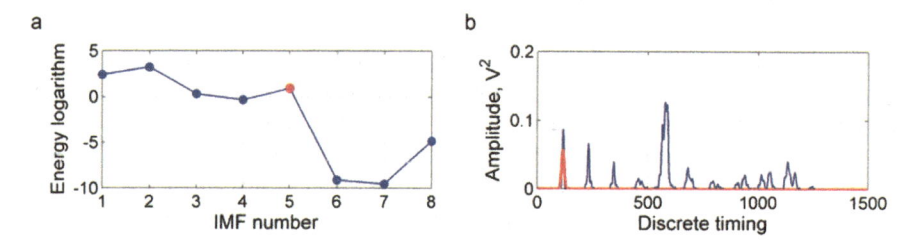

Fig. 3. Determination of an IMF containing the pitch: (a) IMF energy logarithm; (b) Fourier spectrum (voiced section is blue; an IMF containing the pitch is red).

It follows from Fig. 3, that a sharp decrease in the energy logarithm is observed between the fifth informative IMF and the trend sixth IMF. Thus, the fifth mode contains the pitch [9].

To determine the PF value, the method uses the function of the Teager operator or, as it is also called, the function of measuring the instantaneous signal energy. The use of the function of the Teager operator in the processing of speech signals is justified by the efficiency, simplicity of calculations, and good susceptibility to a sharp change in the amplitude of the signal:

$$T(n) = (IMF_{i,PF}(n))^2 - IMF_{i,PF}(n-1) \times IMF_{i,PF}(n+1), \tag{13}$$

where $T(n)$ is the function of the Teager operator; $IMF_{i,PF}(n)$ is the IMF containing the pitch.

Having calculated the values of the Teager operator function, the P period is determined in milliseconds by closely located maxima (through one), and then the PF is determined in Hz:

$$P_0 = \frac{T_{\max}(n+2) - T_{\max}(n)}{f_d} \tag{14}$$

$$f_0 = \frac{1}{P_0}, \tag{15}$$

where P_0 is the pitch; f_0 is the PF; $T_{\max}(n)$, $T_{\max}(n+1)$ are the maxima of the Teager operator function; f_d is the sampling rate.

4   Investigation of the Method

To test the developed algorithm, a group of subjects was formed with the support of K.R. Evgrafov Regional Psychiatric Hospital, and Medical Institute of Penza State University (Penza, Russian Federation). The group comprised 220 male and female subjects, aged 18 to 79 years with signs of psycho-emotional disorders. A control group of 220 conditionally healthy people was formed from the staff and teachers of Penza State University. In accordance with the method developed by the authors, a database of speech signals was registered. To determine the assessment effectiveness of a psycho-emotional state, errors of the first and second kind were used.

Noisy speech signals (signal-to-noise ratio is 30 dB, 20 dB, 10 dB, and 0 dB) were generated by white noise imposing on pure speech signal with different amplitudes.

Software implementation of the method was performed in ©Matlab (MathWorks) package for solving technical computing problems.

Table 1 presents the results of determination of a human psycho-emotional state for pure and noisy speech signals.

Table 1. Results of determination of psycho-emotional states.

Predictable result	Determination result, pers.		Errors, %	
	Pathology	Norm		
Pure speech signal				
Pathology	184	36	1st, α	16.36
Norm	18	202	2nd, β	8.19
Noisy signal, SNR = 30 dB				
Pathology	182	38	1st, α	17.27
Norm	22	198	2nd, β	10.00
Noisy signal, SNR = 20 dB				
Pathology	168	52	1st, α	19.10
Norm	31	189	2nd, β	14.09
Noisy signal, SNR = 10 dB				
Pathology	154	66	1st, α	30.00
Norm	45	202	2nd, β	20.45
Noisy signal, SNR = 0 dB				
Pathology	121	99	1st, α	45.00
Norm	58	157	2nd, β	26.36

5   Discussion and Results

It follows from the obtained results that the values of the 1st (α) and 2nd (β) kind errors for noisy signals are within the acceptable limits (<25% and <20%), including those for high values of noise intensity (SNR is 10 dB and 0 dB). Figure 4 (left column) shows the oscillograms of noisy voiced sections of a speech signal. The reliability of the obtained error values α and β is confirmed by the analysis of the spectral power

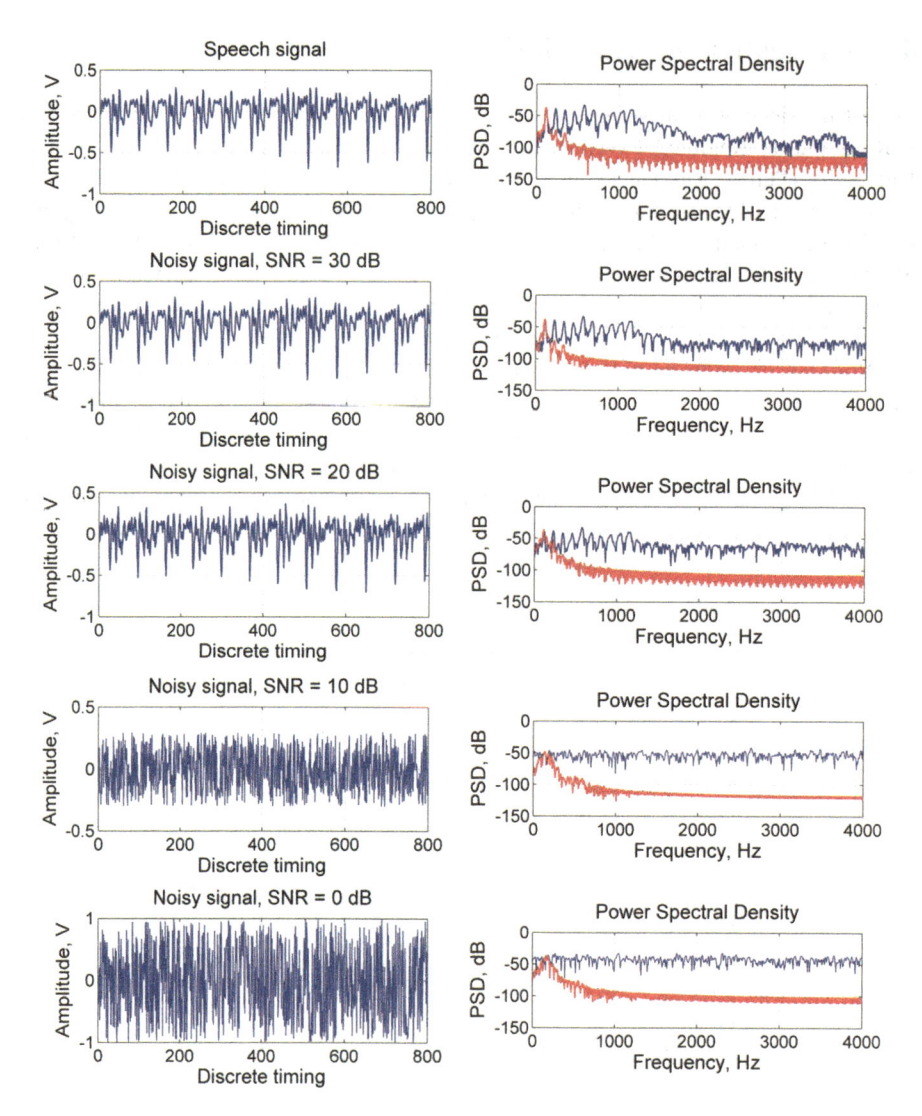

Fig. 4. Oscillograms of noisy voiced sections of the speech signal (left column); power spectral densities (right column) of noisy voiced speech signal sections (blue color), and extracted IMFs containing the pitch (red color).

densities of the IMFs, containing the pitch, obtained as a result of decomposition of pure and noisy speech signals (Fig. 4, right column).

A detailed spectral analysis of the IMFs containing the pitch (Fig. 5) has revealed that the PF difference between the values of pure and noisy speech signals is insignificant, and in some cases, it is completely absent. For the noisiest signal with an SNR of 0 dB, the PF deviation is only 3 Hz. And the PF difference is within 1 Hz for the rest noisy signals, which can be attributed to the error.

Based on the results obtained, we can conclude that the developed noise-robust method for determining the prosodic characteristics of speech signals provides rather good noise robustness even at very high values of the noise intensity. Thus, the developed method can be successfully tested in a modern "aggressive" noise environment in systems for assessing human psycho-emotional state in conditions of free motor activity.

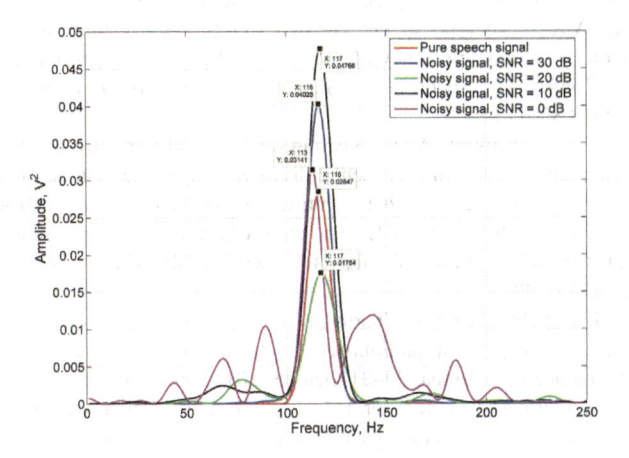

Fig. 5. Fourier spectrum of the IMFs containing the pitch.

Acknowledgments. This work was financially supported by the Russian Science Foundation, the project No. 17-71-20029 "Search for hidden patterns of borderline mental disorders, and the development of a rapid assessment system of human mental health".

References

1. WEVOSYS medical technology GmbH, Baunach, Germany (2018). https://www.wevosys. com/products/lingwaves/lingwaves.html. Accessed 11 Feb 2019
2. Nemesysco Ltd.: Kadima Industrial Park, Kadima, Israel (2019). http://nemesysco.com. Accessed 11 Feb 2019
3. Boll, S.: Suppression of acoustic noise in speech using spectral subtraction. IEEE Trans. Acoust. Speech Sig. Process. **27**(2), 113–120 (1979). https://doi.org/10.1109/tassp.1979. 1163209
4. Berstein, A., Shallom, I.: A hypothesized wiener filtering approach to noisy speech recognition. In: ICASSP, Toronto, 14–17 May 1991, pp. 913–916 (1991)
5. Furui, S.: Cepstral analysis technique for automatic speaker verification. IEEE Trans. Acoust. Speech Sig. Process. **29**(2), 254–272 (1981). https://doi.org/10.1109/tassp.1981. 1163530
6. Viikki, O., Bye, D., Laurila, K.: A recursive feature vector normalization approach for robust speech recognition in noise. In: ICASSP, Washington, 14–15 May 1998, pp. 733–736 (1998)

7. Alimuradov, A.K., Churakov, P.P.: Noise-robust speech signals processing for the voice control system based on the complementary ensemble empirical mode decomposition. In: International Siberian Conference on Control and Communications, Omsk, 21–23 May 2015 (2015). https://doi.org/10.1109/sibcon.2015.7146972

8. Alimuradov, A.K., Tychkov, A.Yu., Ageykin, A.V., et al.: A method to determine cepstral markers of speech signals under psychogenic disorders. In: Ural Symposium on Biomedical Engineering, Radioelectronics and Information Technology, Yekaterinburg, 7–8 May 2018, pp. 128–131 (2018). https://doi.org/10.1109/usbereit.2018.8384568

9. Alimuradov, A.K., Tychkov, A.Yu., Kvitka, Yu.S.: Automation of empirical mode decomposition to increase efficiency of speech signal processing. In: International Russian Automation Conference, Sochi, 9–16 September 2018 (2018). https://doi.org/10.1109/rusautocon.2018.8501732

10. Alimuradov, A.K., Tychkov, AYu., Kuzmin, A.V., et al.: Improved CEEMDAN based speech signal analysis algorithm for mental disorders diagnostic system. Pitch Freq. Detect. Measur. IIJERTCS **10**(1), 22–47 (2019). https://doi.org/10.4018/ijertcs.2019010102

11. Huang, N.E., Zheng, Sh, Steven, R.L.: The empirical mode decomposition and the Hilbert spectrum for nonlinear and non-stationary time series analysis. Proc. Roy. Soc. Lond. Ser. A **454**, 903–995 (1998)

12. Alzamendi, G.A., Schlotthauer, G., Torres, M.E.: Vocal fold activity detection from speech related biomedical signals: a preliminary study. In: VI Latin American Congress on Biomedical Engineering, Parana, 29–31 October 2014, pp. 520–523 (2014)

13. Torres, M.E., Schlotthauer, G., Rufiner, H.L., et al.: Empirical mode decomposition. Spectral properties in normal and pathological voices. In: 4th European Conference of the International Federation for Medical and Biological Engineering, Antwerp, 23–27 November 2008, pp. 252–255 (2008)

14. Schuller, B.W., Batliner, A.M.: Computational Paralinguistics: Emotion, Affect and Personality in Speech and Language Processing. Wiley, New York (2013)

15. Fant, G.K.: Acoustic Theory of Speech Formation. Nauka, Moscow (1964)

16. Colominasa, M.A., Schlotthauera, G., Torres, M.E.: Improved complete ensemble EMD: a suitable tool for biomedical signal processing. Biomed. Sig. Process. **14**, 19–29 (2014)

Automatic Temperature Control System for a Bee Hive

S. V. Oskin[(✉)], N. I. Bogatyrev, and A. A. Kudryavtseva

Kuban State Agrarian University, 13 Kalinina Street,
Krasnodar 350044, Russian Federation
kgauem@yandex.ru

Abstract. The research team has analyzed the physical processes occurring in a hive for the winter aggregation of bees. The primary in-hive processes were analyzed by means of Comsol 5.4. The analysis used the mathematical models of the heat transfer, air flow, and air humidity alteration interfaces. The solutions employed special blocks for simulating the multiphysical relations of the analyzed processes. Visual analysis of the temperature fields proves the bees' outstanding thermal insulation. Simulation-based optimization of the electric heater parameters is proposed. The obtained knowledge of the winter aggregation status of bees has been used to design an electric heating control scheme that makes no use of in-hive temperature feedback. Comparing the thermal field images rendered by the thermal imaging camera as well as by simulation shows a high degree of match. An experiment conducted in Krasnodar Krai shows that the new control system reduces the power consumption by 20 to 25% compared to electric heating controlled by an in-hive temperature sensor.

Keywords: Bee · Hive · Temperature · Ventilation · Model · Automatic control

1 Introduction

Russia's agriculture is on the rise. Beekeeping plays a crucial role in it, as bees are the primary pollinators for many crops. Besides, bees produce such well-known product as honey. Russia is the world's third nation in terms of the total bee population; however, mechanization and automation of the processes in this industry remains underwhelming. Many beekeepers use electricity to heat hives in winter. Wintertime electric heating of hives reduces the amount of bee-consumed honey, improves the internal microclimate, and keeps the bee death toll low. Appropriate use of electric heating also improves early breeding, meaning the family has many more workers by the time of the first forage. However, overheating might have adverse effects, causing the bees to exit hibernation, and the queen to start actively laying the brood too early. This forces the bees to consume a lot of honey and leave the hive in search for food, whereby the shortage of honey and the low outside temperatures may cause multiple deaths. One specific feature of bees is that during winter, they form a specific aggregation referred to as a winter cluster; thus, they survive the winter in a passive state. In winter, bees consume a minimum amount of honey; they keep the cluster temperature at 24 °C to

A. A. Radionov and A. S. Karandaev (Eds.): RusAutoCon 2019, LNEE 641, pp. 827–837, 2020.
https://doi.org/10.1007/978-3-030-39225-3_90

32 °C positive. Some researchers [1–10] have long been trying to understand the biological processes of such winter clusters. Such scientists as V.A. Toboyev, Ye.K. Yeskov, M.S. Tolstov have tried to painstakingly analyze bee aggregation by means of state-of-the-art software [11–16]. This knowledge could be of use for optimizing the heating for bees while also minimizing the power consumption. Today, beekeepers have a variety of electric heaters and special automatic controls. As a rule, at least one temperature sensor must be placed inside the hive. What makes doing so difficult is that the bees try to glue it with propolis; the need for in-hive wiring is another complication. This gives rise to the problem of designing an automatic electric heating control system that uses a mathematical model of the bee family status instead of temperature feedback.

2   Theoretical Studies and Modelling

The primary in-hive physical processes were simulated by means of Comsol 5.4. The equation set comprised the equations contained in this software as well as those the author hereof had earlier derived analytically [17–20]. Thus, the heat processes were simulated in the Heat Transfer interface; the resultant mathematical model could be written as follows:

$$
\begin{cases}
\rho_{air1} \cdot c_{air1} \cdot u_{air1} \cdot \nabla T + \rho_{air2} \cdot c_{air2} \cdot u_{air2} \cdot \nabla T + \\
\rho_{elh} \cdot c_{elh} \cdot u_{air1} \cdot \nabla T + \nabla q_{air1} + \nabla q_{air2} + \nabla q_{wood} \\
+ \nabla q_{hc} + \nabla q_{emptyhc} + \nabla q_{bee} Nu + \nabla q_{elh} = Q_{bee} + Q_{elh} \\
\lambda_{bee} = 0,0076 - 0,0017 \cdot T_0 ; \rho_{n\text{y}} = 243 - 8 \cdot T_0 \\
Q_{bee} = 3,2 \cdot T_0^2 - 20 \cdot T_0 + 922 \\
q_i = -\lambda_i \Delta T \\
Q_{elh} = f(T_o, t)
\end{cases}
\tag{1}
$$

where ρ_{air1}, ρ_{air2} are the incoming air density (Index 1) and the density of the air flowing through the cluster (Index 2); c_{air1}, c_{air2} are the heat capacities for Air Block 1 and Air Block 2 (in-hive air and in-cluster air); u_{air1}, u_{air2} are the Air Block 1/2 velocity fields, m/s; q_{air1}, q_{air2}, q_{wood}, q_{hc}, $q_{emptyhc}$, q_{bee} q_{elh} are the densities of the heat fluxes lost to the thermal conductivity of Air Blocks 1 and 2, wooden elements, honeycombs, empty honeycombs, bee cluster, electric heater, respectively, W/m^2; Nu is the Nusselt number, Q_{bee} is the heat production of the bees, W/m^3; Q_{elh} is the heat produced by the electric heater, W/m^3, T_0 is the outdoor ambient temperature around the hive, °C; q_i λ_i are the lost heat flux density and the heat transfer coefficient of the i-th hive element.

The simulation used the thermophysical properties of materials as recorded in the Comsol library.

Air flow was simulated in the Laminar Flow interface with the following system of equations:

$$\begin{cases} \nabla \cdot (\rho_{air1} u_{air1}) + \nabla \cdot (\rho_{air1} u_{air1}) = 0 \\ 0 = \nabla \cdot (\mu(\nabla u_{air1} + (\nabla u_{air1})^T - \frac{2}{3}\mu(\nabla \cdot u_{air1})I)) \\ \quad + \nabla \cdot (\mu(\nabla u_{air2} + (\nabla u_{air2})^T - \frac{2}{3}\mu(\nabla \cdot u_{air2})I)) \\ u_{in} = \frac{0.0034 \cdot T^2 - 0,0216 \cdot T + 1}{[(0,007 \cdot T + 11,5) - (0,0065 \cdot T^2 + 0,3 \cdot T + 4,03)] \cdot 1,44 \cdot 10^{-3}} \\ -p_1 = -pI + \mu(\nabla u_{air\,in2} + (\nabla u_{air\,in2})^T) - \frac{2}{3}\mu(\nabla \cdot u_{air\,in2})I \\ p_2 = -pI + \mu(\nabla u_{air\,out2} + (\nabla u_{air\,out2})^T) - \frac{2}{3}\mu(\nabla \cdot u_{air\,out2})I \end{cases} \qquad (2)$$

where p is the pressure, Pa; μ is the dynamic viscosity coefficient, Pa·s; I is a unit vector; u_{in} is the hive air inflow velocity, m/s; p_1, p_2 is the air block inlet/outplate plane pressure, N/m^2; $u_{air\,in2}$, $u_{air\,out2}$ is the bee cluster air inflow/outflow velocity, m/s.

Dynamic viscosity coefficients were taken from the materials library, while the cluster-inlet and cluster-outlet air pressure was found iteratively on the basis of this model.

Humidity changes were simulated in the Comsol interfaces with due account of the convection, diffusion, and adsorption processes on the basis of Fick's, Navier-Stokes', and Darcy's laws.

All the mathematical models were interrelated by the Multiphysics block to generate combined solutions when having to adjust this or that parameter.

Simulation was carried out in steady state for several ambient air temperatures: −28 °C, −20 °C, −10 °C, −5 °C, 0 °C, +5 °C, and +15 °C. For example, Fig. 1 shows the temperature fields at an outside temperature of −28 °C. Temperature curves show that regardless of the heating output, the in-cluster temperature was stable at +28 °C to +31 °C. These data prove the model and the assumed constraints adequate. Such simulation can be used to find the temperature at any point in the beehive.

The further task was to optimize the output of the heaters. Review of literature and beekeepers' experience showed that successful overwintering would require minimum fluctuations of the ambient air temperature near the cluster. Any change in the outside temperature disturbs bees, forcing them to worry and to try to cope with it. A drastic increase in temperature provokes bees to leave the hive, causing the cluster to disintegrate. In case the temperature is too low, bees overconsume their food, clogging their intestines.

This is why many beekeepers move beehives into special apiaries set up in basements or dugouts. Apiaries feature minimum temperature fluctuations, as they are well-insulated and have high time constants. This means that heaters must be adjusted in such a way as to keep the inside temperature of beehives constant or nearly constant even if the ambient temperature varies significantly.

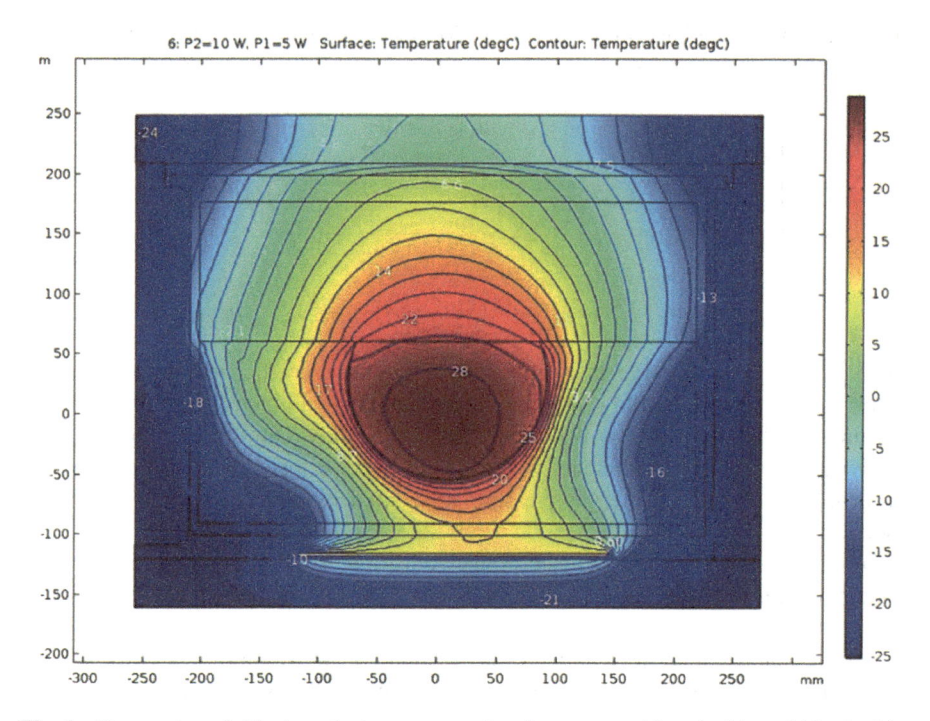

Fig. 1. Temperature fields: bee cluster cross-section, heaters on. Air velocities within the hive shown.

Multiple iterations were run to optimize the heater output. Figure 2 shows the curves of heaters vs bees heat output. As can be seen there, bees have a sustainable heat output of 2.5 W. This is good, meaning that the bees are not disturbed and are kept stable while consuming minimum food. At the same time, changing the temperature from 0 °C to 30 °C negative increases the heater output from 4.8 W to 43.5 W while changing the total in-hive heat power from 7.3 W to 46 W. However, the same curves show the temperature of heaters in the same ambient temperature range; apparently, starting at −17 °C (which corresponds to the heater output of 27 W), the heater temperature goes beyond the acceptable limits; at −30 °C, it reaches +65 °C, which threatens the bees.

This means such heaters are only recommendable as long as the temperature never goes below −17 °C. Further reduction in temperature would require an adjustment in the heater output so that the heater temperature never exceeds +45 °C.

Further simulation produced a heat output value optimized by such criteria as minimum change in the bee-controlled temperature and avoidance of heater overheating. Thus, changing the ambient temperature from 0 °C to −30 °C must cause the total output of the heaters to alter from 4.8 W to 34.5 W with their surface temperature not exceeding 45 °C while the temperature around the bee cluster must thereby change from +5 °C to −1.5 °C.

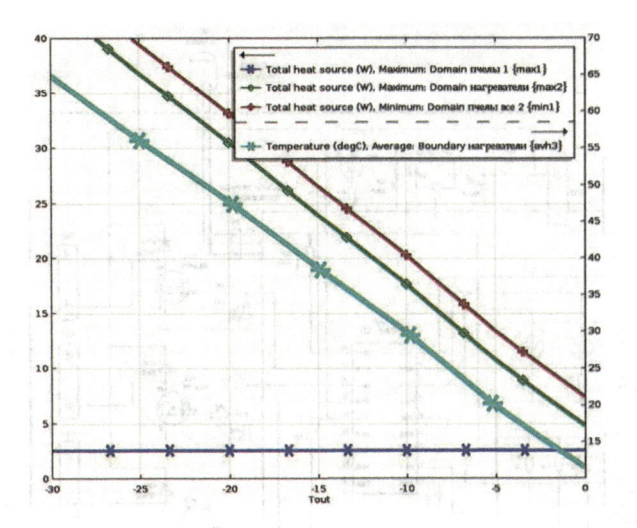

Fig. 2. Bees vs heaters output, heater temperature curve (right axis).

A new electric circuitry has been designed to control three heating elements placed within the hive. Those are controlled by existing algorithms run by a PIC16F1827 microcontroller (made by Microchip, see Fig. 3) [21, 22]. A DS18B20 chip is used as the outside air temperature sensor. This digital thermometer has a 9- to 12-bit conversion resolution and a temperature alarm. DS18B20 communicates with the microcontroller via a 1-Wire interface. Its temperature measurement range is −55 °C to +125 °C. For temperatures from −10 °C to +85 °C, the error is within 0.5 °C. The sensor is connected via the connector X1; the resistor R2 is a pull-up resistor that is necessary for the normal operation of the sensor.

This control unit can operate in several modes. Modes are switchable by the buttons SB2 to SB4; the current mode is indicated by the LEDs VD2 to VD4. In Model 1 (the default mode), the heating elements EK1 to EK3 are only connected to the terminal XS2; their interconnection is as shown in the circuit diagram (EK1 and EK3 are connected in series; EK2 is connected directly to the terminal XS2). In Mode 2, the elements EK1 and EK3 are connected in series and to the terminal XS2 while EK2 is connected to XS3, making the heater control more flexible. In Mode 3, each heating element is connected to their respective terminal (XS2 to XS4), giving the user full control over each such element. In this circuitry, the resistors R10 to R12 are pull-up resistors for the microcontroller inputs; the capacitors C4 to C6 are there to prevent the bounce of the buttons SB2 to SB4.

The heating elements EK1 to EK3 are controlled by high-frequency PWM controls. IRL3705 MOSFET transistors are used for switching; those can control the logic level voltage and accept switching current of up to 63 A. Transistor closures are connected to the microcontroller via the limiting resistors R3, R8, and R13. The diodes VD5 to VD7 are in place to protect the transistors from overvoltage.

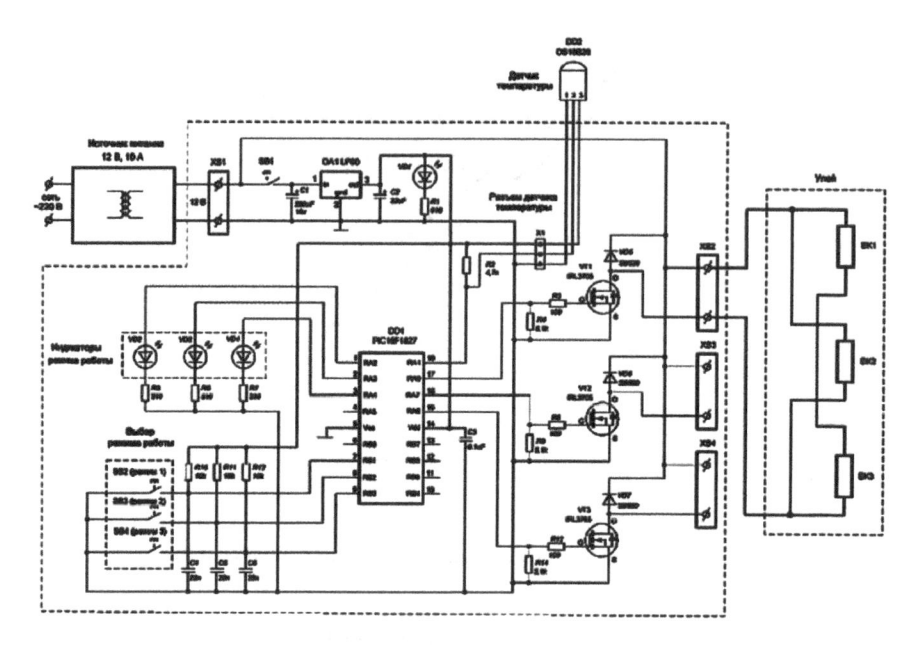

Fig. 3. Beehive heater circuitry.

Temperature sensor readings affect the PWM duty cycle as received by the transistor closures, which in its turn affects the output of the heating elements. Higher ambient air temperature renders the pulses shorter, which reduces such output. The entire operation is controlled by the microcontroller algorithm. The entire circuitry is powered by a stabilized external power supply that features a voltage of 12 V and a minimum permissible load current of 10 A. The power is supplied to the terminal XS1. To power the microcontroller and the digital temperature sensor, the circuitry uses a 5 V LF50-based integrated voltage regulator. Powered-on status is indicated by the LED VD1.

3 Analysis of Results

Using the optimized heater outputs and the 2016/2017 winter weather data for Krasnodar Krai in these models showed that over the entire winter, the bees produced heat at 2.7 W while the heater output peaked at 24 W. Given the calorific value of honey, adaptive heating would have saved 840 g of food over the three winter months (compared to no heating) while consuming 22.1 kWh.

Thermal fields were studied by a thermal imaging camera using a 12-frame Dadant beehive, with 10 frames whereof being populated by a single bee family in fall. Thermal imaging was carried in winter at different outside air temperatures. For example, a thermogram made at −3 °C was compared against the simulated thermograms for the same temperature, see Fig. 4.

Pairwise comparison of the simulated/experimental temperatures at different beehive points showed the difference in such pairs never exceeded 2 °C; such degree of coincidence proves the models adequate for studying the beehive microclimate. The camera-generated thermograms were of the same quality as those published earlier by other researchers.

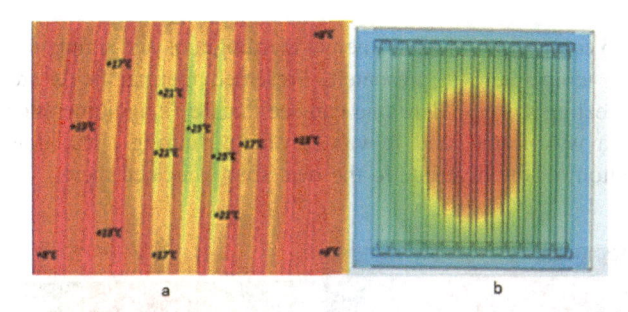

Fig. 4. Thermograms of the beehive surface, uncovered: (a) the experiment; (b) the simulation; note the temperature fields.

Afterwards, validation of modelled thermograms and pictures done by thermal camera in the absence of rare hive wall was carried out. For this purpose, a special bee hive was built – the rare wall was mounted with screws and could be easily removed. Shooting with thermal camera was carried out at the ambient temperature of −8 °C. The same temperature was considered for simulation, and corresponding results were obtained –the thermograms are presented at the Fig. 5. Comparing the pictures, one can see that thermal fields are similar, but in the experimental set up a zone occupied by bees is larger. This peculiarity means that bees occupy more wax frames and, highly likely, bees' agglomeration was spread in the direction parallel to hive's bottom. At the same time, it is visible that temperature magnitude in different points of the hive at the model is equal to the temperature fixed with thermal camera. Modelled heat distribution along the hive is also similar to the real situation.

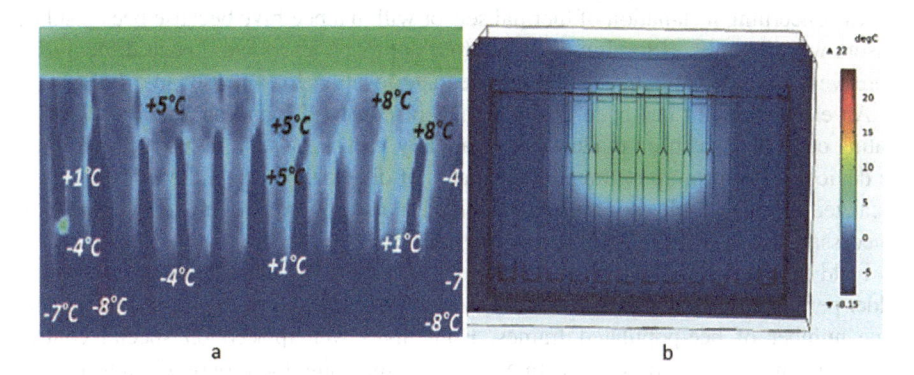

Fig. 5. (a) Thermogram of hive's surface without rare wall in experiment and (b) results of simulation with indication of thermal fields.

Experimental research of thermal fields within a bee's aggregation was done in accordance with the following method. Due to complexity of carrying out such research in winter time – it could be harmful for the bee's family – all experiments were carried out at relatively high ambient temperature +2 °C and with the rapid take out of wax frames for thermal photographing. The remarkable point is the fact that honey has quite high thermal capacity, which gives an opportunity to obtain results with minimal error.

At the Fig. 6 comparison of experimental thermograms and simulation results at the similar ambient temperature is presented. Presented thermal fields are relevant for application of heating system. Here one can see that bee's agglomeration is shifted to the bottom of a wax frame, and it was considered during simulation. Bees move towards the bottom of the hive due to the heat source located there.

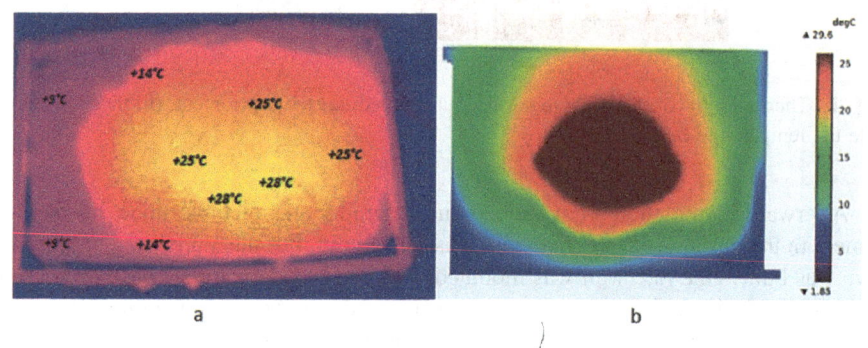

Fig. 6. (a) Thermogram of the wax frame from the center of bees aggregation in experiment and (b) results of simulation with indication of thermal fields.

Analysis of obtained data shows that analytical and empirical data have no significant difference. Thus, empirical honey consumption exceeds simulated values for the case without application of electrical heating system in January. It could be explained by the fact that in January egg rafting starts and first brood comes. While using most popular heating systems bee-keepers mention that there are certain difficulties concerning installation of thermal sensor within a bee-hive because bees insulate sensors with propolis as foreign elements (our approach let to avoid such difficulties). Moreover, it is very difficult to find an adequate location for a sensor in a bee-hive.

An electric circuit has been designed and patented [23], and an adaptive beehive heating controller has been made to automatically keep a constant in-hive temperature; the device can control up to three local heaters of different power in the same beehive. Localized heating enables controlling the temperature field in the beehive, heat the food if necessary while avoiding dangerous overheating.

Field tests were carried out at a 110-beehive yard. Data were recorded during the colder periods of the 2017–2018 winter season. Two beehives were picked that had the same number of bee-populated frames. Both hives were placed on special control scales. One heater was equipped with electric heating using the proposed controls. The same controls were applied to five more beehives of different population. The apiary was located in the Labinsky District, Krasnodar Krai. The ambient air temperature readings were compared to the Labinsk weather data. Virtually no difference in temperature was detected. The electric heaters were connected via an electricity meter, and the readings of the latter were recorded. See Fig. 7 for visualization. The results prove the good coincidence of the theoretical and empirical data.

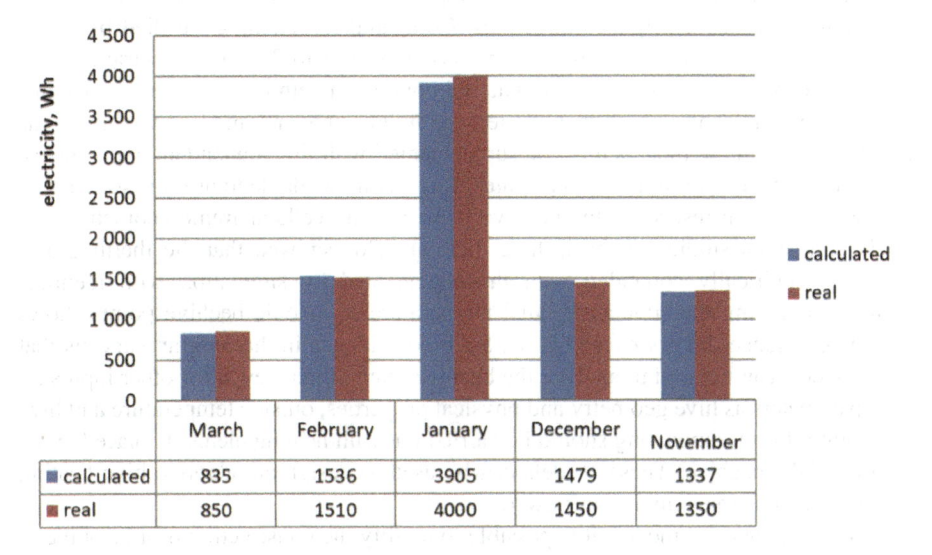

Fig. 7. Electricity consumption in colder season.

4 Basic Results and Conclusions

Equation solutions obtained by the modular simulation of: heat processes, air flow, moisture distribution. The geometric model of bee's aggregation with variable density controlled by temperature is developed. Based on thermal physical and geometrical equations and applying Fourier equation, Newton-Richman equation, Fick laws, Navier-Stokes law, Darcy's law the mathematical model describing basic physical processes inside of bee hive were obtained.

In the systems of differential equations, the following members are included: equations of convective heat transfer, heat conduction of liquid phase, diffusion and adsorption equations, equation of energy characteristic of 15, 000 bees in winter aggregation depending on the temperature of ambient air. Temperature field images prove the bees to have outstanding thermal insulation, showing how drastically the air temperature is different outside the bee cluster. Simulating the microclimate parameters with the bottom heaters on shows the maximum temperatures within the cluster shift towards the bottom; however, the in-cluster temperature never exceeds +31 °C. It has been found that at specific heater output of 74,000 W/m^3, bees reduce their heat output from 3,600 W/m^3 to 1,900 W/m^3, meaning the beekeepers can leave less honey for winter.

Solving the models has enabled the researcher to design the circuitry of an adaptive heater that uses no feedback from inside the beehive. A prototype has been made and proven highly efficient in-field.

An in-situ experiment carried out in Krasnodar Krai showed that the calculated/actual power and honey consumption did not differ significantly, with the relative error not exceeding 12%. Disabling the heaters increased the honey consumption due to the emergence of broods. Compared to heating controlled by an in-hive temperature sensor, the new system consumes 20% to 25% less electricity.

The theoretical and experimental material obtained herein must be used to design a beehive heating control system with remote monitoring and mode switching. The equations used in the models must be supplemented with time-dependent components, then solved in Comsol in non-steady states. It is recommendable to use this software to find solutions with respect to the beehive materials and cellular frame geometry.

Experimental studies of the in-hive thermal fields showed that the thermograms obtained empirically coincided with those generated by simulation with identical inputs. Comparing the simulated/actual temperature at specific beehive points shows that the difference did not exceed 2 °C. Such coincidence in thermograms means that these models can be used to analyze the beehive microclimate even for other inputs by such parameters as hive geometry and physical properties, outside temperature and hive population. Further modeling should be carried out with heating elements placed at the bottom of the beehive. These models can be used to design an adaptive hive heating system that consumes minimum power.

Based on obtained models it is possible to modify the construction of a bee-hive in order to prevent creation of air dead zones where moisture condenses, and overall ventilation efficiency of the bee-hive will be improved. Moreover, constructive materials applied in separate elements of a bee-hive (wax frames, body, bottom, ceiling, lid) could be critically analyzed in order to improve thermal insulation of the latter and facilitate moisture adsorption.

References

1. Korge, V.N.: Basics of Beekeeping. Phoenix, Rostov-on-Don (2008)
2. Lebedev, V.I., Kasianov, A.I.: Thermal mode and energetic of bee families. Beekeeping **2**, 16–19 (2011)
3. Lebedev, V.I., Kasianov, A.I.: Thermogenesis and thermal mode of bee family. Pressa, Rybnoe (2004)
4. Triphonov, A.D.: Feed consumption during the winter. Beekeeping **11**, 21–23 (1990)
5. Triphonov, A.D.: Thermal exchange between a beehive inhabited by bees and environment. Beekeeping **9**, 28–31 (1991)
6. Kasianov, A.I.: Biology of bee hives' heating. Beekeeping **2**, 16–21 (2003)
7. Rybochkin, A.F., Zakharov, I.S.: Computer systems in beekeeping. Kursk State Technological University, Kursk (2004)
8. Triphonov, A.D.: Feed consumption during wintering. Beekeeping **6**, 15–18 (1991)
9. Omholt, S.: Thermoregulation in the winter clusters of honeybee. Apis mellifera. J. Theor. Biol. **128**, 219–231 (1987)

10. Watmough, J., Camazine, S.: Self-organized thermoregulation of honeybee clusters. J. Theor. Biol. **176**, 391–402 (1995)
11. Yeskov, E.K.: Microclimate in beehives. Rosselkhozizdat, Moscow (1983)
12. Yeskov, E.K., Toboev, V.A.: Seasonal dynamics of thermal processes between frames of wintering bees Apis millifera. Zoological J **90**, 335–341 (2011)
13. Yeskov, E.K., Toboev, V.A.: Mathematical modeling of thermal fields distribution in cold aggregation of insects. Biophysics **54**, 114–119 (2009)
14. Toboev, V.A., Tolstov, M.S.: Simulation of thermal processes in aggregation of wintering bees. Phys. Process. Biosyst. 97–102 (2014)
15. Toboev, V.A., Tolstov, M.S.: Simulation of convectional transfer in aggregation of honey bees. Interdiscip. Inst. Sci. Res. **3**, 116–119 (2014)
16. Toboev, V.A.: Contemporary methods of study of thermal homeostasis. Beekeeping **10**, 44–46 (2006)
17. Oskin, S.V., Potapenko, L.V., Blyagoz, A.A.: Necessity of application of electrotechnological means for maintaining of microclimate for bee families. Agrotech. Power Supply **1**, 12–21 (2016)
18. Oskin, S.V., Ovsyannikov, D.A.: Modeling of the main physical processes in a beehive. Biophysics **64**, 153–161 (2019)
19. Oskin, S.V., Potapenko, L.V., Ovsyannikov, D.A., et al.: Adaptive technology of winter heating for bees. Polythemat. Netw. Electron. J. Kuban State Agrar. Univ. **132**, 277–287 (2017)
20. Oskin, S.V., Ovsyannikov, D.A.: Electrotechnological ways and equipment for increasing of labour efficiency in beekeeping at North Caucasus. Kron, Krasnodar (2015)
21. Rashid, M.H.: Power electronics - challenges and trends. In: International Conference on Innovations in Electrical Engineering and Computational Technologies, Karachi, p. 1 (2017). https://doi.org/10.1109/icieect.2017.7916589
22. Panasetsky, D., Tomin, N., Voropai, N., et al.: Development of software for modelling decentralized intelligent systems for security monitoring and control in power systems. In: IEEE Eindhoven PowerTech, Eindhoven, pp. 1–6 (2015). https://doi.org/10.1109/ptc.2015.7232553
23. Oskin, S.V., Bogatyrev, N.I., Potapenko, L.V., et al.: The device for regulation of temperature in a beehive. RU patent 2639324, 21 December 2017 (2017)

Software Application for Determining Comfortable Conditions of the Human-Computer Interaction

A. A. Popov[1(✉)] and A. O. Kuzmina[2]

[1] Plekanov Russian University of Economics, 36 Stremyanny lane,
Moscow 117997, Russian Federation
popov.aa@rea.ru
[2] GMCS, 7 Pokryshkina street, Moscow 119602, Russian Federation

Abstract. The values of ergonomic characteristics of the user interface are influenced by the speed of recognition of visual information on the user interface. The paper considers the influence of three parameters on the perception of visual information: the font size of the character (angular size), the font color of the character, the color of the user interface, on the background of which the character is recognized. The order of actions performed by the software application is considered. The software application is designed to determine the combination of parameter values that affect the perception of visual information which correspond to the best conditions of character recognition (time). The software application can be used as a component of an expert system. The system is designed to configure user interfaces of information systems that provide comfortable working conditions for employees of an enterprise. It also takes into account their current physical and psychological state. For the operation of each component of the expert system, a neural network can be used. The interaction of the components should be carried out using the core of neural networks.

Keywords: Ergonomics · Software application · User interface · Visual information · Characters recognition · Character color · Character size · Background color

1 Introduction

The active improvement of information systems stipulates a more detailed research of the humans and computers interaction. Users, generally, compare the functionality of the software application with those provided by its user interface. If the user interface is user-friendly, then the user, if the necessary functionality is available, can assume that the application meets its requirements. At the same time, more advanced software applications with the same functionality will not be considered a user that meets its requirements due to a non-user-friendly interface. Ergonomic indicators characterize user interfaces. Any user interface that helps the user to avoid errors has better ergonomic characteristics than the user interface, during which the user constantly makes mistakes. The most common indicators characterizing the ergonomics of the user interface are the Schneiderman's indicators [1]:

A. A. Radionov and A. S. Karandaev (Eds.): RusAutoCon 2019, LNEE 641, pp. 838–851, 2020.
https://doi.org/10.1007/978-3-030-39225-3_91

- User speed
- Number of human errors
- Subjective satisfaction
- Speed of learning the skills of operating an interface
- Degree of preservation of the skills of working with the user interface with the long-term non-use of the software application

The values of indicators characterizing the work with user interfaces closely related to the psychophysiological features of the user in the detection, distinction and recognition of visual information. Detection is the stage at which the user selects an object on the user interface from the general background, but cannot decide yet on the form of the object and its attributes. Distinction is the stage at which the user is able to perceive separately two objects on the user interface, located side by side, and highlight their details. Recognition is the stage at which the user selects the significant features of the object on the user interface and determines its purpose.

The difference in the conditions of detection, perception and recognition of objects displayed on the user interface leads to the fact that when performing the actions necessary to solve the same problem by the same user, the different conditions for displaying the characters correspond to different execution time of the actions.

To estimate the execution time of actions, the Fitts' Law [2], Hick's Law [3] or the KLM-GOMS technique [4] are used. In doing so, performing an evaluation of the duration of the action with respect to working with the user interface of the software application implies that the name of the 'target' controls (that is, user interface controls with which the user should work), is recognized immediately. User interface is designed often based on the assumption that users are included in the target group and are well prepared, as developers, to work with information system. If users are not prepared enough to work with information system, user interface, which is designed based on this assumption, probably will not be intuitive and user friendly. Therefore, often there is an impression that developers design information systems 'for themselves', without hesitation about user experience. That is why the information systems, which have wide range of functionality, but at the same time do not have ergonomic user interface, are not successful with users in contrast to the information systems with limited functionality, but with the ergonomic interface. The one way of increasing the ergonomic of the user interface is accounting the results of the ergonomic design of the user interface taking into account the personal qualities of the visual information perception by potential users [5–8] or customization of the operating information system under personal qualities of visual information perception by the employees of the organization. For the research of the personal qualities of visual information perception, could be used specialized software application. One of the problems, solved for the enhancement of ergonomic indicators of Schneiderman, is increasing the speed of working with user interface [9, 10] due to reducing duration and increasing accuracy of physical user actions while working with user interface. Several of practical aspects of using special software package for accounting "human factor" in designing of the user interfaces of the corporate information systems are considered in [9]. Using software application for determining combination of the parameters values, which provide most comfortable conditions for user to determine and to detect characters while working with user interface, are considered in [10]. Speed of working with the user interface could be defined using Fitts' Law [11–13]. Methods to increase the speed of the software applications users, where menu is used, are considered in [14, 15]. In the article [16] is considered using the concept of semantic indication on the elements of the user interface for increasing user speed of work. In the article [17] using Fitts' Law the research of speed of working with the user interface, where as a pointer for gaining 'target' controls is used virtual ball, is rolled over with the incline of device with the built-in accelerometer. As the raw data for research, conducted in [17], ball size, size of 'target' controls and angle of incline are used. In the article [18] was researched advanced Fitts' Law, where considered impact of the four parameters, which characterize 'three-dimensional' interaction with the user interface, on the user speed of work. Questions, associated with using Fitts' Law to evaluate the effect of touching with the fingers (gestures) the touch screen of the device on the speed of working with the user interface, considered in [19–21]. In this article [21] is considered model to determine the user speed of work "Touchless Hand Gesture Level Model" (THGLM), based on model "Key Stroke-Level Model" (KLM) [22] and considering Fitts' Law. In addition, speed of working with user interface could be defined by using Hick's Law [3, 23].

In the article [24] are given particular qualities of work of the software application is developed, based on Hick's Law and intended for testing the speed of work of users of different age. In the article [25] is presented model, which combines Hick's Law and Fitts' Law for predicting the efficiency of using different types of menu and considering element selection probabilities, increasing of the user experience, and adaptability of the user behavior.

In the article [26] is examined the using of Hick's Law for predicting speed of working with the control, which are not displayed at the moment on the user interface, using scrolling lists, elements selection in the hierarchical menu, navigation through the 'arborescent' browsers.

In the article [27] are considered heuristic solutions, based on Hick's Law, for increasing speed of user work depending on the menu structure and catalogues, where selection is carried out. In [28] is presented model based on Hick's Law, usage based on the main memory effects and allow predicting interconnection between number of elements (incentives), presented to human, who passes the test, and time, which is necessary for human to select required elements (incentives).

In [29] is considered opportunity of increasing the user speed of work with touchscreen in case of using 'scrolling' in order to reached the 'target' control, which does not exist in the user's field of view now. In the research of the user performance, which is examined in [29], additionally to the parameters of geometric movement according to Hick's Law (scrolling distance, screen size and width of the 'target' control) two factors are added, which could affect the user performance – scrolling modes (pan and click) and methods of user feedback.

Questions of adaptive kinetic 'scrolling' screen to accelerate access to 'target' information, which is situated out of the user's field of view, in case of working with big data is devoted to article [30], where speed of 'scrolling' is adapted according to data capacity, remaining until the 'target' data achieved. Thus, problem of the increasing the speed of working with the user interface is urgent.

Unfortunately, user recognition of the name of the 'target' control does not always happen at once. Therefore, an additional time component appears to evaluate the execution time of actions, reflecting the amount of time spent recognizing the explanatory text or the icon inside the 'target' control. In this case, the value of the additional time component depends on the individual characteristics of users in recognizing visual information (when different users perform the same operation with the same conditions for displaying information on the user interface, the recognition values may be different).

Therefore, it is relevant to research the influence of parameters characterizing the detection, perception and recognition of information, located on the user interface, on the values of the Schneiderman exponents. The research is aimed at determining the combination of parameter values that affect the detection, perception and recognition of characters on the user interface. In this case, the combination of parameters should provide the user with the best conditions for characters' recognition.

2 Parameters that Affect the Recognition of Objects Displayed on the User Interface

The ability to visually perceive an object on the user interface is determined by the following parameters: the angular dimensions of the user interface objects, the level of adaptive brightness, the contrast between the object and the background, the brightness of the background and user interface objects, the flicker frequency, the image contrast – these are parameters that affect the information recognition time and on the psychological state of the user [31].

In [32, 33], data characterizing the process and conditions of visual perception and affecting the formation of a subjective image for the user of the object. In [34], the methods and results of engineering and psychological research and design of operational and dispatching tools described, based on the structural and psychological concept of synthesis and the multilevel adaptation of information systems to the activities of users of information systems.

Generally, the detection of an object on the user interface performed against a background that is noise in relation to the object (there are almost no ideal cases for detecting objects). Detection occurs if one or another set of features of an object appears and the operational threshold of perception is exceeded (the smallest difference in the signal by some feature of the object, in which the speed and accuracy of detection reach a certain constant value [35]). It should be taken into account that different sources mention different stages of information reception. In [31] three stages are considered (detection, discrimination, identification).

In [35] four stages are considered (detection, discrimination, identification, decoding). In [35], decoding involves interpreting an object, and identification is its identification.

Thus, the content of the 'identification' step [31] includes the content of the 'identification' and 'decoding' steps [35]. Also, the content of the 'distinction' stage differs: in [31], the stage provides for a separate perception of two adjacent objects and the allocation of their common parts that do not allow them to be recognized, and in

[35] the stage provides for the allocation of certain sets of features of the object that do not yet allow it recognize.

Thus, the stage of 'discrimination' in [31] and [35] is generally identical. In a large number of works on engineering psychology, the results of theoretical research and experiments on the processes of detection, distinction and identification of objects along its contours are presented.

In [31] the results of experiments to determine the thresholds for the detection of some objects with elementary forms (straight lines and curved lines of black on a white background), as well as the degree of influence of the length and line thickness on the detection threshold are presented.

The results of the research given in [31] show that when recognizing contours, the threshold for the difference between the line curve and the straight line is 69 arc seconds. If one of the three points displaced relative to an imaginary line passing through these points, then the shift noticed already with a displacement of 60 arc seconds.

In [36], methods are given for quantifying the information content of the perceived object contour, which estimates the contour difference from the circle (its information content is zero). There is a relationship between the information contained in the object and the time it is recognized.

Most often, the apparent magnitude of objects is expressed in angular units (the angular size of the object, which is determined by the linear size of the object and the distance from the object to the user's eye along the line of sight). The results of laboratory studies [31] showed that the threshold value of detection of light lines on a dark background is 3.5 arc seconds, and for black lines on a white background, 9 arc seconds. The value of the sign displayed on the user interface, which provides the most rapid and accurate recognition, is 30–40 angular minutes (in height). If the sign has a larger size, then the reading time is practically unchanged.

It is also established that the smallest permissible sign value is 20 arc min. Such features of information perception matter when selecting a font for characters displayed on the user interface.

An important parameter that affects the recognition of visual information by the user is the color combination used in the user interface. An unsuccessful color combination in the user interface can lead a user to a state where he cannot recognize the name of the 'target' control and perform the necessary action with it. The human eye perceives different combinations of colors in different ways. The best combination is the combination of colors 'blue on white' [37]. As the perception deteriorates, the following color combinations are located: 'black on yellow', 'green on white', 'black on white', 'green on red', 'red on yellow', 'red on white', 'orange on black', 'black on purple', 'orange on white', 'red on green' [37].

The user's perception of visual information is also affected by the color contrast, seven types of which are considered in [38].

The problem of determining the best conditions for perceiving visual information by the user is follows:

$$F : ISH \rightarrow OBN; OBN \rightarrow tobm_{min} \qquad (1)$$

$$F : ISH \rightarrow RASP; RASP \rightarrow trasp_{min} \qquad (2)$$

Using the transformation F, in the role of which the software application acts, and the ISH array containing the combinations ish(i) (i = 1, 2, ..., I) of the M values of font size, J values of character color (element of the SIMV array), K values of the background color (element of the FON array), obtain arrays:

OBN = {obn(i); i = 1, 2, ..., I};
RASP = {rasp(i); i = 1, 2, ..., I},

which provide the best operating conditions for the user of the software application (the smallest value of the detection time tobn or of the recognition time trasp for each value of font size).

Wherein

SIMV = {simv(j), j = 1, 2, ..., J};
J - number of colors selected by the user for the character;
FON = {fon(k), i = 1, 2,..., K};
K - number of colors selected by the user for the background;
I = M × J × K.

Element obn(i) contains font size of the character, the color of the character and the background color and character detection time for element ish(i).

Element rasp(i) contains font size of the character, the color of the character and the background color and character recognition time for element ish(i).

3 Description of the Software Application for Determining the Best Conditions for the Detection and Recognition of the Character

When you start the software application, a dialog box opens with help information. After being acquainted with the features of the software application, you go to the dialog "Setting the background color and the color of the character". The dialog box is used in two modes (for specifying the color of the character and the background color, as well as for detecting and recognizing the character). The operation with the dialog box corresponds to the operator 3 (Fig. 1).

The user selects the color of the character and the background color (user interface) from the proposed list. The color lists for the character and background are the same. In this case, you cannot select the color of the character, which coincides with the background color. The background color is written to the element of the FON array. The color of the character is written to the SIMV array.

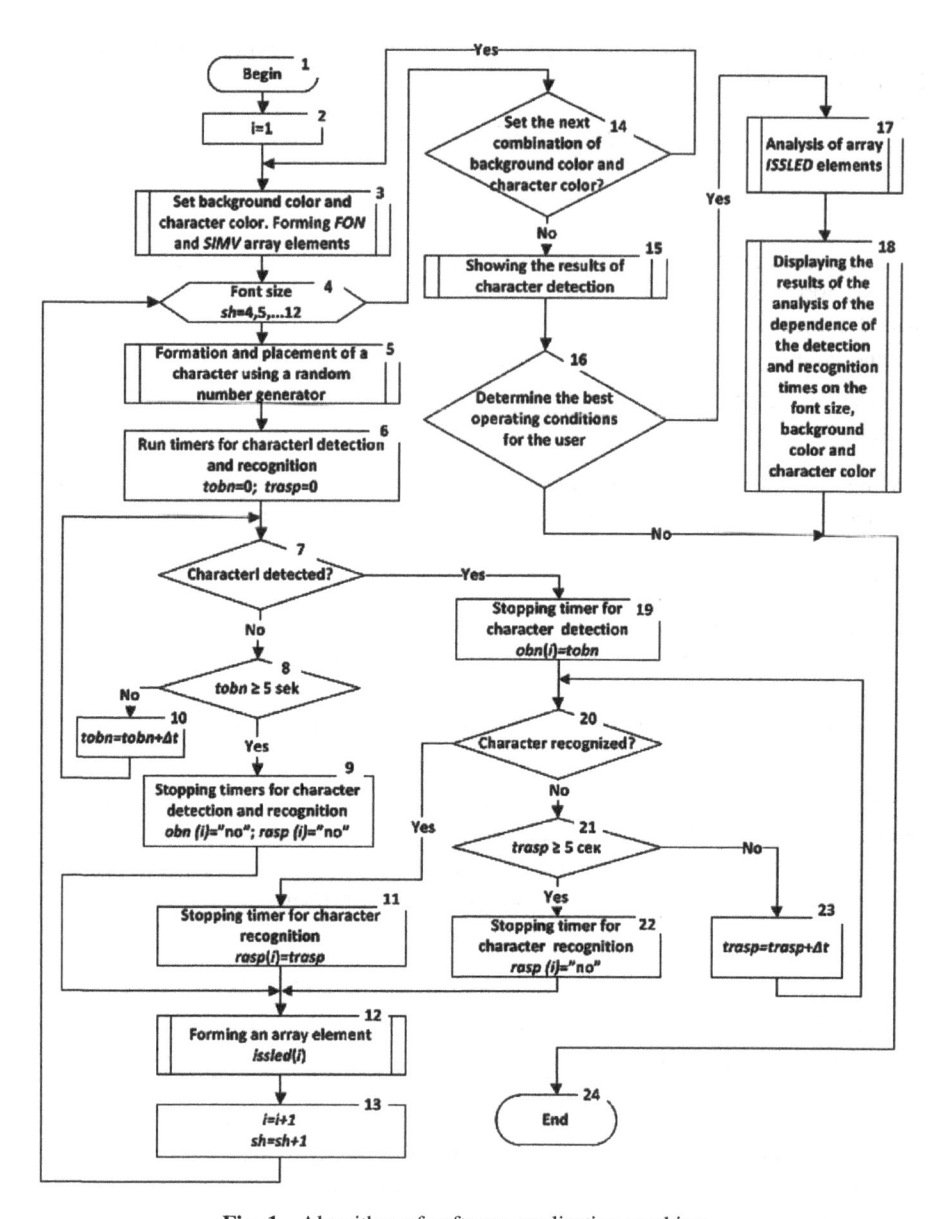

Fig. 1. Algorithm of software application working.

After the background color and the character color are selected, the "Character Detection and Recognition" dialog box appears again, but in the mode for character recognition and detection. The character is formed and placed on the user interface (operator 5, Fig. 1). The character is selected from the list of ASCII characters using a random number generator. The character display in the dialog box is also made using a random number generator.

Verification of detection and recognition by the user of characters begins with a character in the size 4pt. The user must first visually detect the character displayed in the "Character Detection and Recognition" dialog box.

If a character is found, click on the rectangle with the left mouse button with the inscription: "Click here with the left mouse button, if the character is detected", which is located in the lower left part of the dialog box (Fig. 2).

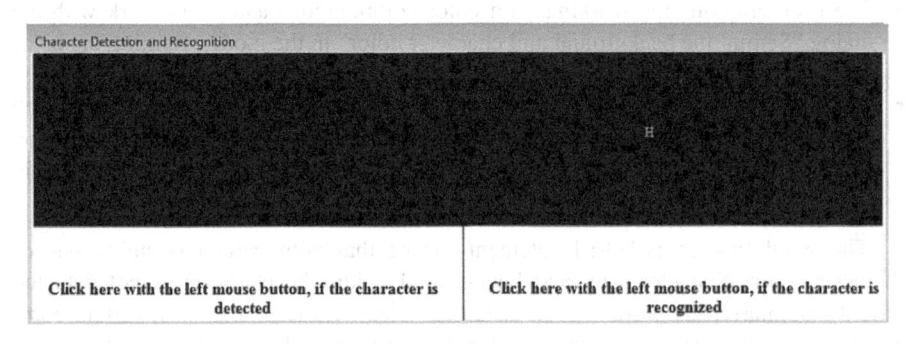

Fig. 2. The "Character detection and recognition" dialog box in detect and recognition mode.

This action corresponds to the operator 7 (Fig. 1). If the user does not detect it within 5 s after the character is displayed, the character is considered not detected and not recognized (operator 9, Fig. 1). In the case where the character is not found, the variables obn(i) and rasp(i) is assigned the value 'no' (operator 9, Fig. 1).

If the character is recognized, then click the left mouse button on the rectangle with the inscription: "Click here with the left mouse button, if the character is recognized", located in the lower right part of the dialog box (Fig. 2). This corresponds to the operators 20, 11 in Fig. 1. If the character is not recognized within 5 s after the character is detected, the character is considered to be detected, but not recognized (operators 20, 21, 22, Fig. 1). In the case where the character is not recognized, the rasp (i) variable is assigned the value 'no' (operators 9, 22, Fig. 1).

After the detection and recognition of the character for a particular font size is completed, the user is given a message stating that the work with a character of a certain size is completed with the given combination of the background color and the character color. After this, the work proceeds to the next character size with the same combination of background and character colors (operators 4, 13, Fig. 1). It returns to the "Character Detection and Recognition" dialog box (Fig. 2) in detection and recognition mode (operator 5, Fig. 1). The font size value varies from 4 to 12pt.

Combinations of detection and recognition results, character size, character color, background color are stored in the ISSLED array for further data processing. Each array element issled(i) (i = 1, 2, ..., I) is represented in the form issled(i) = {issled(i,1), issled(i,2), issled(i, 3), issled(i,4), issled(i,5)}, where

Issled(i, 1) – font size of the character
Issled(i, 2) – background color fon(i)

Issled(i, 3) – the color of the character simv(i)
Issled(i, 4) – the value of the detection time obn(i)
Issled(i, 5) – the value of the recognition time rasp(i)

At end of searching all the font sizes of the character for the next combination of the character color and the background color, application asks the user about his further actions (operator 14, Fig. 1). There are two options:

- The user can continue working with color combinations and go to work with the dialog "Setting the background and character color" in the mode of setting the color of the character and the background color (operator 3, Fig. 1)
- The user can proceed to the end of work with the output of the test results, which involves displaying dialog boxes to display the results of detection and recognition (for example, Fig. 3). Review of the test results corresponds to the operator 15 in Fig. 1

The word 'no' in issled(i,1) element means that with certain combinations of character color, character size, and background color, the character is not detected (Fig. 3). Similarly, the word 'no' in issled(i,2) element means that with certain combinations of character color, character size, and background color, the character is not recognized (operator 9, Fig. 1). After viewing the data (Fig. 3), a dialog box is displayed, with which the order of the user's further actions is selected (operator 16, Fig. 1).

The user can proceed to give recommendations on choosing the best combination for the user "font size of the character - the color of the character - the background color" or finish the work with the software application.

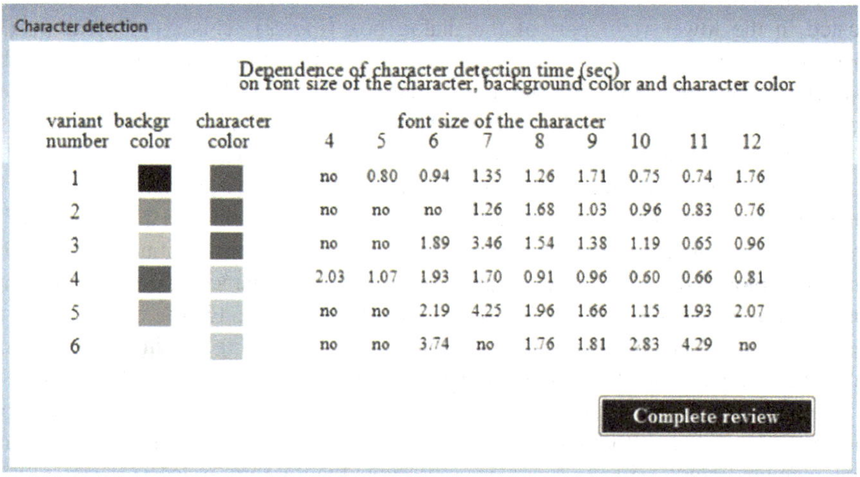

Character detection

Dependence of character detection time (sec)
on font size of the character, background color and character color

variant number	backgr color	character color	font size of the character								
			4	5	6	7	8	9	10	11	12
1			no	0.80	0.94	1.35	1.26	1.71	0.75	0.74	1.76
2			no	no	no	1.26	1.68	1.03	0.96	0.83	0.76
3			no	no	1.89	3.46	1.54	1.38	1.19	0.65	0.96
4			2.03	1.07	1.93	1.70	0.91	0.96	0.60	0.66	0.81
5			no	no	2.19	4.25	1.96	1.66	1.15	1.93	2.07
6			no	no	3.74	no	1.76	1.81	2.83	4.29	no

Complete review

Fig. 3. Dialog box for demonstrating the results of character detection.

In Fig. 3 'background color – character color' ratios were used:

- "ControlText – GrayText" (variant number 1)
- "ControlDark – GrayText" (variant number 2)
- "Scrollbar – GrayText" (variant number 3)
- "GrayText – InactiveCaption" (variant number 4)
- "AppWorkspace – InactiveCaption" (variant number 5)
- "Menu – InactiveCaption" (variant number 6)

In Fig. 4 for the 10pt font, the best combination for detection will be the "InactiveCaption" color character on the background of the "GrayText" color (detection time – 0.6 of second).

After the recommendations for all font sizes have been 'sorted', a transition to saving the issued recommendations in the form that the user chooses (in the form of a diagram with a diagram, in the form of an Excel table) is performed. Then, the program finishes its work.

Combinations for issuing recommendations for detecting characters are contained in the OBN array. Combinations for issuing recommendations for character recognition are contained in the RASP array. Each font size of the character displays its own dialog boxes for detecting and recognizing the character (operators 18, 19, Fig. 1). The analysis results are displayed in the dialog boxes "Recommendations for choosing background and character colors" (Fig. 4).

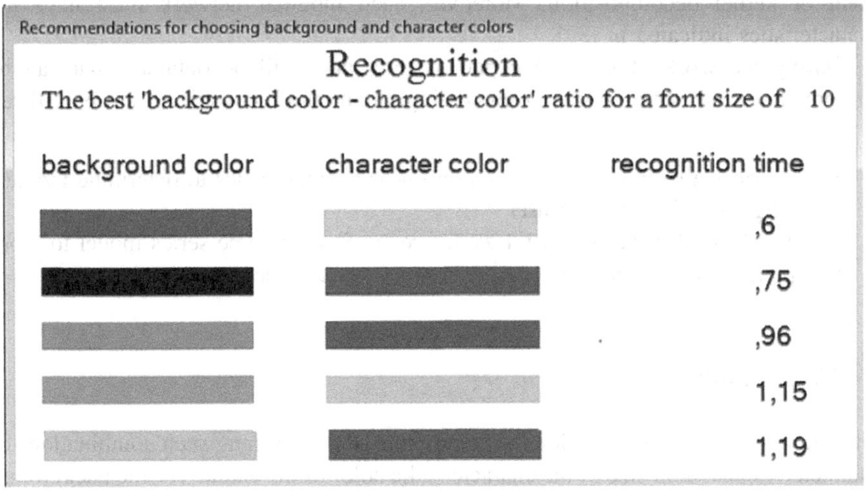

Fig. 4. Dialog box for viewing analysis results.

The developed software application can become a component of the expert system for automatically generating recommendations on the best conditions for recognizing information on the user interface.

Such a system will use neural networks, the initial data for which will be collected from the OBN, RASP arrays as the initial data for training.

Further, during the user's work with the information system, the expert system will analyze the conditions of user interaction with the interface and, taking into account the previously received data, create a user-friendly user interface. Such an expert system will be a hybrid system with interaction. In such a system, several software modules must interact with each other in order to solve a common problem - the formation of the interface most comfortable for the user. Such an expert system should include the following modules, each of which can implement a separate neural network:

- Module for generating initial data for learning a neural network (the software application discussed in this article)
- Module for analyzing the dynamics of user recognition of control objects and characters on the user interface (determining the speed of recognition of control objects)
- Module for determining the dynamics of the user (recognizing the dynamics of manipulation in the performance of tasks)
- Module for determining comfortable working conditions for the user
- Module for automatic generation of the user interface (interacts with the enterprise information system)

For the interaction of such software modules, the core of the modular neural networks must be implemented. In this case, the modular neural networks will be the result of kernel decomposition. Moreover, each modular network must have the characteristics indicated in [39].

During the work of such an expert system, data will be obtained that can be interpreted as time series. For their processing, the methods considered in [39, 40] can be used:

- An integrated method for selecting non-linear characteristics to determine the best operating conditions for the user
- The ANFIS (Adaptive Neuro Fuzzy Inference System) time series model to work with parameters that characterize the recognition of user interface objects

4   Conclusion

The article deals with the solution of the problem of determining such combinations of parameter values "font size of the character - the color of the character - the background color", which provide the best conditions for character recognition on the user interface.

The algorithm of the software application is developed and the sequence of actions for working with it is considered.

With the help of the software application, the user interface of the information system can be customized, taking into account the perception of the visual information of each employee using the information system. In this case, each user is provided with comfortable conditions for working with the user interface.

As a result, the user experience with the user interface is increased. The subjective satisfaction of the user also increases, and the number of user errors caused by character recognition errors is reduced.

A software application can serve as one of the components of an expert system for automatically generating a user interface, taking into account the current conditions of the user's work and his condition.

References

1. Shneiderman, B., Plaisant, C., Cohen, M., et al.: Designing the User Interface: Strategies for Effective Human-Computer Interaction. Pearson, London (2017)
2. Fitts, P.M.: The information capacity of the human motor system in controlling the amplitude of movement. J. Exp. Psychol. Gen. 47(6), 381–391 (1954). https://doi.org/10.1037/h0055392
3. Hick, W.E.: On the rate of gain of information. Q. J. Exp. Psychol. 4(1), 11–26 (1952). https://doi.org/10.1080/17470215208416600
4. Olson, J.R., Olson, G.M.: The growth of cognitive modeling in human-computer interaction since GOMS. Hum.-Comput. Interact. 5, 221–265 (1990). https://doi.org/10.1207/s15327051hci0502&3_4
5. Norman, K.L., Kirakowski, J.: The Wiley Handbook of Human Computer Interaction. Wiley, New York (2018)
6. Cooper, A., Reimann, R., Cronin, D., Noessel, C.: About Face: The Essentials of Interaction Design. Wiley, New York (2014)
7. Tidwell, J.: Designing Interfaces: Patterns for Effective Interaction Sesign. O'Reilly, Sebastopol (2011)
8. Tillman, B., Tillman, P., Rose, R., et al.: Human Factors and Ergonomics Design Handbook. McGraw-Hill Education, New York (2016)
9. Bakanov, A.S.: User interface ergonomics: From design to human-computer interaction modeling. Institute Psychology RAN, Moscow (2011)
10. Chernikov, B.V., Popov, A.A.: Optimization of ergonomic parameters of the information systems interface. News YUFU. Tech. Sci. 3(188), 65–77 (2017)
11. Seow, S.C.: Information theoretic models of HCI: a comparison of the Hick-Hyman law and Fitts' law. Hum. Comput. Interact. 20(3), 315–352 (2005). https://doi.org/10.1207/s15327051hci2003_3
12. MacKenzie, I.S.: Movement time prediction in human-computer interfaces. In: Proceedings of the Conference on Graphics Interface. Morgan Kaufmann Publishers Inc, San Francisco (1992). https://doi.org/10.20380/gi1992.17
13. MacKenzie, I.S., Buxton, W.: Prediction of pointing and dragging times in graphical user interfaces. Interact. Comput. 6(2), 213–227 (1994). https://doi.org/10.1016/0953-5438(94)90025-6
14. Ahlstrom, D.: Modeling and improving selection in cascading pull-down menus using Fitts' Law, the steering law and force fields. In: Proceedings of the SIGCHI Conference on Human Factors in Computing Systems. ACM, New York (2005). https://doi.org/10.1145/1054972.1054982
15. Kobayashi, M., Igarashi, T.: Considering the direction of cursor movement for efficient traversal of cascading menus. In: Proceedings of the 16th Annual ACM Symposium on User Interface Software and Technology. ACM, New York (2003). https://doi.org/10.1145/964696.964706

16. Blanch, R., Guiard, Y., Beaudouin-Lafon, M.: Semantic pointing: improving target acquisition with control-display ratio adaptation. In: Proceedings of the SIGCHI Conference on Human Factors in Computing Systems. ACM, New York (2004). https://doi.org/10.1145/985692.985758

17. MacKenzie, I.S., Teather, R.J.: FittsTilt: The application of Fitts' Law to tilt-based interaction. In: Proceedings of the 7th Nordic Conference on Human-Computer Interaction: Making Sense Through Design. ACM, New York (2012). https://doi.org/10.1145/2399016.2399103

18. YeonJoo, C., RoHae, M.: Extended Fitts' Law for 3D pointing tasks using 3D target arrangements. Int. J. Ind. Ergon. **43**(4), 350–355 (2013). https://doi.org/10.1016/j.ergon.2013.05.005

19. Xiaojun, B., Yang, L., Shumin, Z.: Fitts' Law: modeling finger touch with Fitts' Law. In: Proceedings of the SIGCHI Conference on Human Factors in Computing Systems. ACM, New York (2013). https://doi.org/10.1145/2470654.2466180

20. Sundar, S.S., Bellur, S., Oh, J., Qian, X., Haiyan, J.: User experience of on-screen interaction techniques: an experimental investigation of clicking, sliding, zooming, hovering, dragging, and flipping. Hum. Comput. Interact. **29**(2), 109–152 (2014). https://doi.org/10.1080/07370024.2013.789347

21. Erazo, O., Pino, J.: Predicting user performance time for hand gesture interfaces. Int. J. Ind. Ergon. **65**(5), 122–138 (2018). https://doi.org/10.1016/j.ergon.2017.07.010

22. Card, S., Moran, T., Newell, A.: The keystroke-level model for user performance time with interactive systems. Commun. ACM **23**(7), 396–410 (1980). https://doi.org/10.1145/358886.358895

23. Proctor, R.W., Schneider, D.: Hick's Law for choice reaction time: a review. Q. J. Exp. Psychol. **71**(6), 1–56 (2006). https://doi.org/10.1080/17470218.2017.1322622

24. Hai, Q., Shuping, X.: New Hick's Law based reaction test App reveals "information processing speed" better identifies high falls risk older people than "simple reaction time". Int. J. Ind. Ergon. **58**, 25–32 (2017). https://doi.org/10.1016/j.ergon.2017.01.004

25. Cockburn, A., Gutwin, C., Greenberg, S.: A predictive model of menu performance. In: Proceedings of the SIGCHI Conference on Human Factors in Computing Systems. ACM, New York (2007). https://doi.org/10.1145/1240624.1240723

26. Cockburn, A., Gutwin, C.: A predictive model of human performance with scrolling and hierarchical lists. Hum. Comput. Interact. **24**(3), 273–314 (2009). https://doi.org/10.1080/07370020902990402

27. Rosati, L.: How to design interfaces for choice: Hick-Hyman Law and classification for information architecture. In: Proceedings of Conference: International UDC Seminar. Classification and Visualization. Interfaces to Knowledge, Hague (2013)

28. Schneider, D.W., Anderson, J.R.: A memory-based model of Hick's Law. Cogn. Psychol. **62**(3), 193–222 (2011). https://doi.org/10.1016/j.cogpsych.2010.11.001

29. Zhao, J., Soukoreff, R.W., Ren, X., et al.: A model of scrolling on touch-sensitive displays. Int. J. Hum Comput Stud. **72**(12), 805–821 (2014). https://doi.org/10.1016/j.ijhcs.2014.07.003

30. Jeong, J., Kim, N., Hoh, P.I.: Adaptive kinetic scrolling: kinetic scrolling for large datasets on mobile devices. Appl. Sci. **8**(11) (2018). https://doi.org/10.3390/app8112015

31. Berezkin, B.S., Bakanova, N.M., Volkova, I.M.: Engineering and psychological requirements for management systems. All-Russian NII technical estetique State Committee of science and technique USSR Soviet of Ministers, Moscow (1967)

32. Wickens, C.D., Hollands, J.G.: Engineering and Human Performance. Prentice Hall Inc., New Jersey (2000)

33. Lomov, B.F., Venda, V.F., Zabrodin, J.M.: Psychological problems of mutual adaptation of man and machine in control systems. Science, Moscow (1980)
34. Venda, V.F.: Engineering Psychology and Synthesis of Information Display Systems. Mashinostroenie, Moscow (1975)
35. Litvak, I.I., Lomov, B.F., Soloveichik, I.E.: Basics of Construction of Display Equipment and Automated Systems. Soviet radio, Moscow (1975)
36. Varsky, V.F., Guzeva, M.A.: Dependence of the spatial thresholds of vision on the information content of the contour of a planar figure. Psychol. Questions **2**, 101–115 (1962)
37. Batenkina, O.V.: The Design of the User Interface of Information Systems. OmSTU Publishing, Omsk (2014)
38. Itten, J.: The Art of Color: The Subjective Experience and Objective Rationale of Color. Wiley, New York (1974)
39. Yarushev, S.A., Averkin, A.N.: Time series analysis based on modular architectures of neural networks. Procedia Comput. Sci. **123**, 562–567 (2018). https://doi.org/10.1016/j.procs.2018.01.085
40. Averkin, A.N., Yarushev, S.A.: Hybrid approach for time series forecasting based on ANFIS and fuzzy cognitive maps. In: XX IEEE International Conference on Soft Computing and Measurements (SCM). IEEE, St. Petersburg (2017). https://doi.org/10.1109/scm.2017.7970591

The Control-and-Measuring System Built-in Automatic Control System by the Technical Casting Process with Piezocrystallization

M. Denisov[✉]

Vladimir State University, 87 Gorkogo street,
Vladimir 600901, Russian Federation
denisovmaxim90@mail.ru

Abstract. The research results are reported; It was found that for effective process control under conditions of pressure application, it is necessary to press the casting volume by 8–10% before the crystallization start, and by 2–3% - at the moment of conversion from the liquid to the solid state. The technology based on pressure molding of a liquid metal prior to crystallization can be effectively implemented into production of castings from high-strength heat-hardenable aluminum alloys. The choice of technology parameters is proved, instrumentation, automation devices, control and software is justified. The issues about the choice of technological tooling and special equipment are considered. A process control and measuring system is proposed, a structure scheme for a hydraulic press control system with the possibility of programmatically pressure application on the crystallizing metal is developed. The dependence of formation of casting physical mechanical properties on the applied pressure magnitude is determined.

Keywords: Formation control · Hardness · Aluminum alloys · Measuring system · Control system · Hydraulic press · Crystallization

1 Introduction

For now the processes of crystallization of high-strength aluminum alloys are a complex, up to the present understudied process in which the super duty steel products properties are formed. This process includes heat-mass-exchange, crystalline transformation, chemical reactions, expelling of solution gases and non-metallic compounds. Besides, during the crystallization process the liquation defects, weakness, porousness, gas and shrinkholes occur, which negative effect on such properties as hardness, plasticity and elasticity are removed in the rolling and die forging production.

The study of scientific and technical materials makes the case that the existing automatic process control systems for casting are not effective enough [1–3]. Such control systems are unable to measure in real time, to control and correct the physic-technological process parameters affecting the formation of final product properties such as the time and temperature pattern in the tooling and moulding, the rate of pressure superimposition on the crystallized metal, the value of superimposed pressure.

A. A. Radionov and A. S. Karandaev (Eds.): RusAutoCon 2019, LNEE 641, pp. 852–860, 2020.
https://doi.org/10.1007/978-3-030-39225-3_92

From the public sources [4–6] it will be obvious that the possibility of metal forming before the crystallization remains understudied, therefore the subject matter of appropriate control systems creating remains open.

In this regard, it can be argued that the automation process of the blanks production from high-strength aluminum alloys with enhanced, stable predictable qualitative indicators based on high pressure die casting technology with crystallization using modern control and measurement instruments becomes a current scientific-technical challenge.

2 Body

The forming processes of crystallizing metal, in the context of process control, are discrete-continuous, there is a problem of these technological processes integration in automatic control systems (ACS). The problem consists in that fact that the physical forming processes are not in coincidence with the information processes in time.

At present, the control of technical die-casting process with crystallization consists of the issue commands or of the regulation of parameter values on the "Input" and of the control of certain parameters on the "Output". The progress of the forming process remains unknown.

The use of various models of the studied processes is of practical interest, but does not exclude the necessity of creating of the control instrumentation systems (CIS) that can visualize and transmit the sensors readings on the status and the process progress in time through industrial information networks.

So far as the metals and alloys treatment is concerned, then a certain combination of input parameter values, it means a combination of input system parameters will correspond to the material status and determine its structure and output properties. In this regard, by changing the input parameters values according to programme, it is possible to bring the system to a final (set) state along the path of fixed states.

The proposed approach is used in the present work [7] where the pressure molding of a crystallizing metal under high pressure is considered.

There are three variable parameters in the presented process: the set metal pressure, the metal temperature, holding- pressure time. The pressure molding is carried out as follows: the molten aluminum alloy of type B95 at a temperature of 800–850 °C is poured into the special die cavity heated to 200–250 °C, then with the help of injection plungers moving towards each other, the pressure is applied on the liquid metal according to the program (Fig. 1).

The pressure molding is carried out on a hydraulic press of the original design [8]. The press is equipped with an information system (CIS) including plungers displacement sensors, pressure sensors in the hydraulic cylinder cavity, temperature sensors monitoring the temperature of metal and the die. This system is able to transmit information through the appropriate controllers to the ARM (automated workstation) of the operator. The developed system is a part of the ACS for the process of pressure application and allows to solve the following problems.

- To correct defects and select control modes that ensure consistently high product quality quickly, practically in 3–5 cycles
- To form databases, knowledge bases, on the basis of which to create decision support system

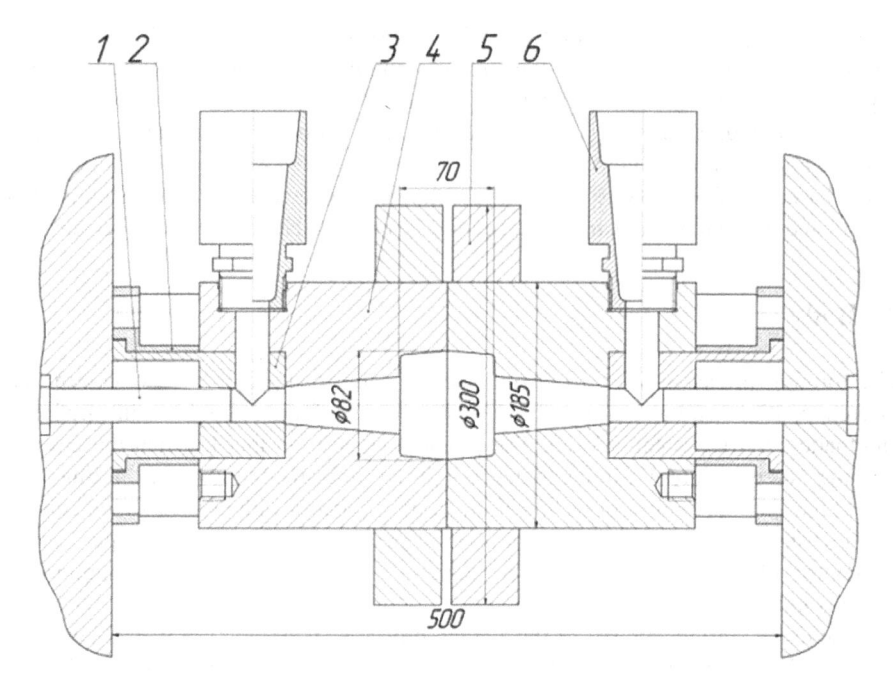

Fig. 1. Scheme of pressure application on the crystallizing metal: 1 – injection plunger; 2 – base; 3 – casting chamber; 4 – matrix; 5 – bandage; 6 – pouring bush.

- To feed information to higher control levels when it is necessary, to form the infomedia of the enterprise

For the unimpaired CIS operation as a part of the ACS, the control system structure for hydraulic press was designed (Fig. 2).

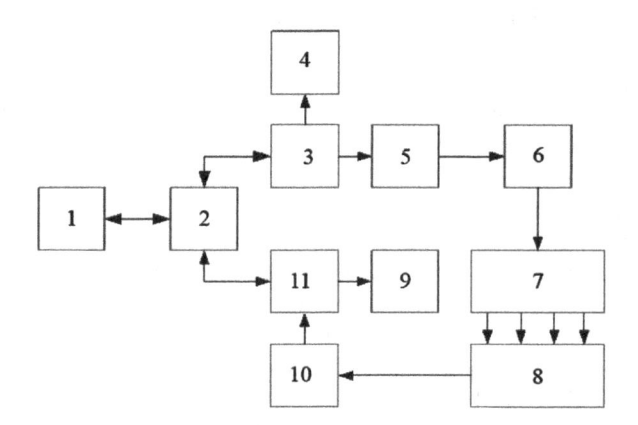

Fig. 2. Structure of the hydraulic press control system: 1 – personal computer; 2 – HUB; 3 – programmable logic controller 1 (PLC1); 4 – operator panel 1; 5 – computer-process interface 2; 6 – stepper motor; 7 – hydraulic press; 8 – sensors; 9 – operator panel 2; 10 – process interface 1; 11 – programmable logic controller 2 (PLC2).

From the earlier researches, it was found [9] that the problem of the control of pressure application process requires making of a formalized model that establishes the dependence of compressibility of the crystallizing metal [10] on the applied pressure value and its application rate. The curve (in the equation system let us denote it by τ) with sections ABC in Fig. 4 taking into account the dependence made in Fig. 3 can be represented by three following Eq. (1), where the output of the system is compressibility and the input the pressure.

The system of equations appears as follows:

$$\begin{cases} \varepsilon = kp(\tau), \ 0 \leq \tau \leq t_1 \\ \varepsilon = k_1 p(\tau) + b, \ t_1 \leq \tau \leq t_2, \\ \varepsilon = k_2 p(\tau), \ t_2 \leq \tau \leq t_3 \end{cases} \tag{1}$$

where k, $k1$, $k2$ – model coefficients, p (τ) – pressure on the sections OA, AB, BC; τ – current time; 0–$t1$, $t1$–$t2$, $t2$–$t3$ – time intervals characterizing the sections OA, AB, BC.

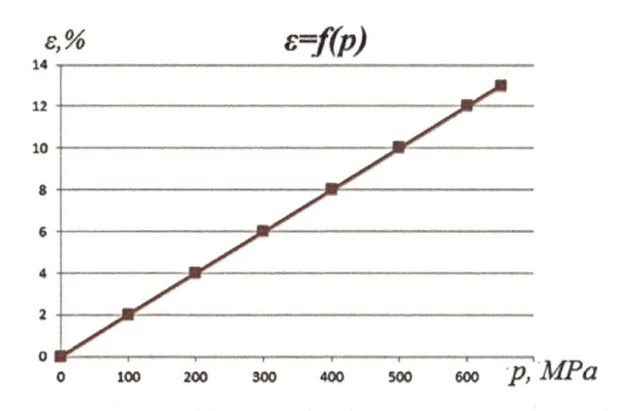

Fig. 3. The dependence of the alloy compressibility on the applied pressure value.

As shown in works [11, 12], the processes of structure and properties formation can be also inferred by the change in the so-called informative parameter. Such a parameter in our work is "compressibility", i.e. relative metal volume reduction under pressure.

All information from the sensors is recorded in graphics. The resulting graphics characterize the behavior of the system during the process of pressure application to the crystallizing metal. The graphics display information about the movement of the injection plungers.

Graphics processing allows us to imagine the progress of the molten metal-forming process in time, taking into account the actual movement of the injection plungers in mm (Fig. 5).

Processing of received information allowed us to make the following conclusions.

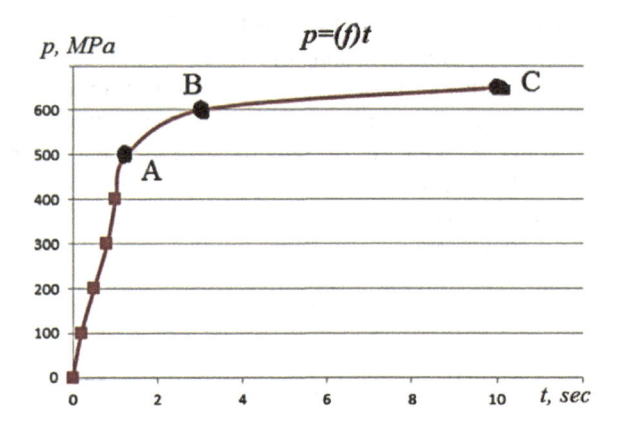

Fig. 4. The pressure change dependence in time.

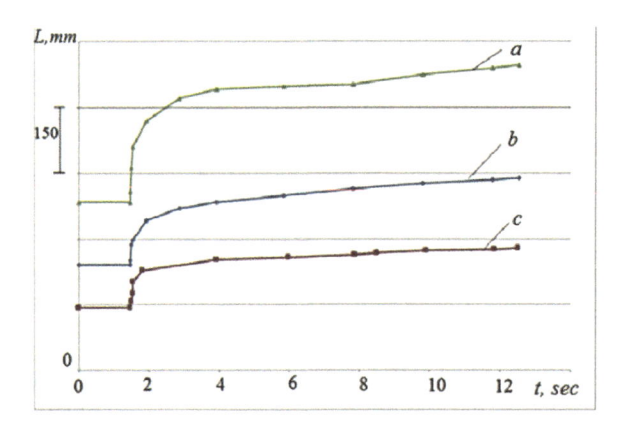

Fig. 5. Recording the injection plungers' movement: a – plunger 1; b – plunger 2; c – summary plungers' movement.

First, the plunger (a) is moving quickly, and then delayed into the casting, whereafter it remains in a constant position, because the corresponding hydraulic cylinder piston comes to stop. The right injection plunger stroke is 150 mm.

The plunger (b), presses the liquid metal at first quickly, and then exponentially in time, approximately, 12 s, its stroke is 125 mm.

In its turn, the dependence characterizing the summary plungers' movement (c) makes it possible to calculate the pressed metal volume inside the casting. With that knowledge in mind, we can determine the relative compressibility, which will be equal to:

$$\varepsilon = \frac{V_{kp}}{V_0} \cdot 100\%, \tag{2}$$

where V_0 – is the total metal volume poured into the mold, V_{cr} – is the metal volume pressed into the casting.

Formula (2) and curves a and b (Fig. 5) provide the information necessary to control the crystallization process, which is represented as the following differential equation system:

$$\frac{dt}{d\tau} = \left(\frac{\delta^2 t}{\delta x^2} + \ldots\right) + LV_{cr} \cdot \rho \tag{3}$$

$$\frac{dp}{d\tau} = \frac{E}{V_0}\frac{dV}{d\tau}, \quad d_p = \varepsilon E \tag{4}$$

$$M\frac{d^2 x}{d\tau^2} = F\Delta p - \lambda\frac{dx}{d\tau} - R \tag{5}$$

$$\frac{dp}{d\tau} = \frac{E}{V}\left(K_y\Delta yl - F\frac{dz}{d\tau}\right) \tag{6}$$

$$Q = \mu yl\sqrt{2(p_h - p_1)} \tag{7}$$

Equation (3) can be used to determine the crystallization rate V_{cr} on condition that the boundary and pre-conditions from experimental data will be described. Equation (4) from the measured relative volume change of the liquid metal under pressure is used to calculate the elastic modulus of the liquid metal ε.

Equations (5)–(7) provide guidance on the hydraulic system operation in dynamic modes.

In the system of Eqs. (3)–(7) the following legends are used: t – temperature; τ – time; x, y, z – coordinates in the volume of the crystallizing metal; L – latent heat of crystallization; ρ – metal density; M – mass of moving parts of the hydraulic cylinder; r – plunger movement; F – piston area; λ – frictional coefficient in the hydraulic actuator; R – the force acting on the liquid metal; p – pressure; V – volume of the liquid metal; R – the force acting on the liquid metal; K_f – force ratio of the pressure regulator; μ – the coefficient of regulator flow; l – moving of controlling element; pH – input pressure in regulator; p_h – pressure on the piston of the hydraulic cylinder; p_1 – backpressure in the hydraulic system.

Based on the researches, the parameters requiring monitoring and controlling were identified, an approach to the equipment automation, including keeping the specified pressure values in the press hydraulic system, controlling of speed and injection plungers moving was developed.

It is assumed that the process control function can be calculated or constructed according to the experimental data. This is a monitored function, and the law of pressure application acts both as a manageable and as a controlled function.

The process control lies in the fact that at first a certain trajectory is set, and then, the path is projected onto the plane by reading the curve coordinates. As a result, a law under which the pressure should be applied is made.

We propose a process in which the forming, crystallization, shrinkage compensation, cancellation of liquation, formation of the required metal products structure and properties proceed consistently in inextricable connection along its predesigned trajectory in accordance with the developed control program.

The control task is to rationalize choosing of such control actions, which ensure the manufacturing of products with required combination of physical, mechanical and structural properties.

In this regard, the problem of choice of control actions on the control object is related to programming, influential expanding the functional capabilities of control systems [13–15] [16].

The developed approach ensures the conversion of melt to the state when the mechanism of the atoms ordering allows the formation of quasi-crystalline structures with the necessary combination of amorphous and crystalline phases. Duc to the stronger and more efficient interatomic linkage, the physical mechanical properties of the final products are enhanced. A fundamentally new approach is proposed for conversion of melt to a quasi-crystalline state -duc to approximation the closest distance of the melt atoms. During all-round compression of the melt it is proposed to assemble the atoms so as to form the associative groups of several atoms having a crystalline structure, while at this time in other parts the atom arrangement would remain chaotic. In such conditions, more efficient interatomic bonds and a general uprating of the physical mechanical properties of the alloy without changing its chemical composition are provided.

The research results show that the uprating of mechanical properties is determined of the structure formation which have no analogues in the alloys used in industry. Samples obtained using the developed process control system for the mold's casting are distinguished by enhanced physical and mechanical properties, namely hardness (Fig. 6).

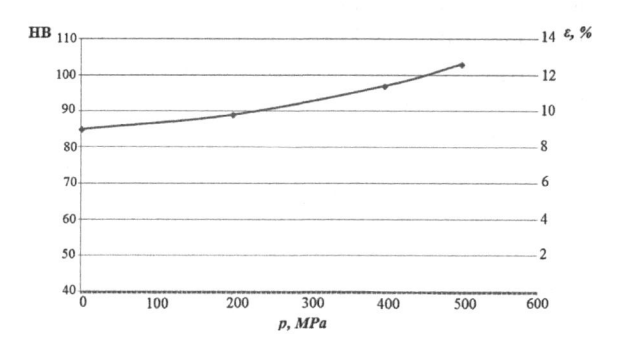

Fig. 6. Change dependency for hardness and compressibility on the pressure in the hydraulic system.

It was established that the physical mechanical properties of the final metal products can be changed depending on the mode of pressure application on the crystallizing metal, a control system is created that adjusts the dynamic properties of the hydraulic actuator in accordance with the behavior of the molten metal under pressure.

Results available from experiments make it possible to identify the main factors that determine the structural and phase transformations of the final products:

- Kinetic factor
- Thermodynamic factor
- Technological factors of pressure

The combined effects of these factors onto the alloy structure provides effective increase of its structural and therefore physical and chemical characteristics.

Consequently, we have confirmed the practicability and perspective of further development of the technology of high pressure application for high-strength aluminum alloys. The proposed experimental technological cycle for the castings production from high-duty aluminum alloy can become basis for industrial implementation of the process, in order to produce the parts and products which are highly competitive in their strength characteristics with leading foreign aluminum alloys of medium and high strength.

The results discussed above were obtained thanks to modern software and hardware, which allow receiving, processing and transmitting the necessary control signals with high speed.

Engineering and implementation of modern automated systems to control the nature and level of interatomic bonds and interactions will allow us to influence, and therefore purposefully change the properties of metal products.

3 Conclusion

It must be noted that automated production equipment used in industry is usually designed for specific technological processes, and automation, if designed, then, basically, as a digital program control.

The concept of flexible, adaptable production suggests the principle of "reverse engineering" when a "product for a resource" is designed. And this resource, in particular, a flexible production line should be quite multifunctional. The higher flexibility degree is, the longer system life cycle is, and a large number of consumers can use it as a basic one.

To this extent, the CIS integrated into the ACS for the technical process does not only reduce the time required for work preparation and production engineering, but also significantly expands opportunities of operators to manage technological processes in optimal modes.

References

1. Denisov, M.S.: Justification of aluminum alloys forming modes in the process of their crystallization. In: Materials of the 13th Russian Annual Conference of Young Scientists and Postgraduates Physics-Chemistry and Technology of Inorganic Materials, Moscow, 18–21 October 2016

2. Denisov, M.S.: Development of technology for formation of quasi-crystalline structures of aluminum alloys by casting with crystallization under pressure. In: Materials of 6th International Conference Functional Nanomaterials and High-Purity Substances. Suzdal, 3–7 October 2016

3. Pilipenko, A.V.: Adaptive control system of non-stationary forming process. Inf. Syst. Technol. 4(46), 115–119 (2011)

4. Korostelev, V.F.: Case- and Volume Hardening of Alloys. Publishing House New Technologies, Moscow (2013)

5. Krstic, M., Wang, H.H.: Stability of extremum seeking feedback for general nonlinear dynamic systems. Automation 36, 595–601 (2000)

6. Parsheva, E.A., Tsykunov, A.M.: Adaptive control of an object with delayed control with a scalar input-output. Autom. Remote Control 1, 142–149 (2001)

7. Denisov, M.S.: Justification of aluminum alloys forming modes in the process of their crystallization. In: Physical Chemistry and Technology of Inorganic Materials of XIII All-Russian Conference of Young Researchers and Graduate Students with International Participation, Moscow, pp. 367–369, 18–21 October 2016

8. Korostelev, V.F.: Theory, Technology and Automation of Casting with the Pressure Application. New Technologies Publishing House, Moscow (2004)

9. Denisov, M.S.: Pressure casting with crystallization as a method of producing metal blanks with nanostructure elements NANO. In: Materials of VI All-Russian Conference on Nanomaterials with Elements of Scientific School for Youth, Moscow, 22–25 November 2016, pp. 342–344 (2016)

10. Denisov, M.S.: The effect of cooling rate on the crystallization of aluminum alloys under pressure. In: 17th International Workshop on New Approaches to High-Technology Nano-Design, Technology, Computer Simulations, pp. 26–28 (2017)

11. Korostelev, V.F., Denisov, M.S.: Crystallization of aluminum alloys under pressure. In: MIST Aerospace 2018: IOP Conference Series: Materials Science and Engineering, vol. 450 (2018). https://doi.org/10.1088/1757-899x/450/3/032012

12. Denisov, M.S.: Modeling of processes of solidification of metals under conditions imposing pressure. Mater. Sci. Forum 945, 603–610 (2019)

13. Rutkovsky, VYu., Glumov, V.M.: Characteristic properties of dynamics of the adaptive control system with a nonlinear reference model. Autom. Telemech. 4, 92–105 (2017)

14. Aseltine, J.A., Mancini, A.R., Sarture, C.W.: A survey of adaptive control systems. IRE Trans. Autom. Control 6(12), 102–108 (1958)

15. Bellman, R., Kalaba, R.: Dynamic programming and adaptive control processes: Mathematical foundations. IRE Trans. Autom. Control 5, 5–10 (1960)

Statistical Simulation and Probability Calculation of Mechanical Parts Connection Parameters for CAD/CAM Systems

S. Skvortsov[1,2], V. Khryukin[1], and T. Skvortsova[2(✉)]

[1] Ryazan State Radio Engineering University, 59 Gagarina Street,
Ryazan 390005, Russian Federation
[2] Academy of the FPS of Russia, 1 Sennaya Street,
Ryazan 3900000, Russian Federation
t.s.skvortsova@yandex.ru

Abstract. The task of calculation procedures formalization for determination of mechanical units connections parameters in CAD/CAM systems are considered. It is shown that in the experimental and single production the task of determining of tolerances, deviations and fits for holes and shafts can be solved with the use of the manufacturing formulas of the Unified System of Tolerances and Fits (USTF). To the calculation of these parameters in large-scale and mass manufacturing, it is proposed to apply the standard ratios of the USTF with correction factors that take into account the distribution laws of size tolerances for mating parts. The purpose of this article is to develop a methodology for calculating the connection tolerance in the most General case under any laws of distribution of tolerance fields of the shaft and hole, as well as in the absence of restrictions on the values of these tolerances. The technique to solve the task of tolerances and fits determining for mating parts based on stochastic simulation taking into account the technological features of manufacturing is developed. The requirements to this stochastic modeling procedure, providing the results with a given accuracy, are formulated. The conditions of complex application in CAD/CAM systems of this technique in combination with known methods taking into consideration technological features of manufacturing are defined.

Keywords: Tolerances and fits · Holes and shafts · Statistical simulation

1 Introduction

In the process of automated design of modern electronic computing tools (ECT), there is often a task of calculating the connection parameters of parts and components of their electromechanical components [1]. During solving such task, the deterministic approach is most often used, which involves the use of the calculation formulas of the Unified System of Tolerances and Fits (USTF) [2]. This system complies with the international standards ISO 286–1: 2010 [3] and ISO 286–2: 2010 [4].

In this approach, deviations from the nominal sizes are considered as limiting, and the tolerance of fit is defined as the sum of the tolerances of the hole and the shaft that make up the connection. Here, "shaft" and "hole" are terms used to refer to the external

A. A. Radionov and A. S. Karandaev (Eds.): RusAutoCon 2019, LNEE 641, pp. 861–870, 2020.
https://doi.org/10.1007/978-3-030-39225-3_93

and internal elements of details, respectively, including non-cylindrical elements. The tolerance of Td shaft size (TD holes) is defined as the difference between the largest and smallest limiting dimensions or the algebraic difference between the upper and lower deviations [5]. In this case, complete interchangeability of products is ensured. However, with this method of calculation, the requirements for dimensional accuracy are tightened, which leads to higher prices for mechanical parts of structures [6].

There is a possibility to reduce the requirements for the accuracy of manufacturing details and components with virtually no deterioration in the parameters of their connections. To solve this task, a theoretical and probabilistic approach is used. It assumes that the tolerance fields obey certain laws of distribution [7–9].

In particular, with the normal distribution of the fields of tolerances of the shaft and the hole, which is typical for mass production of products, the distribution of the tolerance field of the connection will follow the normal law, and with the uniform distribution of the tolerances of the shaft and the hole, typical for individual production of products, – Simpson's law [3].

This allows the use of standard USTF ratios for calculating tolerances and fits based on correction factors. Then, in according to the work [5], the admission of the connection will be determined by the formula:

$$TS = \frac{1}{k_s} \sqrt{k_d^2 \cdot Td^2 + k_D^2 \cdot TD^2}, \tag{1}$$

where k_d, k_D, k_S are correction coefficients for the tolerances of the shaft, hole and their connections, respectively.

During using the above distributions, these coefficients have the following meanings [6]:

- For normal law $k = 1$
- For uniform law $k = 1,73$
- For Simpson's law $k = 1,22$

For example, the required tolerance of connection in mass production can be obtained with an increase of approximately 1.4 times the shaft and hole tolerances compared to tolerances that ensure complete interchangeability. In this case, only for 0,27% of all connections, there is the possibility of a marriage arises [6]. Thus, considering the laws of distribution of tolerances in technological processes can significantly reduce the cost of manufacturing details and assembling components.

2   Formulation of the Problem

However, the use of the coefficient calculation method has constraints. It is applicable only in two cases:

- With the normal distribution of the fields of the tolerances of the shaft and the hole
- With the uniform distribution of the fields of the tolerances of the shaft and the hole and the commensurability of the values of these tolerances

In other circumstances, it is necessary to determine the k_S coefficient, which will not coincide with any of the above values. The same applies to the k_d, k_D coefficients, if the tolerance fields of the shaft or hole do not obey the distribution laws discussed above.

The purpose of this article is to develop a method for calculating the connection tolerance in the most general case, i.e. with arbitrary laws of the distribution of the fields of the tolerances of the shaft and the hole, as well as in the absence of constraints on the values of these tolerances.

Consider the following problem. Let, as a result of the study of the technological process, it is established that the tolerance fields of the mating details (shaft and hole) obey certain laws of distribution with known characteristics.

- Value of TS tolerance
- Limiting clearances (S_{max}, S_{min}) or tightness (N_{max}, N_{min})
- Probability of obtaining clearances P_S or tightness P_N

To achieve this purpose, it is proposed to use the method of statistical simulation, which is universal and is widely used in solving engineering and scientific problems [10, 11].

To solve the task of studying the characteristics of the connections of details, it is proposed to use the procedure of forming a sequence of random numbers that imitate deviations of the shaft and hole sizes from their nominal value. Such procedure can be implemented on the basis of the inverse function method or the elimination method [12–14].

Let the random variables X and Y denote the dimensions of the shaft and the hole, respectively, where $x_1, x_2, \ldots, x_n$ and $y_1, y_2, \ldots, y_n$ are the totality of their values obtained as a result of simulation using one of the above methods. Then it is possible to calculate the differences $\Delta_i = y_i - x_i$ ($i = 1, 2, \ldots, n$) and on the basis of them determine the limit clearance or tightness [15, 16].

For fits with the clearance, where all $\Delta_i > 0$, we get

$$S_{min} = \min \Delta_i; \quad S_{max} = \max \Delta_i; \quad i = 1, 2, \ldots, n$$

and for fits with the clearance, where all $\Delta_i < 0$:

$$N_{min} = \min |\Delta_i|; \quad N_{max} = \max |\Delta_i|; \quad i = 1, 2, \ldots, n.$$

During transitional fits, where Δ_i may have different signs, the maximum clearance and maximum tightness are calculated. The maximum clearance is determined for all $\Delta_i > 0$ by the formula

$$S_{max} = \max \Delta_i; \quad i = 1, 2, \ldots, n \tag{2}$$

The maximum clearance is determined for all $\Delta_i < 0$ by the formula

$$N_{max} = \max \Delta_i; \quad i = 1, 2, \ldots, n \tag{3}$$

The value of connection tolerance will be determined by the ratio:

$$TS = \max \varDelta_i - \max \varDelta_j; \quad i,j = 1, 2, \ldots, n \qquad (4)$$

The probability of obtaining the clearance and tightness can be estimated by the frequency of occurrence of Δ_i values with different signs:

$$P_1 \approx n_1/n_2; \quad P_2 \approx n_2/n_1, \qquad (5)$$

where n_1, n_2 are the number of positive and negative values of Δ_i, respectively, and $n_1 + n_2 = n$.

3 Evaluation of the Reliability of the Simulation Results

During stochastic calculating the parameters of the connection details, the question arises of choosing the characteristics of the law of distribution of the tolerance field [15, 16], such as mathematical expectation (ME) m, variance D or the standard deviation (SD) σ.

For a limited number of values of n random numbers, by which the tolerance is determined, we can only speak of obtaining estimates $\tilde{m}$, $\tilde{D}$ of these characteristics [12]. When m and D are replaced by their estimates $\tilde{m}$, $\tilde{D}$, errors are made. The value of each error depends on n and as $n \to \infty$, the estimate approaches the parameter, and their difference tends to zero. Therefore, the task arises of determining the value of n, at which the error will be less than a given value of ε (estimation accuracy) with probability β (estimation reliability) [17–19].

Find the minimum value of n that satisfies the specified requirements in estimating the mathematical expectation. Assuming that the estimate $\tilde{m}$ is subject to the normal law, it is possible to determine the distribution parameters [12]:

$$M[\tilde{m}] = \tilde{m}; \quad D[\tilde{m}] = \tilde{D}/n; \quad \sigma[\tilde{m}] = \sqrt{\tilde{D}/n} \qquad (6)$$

where $M[\tilde{m}]$, $D[\tilde{m}]$, $\sigma[\tilde{m}]$ are values of mathematical expectation, variance and standard deviation of estimating the mathematical expectation of tolerance field.

The values $\tilde{m}, \tilde{D}$ can be determined approximately, assuming that the distribution of the tolerance field is normal. Then

$$\tilde{m} = \frac{T_d + T_D}{2}; \quad \tilde{\sigma} = \frac{T_d + T_D}{2}; \quad \tilde{D} = \tilde{\sigma}^2 \qquad (7)$$

A more accurate estimate can be made for the first n_0 values of the generated random value (tolerance values TS_i) [20]:

$$\tilde{m} = \frac{\sum_{i=1}^{n_0} TS_i}{n_0}; \quad \tilde{D} = \frac{1}{n_0 - 1} \sum_{i=1}^{n_0} (TS_i - \tilde{m})^2$$

In work [12] was shown that even for $n_0 > 30$ such estimates will be consistent, unbiased, and effective.

The value $\sigma[\tilde{m}]$ in the formula (2) can be determined by assuming that the probability of a deviation of the estimate $\tilde{m}$ from mathematical expectation m less than ε_m, is equal to β_m:

$$P\{|\tilde{m} - m| < \varepsilon_m\} = \beta_m$$

where ε_m, β_m are accuracy and reliability of estimating the mathematical expectation.

By virtue of the above assumption about the normal distribution, this probability $\tilde{m}$ is determined using the Laplace function $\Phi(x)$ as follows:

$$P\{|\tilde{m} - m| < \varepsilon_m\} \approx 2\Phi\left(\frac{\varepsilon_m}{\sigma[\tilde{m}]}\right) \approx \beta_m$$

From here we will have:

$$\sigma[\tilde{m}] = \frac{\varepsilon_m}{\Phi^{-1}\left(\frac{\beta_m}{2}\right)}, \tag{8}$$

where $\Phi^{-1}(x)$ is inverse Laplace function.

The value ε_m is usually given as the fraction Δ_m from the mathematical expectation of tolerance field, but since only the estimate $\tilde{m}$ is known, then $\varepsilon_m = \tilde{m}\Delta_m$.

Then on the basis of (6) and (8) we will get:

$$n = \frac{\tilde{D}}{\sigma^2[\tilde{m}]} = \frac{\tilde{D}\left[\Phi^{-1}(\beta_m/2)\right]^2}{(\tilde{m}\Delta_m)^2} \tag{9}$$

The minimum value n to obtain an estimate of the variance is also calculated under the assumption that $\tilde{D}$ obeys the normal law. From here we can determine the distribution parameters as follows [12]:

$$M[\tilde{D}] = \tilde{D}; \ D[\tilde{D}] = \frac{2}{n-1}\tilde{D}^2; \ \sigma[\tilde{D}] = \tilde{D}\sqrt{\frac{2}{n-1}} \tag{10}$$

The value $\sigma[\tilde{D}]$ can be found similarly, assuming that the probability of the deviation of the estimate $\tilde{D}$ from the variance D less than ε_D is equal to β_D:

$$P\{|\tilde{D} - D| < \varepsilon_D\} \approx 2\Phi\left(\frac{\varepsilon_D}{\sigma[\tilde{D}]}\right) \approx \beta_D$$

where ε_D, β_D are accuracy and reliability of estimating the variance. From here

$$\sigma\left[\widetilde{D}\right] = \frac{\varepsilon_D}{\Phi^{-1}\left(\frac{\beta_D}{2}\right)} \tag{11}$$

From here value ε_D can be also determined as $\varepsilon_D = \widetilde{D}\Delta_D$. Then, considering (10) and (11), we well get:

$$n = \frac{2\widetilde{D}^2}{\sigma^2\left[\widetilde{D}\right]} + 1 = \frac{2\left[\Phi^{-1}(\beta_D/2)\right]^2}{\Delta_D^2} + 1 \tag{12}$$

It is obvious that the required value n will be determined as the maximum of the numbers obtained by formulas (9) and (12):

$$n = \max\left\{\frac{\widetilde{D}\left[\Phi^{-1}(\beta_m/2)\right]^2}{(\widetilde{m}\Delta_m)^2}; \frac{2\left[\Phi^{-1}(\beta_D/2)\right]^2}{\Delta_D^2} + 1\right\}. \tag{13}$$

4 Example

Calculate the parameters of the connection $\varnothing$ 60 H7/m6 or $\varnothing$ 60 $^{+0,03}_{+0,011}$ $^{+0,03}$.

The scheme of tolerance fields in this case will have the form shown in Fig. 1, where the following notation is used: *ES, es* are the upper limit deviations of the hole and shaft from the nominal size; *EI, ei* are the lower limit deviations of the hole and shaft from the nominal size.

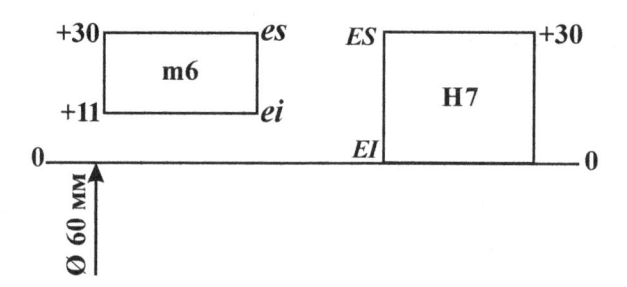

Fig. 1. The scheme of tolerance fields of fit $\varnothing$ 60 H7/m6.

Calculate the parameters of the connection by the method of complete interchangeability, i.e. use the standard ratios of the USTF [5]. As a result, for the connection $\varnothing$ 60 H7/m6 we will get:

- Limiting tightness $N_{max} = es - EI = 30\ \mu m$
- Limiting clearance $S_{max} = ES - ei = 19\ \mu m$
- Tolerance of fit $TS = S_{max} + N_{max} = 49\ \mu m$

Define the parameters of the same connection, assuming that the scattering of the sizes of the holes and shafts, as well as the clearance (tightness), obeys the law of normal distribution. We also assume that the tolerance of the size of each detail is equal to the scattering field, i.e. $T = 6\sigma$. As a result, we will get the tolerance values:

- For hole $\varnothing$ 60 H7: $TD = 6\sigma_D = ES - EI = 30\ \mu m$
- For shaft $\varnothing$ 60 m6: $Td = 6\sigma_d = es - ei = 19\ \mu m$

The tolerance of fit can be calculated by the coefficient method by the formula (1). For the normal distribution, we will get:

$$TS^B = \frac{1}{1,73}\sqrt{1,73^2 \cdot 19^2 + 1,73 \cdot 30^2} = 35,5\ \mu m$$

From here the standard deviation of connection will make

$$\sigma_S = TS^B/6 = 5,92\ \mu m$$

Probabilistic values of the limit tightness N_{max}^B and clearance S_{max}^B can be obtained as deviations $\pm 3\sigma_S$ from the average value of the tightness, which is determined by the average values of the shaft and hole sizes:

$$N_{av} = (es + ei)/2 - (ES - EI)/2 = 5,5\ \mu m$$

As a result, we will get:

$$N_{max}^B = N_{av} + 3\sigma_S = 23,25\ \mu m$$

$$S_{max}^B = -(N_{av} - 3\sigma_S) = 12,25\ \mu m$$

These values are close to the limiting values, since only approximately in three cases out of a thousand connections can the values of clearances (tightness) exceed the calculated ones [6].

The probability of occurrence of tightness (clearance) in the connection is determined by the area under the distribution curve in the range N from 0 to 23.25 μm (Fig. 2).

Calculate this value as the probability of hitting the interval $(0, N_{max}^B)$ by the formula

$$P_N\{0 \leq N \leq N_{max}^B\} = \int_0^{N_{max}^B} f(N)dN$$

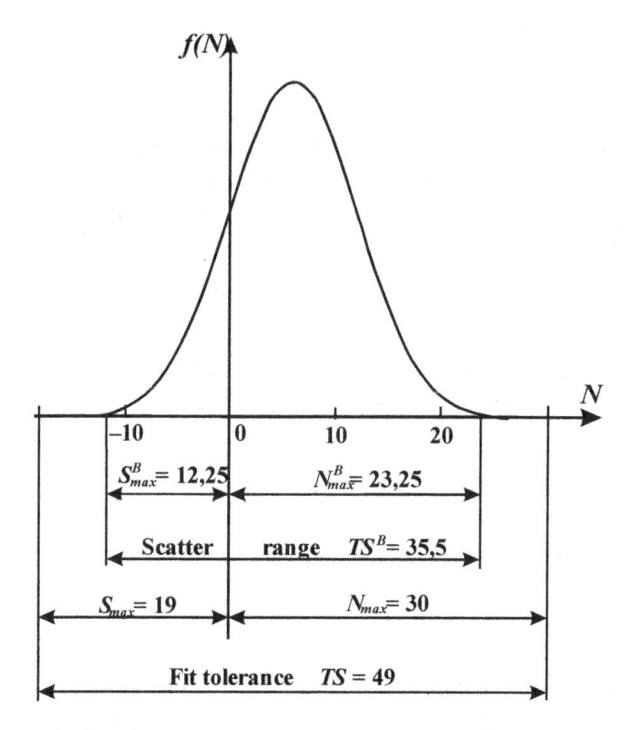

Fig. 2. Tightness distribution function of fit ∅ 60 H7/m6.

Considering the normal distribution law and the values of the mathematical expectation and standard deviation we will get

$$P_N\{0 \le N \le 23,3\} = \int_0^{23,3} \frac{1}{5,92\sqrt{2\pi}} \exp\left[\frac{(N-5,5)^2}{2 \cdot 5,92^2}\right] dN = 0,8186$$

Thus, the probability of obtaining in the connection of tightness will be 81,86%, and the clearance of 18,14%.

Estimate the values of the characteristics of the connection by the method of statistical simulation. First, we calculate the estimated the mathematical expectation and the standard deviation by formulas (7):

$$\widetilde{m} = 24,5 \,\mu\text{m}; \quad \widetilde{m} = 8,1 \,\mu\text{m}$$

Then, let $\Delta_m = \Delta_D = 0,1$; $\beta_m = \beta_D = 0,95$ and based on (13) we will get $n = 835$.

Having formed n pairs $(x_1, y_1), (x_2, y_2), \ldots, (x_n, y_n)$ of random variables with the normal distribution law, using formulas (2–5), we obtain the following values of the desired connection characteristics: the limit clearance is 11,6 µm; limiting tightness is 21,8 µm; fit tolerance is 33,4 µm; the tightness probability in the connection is 81,2%.

Combine the calculation results obtained by different methods and present them in the Table 1.

Table 1. Summary results of calculations connection parameters $\varnothing$ 60 H7/m6.

Connections specifications	Calculation method		
	USTF	Coefficient	Simulation
Limiting clearance, μm	19	12,25	11,6
Limiting tightness, μm	30	23,25	21,8
Fit tolerance, μm	49	35,5	33,4
Tightness probability, %	61,22	81,86	81,2
Clearance probability, %	38,78	18,14	18,8

This table shows that when using standard USTF ratios for calculating the parameters of the detail connections, manufactured in series, the values obtained differ significantly from the practical results.

In particular, the value of the limiting clearance is more than 1, 5 times the value obtained by the coefficient method, considering the technological features of the assembly of details. In addition, the coefficient method shows that the clearance appears in about 19% of cases, while in conditions of complete interchangeability (USTF method) it turns out that this is possible for 39% of connections. Note that the probability of tightness (clearance) for the conditions of complete interchangeability is obtained under the assumption of the uniform distribution of the fit tolerance.

It is also not difficult to notice that the results of statistical simulation are close to the values obtained by the coefficient method. The maximum difference corresponds to the limit tightness and is about 6%, and the probabilities of occurrence of tightness in the connection are almost the same, which confirms the effectiveness of the proposed approach for calculating the characteristics of tolerances and fits.

5   Conclusion

The results obtained allow us to propose the following generalized method for calculating the parameters of connections (tolerances and fits).

- The analysis of the technological processes of manufacturing details and assembly units is carried out
- If it is necessary to ensure complete interchangeability, the calculation is made using standard USTF formulas [5]
- In the absence of the requirement of complete interchangeability, the calculation is performed by the coefficient method [6]. To do this, the laws of the distribution of fields of tolerances of the shaft and the hole must be either normal or uniform. In the latter case, it is also required that the shaft and the hole be performed one by one
- If the specified requirements for the laws of distribution of the fields of the tolerances of the shaft and the hole are not met, the parameters of the connections are calculated by the statistical simulation method

The presented technique is focused on the use in CAD/CAM [1] and can be used in the design of mechanical and electromechanical units [20].

References

1. Koryachko, V.P., Skvortsov, S.V., Taganov, A.I., et al.: The evolution of computer-aided design of electronic computing tools. Radioengineering **3**, 97–102 (2012)
2. Basic norms of interchangeability: Geometrical features. General terms and definitions. http://docs.cntd.ru/document/1200039087. Accessed 15 Jan 2019
3. ISO 286–1:2010: Geometrical Product Specifications (GPS) – ISO code system for tolerances on linear sizes – Part 1: Basis of tolerances, deviations and fits. https://www.iso.org/standard/45975.html. Accessed 15 Jan 2019
4. ISO 286–2:2010: Geometrical Product Specifications (GPS) – ISO code system for tolerances on linear sizes – Part 2: Tables of standard tolerance classes and limit deviations for holes and shafts. https://www.iso.org/standard/54915.html. Accessed 15 Jan 2019
5. Zaycev, S.A., Kurganov, A.D., Tolstov, A.N.: Tolerances, Fits and Technical Measurements in Mechanical Engineering. Academy, Moscow (2004)
6. Yakushev, A.I., Vorontcov, L.N., Fedotov, N.M.: Interchangeability, Standardization and Technical Measurements. Engineering, Moscow (1987)
7. Matveev, V.V., Tverskoy, M.M., Boykov, F.I.: Dimensional Analysis of Technological Processes. Engineering, Moscow (1982)
8. Kazantseva, N.K.: Interchangeability and Standardization of Accuracy. Publishing House of USU, Yekaterinburg (2015)
9. Publishing house of standards: Dimensional chains. The basic concepts. Methods for calculating linear and angular chains. Guidelines RD 50–635–87, Moscow (1987)
10. Ryzhikov, Y.I.: Simulation. Theory and Technology. Corona Print, St. Petersburg (2004)
11. Knut, D.: The Art of Computing Programming. Vol. 2. Seminumerical Algorithms. Williams, Moscow (2001)
12. Venttcel, E.S., Ovcharov, L.A.: Probability Theory and its Engineering Applications. High School, Moscow (2000)
13. Lebedev, A.N., Kupriyanov, M.S., Nedosekin, D.D., et al.: Probabilistic Methods in Engineering Problems. Energoatomizdat, Moscow (2000)
14. Korshunov, Y.M.: Mathematical Foundations of Cybernetics. High School, Moscow (1995)
15. Dunaev, P.F.: Designing Units and Parts of Machines. Academy, Moscow (2006)
16. Skvortsov, N.V., Skvortsov, S.V., Khryukin, V.I.: Development of software for theoretical and probabilistic calculations in the unified system of tolerances and fits. In: Information Technology in Scientific Research, Interuniversity Collection, pp. 134–136. Ryazan State Radio Engineering University, Ryazan (2012)
17. Skvortsov, S.V., Khryukin, V.I.: Probability technique of standard fits selection. In: Informatics and Applied Mathematics, Interuniversity Collection 23, pp. 110–113. Ryazan State University, Ryazan (2017)
18. Skvortsov, S.V., Khryukin, V.I.: Probability technique of tolerances and fits calculations. In: Informatics and Applied Mathematics, Interuniversity Collection 21, pp. 110–113. Ryazan State University, Ryazan (2017)
19. Skvortsov, S.V., Skvortsova, T.S., Khryukin, V.I.: Probability technique of tolerances and fits calculations for CAD/CAM systems. Bull. Ryazan State Radiotech. Univ. **48**, 92–97 (2014)
20. Skvortsov, S.V., Skvortsova, T.S., Khryukin, V.I.: Selection of standard fits with random nature of sizes of holes and shafts for CAD/CAM systems. Bull. Ryazan State Radiotech. Univ. **61**, 32–40 (2017). https://doi.org/10.21667/1995-4565-2017-61-3-32-40

Evaluation of the Harmonic Locus of the Milling Technological System Based on the Analysis of the Vibro-Acoustic Signal

R. M. Khusainov[✉], A. R. Sabirov, and D. D. Safin

Naberezhnye Chelny Institute of Kazan Federal University, 68/19 Mira street,
Naberezhnye Chelny 423812, Russian Federation
rmhusainov@gmail.com

Abstract. The article deals with the use of acoustic signals of noise caused by vibrations of the technological system, consisting of technological equipment, tools and blanks, in order to obtain the harmonic locus of the dynamic system of equipment. Vibrations are one of the factors that prevent productivity increase in milling machines. Therefore, vibrations prevent the efficient use of such machines. For effective use of technological systems, it is necessary to determine experimentally the amplitude-phase frequency response. The analysis of vibroacoustic signals is the easiest way to obtain such a characteristic experimentally. The article gives the principles of mathematical processing of vibroacoustic signals, which allows obtaining the required characteristics. The experimental part is presented, which describes the acquisition of signals during the processing of the workpiece on the machine and obtaining its amplitude phase-frequency characteristics. The practical application of this technique for the analysis of the dynamic stability of the technological system and for the selection of optimal cutting conditions is considered.

Keywords: Harmonic locus · Vibroacoustic signals · Technological system

1 Introduction

The implementation of modern requirements of "Industry 4.0" in metalworking production involves accelerating the process of production of new products, and especially new modifications of products. One of the essential conditions for solving this problem is to increase the machining rates of machining operations. However, the solution of this problem is prevented by increasing vibrations of the technological system. It is safe to say that vibrations are the main factor preventing the increase of productivity. In addition, vibrations have a significant negative impact on many aspects of production [1–3], including processing accuracy [4]. In this regard, such a function is necessary, which would allow to assess the vibration resistance of the technological system, as well as to select machining rates, providing high-performance cutting without significant vibrations [5, 6].

A. A. Radionov and A. S. Karandaev (Eds.): RusAutoCon 2019, LNEE 641, pp. 871–878, 2020.
https://doi.org/10.1007/978-3-030-39225-3_94

2 The Relevance of the Study

2.1 Research Overview

In the earliest studies of the dynamics of technological equipment [7, 8] it was found that the main functions to assess the vibration resistance are the amplitude-frequency response and amplitude phase-frequency response. The first is mainly used for the study of resonance phenomena, the second - to assess the dynamic stability. Dynamic stability is understood as the absence of self-oscillations during cutting [9]. According to the Nyquist criterion, the dynamic system is stable if the Nyquist plot does not cover the point "−1" on the real axis. Thus, as the desired function, you can take the transfer function of the technological system, the graph of which is the Nyquist plot. For a dynamic system, conventionally consisting of an elastic system, including technological equipment, device, tool and workpiece, and the cutting process, the expression of the transfer function is as follows [9]:

$$W(j\omega) = \frac{K_e \cdot K \cdot b}{(1 - T_1^2\omega^2 + T_2 j\omega)(1 + \frac{m}{n} \cdot \frac{a_0 \xi_0}{v} j\omega)},$$ (1)

where K_e – is the dynamic compliance of the elastic system; a_0 – thickness of cutting layer; b – width of the cutting layer; K – specific cutting force; T_1 – inertial time constant; T_2 – time constant of damping; ξ_0 – average chip shrinkage; m/n – some constant coefficient; v – cutting speed; j – unit imaginary number.

Using (1), you can solve two problems:

- Direct – the estimation of dynamic stability under given cutting conditions
- Reverse – the selection of machining rates

In both cases, to obtain the amplitude phase-frequency response of a particular technological system, it is necessary to perform an experimental study, which requires expensive experimental equipment [10–13]. In addition, there are problems with the installation of sensors on a technological equipment. On the other hand, the method of vibroacoustic diagnostics of machines is widely used, with the use of which it is possible to estimate the dynamic state, for example, of technological equipment [14]. This method is simpler, does not require analog-to-digital converters for signal processing, the signal can be recorded and processed directly on the computer using universal software. However, in this case, the mathematical processing of signals is a known complexity [15]. First of all, this is due to the fact, that vibroacoustic signals are generated by waves of elastic vibrations of the elements of the technological system, which in turn are the product of disturbing forces acting on this system. Directly vibroacoustic signals do not characterize the amplitude phase-frequency response of the dynamic system.

2.2 Work Statement

Thus, to ensure the performance of the technological system, it is necessary to assess its dynamic stability, and for this, it is necessary to build its Nyquist plot. Moreover, this task must be performed in a production environment, that is, quickly, without creating additional downtime, and with minimal production costs [16].

3 Theoretical Part

The most suitable method for solving this problem is the analysis of vibroacoustic signals. However, in order to use this method to evaluate the transfer function of the dynamic system of the equipment, it is necessary to perform subsequent mathematical processing of such signals.

This processing is based on the analysis of the expression for the amplitude of vibroacoustic signals [17]:

$$L(j\omega) = \psi_j |\lg| \frac{\Gamma - \Gamma_\Pi}{v} |c_*| + \Delta L_j, \tag{2}$$

where ψ_j – an empirical coefficient; Γ – the flow of vibrational energy; v – the speed of the flow of vibrational energy; c_* – the speed of sound; Γ_n – the flow of energy to overcome the inelastic resistance; ΔL_j – amendment to level L.

It is difficult to determine the exact numerical values of the components included in the (2). However, on the other hand, it follows from this expression, that the amplitude of the vibroacoustic signal is a function of the amplitude of elastic vibrations and disturbing forces. Thus, it is possible to present an expression for the amplitude of the vibroacoustic signal in the following form:

$$L(j\omega) = f[A(j\omega), P(j\omega)], \tag{3}$$

where $A(j\omega)$ – a function of oscillation amplitude; $P(j\omega)$ – function of disturbing forces.

This expression can be rewritten as a polynomial:

$$L(j\omega) = a_0 + a_1 A(j\omega) + a_2 P(j\omega), \tag{4}$$

where a_0, a_1, a_2 – unknown coefficients.

Performing the transformations, we obtain:

$$L(j\omega) = a_0 \cdot A(j\omega)^{a_1} \cdot a_2 P(j\omega)^{a_2} \tag{5}$$

On the other hand, the expression for the transfer function of a dynamic system can be written as follows:

$$W(j\omega) = R(\omega)e^{j\varphi(\omega)} \tag{6}$$

where $R(\omega)$ – amplitude-frequency response, $\varphi(\omega)$ – phase-frequency response.

The amplitude-frequency response of the dynamic system is the ratio of the amplitude of oscillations to the amplitude of the disturbing force:

$$R(\omega) = A(j\omega)/P(j\omega) \tag{7}$$

Substituting this expression in (6), we obtain:

$$W(j\omega) = [A(j\omega)/P(j\omega)] \cdot e^{j\varphi(\omega)} \tag{8}$$

Now in (5) we divide both parts by $P(j\omega)^{a1}$ and multiply by $e^{j\varphi(\omega)}$, we obtain:

$$e^{j\varphi(\omega)} \frac{L(j\omega)}{P^{a_1}} = \frac{a_0 \cdot A(j\omega)^{a_1} \cdot P(j\omega)^{a_2}}{P(j\omega)^{a_1}} e^{j\varphi(\omega)} \tag{9}$$

Given (7), we obtain:

$$\frac{L(j\omega)}{P(j\omega)^{a_1}} e^{j\varphi} = a_0 \cdot P(j\omega)^{a_2} \cdot [W(j\omega)]^{a_1} \tag{10}$$

By converting this expression, you can obtain the following expression for the amplitude function of the vibroacoustic signal:

$$L(j\omega) = \frac{a_0 \cdot P^{(a_1 + a_2)} \cdot [W(j\omega)]^{a_1}}{e^{j\varphi(\omega)a_1}} \tag{11}$$

In this expression, the formula of the transfer function of the dynamic system (1) is used, in which the K_e – is the dynamic compliance of the elastic system; a_0 - thickness of cutting layer; b – width of the cutting layer; K – specific cutting force; T_1 – inertial time constant; T_2 – time constant of damping; ξ_0 – average chip shrinkage; $m/n.$ – some constant coefficient; v – cutting speed; j – unit imaginary number. These values characterize the current state of the elastic system of the machine and the state of the cutting process, and the analytic expression for them does not exist.

The phase frequency response can be expressed as follows. The phase-frequency response is determined by the expression:

$$\varphi(\omega) = arctg[Jm(\omega)/Re(\omega)], \tag{12}$$

where $Re(\omega)$ – real component; $Jm(\omega)$ – imaginary component. On the other hand, the phase-frequency characteristic of a dynamic system is the sum of two characteristics – the characteristic of the elastic system and the characteristic of the cutting process:

$$\varphi_d(\omega) = \varphi_e(\omega) + \varphi_c(\omega), \tag{13}$$

where $\varphi_e(\omega)$ – phase-frequency response of elastic system, $\varphi_c(\omega)$ - phase-frequency response of the cutting process.

The phase-frequency characteristic of the elastic system can be obtained from the expressions for the real and the imaginary components:

$$Re_e(\omega) = \frac{K_e(1 - T_1^2\omega^2)}{(1 - T_1^2\omega^2)^2 + T_2^2\omega^2} \tag{14}$$

$$Jm_e(\omega) = \frac{K_eT_2\omega}{(1 - T_1^2\omega^2)^2 + T_2^2\omega^2} \tag{15}$$

Substituting in the (12), we obtain:

$$\varphi_e(\omega) = arctg\left(-\frac{T_2\omega}{1 - T_1^2\omega^2}\right) \tag{16}$$

The phase-frequency characteristic of the cutting process can be obtained from the expressions for the real and the imaginary components:

$$Re_c(\omega) = \frac{K_p}{1 + T_p^2\omega^2} \tag{17}$$

$$Jm_c(\omega) = -\frac{K_pT_p\omega}{1 + T_p^2\omega^2}, \tag{18}$$

where K_p – the stiffness of the cutting process, T_p – inertial component of the cutting process.

Substituting in the (12), we obtain:

$$\varphi_c(\omega) = arctg(-T_p\omega) \tag{19}$$

Substituting (16) and (19) into (13), we obtain the expression of the phase-frequency response of the dynamic system:

$$\varphi_d(\omega) = arctg\frac{-T_2\omega - T_p\omega\left(1 - T_1^2\omega^2\right)}{1 - T_1^2\omega^2 - T_pT_2\omega^2} \tag{20}$$

Perturbing forces, that are generating oscillations, are usually periodic in nature. However, the function (11) is considered in the frequency domain. In this case, it is better to model the disturbance in the form of a Dirac delta function, for example, in the form of:

$$P(\omega) = \sum_i \frac{P_i}{\sqrt{\pi}} e^{-P_i^2(\omega - n_i)^2}, \tag{21}$$

where P_i – the amplitudes of individual disturbing forces, n_i – their frequencies.

Thus, the unknown quantities are the values, specified in the signature to the (11), as well as the coefficients a_0, a_1, a_2, amplitudes of individual disturbing forces P_i. These values can be found by the least squares method. Finding them, you can get the full expression for the transfer function of the dynamic system. You can use this expression to perform the following dynamic exploration tasks:

- Assessment of the dynamic stability of the technological system
- Finding the machining rates, that ensure the dynamic stability of the process system

4 Experimental Part

Using this technique, a study of the dynamic stability of milling on the machine JMD3CNC was performed. Milling of the ledge on a workpiece of steel 45 with an end mill of high-speed steel P6M5 with a diameter of 10 mm, the number of teeth 4 [18], was performed. The machining rates during processing are presented in the Table 1.

Table 1. Cutting conditions.

Machining rates	Cutting width (b), mm	Cutting depth (t), mm	Feed (sz), mm/tooth	Cutting speed (v), m/min
Values of machining rates	1	1	0.09	25

The time realization of the vibroacoustic signal is shown in Fig. 1a, the frequency spectrum of the vibroacoustic signal is shown in Fig. 1b. As a result of solving the problem by the above method, we have an expression for the transfer function, the graph of which is shown in Fig. 2.

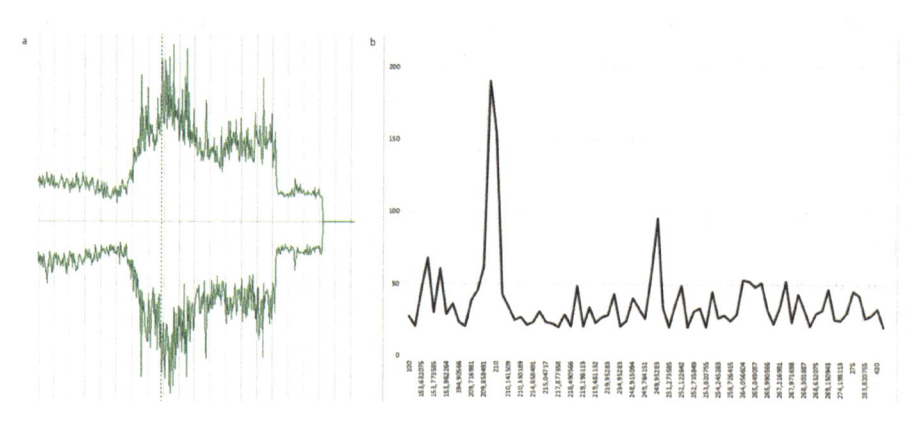

Fig. 1. The time realization (a) and the frequency spectrum (b) of the vibroacoustic signal.

$$W(j\omega) = \frac{2093 \cdot 8.06 \cdot 10^{-6} \cdot b}{(-0.268 \cdot 10^{-3}\omega^2 + 0.426 \cdot 10^{-4}j\omega + 1)(1.2 \cdot \frac{a_0 \cdot 0.85}{v} j\omega + 1)} \quad (22)$$

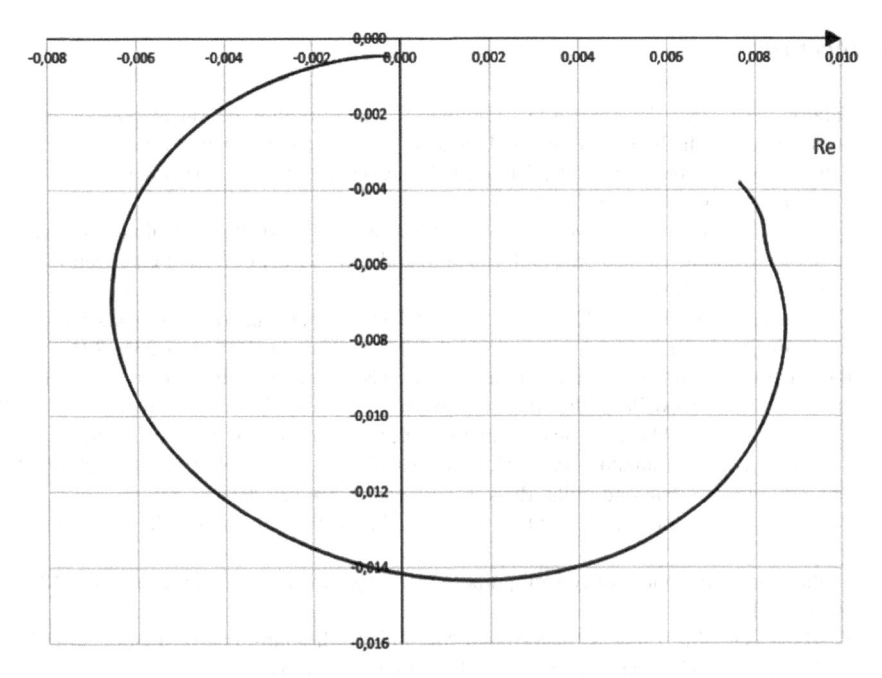

Fig. 2. The graph of the transfer function.

This graph shows that it does not cover the point "−1" on the real axis, therefore, processing in the cutting conditions of this technological system is dynamically stable.

5 Conclusion

The proposed method is characterized by simplicity in the execution of the experiment and in the mathematical processing of the results.

The use of this technique makes it possible to perform mathematical modeling of a dynamic system, which makes it possible to solve the following problems [19, 20]:

1. Assess the dynamic stability of the technological system in the technological process planning;
2. Determine the optimal cutting conditions, taking into account a given margin of dynamic stability;
3. To make diagnostics of the state of technological equipment.

References

1. Safarov, D.T., Kondrashov, A.G., Glinina, G.F.: Algorithm of calculation of energy consumption on the basis of differential model of the production task performed on machines with computer numeric control (CNC). In: IOP Conference Series: Materials Science and Engineering, vol. 240, p. 012060 (2017)
2. Govorkov, A., Lavrenteva, M., Fokine, I.: Mathematical modeling of making mechanical engineering products based on an information model. In: MATEC Web of Conferences, vol. 224, p. 02022 (2018)
3. Gavariev, R.V., Savin, I.A.: Research of the mechanism of destruction of compression molds for casting under pressure of color alloys. Solid State Phenom. **284**, 326–331 (2018)
4. Balabanov, I.P., Simonova, L.A., Balabanova, O.N.: Systematization of accuracy indices variance when modelling the forming external cylindrical turning process. In: IOP Conference Series: Materials Science and Engineering, vol. 86, p. 012010 (2015)
5. Ryabov, E.A., Khisamutdinov, R.M., Grechishnikov, V.A.: Selection of surfaces for endurance tests of ball end mills. Russ. Eng. Res. **38**, 10–12 (2018)
6. Balla, O.M., Zamashchikov, Yu.I., Livshits, O.P.: Mills and milling. ISTU Publishers, Irkutsk (2006)
7. Method for testing mid-size general-purpose lathes for vibration stability in cutting ENIMS, Moscow (1961)
8. Zharkov, I.G.: Vibrations in Blade-Based Machining. Mashinostroeniye, Leningrad (1986)
9. Kudinov, V.A.: Machine Dynamics. Mashinostroeniye, Moscow (1967)
10. Gorin, Y.Yu., Kryazhev, A.Yu., Tatarkin, Y.Yu., et al.: Improving the vibration stability in end milling. Polzunovsky Bull. **2**, 43–48 (2015)
11. Kozochkin, M.: Particularities of vibrations in metal cutting. STIN **1**, 25–29 (2009)
12. Yakovlev, E.Yu.: Improving the quality of metal finishing in CNC end milling by neural-network modulation of cutting parameters. Inf. Secur. Questions **2**, 65–69 (2009)
13. Kozochkin, M.P.: Cutting process stability. Bull. Eng. **2**, 77–81 (2013)
14. Podurayev, V.N.: Technological Diagnosis of Cutting by Acoustic Emission. Mashinostroeniye, Moscow (1988)
15. Lavrentyieva, M., Govorkov, A.: Identifying the objects in the structure of an e-model by means of identified formal parameters in the design and engineering environment. In: MATEC Web of Conferences, vol. 129, p. 03002 (2017)
16. Serebrenitsky, P.P.: Some Specific Features of High-Speed Machining, vol. 4, pp. 6–15. Metalloobrabotka (2007)
17. Poryadkov, V.I.: Designing Low Noise Mechanisms. Mashinostroeniye, Moscow (1991)
18. Golovko, A., Petrov, S.: Definition of the initial information for selecting the cutting tool for automated production preparation. In: MATEC Web Conference, vol. 224, p. 0110312 (2018)
19. Shastin, V.I., Kargapoltcev, S.K., Gozbenko, V.E.: Results of the complex studies of microstructural, physical and mechanical properties of engineering materials using innovative methods. Int. J. Appl. Eng. Res. **12**, 15269–15272 (2017)
20. Averyanova, I.O., Shestakov, N.A.: Analysis of chipping in cutting. Bull. Eng. **2**, 74–78 (2013)

An Adaptive Speech Segmentation Algorithm to Determine Temporal Patterns of Human Psycho-Emotional States

A. K. Alimuradov[(⊠)], A. Yu. Tychkov, and A. V. Ageykin

Penza State University, 40 Krasnaya street, Penza 440026, Russian Federation
alansapfir@yandex.ru

Abstract. An algorithm for adaptive segmentation to determine temporal patterns of speech signals reflecting human psycho-emotional states has been developed. The nature of the algorithm is the segmentation of speech into informative areas using adaptive decomposition and energy analysis of intrinsic mode functions, and determining patterns of temporal intervals of voiced, unvoiced, and pause sections. A research using the formed base of speech signals of 100 subjects, experiencing natural positive and negative emotions, was conducted. The research results were evaluated in comparison with the known methods of segmentation, followed by the determination of temporal patterns. In accordance with the results, the percentage of false assignments of the "normal" status to speech signals uttered by subjects under psycho-emotional excitement is greater in the known analogues than in the developed algorithm. This is also observed in false assignments of the "pathology" status to speech signals pronounced by subjects in a neutral state. Based on the determination results of psycho-emotional states, it was concluded that the developed adaptive algorithm more accurately determines the boundaries of informative areas, due to the advantages of the energy analysis of the modes obtained by the method of adaptive decomposition.

Keywords: Speech processing · Segmentation · Adaptive decomposition · Temporal patterns · Psycho-emotional states

1 Introduction

Speech reproduction is one of the most complex human skills acquired throughout life. The speech apparatus is extremely sensitive to disorders of the nervous system [1]. For years, the assessment of the instability of the motility of the speech apparatus under psycho-emotional disorders was limited to laboratory analyzes and tests of direct mental perception. Today, this problem is successfully solved by the methods based on the analysis of speech signals [2].

A human psycho-emotional state is encrypted in certain patterns of speech, reflecting the type and degree of manifestation of the state. The basic concepts that characterize speech patterns associated with the form, size, dynamics of changes in the speech apparatus and describing the psycho-emotional state of a person can be divided

© Springer Nature Switzerland AG 2020
A. A. Radionov and A. S. Karandaev (Eds.): RusAutoCon 2019, LNEE 641, pp. 879–890, 2020.
https://doi.org/10.1007/978-3-030-39225-3_95

into three groups of objective features that allow to distinguish speech patterns: spectral-temporal, cepstral, and amplitude-frequency [3].

Spectral-temporal patterns characterize a speech signal in its physical and mathematical essence based on the presence of three types of components: voiced sections, unvoiced sections, and pause sections.

The spectral-temporal patterns quite well reflect the peculiarity of the form of the time series and the spectrum of voice impulses, as well as the uniqueness of the filtering functions of the speech apparatus during psycho-emotional excitement of a person.

Temporal patterns characterize the features of the speech flow associated with the dynamics of the restructuring of the speaker's articulation organs, and are integral characteristics reflecting the relationship of movements of the articulation organs of the speaker [4]. Currently, there are many various methods for determining the temporal patterns of speech [5–7]. Each of the techniques has a number of advantages and disadvantages. The accuracy of determining temporal patterns in the presented approaches depends on the effectiveness of speech segmentation into informative areas, characterized by a certain duration.

This paper presents an algorithm for adaptive segmentation of speech into voiced, unvoiced, and pause sections for determining temporal patterns of psycho-emotional states. The essence of the algorithm lies in the use of adaptive decomposition for speech signal segmentation.

The article is a continuation of the published works of the authors [8, 9], and is structured as follows. The second part of the article briefly presents information on adaptive decomposition and temporal patterns of speech. The developed algorithm is briefly described and investigated in the third and fourth parts. The fifth part presents the research results and conclusions.

2 Materials and Methods

2.1 Adaptive Decomposition

An important condition for adaptive decomposition is the formation of a basic decomposition function, functionally dependent on the internal structure of the original speech signal. This condition can be fulfilled with the use of empirical mode decomposition (EMD) [10]. The classical EMD method was developed by Norden Huang in 1998, and was intended to decompose non-stationary signals arising in nonlinear systems. The EMD provides the decomposition of a non-stationary signal into high-frequency and low-frequency components, called intrinsic mode functions (IMFs).

During decomposition, the signal model is not specified in advance, and the IMFs are calculated during the sifting procedure taking into account local features (such as extremes and signal zeros), and the internal structure of each particular signal. Thus, IMFs do not have a rigorous analytical description, but must satisfy two conditions that guarantee the process adaptability, as well as a certain symmetry and narrowband basic functions:

- The total number of extremes of the signal function should be equal to the number of zeros of the function with an accuracy of one

- The average value of the upper and lower envelopes, interpolating local maxima and minima of the signal function, should be approximately equal to zero

Analytically, the EMD has the following expression:

$$x(n) = \sum_{i=1}^{I} IMF_i(n) + r_I(n), \tag{1}$$

where $x(n)$ is the original signal; $IMF_i(n)$ is an IMF; $r_I(n)$ is the final residue; $I = 1, 2, \ldots, I$ is the IMF number; n is discrete timing.

In addition to adaptability, the EMD technology has some other important properties:

- Locality, that is the ability to take into account local features of the signal
- Orthogonality, that is ensuring signal recovery with a certain accuracy
- Completeness, that is a guarantee of a finite number of basic functions at a finite signal duration

The uniqueness of the decomposition lies in the fact that the functions of the decomposition basis are extracted directly from the internal structure of the original signal. This allows to take into account the features of the original signal, such as frequency and amplitude modulation, the concentration of the energy with a certain frequency range, and etc.

Since the appearance of the classical EMD, lots of decomposition types have been developed. The most adaptive one to non-stationary speech signals of a complex form is an improved complete ensemble empirical mode decomposition with adaptive noise (CEEMDAN). The emergence of the improved CEEMDAN has made it possible to solve a number of problems inherent in other types of decomposition [11]:

- Mixing of IMFs incommensurable in amplitude and frequency scales (due to the overlap of the large-scale energy spaces of modes)
- Minimum residual noise in the IMF
- Parasitic IMFs at early stages of decomposition

A distinctive feature of the improved CEEMDAN is the addition of controlled adaptive noise to the original signal to create new extremes:

$$x_j(n) = x(n) + w_j(n), \tag{2}$$

where $x_j(n)$ are the noise copies of the speech signal; $w_j(n)$ is the implementation of white noise,

$$x_j(n) = \sum_{i=1}^{I} IMF_{ji}(n) + r_{jI}(n); \tag{3}$$

$$IMF_i(n) = \sum_{j=1}^{J} \frac{IMF_{ji}(n)}{J}; \tag{4}$$

$$r_I(n) = \sum_{j=1}^{J} \frac{r_{jI}(n)}{J},$$

(5)

where $j = 1, 2, \ldots, J$ is the amount of white noise implementations.

Different types of decomposition, their advantages and disadvantages in speech processing are described in detail in [12–14].

2.2 Temporal Patterns

A review of the informative parameters of speech signals [15] has revealed the following temporal patterns relevant to psycho-emotional states:

- Distribution rate of time intervals of voiced, unvoiced, and pause sections (rate of speech timing, RST)
- Distribution acceleration of time intervals of voiced, unvoiced, and pause sections (acceleration of speech timing, AST)
- Distribution entropy of time intervals of voiced, unvoiced, and pause sections (entropy of speech timing, EST)
- Duration of pause intervals (DPI)

3 Description of the Method

Figure 1 shows a block diagram for the algorithm of adaptive speech segmentation to determine temporal patterns of psycho-emotional states. The first stage of the algorithm consists in speech segmentation into informative areas using adaptive decomposition and energy analysis of IMFs. The second stage consists in determining the temporal patterns of voiced, unvoiced, and pause sections, reflecting motility disturbance of the speech apparatus, caused by psycho-emotional disorders. Let us consider some stages of the algorithm in more details.

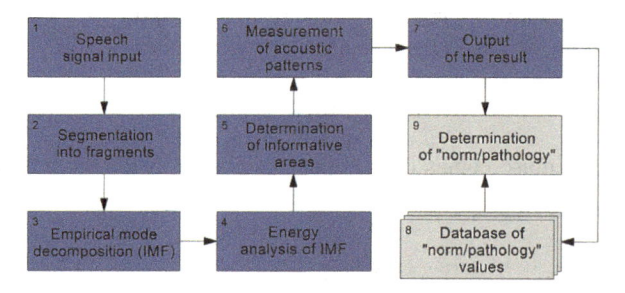

Fig. 1. Simplified block diagram for the algorithm of adaptive speech segmentation to determine temporal patterns of psycho-emotional states.

Speech is a process whose spectrum remains relatively unchanged for a short period of time. This allows to divide a speech signal into equal fragments of 10 ms, within which a signal can be considered as conditionally stationary. After segmentation, the signal is a set of fragments, and the further operation of the algorithm is carried out with each fragment separately.

Signal decomposition is carried out by the improved CEEMDAN method. The decomposition parameters were set in accordance with an automated method for determining the optimal values [16]. Figure 2 presents the results of the decomposition of fragments of voiced and unvoiced speech sections with a duration of 100 ms.

As can be seen from Fig. 2, voiced and unvoiced sections were decomposed into eight IMFs. The first IMFs for both types of sections are high-frequency, and the subsequent modes are low-frequency (descending). The first five IMFs for the voiced section are informative (concentrating the main signal energy), and the last three IMFs are compensating trend ones. It is impossible to unambiguously determine the informative and compensating components for the unvoiced section. This is due to the fact that unvoiced speech signal is similar in characteristics to noise.

The change in the level of the speech signal in time is characterized by an important informative parameter, namely, amplitude distribution. The distribution of the amplitude of the signal in time is adequately described by the logarithm function of the short-term energy. In accordance with the functionality of the auditory apparatus, a person perceives speech nonlinearly, determining the difference between the energies of the informative areas of speech. Approaching the operation of the algorithm to the functionality of the auditory apparatus, to compress the amplitude of the signal in a large dynamic range, energy logarithmization is used:

$$LE_{s,i}(n) = \log_2\left(\sum_{n=1}^{N}\left(IMF_{s,i}(n)\right)^2\right), \tag{6}$$

where $LE_{s,i}$ is the IMF energy logarithm of a speech signal fragment; s is the fragment number.

In the developed algorithm, the segmentation is carried out on the basis of the energy analysis of the IMF fragments of the speech signal in a sliding window with a duration of 10 ms.

In accordance with the physiological aspect of speech formation, a person makes an initial short pause before pronunciation, which does not contain speech and corresponds to silence. Typically, the duration of the initial pause is 200–500 ms. Using the averaged values of the energy of the IMF fragments of the initial pause, one can determine the threshold values of the energy logarithms for the speech stream segmentation into a useful signal (voiced and unvoiced sections), and pauses. The determination of the threshold values of the IMF energy logarithms is carried out according to the formula:

$$LE_{thres.,i}(n) = \frac{1}{S}\sum_{s=1}^{S} LE_{IMFs,i}, \tag{7}$$

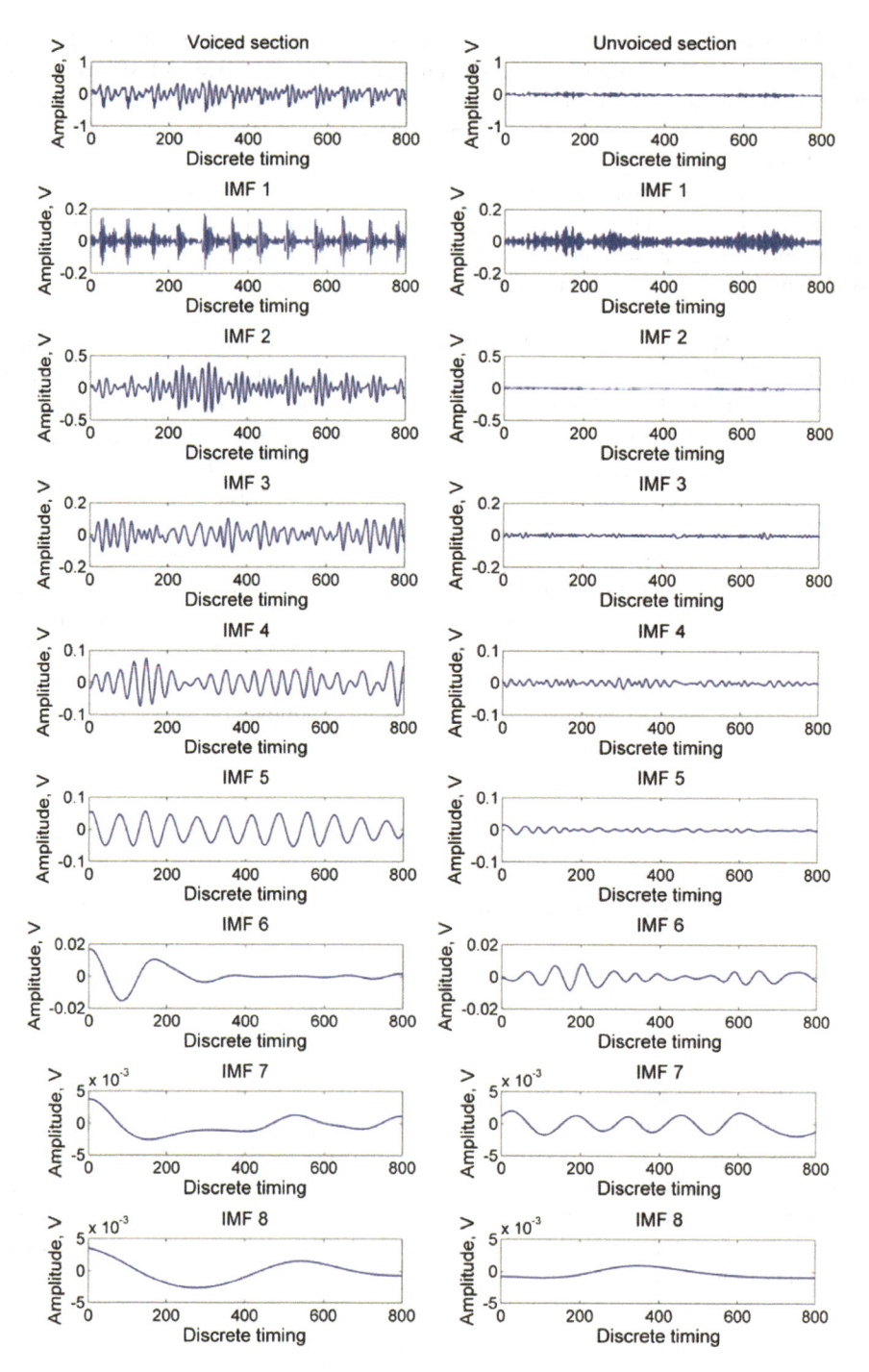

Fig. 2. The result of the decomposition of voiced (left column) and unvoiced (right column) fragments of speech sections using the improved CEEMDAN method.

where $LE_{thres.,i}$ is a threshold value of the IMF energy logarithms; S is the number of fragments corresponding to the initial pause; $LE_{IMFs,i}$ is the IMF energy logarithm.

Figure 3a shows a graphical interpretation of determination of threshold values for five modes.

Red dashed lines in Fig. 3a and b show the values of the IMF energy logarithms of the initial pause fragments. The averaged threshold values of the IMF energy logarithms are marked with a thick solid black line. After determining the threshold values, the first stage of speech signal segmentation into a useful signal and pauses is performed. Figure 3b presents a graphical interpretation of the threshold processing for the first five IMFs. A thick blue solid line indicates the IMF energy logarithms of the voiced speech fragment. A thick red line marks the IMF energy logarithms of the unvoiced speech fragment.

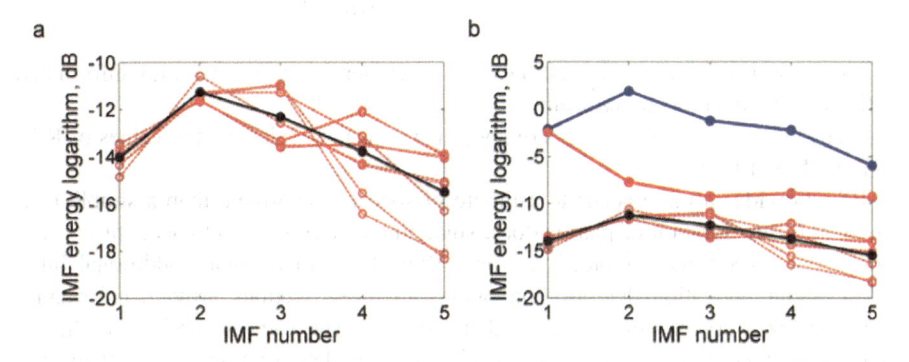

Fig. 3. Energy analysis of the first five IMFs: (a) determination of threshold values; (b) threshold processing during speech segmentation.

The next stage of segmentation is to divide the useful signal into voiced and unvoiced sections. Segmentation is also carried out on the basis of the IMF energy analysis in a sliding window of 10 ms, and the additional calculation of the following parameters for each fragment:

- Fragment energy/power (PWR):

$$PWR = \frac{1}{N} \sum_{n=1}^{N} x^2(n) * h(n), \tag{8}$$

where $x(n)$ is a speech signal in a window of N countings; $h(n)$ is the Hamming window

- Autocorrelation function (ACR):

$$R_x(k) = \frac{1}{N * \sigma_x^2} \sum_{n=1}^{N} (x(n) - \mu_x) * ((n+k) - \mu_x); \tag{9}$$

$$ACR = \frac{1}{N-1} \sum_{n=1}^{N} \left(R_x(k) - \overline{R_x} \right), \tag{10}$$

where R_x is the autocorrelation function; σ_x is the signal standard deviation; μ_x is the signal mean value

- Zero-crossing rate (ZCR):

$$ZCR = \frac{1}{N-1} \sum_{n=1}^{N-1} |sign(R_x(n+1)) - sign(R_x(n))|; \tag{11}$$

$$sign(R_x(n)) = \begin{cases} 1, & x(n) \geq 0 \\ -1, & x(n) < 0 \end{cases} \tag{12}$$

Figure 4 shows the interpretation of the segmentation process into informative sections in the developed algorithm.

To determine temporal patterns of speech signals is to calculate the values of RST, AST, DPI, and EST.

RST provides a more accurate estimate of speech impairment than a simple measurement of the duration of pauses does, since this pattern takes into account not only pauses, but voiced and unvoiced sections. Voiced sections provide additional information about phonation deterioration, while unvoiced sections provide information about the fuzzy articulation of the speaker's organs. The value of RST as a whole is approximately equal to the speed of speech, since the deterioration of speech speed is associated with defects in all elements of speech signals. Each voiced, unvoiced and pause section is described by the time of its occurrence, defined as the mean time between the beginning of the section and its end.

AST determines the degree of time acceleration. Each analyzed fragment of the speech signal is divided into two parts with an overlap of 25%, which ensures a smooth transition between the parts.

The AST value is calculated as the difference between the RST values of both parts divided by the total duration of the speech signal fragment.

EST describes the ordering (predictability) of speech signals, including voiced, unvoiced, and pause sections. Accordingly, a decrease in entropy is equivalent to a violation of speech motor.

To determine the EST, the number of all intervals of voiced sections kv, unvoiced sections ku, pause sections kp, and the total number of sections kt is calculated. The EST value is defined as follows:

$$EST = -\frac{kv}{kt} \cdot \log_2\left(\frac{kv}{kt}\right) - \frac{ku}{kt} \cdot \log_2\left(\frac{ku}{kt}\right) - \frac{kp}{kt} \cdot \log_2\left(\frac{kp}{kt}\right). \tag{13}$$

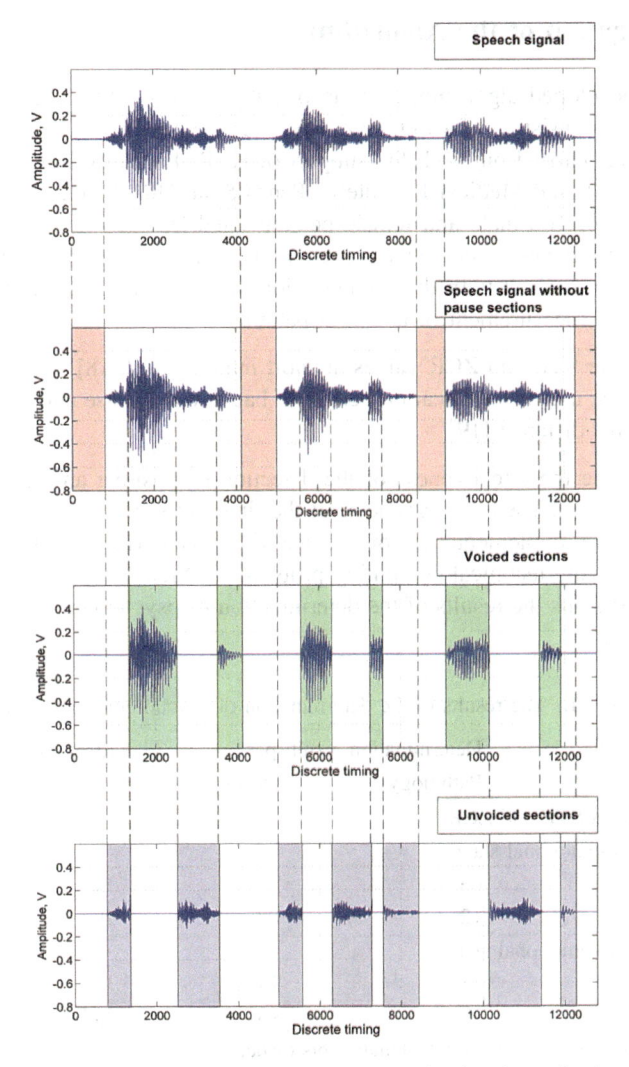

Fig. 4. Speech signal segmentation into pause, voiced and unvoiced sections.

DPI defines the ability of the speaker to start speech reproduction. The instability of the motility of the speech apparatus under psycho-emotional disorders can cause difficulties in reproduction, which result in an increase in the duration of pauses. The DPI value is calculated as the average duration of all sections of pauses.

4 Investigation of the Algorithm

To test the developed algorithm, a group of subjects was formed, and a database of speech signals, consisting of 1000 records was registered. The formation of the group of subjects was carried out with the support of K. R. Evgrafov Regional Psychiatric Hospital (Penza), and Medical Institute of Penza State University. The group of subjects consisted of 100 male and female persons aged from 18 to 79 years old, experiencing natural positive and negative emotions. The segmentation efficiency was evaluated in comparison with the segmentation methods that are popular in practice, followed by the measurement of temporal patterns:

- Based on the STE and ZCR values at short intervals [17, 18]
- Based on the use of statistical properties of background noise and one-dimensional Mahalanobis distance [19]

To determine the effectiveness of the detection of positive and negative psycho-emotional states, we used the parameter of the first and second kind error.

The software implementation of the algorithm is performed in an application package for solving technical computing problems © Matlab (MathWorks).

Table 1 presents the results of the determination of psycho-emotional positive and negative states.

Table 1. The results of the determination of psycho-emotional states.

Predictable result	Determination result, pers.		Error, %	
	Pathology	Norm		
Method based on STE and ZCR				
Positive psycho-emotional state				
Pathology	63	37	1st	37
Norm	12	88	2nd	12
Negative psycho-emotional state				
Pathology	81	19	1st	19
Norm	7	93	2nd	7
Method based on one-dimensional Mahalanobis distance				
Positive psycho-emotional state				
Pathology	71	29	1st	29
Norm	8	92	2nd	8
Negative psycho-emotional state				
Pathology	84	16	1st	16
Norm	6	94	2nd	6
Developed algorithm				
Positive psycho-emotional state				
Pathology	88	12	1st	12
Norm	5	95	2nd	5
Negative psycho-emotional state				
Pathology	89	11	1st	11
Norm	4	96	2nd	4

5   Results and Conclusion

The table shows that the percentage of false assignments of the "normal" status to speech signals uttered by subjects in a state of psycho-emotional arousal in methods based on STE, ZCR and one-dimensional Mahalanobis distance is higher by 25% and 17% for negative psycho-emotional states, by 7% and 4% for a positive psycho-emotional state, respectively, than the developed algorithm has. The same can be said about false assignments of the "pathology" status to speech signals pronounced by subjects in a neutral state: by 7% and 3%, 3% and 2%, respectively. The lowest values of the errors of the 1st and 2nd kind were achieved by the developed algorithm: only 12% and 5% for positive emotions, 11% and 4% for negative emotions. Based on the results, we can conclude that the developed algorithm more accurately defines the boundaries of voiced, unvoiced sections and pauses. This is achieved due to the advantages of the IMF energy analysis obtained by the method of the improved CEEMDAN for each analyzed fragment.

Acknowledgments. This work was financially supported by the Russian Foundation for Basic Research according to the research project No. 18-37-00256.

References

1. Schuller, B.W., Batliner, A.M.: Computational Paralinguistics: Emotion, Affect and Personality in Speech and Language Processing. Wiley, New York (2013)
2. Trigeorgis, G., Ringeval, F., Brueckner, R., et al.: Adieu features? end-to-end speech emotion recognition using a deep convolutional recurrent network. In: IEEE International Conference on Acoustics, Speech and Signal Processing (ICASSP), 20–25 Mar 2016, pp. 5200–5204 (2016)
3. Huang, X., Acero, A., Hon, H.-W.: Spoken Language Processing. Guide to Algorithms and System Development. Prentice Hall, New Jersey (2001)
4. Fant, G.K.: Acoustic Theory of Speech Formation. Science, Moscow (1964)
5. Whitehead, R.L., Schiavetti, N., Metz, D.E., et al.: Temporal characteristics of speech produced by inexperienced signers during simultaneous communication. J. Commun. Disord. **32**(2), 79–95 (1999)
6. Bona, J.: Temporal characteristics of speech: the effect of age and speech style. J. Acoust. Soc. Am. **136**(2), 116–121 (2014). https://doi.org/10.1121/1.4885482
7. Bakaev, A.V.: Spectral and temporal characteristics of vocal speech in the emotional aspect. Almanac Mod. Sci. Educ. **4**(83), 28–32 (2014)
8. Alimuradov, A.K., Tychkov, A.Yu., Ageykin, A.V. et al.: Speech/pause detection algorithm based on the adaptive method of complementary decomposition and energy assessment of intrinsic mode functions. In: XX IEEE International conference on soft computing and measurements, St. Petersburg, 24–26 May 2017, pp. 610–613 (2017)
9. Alimuradov, A.K., Tychkov, A.Yu., Kuzmin, A.V. et al.: Measurement of speech signal patterns under borderline mental disorders. In: 21st Conference of Open Innovations Association FRUCT, Helsinki, 6–10 November 2017, pp. 26–33 (2017)
10. Huang, N.E., Zheng, S., Steven, R.L.: The empirical mode decomposition and the Hilbert spectrum for nonlinear and non-stationary time series analysis. Proc. R. Soc. Lond. **A454**, 903–995 (1998)

11. Colominasa, M.A., Schlotthauera, G., Torres, M.E.: Improved complete ensemble EMD: a suitable tool for biomedical signal processing. Biomed. Sig. Process. **14**, 19–29 (2014)
12. Z, W., Huang, N.E.: Ensemble empirical mode decomposition: a noise-assisted data analysis method. Adv. Adapt. Data Anal. **1**(1), 1–41 (2009)
13. Yeh, J.-R., Shieh, J.-S., Huang, N.E.: Complementary ensemble empirical mode decomposition: a novel noise enhanced data analysis method. Adv. Adapt. Data Annal. **2**(2), 135–156 (2010)
14. Torres, M.E., Colominas, M.A., Schlotthauer, G., et al.: A complete ensemble empirical mode decomposition with adaptive noise. In: Proceedings of IEEE International Conference on Acoustics, Speech and Signal Processing, Prague, 22–27 May 2011, pp. 4144–4147 (2011)
15. Hlavnicka, J., Cmejla, R., Tykalova, T., et al.: Automated analysis of connected speech reveals early biomarkers of Parkinson's disease in patients with rapid eye movement sleep behavior disorder. Sci. Rep. **7**(12), 13 (2017)
16. Alimuradov, A.K., Tychkov, A.Yu., Kvitka, Yu.S.: Automation of empirical mode decomposition to increase efficiency of speech signal processing. In: 2018 International Russian Automation Conference, Sochi, 9–16 September 2018 (2018). https://doi.org/10.1109/rusautocon.2018.8501732
17. Bachu, R.G., Kopparthi, S., Adapa, B., et al.: Separation of voiced and unvoiced using zero crossing rate and energy of the speech signal. In: American Society for Engineering Education Zone Conference, pp. 1–7 (2008)
18. Moattar, M.H., Homayounpour, M.M.: A simple but efficient real-time voice activity detection algorithm. In: 17-th European Signal Processing Conference, Glasgow, 24–28 August 2009, pp. 2549–2553 (2009)
19. Saha, G., Sandipan, C., Senapat, S.: A new silence removal and endpoint detection algorithm for speech and speaker recognition applications. In: 11th National Conference of Communications, India, 28–30 January 2005, pp. 291–295 (2005)

Development of Algorithms for the Correct Visualization of Two-Dimensional and Three-Dimensional Orthogonal Polyhedrons

V. A. Chekanin$^{(\boxtimes)}$ and A. V. Chekanin

Moscow State University of Technology «STANKIN», 3a Vadkovsky Lane,
Moscow 127055, Russian Federation
vladchekanin@rambler.ru

Abstract. The article is devoted to the two-dimensional and three-dimensional orthogonal polyhedrons outlines obtaining algorithms. An orthogonal polyhedron is a geometric figure representing the union of non-overlapping orthogonal objects (rectangles or parallelepipeds in the two-dimensional or three-dimensional case, respectively) with a fixed position relative to each other, considered as a single whole object. The need to use the orthogonal polyhedrons as individual objects arises, in particular, during solving certain resource allocation problems. The popular application program interfaces for rendering graphic images such as OpenGL and DirectX do not provide the ability to visualize only the outline of the union instead of all the edges of its objects. The developed algorithms are based on the idea of searching and cutting off segments located on edges belonging to several orthogonal objects belonging to the considered orthogonal polyhedron. The described algorithms provide the possibility of visualization of arbitrary orthogonal polyhedrons, including those containing any holes. These algorithms are implemented in the developed applied software intended to optimize the solution of resource allocation problems of any dimension, including the orthogonal packing and rectangular cutting problems.

Keywords: Orthogonal polyhedron · Visualization · Outline · Orthogonal object · Rectangular cutting · Packing problem

1 Introduction

An orthogonal polyhedron is a geometric figure formed by combining non-overlapping orthogonal objects [1, 2] (rectangles or parallelepipeds in the two-dimensional or three-dimensional case, respectively) with a fixed position relative to each other, considered as a single whole object [3, 4]. The need to work with orthogonal polyhedrons used as separate objects arises, in particular, when solving a number of resource allocation problems [5–9], many of which are NP-hard [10, 11] and require the use of heuristic [12, 13] and metaheuristic optimization algorithms [14–16]. Most often, the problem of packing the orthogonal polyhedrons takes place when solving problems of industrial cutting of cardboard and plywood [5, 17]. The problem of constructing and packing two-dimensional orthogonal polyhedrons in the form of polymino is relevant in the

A. A. Radionov and A. S. Karandaev (Eds.): RusAutoCon 2019, LNEE 641, pp. 891–900, 2020.
https://doi.org/10.1007/978-3-030-39225-3_96

design of wideband phased arrays [18]. Mathematical models using orthogonal poly-hedrons of various dimensions may be using for solving a number of optimization problems in the field of functional voxel modeling [19].

Algorithms for the formation of orthogonal polyhedrons from sets of orthogonal objects are described in detail in [3].

We will consider a N-dimensional ($N = 2, 3$) orthogonal polyhedron O, consisting of m orthogonal objects in the form of rectangles or parallelepipeds (for the two-dimensional and three-dimensional case, respectively) o_k, $k \in \{1,\ldots,m\}$ with the overall dimensions $\{w_k^1; w_k^2; \ldots; w_k^N\}$ (the superscript in formulas means the number of the coordinate axis), the position of which relative to each other is specified with vectors $\{z_k^1; z_k^2; \ldots; z_k^N\}$ containing the coordinates of orthogonal objects [20, 21] in coordinate system associated with the considered orthogonal polyhedron O.

We will use three intermediate coordinate axes d_1, d_2 and d_3, each of which matches with one of the local coordinate axes of the considered orthogonal polyhedron. The numbers of the coordinate axes d_2 and d_3 are set depending on the value d_1. For a two-dimensional orthogonal polyhedron the coordinate axis $d_3 = 0$, and the coordinate axis d_2 is selected as follows:

$$\begin{cases} d_2 = 2 & for \quad d_1 = 1; \\ d_2 = 1 & for \quad d_1 = 2. \end{cases} \tag{1}$$

For a three-dimensional orthogonal polyhedron the coordinate axes d_2 and d_3 are selected as follows:

$$\begin{cases} d_2 = 2, & d_3 = 3 & for \quad d_1 = 1; \\ d_2 = 3, & d_3 = 1 & for \quad d_1 = 2; \\ d_2 = 1, & d_3 = 2 & for \quad d_1 = 3. \end{cases} \tag{2}$$

The origin of each coordinate axis d_1, d_2 and d_3 matches with the position of the beginning of the local coordinate system of the orthogonal polyhedron.

Each edge r of an orthogonal object o_k is described by a set of four numbers $\{z, y; x_1; x_2\}$, where z is the position of the edge on the coordinate axis d_3 (for a two-dimensional orthogonal polyhedron $z = 0$), y is the position of the edge on the coordinate axis d_2, x_1 and x_2 are the coordinates of the start and end point of the edge, measured along the coordinate axis d_1, respectively.

Under the segment t will be understood some part of the edge of an orthogonal object. Each segment is characterized by a set of two numbers $\{t_1; t_2\}$ defining the coordinates of its beginning and end (respectively), measured along the coordinate axis onto which it gives a nonzero projection.

The popular application program interfaces for rendering graphic images such as OpenGL and DirectX do not provide the ability to visualize only the outline of the union instead of all the edges of its objects. The developed and described in article algorithms provide obtaining the outlines of orthogonal polyhedrons as a set of segments used for the subsequent correct visualization of these orthogonal polyhedrons.

2 Algorithm of Deleting Overlapping Segments of Edges

This algorithm is used for both two-dimensional and the three-dimensional orthogonal polyhedrons. The algorithm for deleting overlapping segments of edges from a set A has the following steps.

Step 1. Create a new set of edges B.

Step 2. Select the first edge r from the set A as the current edge.

Step 3. Create an empty set of segments $C = \varnothing$.

Put in the set C the segment $\{t_1; t_2\}$ the coordinates t_1 and t_2 of which are equal to the coordinates x_1 and x_2 (respectively) of the current edge r.

Create an intermediate set of edges A': $\forall r' \in A'$ the coordinate y of the edge r' is equal to the coordinate y of the current edge r.

Put segments $\{t_1; t_2\}$ in the set C, for each of which the coordinates t_1 and t_2 are equal to the coordinates x_1 and x_2 the corresponding edges r'.

Sort the set C in ascending order of the coordinate t_1 of all segments contained in it.

Step 4. Create a set D for storing the unique coordinates p of points located at the beginning and end of each considered segment.

Put in the set D the coordinates t_1 and t_2 of each segment from the set C.

Reduce the set D by removing duplicate values from it.

Step 5. Divide all the segments from the set C into smaller segments according to the coordinates from the set D. For this, for each segment $t \in C$ with coordinates $\{t_1; t_2\}$ is created a set $D' \subset D$ of points, such that an inequality $t_1 \leq p \leq t_2$ holds for each coordinate p from the set D'. For a given set of coordinates $\{p_1; p_2; \ldots; p_{|D'|}\}$ (where $p_1 = t_1$, $p_{|D'|} = t_2$) from the set D', the set of segments $\{p_1; p_2\}$, $\{p_2; p_3\}$, $\ldots$, $\{p_{|D'|-1}; p_{|D'|}\}$ is created when any original segment $t \in C$ is being divided.

Put all created segments in a new set C'.

Sort the set C' in ascending order of the coordinate t_1 of all segments contained in it.

Step 6. Create a new set of segments C''. Put in the set C'' all segments found only once in the set C'.

Step 7. On the basis of each segment $t \in C'$ with the coordinates $\{t_1; t_2\}$, create a new edge $\{z, y; t_1; t_2\}$, for which the coordinates z and y are equal to the coordinates z and y of each edge from the initial set of edges A. All created edges put in the set B.

Step 8. If the set A contains any edges with a coordinate y greater than the coordinate y of the current edge r, then select the next edge in order (with a larger coordinate y) as the current edge r, and go to step 3.

As a result of this algorithm, a set B of non-overlapping edges will be created.

For each coordinate y and some current axis of the orthogonal polyhedron, all edge fragments (segments) that are also found on other its edges are deleted. As an example, we consider the set of segments C formed in step 3 which is presented in Table 1 and shown on Fig. 1. In this case, after performing step 4, a set D of unique coordinates will be created: $D = \{\ 0, 10, 20, 30, 35, 40, 50, 60, 65, 70, 75, 80\ \}$.

Table 1. Parameters of original segments (set C).

Segment number	Coordinate t_1	Coordinate t_2
1	0	20
2	10	35
3	30	65
4	40	60
5	50	70
6	75	80

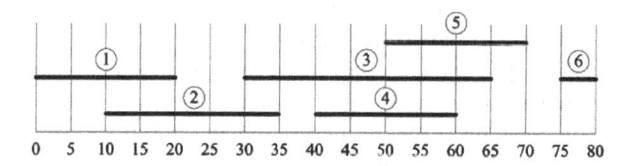

0   5   10 15 20 25 30 35 40 45 50 55 60 65 70 75 80

Fig. 1. Original set of segments.

Table 2. Segments obtained after dividing (set C').

Segment number	Number of original segment from the set C	Coordinate t_1	Coordinate t_2
1	1	0	10
2	1	10	20
3	2	10	20
4	2	20	30
5	2	30	35
6	3	30	35
7	3	35	40
8	3	40	50
9	4	40	50
10	3	50	60
11	4	50	60
12	5	50	60
13	3	60	65
14	5	60	65
15	5	65	70
16	6	75	80

The set of segments C' which will be obtained in step 5 is presented in Table 2. The set of segments C'' which will be obtained in step 6 is presented in Table 3 and shown on Fig. 2.

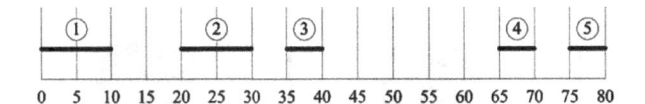

Fig. 2. Resulting set of segments.

Table 3. Segments obtained after deleting all duplicates (set C'').

Segment number	Coordinate t_1	Coordinate t_2
1	0	10
2	20	30
3	35	40
4	65	70
5	75	80

3 Algorithm of Creation the Outline of a Two-Dimensional Orthogonal Polyhedron

For the formation the set of edges belonging to the outline of a two-dimensional orthogonal polyhedron, it is necessary for each variant i of the coordinate axis $d_1 = i$, $i = 1,\ 2$ perform steps 1–5.

Step 1. Determine the value d_2 by (1) for the given value d_1.

Step 2. Create an empty set of edges $A^i = \varnothing$.

Step 3. For each orthogonal object o_k, $k \in \{1, \ldots, m\}$, select a pair of edges, giving a non-zero projection on the coordinate axis d_2: $\left\{0;\ z_k^{d_1};\ z_k^{d_2};\ z_k^{d_2} + w_k^{d_2}\right\}$ and $\left\{0;\ z_k^{d_1} + w_k^{d_1};\ z_k^{d_2};\ z_k^{d_2} + w_k^{d_2}\right\}$.

Include this pair of edges in the set A^i.

Step 4. Arrange the set A^i in ascending order of the value y of all edges contained in it.

Step 5. For the set of edges A^i, execute the algorithm of deleting overlapping segments of edges which will give a set B^i of edges belonging the outline the orthogonal polyhedron obtained by analyzing a variant i of the coordinate axis $d_1 = i$.

As a result of the application of this algorithm, two sets of edges (B^1 and B^2) will be formed, the visualization of which will present the outline of the considered two-dimensional orthogonal polyhedron including all its holes.

On Fig. 3a as an example, a two-dimensional orthogonal polyhedron is shown, for which are displayed all the edges of its orthogonal objects. On Fig. 3b are displayed only the edges belonging to the outline of this orthogonal polyhedron that were obtained by applying the described algorithm.

On Fig. 4 are presented the examples of packings of two-dimensional polyhedrons displayed with different visualization modes.

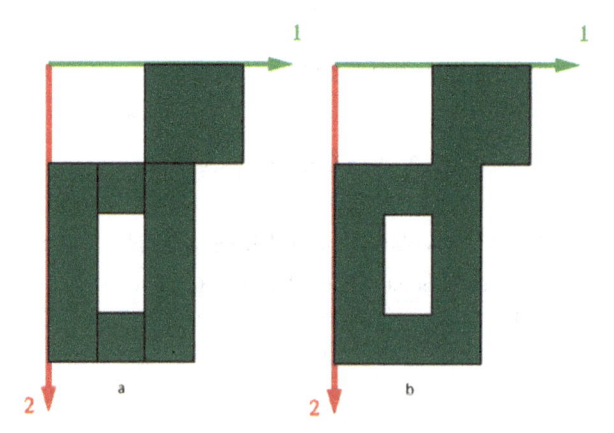

Fig. 3. (a) visualization all the edges of objects belonging to the two-dimensional orthogonal polyhedron; (b) the outline of the two-dimensional orthogonal polyhedron.

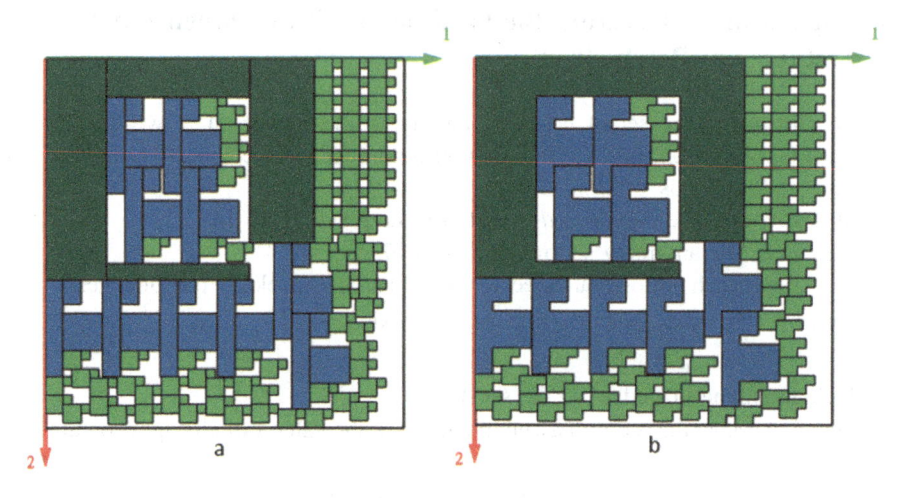

Fig. 4. (a) visualization all the edges of objects belonging to the orthogonal polyhedrons in the two-dimensional packing problem; (b) visualization the outlines of orthogonal polyhedrons in the two-dimensional packing problem.

4   Algorithm of Creation the Outline of a Three-Dimensional Orthogonal Polyhedron

The algorithm for forming the edges belonging to the outline of a three-dimensional orthogonal polyhedron based on the edges of its orthogonal objects is a two-pass.

At the first iteration of the algorithm, the edges of an orthogonal polyhedron are formed when some cutting plane moves from the origin of the orthogonal polyhedron to its most distant edges (the so-called direct view).

At the second iteration, the cutting plane moves backward in the direction to the origin of the orthogonal polyhedron (the so-called reverse view). The direction of viewing the faces of the orthogonal polyhedron will be determined by the parameter λ (viewing of the front faces at $\lambda = 1$ and viewing of the back faces at $\lambda = 2$).

Step 1. Create empty sets of edges $B^1 = \varnothing$, $B^2 = \varnothing$ and $B^3 = \varnothing$.

For each variant of the coordinate axis i, perform steps 2–7.

Step 2. Determine the values d_2 and d_3 by (2) for the given value d_1.

Step 3. Create a set F for storing unique coordinates of points defining the position of the faces perpendicular to the coordinate axis d_3. The coordinates $z_k^{d_3}$ and $z_k^{d_3} + w_k^{d_3}$ of each orthogonal object o_k, $k \in \{1, \ldots, m\}$ of the considered orthogonal polyhedron are included in the set F.

Reduce the set F by removing duplicate values from it.

Step 4. Set $\lambda = 1$ (the direct view).

Step 5. For each coordinate p form the set F perform step 6.

Step 6. Create empty sets of two-dimensional orthogonal (rectangular) objects $C^+ = \varnothing$ and $C^- = \varnothing$, which will be used in addition and subtraction operations (respectively) of orthogonal objects [3] with the aim to form a two-dimensional orthogonal polyhedron O' defining a set of visible fragments of faces of the original three-dimensional orthogonal polyhedron that are in a cutting plane perpendicular to the coordinate axis d_3 and located at a distance p from the origin of the considered orthogonal polyhedron. The set C^+ will include new rectangular objects created by the faces of orthogonal objects lying in the cutting plane. The set C^- will include new rectangular objects created by the faces of orthogonal objects located on the path from the beginning of the cutting plane to its current position p.

For each orthogonal object o_k, $k \in \{1, \ldots, m\}$, define a set of faces, on the basis of which new rectangular objects will be created, included in the sets C^+ and C^-. To do this, under the conditions $z_k^{d_3} \leq p$ and $z_k^{d_3} + w_k^{d_3} \geq p$ create a new two-dimensional orthogonal object o' with the overall dimensions $\{w_k^{d_1}; w_k^{d_2}\}$, the position of which is determined by the vector $\{z_k^{d_1}; z_k^{d_2}\}$. An orthogonal object o' is put in a set C^+ when fulfilled simultaneously two conditions $\lambda = 1$ and $z_k^{d_3} = p$ or the conditions $\lambda = 2$ and $z_k^{d_3} + w_k^{d_3} = p$. Otherwise, the object o' is put in the set C^-.

Perform the operations of addition and subtraction of orthogonal objects from sets C^+ and C^- according to the algorithm given in article [3]. As a result, a set C of orthogonal objects forming a two-dimensional orthogonal polyhedron O' will be created.

If $|C| \neq 0$, then perform steps 6.1–6.4.

Step 6.1. Create an empty set of edges $A' = \varnothing$.

Step 6.2. For each orthogonal object $o'_{k'} \in O'$ ($k' \in \{1, \ldots, |C|\}$), select a pair of edges: $\{p; z_{k'}^1; z_{k'}^2; z_{k'}^2 + w_{k'}^2\}$ and $\{p; z_{k'}^1 + w_{k'}^1; z_{k'}^2; z_{k'}^2 + w_{k'}^2\}$.

Include this pair of edges in the set A'.

Step 6.3. Arrange the set A' in ascending order of the value y of all edges contained in it.

Step 6.4. For the set of edges A', execute the algorithm of deleting overlapping segments of edges which will give a set of edges B' which is necessary to put in the set B^{d_1}.

Step 7. Set $\lambda = 2$ (the reverse view) and perform steps 5–6.

As a result of the application of this algorithm, three sets of edges (B^1, B^2 and B^3) will be formed, the visualization of which will present the outline of the considered three-dimensional orthogonal polyhedron including all its holes.

On Fig. 5a as an example is presented a three-dimensional orthogonal polyhedron rendered with the OpenGL without using the developed algorithms (on the image are displayed all visible edges of orthogonal objects). On Fig. 5b is presented the same polyhedron without overlapping segments of edges, obtained using the described algorithm of creation the outline of a three-dimensional orthogonal polyhedron.

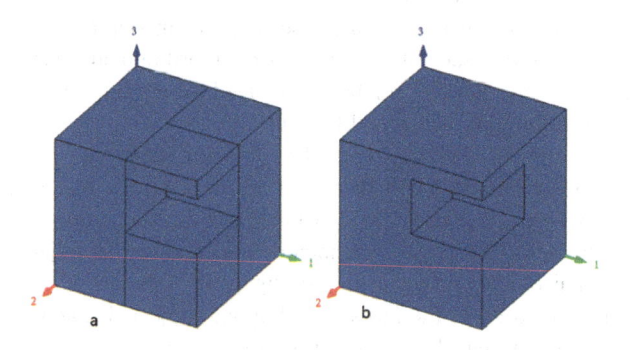

Fig. 5. (a) visualization all visible edges of objects belonging to the three-dimensional orthogonal polyhedron; (b) the outline of the three-dimensional orthogonal polyhedron.

On Fig. 6a is presented an example of a complex three-dimensional orthogonal polyhedron visualized without using the described algorithms. On Fig. 6b is presented the obtained outline of the considered three-dimensional orthogonal polyhedron.

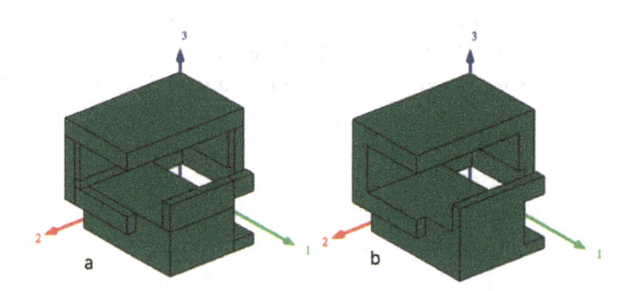

Fig. 6. (a) visualization of all edges of orthogonal objects of the three-dimensional orthogonal polyhedron; (b) the outline of the three-dimensional orthogonal polyhedron.

On Fig. 7a are shown the edges belonging to the outline of the considered orthogonal polyhedron obtained at the first iteration algorithm (the direct view). Figure 7b shows edges of the contour obtained at the second iteration of this algorithm (the reverse view).

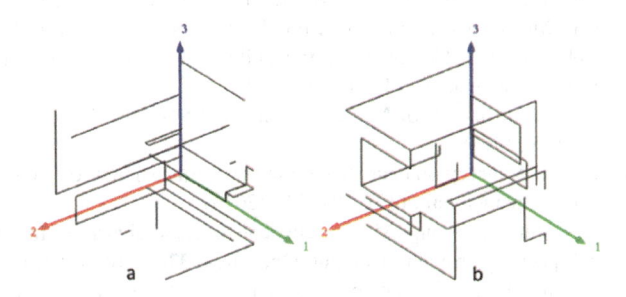

Fig. 7. (a) edges belonging to the outline of the three-dimensional orthogonal polyhedron obtained on the first iteration; (b) edges belonging to the outline of the three-dimensional orthogonal polyhedron obtained on the second iteration.

5 Conclusion

The algorithms to obtain a set of edges belonging to the contour of a two-dimensional or a three-dimensional orthogonal polyhedron are developed. The algorithms are based on the idea of searching and deleting all segments of edges belonging simultaneously to several orthogonal objects from the considered orthogonal polyhedron. The developed algorithms provide the possibility of correct visualization of arbitrary orthogonal polyhedrons, including those containing holes. The algorithms presented in the article are implemented in the developed application software [22, 23], designed to optimize a number of resource allocation problems of various dimensions, including the orthogonal packing and rectangular cutting problems.

References

1. Alvarez-Valdes, R., Carravilla, M.A., Oliveira, J.F.: Cutting and packing. In: Marti, R., Panos, P., Resende, M. (eds.) Handbook of Heuristics, pp. 1–46. Springer, Cham (2018)
2. Chekanin, V.A., Chekanin, A.V.: An efficient model for the orthogonal packing problem. Adv. Mech. Eng. **22**, 33–38 (2015)
3. Chekanin, V.A., Chekanin, A.V.: Algorithms for the formation of orthogonal polyhedrons of arbitrary dimension in the cutting and packing problems. Vestnik MGTU «Stankin» **3**, 126–130 (2018)
4. Aldana-Galvan, I., et al.: Beacon coverage in orthogonal polyhedra. In: 29th Canadian Conference on Computational Geometry, Ottawa, pp. 156–161 (2017)
5. Wascher, G., Haubner, H., Schumann, H.: An improved typology of cutting and packing problems. EJOR **183**(3), 1109–1130 (2007)
6. Bortfeldt, A., Wascher, G.: Constraints in container loading—a state-of-the-art review. EJOR **229**(1), 1–20 (2013)

7. Boschetti, M.A.: New lower bounds for the finite three-dimensional bin packing problem. Discrete Appl. Math. **140**, 241–258 (2004)
8. Chekanin, V.A., Chekanin, A.V.: Algorithms for management objects in orthogonal packing problems. ARPN J. Eng. Appl. Sci. **11**(13), 8436–8446 (2016)
9. Chekanin, V.A., Chekanin, A.V.: Deleting objects algorithm for the optimization of orthogonal packing problems. In: Evgrafov, A. (ed.) Advances in Mechanical Engineering. Lecture Notes in Mechanical Engineering, pp. 27–35. Springer, Cham (2017)
10. Garey, M., Johnson, D.: Computers Intractability: A Guide to the Theory of NP-Completeness. WH Freeman, San Francisco (1979)
11. Johnson, D.S.: A Brief history of NP-completeness, 1954–2012. Doc Math. Extra volume ISMP, pp. 359–376 (2012)
12. Oliveira, J.F., et al.: A survey on heuristics for the two-dimensional rectangular strip packing problem. Pesquisa Operacional **36**(2), 197–226 (2016)
13. Chen, D., Liu, J., Fu, Y., Shang, M.: An efficient heuristic algorithm for arbitrary shaped rectilinear block packing problem. Comput. Oper. Res. **37**(6), 1068–1074 (2010)
14. Gao, Y., et al.: A multi-objective ant colony system algorithm for virtual machine placement in cloud computing. J. Comput. Syst. Sci. **79**(8), 1230–1242 (2013)
15. Chekanin, V.A., Chekanin, A.V.: Development of the multimethod genetic algorithm for the strip packing problem. Appl. Mech. Mater. **598**, 377–381 (2014)
16. Chekanin, V.A., Chekanin, A.V.: Design of library of metaheuristic algorithms for solving the problems of discrete optimization. In: Evgrafov, A. (ed.) Advances in Mechanical Engineering. Lecture Notes in Mechanical Engineering, pp. 25–32. Springer, Cham (2018)
17. Crainic, T.G., Perboli, G., Tadei, R.: Recent advances in multi-dimensional packing problems. In: Volosencu, C. (ed.) New Technologies–Trends, Innovations and Research, pp. 91–110 (2012)
18. Rocca, P., Mailloux, R.J., Toso, G.: GA-based optimization of irregular subarray layouts for wideband phased arrays design. IEEE Antennas Wirel. Propag. Lett. **14**, 131–134 (2015)
19. Caulkin, R., et al.: Impact of shape representation schemes used in discrete element modelling of particle packing. Comput. Chem. Eng. **76**, 160–169 (2015)
20. Chekanin, V.A., Chekanin, A.V.: Multilevel linked data structure for the multidimensional orthogonal packing problem. Appl. Mech. Mater. **598**, 387–391 (2014)
21. Chekanin, V.A., Chekanin, A.V.: Improved data structure for the orthogonal packing problem. Adv. Mater. Res. **945–949**, 3143–3146 (2014)
22. Chekanin, V.A., Chekanin, A.V.: Design of software for orthogonal packing problems. In: Proceedings of the International Conference on Advanced Materials, Structures and Mechanical Engineering, Incheon, South Korea, 29–31 May 2015, pp. 277–280. CRC Press (2016)
23. Chekanin, V.A., Chekanin, A.V.: Implementation of packing methods for the orthogonal packing problems. J. Theor. Appl. Inf. Technol. **88**(3), 421–430 (2016)

Long-Term Digital Documents Storage Technology

A. V. Solovyev[✉]

Federal Research Center "Computer Science and Control" of Russian Academy of Sciences, 44/2 Vavilova Street, Moscow 119333, Russian Federation
soloviev@isa.ru

Abstract. This study is devoted to the creation of technology of long-term digital documents storage. The article presents an overview of the problem and the existing solutions, as well as formally sets the task of ensuring long-term keeping. It describes the problems arising in the organization of long-term keeping. It describes the creation of mathematical models for estimating the parameters of keeping, which are, in essence, an assessment of parametric perturbations of the digital document storage environment. It justifies and describes the algorithms that allow monitoring and organizing the keeping of digital documents. Examples of practical implementation of the developed technology in the framework of large information systems are given. The advantage of the presented technology is that it allows you to comprehensively solve the scientific and technical problem posed by the organization of long-term keeping of digital documents. The technology development plans are defined to solve the problem of long-term keeping of big data and systems built using distributed registry technologies.

Keywords: Long-term keeping · Digital document · Authenticity · Interpretability · Stability · Reliability

1 Introduction

The program of the digital economy of the Russian Federation [1], adopted by the Government of the Russian Federation on July 28, 2017, contains the statement that "data in digital form" is within the framework of the digital economy (DE) "a key factor of production in all areas of socio-economic activity". Thereby, very high requirements must be placed on the keeping of data in digital form (DiDF). These requirements are especially high if DiDFs are of high value and need to be stored for years or decades. However, DiDF in the conditions of DE: (1) are in a dynamically changing digital platform (DP), (2) must be interpretable, authentic, available during the entire storage period. This contradictory situation determines the need to solve an important scientific and technical problem of ensuring the long-term keeping of DiDF.

However, it is still necessary to understand what the DiDF will be transmitted via digital channels between the state and the DE branches, the state and the citizens. The answer to this question provides an analysis of the current state of business processes in

© Springer Nature Switzerland AG 2020
A. A. Radionov and A. S. Karandaev (Eds.): RusAutoCon 2019, LNEE 641, pp. 901–911, 2020.
https://doi.org/10.1007/978-3-030-39225-3_97

the economy. For any elementary business process, inputs and outputs are streams of information representing digital documents (DD). Why exactly DD and what is it?

For example, DiDF is stored as "35000". What it is? Distance? Time? Money? When DiDF have the form "salary 35000", then there is no doubt what it is. Then you can define DD as DiDF, which have semantics. Below in the article will be given a more precise definition. So, consider the task of ensuring the long-term keeping of DD in a dynamically changing DP environment.

2   Literature and Problem Review

Attempts to solve the problem of keeping DD are actively undertaken both in the Russian Federation and abroad. Consider the set of existing methods and means of storing DD in the world to confirm the relevance of the problem posed, as well as to identify factors hindering its solution.

The experience of the National Archives and Records Administration (NARA), which is the world's largest repository of DD, is of the greatest interest from the point of view of the organization of long-term storage. DD was first accepted for storage at NARA in 1970 [2]. The US Department of State and the Pentagon annually transfer tens of millions of DDs to NARA for long-term storage [3].

The extensive experience of NARA allowed us to create the following storage standards [4]: (1) MARC bibliographic format, (2) DD storage and conversion format on ODISS optical discs, (3) GSA 6710 A archive descriptions format, (4) DD storage format in ASCII and EBCDIC encodings on magnetic tapes. In 2013, NARA developed a draft open standard "Encoded Archival Description" of DD (EAD) based on XML. Also, a draft bulletin has been posted for public discussion on the acceptable file formats of DDs transferred for archiving [5]. In 2017, NARA released the "Universal Electronic Records Management Requirements" document. These requirements can be used by employees of federal agencies when writing technical assignments for tools or DD management services [6].

Since 2005, NARA has been developing ERA (Electronic Records Archives) for long-term storage of DD. ERA should have provided the ability to save DDs in the network of NARA-controlled servers located throughout the USA in at least 20 clusters. This will protect the DD from possible loss and forgery. In 2012, the ERA project was suspended and the limited possibilities for using ERA were announced [7]. It is assumed that full-text search DD is possible no earlier than 2021 [8]. The reason for this was a huge variety of computer formats, DD, the use of which entails the risk of not interpreting DD after decades. Storage in a single ASCII encoding, as this was done in NARA before, has become impossible. Thus, it is clear that the problem of interpretability at the beginning of development was not considered at all, which led to significant difficulties in creating an ERA. In addition, problems of technological and technical aging of information carriers were revealed, as well as a lack of understanding of what DD is, what kind of information and in what form should be preserved [9]. However, NARA continues to accumulate DD. D 2017, NARA reported on the creation of a new model for the storage and accessibility of presidential DDs, which received more than 250 TB in a year [10]. In order to partially solve the problem of interpretability, in 2017, NARA made a decision after December 31, 2022 to accept DD in digital format with descriptive metadata [11].

The problem of storing digital data is also handled by the National Archives of the United Kingdom. In 2017, the National Archives introduced a strategy, according to which a solution should be created, allowing to ensure the long-term keeping of data transferred from government agencies of the DD [12]. This, however, is not the first attempt to create such a decision, since as early as 2013 a similar strategy was adopted [13], which involved the creation of such a decision by 2017; however, it was never created. Although the result of the efforts to create a solution was the normative documents, the most famous of which [14] regulates the access control procedure, including audit issues, DD movements, logging of operations, and backup management. Thus, there is an understanding that, in addition to the text of the document, DD data should also be used for long-term storage. Although the DD model is not represented, but the composition of the information of digital documents is approximately defined in [14].

In the Russian Federation, relatively recently, they began to address the issue of the long-term keeping of DD. Currently, the Rosarchiv, together with The All-Russian Scientific and Research Institute for Records and Archives Management (VNIIDAD), is developing documents that regulate the organization of DD storage in an advisory manner [15, 16]. In particular, these recommendations prescribe DD normalization during long-term storage in PDF/A-1 format. Which, however, immediately reduces the number of storage formats to one, and a strictly defined version.

Bundesamt für Sicherheit in der Informationstechnik (BSI, BRD) prepared the document: Technical Guide BSI TR-03125 [17]. The document provides recommendations on the long-term storage of DD, certified by an digital signature (DS), in order to ensure their authenticity, integrity, confidentiality. Protection using DS, however, contains vulnerabilities associated with changing cryptographic protection technologies, as well as the fact that the duration of a DS is a maximum of 5 years. Other countries are also engaged in solving the problem of long-term keeping, but the format of the article does not allow for a detailed review.

However, the following conclusions can be made even from the above brief review: (1) the problem is extremely urgent, (2) the problem is still far from being resolved, because there is no universal replicable software/hardware solution for the keeping of DD. The lack of solution is due to several reasons:

- The problem is not purely technical, but is interdisciplinary
- There is no systematic approach to the problems of long-term storage of DD
- Not all problems of keeping DD are fully studied and systematized
- There is no methodology for monitoring the parameters of keeping DD

3 Setting of the Problem

We define long-term storage as DD storage for at least 10 years. Keeping - as the DD property to exist as an accessible and authentic evidence at an arbitrary point in time.

The life cycle of the storage DD exceeds the terms of operation of the equipment and software. For example, documents on personnel must be kept for 50 years. By comparing this period with the average period of technological storage of equipment (usually 5–10 years), one can easily make sure that during such a period of storage both the system software and the storage media will be updated several times.

Statement 1. DD is an object of management, and the task of long-term keeping is the problem of optimal control of DD under conditions of parametric perturbations [18, 19]. And for optimal control, it is necessary to learn how to control the parameters of the storage environment and develop disturbance compensation algorithms to stabilize the control object [20].

The problems of long-term storage were systematized, which made it possible to formulate a mathematical formulation of the task of keeping in the following form. Given:

1. The set of DD $D = \{d_i\}$;
2. The set of parametric perturbations of the storage medium $E = \{\varepsilon_q\}$: (a) ε_1 – authenticity violation D, (b) ε_2 – interpretability violation D, (c) ε_3 – changes in software and hardware environment D, (d) ε_4 – reliability violation of storage D, (e) ε_5 – information security breach D;
3. The set of information systems $IS = \{IS_k\}$ perturbed by E and processing, transfer and storage of D;
4. The set of information security requirements $T = \{NTD_l\}$.

Find:

1. The set of mathematical models $\mu = \{\mu_r\}$, who control the parameters of keeping including reliability, authenticity (immutability), interpretability (readability), compliance with information security requirements for information exchange D between IS;
2. The set of algorithms $A = \{A_r\}$, providing the keeping D at baseline SV_0: $A_r(d_i) = SV_0$.
3. The set of software and hardware solutions $R = \{R_r\}$, providing the keeping D at baseline SV_0: $R_r(d_i) = SV_0$.

4   Mathematical Models of Control of Keeping Parameters

Statement 2. The task of monitoring the keeping of DD as objects of management is the task of optimal selection by many criteria.

Indeed, let keeping of d_i is characterized by a function $\mu(t)$, whose value is the probability of keeping d_i at baseline SV_0 at an arbitrary time t. Then it can be argued that the task of keeping d_i is formulated as the task of achieving the maximum of the function $\mu(t)$ in an arbitrary time interval t.

I.e. $M = \max_{t \in [t0, \infty]} \mu(t)$, where $t0$ – is the instant of creation d_i in some IS_0.

Then $\mu(t) = \alpha(t)^{\omega 1} \zeta(t)^{\omega 2} \varphi(t)^{\omega 3} \rho^{\omega 4} \sigma(t)^{\omega 5}$, where $\sum \omega_i = 1$, $\omega_i > 0$, $i = [1, 5]$. $\alpha(t)$, $\zeta(t)$, $\varphi(t)$, ρ, $\sigma(t)$ – according to the probability of preserving authenticity, interpretability, stability to changes in software and hardware storage environment, reliability, stability to threats of breach of information security D in an arbitrary time interval t.

As a result of the research, the following mathematical models were developed to control the parameters of keeping. Unfortunately, the format of the article does not allow for the complete derivation of the obtained mathematical formulas, therefore, we present only the results of the research.

4.1 Evaluation of the Preservation of Authenticity

Preserving the authenticity (immutability and integrity) of DD during the entire storage period is the most important problem in the organization of long-term keeping. It is necessary to find a way to confirm and control the preservation of authenticity.

Suppose we have found a way to validate DD authenticity with DS. Further it will be shown that such a method was found. As a result of the study, the DS inventory algorithm was developed. Then it is necessary to estimate the probability of not breaking DS ($\alpha 1(t)$), as well as the probability that DS is still relevant ($\alpha 2(t)$). Then the assessment of the preservation of authenticity can be represented by the following relations:

$$\alpha 1(t) = \frac{1}{N_{DS}} \sum_{i=1}^{N_{DS}} \left(\frac{1}{2} - 2^{n-1} \prod_{ii=1}^{n} \left(p_{ii}(t) - \frac{1}{2}\right)\right) \tag{1}$$

$$\alpha 2(t) = \frac{1}{N_{DS}} \sum_{i=1}^{N_{DS}} \left(Vf_i(t)\left(\frac{N_{sert\_dbi}(t)}{N_{serti}} \sum_{j=1}^{N_{serti}} S_{true\_ij}(t)\left(\frac{T_{dij} - T_{dcij}(t)}{T_{dij}}\right)\right)\right) \tag{2}$$

$$\alpha(t) = min(\alpha 1(t), \alpha 2(t)) \tag{3}$$

where N_{DS} – is the number of DS, $p_{ii}(t)$ – is the probability of selecting the correct combination of an arbitrary bit of a key – ciphertexts with a length of n, T_{dij} – is the time of the j-th certificate of the i-th DS, $T_{dcij}(t)$ – is the current time from the beginning of the validity period of the DS. $S_{true\_ij}(t)$ – the function of the validity of the j-th certificate of the i-th DS at time t, $N_{sert\_dbi}(t)$ – the number of downloaded certificates of the i-th DS at time t, N_{serti} – the required number of certificates for the i-th DS, $Vf_i(t)$ – is the result of the check (1 or 0) of the i-th DS at time t.

4.2 Evaluation of the Preservation of Interpretability

When ensuring the long-term keeping of DD, there can be a problem of interpretability and data mapping in new informational conditions. In other words, it should be possible to decode the stored DD format in decades and display the DD on the screen, print, etc. In addition, along with DD, it is necessary to store metadata, which must also be read after years.

Thus, it is necessary: (1) to control the preservation of interpretability by considering risks and checking the readability of DD; (2) to ensure interpretability by all available means up to the normalization of DD into a suitable format for long-term storage.

In addition, you need to determine what is DD with long-term keeping. The following definitions have been introduced. Document - structured information, which is a set of interrelated semantic blocks. DD is a document whose semantic blocks and relationships between them are represented in digital form. Semantic blocks - some fragments of DD, highlighted in semantic content.

To control the interpretability (IPT) of DD, a long-term DD model has been developed (4) and a control model (5) of the probability of IPT ($\zeta(t)$):

$$d_i = MD \bigcup OrD \bigcup OdfD \bigcup FTI \bigcup CLI \bigcup LDI \bigcup OpD \qquad (4)$$

$$\zeta(t) = \zeta OdfD(t)^{k1} \zeta OrD(t)^{k2} \zeta MD(t)^{k3} \zeta CLI(t)^{k4} \zeta OpD(t)^{k5} \zeta FTI(t)^{k6} \zeta LDI(t)^{k7} \qquad (5)$$

where k_i – weighting coefficients of importance (appointed by experts) $\sum k_i = 1$, $k_i > 0$, $i = [1, 7]$. On the basis of practical application, it is recommended to assign the following values $k_1 = k_2 = 0,1$, $\sum_{I = [3, 7]} k_i = 0,8$, since the IPT of original (OrD) DD (d_i) and the normalized copy ($OdfD$) are the most critical. MD – DD metadata, OrD – original of DD, $OdfD$ – normalized copy of the original of DD, FTI – normalized text of the original of DD, CLI – classifiers of DD, LDI – vector of relations with other DD, OpD – log of operations with DD. $\zeta OdfD(t)$ – probability IPT of $OdfD$, $\zeta OrD(t)$ – probability IPT of OrD, $\zeta MD(t)$ – probability IPT of MD, $\zeta CLI(t)$ – probability IPT of CLI, $\zeta OpD(t)$ – probability IPT of OpD, $\zeta FTI(t)$ – probability IPT of FTI, $\zeta LDI(t)$ – probability IPT of LDI.

4.3   Evaluation of Stability to Changes in Software and Hardware Storage Environment

As with the aging of the media, and when updating the software and hardware environment there is a need for the correct migration of DD. A significant risk is that there is a high probability of losing DD due to negligence or malicious intent. During migration, it is important to pay attention to the likelihood of DD metadata transfer, classifiers, transaction logs, links to other documents, in order to avoid loss of valuable information.

Based on the DD (4) model, an algorithm for controlling the storage environment was developed, as well as a mathematical model for assessing stability (6).

$$\varphi(t) = \varphi OdfD(t)^{k1} \varphi OrD(t)^{k2} \varphi MD(t)^{k3} \varphi CLI(t)^{k4} \varphi OpD(t)^{k5} \varphi FTI(t)^{k6} \varphi LDI(t)^{k7} \qquad (6)$$

where k_i – are the weighting coefficients of importance assigned by experts $\sum k_i = 1$, $k_i > 0$, $i = [1, 7]$. On the basis of practical application, it is recommended to assign the following values $k_1 = k_2 = 0,1$, $\sum_{i = [3, 7]} k_i = 0,8$, since the stability (STB) of the original (OrD) DD (d_i) and the normalized copy ($OdfD$) are the most critical. $\varphi OdfD(t)$ – probability STB of $OdfD$, $\varphi OrD(t)$ – probability STB of OrD, $\varphi MD(t)$ – probability STB of *MD*, $\varphi CLI(t)$ – probability STB of *CLI*, $\varphi OpD(t)$ – probability STB of *OpD*, $\varphi FTI(t)$ – probability STB of *FTI*, $\varphi LDI(t)$ – probability STB of *LDI*.

4.4   Evaluation of Reliability

DD is inseparable from software and hardware means of storage, interpretation, control. Therefore, it is important in the organization of long-term storage is to control the reliability of the software-hardware storage environment.

The reliability model is described in sufficient detail in [21]. Here we can briefly say that the author proved that the IS reliability scheme can always be represented as a tree or forest.

The general reliability model scheme for a three-level hierarchical distributed IS [21] is as follows (7):

$$\rho = K_{r\_IS} = K_{r\_tl} \sum_{i=1}^{N_{ml}} \left(b_i K_{r\_cc\_ml\_i} K_{r\_ml\_i} \left(\sum_{j=1}^{N_{lli}} a_{ij} K_{r\_cc\_ll\_ij} K_{r\_ll\_ij} \right) \right) \tag{7}$$

where $K_{r\_tl.}$ – reliability parameter for IS top-level objects (non-detailed object reliability model for objects like data processing centers, central servers, software of central database etc.); $b_i = Ob_i / \sum_{i=1}^{N_{ml}} Ob_i$ – share of objects serviced by i- node of the middle level IS, Ob_i – a number of reliability elements serviced by i-node of the middle level IS ($\sum_{i=1}^{N_{ml}} b_i = 1$); N_{ml} – a number of elements of the middle level IS; $K_{r\_cc\_ml\_i}$ – indicator of communication channel availability between top and middle level; $K_{r\_ml\_i}$ – indicator of i-node availability of the middle level IS; $a_{ij} = Ob_{ij} / \sum_{j=1}^{N_{lli}} Ob_{ij}$ – share of objects serviced by j-node of the IS related with the middle level i ($\sum_{j=1}^{N_{lli}} a_{ij} = 1$) against general number of low level objects connected with i-node of the middle level IS; N_{lli} – number of low-level nodes connected with i-node of the middle level; $K_{r\_cc\_ll\_ij}$ – indicator of communication channel availability of the middle level i-node – low level j-node; $K_{r\_ll\_ij}$ – availability indicator of the low level j-node connected with the middle level i-node.

4.5   Evaluation of Stability to Threats of Breach of Information Security

We can define stability as the probability of preserving authenticity, reliability and interpretability of DD in the case of an external impact (changes in software and hardware storage environment, the realization of information security threats).

Then the mathematical model for estimating the probability of preserving resistance to threats to information security breaches ($\sigma(t)$) can be represented as follows (8):

$$\sigma(t) = p_{isc}(t)^{v1} (1 - p_{cr}(t))^{v2} p_{hra}(t)^{v3}, \tag{8}$$

where $\sum v_i = 1$, $v_i > 0$, $i = [1, 3]$. $p_{isc}(t)$ is the probability of implementing measures to counter information security threats, $p_{cr}(t)$ is the probability of losing critical information, $p_{hra}(t)$ is the assessment of the influence of the human reliability assessment (HRA).

5   Algorithms for Long-Term Keeping

Of course, only the control of the parameters of keeping the problem of ensuring cannot be solved. Therefore, algorithms have been developed to ensure long-term preservation.

Together with mathematical models for monitoring the parameters of long-term safety, they became the basis of the established technology of long-term storage of DD.

Unfortunately, the article format does not allow to provide a fully detailed description of these algorithms. Therefore, we confine ourselves to a brief description.

5.1   DS Inventory Algorithm

When implementing the algorithm, it is assumed that DS is used as an independent means of ensuring the authenticity (immutability) of DD. The algorithm involves the periodic re-DD with the new DS. In this case, information about the authorship of the old DS should be preserved. In addition, the authenticity of the DS must be confirmed by an electronic notary.

Brief description of the algorithm:

- The probability of authenticity violation is estimated using a mathematical model (3)
- If the probability value $\alpha(t)$ is less than a certain threshold value α_T, the violation is recorded in the authenticity audit log
- Means of cryptographic protection checks the existing DS
- In the case of DS authentication, the DS time stamp is checked
- In the case of confirmation of the verification of the time stamp, the fact of the positive DS verification is recorded in the audit log of authenticity
- The audit log is certified by the DS indicating the authorship of the verifier
- The author's data is extracted from the certificate of the current DS
- DD is authenticated with a new DS, the data about the author of the current DS is recorded
- The fact of assurance "new" DS is recorded in the audit log of authenticity
- The audit log is certified by the DS indicating the authorship of the reviewer
- If at least one check did not give a positive result, the fact of keeping is recorded in the audit log of authenticity. In this case, a detailed investigation of the causes of authenticity violations is needed.

5.2   Algorithm Inventory of Media

In addition to software obsolescence, there are a number of problems of long-term keeping. Currently there are no carriers capable of functioning without problems for decades. In addition, there is still the problem of technological aging, when an absolutely safe carrier cannot be read over time due to the absence of reading devices. The reason for this is that over time, it becomes economically unprofitable for manufacturers of IT technologies to maintain old, poorly demanded technologies.

The solution to these problems lies, firstly, in the duplication of information, and secondly, in regular verification and transfer of information to new data carriers. The authors proposed an algorithm for the inventory of information carriers to solve the problem of changing the software and hardware environment.

Brief description of the algorithm:

1. A check is performed on the health of the media;
2. If the media is faulty, go to step (3), if it's OK, go to step (5);
3. Search for backup media;
4. If the media is not found, fixation in the audit log of the media that DD is lost, the end of the algorithm, if found, go to step (1);
5. The DD is checked for the possibility of migration (DD stability) according to the model (6);
6. If migration is possible, go to step (7), otherwise - to step (3). Fixation of audit result in the audit log media;
7. DD interpretation interpretability is performed by model (5);
8. If DD is interpretable, go to step (9), otherwise go to step (3). Fixation in the audit log media audit result;
9. DD authentication is being verified using the DS inventory algorithm;
10. If DD is authentic, go to step (11), otherwise go to step (3). Fixation in the audit log media audit result;
11. Creates the necessary number of copies of DD on new media;
12. Conduct interpretability, authenticity and robustness checks after creating copies. Fixation of audit result in the audit log media.

5.3 Others Algorithms

In addition to those described above, the following algorithms were also developed as part of the created technology for ensuring long-term security.

Algorithm for assessing resistance to threats to information security breaches. The algorithm includes steps for analyzing the software and hardware DD environment, identifying and ranking threats, developing countermeasures for each threat, and performing a sustainability assessment using the model (8).

The algorithm for assessing the impact of HRA on the keeping of DD. The algorithm includes the following steps: ranking of operations with DD according to the degree of criticality, determining the degree of congestion of a human operator in each operation, determining the composition of possible operator errors, evaluating the impact of HRA on the reliability and robustness indicators of DD to external influences, determining opportunities to reduce the impact of HRA.

6   Implementation of Long-Term Storage Technology

On the basis of the presented technology for ensuring long-term security, software solutions have been created, which are implemented in several electronic archive projects. This is an electronic archive for the Pension Fund of the Russian Federation, which has been functioning for more than 15 years in 80 regions of the Russian Federation and allows you to organize secure long-term storage of DD. The introduction of an electronic archive allowed a 10-fold increase in the efficiency of the DD search and the organization of reliable storage of more than 500 million DD (over 50% of the fund's documents). No violation of the authenticity and interpretability of DD was noted. The rented space for storage of documents has been reduced several times.

Based on the presented technology, electronic archives have been created for Gazprombank JSC, Law Firm Gorodissky and Partners LLC, Cognitive Technologies LLC.

Besides, the suggested here technology was used at the development of Skolkovo "Smart city" concept [22]. The suggested approach was implemented during reliability assessment of the Russian Federation State Automated System "Vybory" [23, 24].

7 Conclusion

This article discusses the technology of organizing the long-term storage of DD, which was brought to the level of existing software and methodological solutions. These solutions are implemented and applied in industrial operation within the framework of the two largest information systems of the federal level and several IS of large companies.

The advantage of the presented technology is that it allows you to comprehensively solve the scientific and technical problem posed by the organization of long-term keeping of DD. On the basis of the development of technology, it is possible to build software solutions capable of conducting an automated quantitative assessment of the parameters of keeping DD. The technology makes it possible to comprehensively solve such problems of keeping as: ensuring the authenticity, interpretability, reliability and stability of DD.

In the future, it is planned to develop a mathematical and algorithmic apparatus of technology for solving the problem of long-term keeping of big data and systems based on distributed registry technologies.

References

1. Program Digital Economy of the Russian Federation: Approved by the order of the Government of the Russian Federation at 28 July 2017, № 1632-r (2017)
2. Ryskov, O.I.: On the main activities of foreign archival bodies in the field of research and regulatory work with electronic documentation. Secretarial Bus. **3**, 76 (2005)
3. Afanasyeva, L.P.: Automated archive technologies. Russian State University for the Humanities, Moscow (2005)
4. Ryskov, O.I.: The main activities of the national archives of the United States and the United Kingdom of Great Britain and Northern Ireland in the field of management of electronic documents of government agencies. Russian archives 3, Moscow (2004)
5. US National archives blog (2013). http://blogs.archives.gov/records-express/2013/11/01/opportunity-for-comment-transfer-guidance-bulletin. Accessed 22 Mar 2019

6. Universal Electronic Records Management (ERM) Requirements. U.S. National Archives and Records Administration (2017). https://www.archives.gov/records-mgmt/policy/universalermrequirements. Accessed 20 Mar 2019
7. Miller, J.: NARA to suspend development of ERA starting in 2012. FederalNewsRadio.com (2012). http://www.federalnewsradio.com/?sid=2204570&nid=35. Accessed 25 Mar 2019
8. Lipowicz, A.: NARA officials defend searchability of electronic archive. Federal computer week (2011). http://fcw.com/articles/2011/11/01/nara-officials-defending-searchability-of-electronic-archive.aspx. Accessed 22 Mar 2019
9. Carlstrom, G.: Is DoD's new pay system fair? FederalTimes.com (2014). http://federaltimes.com/index.php?S=3502888. Accessed 24 Mar 2019
10. National archives announces a new model for the preservation and accessibility of Presidential records. U.S. National Archives and Records Administration (2017). https://www.archives.gov/press/press-releases/2017/nr17-54. Accessed 24 Mar 2019
11. Draft National archives strategic plan. U.S. National Archives and Records Administration (2017). https://www.archives.gov/about/plans-reports/strategic-plan/draft-strategic-plan. Accessed 24 Mar 2019
12. Suvorovtseva, N.G.: Storage of electronic documents: foreign experience. Bull. Cult. Art **4** (52), 17–23 (2017)
13. Open government partnership UK National action plan 2013 to 2015, London SW1A 2AS (2013)
14. Typical requirements for automated electronic document management systems. Specification MoReq. Office for Official Publications of the European Communities as INSAR Supplement VI (2001)
15. Solovyev, A.V., et al.: Recommendations for the acquisition, accounting and organization of storage of electronic archival documents in the archives of organizations. The all-Russian scientific and research Institute for Records and Archive Management, Moscow (2013)
16. Solovyev, A.V., et al.: Recommendations for the acquisition, accounting and organization of storage of electronic archival documents in state and municipal archives. The all-Russian scientific and research Institute for Records and Archive Management, Moscow (2013)
17. Preservation of evidence of cryptographically signed documents. BSI Technical Guideline TR-03125 – Version 1.2 – Federal Office for Information Security, Germany (2015)
18. Emelyanov, S.V.: Automatic control systems with variable structure. Science, Moscow (1967)
19. Emelyanov, S.V., Kostileva, N.E., Matich, B.L., et al.: System design automation. Mashinostroeniye, Moscow (1978)
20. Emelyanov, S.V.: New types of feedback. Fizmatlit, Moscow (1997)
21. Akimova, G.P., Solovyev, A.V., Tarkhanov, I.A.: Reliability assessment method for geographically distributed information systems. In: AICT 2018: The IEEE 12th International Conference on Application of Information and Communication Technologies, Almaty, 17–19 October 2018, pp. 188–191 (2018)
22. Soloviev, A.V.: Concept Smart city Skolkovo for Skolkovo innovation center and set of measures for its implementation, vol. 3. Logic-mathematical model of Smart City: research report. ISA RAS, Moscow (2012)
23. Akimova, G.P., Pashkina, E.V., Soloviev, A.V.: Analysis of the assessment of the effectiveness of a hierarchical geographically distributed information system by the example of GAS "Vybory". Proc. Inst. Syst. Anal. Russ. Acad. Sci. **58**, 25–38 (2010)
24. Soloviev, A.V.: Methodological support of reliability in the field of electronic documents storage. In: Proceedings of the XXIII International Conference Documentation in the Information Society: Archival Studies and Documentation in the Modern World, pp. 321–331. Rosarchive, Moscow (2017)

Automation of the Process a Comprehensive Assessment of Educational Organization

L. A. Ponomareva[✉], O. N. Romashkova, and E. N. Pavlicheva

Moscow City University, 4/1 Vtoroy Selskohozyaystvenniy Avenue,
Moscow 129226, Russian Federation
lmgpu@yandex.ru

Abstract. The purpose of this research is the development of an information system that would allow to carry out quality management in terms of the results of the learning process. The work is of practical importance, as algorithm for assessing the learning process is proposed, which is implemented and integrated into the information system of monitoring, evaluation, correction of the learning process for any discipline. With the help of the developed module it is possible to evaluate the degree of mastering competences by students and to carry out long-term planning of the educational process. In the work, the process of teaching students is modeled on the basis of studies of general information processes in educational environments. A dynamic model is constructed in the notation of colored hierarchical Petri nets. The simulation model of one of the stages of mastering the discipline of the curriculum is analyzed. Methods of statistical analysis developed an algorithm for rating the work of the department. The equation of the discriminant function was obtained, which served as a rating for real departments of the Moscow City University.

Keywords: Model · Assessment of the department · Educational process · Rating · Information system

1 Introduction

Relevance of the topic: a specific indicator of the development of national Russian higher education is the position of universities in world university rankings [1]. The popularity of various systems for assessing the activities of a higher educational institution has determined their recognition and widespread usage [2, 3]. One of the indicators of a high rating of the university is the quality of education [4].

The key structural element that is responsible for the educational process is the academic department (department). Evaluation of the results of the department is an important indicator of the university's assessment in the educational services market in general [5].

Effective management of the university is impossible without a corporate information system, which is an integrated system consisting of many subsystems [6]. One of these subsystems is the module for managing the activities of the department.

A. A. Radionov and A. S. Karandaev (Eds.): RusAutoCon 2019, LNEE 641, pp. 912–922, 2020.
https://doi.org/10.1007/978-3-030-39225-3_98

The task of the authors is to modernize the module by implementing the proposed algorithms for evaluating the activities of the department and for planning the educational process on the basis of its analysis.

The objects of the research are the informational processes taking place in the structural subdivision of the university aimed at teaching students.

The subject of the study is the process of developing an information system unit for evaluating and adjusting learning process of students of a higher education institution.

2   Problem Statement

Analyze the existing processes associated with the study of students of a certain discipline;

- Construct a dynamic model of the learning process
- Carry out a research into the activities of the Moscow University Department
- Build a mathematical model of the rating evaluation of the department
- Formulate the requirements for the module being developed
- Develop a prototype of the information system module for strategic planning of the educational process.

3   Modeling the Education Process

The study of the process of education is a very important task for the management of educational institutions, and consequently, an improvement of training quality of future specialists [7]. The construction of mathematical models makes it possible to study the regularities of the process of the educational process, which is part of the educational process.

A change in the state of an object is called a process [8]. Educational process can be discussed in relation to students as studied objects. The educational process is part of the educational system [9]. The differences are presented in Table 1.

Table 1. Comparison of educational process and educational system [10].

	Educational system	Educational process
Objectives	The main goal - the development of individual characteristics of the student, the promotion of his cognitive independence through subject knowledge, skills and habits	The main goal is the assimilation of subject knowledge, skills
Content	On the forefront are general cultural values	The content is determined by the curricula
Process	The activity of the teacher is the unity of education and upbringing	Training is conducted with the teacher's dominant role

To construct a mathematical model, the authors proposed the definition of an informational object - an educational process that reflects its formalized structure.

The educational process (UE) is a continuous, dynamic, deterministic process with rigidly defined, limited resources, consisting of elements:

- Lecture forms of training
- Laboratory forms of training
- Seminar classes
- Practical lessons
- Technological practice
- Independent work
- Internship
- Knowledge control

Resources for the implementation of the educational process:

- Faculty
- The audience
- Various benefits
- Computer and multimedia audiences
- Corporate networks, etc.

The authors have studied information flows that occur during training [10, 11].

A student can be represented as a finite automaton, with a finite set of states that depend on external influences. Therefore, it was decided to construct a dynamic model in the notation of Petri nets. To implement the dynamic model, the CPN Tools software [12] is used, which is freely distributed for non-profit organizations.

Each network position corresponds to the state of the learning process. Transition is the study of any topic: tests, completion of laboratory work, exams, course projects, etc. Transition triggering corresponds to the successful completion of the study of the topic. Each token is added a color that stores the attributes: either "passed" or "failed". Following the instructions, the chip moves along the network (Fig. 1).

To evaluate the performance of the model, the matrix method of network analysis was applied [13, 14]. A necessary condition for the attainability of any network marking was fulfilled. The analysis showed that only two combinations of transitions are suitable: t3, t2, t1 and t1, t2, t3. This conclusion is confirmed by Fig. 2.

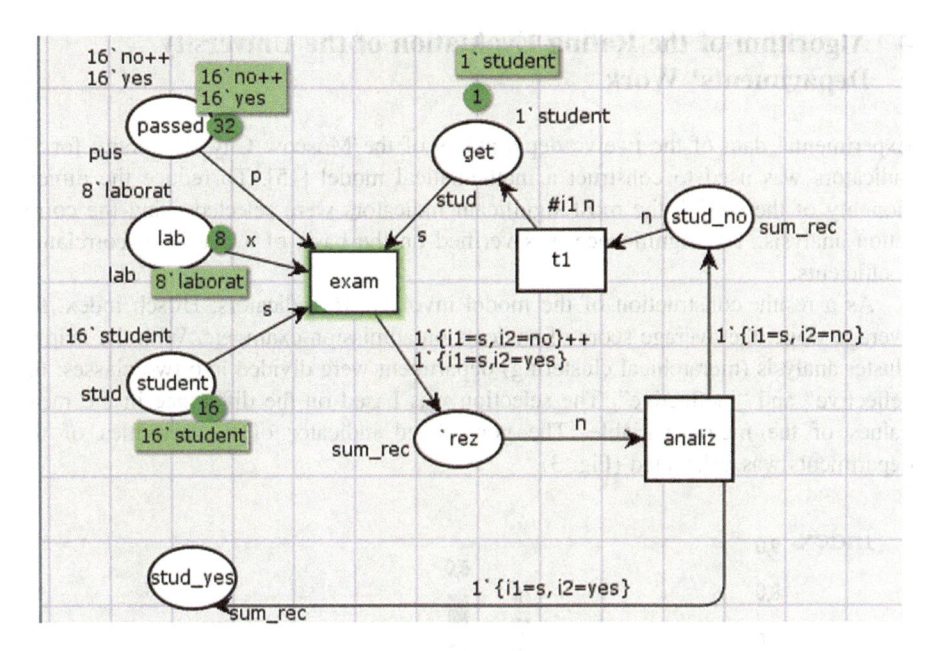

Fig. 1. Fragment of a colored Petri net learning process.

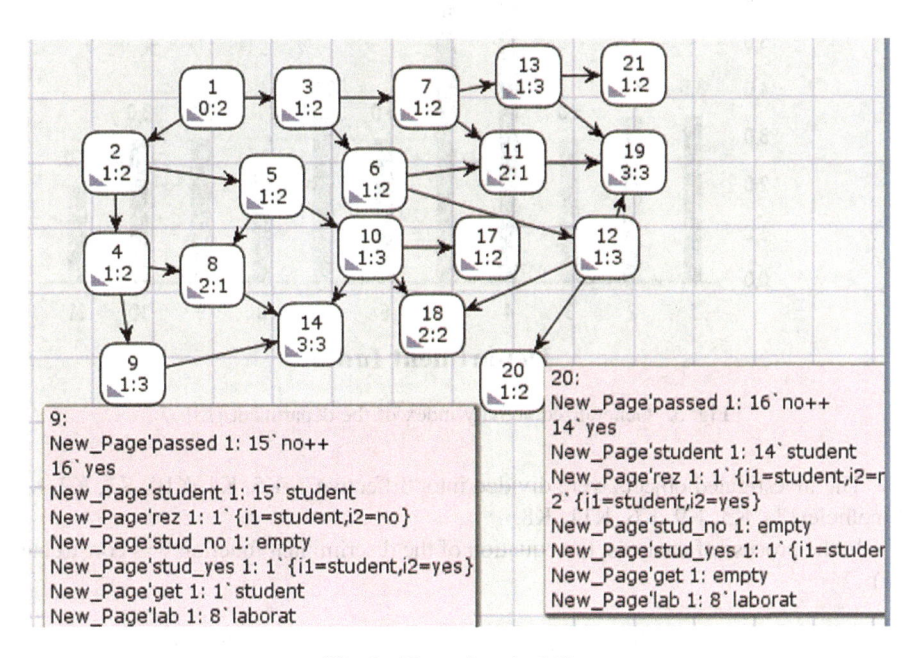

Fig. 2. Tree of attainability.

4 Algorithm of the Rating Evaluation of the University Departments' Work

Experimental data of the twelve departments of the Moscow City University for 25 indicators was used to construct a mathematical model [15]. To reduce the dimensionality of the model, the most significant indicators were selected using the correlation analysis. The significance was verified on the basis of a matrix of correlation coefficients.

As a result, construction of the model involved 16 indicators: Hirsch index, the average wage, the average score of students on admission exam, etc. With the help of cluster analysis (hierarchical clustering) department were divided into two classes: the "effective" and "ineffective". The selection was based on the difference in the mean values of the metric variable. The generalized indicator of the activities of the departments was calculated (Fig. 3).

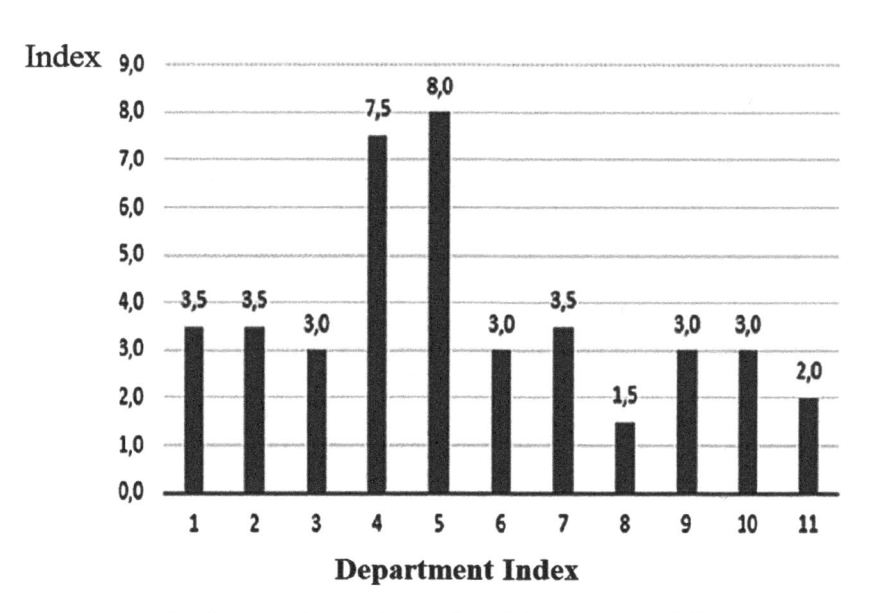

Fig. 3. Generalized activity index of the department [15].

The investigated objects were divided into "Effective" - K5, K4, K10, K1, K2, K7; "Inefficient" - K3, K9, K6, K11, K8.

In the process of analysis, the equation of the discriminant function was constructed (1)

$$D = -7,485 \cdot x_1 + 5,76 \cdot x_2 + 3,481 \cdot x_3 + 30,173 \cdot x_4 - 16,599 \cdot x_5 + 8,190 \cdot x_6 \\ - 11,449 \cdot x_7 + 6,867 \cdot x_8 - 3,367 \cdot x_9$$

$$(1)$$

where xi is the indicator of the department.

Using the constructed model with a probability of 0.999, it is possible to predict the evaluation of the work of structural units without special studies of their indicators.

The prototype of the information system implementing the algorithms described above was developed on the "1C: Enterprise" platform [16–18].

5 Basic Concepts Used in the Information System Module

The performance indicators of the department and the results of students' training are assessed with points for each unit of work. Objects of accounting are various types of work performed by the department or students in the learning process. Achievements are indicators for a certain period of time. The rating of the department is the sum of all points scored for the whole period of work. Accounting rules are automated summation [19, 20].

Processes that automate the module of the information system [21–23]:

- Definition of the system of indicators for evaluating the work of the department
- Determination the success of student learning
- Determination of rules for calculating the rating of the department and the performance of each student
- Formation of recommendations on the prospective planning of the educational process
- Construction of rules for the success of the department
- Creation of a variety of customizable reports

Functional requirements for the information system module [24, 25]:

- Ensure the collection of data on the learning process in the system
- Ensure the storage in a single database of all information about the activities conducted, the type of employment, the completed, unsuccessful tasks, tests, the number of students who have passed and not passed control, various types of activities of the department
- Provide preprocessing of data using methods of statistical data analysis
- Automate the construction of a learning process model based on data on the learning process in the electronic learning system and its editing on the basis of a hierarchical colored Petri net
- Automate the process of creating various reports

The database stores list of students and lists of the faculty members of the department (Fig. 4).

To account for the progress of students, scores are added for each reporting period for each subject in the schedule (Fig. 5).

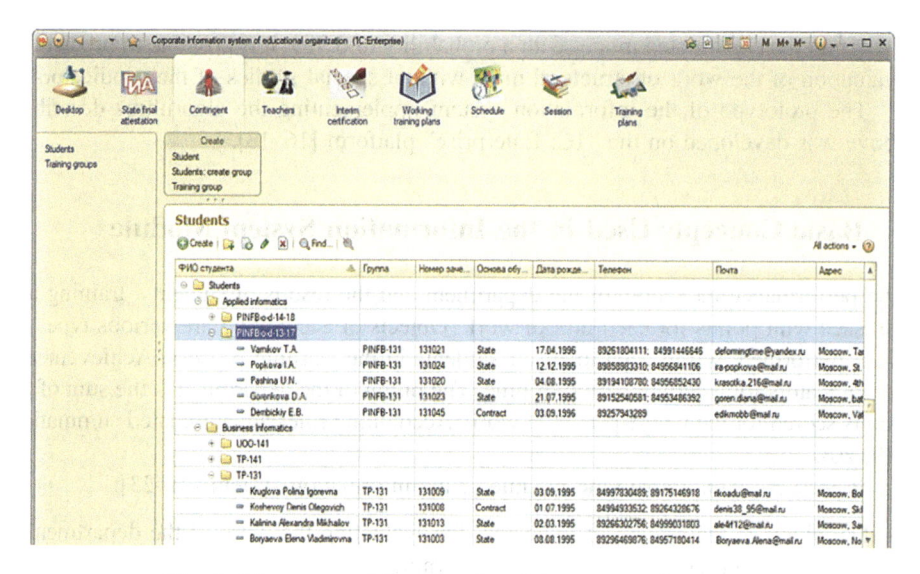

Fig. 4. The command interface list of the catalog «Students».

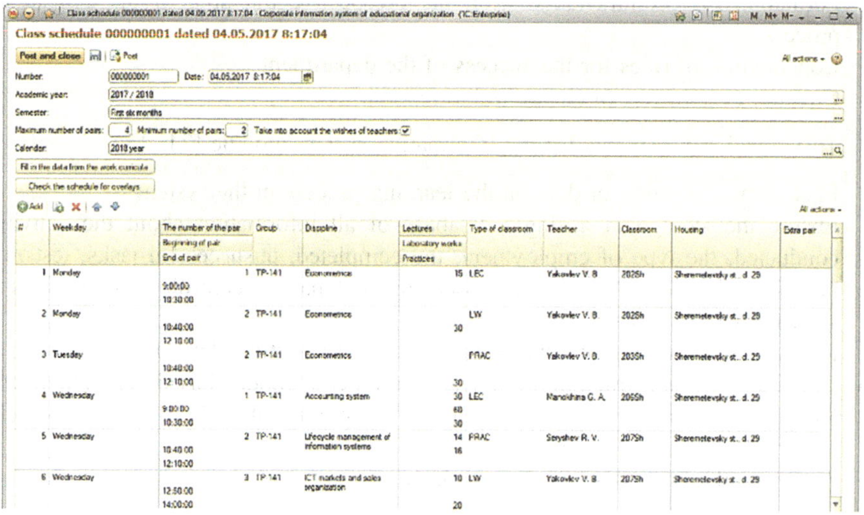

Fig. 5. The main form of the document «Class schedule».

In general, the repository information system of the department are various test and verification tasks. The results of the checks are sent to the developed module, where the rating of the student is evaluated (Fig. 6).

Examples of summary reports of the results of the department are presented in Figs. 7 and 8.

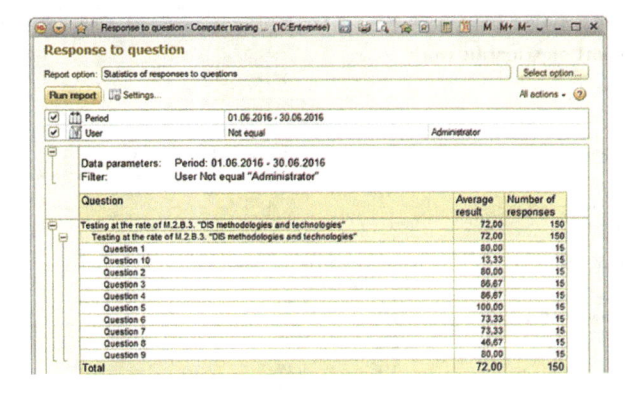

Fig. 6. Report form with test results.

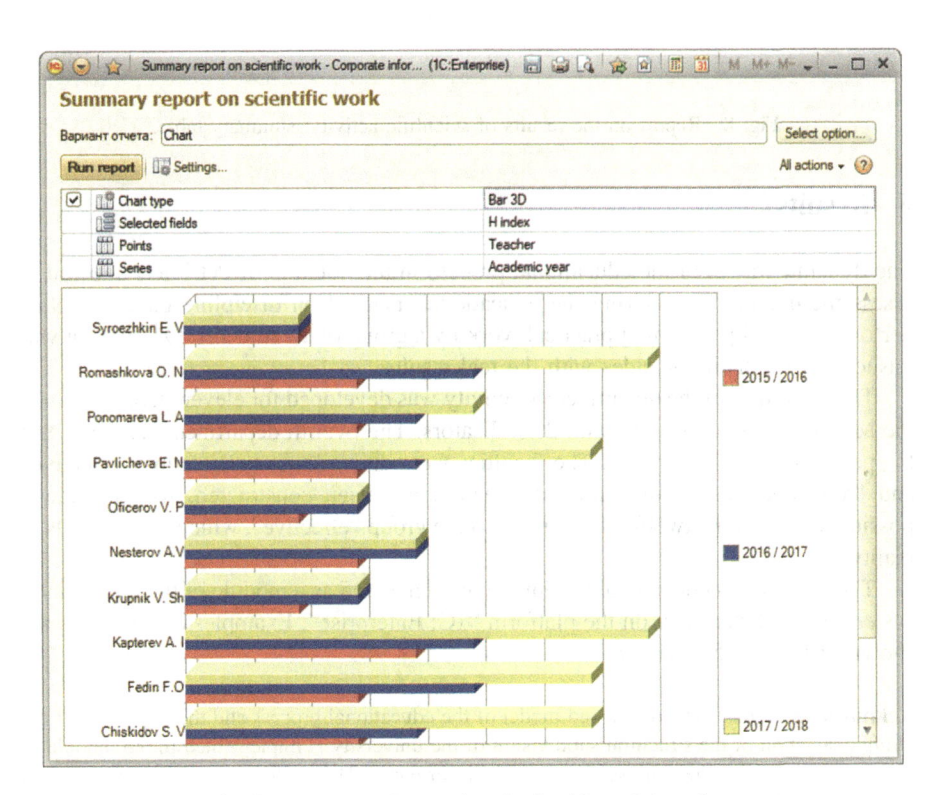

Fig. 7. Report on the results of scientific activity: chart.

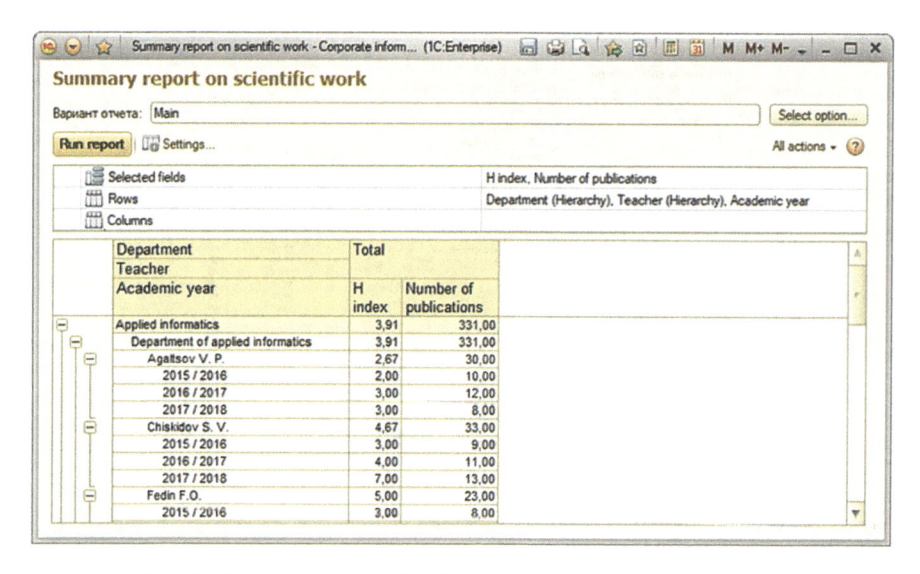

Fig. 8. Report on the results of scientific activity: summary table.

6 Results

The dynamic model of the educational process in the notation of Petri nets allowed to assess the degree of mastering the educational material on discipline based on class performance and performing practical work by a group of 16 students. One person was unsuccessful, which coincides with the real results.

The algorithm of the department's activity was developed for eleven departments of the Moscow City University with 25 indicators. The twelfth department served to test the performance of the model. The faculties were divided into two groups: "effective" and "not effective." A discriminant equation has been constructed, which made it possible to assign the twelfth department to the group "effective", which corresponds to reality.

The proposed model and algorithm are implemented as a module of the information system of the department on the platform "1C: Enterprise". Examples of interfaces are shown in Figs. 4, 5, 6 and 7.

Acknowledgements. The developed model of the educational process and the algorithm of the rating evaluation of the structural subdivision of the university is implemented in the module of the information management system of the department. This will improve the quality and effectiveness of university management in general, as based on reliable and objective methods of evaluation and prospective planning. The implementation of the module will simplify the collection of data on the work of departments and the quality of the educational process and assist with formation of various types of reporting.

References

1. Lee, J.Y., Lee, S.H., Kim, J.H.: A review of the curriculum development process of simulation-based educational intervention studies in Korea. Nurse Educ. Today **64**, 42–48 (2018). https://doi.org/10.1016/j.nedt.2018.01.029
2. Svetsky, S., Moravcik, O.: The empirical research on human knowledge processing in natural language within engineering education. Adv. Intell. Syst. Comput. **627**, 3–15 (2018). https://doi.org/10.1007/978-3-319-60937-9_1
3. Shaydurov, A.A., Chenushkina, S.V., Shaydurova, T.Y.: Technology of implementation of the software module 1C: University in activities of primary labour union employee organization of a higher education institution. Espacios **39**(5), 12 (2018)
4. Gottipati, V., Shankararaman, S.: Competency analytics tool: analyzing curriculum using course competencies. Educ. Inf. Technol. **23**(1), 41–60 (2018). https://doi.org/10.1007/s10639-017-9584-3
5. Kaur, A., Kaur, K., Chopra, D.: An empirical study of software entropy-based bug prediction using machine learning. Int. J. Syst. assur. Eng. Manage. **8**, 599–616 (2017). https://doi.org/10.1007/s13198-016-0479-2
6. Kumar, S.S., Blaga, F., Vesselenyi, T.: Automated production-based model of flexible cell using timed colored Petri Nets. In: MATEC Web of Conferences, vol. 126, p. 02002 (2017). https://doi.org/10.1051/matecconf/201712602002
7. Gottipati, S., Shankararaman, V., Goh, M.: Mining capstone project wikis for knowledge discovery, pp. 371–380 (2017). https://doi.org/10.1109/COMPSAC.2017.169. 8029631
8. Kushik, N., Yevtushenko, N., Evtushenko, T.: Novel machine learning technique for predicting teaching strategy effectiveness. Int. J. Inf. Manage. **2**, 231 (2015). https://doi.org/10.1016/j.ijinfomgt.2016.02.006
9. Ponomareva, L.A., Golosov, P.E.: Development of a mathematical model of the educational process at the University to improve the quality of education. Fundam. Res. **2**, 77–81 (2017)
10. Romashkova, O.N., Ermakova, T.N.: Monitoring the quality of education in secondary organizations with the use of modern means of informatization. Bull. Russ. Univ. Friendship Peoples Ser. Informatizion Educ. **4**, 10–17 (2014)
11. Ovchinnikova, E.V., Chiskidov, S.V.: Problems of development and application of interactive educational modules in learning. In: In the Collection: Science, Education, Society: Trends and Perspectives the Collection of Scientific Works on Materials of the International Scientific-Practical Conference, pp. 80–85 (2014)
12. Ponomareva, L.A., Kodanev, V.L.: Development module of the corporate information system. Educational environment of the University based on cloud technologies. In: Computer Science: Problems, Methodology, Technology the Collection of Materials of XVII International Scientific Conference, pp. 393–398 (2017)
13. Romashkova, O.N., Morgunov, A.I.: Information system for evaluation of results of activity of educational institutions of Moscow. Bull. Russ. Univ. Friendship Peoples Ser. Informatizion Educ. **3**, 88–95 (2015)
14. Orlov, Y., Zenyuk, D., Samuylov, A., et al.: Time-dependent SIR modeling for D2D communications in indoor deployments. In: Proceedings of the 31st European Conference on Modelling and Simulation, pp. 726–731 (2017)
15. Drozdova, A.A., Guseva, A.I.: Modern technologies of e-learning and its evaluation of efficiency. Procedia Soc. Behav. Sci. **237**, 1032–1038 (2017)
16. Kireev, V.S.: Development of fuzzy cognitive map for optimizing e-learning course. Commun. Comput. Inf. Sci. **706**, 47–56 (2017)

17. Kireev, V., Silenko, A., Guseva, A.: Cognitive competence of graduates, oriented to work in the knowledge management system in the state corporation Rosatom. J. Phys. Conf. Ser. **781**(1), 012060 (2017). https://doi.org/10.1088/1742-6596/781/1/012060
18. Romashkova, O.N., Chiskidov, S.V., Frolov, P.A.: Improvement of information technologies of the decision of problems of management in economic systems. Mod. High Technol. **10**, 63–67 (2017)
19. Romashkova, O.N., Ponomareva, L.A.: Model of educational process in high school using Petri nets. Mod. Inf. Technol. IT Educ. **13**(2), 131–139 (2017). https://doi.org/10.25559/SITITO.2017.2.244
20. Ponomareva, L.A., Litvinova, K.R., Gorelov, V.I.: Comparative analysis of the Russian rating systems of the University assessment. In: Methods, Mechanisms and Factors of International Competitiveness of National Economic Systems Collection of Articles of the International Scientific and Practical Conference, pp. 55–58 (2017)
21. Romashkova, O.N., Ponomareva, L.A.: Model of effective management of the United educational system (structure). In: New Information Technologies in Scientific Researches Materials of the XXI All-Russian Scientific and Technical Conference of Students, Young Scientists and Specialists, pp. 16–18 (2017)
22. Ponomareva, L.A., Kodanev, V.L., Chiskidov, S.V.: Model of management of process of development of competences in educational organizations. In: New Information Technologies in Scientific Research Materials of the XXII All-Russian Scientific-Technical Conference of Students, Young Scientists and Specialists, pp. 20–22 (2017)
23. Ponomareva, L.A., Romashkova, O.N., Vasilyuk, I.: Conceptual model of changing the rating assessment of the University. In: Methods, Mechanisms and Factors of International Competitiveness of National Economic Systems. Collection of Articles of the International Scientific-Practical Conference, pp. 75–77 (2017)
24. Ponomareva, L.A., Golosov, P.E., Mosyagin, A.B., et al.: Method of effective management of competence development processes in educational environments. Mod. Sci. Actual Probl. Theor. Pract. Nat. Tech. Sci. **9**, 48–53 (2017)
25. Ponomareva, L.A., Kochergina, G.M., Perelygina, E.N.: The use of information and communication technologies in the study of banking in College. In: Theoretical and Applied Issues of Science and Education. Collection of Scientific Works on the Materials of the International Scientific-Practical Conference, pp. 104–107 (2015). https://doi.org/10.17117/na.2015.02.083

Fuzzy Modeling of the Assessment of Using an Educational Audience in Order to Improve the Quality of Training of the Educational Process

R. U. Stativko and A. I. Rybakova[✉]

Belgorod State Technological University, 46 Kostyukova Street, Belgorod 308012, Russian Federation
aribakova@intbel.ru

Abstract. This paper emphasizes the importance of education for a person in a developed society. There is a need to assess the effectiveness of using classrooms, laboratories and software in the educational process of higher educational institution in order to improve the quality of training. Emphasis is placed on the importance of intelligent technologies as a tool for assessing the quality of the educational process, the use of which improves the quality of training. A brief description of the main provisions of the theory of fuzzy sets is presented. The features of an educational and laboratory fund of a higher educational institution (area, number of working places, software available) that must be taken into account in the educational process are listed. It is shown that all features have a different nature, which led to the need to use the apparatus of fuzzy sets. An example of the description of the input linguistic variable with the original base set and term-sets is given. This work allows us to expand the scope of application of the apparatus of fuzzy sets. Improving information communication technologies (creating local and global networks, databases and knowledge, as well as expert systems) forms a specific educational information computer area of a higher educational institution that enriches traditional forms of education.

Keywords: Quality of training · The role of information · The apparatus of fuzzy sets · The input linguistic variable · Fuzzy rules

1 Introduction

In a developed society, education is treated as an important resource of the human person. It is a resource that allows the consumer of educational services to have an advantage, both in the labor market and in the social sphere.

The rapid development of information communication technologies makes it possible to implement the main principles of the existing training system: the principle of accessibility and the principle of continuity i.e. use of online technology. While observing the principles of accessibility and continuity in the educational process of a technical higher education institution, there is a steady trend towards the implementation

A. A. Radionov and A. S. Karandaev (Eds.): RusAutoCon 2019, LNEE 641, pp. 923–932, 2020.
https://doi.org/10.1007/978-3-030-39225-3_99

and application of modern information technologies in education. In recent years, there is a continuous trend towards informatization of the education sector.

Russia's entry into the Bologna process led to dramatic changes in the field of higher professional education. These changes contributed to the emergence of competition in the market of educational services between higher educational institutions e.g. competition for resources (material, human, etc.). To obtain a successful position in the educational services market, an institution of higher education needs to make the most effective use of its own capabilities, in particular, its own tangible assets, educational and methodological complexes, etc. Productive use of classrooms and laboratories is required. Therefore, approaches are needed in assessing the effective use of classrooms.

The educational process in higher educational institution is actively accompanied by the use of innovative educational technologies. The latest educational technologies allow combining the advantages of classic technologies and the advantages of innovative ones.

The latest educational technologies include the use of multimedia technologies, online technologies. This combination allows the consumer of educational services, in various areas, to improve their educational level in the development of the required disciplines. Also, timely adjustment of the educational trajectory becomes available [1–5].

Currently, the educational process should be aimed at the personality of the student with the development of his creative abilities, using information and communication educational technologies.

Effective use of new information technologies in the educational process is designed to improve the quality of training. An important component of quality assurance is the use of a classroom fund. The classroom fund of a higher educational institution consists of educational audiences of various purposes. Each classroom is characterized by the area, the number of working places for students and related equipment. Depending on the purpose of use, the classroom can be considered as a laboratory room or a computer class. After analyzing the characteristics of classrooms, it can be concluded that all characteristics have a different nature. For example, the number of working places is an indicator of numerical nature. Studying a certain discipline in a computer classroom requires appropriate software. The degree of compliance is difficult to determine using standard mathematical methods. The task of assessing the effectiveness of the use of an educational audience can be solved using the apparatus of fuzzy sets [6–8].

2 Basic Provisions of the Theory of Fuzzy Sets. Fuzzy Variable, Linguistic Variable

To describe fuzzy (blurry) knowledge in the early 70s, the American mathematician Lofti Zadeh developed a formal apparatus for fuzzy mathematics and fuzzy logic (L.A. Zadeh, Fuzzy Sets, Information and Control, 8 (1965) pp. 338–353). The proposed theory of L. Zadehis based on a subjective fact i.e. the subjective judgments of a person about a goal are always fuzzy. But he also takes the next step - he believes that all assessments of the subject and the limitations with which he works are also, as a rule, fuzzy, and sometimes even lack quantitative characteristics in their initial form. L. Zadeh introduced one of the main concepts in fuzzy logic - the concept of a linguistic variable.

With the help of fuzzy sets, one can formally define inaccurate ambiguous concepts, such as "the correspondence of the educational audience to the subject being conducted", "the large number of students", "the average correspondence of the profile software", "the large classroom fund", etc. [9–15]. Before formulating the definition of a fuzzy set, it is necessary to define the so-called universe of discourse. In the case of the ambiguous concept of "large number of students," one number will be recognized as a large number, if we confine ourselves to the range [1..25 people] and quite another - in the range [10..100 people]. The area of reasoning, hereinafter referred to as space or set, is most often denoted by the symbol X. It must be remembered that X is a clear set.

The fuzzy set A in some (non-empty) space X is the set of pairs:

$$A = \{(x, \mu_A(x); \; x \in X\}, \tag{1}$$

$$\mu_A : X \to [0, 1], \tag{2}$$

where the membership function of the fuzzy set A. This function assigns to each element $x \in X$ the degree of its belonging to the fuzzy set A, and three cases can be distinguished: $\mu_A(x) = 1$ means the full belonging of the element x to the fuzzy set A, i.e. $x \in A$; $\mu_A(x) = 0$ means the absence of x belonging to the fuzzy set A, i.e. $x \notin A$; $0 < \mu_A(x) < 1$ means the partial belonging of the element x to the fuzzy set A.

One of the descriptions used in literature is a symbolic description of fuzzy sets. In this description, X is a space with a finite number of elements, i.e. $X = \{x_1, \ldots, x_N\}$, the fuzzy set $A \subseteq X$ is written as

$$A = \frac{\mu_A(x_1)}{x_1} + \frac{\mu_A(x_2)}{x_2} + \ldots + \frac{\mu_A(x_n)}{x_n} = \sum_{i=1}^{n} \frac{\mu_A(x_i)}{x_i}, \tag{3}$$

The formal definition of a fuzzy set does not impose any restrictions on the choice of a particular membership function for its representation [16–20]. However, in practice it is convenient to use those that allow an analytical representation in the form of some simple mathematical function. This simplifies not only the corresponding numerical calculations, but also reduces the computational resources required to store the individual values of these membership functions. The need for typing of individual membership functions is also due to the presence of implementations of the corresponding functions in the instrumental means used.

A fuzzy variable is characterized by a triple $(\alpha, X, A (\alpha; x))$, where α is the name of the variable, X is the universal set (finite or infinite), x is the common name of the elements of the set X, $A (x \in X)$ is a fuzzy subset of the set X, which is a fuzzy restriction on the values of the variable x, due to X.

A linguistic variable (LP) is described by a tuple of the form <β, T, U, G, M>, where β is the name of the linguistic variable, T is the set of its values (term set), which are the names of fuzzy variables, the scope of each of which is the set U. The set T is called the base term-set of the linguistic variable. G is a syntactic procedure describing the process of the formation of the values of the linguistic variable (term) meaningful for this task from the elements of the set T of new ones. M is a semantic procedure that allows you to turn each new value of LP, formed by procedure G, into a fuzzy variable, i.e. form the corresponding fuzzy set.

The mechanism of fuzzy inference (Fig. 1) is based on the knowledge base, defined by subject matter experts in the form of a set of fuzzy rules of the form:

P1: if x is A1, then y is B1,
P2: if x is A2, then y is B2,
...
Pn: if x is An, then y is Bn,

where x is the input variable, y is the output variable, A and B are the membership functions defined on x and y, respectively.

The knowledge of expert $A \rightarrow B$ reflects a fuzzy causal relation of premise and conclusion; therefore, it is called a fuzzy relation:

$$R = A \rightarrow B,$$

where "$\rightarrow$" is a fuzzy implication.

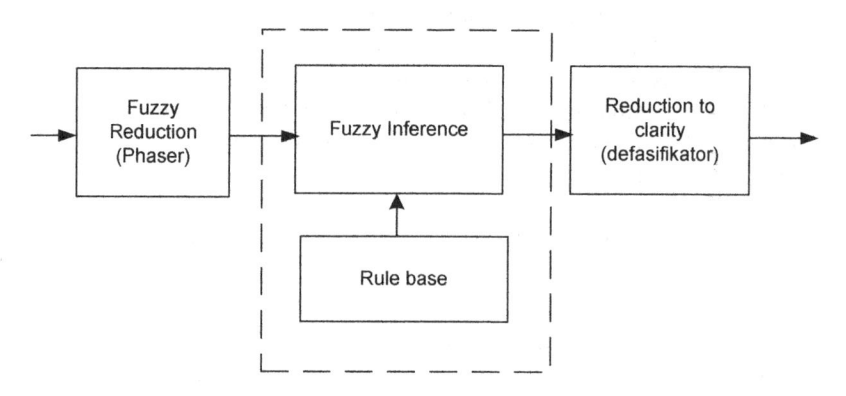

Fig. 1. Diagram of fuzzy output.

Let's describe briefly the stages of fuzzy inference:

- Fuzzification stage. At this stage we should determine the type of membership functions for the input variables according to the requirements of the subject area. The membership functions are applied to the actual values of the input variables to determine the degree of truth of each premise of each rule
- Logical inference phase. For activated rules, the truth value for the premises is determined and applied to the conclusions of each rule. As a result, one fuzzy set is formed, which determines the value of the output variable for each rule

- Stage of composition. The fuzzy subsets assigned to each output variable (in all rules) are combined together to form one fuzzy subset for each output variable. With such a union, max operations are usually used
- Stage of defuzzification - bringing to clarity. It is a conversion a fuzzy set of outputs to a number

3 Development of a Fuzzy Model for Evaluating the Use of Training Audience

In this paper, we focus on the development of a fuzzy model for assessing the effectiveness of the use of an educational audience. Define the linguistic variable Y "audience utilization efficiency" by the following tuple:

$$Y = <S, C, ND>, \tag{4}$$

where S is the classroom area; C is the number of students; ND - the availability of special equipment in the audience (laboratory, software, for the discipline).

We define a fuzzy model for the linguistic variable Y as a system with three inputs and one output (Fig. 2).

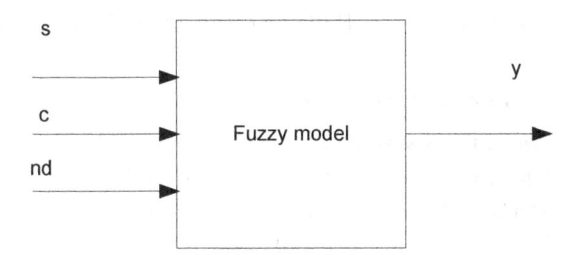

Fig. 2. Fuzzy model for linguistic variable Y.

We define the parameters specified in expression 4 as input linguistic variables. The variable Y is the output linguistic variable. The value of the output linguistic variable will be obtained using simplified fuzzy inference. Let's perform an assessment of the effectiveness of the use of an educational audience, designed to conduct classes with software. For fuzzification of input linguistic variables, we will use a triangular membership function of the type 5.

$$\mu(x; a, b, c) = \begin{cases} 0 & for \quad x \leq a, \\ \frac{x-a}{b-a} & for \quad a \leq x \leq b, \\ \frac{c-x}{c-b} & for \quad b \leq x \leq c, \\ 0 & for \quad x \geq c. \end{cases} \tag{5}$$

This membership function can be used to describe uncertainties of the type: "approximately", "roughly". The parameters of the function a, b, c are determined by an expert for each term of the linguistic variable, respectively.

As an example of assessing an educational audience, let's consider the use of a computer audience to conduct classes that require special software.

To form an assessment of the choice of a computer room, a procedure is used in which the characteristics of different criteria are compared. The system works with indicators that characterize a computer audience. It is based on the idea that determines the choice of a computer room in accordance with the requirements for the educational process. Requirements are determined with the involvement of experts.

According to the expert opinion, the values of the membership function will be chosen in such a way as to satisfy the previously formulated conditions. Expert information is only the initial information for further processing [17–27].

For all input indicators (classroom size, number of students, software availability) a triangular membership function (trimf) was chosen. We use three types of assessment (small, medium, high).

We describe the input linguistic variables:

- S is the training area. The training audience is characterized by state standards (ORDER. RUSSIAN FEDERATION of September 1, 2009 N 390) [27]. According to expert estimates, the size of the classroom can be estimated (Table 1). Linguistic variable S = <area of the classroom, TS, [20-35], GS, MS>, where: TS is a term set with the following values; TS = {little (A1), middle (A2), big (A3)}

Table 1. Expert estimates of the audience area.

Size of the classroom	20	22	24	28	30	32	35
Little (A1)	1	0,9	0,5	0,3	0,2	0	0
Middle (A2)	0,1	0,3	0,5	1	0,5	0,3	0,1
Big (A3)	0	0	0,2	0,3	0,5	0,9	1

- C is the number of students. Student groups may include a different number of students. Expert estimates of the number of students are presented in Table 2

Linguistic variable C = <number of students, T, [10–30], GC, MC>, TC = {small (B1), medium (B2), large (B3)}

Table 2. Expert estimates of the number of students.

Number of students	10	14	18	18	24	27	30
Little (B1)	1	0,9	0,5	0,3	0,2	0	0
Middle (B2)	0,1	0,3	0,5	1	0,5	0,3	0,1
Big (B3)	0	0	0,2	0,3	0,5	0,9	1

- ND - the degree of software security for the subject being taught. The use of software in the classrooms is determined by the availability of a license in a higher educational institution. The indicator, unlike the previous indicators, does not have a quantitative base set. It was proposed to estimate this indicator using a numerical scale from 1 to 100. ND = <degree of security, T, [10–100], GND, MND>, TND = {acceptable (Z1), average (Z2), full (Z3)} (Table 3).

Table 3. Expert software conformity assessments.

Degree of security	10	20	40	60	80	90	100
Acceptable (Z1)	1	0,9	0,5	0,3	0,2	0	0
Middle (Z2)	0,1	0,3	0,5	1	0,5	0,3	0,1
Full (Z3)	0	0	0,2	0,3	0,5	0,9	1

For the output linguistic variable Y (audience efficiency), a triangular membership function (trimf) was also chosen. For the output linguistic variable, three types of assessment are used (low efficiency, medium efficiency, high efficiency).

Let's illustrate a fragment of fuzzy rules for the operation of the system:

1. If ("classroom size" is small) and ("number of students" is small) and ("degree of security" is acceptable) then ("audience use efficiency" is low efficiency) (1)
2. If ("classroom size" is high) and ("number of students" is high) ("degree of security" is high) then ("audience efficiency" is high efficiency) (1)
3. If ("classroom size" is average) and ("number of students" is average) and ("degree of provision" is average) then ("audience efficiency" is average efficiency) (1)
4. If ("classroom size" is high) and ("number of students is high) and ("degree of security "" is full) and ("audience efficiency" is high efficiency) (1)
5. If ("classroom size" is high) and ("number of students" is high) and ("number of students in a group" is average) and ("degree of sufficiency" is average) then ("audience efficiency" is average efficiency) (1)

To implement the developed fuzzy model, we use MatLab (an interactive development environment for algorithms, a powerful data analysis tool). For example, the description of the input linguistic variable S (size of the classroom) with term-sets and triangular membership functions is presented in Fig. 3.

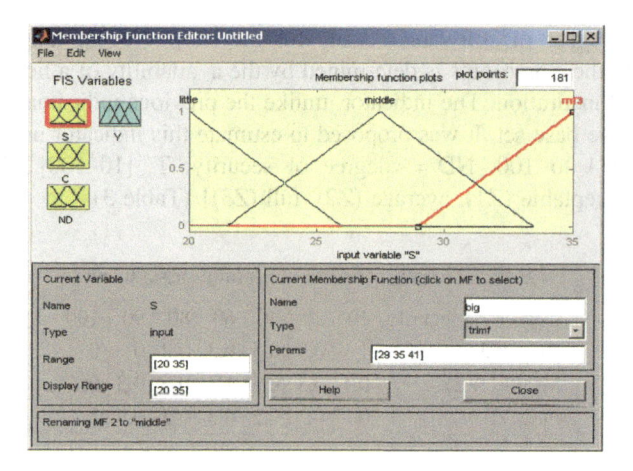

Fig. 3. Linguistic variable S (size of the classroom).

Other input and output linguistic variables are described by analogy. In Fig. 4 a list of rules is given.

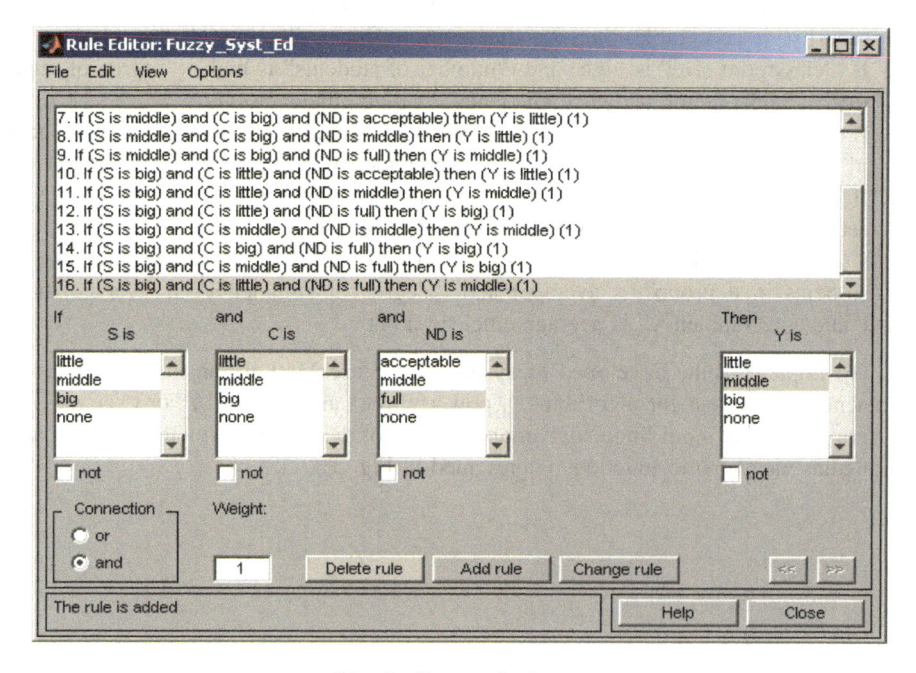

Fig. 4. Fuzzy rules base.

Let us demonstrate the operation of a fuzzy system, setting the initial values, for example: S = 25, C = 20, ND = "middle"/The results are shown in Fig. 5.

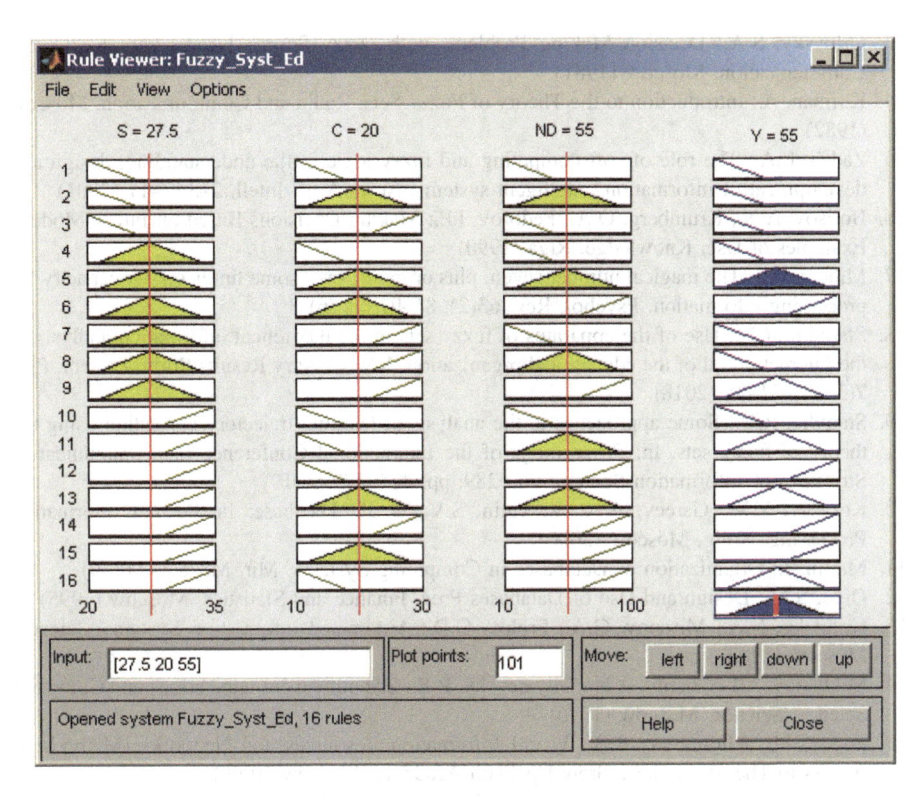

Fig. 5. The results of the fuzzy system.

4   Conclusion

The presented work contains a brief description of the positions of the apparatus of fuzzy sets. The theory of fuzzy sets allows you to perform an assessment of indicators that have in their composition indicators of a different nature. The proposed method allows monitoring and evaluation of the effectiveness of the use of the classroom fund of a higher educational institution. Timely assessment will improve the quality of the educational process in higher education. The work may be useful to the management of the higher echelon of the educational institution.

Acknowledgements. The work was performed as part of the development program of the Belgorod State Technological University named after V.G. Shukhov for the period of 2017–2021.

References

1. Zade, L.: The Concept of a Linguistic Variable and Its Application for Making Approximate Decisions. Mir, Moscow (1976)
2. Ryzhov, A.P.: Elements of the Theory of Fuzzy Sets and Fuzziness Measurements. Dialogue-MSU, Moscow (1998)

3. Orlovsky, S.A.: Decision Making Problems with Fuzzy Source Information. Radio and Communication, Moscow (1981)
4. Kofman, A.: Introduction to the Theory of Fuzzy Sets. Radio and Communication, Moscow (1982)
5. Zadeh, L.A.: The role of soft computing and fuzzy logic in the understanding, design and development of information, intelligent systems. News Artif. Intell. **2–3**, 7–11 (2001)
6. Borisov, A.N., Krumberg, O.A., Fedorov, I.P.: Making Decisions Based on Fuzzy Models: Examples of Use. Knowledge, Riga (1990)
7. Miller, G.A.: The magical number seven, plus or minus two: some limits on our capacity for processing information. Psychol. Rev. **63**(2), 81–97 (1956)
8. Stativko, R.U.: Use of the apparatus of fuzzy sets in the theoretical information analysis of the Internet portal of the educational organization. XXI Century Results Probl. Present. Plus **7**(3–43), 31–35 (2018)
9. Stativko, R.U.: Some approaches to the analysis of learning trajectory correction using the theory of fuzzy sets. In: Proceedings of the International Conference on Communicative Strategies of Information Society, vol. 289, pp. 474–479 (2019)
10. Korncev, V.V., Gareev, A.F., Vasyutin, S.V., et al.: Database: Intellectual Information Processing. Nolig, Moscow (2000)
11. Martin, J.: Organization of Databases in Computing Systems. Mir, Moscow (1999)
12. Digo, S.M.: Design and Use of Databases Proc. Finance and Statistics, Moscow (1995)
13. Krinitsky, N.A., Mironov, G.A., Frolov, G.D.: Automated Information Systems. Science, Moscow (1982)
14. Mikhailov, A.I., Cherny, A.N., Gilyarevsky, R.S.: Scientific Communications and Computer Science. Science, Moscow (1976)
15. Popov, I.I., Khramtsov, B.B.: World Information Resources and Networks (Methods of Access to Them): Studies. Grew Up Econ Academy, Moscow (1999)
16. Revunkov, G.A., Samokhvalov, E.N., Chistov, V.V.: Databases and Data Banks and Knowledge. High School, Moscow (1992)
17. Romanenko, A.G., Samoylyuk, O.F.: Information Retrieval Systems. RGGU, Moscow (1998)
18. Larichev, O.I., Moshkovich, E.M.: Qualitative Decision-Making Methods: Verbal Analysis of Solutions. Fizmatlit, Moscow (1996)
19. Aizerman, M.A., Aleskerov, F.T.: The Choice of Options (Basic Theory). Science, Moscow (1990)
20. Andreichikov, A.V., Andreichikova, O.N.: Intellectual Information Systems: Textbook. Finance and Statistics, Moscow (2004)
21. Aylamazyan, A.K., Stas, E.V.: Computer Science and Development Theory. Science, Moscow (1989)
22. Larichev, O.I.: Science and the Art of Decision Making. Science, Moscow (1979)
23. Keeney, R.L., Raifa, H.: Decision Making Under Many Criteria: Preferences and Substitutions. Radio and Communication, Moscow (1981)
24. Robinson, J.: The Economic Theory of Imperfect Competition. Progress, Moscow (1986)
25. Aizerman, M.A., Malishevsky, A.V.: Some aspects of the general theory of choosing the best options. Autom. Remote Control **2**, 65–83 (1982)
26. Batyrshin, I.Z.: To analysis of preferences in decision-making systems. Proc. MEI **533**, 57–62 (1981)
27. Ministry of regional development of the Russian Federation: Order N 390 dated 1 September 2009. On Amendments to SNiP 2.08.02-89 Public buildings and structures (2009)

Process Modeling for Energy Planning of Technological Systems

A. Sychugov, Yu. Frantsuzova, and V. Salnikov[✉]

Tula State University, 95 Lenina Avenue, Tula 300026, Russian Federation
vladimirsalnikov95@yandex.ru

Abstract. The paper proposes an approach to the modeling of technological systems and the technological processes within them. The technological system structure represents in the following form of connected directed graph (tree), the state of the elements of which is given by the vector function of controlling and disturbing stimuli. Technological process, implemented in a specific technological system, represents as a sequence of operations for changing the finite geometric, physical and mechanical properties from workpiece to the finished part. The article also presents a method of estimating energy costs depending on the labor intensity of manufacturing of specific products. Production time norms are mainly set in each production system and are reflected in flow process charts. This model is based on the concept of labor-intensity – the time spent on manufacturing of products. To convert labor intensity into energy intensity, we use coefficients of normative energy intensity which depend on the existing production capacity.

Keywords: Technological system · Efficiency modeling · Energy · Energy consumption

1 Introduction

Modern industries are based on technological systems that have a complex structural and functional organization. As a rule, specific technological processes are kept under control in these systems. If we abstract from a specific type of technological process, then any technological process can be represented as a set of actions, conditions and connections. Any production consists of steps (stages) and each of them produces a certain effect on material flow and energy conversion. The sequence of stages is usually described with the help of a flow process chart, with each element corresponding to a specific technological process. The connections between the elements of the flow chart reflect the material and energy flows in the system [1–3]. The system is characterized by a functioning algorithm aimed at achieving a specific goal.

From the standpoint of the systematic approach, technological process is a complex dynamic system with the following interacting elements: equipment, control and management tools, auxiliary and transport devices, processing tools or environments that are constantly moving and changing, production objects and people who carry out and control the process. For the purpose of analysis, a complex technological process can be divided into subsystems of various levels. The decomposition of the system into

A. A. Radionov and A. S. Karandaev (Eds.): RusAutoCon 2019, LNEE 641, pp. 933–943, 2020.
https://doi.org/10.1007/978-3-030-39225-3_100

subsystems leads to finding the hierarchy of the structure and viewing the system at different levels of detail [3].

2   Mathematical Model of the Technological System

Technological system (TS) is a set of functionally interrelated elements of enterprise's manufacturing structure, technological equipment and performers that carry out the technological processes of production of goods in a regulated environment, in accordance with the requirements of regulatory and technical documentation [4].

The model of the TS structure element is given by the sets of input, disturbing and output variables [5–7]:

$$u = (u_1, u_2, \ldots, u_m) \tag{1}$$

$$z = (z_1, z_2, \ldots, z_l) \tag{2}$$

$$x = (x_1, x_2, \ldots, x_n) \tag{3}$$

where $u_1, u_2, \ldots, u_m$ are input variables; $z_1, z_2, \ldots, z_l$ are stimuli; $x_1, x_2, \ldots, x_n$ are output variables.

The state of an element is characterized by its output and at any moment t it can be presented as follows [5–7]:

$$x(t) = \phi\{ u(t), \quad z(t) \} \tag{4}$$

where $\varphi\{\cdot\}$ is a vector-function of controlling and disturbing stimuli.

According to the concept of electronic description of technological systems, their model during their life cycle should be presented in the computer environment in the form of a hierarchy of information models, comprising a single unit and having subordination. Each subsequent level of detail contains additional information [8].

Then the TS structure can be represented in the following form of connected directed graph (tree):

$$GTS = (V, E) \tag{5}$$

where $V = \{v_i^k\}$, $i = 1, 2, \ldots, N$ is a set of structural elements of the TS (nodes of the tree); N – is the number of structural elements in the TS; k – depiction of $V \rightarrow K$, $K = \{k_l\}, l = 0, I, II, \ldots, L$ is a set of sequence numbers of TS hierarchy levels; L is the number of levels (tree height); E is the depiction of the set V in V that can be expressed with the following rule:

$$\Gamma(vi) = \{ vj : \exists ark(vi, vj) \in A\} \tag{6}$$

where A is a set of connections between TS elements (tree branches GTS).

Figure 1 shows the graphical representation of the structure of a TS.

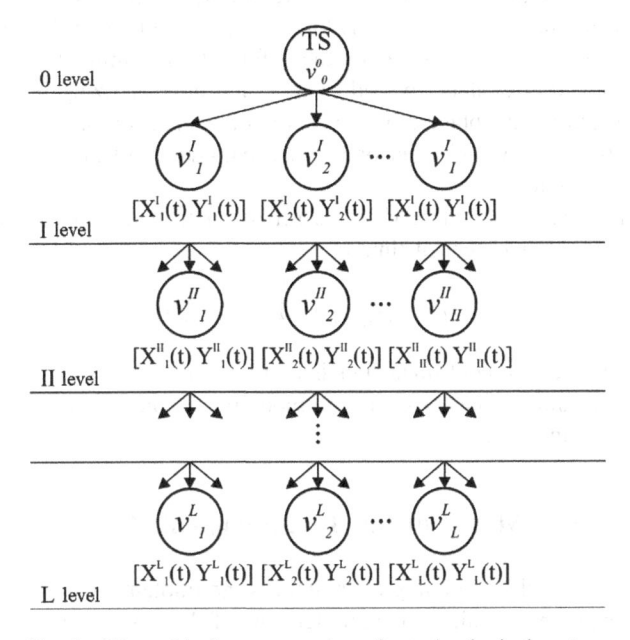

Fig. 1. Hierarchical representation of a technological system.

Each structural element vki is characterized by parameters P needed to organize the modeling:

$$P_i^k = \{p_{i1}^k, p_{i2}^k, \ldots, p_{iM}^k\} \tag{7}$$

where M – the number of parameters of i-th structural element on k-th level of TS hierarchy.

This set characterizes the impact of the element (behavior) on the information that comes at its inputs.

The parameter can be represented as a simple number, as a vector or in the form of inequality constraint.

$$p_{i1} = const; p_{i1} = \begin{bmatrix} p_{ij}^1 \\ p_{ij}^2 \\ \vdots \\ p_{ij}^A \end{bmatrix}; p_{i1} = \begin{bmatrix} p_{ij}^{11} & \cdots & p_{ij}^{1B} \\ \vdots & \ddots & \vdots \\ p_{ij}^{A1} & \cdots & p_{ij}^{AB} \end{bmatrix} \tag{8}$$

For example, a parameter can be represented by the position of the element with respect to the basing point of the technological system, which will allow to assess the characteristics of transport operations and to consider them during the simulation of the whole system; the list and characteristics of energy consumers (equipment, light and heating sources, etc.) to assess the energy efficiency of the TS; the list of element's operations and their valid characteristics.

Almost all modern technology systems have extensive diagnostic system structure. Diagnostic system consists of methods of determining the technical state of the node, device without its disassembly, i.e. it is the system for obtaining diagnostic information, which is a logical processing of the information coming from the operating equipment during the regulated period of time and characterizing its condition. The diagnostic information is obtained by the survey of system sensors, which characterize the electrical, mechanical and other operational parameters of the node, the device or the system as a whole.

Then each structural element v_i^k has a set of controlled signals S, describing its condition in a certain moment of time t:

$$S_i^k(t) = \{s_{i1}^k(t), s_{i2}^k(t), \ldots, s_{iM}^k(t)\} \tag{9}$$

where L is the number of controlled signals in the element v_i^k. Usually there are tens or hundreds of thousands of these signals in a modern TS. They are used for monitoring, diagnosing and managing.

3   Mathematical Model of the Technological Process

The main purpose of the technological system is the implementation of technological processes by engaging certain structural elements of the system. According to GOST 14.004-83 [9] the production process is the set of all the actions of workers and tools required to manufacture and repair the products. According to GOST 3.1109-82 [10] the technological process is a part of the production process containing targeted actions for changing and (or) the determining the state of the subject of labor. The subject of labor is a thing or a complex of things to which man's labor is applied in the process of production by means of tools for the production of material goods. Together with the instruments of labor they comprise production tools. The instruments of labor include tools, industrial buildings, means of transportation of goods, and land.

Technological process, implemented in a specific TS, can be represented as a sequence of operations $O = \{o_1, o_2, \ldots, o_U\}$ for changing the finite geometric, physical and mechanical properties $C = \{c_1, c_2, \ldots, c_B\}$ from workpiece to the finished part, where U is the number of operations in the technological process; B is the number of workpiece's properties.

Assuming that the technological process is implemented on the previously defined structure of the TS, it takes the form of a directed graph, where the initial vertex is the subject of labor (material, workpiece, etc.), fixed for the structural element of the TS, the final vertex is the finished item, and the directed edges of the graph show the transfer (change) of properties of the item in the transition from one operation to the next (from one item to the other), which corresponds to the principle of technological heredity.

In this case, the first step in the modeling of the technological process is to define the operators of transformation of workpiece's properties during the transition from one element to the other. The choice of the operator depends on the parameters of the performed operations.

Technological system can provide the optimal technological process according to performance, energy efficiency, costs, etc. The choice depends on the internal and external conditions of functioning of the TS. Figure 2 shows the graphic representation of the pre-modeling of technological process.

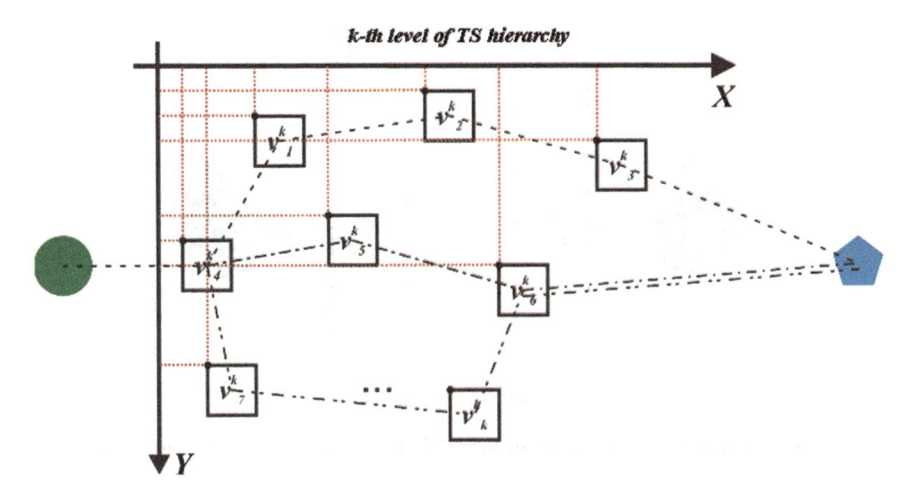

Fig. 2. Possible options of technological process organization on k-th level of the TS hierarchy.

The formed technological process can be represented as a chain of operations, each of which is fixed for certain structural element v_i^k of the TS. The steps of the technical process are usually depicted in the form of flow process chart.

The technological process regulates the mode of each element involved in its implementation. The mode can be represented as a set of parameters R:

$$R_i = \{r_1, r_2, \ldots, p_W\}; i = 1, 2, \ldots, U \tag{10}$$

where W the number of parameters characterizing the mode of equipment. At the same time, as a rule, mode R is a subset for the set of equipment parameters P. Figure 3 illustrates the introduced values.

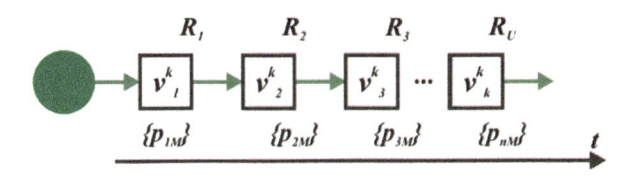

Fig. 3. Representation of technological process as a chain of operations.

Green arrow means the transfer of the changed properties of the workpiece from one operation to the other. These changes are determined by a combination of technological factors for each operation of the technological process. Figure 4 demonstrates the graphic representation of the transfer of properties.

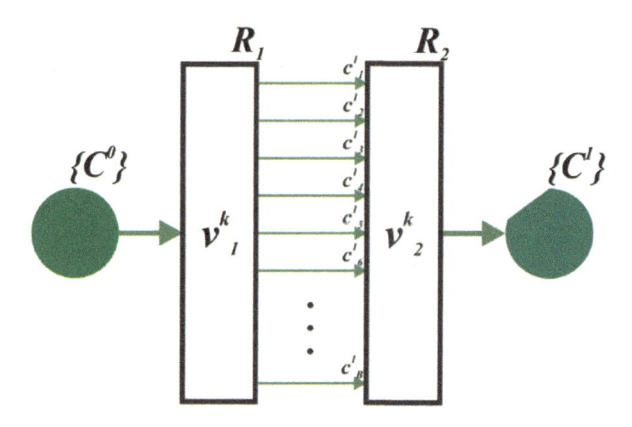

Fig. 4. The transfer of properties of the workpiece from one operation to the other.

As a rule, transformation of the set of the properties of the workpiece is formed with explicit mathematical description of physical processes occurring in the real object in the form of a set of differential, algebraic and logical equations:

$$
\begin{cases}
c_1^i(t) = f(p_1^{i-1}(t), t) \\
c_2^i(t) = f(p_2^{i-1}(t), t) \\
\quad \vdots \\
c_B^i(t) = f(p_B^{i-1}(t), t)
\end{cases}
\tag{11}
$$

Figure 5 shows the examples of possible transfer of the properties of the workpiece: a – no change of the parameter during the operation; b – linear change of the parameter; c – dynamic change of the parameter.

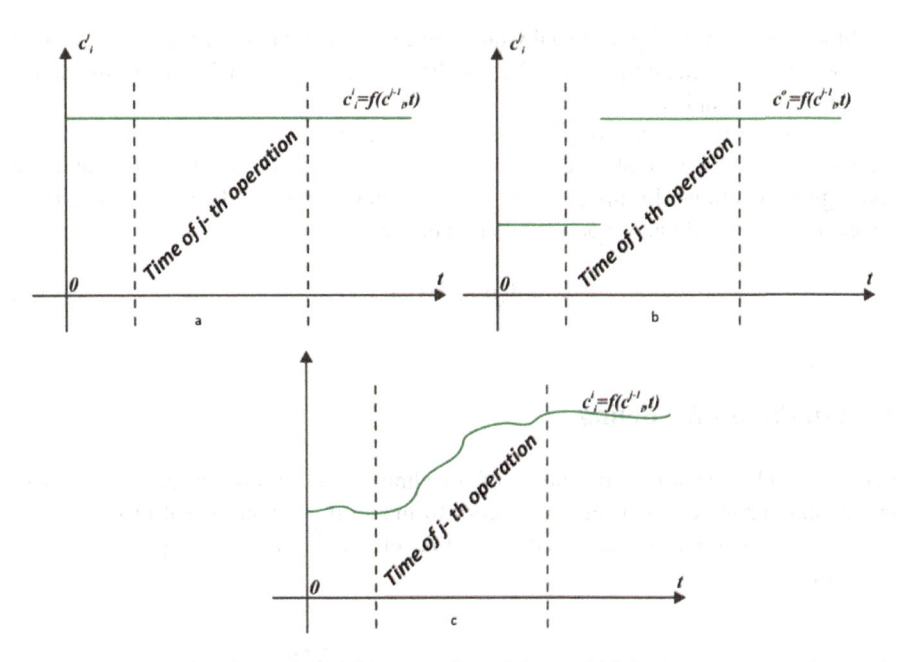

Fig. 5. The example of transfer of the properties of the workpiece from one operation to the other.

Previously it was proposed to find a set of controlled signals Y among parameters V that can be compared with the sensors in real technological system. Figure 6 graphically represents the formation of data signal.

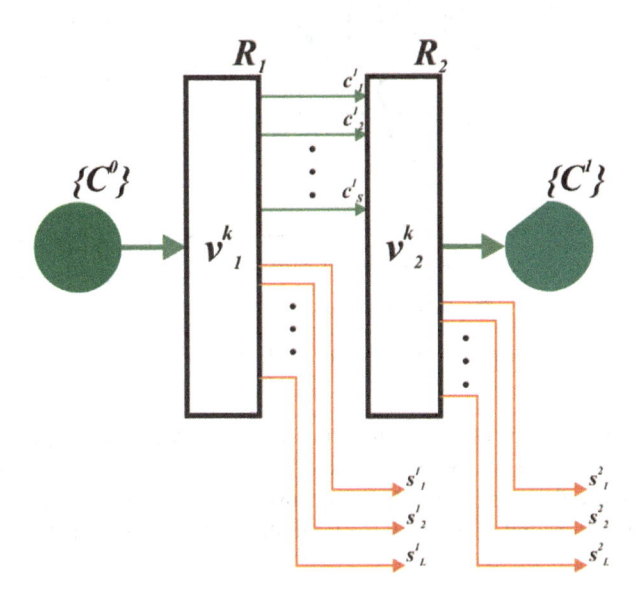

Fig. 6. Generation of diagnostic signals during the technological process.

Modeling of these signals usually takes place on the lower hierarchical levels of the TS consisting of operating tools. During the transition to a higher level the signals become more complex.

The modeling of the controlled signals depends on the specific technological process and, in the general case, can be described with a set of differential, algebraic and logical equations. In this case the equations must consider the current state of the TS element, its mode and workpiece's parameters

$$y_j^i = f(p(t), r(t), c(t), t) \tag{12}$$

4   Simulation Modeling

Let us consider the following model of a technological system, based on one of the private enterprises of Tula region in order to model the technological process and to obtain the parameters of energy consumption. Figure 7 gives the graphical representation of the model.

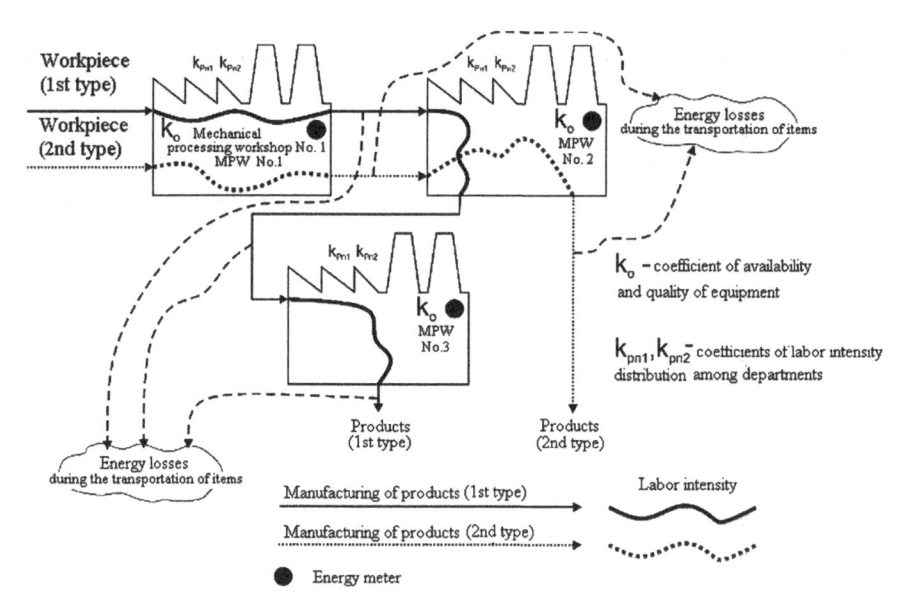

Fig. 7. Technological system model for modeling the energy consumption plan.

This system consists of three structural units - three workshops of mechanical processing. It is aimed at the production of two types of products. Manufacturing technology for the first type of products is considered to be more time-consuming than the production of second type of products. It should be noted that the second type of production involves only two of the structural units of the production system, which significantly reduces the cost of transportation of the processed workpieces between the workshops than in the case of the first type of products.

Level of detail of the technological system depends on the purposes of modeling. For this example it is sufficient to distinguish the following levels: workshop s^I, machine s^{II} and tool s^{II}. At the same time each level of the hierarchy is a set: a set of workshops, machines, or tools. Let us assume that for each workshop S_i^I the parameter X_{i1}^I is the average deviation of the actual run-time of the processing from the planned one. The modes of mechanic processing are previously known from the technological process, then for each machine S_i^{II} the parameter X_{i1}^{II} includes a number of limitations characterizing the possibility of a particular technology. In addition to the main geometric parameters each tool S_i^{II} is characterized by durability X_{i1}^{III} and working time X_{i2}^{III}. Input S_i^{II} receives information about the workpiece's geometry, quality of its original surface, and technical requirements, i.e. the flow chart and the workpiece processing mode, etc. Mechanical processing generates controlled signals at all levels of the hierarchy Y_i^K, that characterize the state of the structural element. At the tool level the signal $Y_i^1(t)$, characterizing the state of the tool, is generated. In certain cases, the tool is replaced or sharpened, which must be considered when assessing the time parameters.

The production is characterized by labor intensity – time required for the production. To calculate labor intensity of a particular product for each structural unit (SU) we suggest using the parameters p_{l1}, p_{l2}, p_{l3}. Zero equality says that this workshop is not involved into the production.

To convert labor intensity into of the energy consumption we introduce parameter p_e for each SU, with the value depending on the capacity of the SU. The higher the value of this coefficient, the higher the energy cost of the production.

The introduced coefficients allow us to assess the cost of energy for the manufacturing of products. For example, the expected cost of energy to produce the first type of items is described with the following expression:

$$E_{i1} = ((T \cdot p_{l1}) \cdot p_{e1} + (T \cdot p_{l2}) \cdot p_{e2} + (T \cdot p_{l3}) \cdot p_{e3}) + F \tag{13}$$

where E_{i1} is the cost of energy; T – labor intensity,. p_{l1} – the share of total labor intensity for the first workshop in the manufacturing of the first type of products; p_{e1} – coefficient for the conversion of labor intensity into the costs of energy for the first workshop; p_{l2} – the share of total labor intensity for the second workshop in the manufacturing of the first type of products; p_{e2} – coefficient for the conversion of the labor intensity into the costs of energy for the second workshop; p_{l3} – the share of total labor intensity for the third workshop in the manufacturing of the first type of products, p_{e3} – coefficient for the conversion of the labor intensity into the costs of energy for the third workshop, F – energy consumption during transportation.

The technological system works as follows: for three days the system works to manufacture the products of the first type, and then for the next three days it produces the second type of items.

5   The Performance of the Experiment

Modeling results can be considered correct, as they correspond to the production in a typical production system – the work is only performed during the working days and during the working time. The modeling was carried out using the production function, peaking in the middle of a working day, which also corresponds to a typical production. We have also taken into account the fact that the production system has permanent costs that do not depend on the working mode (the cost of lighting). It is illustrated on the chart (Fig. 8) by a change in energy consumption during non-working time and weekends.

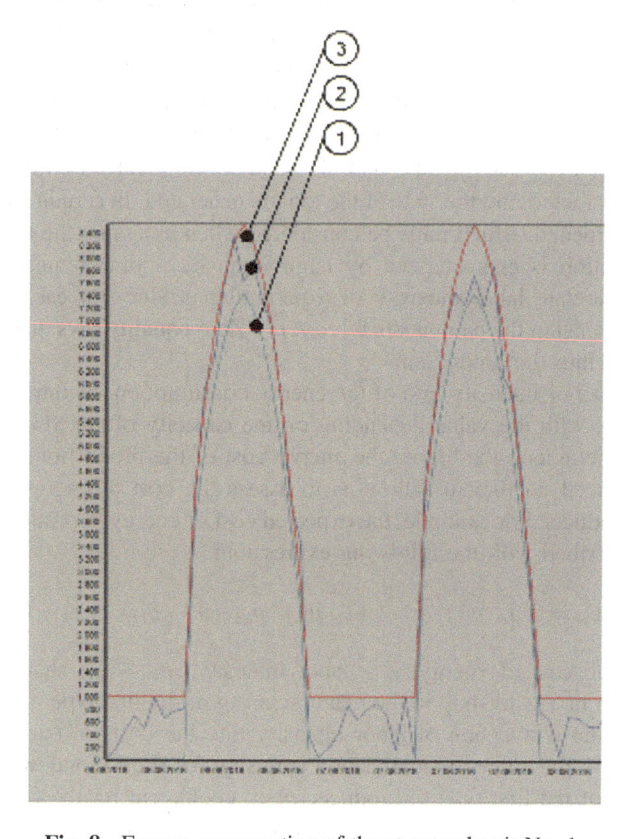

Fig. 8. Energy consumption of the structural unit No. 1.

One of the curves (number 1 in Fig. 8) shows the expected energy consumption, which is calculated according to the labor intensity of the production and the coefficients of regulatory energy intensity. The curve number two shows the real power consumption for the period of time. The third curve shows the acceptable deviation in power consumption during the day.

6 Result and Discussion

The results allow the end-user to evaluate the effectiveness of energy consumption based on the information about the production. The comparison of graphs of the expected and real consumption makes it possible to adequately analyze the reasons for excess energy consumption.

The difference between the real consumption and the planned one can be explained by non-technological energy consumption (lighting, heating, etc.). If at some time period the real consumption exceeds the permissible value, it can be concluded that the system's normal technological process fails due to the failure of the equipment, machine, tools or the structure of the treated material; or due to unauthorized energy theft or connection of unaccounted equipment in the unit.

This paper presents the method of modeling of technological processes and the example of practical implementation of such models which demonstrate a huge potential of the modeling in traditional productions, as well as in other processes of social and economic nature. After a relatively small revision, the proposed method of process modeling can be applied in various areas, for example, for the effective control and management of energy consumption of technological systems.

Acknowledgments. The reported study was funded by RFBR according to the research project № 19-07-01107.

References

1. Hartmann, T., Sovietin, F.: Analytical review of modern packages of modeling programs for computer modeling of chemical-technological systems. Adv. Chem. Chem. Technol. **26** (140), 117–120 (2012)
2. Emelyanov, S., et al.: Information technologies of regional management. Editorial URSS, Moscow (2004)
3. Malkov, M., Oleynik, A., Fedorov, A.: Modeling of technological processes: methods and experience. Proc. Kola Res. Cent. RAS (3), 93–101 (2010)
4. GOST 27.004-85: Reliability in technology (SSNT). Technological systems. Terms and definitions. Standartinform, Moscow (1985)
5. Smilansky, G., Amlinsky, L., Baranov, V., et al.: Directory of designer APCS. Mashinostroenie, Moscow (1983)
6. Kafarov, V., Dorokhov, I.: System analysis of chemical technology processes. Nauka, Moscow (1979)
7. Avramchuk, E., Vavilov, A., Emelyanov, S.: System modeling technology. Mashinostroenie, Moscow (1988)
8. Kuritsyna, V., Liokumovich, D., Siluyanova, M.: Tools Matlab Simulink in the system of technological quality management of precision engineering. Electr. Inf. Syst. Syst. **8**(1), 22–31 (2012)
9. GOST 14.004-83: Technological preparation of production. Terms and definitions of basic concepts. Standartinform, Moscow (1983)
10. GOST 3.1109-82: Unified system of technological documentation. Terms and definitions of basic concepts. Standartinform, Moscow (1982)

Problems and Solutions of Automation of Magnetron Sputtering Process in Vacuum

S. V. Sidorova[✉], A. D. Kouptsov, and M. A. Pronin

Bauman Moscow State Technical University, 5 Vtoraya Baumanskaya Street,
Moscow 105005, Russian Federation
sidorova@bmstu.ru

Abstract. The problems of operation of the magnetron sputtering module in vacuum are considered. Elimination of these disadvantages was done by development of new shutter design, which will help to protect wafer from particles from the target. A variant of automation with the aim of improving the quality of the coatings is introduced. In addition to the presence of defects on the target, interfering with the film growth process, failures can occur in the magnetron sputtering module itself. Plasma failures are of several types: pollution in the form of dust particles, metal shavings cause dielectric inclusions, which block the power to the target. Spraying becomes not high-volt, but variable. Chosen motor was implemented to the control system of the described magnetron sputtering module of the machine. Base experimental researches were performed to check quality and efficiency of to be performed modification of design and methods of magnetron sputtering system. Basic experimental studies have confirmed the assumption of obtaining a thin film of copper with good adhesion. The main purpose of the research is to investigate current behavior while film growth. Ampermeter allowed to write to its memory series of measurements. Stored measurement results could be transferred by RS interface to the PC. Using special software ExceLINX this measurement results could be checked. Separate PC is used for this. It is still necessary to evaluate the composition of the obtained thin-film coatings to verify the reduction of impurity oxides in the coating.

Keywords: Magnetron · Plasma · Automation · Stepper motor · Vacuum · Experiment · Process control

1 Introduction

There is a need for metal and dielectric thin films application on various surfaces at the manufacturing enterprises of the microelectronic industry. Thin films are widely used as hardening, reflective, conductive and dielectric coatings. Most often these types of work are carried out to obtain and researching new promising materials for the microelectronic industry, and for micro- and nanoelectronics devices, optics, photonics, and other branches of technology.

The simplest use of thin films is decorative - creating mirrors and coatings for jewelry. However, in general coatings of small thickness are used to study the electrical properties of new materials during the formation of contacts; applying resistive and

A. A. Radionov and A. S. Karandaev (Eds.): RusAutoCon 2019, LNEE 641, pp. 944–952, 2020.
https://doi.org/10.1007/978-3-030-39225-3_101

conductive coatings in industry and in the manufacture of elements of integrated circuits in microelectronics; in the creation of light filters, reflective and light-conducting optoelectronic coatings; modern lithographic processes.

For application of thin-film coatings of a wide range of purposes from various materials, both conductive and non-conductive, the method of magnetron sputtering is mainly used. This method has several advantages compared with other coating methods [1–3]:

- High deposition rate
- High uniformity of coatings
- Almost complete absence of overheating of the part surface
- Low degree of contamination of the films
- Ability to automate the process

The disadvantages of this method of coating are in the presence of the porosity of the coating, low adhesion of the coating, the need for additional heating of the substrate. The elimination of these deficiencies of the obtained thin-film coatings is possible if one tries to identify and solve the problems of the operation of the technological module - a magnetron source.

The authors consider the problems of the magnetron source and offer options for eliminating the identified disadvantages.

2 Identify the Problems of the Method of Magnetron Sputtering in Vacuum

It is important to ensure the precision of a number of factors and limit the various external harmful interventions to the process to obtain the require structure and topology. Concerning vacuum hygiene requirements are applied to operators. For example, the use of bathrobes, change of clothes, gloves on hands, masks for the mouth and hats for hair minimizes the ingress of external contamination to samples [4, 5].

In addition, there are devices and methods for vacuum equipment. For example, the method of training the target before applying thin-film coatings by ion-plasma spraying allows you to get rid of various contaminants, oxides and other inclusions from the surface of the target without applying a coating layer to the sample. View of the target before and after training is presented in Fig. 1.

Figure 1 shows that the target has characteristic zones. The erosion zone is a smooth pink tint; chipping zone - recess in the center of the target disk; the blank zone of the target is the layer not affected by spraying. In the case when the equipment is not used for some time, this target is oxidized, which adversely affects the quality and characteristics of the resulting coating. And after training the target, the zone of erosion becomes clean.

In addition to the presence of defects on the target, interfering with the film growth process, failures can occur in the magnetron sputtering module itself. Plasma failures are of several types: pollution in the form of dust particles, metal shavings cause dielectric inclusions, which block the power to the target. Spraying becomes not high-volt, but variable. This is unacceptable because the type of signal is constant with large

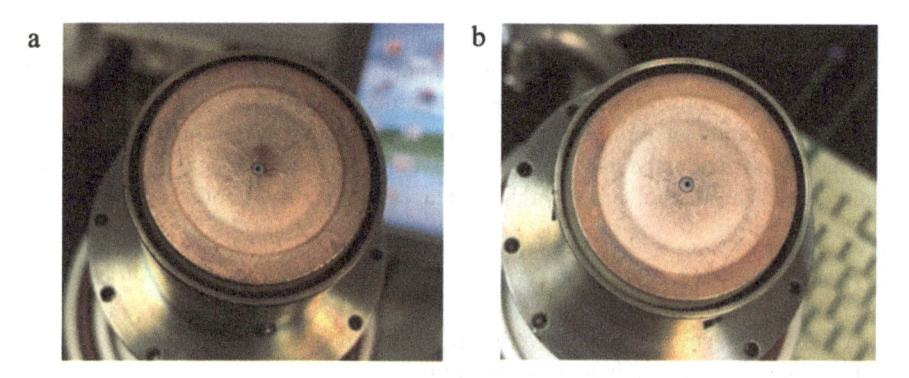

Fig. 1. Target of module of magnetron sputtering before (a) and after (b) the training.

voltage values. This type of malfunction is the most dangerous, since, when a dielectric element is formed, the power supply from overload ceases to give power and the sputtering process stops. Then the power supply unit re-energizes the system. If such a cycle is repeated several times, then the system becomes unstable for the system, because it is not intended for operation in such pulsating modes and begins to overheat.

In addition, the plasma displacement along the height, or its linear displacement, can be caused by insufficient vacuum purity, or a failure of the magnetic system. Plasma displacement in height is the most common defect that affects the thickness of the resulting coating.

Dielectric contamination and plasma displacement are presented in Fig. 2.

All of the above defects adversely affect the properties, characteristics of the resulting films. Equipment also suffers and the process time increases, which is unacceptable for industrial production. Protection from harmful factors can serve as a simple device - valve. It covers the target, or the sample and everything that is sprayed from the target does not fall on the substrate. Also, this device allows in one vacuum cycle to reach the mode of the working sputtering of the target. Working spraying is called spraying, when the plasma burns continuously and without fail.

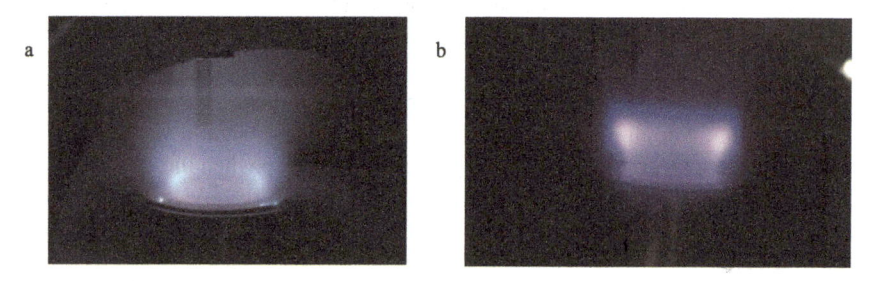

Fig. 2. Various of incorrect work of plasma because of dielectric pollution (a) and shift of plasma (b).

3 Development of the Input Rotation of the Damper for the Magnetron

Forms of valves, methods of their rotation and implementation are different. For the installation of the UVN-1M and the magnetron sputtering module located at the Department of Electronic Technologies in Mechanical Engineering of Bauman Moscow State Technical University, a gate was developed and its movement was introduced to implement the subsequent process automation (target development and coating in a single cycle, eliminating contamination products) [6–12].

Considering the size and shape of the vacuum chamber, 5 forms and the location of the axis of rotation of the valve were selected for analysis. Each option has its advantages and disadvantages. The resulting factor when choosing the final option was the area of the open part of the target in the position of the valve "open". The best indicator showed the option located on the bottom right.

After the choice of the damper form factor and its manufacture, the next question becomes for the operator and the designer. How to ensure the rotation of this element? There are two simple solutions: turning using a hand drive (the operator himself rotates the flap) and using a drive from a stepper motor. A stepper motor (hereinafter referred to as the SM) will increase the accuracy of the flap rotation, which will positively affect the quality of the coatings obtained during the spraying process.

Also, an analysis of important criteria for selecting the SM was carried out.

3.1 Angle of Rotation and Speed

SM can be controlled with high accuracy and efficiency. In the case of a five-phase step-by-step control, control is carried out by means of an electric pulsed (digital) signal. Each such signal sets a certain angle of rotation of the rotor, which is called pitch. Currently used angles: $90°$, $45°$, $15°$, $7.5°$, $1.8°$, $0.9°$, $0.45°$.

3.2 High Torque and Low Response Time

SM – compact, lightweight and provides high torque. Moreover, due to the rapid acceleration and high stopping and starting points, it is possible to immediately start/stop and change the direction of rotation of the rotor.

3.3 Highest Resolution and Positioning Accuracy

The five-phase SM has a pitch of $0.72°$ and a high resolution of $0.009°$ in the microstepping mode. The stopping accuracy is ± 3 min ($0.05°$ without load) when moving in $0.72°$ increments.

3.4 Holding Moment

In the case of stopping the five-phase SM with the power on, the rotor position will be held due to the high holding torque. For this reason, it is not necessary to apply a mechanical brake or constantly send a control signal to maintain the stopped rotor.

3.5   Low Installation Time and No Hesitation After Shutdown

When stopping after rotation in any direction, the time from the beginning of the step to the moment when the rotor remains in a given position is extremely short. Moreover, after the installation time, i.e. after stopping, no rotor vibrations.

Since the SM should be installed directly on the vacuum chamber (chamber flange), taking into account the small dimensions of this chamber - (inner diameter of the chamber is 87 mm), the basic requirements were specified in advance before choosing the SM: the SM should have a built-in gearbox, the engine size should not be large.

The following companies were considered during the analysis: Stepmotor, Electroprivod, Autonics, B & R, Purelogic R & D.

Table 1 presents the analysis for the selection of the SM, as well as the criteria for the selection were assigned weights to identify more significant parameters of the analysis.

Table 1. The choice of SM for the magnetron sputtering valve with a WF.

Company	Model	Sizes (dxl), mm	Weight, gr	Rotation moment, H·m	Step angel, deg	Numbers of points	Place
Pure-logic R&D	PL57GH76-50D8	2	2	1	2	1.75	4
B&R	80MPF5.250S113-01	1	1	2	2	1.35	5
Auto-nics	A10K-S545-G5	3	3	5	1	3.3	3
Elec-tropri-vod	FL28STH45-0956A	4	5	3	2	3.7	2
Step-motor	PMG3530-01	5	4	4	3	4.4	1
Weighting factor		0.5	0.15	0.25	0.1		

Thus, for the refinement of the magnetron sputtering module and the manufacture of the input of the rotation of the damper, a Stepmotor stepping motor was chosen.

4   Experimental Basis

Base experimental researches were performed to check quality and efficiency of to be performed modification of design and methods of magnetron sputtering system [13–19].

The experiments were performed by using of mini laboratory tool UVN-1 M (Fig. 3). The tool is made by using of modern components of vacuum and electrical blocks. Dry spiral pump and hybrid turbo pump are used for oil free pumping of the system [15, 16].

The set of samples of sitall with previously formed contact pads of Cu were used to investigate the process of thin films formation.

Cu deposition of contact pads were made using magnetron sputtering. Monitoring of thin film continuous coating was performed by through film current measurement (on earlier phases the current is tunnel).

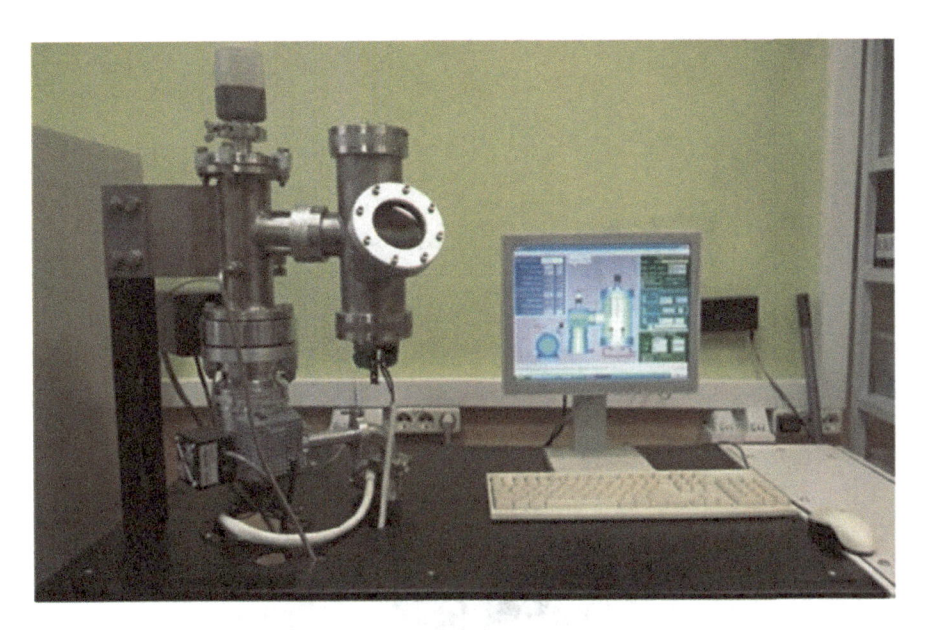

Fig. 3. General view of the tool UVN-1M.

The power supply which was used (Fig. 4) allowed to limit maximum current value (short circuit current). It makes possible to protect picoampermeter. This power supply provides DC voltage of the 0...18 V range and current of 0...3 A range. The accuracy of output voltage measurement was less then 1% and the accuracy of output current measurement was less then 2%. Monitoring of nanorange current was performed by using of picoampermetter Keithley 6485 [15], which is designed for high rate and high-resolution measurement of low currents (less than 2 nA).

The substrate with contact pads is fixed on the workholder. The Cu wires are connected to the contact pads.

The main purpose of the research is to investigate current behavior while film growth. Mentioned above ampermetter tool allowed to write to its memory series of measurements. Stored measurement results could be transferred by RS interface to the PC. Using special software ExceLINX this measurement results could be checked. Separate PC is used for this.

Vacuum chamber was pumped down to pressure $6,3 \cdot 10^{-5}$ mbar. Operator start power supply and supply preset step-by-step current value which allowed to monitor real time formation of thin film. Picoampermeter starts recording simultaneously with process start. Current growth up to power supply preset short current value. After that power supply starts lowing the voltage with keeping current on the same level. While chamber is vented measurement results are uploaded to the PC and are imported to the ExceLINX software. After venting is finished the substrate is taken out for further investigations of topology using atomic force microscopy. Uploaded measurement results are shown on the Current - Measurement number diagram [15].

Fig. 4. Precision DC power supply QJ1803C.

To measure topology and sizes of the structures of the coverage atomic force microscope [20] was used. Investigations of the coating shows thickness of this coating (Fig. 5). This result allowed to investigate correct process parameters.

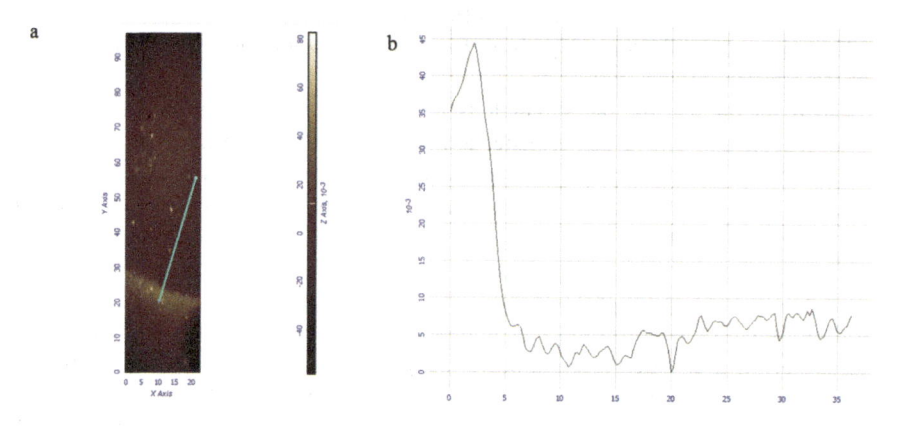

Fig. 5. AFM-image (a) of the thin film and the result of thickness measurement (b).

5 Conclusions

While investigation of existing problems of magnetron sputtering module main disadvantages of current module was shown. Elimination of these disadvantages was done by development of new shutter design, which will help to protect wafer from particles from the target. Also, motor (stepping motor) model was chosen to arrange rotation feedthrough of the shutter. Chosen motor was implemented to the control system of the described magnetron sputtering module of the machine.

Basic experimental studies have confirmed the assumption of obtaining a thin film of copper with good adhesion. Judging by the measurements carried out on the atomic force microscope, the obtained thin-film coatings were obtained with a thickness of up to 45–50 nm, which shows a better result than before the modernization of the module.

It is still necessary to evaluate the composition of the obtained thin-film coatings to verify the reduction of impurity oxides in the coating.

References

1. Sidorova, S.V.: Methods of thin film formation: initial stage of formation. Eng. J. App **9**, 13–17 (2011)
2. Kouptsov, A.D.: Metal thin-film coatings for solar panels. In: XXV Scientific and Technical Conference with the Participation of Foreign Experts Vacuum Science and Technology, Sudak, 16–22 September 2018, pp. 187–192 (2018)
3. Nikitin, M.M.: Vacuum deposition technology and equipment. Metallurgy **8**, 110 (1992)
4. Laser MEASURER LAES (2018). http://www.spectrosystems.ru/analytical/laes/laes_measurer.shtml. Accessed 16 Dec 2018
5. Efremov, A.M., Svettsov, V.I., Rybkin, V.V.: Vacuum Plasma Processes, and Technology. ISCTU, Ivanovo (2006)
6. How stepper motors work (2018). http://robotosha.ru/electronics/how-stepper-motors-work.html. Accessed 26 Sept 2018
7. Stepper motor control (2018). http://www.cnccontrollers.ru/shagovye_dvigateli.html. Accessed 26 Sept 2018
8. Stepper motor for CNC machine tools (2018). https://stepmotor.ru/elektrodvigateli/shag. Accessed 26 Sept 2018
9. Development and production of stepper motors (2018). https://electroprivod.ru/st_motor.htm. Accessed 26 Sept 2018
10. Six-speed stepper motors (2018). https://www.autonics.com/series/3000665. Accessed 26 Sept 2018
11. Stepper motor: products of B&R (2018). https://www.br-automation.com/ru/products/motion-control/stepper-motors-80mp. Accessed 26 Sept 2018
12. Stepper motors with gear (stepper motors): products of B&R (2018). https://purelogic.ru/catalog/elektroprivod/shagovye_dvigateli_i_aksessuary/shd_s_tsilindricheskim_reduktorom. Accessed 26 Sept 2018
13. Panfilov, Y.V., Kolesnik, L.L., Ryabov, V.T., et al.: Research and development complex with remote access. J. Phys. Conf. Ser. **872**, 012010 (2017)
14. Wong, P.S.: Fabrication and characteristics of broad-area light-emitting diode based on nanopatterned quantum dots. Nanotechnology **20**, 035302–035306 (2009)

15. Sidorova, S., Pronin, M., Isaeva, A.: Automated unit for control of initial stages of metal islands thin films and nanostructures growth. In: International Russian Automation Conference, vol. 1, pp. 1–4 (2018)

16. Isaeva, A.A., Pronin, M.A., Sidorova, S.V.: Stand for control of the initial stages of growth of metal island thin films and nanostructures and development of their formation modes. In: Vacuum Science and Technology. Materials of the XXV Scientific and Technical Conference with the Participation of Foreign Experts, pp. 181–186 (2018)

17. Accuracy class: stepper motor ST42–40 (2018). https://darxton.ru/catalog_item/shagovyy-dvigatel-st42-40. Accessed 4 Dec 2018

18. MakerGears: couplings (2018). https://makergears.ru/category/mufty-soedinitelnye. Accessed 4 Dec 2018

19. Sarkar, D.K.: Growth of self-assembled copper nanostructure on conducting polymer by electrodeposition. Solid State Commun. **125**, 365–368 (2003)

20. Laucht, A.: Electrical control of spontaneous emission and strong coupling for a single quantum dot. New J. Phys. **11**, 023034 (2009)

Information and Analytical Support for the Protection of Important Critical Information Infrastructure Objects

V. Berdyugin$^{(\boxtimes)}$ and L. Dronova

South Ural State University, 76 Lenina Avenue,
Chelyabinsk 454080, Russian Federation
berdiuginvi@susu.ru

Abstract. This article considers issues related to information and analytics support of organization and management activities for the protection of critical information infrastructure (CII). The covered problem is of interest for large industrial enterprises (subjects of CII), therefore, there is a need to accumulate and process large data arrays that are of primary importance for ensuring information security. To meet the information needs arising while protecting CII, it is proposed to use the information-analytical system (IAS), created on the basis of the integrated database management system (IDMS) "CronosPro". The scientific novelty of the work is that the newly created IAS allows to formalize and ensure the information needs of the security administrator at all stages of the protection of CII objects. An example of information and analytical support for activities related to the formation and operation of databases is given to illustrate the full range capabilities of IAS: the categorization of CII object, confidential information storage devices, and persons allowed to the confidential data. These databases greatly facilitate the definition of the actual awareness of employees of CII subject in the confidential information, as well as conducting checks of confidential paperwork. Since all the listed databases consist of similar and interrelated information objects ("Organization", "CII object", "Person", "Action", "Data Storage", "Document"), IAS allows you to decide not only knowledge management but also information and logical tasks as well as to identify non-obvious connections (second-level connections) between information objects, which is especially important in the process of investigating computer incidents.

Keywords: Information security · CII · IAS · Computer incident · Data storage · CronosPro

1 Introduction

Nowadays, the development of global cyberspace is occurring at a high rate, which significantly increases the danger of the realization of such threats and determines the importance of ensuring the security of all its components. Almost all developed and developing countries have adopted or are working on cyber-security strategies. According to the international standard ISO/IEC 27032:2012 [1] the term "cybersecurity" is defined as a generalizing one and includes the issues of ensuring the security of systems,

© Springer Nature Switzerland AG 2020
A. A. Radionov and A. S. Karandaev (Eds.): RusAutoCon 2019, LNEE 641, pp. 953–964, 2020.
https://doi.org/10.1007/978-3-030-39225-3_102

communications, and objects that belong to the cyberspace: information security, network security, Internet security, protection of critical information infrastructure.

According to the article of the № 390-FZ Federal Law [2] called "About safety", one of the basic principles of ensuring security is consistency and complexity.

At the same time, consistency means the need to use system analysis and synthesis in each management decision. A wrong management decision [3] can nullify all the activities of the system and lead to its destruction. The complexity of management means the need for all-sided coverage of the entire controlled system, considering all directions, aspects of the activity, and properties.

Forms and methods of any organization and management activity are applied in a certain sequence (cycling), dictated by the interests and goals of the preparation, adoption, and execution of management decisions [4]. The stages of management activity have a logical connection and set up a cycle of management actions [5]. Each stage needs information and analytics support [6].

This article addresses the problem of organizing information and analytics support for the protection of critical information infrastructure (CII) subjects, which, according to 187-FZ Federal Law [7] called "About safety of critical information infrastructure of the Russian Federation", include: public authorities, government agencies, Russian legal entities and (or) individual entrepreneurs owning CII objects based on the right to ownership or lease or any other legitimate grounds.

CII objects are information systems, information and telecommunication networks, or automated management systems operating in the field of health, science, transport, communications, energy, banking and other areas of the financial market, fuel and energy complex, in the field of nuclear energy, defense, rocket and space, mining, steel and chemical industries.

2   Choice of Database Management System

The information and analytics support for the process of organizing the protection of CII subject should be understood as a set of organization and management measures for studying and evaluating information that characterizes: the state of CII object, the results of tasks being accomplished by information security units as well as the conditions in which these tasks are solved.

Based on the list of measures ensuring the protection of CII objects, specified in the orders of the Federal Service for Technical and Export Control of Russia [8, 9], CII subjects need to be provided with:

- Categorization of CII objects
- Maintaining the current status of the documentation of the security system of CII object
- Analysis of vulnerabilities of CII objects
- Accounting of confidential data storages
- Accounting the awareness of employees operating and maintaining CII objects
- The accumulation of information about the training and instruction of employees
- Consideration of information security tools that provide technical protection of CII objects

- Fixing interactions of CII subject with third-party institutions (legal entities), including regulatory authorities
- Documenting the results and procedures for monitoring the safety of CII object
- Conducting investigations of computer incidents

The solution of each of these tasks involves the accumulation and processing of large amounts of information even in relation to a single CII object. If there is a large CII subject where dozens of CII objects are simultaneously operating, it becomes necessary to create special databases to automate the compliance with the legislation requirements of the Russian Federation. To ensure the above activities, it is proposed to use information-analytical system (IAS), which must meet the following requirements [10]:

- Correct selection of primary data
- Systematization and classification of information
- The absence of duplication of information
- Control of the correctness of information
- Conversion of the processed information to a common format

The main functions of IAS are the following:

- Loading and extracting data from storage and its further conversion
- Data storage
- Solving statistical and logical problems
- Identification of hidden links between information objects

When it comes to ensuring information and analytics work on the protection of CII objects, the choice of a database management system (DBMS) plays an important role, which should not only provide the solution of the above tasks but also comply with the legislative framework in the field of information security.

A comparative analysis of the following DBMSs was conducted: Linter, Cronos-Pro, Oracle MySQL [11], which have a set of corresponding tools given in Table 1.

Table 1. Comparative analysis of DBMSs.

Function	DBMS		
	Linter	CronosPro	Oracle MySQL
Prompt satisfaction of information consumers' needs	+	+	+
Continuity of information selection and processing	−	+	+
Lack of information duplication	+	+	+
Control of information correctness	−	+	+
Uniform formatting of the processed information	−	+	+
Filtration, aggregation, and updating of information	−	+	+
Price	From 25 000 rubles	8 500 rubles (for 10 years)	130 000 rubles (for 1 year)
Certificates	Ministry of Defense of the Russian Federation, FSTEC	FSTEC	−

As a result, DBMS called "CronosPro" was chosen [12] as the most appropriate means of solving the issue under consideration, since it implements:

- Use of the network model of data organization [13]
- Dynamic compression of data banks, use of records and fields of variable length
- Identification of database records by user-defined criteria
- Creation, storage, and launch of customizable code
- The option of entering/correcting data using customizable forms or standard tools [14]
- The ability to correct information in batch mode (mass correction)
- Visualizing the construction of complex queries using different criteria and conditions, including several related databases
- The ability to create, store and use query patterns
- The option of a report generator that allows the use of the built-in formula language
- Visualization and analysis of the relationships between objects
- Statistical data analysis with the tool of exporting results in MS Excel
- Compatibility with external data formats (MS Access, MS Excel, Oracle, XML, etc.)

3   Structure of Information-Analytical System

To let the information security specialists, solve information and analytics tasks, a data bank structure was created within this study. The structure includes the following interrelated databases:

- "Person" (the database contains information about the employees of the organization, and other persons interacting with CII object)
- "Organization" (contains information about CII subject, as well as about enterprises, institutions, organizations that interact with CII subject)
- "CII Object" (contains information about the object of protection)
- "Action" (records the actions of persons associated with CII object, including the facts of unauthorized access to CII object, as well as the actions of information security specialists of the organization in the process of investigating computer incidents)
- "Data Storage" (contains information on all types of protected information media, confidentiality neck, date of the entry and decommissioning)
- "A Means of Protecting" (contains information on all used means of protection of CII object)

The above-mentioned information objects are the places for:

- The work of "Persons"
- The performance of various kinds of "Actions"
- The use of "Data Storages" and "Means of Protection"

In order to maximize the potential of the DBMS, data input forms, query library for solving typical information-logical tasks and user guidelines for the data bank were developed.

As an example, the process of CII object categorizing is considered below.

The following abbreviations and symbols that refer to the developed algorithm are used in the text: R – responsible; E – executor; P – participant; Dir – organization's structural subdivision director; Emp – organization's employee; Exp – organization's expert in information security. The symbols on Fig. 1. are also used in the algorithm.

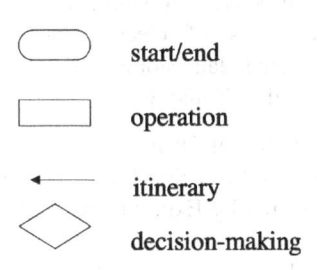

Fig. 1. Symbols used in the algorithm.

A brief sequence of actions is presented in Table 2.

1. Exp addresses Dir with the proposal to establish the CII Commission and determines its composition.
2. Dir decides on the establishment of the CII Commission (Emp 1, Emp 2). Data is entered into the database.
3. The Commission identifies critical processes, determines CII objects and makes a list of CII objects to be categorized. The result is entered into the database.
4. Exp and Dir agree on the list of CII objects, which the commission has determined, and send it for agreement with the industry regulator. In response, a Coordinated List of CII objects is received. Information is entered into the database.
5. Exp sends the list of CII objects for approval to the FSTEC and in response receives FSTEC act regarding the CII Object Lists. Exp saves the result in the database.
6. The Commission conducts a survey of potential CII objects. The obtained data will be the initial data for categorizing CII objects.
7. The Commission analyzes information security threats and vulnerabilities that may lead to computer incidents and develops a Security Threat Model, mentioned in the database as well.
8. The Commission categorizes CII object and draws up a notification on the results of assigning (or not assigning) CII object one of the categories of the importance.
9. Dir makes approves (or rejects) the decision of the categorization Commission.

10. Exp sends to the FSTEC the results of assigning CII object one of the categories of the importance.
11. After checking the information, the FSTEC notifies CII subject on its entering the CII Registry. Information is entered into the database.

The main feature of this algorithm is that the "Actions" and "Persons" databases have connections with several organizations at once. In the algorithm presented above, it is well exemplified that CII subject, its industry regulator, and the FSTEC of Russia are involved in the categorization process. This means that with the help of the IAS one can easily trace the connection between different organizations.

As another example, the process of selecting and systematizing information in a databank related to the protected data storages information is considered. A brief sequence of actions is presented in Table 3.

1. Emp turns to Dir with a proposal for the issuance of the Data Storage.
2. Dir, based on a proposal from Emp, decides on the registration of the Data Storage.
3. The Data Storage is registered by Exp, who also adds information about the Data Storage to the database.
4. Exp provides Emp with the Data Storage, the fact of the issue is recorded in the database.
5. When it is necessary to transfer the Data Storage from Emp 1 to Emp 2, Emp 1 refers to Dir.
6. Dir makes the decision on the transfer of the Data Storage to Emp 2 on the basis of the Emp 1 proposal.
7. After obtaining permission from Dir, Emp 1 gives the Data Storage to Exp, who issues it to Emp 2. This issue is fixed in the database.
8. If the destruction of the Data Storage is needed Exp comes out with the appropriate proposal to Dir.
9. Dir decides on the destruction of the Data Storage.
10. Emp gives the Data Storage to Exp. A commission is established, and an act of destruction to be approved by Dir is drawn up.
11. The Commission destruct the Data Storage, Exp makes a note in the database.

In the considered algorithm, all events take place in one "Organization", within one of the "CII Objects". Information about the object's director, specialist, and staff is accumulated in the database "Person". At the same time, the DBMS allows establishing connections between databases, for example, between the "Person" and "Action". The last one accumulates information on registration, issue, receipt, transfer, and destruction of media. In turn, the "Data Storage" database containing information about paper and electronic data storages (documents, files, CDs, DVDs, Blu-ray disks; flash memory, SSD disks, floppy disks, hard disks) will be connected with "Person" and "Action" databases. In other words, persons are qualified as participants of actions that are carried out with data storages and documents assigned to them [15].

Table 2. Algorithm of categorizing CII object.

operation	R	E	P
Establishment of a commission for CII objects categorizing (1,2)	Dir	Exp	Emp1 Emp2
Creation of the list of CII objects (3)	Dir	Exp Emp1 Emp2	-
Coordination of the list with the industry regulator (4)	Dir	Exp	Emp1 Emp2
Coordination of the list with FSTEC (5)	Dir	Exp	-
Collection of data for categorization (6)	Dir	Exp Emp1 Emp2	-
Security Threat Analysis (7)	Dir	Exp Emp1 Emp2	-
Categorization of CII objects (8)	Dir	Exp	Emp1 Emp2
Approval of categorization acts (9)	Dir	Exp Emp1 Emp2	-
Verification of categorization information by FSTEC of Russia (10)	Dir	Exp	-
Receipt of notification on entering the Registry of CII (11)	Dir	Exp	Emp1 Emp2

At the same time, the DBMS allows the user to associate the "Person" database object with an almost unlimited number of "Action" and "Data Storage" objects, and several persons may be participants of the same action or have relation to one particular information storage.

Thus, IAS can be considered as continuously revolving structured information, which will help solve the following information-logical tasks [16]:

- Promptly receive information on the number of confidential information storages
- Compile reports on the presence of data storages in the organization

- Receive information about the number of employees worked with a particular data storage or document and, conversely, full list of data storages or documents each employee worked with
- Control the process of transferring media from employee to employee
- Timely detect the facts of data storages loss

Table 3. Algorithm of data storages accounting.

Operation	R	E	P
Decision on the data storage registration (1,2)	Dir	Exp	Emp
Data storage registration (3)	Exp	Exp	Emp
Data storage issue (4)	Exp	Exp	Emp
Decision on the data storage transfer (5,6)	Dir	Emp 1	-
Reception of the data storage (7)	Exp	Emp 1	-
Data storage issue (7)	Exp	Emp 2	-
Decision on the data storage destruction (8,9)	Dir	Emp	Emp
Data storage destruction (10,11)	Dir	Exp	Emp 1 Emp 2

The accumulation and processing of information in respect of persons with access to confidential information proceed in the same way. A brief algorithm of actions is presented in Table 4.

Thus, IAS can be considered as continuously revolving structured information, which will help solve the following information-logical tasks [16]:

- Promptly receive information on the number of confidential information storages
- Compile reports on the presence of data storages in the organization
- Receive information about the number of employees worked with a particular data storage or document and, conversely, full list of data storages or documents each employee worked with
- Control the process of transferring media from employee to employee
- Timely detect the facts of data storages loss

1. Exp address to Dir with a proposal to issue a pass for Emp.
2. Dir decides on the start of the security clearance process and providing information access for Emp.
3. Once the decision is made by Dir, Exp conducts training with the Emp who received access to confidential information. Information is entered into the database.
4. To check the knowledge received by Emp within the training, Exp conducts testing of Emp, the results are recorded in the database.
5. If the test has been passed, then Dir decides on granting Emp with access to the information.
6. Exp provides Emp with access, sets a personal password. The corresponding record is made to the database.
7. If access is needed to be extended, Emp comes out with the proposal to Dir.
8. Dir decides on expanding Emp's access.
9. Exp provides additional rights. The corresponding record in the database is made.
10. If the access to the information needs to be suspended, Dir makes the appropriate decision, Exp revokes Emp's access to the information and leave a note in the database.

Accounting of information on the form of access of an employee (in the "Person" database), and the media confidentiality neck (in the "Data Storage" database) allows user to quickly establish actual employee awareness of confidential information, as well as identify possible violations related to confidentiality stamp, not corresponding to the form of their access [17].

In the "Action" database, the user can not only record the fact of instructing the employee on the rules for working with confidential information but also the degree of assimilation of these rules by each employee.

In the case of investigation of computer incidents [18], an employee of the information security subdivision will actively use the information accumulated in the IAS during the execution of the previous algorithms, as well as introduce additional information characterizing the employees of the organization.

As a result, a file will be formed for each employee gradually, which will provide irreplaceable assistance in deciding on whether to expand employee access or issue confidential data storages.

Table 4. Algorithm of accounting of allowed persons.

operation	R	E	P
Decision on the start of providing access to the information (1,2)	Dir	Exp	Emp
Training (3)	Exp	Exp	Emp
Testing (4)	Exp	Exp	Emp
Decision on granting access to the information (5)	Dir	Exp	Emp
Providing the access to the information (6)	Exp	Exp	Emp
Decision on expanding the access (7,8)	Dir	Exp	Emp
Expanding the access to the information (9)	Exp	Exp	Emp
Suspension of access (10)	Dir	Exp	Emp

When investigating incidents [19], information from previously described databases will be of great interest. In particular, a suspected employee of CII object may be associated with a competing organization, or have information about the means of protecting CII object.

4   Conclusion

From the above examples, it follows that the DBMS allows the implementation of algorithms for information and analytical support at all stages of the creation of the CII object protection system. This is the scientific novelty of the study.

The IAS allows not only to accumulate, systematize and provide the information necessary to create an integrated CII object protection system, but also to identify implicit links between information objects. For example, detection of the fact that confidential information was leaked after one of the former employees of CII object (previously granted access to the information stolen or having connections among existing employees of CII object) has been hired by a third-party organization [20].

The most important advantage of using the IAS is a guarantee of continuity in ensuring the safety of CII objects. The accumulated information does not disappear if the security specialist leaves the company or is transferred to another subdivision. It continues to be supplemented, systematized and stored in a convenient form for consumers.

In view of the above, the proposed IAS has the prospect of practical application on large CII subjects.

References

1. ISO/IEC 27032:2012 (2012) ISO. https://www.iso.org/standard/44375.html. Accessed 8 May 2019
2. State Duma of the Russian Federation. Federal law № 390-FZ of 28 December 2010 (2010). http://www.consultant.ru/document/cons_doc_LAW_108546/. Accessed 8 May 2019
3. Taylor, F.: Information systems management, pp. 105–115 (2018)
4. NIST: Managing information security risk: organization, mission, and information system view (2011)
5. Science and Engineering Research Support Society. Int. J. Database Theory Appl. (2016)
6. Burney, S., Burney, S.: Security and Frontend Performance. O'Riley, New York (2017)
7. State Duma of the Russian Federation. Federal Law N 187-FZ of 26 July 2017 (2017). http://www.consultant.ru/document/cons_doc_LAW_220885/. Accessed 8 May 2019
8. FSTEC of Russia: Requirements for arrangement of security systems at significant facilities of critical information infrastructure of the Russian Federation, for provision of their functioning. FSTEC of Russia order no 235 of 21 December 2017 (2017)
9. FSTEC of Russia: On Approval of requirements for ensuring the security of significant facilities of the information infrastructure of the Russian Federation: order 239. Federal service for technology and export control of 25 December 2017 (2017)
10. Hellerstein, J.M., Stonebraker, M., Hamilton, J.: Architecture of a database system. Found. Trends Databases 1(2), 141–259 (2007)
11. Korotkevich, D.: Pro SQL Server Internals. Apress, New York (2016)
12. Cronos Inform: IDBMS CronosPRO (2018). http://www.tadviser.ru/index.php/%D0%9F%D1%80%D0%BE%D0%B4%D1%83%D0%BA%D1%82. Accessed 20 Mar 2019
13. Bejtlich, R.: The Practice of Network Security Monitoring: Understanding Incident Detection and Response. No Starch Press, San-Francisco (2013)
14. Checkpoint: SQL slammer comeback (2017). https://blog.checkpoint.com/2017/02/02/sql-slammer-comeback. Accessed 20 Mar 2019
15. IBM: Process pseudo-file system and system pseudo-file system support (2019). https://www.ibm.com/support/knowledgecenter/en/SSB23S_1.1.0.13/gtpd1/d1procfs.html. Accessed 20 Mar 2019
16. Andreeva, O., Gordeychik, S.: Industrial control systems Vulnerabilities statistics. Kaspersky Lab, Moscow (2016)
17. NIST: Digital data acquisition tool specification, May 2017. https://www.nist.gov/sites/default/files/documents//05/09/pub-draft-l-dda-require.pdf. Accessed 20 Mar 2019

18. Ligh, M.H., Case, A., Levy, J., Walters, A.A.: The Art of Memory Forensics, Detecting Malware and Threats in Windows, Linux, and Mac Memory. Wiley, New York (2014)
19. Barkly: 10 Must-know cybersecurity statistics for 2018 (2018). https://blog.barkly.com/2018-cybersecurity-statistics. Accessed 20 Mar 2019
20. Positiv research: Digest of studies on practical security (2016). https://www.ptsecurity.com/upload/ptru/analytics/Positive-Research-2016-rus.pdf. Accessed 1 Apr 2019

The Artificial Neural Network Application for Service-Oriented Evaluation of the Used Cars

A. N. Guda and A. N. Tsurikov[(⊠)]

Rostov State Transport University, 2 Rostovskogo Strelkovogo Polka Narodnogo Opolcheniya Square, Rostov-on-Don 344038, Russian Federation
tsurik7@yandex.ru

Abstract. The article is devoted to the development of a system of the service-oriented valuation to the used cars' cost. "Rapid" produced by the famous Czech manufacturer "Skoda" was chosen as the target brand and model of the car. The paper presents a mathematical formulation of the determining problem of the value of the car in the form of a classification problem. To solve the problem, the authors propose to use the technology of the artificial neural networks (ANN). The universal classifier of the "multilayer perceptron" type with two hidden layers was used as the ANN to be trained. The "NeuroSolutions" neuropackage program is used as the neural network modeling tool. The input and output parameters of the trained ANN are determined. The statistics collection for training the ANN was collected on Russian websites specializing in the sale and purchase of the used cars, in particular, on "auto.ru" and "drom.ru". As a result, the created ANN is able to solve the presented problem with a sufficient degree of accuracy. The article discusses the prospects for the research results' implementation.

Keywords: Classification problem · Machine learning · Multilayer perceptron · Neural network · Neural training · Service-oriented evaluation · Used cars

1 Introduction

The technology of the artificial neural networks (ANN) today is widespread [1, 2]. Almost every day, there are reports about new areas of successful application ANN: the creation of the software products and devices that are based on them.

In this regard, the study of the possibilities for using ANNs and machine learning are of great interest when create modern service-oriented software products, especially if the solving problems can emphasize the classification problems, as we know, the ANN can successfully cope with them [3, 4].

One of the main problems is the task of evaluating used cars, namely, determining price of a specific vehicle instance. The price is the money amount in exchange for which the seller agrees to sell the product [5].

A. A. Radionov and A. S. Karandaev (Eds.): RusAutoCon 2019, LNEE 641, pp. 965–975, 2020.
https://doi.org/10.1007/978-3-030-39225-3_103

2 The Purpose of the Study

The purpose of this paper is to study the possibility of using the artificial neural network in software services designed to approximate price of the used cars of a selected brand and model with a number of the characteristics.

Similar services (using elements of machine learning) are still few and are under development or test launch [6]. The topic is of high importance due to the expansion of the number of the relevant Internet sites; the transactions' increase with used cars and decrease in demand for new cars due to the difficult situation in the economy are obtained.

3 Materials and Research Methods

To carry out the study, it is necessary to determine the appropriate brand and model of the car, to select the data source for the ANN's training, to choose the model allowing to train the ANN, to carry out the mathematical formulation of the problem and to determine the ANN's architecture with a number of the layers and neurons.

3.1 The Choice of a Model and a Car Brand

It was decided to choose as a target brand and a car model "Rapid" of the famous Czech manufacturer Skoda (Fig. 1); it is a part of the German concern Volkswagen.

This choice has been done for several reasons. The model is quite popular in the automobile market, it appeared relatively recently (2012–2014 year), but at the same time it has been released so far that a sufficient number of ads can appear on the sale sites for used cars.

Previously, this popular car model was not produced (models of 1935 and 1984 can be ignored as we couldn't find any at the automobile market). During the production (6 years) Rapid there was no change of the generations, only a number of restyling was carried out which did not significantly affect the outside appearance and technical equipment of the car [7].

The cars assembled at different factories (Mlada Boleslav in the Czech Republic, Kaluga in Russia, Ukraine and Kazakhstan) do not have any fundamental differences (for Russia it is improved the suspension mount). The external part of the car is presented in a single version, it is liftback, and other options are practically absent.

Technically, all cars have only three variants of the gearboxes (manual, automatic, robotized DSG (direct-shift gearbox)) and engines (atmospheric – 1.6 L volume (90 hp or 110 hp) as well as turbocharged – 1.4 L volume (125 hp). In addition, it was previously produced cars with engines of 105 and 122 hp, which are presented in the secondary market [8].

So, the Rapid is an ideal option for classification at the initial stage for applying the ANN, therefore this car will be used in our work.

Fig. 1. The appearance of Skoda Rapid.

3.2 The Data Source for Training ANN

Currently, the Internet can find a large number of the sites specializing in sale the used cars. We've studied the following resources in Russia: auto.ru, drom.ru, car.ru, genser. ru and the avito.ru website section.

As a result, we will focus on the oldest and largest auto.ru website and on the drom. ru portal in Russia. It is collected the statistics of the car studied model, its characteristics and associated price.

Nowadays the site auto.ru placed over 500 thousand sale advertisements of the used cars. The website experts ring up the ads and check their accuracy [9]. According to the collected information, the adverts about the sale "Skoda Rapid" are about 3 thousand in all Russian regions.

The statistics were collected from sites manually for a week in November 2017 and among advertisements it was published a number of geographically close regions in Southern Russia (city: Rostov-on-Don + 500 km). Obviously, the possible price fluctuations caused by differences in the economic development of the regions and changes in exchange rates over long time periods were nearly eliminated. The results were processed and saved in a file using the Microsoft Excel spreadsheet processor.

3.3 Tools for Modeling and Training ANN

The use of the neuropackage program and a software product that allows you to create and train ANN's emulating their behavior on the ordinary personal computer is chosen as the ANN simulation tool.

The following tools were considered as alternatives: NeuroSolutions, Process Advisor, NeuralWorks Professional II/Plus, BrainMaker Pro, NeuroShell, MatLab + Neural Network Toolbox.

Each of them has its own advantages and disadvantages, which are described in detail in the articles [10, 11]. As a result, NeuroSolutions was chosen for this work as the development environment of the neural networks (neuropackage) from the NeuroDimension.

3.4   Mathematical Formulation of the Problem

The mathematical formulation of the problem for determining cost is almost similar to the formulation of the classification problem in general form [12]. The task is formulated as a classification task, where G = [mileage, age, engine power, gearbox type] is the set of the parameters for an individual vehicle, $Q = \{A, B, C, D, E, F\}$ is the range of the price changes. There is an unknown mapping $g^*: G \to Q$ its values are known only on the objects of the final sample $G_j = \{(g_1, q_1), \ldots, (g_j, q_j)\}$. It is required to find the algorithm $\beta: G \to Q$, which is able to correctly classify an arbitrarily taken instance of the car $g \in G$.

3.5   Definition of the Input and Output Parameters

We define the basic parameters by use the sale site for used cars auto.ru (Fig. 2). You can see that the price is mainly determined by four main parameters: mileage, engine power, type of a gearbox and production year (age of the car). We will select them as ANN inputs.

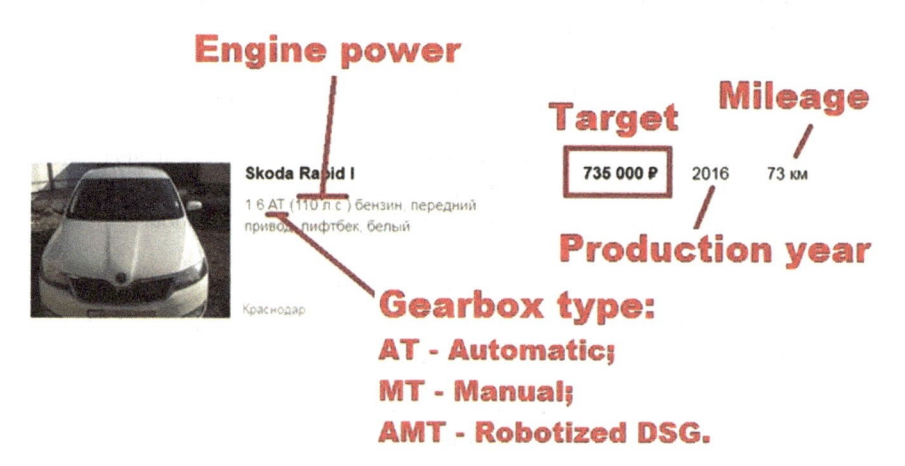

Fig. 2. Skoda Rapid on the website auto.ru.

Firstly, it is selected the range of an input data for each parameter of the vehicle and their data type as well as the scale of the measuring parameters.

In our case, the following ranges of the input variables are defined. Mileage is from 100 to 750,000 km. The engine power is 75, 90, 105, 110, 122 and 125 hp. Type of gearbox is AMT, AT, MT. In 2018, the automobile from 2014 to 2017 are on sale in the auto markets as the car age is calculated from 1 year to 4 years.

Since the work is aimed at determining the price of a used car, the output variable will be its price. To set the parameters for working with a neural network correctly, it is necessary to determine the price range. As the Skoda Rapid brand is investigated, it is necessary to use data on it and its characteristics.

After examining sites used in the work, we should note that the minimum price for the Skoda Rapid is approximately 400,000 rubles. The maximum price on Rapid is approximately 1 million rubles.

It follows that the range of the price changes for this car is from 400,000 to 1,000,000. We divide this range into 6 conditional segments of 100,000, each that will be the output parameters.

For getting the input feature vector consisting of four vehicle parameters G = [mileage, engine power, gearbox type, car age], trained by the ANN should classify an arbitrary price $g \in G$ by assigning it to one of 6 price ranges Q = {A, B, C, D, E, F}, where A = 400,000–500,000; B = 500,000–600,000; C = 600,000–700,000; D = 700,000–800,000;   E = 800,000–900,000;   F = 900,000–1,000,000   (price in rubles).

3.6   Architecture and the ANN's Type

The universal classifier of the "multilayer perceptron" type [12–14] will be used as an ANN that will be trained. The number of the feature vector parameters is four; there should be 4 neurons in the input layer. The number of the neurons in the output layer, which is equal to the number of the possible classes, for our example, is 6. In such a perceptron, the number of the hidden layers is two.

It is possible to determine the number of the neurons in the hidden layer of the ANN, provided that there is only one hidden layer [12]. This is possible according to the Hecht-Nielsen equation:

$$N_C = 2 \cdot N_x + 1 \tag{1}$$

where N_x is the number of the neurons in the input layer, N_C is the number of the neurons in a single hidden layer.

In our case, $N_x = 4$, and we get: $N_C = 2 \cdot 4 + 1 = 9$.

In the case of two hidden layers in a perceptron, there is no strict mathematical formula and the number of the neurons can be determined empirically [15]. We will place 8 and 4 neurons in the first and second hidden layers, respectively, which is totally more than the estimate obtained by (1).

We will apply a two-layer perceptron with two hidden layers, which will have 4 input neurons and 6 output (Fig. 3).

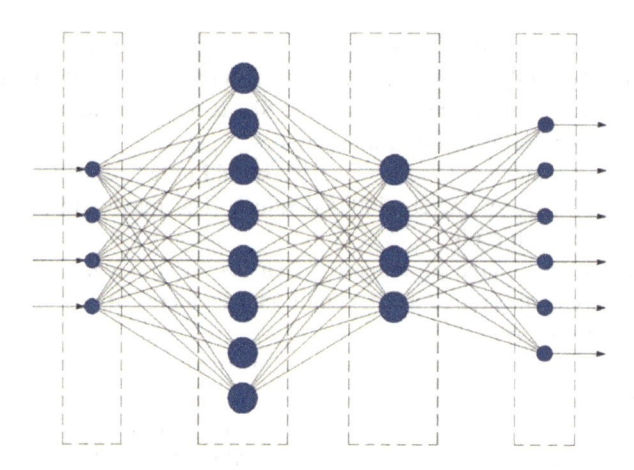

Fig. 3. Structure of the double-layer perceptron.

As the activation function of the neurons in hidden layers S_{neuro}, the nonlinear function of the hyperbolic tangent is used, it is given by the equation:

$$S_{neuro}(Ax_{lin}) = th(Ax_{lin}) = \frac{Sh(Ax_{lin})}{Ch(Ax_{lin})} = \frac{e^{Ax_{lin}} - e^{-Ax_{lin}}}{e^{Ax_{lin}} + e^{-Ax_{lin}}}, \tag{2}$$

where x_{lin} is the signal of the linear part of the neuron, A – function shift.

3.7   The Training Vectors

To determine the required number K of the training examples for correct training of an ANN using the "Supervised learning" method, we use the equation:

$$K \geq \frac{w_C}{e}, \tag{3}$$

where w_c – is the number of free ANN parameters, $e \leq 1$ is a valid classification error.
The value of K is determined with a margin of at least 5% of the calculated.
In our case, we use the following values: $e \approx 0,15$, $w_c \approx 90$ and we get $K \geq 600$. We increase the calculated value by 5%: $K = 630$ (pieces).
The required number of training examples from the sites auto.ru and drom.ru will be entered into the Excel. Next, we will transform the table into a text document, which is necessary for the correct operation of the NeuroSolutions program, by performing the necessary data transformations.
Now the data of training vectors is full and ready to use, we can proceed to the next stage it is training the ANN.

4   The ANN Training and Analysis of the Results

We will perform ANN modeling in the NeuroSolutions package in the Neuroexpert mode using the structure shown in Fig. 3.

The number of the training examples reserved for test and exam datasets was 20% of the vectors' total number. The test set (Cross validation) is used to prevent the "overtraining" of the ANN during training; the examination set (Testing) is to assess the quality after training.

Having set all the necessary parameters of the ANN, we complete the construction of the network, after which a new project "NeuroSolutions" is opened (Fig. 4).

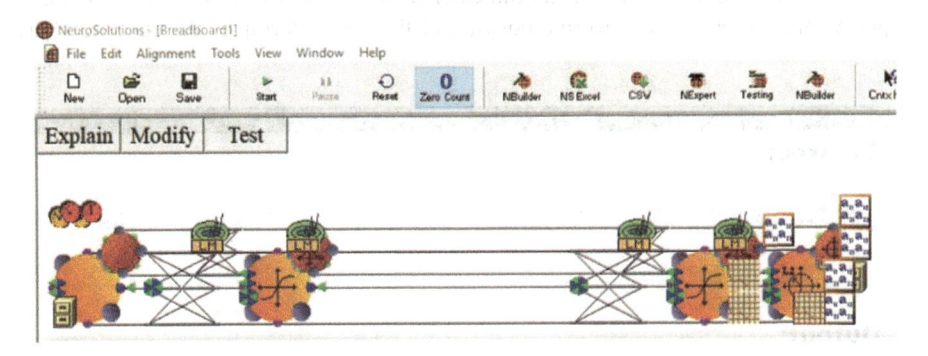

Fig. 4. The project "NeuroSolutions" with created ANN.

Further, we start the training process of the neural network. In total, we use 1000 epochs of training. As a result, the learning curve takes the form shown in Fig. 5, where T is the mean square error on the training data, CV is the mean square error on the test data set.

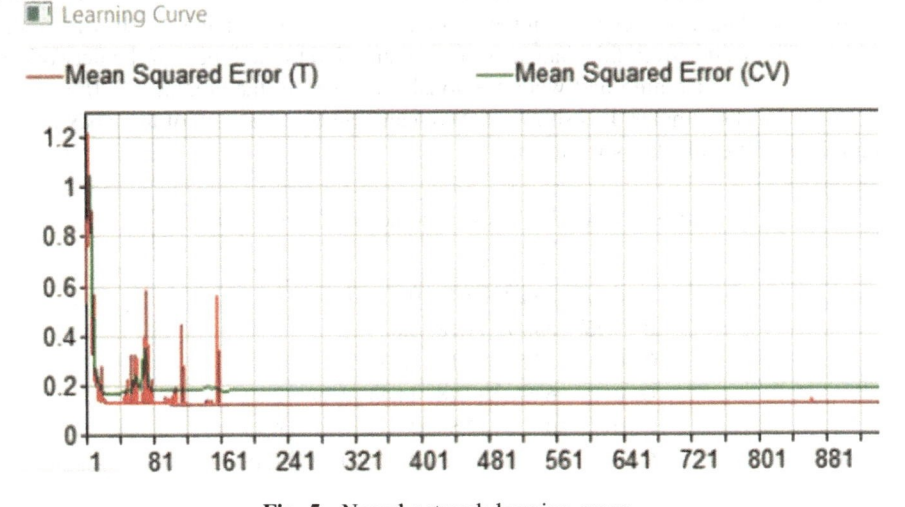

Fig. 5. Neural network learning curve.

It can be seen from the graph that both errors gradually decrease as the ANN learns. You may notice that the learning curve takes a horizontal view at a level of just under 0.2 after 170 epochs of learning. Therefore, the ANN is trained and can solve the problem presented to it with a mean-square accuracy of about 80%.

5 Research Results and Discussion

We have tested the ANN on test records and note that in most cases the ANN successfully copes with the determination of the price range (Fig. 6). The output vector Q consists of the numbers approximately equal to zeros, and one number approximately equals to one. The unit is an indicator of the class which the current vector of input parameters gi belongs to (the position of the unit determines the class number).

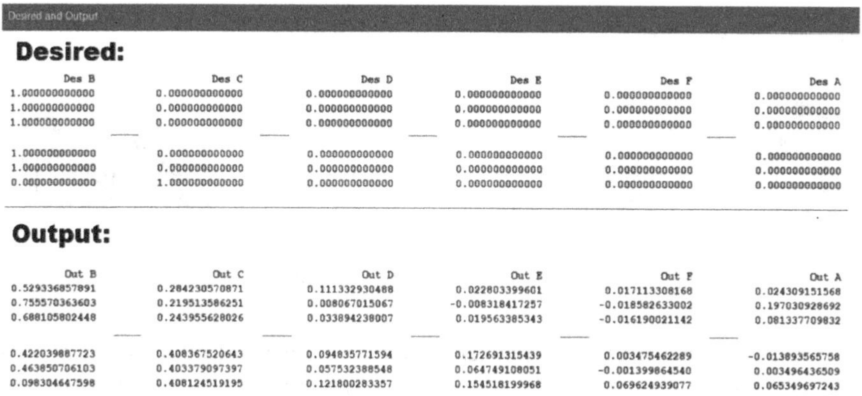

Desired and Output

Desired:

Des B	Des C	Des D	Des E	Des F	Des A
1.000000000000	0.000000000000	0.000000000000	0.000000000000	0.000000000000	0.000000000000
1.000000000000	0.000000000000	0.000000000000	0.000000000000	0.000000000000	0.000000000000
1.000000000000	0.000000000000	0.000000000000	0.000000000000	0.000000000000	0.000000000000
1.000000000000	0.000000000000	0.000000000000	0.000000000000	0.000000000000	0.000000000000
1.000000000000	0.000000000000	0.000000000000	0.000000000000	0.000000000000	0.000000000000
0.000000000000	1.000000000000	0.000000000000	0.000000000000	0.000000000000	0.000000000000

Output:

Out B	Out C	Out D	Out E	Out F	Out A
0.529336857891	0.284230570871	0.111332930488	0.022803399601	0.017113308168	0.024309151568
0.755570363603	0.219513586251	0.008067015067	-0.008318417257	-0.018582633002	0.197030928692
0.688105802448	0.243955628026	0.033894238007	0.019563385343	-0.016190021142	0.081337709832
0.422039887723	0.408367520643	0.094835771594	0.172691315439	0.003475462289	-0.013893565758
0.463850706103	0.403379097397	0.057532388548	0.064749108051	-0.001399864540	0.003496436509
0.098304647598	0.408124519195	0.121800283357	0.154518199968	0.069624939077	0.065349697243

Fig. 6. Testing the neural network after training.

NeuroSolutions allows to calculate the overall results of a trained ANN using an examination data set (Testing) and present them as a percentage as shown in the Table 1. The values from this data set were not supplied to the inputs of the network in the process of its learning and were previously reserved in the amount of 20% of the total number of examples. It is shown that we can simulate the work of the ANN in real conditions and evaluate its results.

Table 1. The ANN's results at the examination data set.

	A	B	C	D	E	F
A	61.77	35.29	2.94	0	0	0
B	3.73	84.47	11.18	0.62	0	0
C	0	32.5	59.17	5.83	2.5	0
D	0	6.67	37.78	44.44	8.89	2.22
E	0	0	12.5	25	56.25	6.25
F	0	0	0	0	0	100

As can be seen from the Table 1, the ANN achieved the best results when it classified the automobiles to the price ranges (classes) F and B (100% and 84.47%, respectively). There were the worst things with the range D, where the percentage of correct answers was only 44.44%. At the same time, a significant proportion of the cars in this category (37.78%) attributed the neural network to the nearest (cheaper) price range C.

On the one hand, this may indicate that some car owners incorrectly estimate the value of their cars when the sale price is determined. This is due to the fact that the prices which the neural network relies on are represented not only by experts, but also by ordinary users of the services.

On the other hand, some other factors that we did not consider in the study and did not apply to the ANN for the real evaluation of the car may exist. For example, the presence of the hidden defects (the consequences of the accidents, damage to the paintwork and so on) or problems with paperwork (frequent change of the ownership, sale by proxy or duplicate documents, etc.).

To improve the results, some of these parameters can be added as additional ANN inputs, but there is a danger of complicating the system with a slight increase in the classification accuracy. Adding too many features it leads to an exponential growth [4, 16] of the reference points numbers that are required to describe space (the so-called "curse of a dimension").

Of course, the errors in determining prices will always occur even if the experienced experts work. In general, the presented result of the work of the ANN can be considered acceptable. The results of the neural network can be based on a rough estimate of the cost of a used Skoda Rapid.

6 The Possibility of an Implementation

Using the built master into NeuroSolutions, a dynamic-link library DLL-module was created that contains a trained ANN and allows it to be reused by various software applications. To work with the DLL-module, a software application can be written in C++ (Visual Studio) with a user interface that allows to control the necessary actions with the ANN, enter and output the results in a convenient form.

The program has two working modes with the ANN: receiving results (Output) and training (Training). In the "Output" mode, the user enters data about a specific vehicle instance, clicks "Calculate" and gets the result; it is one of the possible price ranges. In the "Training" mode, it is possible both the additional training of the ANN on the newly obtained examples and the zeroing of the synaptic weights with the launch of a new learning process. The presence of such a mode allows to increase the accuracy of the ANN's classification when new data are received.

The ability to work with the ANN to estimate the cost of a car can be realized using similar software products both on desktop computers running Windows and on mobile devices running on Android or iOS operating systems.

At the same time, the implementation of a service using an artificial neural network should be considered as the preferred of its implementation on popular Internet sites for the sale used cars (auto.ru, drom.ru, etc.) as a Web service available to users online, it is one of the sections of the site through the browser.

7 Conclusion

As a result of the study, the possibility of using ANN in software services designed to estimate the value of the used cars for a number of some characteristics was studied.

Skoda Rapid was chosen as the target car brand and model. The required number of the training examples was collected from auto.ru and drom.ru in the amount of 630 pcs. NeuroSolutions neuropackage is selected as the ANN modeling tool. The mathematical formulation of the price determination problem as a classification problem is presented. The ANN input and output parameters are defined. As the ANN, a "multilayer perceptron" with two hidden layers was used.

It is produced a simulation of the ANN in NeuroSolutions and its training on the collected examples. The ANN has been trained and can solve the presented problem with a standard error of 20%. The results of the work of the ANN are analyzed, the ways to improve the quality of the neural network are considered.

A dynamic-link library (DLL-module) has been generated that contains a trained ANN. The user interface of the software application was created in the Visual Studio environment, which allows performing actions with the ANN's inputting and outputting results in a convenient form.

It is analyzed the possibility of the implementing results. A trained neural network can be used to save time and to provide user with convenience. The study of this problem requires further research.

Acknowledgements. This research was supported by the Russian Foundation for Basic Research (RFBR); the projects are 18-08-00549-a, 17-07-00620-a, 19-01-00246-a, 19-07-00329-a. The authors express their acknowledgment to Anastasia Alekseevna Gamagina for her help in collecting data for artificial neural network training.

References

1. Haykin, S.: Neural Networks: A Comprehensive Foundation. Prentice Hall, New Jersey (1999)
2. Vereskun, V.D., Tsurikov, A.N.: Information-operating systems in research and production. Rostov State Transport University, Rostov-on-Don (2016)
3. Tsurikov, A.N., Guda, A.N., Karsyan, A.Z.: Theoretical foundations of intellectualization solving classification problems in weakly formalizable areas. Sci. Tech. Bull. Povolzhye **1**, 90–93 (2016)
4. Shukla, N., Fricklas, K.: Machine Learning with TensorFlow. Manning Publications, Shelter Island (2018)
5. Vianello, F.: Natural (or normal) prices: some pointers. Polit. Econ. Stud. Surpl. Approach **5**, 89–105 (1989)

6. Khabibrakhimov, A.: Avto.ru launched a free service for estimating the cost of a car (2019). https://vc.ru/transport/29899-avto-ru-zapustil-besplatnyy-servis-dlya-ocenki-stoimosti-avtomobilya. Accessed 12 Feb 2019
7. Skoda Rapid: Electronic resource (2012). https://en.wikipedia.org/wiki/%C5%A0koda_Rapid_(2012). Accessed 14 Mar 2018
8. Skoda Auto Russia: The official site of the brand Skoda (2018). http://www.skoda-avto.ru. Accessed 15 Mar 2018
9. Auto.ru: Electronic resource (2019). https://ru.wikipedia.org/wiki/Auto.ru. Accessed 10 Feb 2019
10. Gamagina, A.A., Tsurikov, A.N.: Software for simulation of artificial neural networks: advantages and disadvantages. In: The Actual Problems and Challenge Developments of the Transportation, Industry and Russian Economy. Collection of Scientific Papers, Part 1: Engineering Sciences, Rostov State Transport University, Rostov-on-Don, 1–2 March 2018 (2018)
11. Tsurikov, A.N.: The use of the neuropackage software for solving practical problems. Engineer **5**, 6 (2012)
12. Principe, J.C., Euliano, N.R., Lefebvre, W.C.: Neural and adaptive systems: fundamentals through simulations. Wiley, New York (2000)
13. Tsurikov, A.N., Guda, A.N.: Practical application of the original method for artificial neural network's training. In: Abraham, A., Kovalev, S., Tarassov, V., et al. (eds.) IITI 2017: Proceedings of the Second International Scientific Conference on Intelligent information technologies for industry, Varna, 14–16 September 2017. Advances in Intelligent Systems and Computing, vol. 679, pp. 84–93. Springer, Cham (2017). https://doi.org/10.1007/978-3-319-68321-8_9
14. Tsurikov, A.N., Guda, A.N.: Method, algorithm and device for training of artificial neural network: theoretical foundations and practical realization. In: 2nd International Conference on Industrial Engineering, Applications and Manufacturing, Proceedings 2, 19–20 May 2016. IEEE Xplore, Chelyabinsk (2016). https://doi.org/10.1109/icieam.2016.7911613
15. Tsurikov, A.N.: Application of artificial neural network for identification of stability of bottom layer of atmosphere. In: Proceedings of the 2nd International Academic Conference on Applied and Fundamental Studies, St. Louis, March 2013 (2013)
16. Callan, R.: The Essence of Neural Networks. Prentice Hall Europe, Harlow (1999)

Analysis of the Problems of Industrial Enterprises Information Security Audit

I. I. Barankova, U. V. Mikhailova$^{(\boxtimes)}$, and O. B. Kalugina

Nosov Magnitogorsk State Technical University, 38 Lenina Avenue,
Magnitogorsk 455000, Russian Federation
ylianapost@gmail.com

Abstract. The article analyzes the statistics of cyber attacks on industrial enterprises in 2018. The relevance of the information security audit is determined by the increased dependence of the success of the enterprise's activities on the corporate information protection system as a whole. In order to protect from such attacks risks in the future, management of enterprises needs to take measures to ensure information security of production. This article describes the basic principles, goals and objectives of the audit of enterprises information security. In particular, it describes the general approach to conducting of IT audit at an enterprise. It is prove the importance of correct analysis of the enterprise's belonging to the objects of critical information. In the course of the experiment, a security audit was conducted of one of the industrial enterprises of our region. The main problems arising during the audit of information security are listed on the example of the enterprise "Machine-Building Plant". The results of the information security audit were presented and deficiencies in the provision of information security were shown.

Keywords: Information system · Information protection · Security audit · Enterprise security · Information security

1 Introduction

Information security audit is a systematic process of obtaining objective qualitative and quantitative assessments of the current state of an enterprise information security in accordance with certain criteria and safety indicators. The information security audit significance is caused by the increased dependence of the enterprise's activities success on the corporate information system protection and an increase in the volume of vital data for the enterprise processed in the corporate information system.

According to Kaspersky Lab, about half of the computers in Russian industry ran into cyber threats in 2018 [1]. Cybercriminals, in the opinion of Laboratories experts, as the matter of fact were simply forced to pay attention to the industry, as banks and financial institutions constantly strengthened their security systems.

In percentage terms, about 48% of industrial computers were attacked. First of all, these are automated process control systems. About a third of the systems, the source of threats was the Internet, 5% were attacked through removable storage media, and 2% through email programs of various kinds.

© Springer Nature Switzerland AG 2020
A. A. Radionov and A. S. Karandaev (Eds.): RusAutoCon 2019, LNEE 641, pp. 976–985, 2020.
https://doi.org/10.1007/978-3-030-39225-3_104

Modern IT technologies, on the one hand, increase an enterprises efficiency and, on the other, make them vulnerable to computer attacks, says Kaspersky Lab expert Kirill Kruglov. Historically, automated process control systems were built without taking into account possible cyber interference, therefore not all incidents can be observeed. Head of IT Infrastructure Protection at Norilsk Nickel Andrei Kulpin, generally agreed with the findings of Kaspersky Lab.

A cyber attack on the enterprise information system can stop the production process so that it will take several months to restart, and another company will receive profitable orders, says Dmitry Darrensky, head of industrial cybersecurity practices for Positive Technologies (security analysis of IT systems). This could be, for example, disabling gas turbines or shutting off the power supply at an aluminum plant with subsequent solidification of aluminum in electrolysis baths, which would require replacing equipment worth of million dollars.

Most Russian industrial companies spend on information security less than 50 million rubles. per year, ascertained Positive Technologies in the course of the study "How much does safety cost?" At the same time, 27% of respondents estimated the same amount of loss for one day of infrastructure downtime due to a cyber attack. One third of industrial organizations estimated the possible damage from the failure of corporate infrastructure in one day at 0.5–2 million rubles, 13% from 2 million to 10 million and 17% from 10 million to 50 million [1].

For effective protection against attacks enterprises need an objective IT security assessment. Conducting an IT security audit will help to keep of enterprise information resources security [2].

The main objectives of the audit of enterprise information security include the following [3]:

- Obtaining an independent objective assessment of the current state of enterprise information resources protection
- Get the most out of investition into creation of an information security system
- Assessment of possible damage from unauthorized activity
- Requirements engineering for information security systems planning
- Determination of responsibility areas of employees
- Calculation of necessary resources
- Development of the procedure of information security system implementation

There are main types of audit:

- Expert audit, which resulted in the identification of deficiencies in the system of information protection measures based on the experience of experts participating in the survey procedure
- Evaluation of compliance with the recommendations of the international standard ISO 17799, and the requirements of the governing documents of the federal service for technical and export control
- Instrumental analysis of the security of the information system, which aims to identify and eliminate vulnerabilities of software and hardware systems
- Comprehensive audit, includes all of the above forms of survey

Considered types of security audits can be conducted individually or in combination, depending on the enterprise tasks [4]. The object of an audit can be either the organization's information system as a whole or its certain segments, which processed information to be protected.

Irrespective of the audit form, the security audit consists of:

- The development of the audit regulations
- Collection of source data
- Analysis of the data to assess the current level of security
- Development of recommendations to improve the level of information system security

2   Objectives and Methods of Researching Enterprise Information Security

The main task of the regulation is to determine the survey scope in order to avoid mutual claims at the end of the audit, since the regulation establishes clear responsibilities for both parties. At the next stage, in accordance with the agreed regulations, the source data is collected. Methods of collecting the necessary information include interviewing the customer's employees, filling out questionnaires, analyzing the provided organizational, administrative and technical documentation, using specialized tools. After collecting the initial data, they are analyzed to assess the level of information security. In the course of the analysis, the IS risks to which the organization is subjected are determined. At the last stage of the IS audit, recommendations are developed to improve the organizational and technical support of security at the enterprise [5, 6].

The most obvious type of audit of IT infrastructure security is scanning for vulnerabilities using special software [7, 8]. This type of audit allows you to identify most of the known vulnerabilities in information resources and get detailed recommendations on how to fix them.

The search for vulnerabilities is one of the testing stages for the exploration of unauthorized entry possibility. At this stage, we imitate the intruders penetrating actions into a corporate system.

There are external penetration testing, in which specialists try to penetrate the corporate network via the Internet, internal, when the actions of an attacker who has physical access to the company's network are imitated [9–11].

During inspections, specialists use software and techniques used by real attackers to hack systems [12, 13]. Testing is carried out at the network, system and application levels and allows us to identify not only the majority of vulnerabilities, but also to identify among them those that really compromise the information security system.

The analysis of the information systems security settings is most laborious stage, which is carried out using special checklists containing a description of the system configuration recommended by information security professionals. The main advantages of such analysis are the high reliability of vulnerabilities intelligence and minimal impact on the system's performance.

The security audit, which includes all types of safety checks is most effective, since it allows a comprehensively assessment the security of the company in the face of information security threats. It is best when such an audit is conducted professionally and independently [14].

To conduct an information security audit of a company, external companies that provide consulting services in the field of information security are involved. They are carried out by a group of experts whose strength depends on the goals and objectives of the survey, as well as on the complexity of the object being assessed [15, 16].

When developing of an audit plan, it is necessary to take into account the belonging of the enterprise to critical information infrastructures (CII), since in this case it is impossible to speak about some single "abusers", but it is necessary to consider such "cyber-forces" as a highly developed and technically trained attackers. They will conduct continuous sophisticated attacks in the information space [17–19]. Most of the managers of enterprises do not believe that their objects belong to the CII, this leads to insufficient protection of the data processed in the enterprise information system [20, 21].

The first stage of the audit is the collection of basic data on the enterprise's information management system. This is such data as an available documentation on the protection of information resources and data of the existing protection systems [22]. As practice shows, this stage is the most problematic [23].

3   Enterprise Information Security Survey

The main problems encountered in the audit process, on the example of "Machine-building plant" are:

- The lack of an information security policy, which describes the objectives, the audit plan and the designated responsible specialist
- Lack of support and company top management understanding of the in matters of information security and auditing, in particular
- Chefs of head enterprises are not always ready to hand over documents related to information security to branches
- The absence of an information security specialist at the enterprise. In most organizations, its tasks are carried out by system administrators or specialists from other areas who are not aware of the specifics of the work, the risks, and the possible damage [24]

After collecting the data, they were analyzed, which allows us to assess the level of information security. In the course of the analysis, the risks of information security are determined by which the enterprise is subjected. For the analysis, it was chosen the classification code of 2005 risk assessing methodology from Digital Security [25]. At the first stage, the threat level for vulnerability Th was calculated based on the criticality and probability of the threat occurring through this vulnerability. The threat level shows how critical the impact of this threat is on a resource, taking into account the probability of its realization:

$$Th = \frac{P(V)}{100} \cdot \frac{ER}{100},$$ (1)

where ER is the criticality of the threat (indicated as a percentage); $P(V)$ is the probability of the threat being realized through this vulnerability (indicated as a percentage).

To calculate the threat level across all vulnerabilities (CTh), through which it is possible to implement this threat on a resource, it is necessary to sum the obtained threat levels through specific vulnerabilities using formula "(2)".

$$CTh = 1 - \prod(1 - Th),$$ (2)

where Th is the threat level by certain vulnerability.

Next, the total threat level is calculated using the $CThR$ resource according with the formula "(3)":

$$CThR = 1 - \prod(1 - CTh),$$ (3)

where CTh is the threat level for all vulnerabilities.

The risk of the resource R is calculated by the formula "(4)":

$$R = CThR \cdot D,$$ (4)

where D is the criticality of the resource (specified in money units or by levels); $CThR$ is the overall threat level for a resource.

4   Results and Discussion

The information system of enterprise management processes the following types of information:

1. Financial and accounting data:

   - Budget plans of the enterprise
   - Operations data and cash flow
   - The result of analytical audit reports
   - Information on the main economic indicators

2. Marketing data:

   - Systematic data about contractors, shippers, middlemen and customers
   - A plans for the submission of the tender commission
   - Strategic plans for the enterprise development

3. Personal data:

   - Personal files of employees
   - Personal official salaries, allowances and amounts of remuneration to employees of the enterprise
   - A personnel and special checks information

4. Organizational and administrative information:

- Job descriptions
- Internal instructions, orders, regulations; minutes of meetings and negotiations

5. Profile information (by type of enterprise activity):

- Technological plans, schemes, technical documentation
- Information on the inventory of equipment and technology
- Business correspondence

6. Information for internal use:

- Surveillance data

Table 1 presents an expert evaluation of the threats realization probabilities on the investigated enterprise.

Table 1. The probability of threats.

Treatment	P, %	Server 1	Server 2	Server 3
1. Exposure on the resource				
1.1. Reading valuable information from paper and PC screens	60	50	65	30
1.2. Destroy or damage media with valuable information	50	30	40	10
1.3. Theft of data storage device with valuable information	50	35	50	20
2. Exposure on OS				
2.1. User privilege escalation when implementing local vulnerabilities using OS administration errors	60	60	70	5
2.2. Extension of user privileges in the implementation of local vulnerabilities that use OS design errors	40	40	40	20
2.3. Extending remote user privileges when implementing vulnerabilities that exploit network service development or administration errors	35	60	60	20
2.4. Denial of service OS	15	30	30	15
2.5. Substitution of system configuration files/application software	35	50	70	55
3. Exposure on data				
3.1. Unauthorized modification of electronic documents/information in the database	20	60	70	50
3.2. Unauthorized reading of confidential information in the database/in electronic documents	30	50	60	10
4. Exposure on network services				
4.1. Denial of service for network service (internal software failure)	35	40	50	20
4.2. Extending remote user privileges when implementing vulnerabilities that exploit network service development or administration errors	15	70	70	30
4.3. Pick up of user authentication data	15	50	60	10
4.4. Interception of network services data using vulnerabilities of data transfer protocols	40	60	30	10

Next, we calculated the threat level for a certain vulnerability (*Th*) and threat level for all vulnerabilities (*CTh*) for each resource (*ER* is server) using the formulas "(1–2)".

The results of the calculations of threat level for all vulnerabilities (*CTh*) for each resource (*ER*) and the threat level for a certain vulnerability (*Th*) are shown in Table 2. The list of threats in Table 2 is given in accordance with their numbering in Table 1.

Table 2. Threat level for certain vulnerability and for all vulnerabilities.

Threat/Vulnerability	Server 1		Server 2		Server 3	
	Th	*CTh*	*Th*	*CTh*	*Th*	*CTh*
1/1	0,3	0,509	0,39	0,064	0,18	0,299
1/2	0,15		0,2		0,05	
1/3	0,175		0,25		0,1	
2/1	0,36	0,666	0,42	0,910	0,03	0,341
2/2	0,16		0,16		0,08	
2/3	0,21		0,21		0,07	
2/4	0,045		0,045		0,02	
2/5	0,175		0,245		0,19	
3/1	0,12	0,132	0,14	0,155	0,1	0,127
3/2	0,15		0,18		0,03	
4/1	0,035	0,192	0,175	0,081	0,07	0,165
4/2	0,105		0,105		0,05	
4/3	0,075		0,09		0,015	
4/4	0,24		0,12		0,04	

The maximum damage, expressed in thousand rubles, which can be subjected to a resource (server) as a result of the realization of threats for all vulnerabilities is presented in Fig. 1.

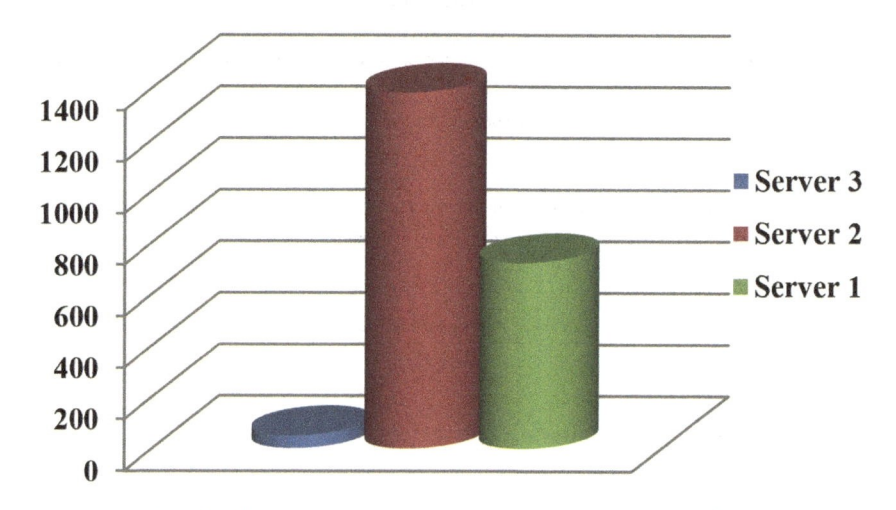

Fig. 1. Maximum damage to the resources, thousand rubles.

The calculating risk results are presented in Table 3. The total risk for 3 servers enterprise information system merely is 2,018,610 rubles.

Table 3. Resource risk assessment.

Resource	Threat level for all vulnerabilities, %	Resource risk, rub
Server 1	0,885	637 200
Server 2	0,977	1 348 260
Server 3	0,663	33 150

5 Conclusion

As a result of the information system security audit, the following deficiencies in the provision of information security were identified:

1. Organizational and legal:

   - Lack of an information security specialist
   - Lack of information security policies
   - There is not training of personnel in the field of information security
   - Disciplinary of non-compliance with information security are not approved
   - Inadequate protection of server room equipment

2. Software and hardware:

   - Weak password protection
   - Not all users comply with the rules for using passwords (storing a password, actions to compromise, regular password changes)
   - Not all workstations have uninterruptible power supplies

According to IT analysts, domestic automated process control systems were initially created without an external intervention prediction, thus, the protective mechanisms of such systems were not adapted to repel cyber-attacks.

On the basis of the information security audit only, after identified deficiencies in the process implementation and timely remedied of them, we can create effective and reliable information security systems, including at the enterprises informatization facilities.

References

1. Hacker attacks underwent half of the computers of Russian industry. Vedomosti (2019). https://www.vedomosti.ru/technology/articles/2018/12/05/788447-haerskim. Accessed 11 Jan 2019

2. Barankova, I.I., Mikhailova, U.V., Barankov, V.V., et al.: Experience of developing cloud service for accounting sales in installments. In: Journal of Physics: Conference Series, vol. 1015, p. 042004 (2018)
3. Types of information security audits. The art of information security management (2019). http://www.iso27000.ru/chitalnyi-zai/audit-informacionnoi-bezopasnosti/vidy-audita-informacionnoi-bezopasnosti. Accessed 15 Jan 2019
4. Kudryavtsev, M.E., Kalugina, O.B.: Development of information security threat modeling tool based on graph theory. In: Security of the Information Space Collection of the XVII All-Russian Scientific and Practical Conference of Students, Postgraduates and Young Scientists, pp. 133–138 (2018)
5. Kong, H.K., Kim, T.S., Kim, J.: An analysis on effects of information security investments: a BSC, perspective. J. Intell. Manuf. 23(4), 941–953 (2012)
6. European Commission: Cybersecurity strategy of the European Union: An open, safe and secure cyberspace JOIN. European Commission, Brussels (2013)
7. Likhonosov, A.G., Denisov, D.V.: Fundamentals of Information Security Audit: Study Guide. MFPA, Moscow (2010)
8. Barankova, I.I., Mikhailova, U.V., Lukyanov, G.I.: Prediction of local and external threats to enterprise information servers. Actual Probl. Mod. Sci. Technol. Educ. 1, 217–220 (2017)
9. Hang, A.C., Fung, W.S.L.: Knowledge Audit Model For Information Security. IKMAP, Kobe (2016)
10. Garber, L.: Have java's security issues gotten out of hand? Computer 45(12), 18–21 (2012)
11. Kemshall, A.: The RSA security breach - 12 Months down the technology turnpike. Database Netw. J. 4(2), 21 (2012)
12. Barankova, I.I., Mikhailova, U.V., Lukyanov, G.I.: DLP system: protection against information leakage. Analysis of the search WORDSEARCH. Actual Probl. Mod. Sci. Technol. Educ. 1(1), 187–191 (2016)
13. Barankova, I.I., Mikhailova, U.V., Lukyanov, G.I.: Approach to designing a network of enterprises in a protected version. Bulletin of the ural federal district. Inf. Secur. 1(27), 24–28 (2018)
14. Mikhailova, U.V., Ershov, V.A.: Methods of organization and methods of countering DOS/DDOS – attacks. In: Security of the Information Space: A Collection of Works of the XIII All-Russian Scientific and Practical Conference of Students, Postgraduates and Young Scientists, pp. 73–79 (2015)
15. Barankova, I.I., Mikhailova, U.V.: Features of the formation of evaluative means for the formation of competencies of an information security specialist. Inf. Countering Threats Terrorism 2(25), 26–30 (2015)
16. Mikhailova, U.V., Barankova, I.I., Lukyanov, G.I.: Automated control system of the ZIGBEE factory railroad transport route. In: 2nd International Conference on Industrial Engineering, Applications and Manufacturing Proceedings IEEE 16838849, p. 7910923 (2016)
17. Uriev, V.N., Erman, S.A.: Theoretical probabilistic model for assessing information security risks of an enterprise. Scientific and technical statements St Petersburg University. Econ. Sci. 4(199), 188–194 (2017)
18. Antokhina, V.A.: Management situation and risks. Scientific and technical statements of the St Petersburg University. Economics 61(185), 287–291 (2013)
19. Kozminykh, S.I., Kozminykh, P.S.: Information security audit. Bulletin of the Moscow University of the Ministry of Internal Affairs of Russia, vol. 1, pp. 181–186 (2016)
20. Miloslavskaya, N.G.: Verification and evaluation of information security management activities. Hotline - Telecom, Moscow (2012)

21. Barankova, I.I., Nosova, T.N., Permyakova, O.V.: Use of high technologies in the management of the quality of knowledge in higher education in the discipline Computer Science. Bulletin of Nosov Magnitogorsk State Technical University, vol. 3, no. 15, pp. 14–15 (2006)
22. Pantyukhina, I.S., Zikratova, I.A.: Methods of conducting postincedent internal audit of computer equipment. Sci. Tech. J. Inf. Technol. Mech. Opt. **17**(3), 467–474 (2017)
23. Barankova, I.I., Mikhailova, U.V., Lukyanov, G.I.: Company railway transport control automation. In: International Conference on Industrial Engineering, Applications and Manufacturing IEEE 17284001, p. 8076138 (2017)
24. Mikhailova, U.V., Saigushev, N.Ya., Vedeneeva, O.A., et al.: Information systems at enterprise design of secure network of enterprise. In: Journal of Physics: Conference Series, vol. 1015, p. 042054 (2018)
25. International organization for standardization: ISO/IEC 27001 Information technology. Security techniques. Information security management systems. Requirements (2017)

The Method of Automated Configuration Objects of the WinCC Project for the Oil and Gas Industry

Sh. Khuzyatov and R. Valiev[✉]

Naberezhnye Chelny Institute, Kazan Federal University,
68/19 Mira Street, Naberezhnye Chelny 423810, Russian Federation
rustvali@mail.ru

Abstract. The paper suggests method of automated configuration of WinCC project objects for the development of supervisory control system software in the case of a large number of similar sensors and actuators. The method is based on template projects with predefined structure and functionality of the software. The system implemented on the basis of the proposed method automatically configures the static and dynamic properties of the process screen objects, creates event handlers for graphic objects, as well as determines archive tags and conditions for the appearance of emergency messages. The use of independent input tables for configuration allows for the group formation and editing of all properties of project objects. The automated configuration system exempts the developer from performing routine, single-type work, and significantly reduces the development and implementation time of the software, which ultimately reduces the cost of the project. This system is used for configuration of projects of automated control systems of oil production processes. The universal nature of the proposed method makes it possible to implement it also in other industries, where the same type of sensors and actuators are widely used.

Keywords: Automation · Supervisor control · Pattern design · Configuration · Oil production

1 Introduction

In the oil and gas industry to separate oil, gas, condensates, water and various contaminants from the oil emulsion, various installations (booster pumping stations, water pre-discharge units, oil and gas separators, etc.) are used. Automation of these installations involves a number of features that will appear in the development of control system software [1, 2].

The uniformity of sensors and actuators in these installations facilitates the process of designing an automated process control system. But at the same time, the use of a large number of sensors and actuators requires a lot of time for software development [3–5]. It should be noted that information processing and data exchange over the network between the user program of the controller and the software of the human-machine interface occurs in real time [6–9].

A. A. Radionov and A. S. Karandaev (Eds.): RusAutoCon 2019, LNEE 641, pp. 986–993, 2020.
https://doi.org/10.1007/978-3-030-39225-3_105

The use of modern software development tools for controllers and SCADA systems simplifies the process of developing control system software [10–12]. Moreover, in cases where control systems differ from each other only by the number of sensors and actuators, further simplification of the process can be achieved by using the design pattern technique [13–15].

Pattern design allows to develop template projects in which the structure and functionality of the software of automated process control systems are predetermined [16, 17]. The proposed structure of template projects allows to adapt them to the requirements of a specific control system, which greatly simplifies the process of developing Supervisory control system software [18, 19].

2 Statement of the Problem

In accordance with the design pattern technique, similar project objects are accepted as elements of the lower hierarchical level, which have the same algorithms for data processing and calculation of control actions [20]. The same type of objects of automated control systems of crude oil separation plants are analog and discrete sensors, valves, pumps and compressors, gas meters, etc. [21–23].

The upper level software is a hierarchical system consisting of the main process screen (object screen) and the subordinate process screens (status screens). The objects screen contains graphic objects of typical sensors and actuators, for example, analog signals, latches, pumps (Fig. 1). On the basis of these graphical objects, a process flow diagram is created, where all the objects of the control system and their main characteristics are displayed.

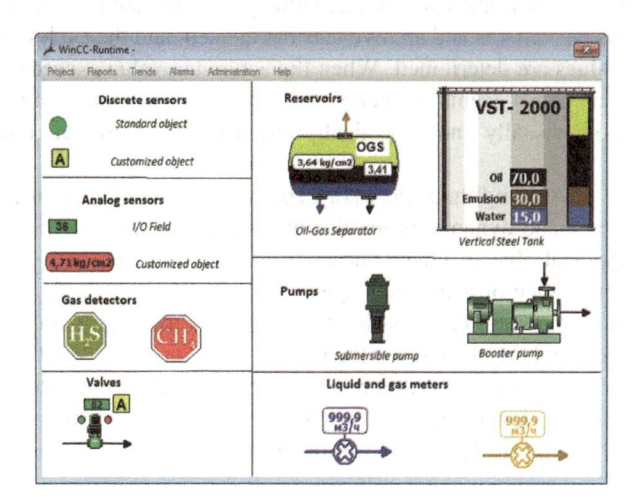

Fig. 1. The screen of typical graphic objects.

The process flow diagram is the main part of the Supervisory control system software. After designing a process flow diagram, you should configure the static and dynamic properties of its graphic objects. As an example, we present some static and dynamic properties of the analog signal pictures on the process flow diagram – the output values of the temperature, pressure, liquid level and valve position sensors (Table 1).

Table 1. Basic properties of the analog signal pictures on the process flow diagram.

Graphic object properties	Analog signals		
Name	An [1]	An [2]	An [3]
Control object	Valve	Pump	Separator
Tooltip text	Position	Pressure	Level
Unit of measure	%	atm	m
Refresh rate (ms)	200	250	1000
Archive tag	+	–	+
Control of alarm levels	–	+	+

For each type of sensor or actuator of the object screen, subordinate process screens are created - status screens that display the values of the input and output signals of the object in real time, trends in signal changes and alarm messages.

The status screen also allows to configure the parameters of the selected object in the process of commissioning, when diagnosing the process and in the supervisory control. The configuration of the status screen to display the parameters of a specific graphical object of the process flow diagram is carried out using the event handler of the graphical object, which is implemented as macros and scripts.

In addition to configuring static and dynamic properties and creating event handlers for graphic objects, archives and archive tags are created, and the cycles for polling and archiving tag values are determined. When the actual tag value exceeds or falls below the limit value an alarm events are generated. The message system indicates detected alarm events both visually and acoustically and archives them electronically and on paper.

Thus, when developing the supervisory control system software, it is necessary to configure the static and dynamic properties of the process screen objects, create event handlers of graphical objects, create and configure archive tags, as well as the conditions for the appearance of alarm messages.

The project configuration process is complicated by the large number of objects of the control system and the need to use the various components of WinCC. On the other hand, typical WinCC project objects are characterized by the same settings. Therefore, in this case, the solution to the problem is the automated configuration of objects developed on the basis of the WinCC template project.

3   Results and Discussion

The process flow diagram of the crude oil separation process is created by placing graphical objects on the main screen of the process and connecting them to each other by pipelines. These graphical objects for typical sensors and actuators are contained in the screen objects of the WinCC template project.

The names of the same type of graphic objects are formed in the form of array elements. This allows to establish a correspondence between the process flow diagram and the program code, and to configure graphical objects using a loop in the program. For the process under consideration, the names of the process screen objects can be designated as An[1], An[2], An[3] for analog signals, Valve [1], Valve [2], Valve [3] for valves, M[1], M[2], M[3] for pump motors, etc.

The transfer of the characteristics of sensors and actuators to the program cycle is organized using property tables. A separate table is created for each type of graphic object, since they are characterized by their own list of properties and event handlers. Saving the characteristics of all the same type of project objects in a single table also provides an additional opportunity to use these tables when drawing up the project specification and ordering project components.

At the initial stage, a correspondence is established between the process flow diagram and the tables by the names of graphic objects based on filtering the objects of the collection ActiveDocument.HMIObjects. This collection contains pointers to all graphic objects of the current screen of the process and provides programmatic access to them.

At the next stage, these tables are filled with the values of static and dynamic properties of graphic objects. Based on the established correspondence, the reverse assignment of the values of static properties to graphic objects of the process flow diagram is performed. To create dynamic properties of a graphic object, the method CreateDynamic of a graphic object and then determine the frequency of updating the property value is used. To define an event handler for a graphic object, first an event from the collection Events is selected and then the method AddAction to add a VB script or C script to the Actions event handler collection is used.

Configuring static and dynamic properties, as well as defining event handlers are implemented by separate modules of the configurator (Fig. 2).

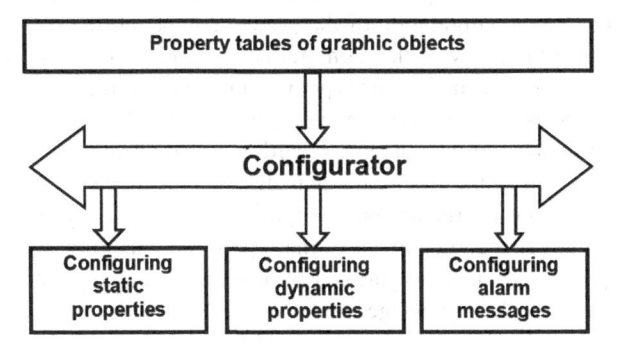

Fig. 2. Configuring the same type of graphical object of the main process screen.

On the other hand, in addition to configuring static and dynamic properties, defining event handlers for graphic objects, archives and archive tags are created and configured, as well as conditions for the appearance of alarm messages. Archive tags and alarm messages are configured using the Tag Logging and Alarm Logging editors. Functions for accessing the objects of these editors are contained in the HMIGO class.

As in the case of graphic objects, the list of archive tags and conditions for the appearance of alarm messages are formed in the form of separate tables. The filled tables are further processed by the corresponding modules of the Configurator, which leads to the creation and configuration of alarm tags and alarm conditions.

4 Practical Use

The proposed method for automated configuration of the WinCC project objects is implemented as an application. It is developed using the built-in VBA programming language, which allows to extend the functionality of WinCC and automate configuring when creating industry-specific projects. The input data of this application are tables that contain properties of the same type objects of the WinCC project.

Compliance with the rules of this method ensures the correct operation of the developed Configurator. According to the rules of this WinCC project configuration system, a process screen is first created in the form of a process flow diagram. The Graphics Designer of the WinCC system for creating process screens provides various toolbars that allow you to implement the following types of tasks: displaying static objects and operator-controllable objects, such as texts, graphics or buttons; creating a dynamic objects, e.g. modifies the length of a bar graph in relation to a process value; creating objects for operator input, e.g. the clicking of a button, or the entry of a text in an input field.

This configuration step requires the developer to present the process objects in a convenient form for viewing. All design work associated with the creation of process flow diagrams is performed in an interactive mode. The same type of graphic objects are assigned the same type of names, such as array elements.

In this implementation of the VBA application, the data of the same type of graphics objects is generated in Excel format, which contain the names of these graphics objects. Then, in Excel, these tables are populated with the corresponding characteristics of the WinCC project objects. After that, the VBA application is run again for step-by-step configuration of the project components.

The three-level hierarchy of the Configurator windows allows to perform the following types of work: selection of the type of configurable project objects; selection of typical project objects for configuration; creating property tables for project objects and their filling; configuring project objects according to the tables.

The user interface of the Configurator is designed as a "guide" for the developer, which is a step-by-step instruction on configuring the project objects. The main window of the Configurator allows to select the type of actions to be performed: configuration of graphical objects, configuration of archive tags, configuration of conditions for the appearance of alarm messages.

Configuration of graphic objects of the main process screen is carried out by types of project objects. For automated control systems of the oil production processes, such typical objects of the project are discrete signals from gas contamination sensors and limit switches of valves, analog signals from temperature, pressure, level sensors, as well as pumps, valves, intelligent liquid and gas flow sensors.

The choice of types of graphic objects of the project is carried out in the window of the second level of the hierarchy of the Configurator. After selecting a specific type of graphic object, the system goes to the third level window, which allows to configure the static and dynamic properties of the graphic object, as well as to determine the event handlers according to the property tables.

To create archives and archive tags, the TagLogging editor of the WinCC system is used. To automatically configure objects of this TagLogging system, the Configurator also creates a property tables, which are then populated in Excel. Archives and archive tags are also created and configured based on the property tables. The configuration parameters of all archive tags are located in a single table, which allows group generation and editing of all archive tags.

To create and configure alarm messages, the AlarmLogging editor of the WinCC system is used. During alarm messages configuration, the text and class of the message, the acknowledgement model and the archiving of the message are determined. Messages are intended to alert the installation operator of non-periodic events occurring in the process. The alarm configuration module of the Configurator allows you to configure these properties of the AlarmLogging system objects based on the alarm properties table.

The use of independent tables for configuring project objects has another advantage: when changing the composition of the project, it is enough to load a table with new data through the Configurator. Fixing the date and time in the project allows to control these changes.

5 Conclusion

The method of automated configuration of the WinCC project objects is advisable in the development of Supervisory control system software, where a large number of similar sensors and actuators are used. In this case, template projects are used in which the structure and functionality of the software are predetermined.

The Configurator, developed on the basis of the proposed method, configures the static and dynamic properties of objects on the process screen, creates event handlers for graphic objects, as well as archive tags and conditions for the appearance of alarm messages.

The advantage of the proposed method is also the use of independent tables for configuration, which allows group formation and editing of all properties of the project objects.

The use of this automated configuration system exempts the developer from performing routine tasks of the same type, can significantly reduce the time of development and implementation of the automated process control system software, and requires less development skills, which ultimately reduces the cost of the project.

This configuration system is used for automated configuration of the Supervisory control systems software of oil production facilities. The proposed method is universal, which allows its use in the automation of similar objects of the enterprises of another industry.

References

1. Tynchenko, V.S., Kukartsev, V.V., Tynchenko, V.V., et al.: Automation of monitoring and management of conveyor shop oil-pumping station of coal industry enterprise. IOP Conf. Ser.: Earth Environ. Sci. **194**(2). https://doi.org/10.1088/1755-1315/194/2/022044
2. Bukhtoyarov, V.V., Tynchenko, V.S., Petrovskiy, E.A., et al.: Improvement of the methodology for determining reliability indicators of oil and gas equipment. Int. Rev. Model. Simul. **11**(1), 37–50 (2018). https://doi.org/10.15866/iremos.v11i1.13994
3. Ibragimov, N.G., Zabbarov, R.G., Idiatova, V.R.: Operational supervision and oil production control system based on monitoring of telemetry-controlled well performance in ARMITS corporate information system. Oil Ind. **4**, 106–109 (2014)
4. Sobolev, S.A., Fattakhov, R.B.: Coordination of booster pump stations operating modes. Oil Ind. **6**, 122–125 (2013)
5. Khaziev, E.L., Khaziev, M.L.: Intelligent diagnostic system for hydraulic actuator. In: International Conference on Industrial Engineering, Applications and Manufacturing, p. 8742779 (2019)
6. Khuzyatov, S.S., Valiev, R.A.: Organization of data exchange through the Modbus network between the SIMATIC S7 PLC and field devices. In: Proceedings of International Conference on Industrial Engineering, Applications and Manufacturing, pp. 1–3 (2017). https://doi.org/10.1109/icieam.2017.8076369
7. Tcaciuc, S.A.: A solution for the uniform integration of field devices in an industrial supervisory control and data acquisition system. Int. J. Adv. Comput. Sci. Appl. **9**(3), 319–323 (2018)
8. Tcaciuc, S.A.: Performances analysis of a SCADA architecture for industrial processes. Int. J. Adv. Comput. Sci. Appl. **8**(11), 456–460 (2017). https://doi.org/10.14569/ijacsa.2017.081155
9. Biktimirov, R.L., Valiev, R.A., Galiullin, L.A., et al.: Automated test system of diesel engines based on fuzzy neural network. Res. J. Appl. Sci. **9**(12), 1059–1063 (2014)
10. Chen, X., et al.: PLC and configuration software based supervisory and control system for oil tanks area. In: 3rd International Conference on Power Electronics Systems and Applications, p. 83, May 2009
11. Butta, R.: An overview of oil drilling and production monitoring system using SCADA automation in oil and natural gas corporation Ltd. In: International Conference on Electrical, Electronics, Signals, Communication and Optimization, pp. 1–4, September 2015. https://doi.org/10.1109/eesco.2015.7253920
12. Reeser, J., Jankowski, T., Kemper, G.M.: Maintaining HMI and SCADA systems through computer virtualization. IEEE Trans. Ind. Appl. **51**(3), 2558–2564 (2015). https://doi.org/10.1109/TIA.2014.2384132
13. Khalil, M.: Pattern libraries guiding the model-based reuse of automotive solutions. In: Khendek, F., Gotzhein, R. (eds.) System Analysis and Modeling. Languages, Methods, and Tools for Systems Engineering. LNCS, vol. 11150. Springer, Cham (2018)
14. Khuzyatov, S.S., Valiev, R.A.: Designing of automated control systems based on pattern methods. Sci. Tech. Bull. Volga Region **2**, 215–218 (2015)

15. Valiev, R.A., Khuzyatov, S.S.: Pattern-design software of automated control systems. In: 2nd International Conference on Industrial Engineering, Applications and Manufacturing, May 2016. https://doi.org/10.1109/icieam2016.7910942

16. Iliukhin, A.N., Khuzyatov, S.S., Valiev, R.A.: Unified approach to software development of automated control systems for oil equipment. HELIX 8(1), 2455–2459 (2017). https://doi.org/10.29042/2018-2455-2459

17. Zubkov, E.V.: Computer modeling of the automated tests of diesel engines various conditions of their operation. In: Proceedings of 2nd International Conference on Industrial Engineering, Applications and Manufacturing, p. 7911586. https://doi.org/10.1109/ICIEAM.2016.7911586. ISBN: 9781509013227

18. Iliukhin, A.N., Khuzyatov, S.S., Valiev, R.A.: Methodology for the development of a dispatch control system for oil production facilities. Sci. Tech. Bull. Volga Region 5, 215–218 (2018)

19. Belousov, A., et al.: An approach of lower-level communication line implementation of automated dispatch control systems of distributed facilities. In: International Multidisciplinary Scientific GeoConference Surveying, Geology and Mining Ecology Management, vol. 17, no. (21), pp. 183–190. https://doi.org/10.5593/sgem2017/21/s07.024

20. Berger, H.: Automating with Simatic: Controllers, Software, Programming, Data Communication, Operator Control and Process Monitoring. Publicis Publishing, Erlangen (2013)

21. Wang, Q.C., Hu, L.K.: Research on WinCC-based SCADA software for acrylic fibres filature. Control Instr. Chem. Ind. 1, 35–38 (2006)

22. Liao, R., Chan, C.W., Huang, G.G.: A fuzzy logic controller for an oil separation process. In: Canadian Conference on Electrical and Computer Engineering, pp. 1045–1048 (2005). https://doi.org/10.1109/CCECE.2005.1557155

23. Liao, R., Chan, C.W., Hromek, J., et al.: Fuzzy logic control for a petroleum separation process. Eng. Appl. Artif. Intell. 21, 835–845 (2008). https://doi.org/10.1016/j.engappai.2007.09.006

Optimization the Process of Catalytic Cracking Using Artificial Neural Networks

E. Muravyova$^{(\boxtimes)}$

Ufa State Petroleum Technological University,
2 Oktyabrya Avenue, Sterlitamak 453120, Russian Federation
muraveva_ea@mail.ru

Abstract. Industrial production is one of the promising areas of application of artificial neural networks (ANN). There is a tangible trend towards manufacturing modules with a high level of automation in this area, which requires an increase in the number of intelligent self-regulating and self-adjusting objects. However, industrial processes are characterized by a large variety of dynamically interacting parameters, which complicate the creation of adequate analytical models. Modern industrial production is constantly becoming more complicated. This slows down the introduction of new technological solutions. In this regard, there is an increasing interest in alternative approaches to modeling industrial processes using ANN, which provide the possibility to create models that operate in real time with small errors that can be trained in the process of use. The advantages of neural networks make their use attractive for solving problems such as: forecasting, planning, designing of automated control systems, quality management, manipulator and robotics management, process safety management: fault detection and emergency situations prevention, process management: optimization of industrial process regimes, monitoring and visualization of supervisory reports. Neural networks can be useful in industrial production, for example, when creating an enterprise risk management model, planning a production cycle. Modeling and optimization of production is characterized by high complexity, a large number of variables and constants, defined not for all possible systems. Traditional analytical models can often be built only with considerable simplification, and they mostly have evaluative nature. While the ANN is trained on the basis of data from a real or numerical experiment.

Keywords: Catalytic cracking · Artificial neural network · Neuron · Gasoline fraction

1 Introduction

The object of research in this article is the process of catalytic cracking. The influence of the process parameters on the final yield of the NK-200 gasoline fraction and the possibility of efficiency upgrading with the help of a new and promising direction-artificial neural networks is studied [1].

Catalysis is an exceptionally efficient method of implementing chemical transformations in the industry. Currently, up to 90% of all chemical products in the world are

A. A. Radionov and A. S. Karandaev (Eds.): RusAutoCon 2019, LNEE 641, pp. 994–1004, 2020.
https://doi.org/10.1007/978-3-030-39225-3_106

manufactured by catalytic methods. The technical progress of chemical, petrochemical, oil refining and other industries largely depends on the development of catalysis. The process of catalytic cracking is one of the most common large-scale processes of deep oil processing and largely determines the technical and economic performance of modern and prospective refineries.

Table 1. Mass fraction of gasoline fraction NK-200 at the output, depending on process conditions.

Thermolysis conditions		Content of fractions at the output, % mass
Temperature, °C	Duration, min	NK-200
400	60	0
450	40	0,6
450	60	1,2
450	80	2,4
450	100	7,7
500	60	12,3
500	80	12,6
500	100	17,5

The main purpose of catalytic cracking is production with the highest possible yield (up to 50% or more) of high-octane gasoline, light gas oil and unsaturated fatty gases. The following factors affect the yield and quality of cracking products: the type of raw materials, the composition and activity of the catalyst, temperature and pressure of the process, the volumetric feed rate to the reactor and the duration of continuous cracking without catalyst regeneration. When carrying out catalytic cracking, catalysts of different composition and method of preparation are used. The direction of the chemical transformation of the fuel depends on the quality of the catalyst used, as well as on the technological regime of the cracking process. To improve this process, it is necessary to more fully study the influence of process parameters on the quality indicators of the final product. This task is complicated by the fact that a complex mixture of hydrocarbons is used as the raw material, as well as by a multitude of reactions taking place in the catalytic cracking process. The use of modeling using ANN will instantly reproduce complex dependencies of the catalytic cracking process [2, 3].

Control and regulation of input parameters in multidimensional technological processes is the most important part of the process, since the final quality of the products, efficiency, amount of waste, etc. depend on this.

In this paper, the process of producing gasoline fraction NK-200 under catalytic cracking is considered. In order to achieve the maximum output of cracking, it is necessary to select the optimum thermolysis conditions. It follows that for the neural network, the input parameters are the temperature and duration of the treatment, and the output parameters are the content of the gasoline fraction NK-200 at the output (Table 1).

2 Experimental Results and Discussion

The Neural Network Toolbox of the MATLAB environment has annstart [4, 5] tool that allows to design an artificial neural network in a graphical environment. You need to type nnstart in command line mode to start (Fig. 1).

You should select one of the available types of neural network in the starting panel of the Neural Network tool. In our case this is "input output and curve fitting" (Fig. 2).

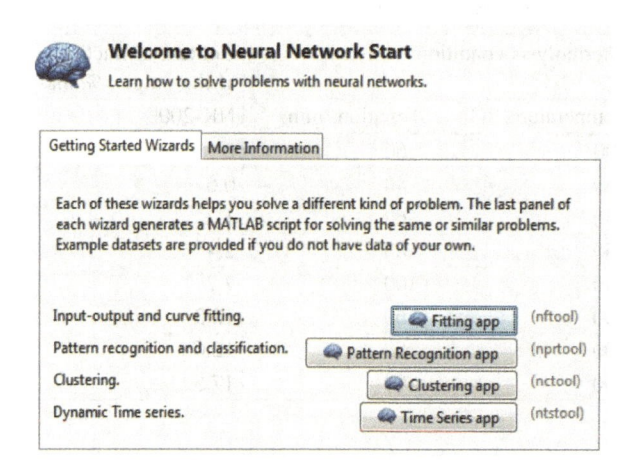

Fig. 1. Annstart tool.

The Network Architecture panel (Fig. 3) presents a standard network used for approximation and is a two-layer direct propagation network with sigmoid activation function in the first layer and a linear activation function in the output layer [6, 7]. By setting Number of Hidden Neurons parameter, we set the number of neurons in the hidden layer to 3, this value was chosen empirically and corresponds to the maximum accuracy of the output data. The number of neurons in the output layer is 1.

The fitting panel (Fig. 4) indicates that a two-layer feed-forward network with sigmoid hidden neurons and a linear output neurons will be used. Such a network allows you to arbitrarily accurately solve the problem of multidimensional approximation, provided that the data is consistent and a sufficient number of neurons in the hidden layer. The network will be trained by the method of back propagation with Levenberg-Marquardt algorithm [7, 8].

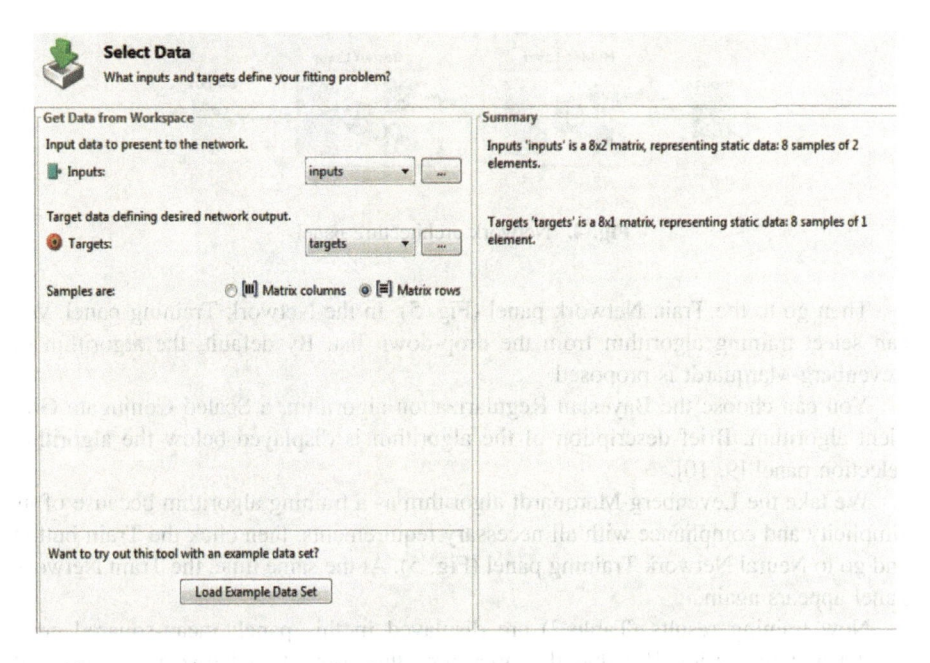

Fig. 2. Input and targets data.

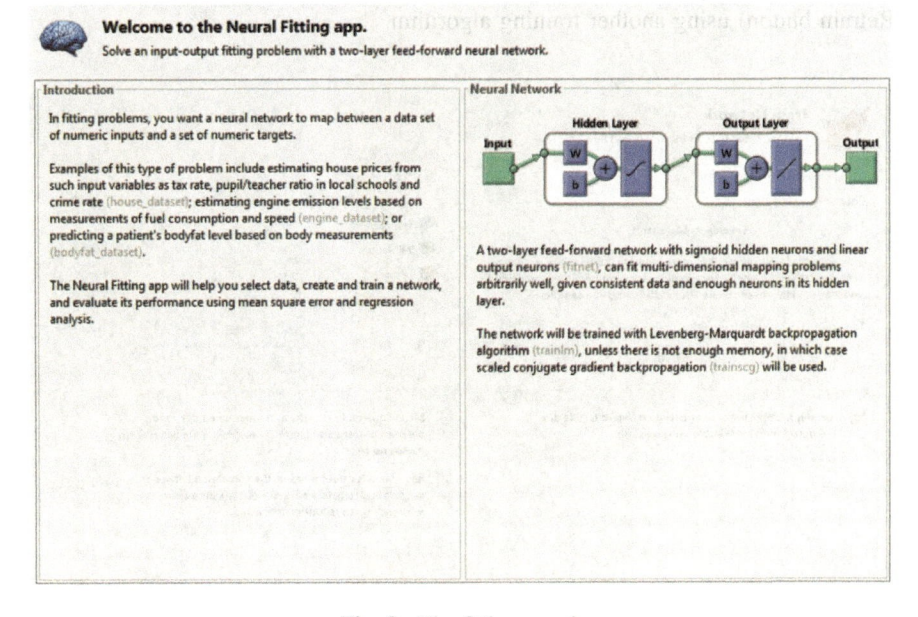

Fig. 3. The fitting panel.

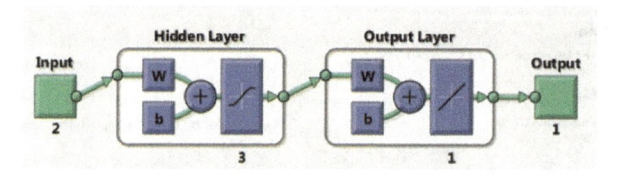

Fig. 4. Network architecture panel.

Then go to the Train Network panel (Fig. 5). In the Network Training panel, you can select training algorithm from the drop-down list. By default, the algorithm of Levenberg-Marquardt is proposed.

You can choose the Bayesian Regularization algorithm, a Scaled Conjugate Gradient algorithm. Brief description of the algorithm is displayed below the algorithm selection panel [9, 10].

We take the Levenberg-Marquardt algorithm as a training algorithm because of its simplicity and compliance with all necessary requirements, then click the Train button and go to Neural Network Training panel (Fig. 5). At the same time, the Train Network panel appears again.

Now training results (Table 2) are displayed in this panel: mean-squared error (MSE) and regression R value that measures the correlation between outputs and targets (for the training, validation and testing sets) [11, 12]. You can display plots characterizing the quality of training in this panel (Fig. 6). You can repeat the training (Retrain button) using another training algorithm.

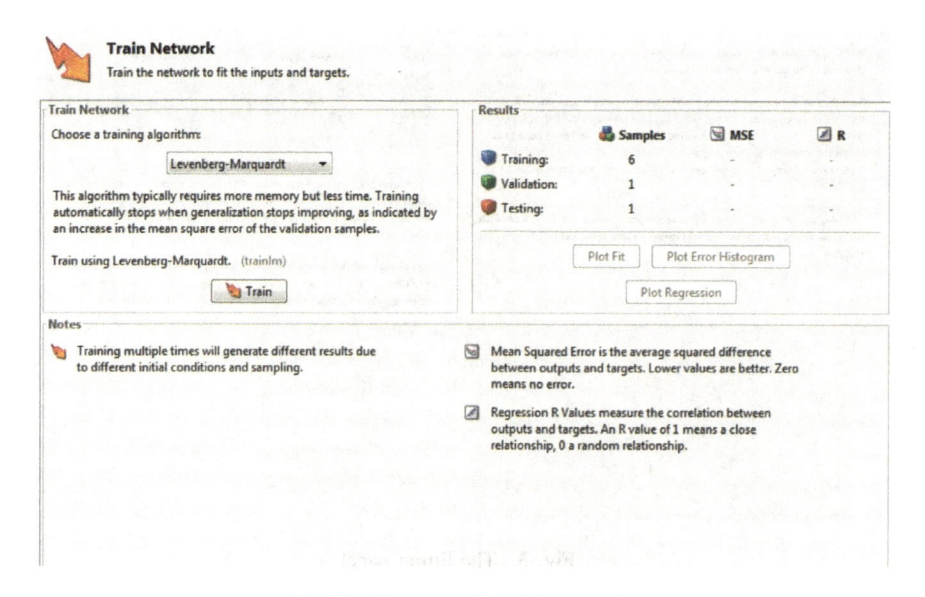

Fig. 5. The network training panel.

Table 2. Train panel after completion of the training process of NW.

	Samples	MSE	R
Training	6	2.90105·e−25	1
Validation	1	6.91368·e−1	0
Testing	1	218.78074	0

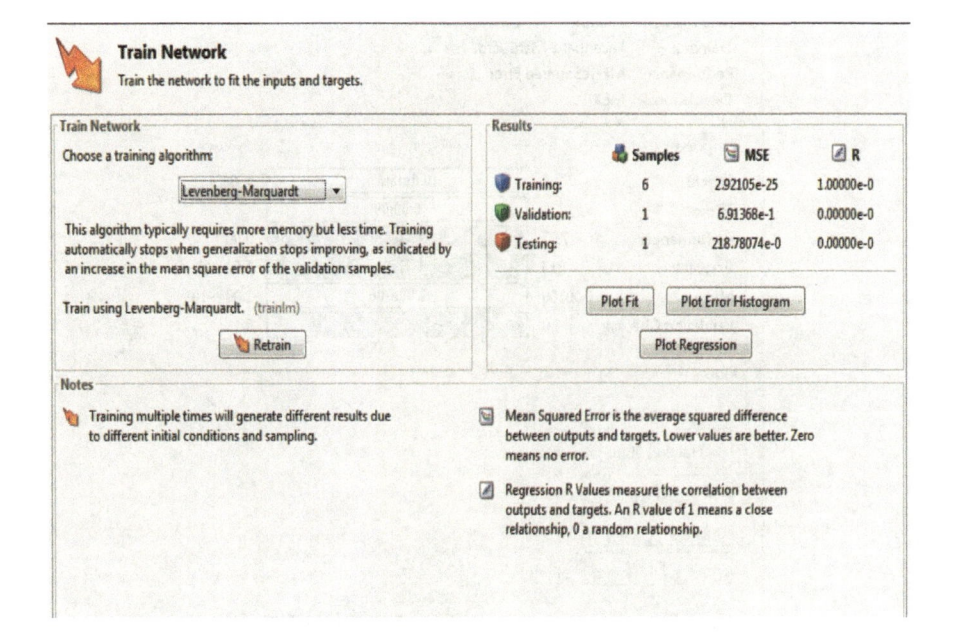

Fig. 6. The network training plot.

In the neural network training process panel the Performance button (Fig. 7) shows the network training plot (Fig. 8), performing the behavior of the training error. It can be seen from the plot that for 5 epochs the mean square error 0.002199 has been reached. Training function uses training with early stoppage as a means of combating retraining. It can be seen from the plot that training is stopped when the error on the test set has ceased to decrease [13].

Training State plots are shown in Fig. 9. Gradient plot shows the change in the gradient of the training error function by the network weights. Mu plot reflects the variation of the regularization parameter (m) of the Levenberg-Marquardt method designed to add additional information to the condition in order to solve an incorrectly posed problem or prevent retraining [14]. Val fail plot shows the change in the error on the control set. It can be seen that after the epoch 5 the error begins to grow.

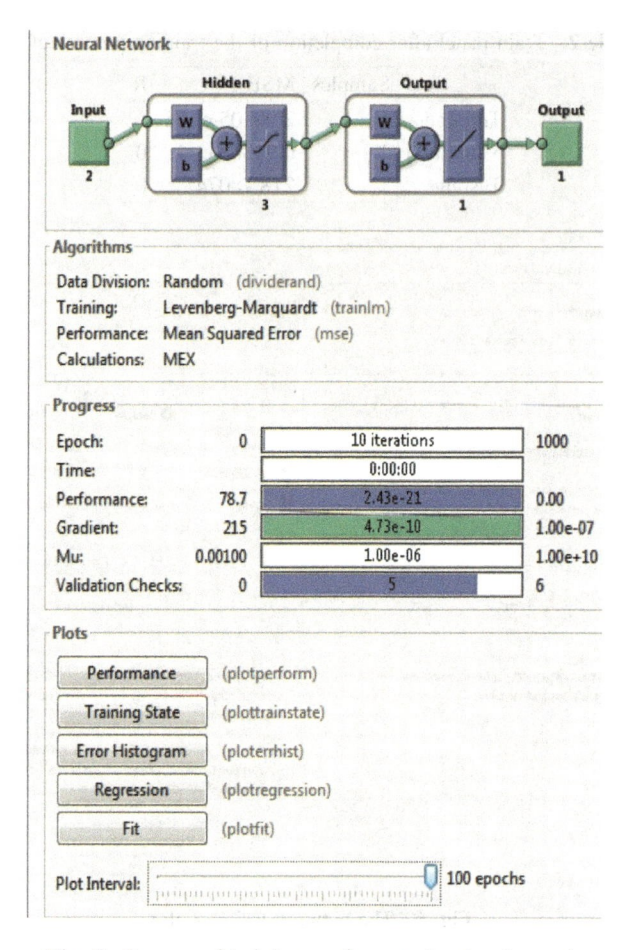

Fig. 7. Process of training on the neural network panel.

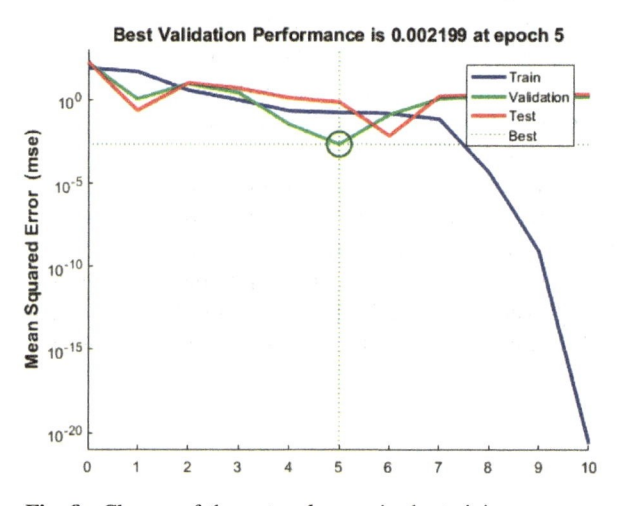

Fig. 8. Change of the network error in the training process.

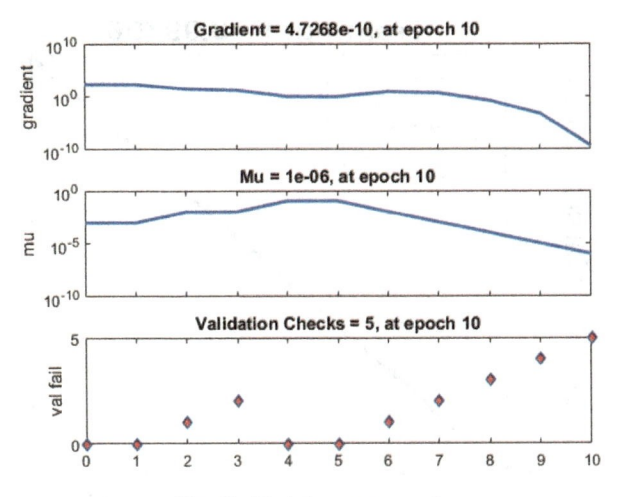

Fig. 9. Training state panel.

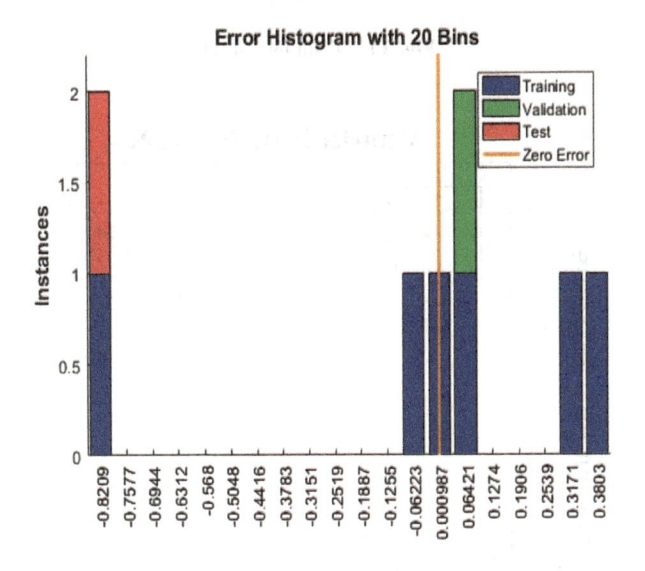

Fig. 10. Error histogram.

Error Histogram (Fig. 10) shows on how many instances the network gives this or that error. Error is calculated as the difference between target value and network output. The plot shows errors for the training, validation and testing sets [15]. It can be seen from the histogram that most of errors lie in the range from −0.06223 to 0.0641.

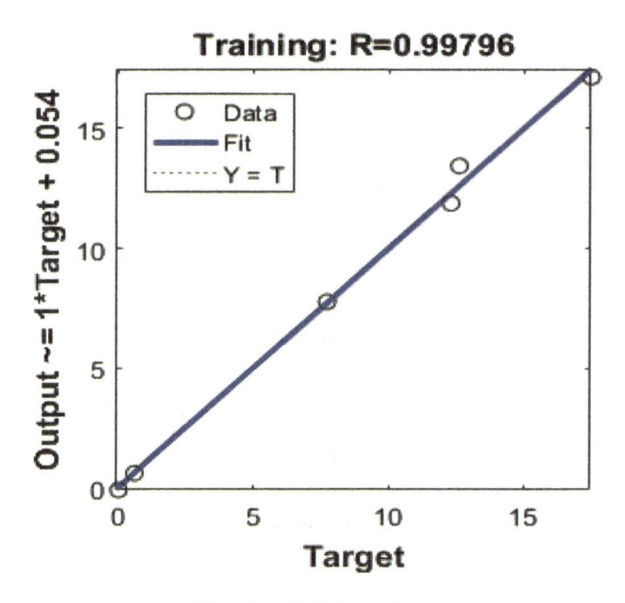

Fig. 11. Training subset.

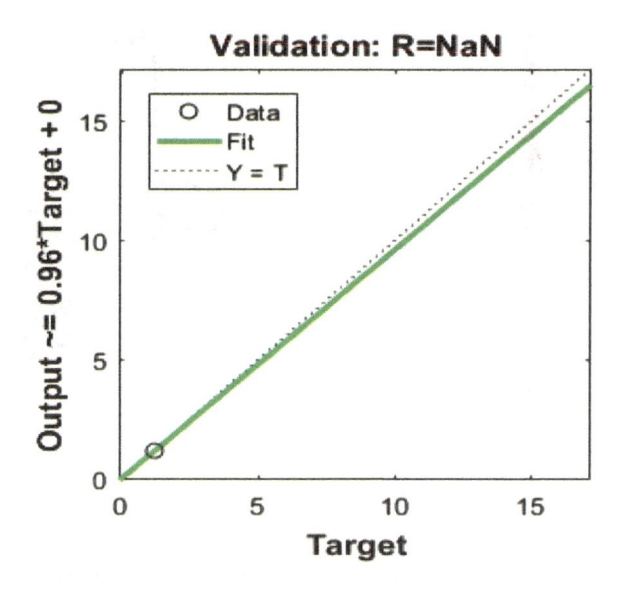

Fig. 12. Validation subset.

The Regression plot shows a linear regression of training outputs of the network on the three considered subsets: training (Fig. 11), validation (Fig. 12) and test (Fig. 13) and on all (Fig. 14) the sets [16–20]. For each result, the correlation coefficient R is calculated, a plot is constructed and the regression equation is derived as

Output = $a \times$ Target + b (Fig. 9). When the outputs of the network coincide with the target values $R = 1$, $a = 0$, $b = 0$. Figure 8 shows that network almost perfectly approximates the function.

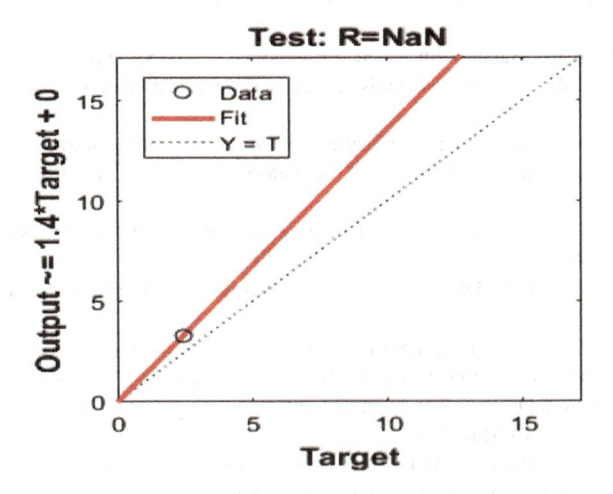

Fig. 13. Test subset.

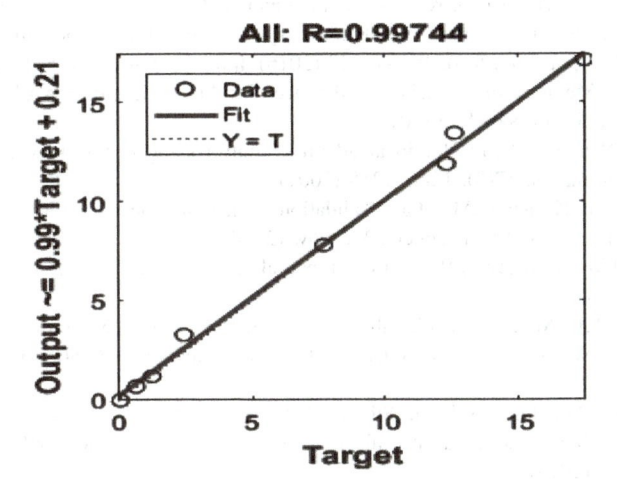

Fig. 14. All subsets.

3   Conclusion

A neural network was created simulating the process of catalytic cracking and predicting the yield of gasoline fraction NK-200, depending on conditions of thermolysis: temperature and duration of process. As a result of the tests carried out, we can conclude that the model is adequate, since the values obtained are in agreement with the experimental data.

References

1. Muravyova, E., Sagdatullin, A., Sharipov, M.: Modelling of fuzzy control modes for the automated pumping station of the oil and gas transportation system. IOP Conf. Ser. Mater. Sci. Eng. **132**(1), 012028 (2016)
2. Muravyova, E.A., Sagdatullin, A.M., Emekeev, A.A.: System-integrative approach to automation of the oil and gas fields design and development control. Oil Ind. **3**, 92–95 (2015)
3. Spitsyn, V.G., Tsoi, Yu.R: Application of artificial neural networks for information processing: methodical instructions to laboratory works. TPU Publishing House, Tomsk (2007)
4. Burkov, M.V.: Neural networks and neurocontrollers: Textbook. Allowance, GUAP, St. Petersburg (2013)
5. Akhmetov, B.S., Gorbachenko, V.I.: Neural networks. Laboratory practical work, KazSTU, Almaty (2015)
6. Muravyova, E.A.: Two fuzzy controller synthesis methods with the double base of rules: reference points and training using. In: Industrial Engineering, Applications and Manufacturing. International Conference on Industrial Engineering, Applications and Manufacturing, St. Petersburg, 16–19 May 2017
7. Muravyova, E.A., Sharipov, M.I.: Method of fuzzy controller adaptation. In: Proceedings of the International Conference Actual Issues of Mechanical Engineering, November 2017. https://doi.org/10.2991/aime-17.2017.82
8. Amarendra, Ch., Harinadha, K.: Investigation and analysis of space vector modulation with matrix converter. Determined based on fuzzy C-means tuned modulation indexes. Int. J. Electr. Comput. Eng. **6**(5), 1939–1947 (2016). https://doi.org/10.11591/ijece.v6i5.11378
9. Bonato, J.: Methods of artificial intelligence. In: Fuzzy Logic. DAAAM International Scientific Book, pp. 849–856 (2013)
10. Chen, L., Narendra, K.S.: Nonlinear adaptive control using neural networks and multiple models. Automation **37**(8), 1245–1255 (2001)
11. Chen, M., Qin, K., Ku, H.M., et al.: Validation at the system level. High-level modeling and testing management. Technosphere, Moscow (2014)
12. Eremin, D.M., Garceev, I.B.: Artificial neural networks in intelligent control systems. MIREA, Moscow (2004)
13. Galushkin, A.I.: Neural networks: the basis of the theory. ReS, Moscow (2015)
14. Haykin, S.: Neural Networks: A Comprehensive Foundation. Prentice Hall PTR, Upper Saddle River (2006)
15. Maltarollo, V.G., Honorio, K.M., Silva, A.B.F.: Application of artificial neural networks in chemical problems, artificial neural networks. Architectures and applications, InTech, pp. 203–223 (2013)
16. Pegat, A., Pendvesovsky, A.G., Tyumentsev, Yu.V.: Fuzzy modeling and control. Binom, Moscow (2013)
17. Tadeusewicz, R.: An elementary introduction to the technology of neural networks with examples of programs. Peter, St. Petersburg (2011)
18. Wang, F.Y., Bahri, P.L., Lee, I.T., et al.: A multiple model, state feedback strategy for robust control of non-linear processes. Comput. Chem. Eng. **31**(5), 410–418 (2007)
19. Muravyova, E., Sagdatullin, A., Emekeev, A.: Intellectual control of oil and gas transportation system by multidimensional fuzzy controllers with precise terms. Appl. Mech. Mater. **756C**, 633 (2015)
20. Muravyova, E., Bondarev, A., Kadyrov, R., et al.: The analysis of opportunities of construction and use of avionic systems based on cots-modules. ARPN J. Eng. Appl. Sci. **11**, 78–92 (2016)

Simulation of a Multi-connected Process in iThink Program

E. Muravyova$^{(\boxtimes)}$ and Y. Stolpovskaya

Ufa State Petroleum Technological University,
2 Oktyabrya Avenue, Sterlitamak 453120, Russian Federation
`muraveva_ea@mail.ru`

Abstract. Quite often, multiply connected processes with parameters that affect each other can be observed in operation at many industrial facilities. In the article, the item to be automated is a reactor for the synthesis of a ready product, in which there are parameters that have a mutual influence on each other. To build an effective control system, the use of classical PID controller for this object is not possible. In the course of the research, a control system for the electrical heating of the reactor was developed, taking into account the mutual influence of the control loops on each other. The article describes the development of a control system for electrical heating of the reactor with the joint use of a fuzzy controller and neural networks in the Matlab environment to compensate for the mutual influence of control loops. The paper presents a software implementation of the reactor electric heating control system, which can, depending on the given input parameters, monitor and control the reactor electric heating process using neural networks, displaying the dynamics in form of parameters with respect to time. The control system can be implemented for reactors of different applications. It is also possible to use similar solutions for different kinds of complex objects.

Keywords: Simulation · Reactor · Electric heating · iThink program · Temperature

1 Introduction

The simulation model of electric heating control system of a reactor which allows to simulate the process of electric heating of a reactor depending on the given input parameters, with displaying the dynamics of work in the form of time-based diagrams is presented in this paper.

The model is developed in iThink simulation modeling package using cognitive maps. It includes 2 submodels: "Upper part of the reactor" and "Lower part of the reactor". The temperature control of the heating element and the temperature control of the product in the reactor were simulated using set values of the heating element power and the heating rate.

A. A. Radionov and A. S. Karandaev (Eds.): RusAutoCon 2019, LNEE 641, pp. 1005–1018, 2020.
https://doi.org/10.1007/978-3-030-39225-3_107

2   Problem Definition

At present, in the technology of automation of complex objects and processes, multidimensional systems with mutually influencing of controlled parameters are most often used [1]. Multidimensional systems are characterized by a number of specific features, the main of which is a significant mutual influence of the control loops while maintaining the values of process parameters in the required range. In particular, in the control system of electric heating there is a mutual influence of the control loops on each other. In this article, we consider the development of a model for the control system for the electric heating of a reactor with compensation of the mutual influence of control loops using the software product for modeling iThink [2–5].

3   Description of Technological Process of Product Synthesis

A chemical substance "Polytryl" production process served as a process for simulation. The reactor was loaded with all the necessary ingredients.

In the course of the study a model of electric heating of the reactor with the synthesis of the product has been developed. It includes the following stages:

- Synthesis process of the chemical substance "Polytryl"
- Precipitation and separation from morpholine by filtration
- Sediment dissolution and decontamination
- Recovery of the finished product

The raw materials required for the production of the "Polytryl" chemical substance include morpholine, N-Et-CIK (semi-finished product), caustic soda and acetic acid.

During the synthesis of the product, morpholine is loaded into the first reactor. Then dry N-Et-CIK is loaded with mixer being in operation. The temperature is brought to 108–110 °C using electric heater of the reactor and the synthesis process begins. After 10–12 h the temperature of the obtained mass decreases to 60–80 °C, and it is drained into the second reactor.

To separate the sediment, cold water is supplied and the mixer of the second reactor starts. After that, the formed suspension is drained from the reactor to the nutsch filter and filtration takes place. The obtained filtrate is directed for recovery, and the sediment is discharged for subsequent dissolution and decontamination.

After that, water, sediment and caustic soda are mixed in the reactor. The sediment dissolves at a temperature 55–60 °C within an hour. Then the mass cooled to 20–30 °C is drained onto the nutsch filter. The obtained filtrate is directed to separation the finished product.

Then the filtrate is pumped into the reactor. With operating mixer the mass is cooled to 15–20 °C, the process of precipitation of Polytryl takes place. Then acetic acid is poured in portions to acidify the medium. When the pH of the medium reaches a value of 6 the process of Polytryl separation is considered to be complete.

In Fig. 1, it is shown that two heating elements installed in the lower and upper parts of the reactor are used. Temperature sensors of the heating element and of the product are installed for both of them. Therefore lower and upper parts of the reactor are introduced in the model.

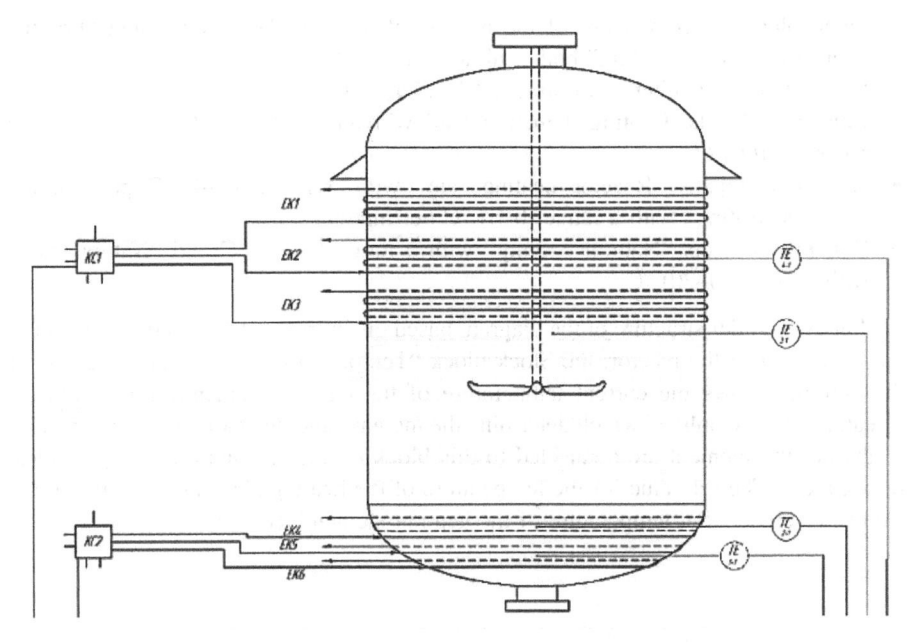

Fig. 1. Reactor.

4 "Lower Part of the Reactor" Model Development

The initial model of the lower part of the reactor is shown in Fig. 2.

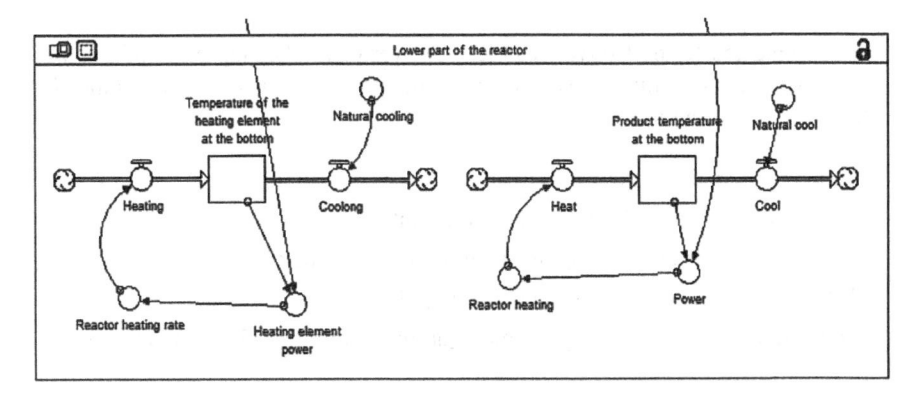

Fig. 2. "Lower part of the reactor" model with entered data.

This model is developed using the Sector frame object. In order to display 2 groups of heating elements, the model is divided into 2 sectors: "Lower part of the reactor" and "Upper part of the reactor".

The "Lower reactor of the reactor" model is developed on the Model tab. The following elements were used to develop it:

- Stock blocks "Temperature of the heating element at the bottom" and "Product temperature at the bottom" both with a value of 20 °C
- Flow objects "Heating", "Cooling", "Heat" and "Cool"
- Converter objects "Heating element power" with a range (0–100)%, "Power" with a range (0–100)%
- Converter objects "Reactor heating rate" with a range (0–4) °C per minute, "Reactor heating" with a range (0–4) °C per minute;
- Converter objects "Natural cooling" with a range (0–20) °C and "Natural cool" with a range (0–20) °C

Let us consider structure of the diagram based on example of the heating element in the lower part of the reactor: the Stock block "Temperature of the heating element at the bottom" shows the current temperature of the heating element. "Flow" objects "Heating" and "Cooling" which determine the increase and decrease of the temperature of the heating element are connected to this block on the left and on the right sides, respectively. The set value for the temperature of the heating element is entered in the Converter object "Set temperature of the heating element" (Fig. 3).

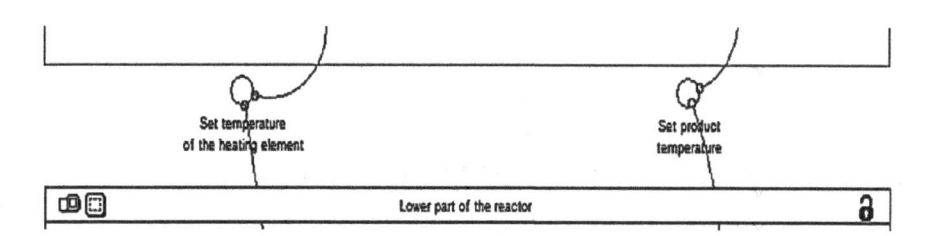

Fig. 3. Set temperature of the heating element and set product temperature.

It is connected to the Converter object "Power of the heating element", in which this parameter can be adjusted. The power of the heating element is given in the object in the form of a formula:

Heating_element_power =

IF(Set_temperature_of_the_heating_element−

Temperature_of_the_heating_element_at_the_bottom < 13)   (1)

THEN((Set_temperature_of_the_heating_element−

Temperature_of_the_heating_element_at_the_bottom + 12)/25)ELSE(1)

This object is connected to the Converter object "Reactor heating rate" to control the heating rate of the reactor. The power of the heating element and the heating rate of the reactor are set in these objects as formulas:

$$\text{Reactor\_heating\_rate} = 4 \cdot \text{Heating\_element\_power} \qquad (2)$$

Since the mass should be heated gradually within an hour, the heating rate of 4 ° C/min is set in the formula (2).

The formed chain of the Converter objects "Set temperature of the heating element", "Heating element power" and "Reactor heating rate" (Figs. 2 and 3) is designed to read data on temperature increase of the heating element so the last object of this chain is connected to the Flow object "Heating".

The Converter object "Natural cooling" which is connected to the Flow object "Cooling" is responsible for the free cooling of the heating element. It is set by a dependence of temperature on time (Fig. 4) as it will vary depending on the disturbances in this case presented by the ambient temperature.

Similarly, the product temperature control loop is arranged.

To produce Polytryl, the heating element should be heated to 275 °C, and the initial mass to 150 °C. The set temperature of the heating element and set product temperature are input as constants in Converter objects "Set temperature of the heating element" with a value of 275 °C and "Set product temperature" with a value of 150 °C, respectively. These quantities on the diagram are located outside of the "Lower part of the reactor" block since these temperatures are maintained throughout the reactor (Fig. 3).

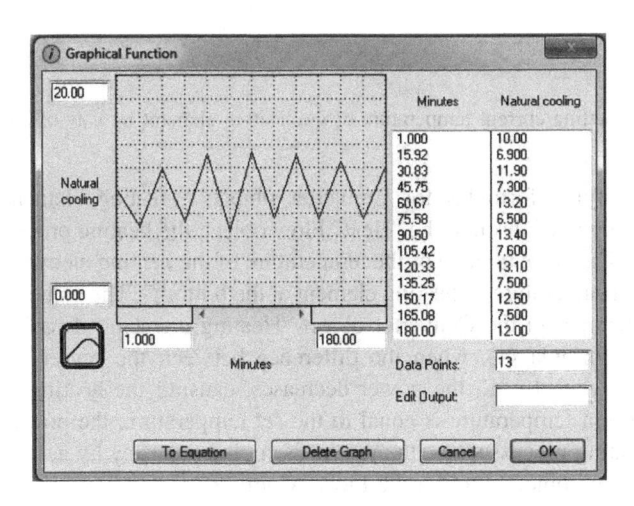

Fig. 4. Setting natural cooling in the form of dependency diagram.

The current temperature of the heating element and the current product temperature are set in the Stock blocks "Temperature of the heating element at the bottom" and "Product temperature at the bottom" as constants. Figure 5 shows the Stock block "Temperature of the heating element at the bottom" with the initial temperature of the heating element set to 20 °C.

The operation of this model is as follows: the desired temperature value of 275 °C for the heating element in the Converter object "Set temperature of the heating element" is set, temperature value of 150 °C for the product in the Converter object "Set temperature of the product", and their initial temperature 20 °C in Stock blocks "Temperature of the heating element at the bottom" and "Product temperature at the bottom" (Fig. 5).

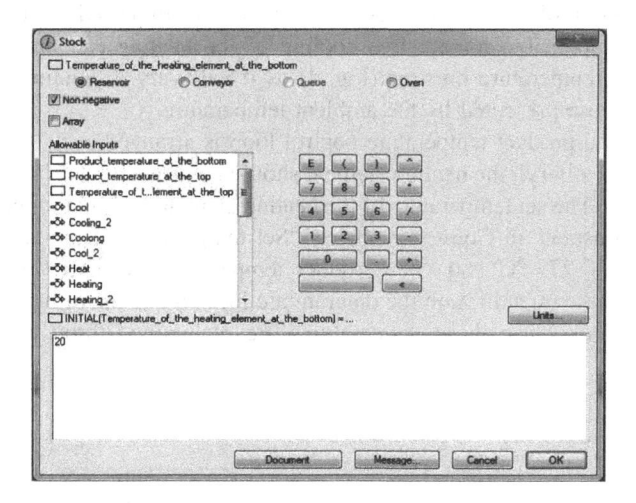

Fig. 5. Setting current temperature of the heating element by way of a constant.

After specifying formulas for Converter objects "Heating element power" and "Reactor heating rate" through the "Heat" Flow object, the heating process is simulated at full power. The current value of the temperature of the heating element is fixed in the Stock block "Temperature of heating element at the bottom". The value of the power of the heating element in the Converter object "Heating element power" depends on it. According to formula (1), when the difference between the preset and the current temperature reaches 13 °C, the power decreases, causing the heating to slow down. When the current temperature is equal to the set temperature, the power tends to 0.

As the heating process stops, the reactor is cooled naturally by using the Converter "Natural cooling" object through the Flow "Cooling" object (Fig. 4), the temperature drops. Therefore, if the temperature drops by more than 5 °C, the model increases the power, the temperature rises again. Thus, the temperature in the lower part of the reactor is controlled.

5 Creating the "Upper Part of the Reactor" Model and Entering the Initial Data

To create the "Upper part of the reactor" model, the Sector frame blocks were also used. In all the initial data, the range is the same as in the "Lower part of the reactor" model.

In this part, the initial elements used are the same as in the "Lower part of the reactor" model, only the digit 2 is added.

As the temperature of the heating element and the product is maintained throughout the reactor in the ranges (20–300) °C and (20–200) °C, respectively, the sector model "Upper part of the reactor" therefore will be identical to the "Lower part of the reactor" sector model (Fig. 6).

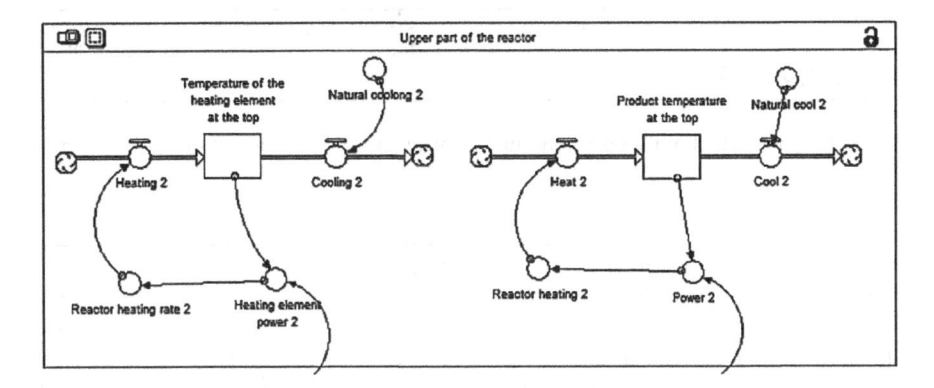

Fig. 6. Upper part of the reactor" model with entered initial data.

6 Analysis of the Obtained Model

The dependence of the current temperature of the heating element on time is plotted, as well as the dependence of the current product temperature on time in the lower part of the reactor in order to detect the mutual influence of the temperature control loops (Figs. 7 and 8).

Analysis of the graphs obtained showed that the heating of the reactor initially takes place at full power. When the current temperature approaches the set value, the power starts to fall. There is a slowdown in temperature increase. When the power value approaches 0, there is a natural drop in temperature, therefore the power again increases. Thus, the temperature control simulation process takes place.

Then, similar graphs for the upper part of the reactor are plotted (Figs. 9 and 10).

On these graphs, you can also trace the temperature control simulation process after approaching the current value to the set value.

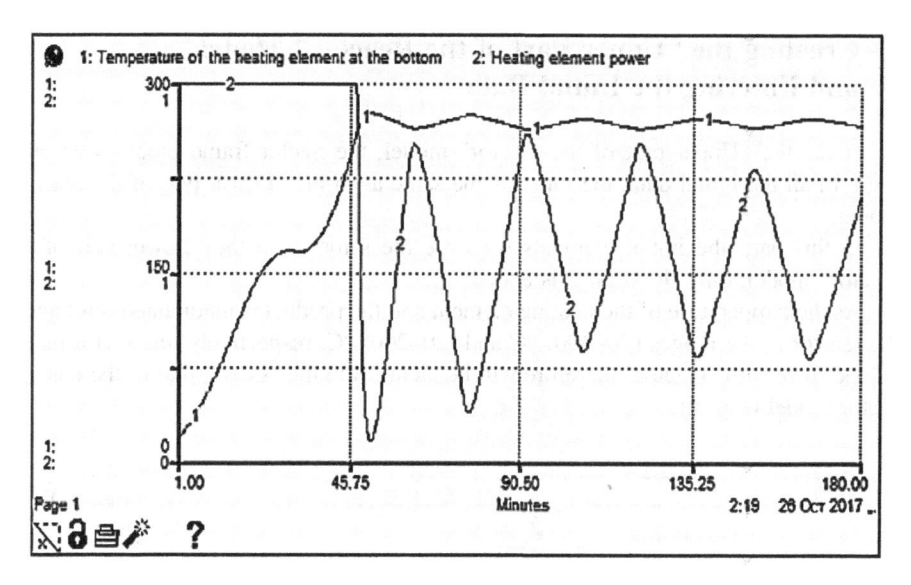

Fig. 7. Graphical chart of the current heating element temperature versus time in the lower part of the reactor.

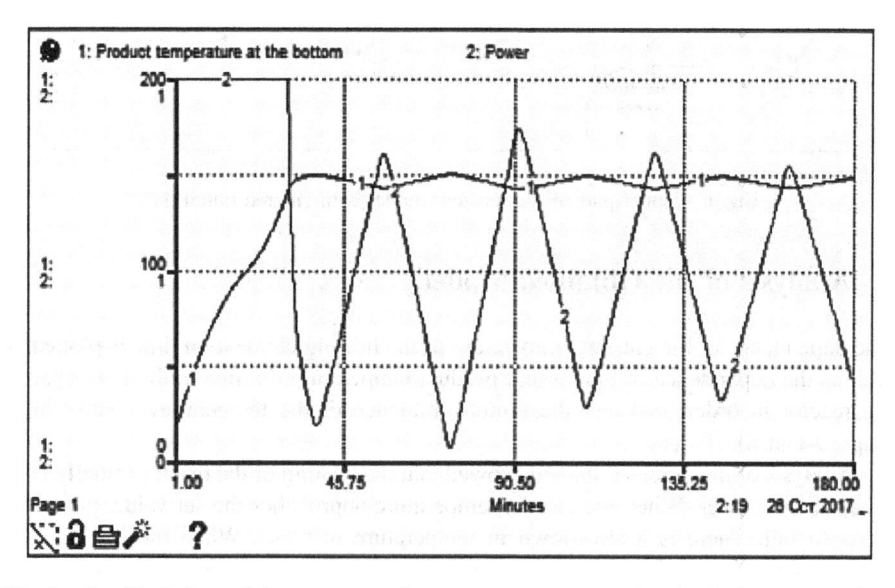

Fig. 8. Graphical chart of the current product temperature versus time in the lower part of the reactor.

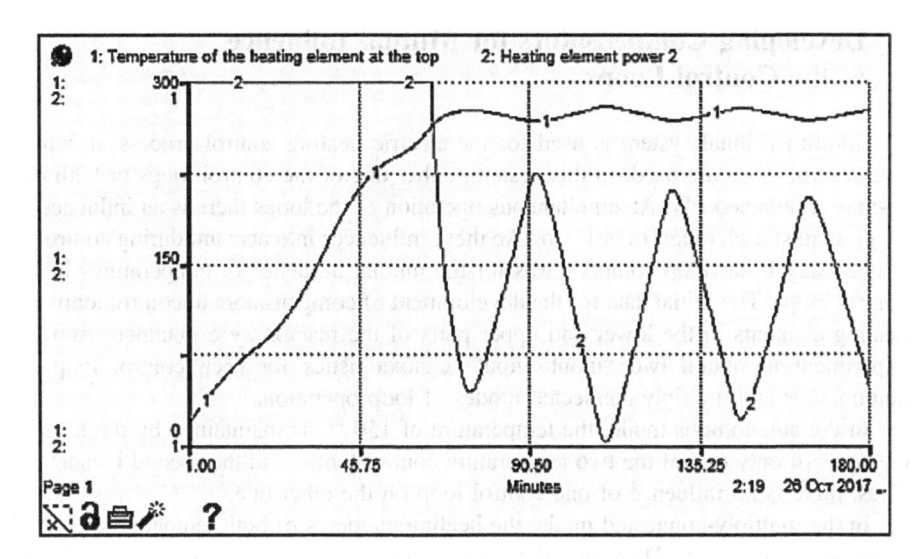

Fig. 9. Graphical chart of the current heating element temperature versus time in the upper part of the reactor.

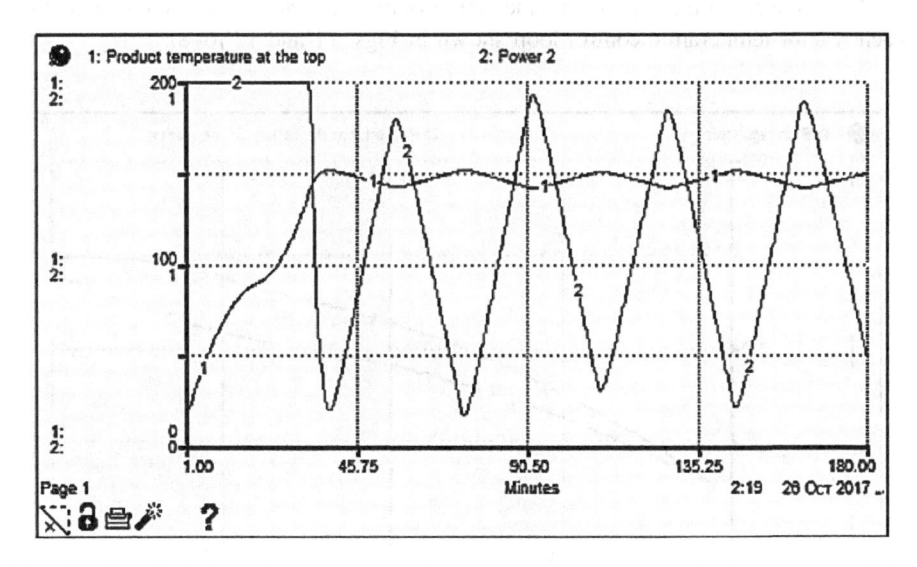

Fig. 10. Graphical chart of the current product temperature versus time in the upper part of the reactor.

7 Developing Compensators for Mutual Influence of the Control Loops

A multidimensional system is used for the electric heating control process, in which two control loops are used. In this system, either one of the control loops or both can operate simultaneously. At simultaneous operation of the loops there is an influence of loops against each other. In order to take these influences into account during control, it is necessary to develop compensators for the mutual influence of temperatures in the control loops. The initial data for the development of compensators in control loops for heating elements in the lower and upper parts of the reactor were obtained from the experiment to obtain two "input-output" characteristics for each control loop: in autonomous and multiply-connected modes of loop operation.

In the autonomous mode, the temperature of 150 °C is maintained by the heating elements of only one of the two temperature control loops, and the second is inactive. Thus, there is no influence of one control loop on the other one.

In the multiply-connected mode, the heating elements of both temperature control loops are in operation. Therefore, the "input-output" characteristic of the control loop under consideration in these conditions will take into account the influence of the other loop.

In the course of the experiment, the "input-output" characteristics were obtained for each reactor temperature control loop shown in Figs. 11 and 12 [6–8].

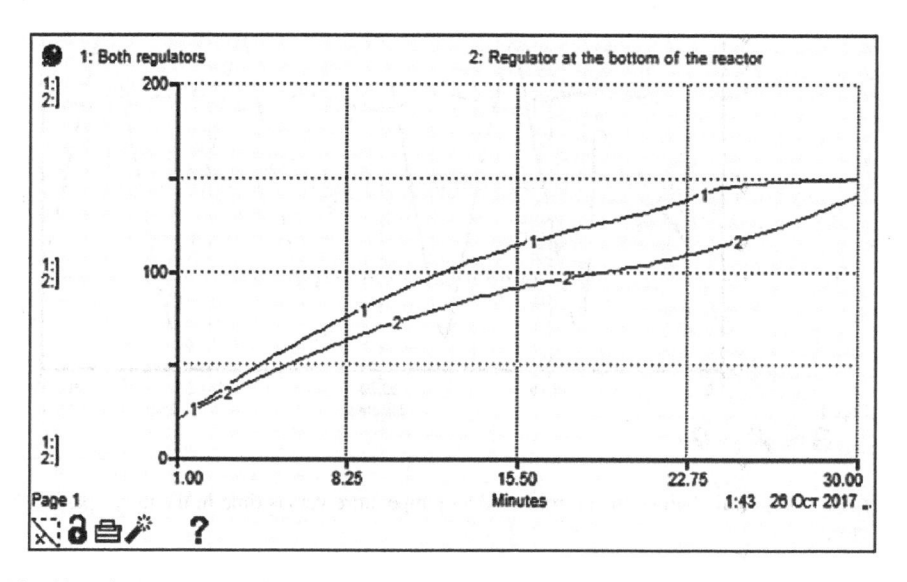

Fig. 11. The "Input-output" characteristic for the temperature control loop at the bottom of the reactor in autonomous (graph 2) and multiply-connected (graph 1) modes.

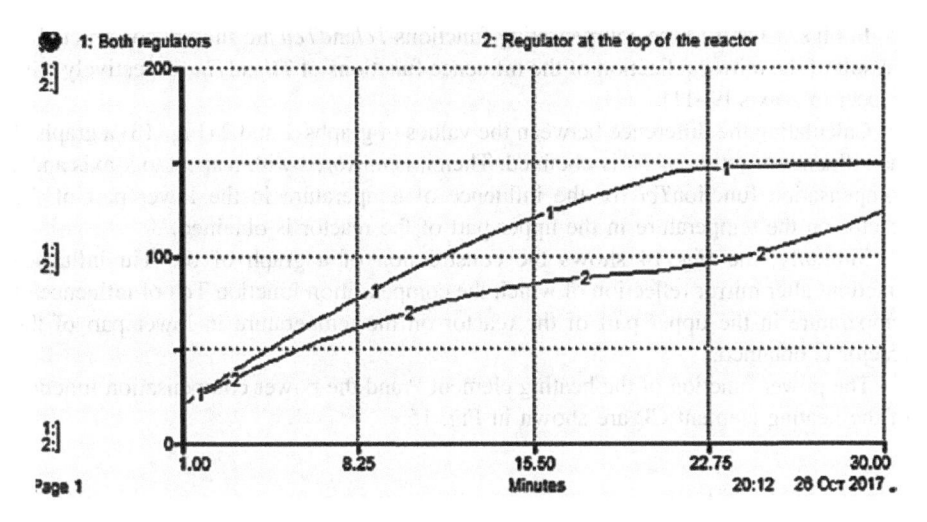

Fig. 12. The "input-output" characteristic for the temperature control loop in the upper part of the reactor in autonomous (graph 2) and multiply-connected (graph 1) modes.

In each of the Figs. 11 and 12 there are graphs for the autonomous and multiply-connected operating modes of the temperature control loop. For example, to find function of the loop influence in the lower part of the reactor Til, in Fig. 11, it is necessary to subtract the values of graph 2 from the values of graph 1.

Further, to develop compensators for mutual influence of loops, it is necessary to use the influence functions of temperature control loops of the reactor in the lower *Til* and upper *Tiu* parts of the reactor. The graphs of these functions are shown in Figs. 13 and 14. The idea of compensation itself was in generating of a function in operating environment of each control loop which is a mirror image of the functions *Til* and *Tiu* with respect to x-axis.

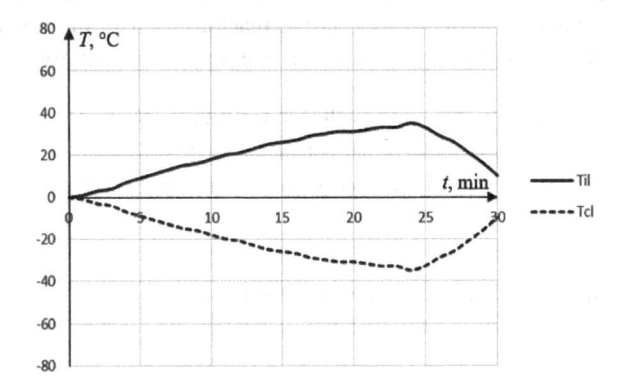

Fig. 13. The influence function Til and the compensation functions Tcl for the temperature controller in the lower part of the reactor.

In Figs. 13 and 14 the compensating functions *Tcl* and *Tcu* are shown, constructed as a result of the mirror reflection of the influence functions of *Til* and *Tiu* respectively with respect to x-axis [9–11].

Calculating the difference between the values of graphs 1 and 2 (Fig. 13) a graph of the influence function of *Til* is obtained. Then it is mirrored with respect to x-axis and a compensation function *Tcl* for the influence of temperature in the lower part of the reactor on the temperature in the upper part of the reactor is obtained.

Similarly, the Fig. 14 shows the construction of a graph of the Tiu influence function, after mirror reflection of which the compensation function Tcu of influence of temperature in the upper part of the reactor on the temperature in lower part of the reactor is obtained.

The power function of the heating element P and the power compensation function of the heating element CP are shown in Fig. 15.

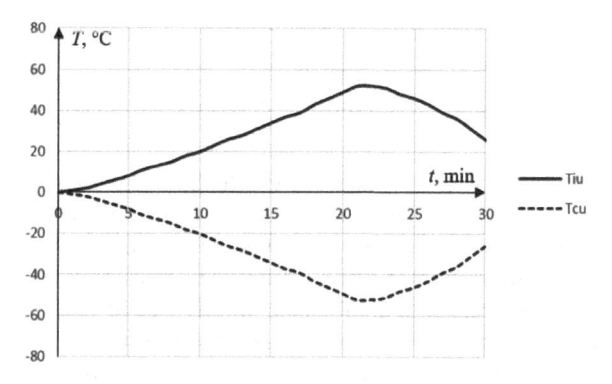

Fig. 14. The influence function Tiu and compensation function Tcu for the temperature controller in the upper part of the reactor.

The introduction of compensation functions changed the structure of the model executing the temperature control loop in the reactor [12, 13], the function of power compensation of the heating element CP was introduced into the formula (1):

Compensation_of_heating_element_power $=$ IF(Set_temperature_of_the_heating_element$-$
Temperature_of_the_heating_element_at_the_bottom < 13)THEN(CP $*$ (
Set_temperature_of_the_heating_element_Temperature_of_the_heating_element_at_the_bottom $+$
12)/25)ELSE(-1)

Thanks to these formulas, the mutual influence of the control loops is taken into account, and the quality of the temperature control in the reactor at product synthesis stage is increased [14, 15].

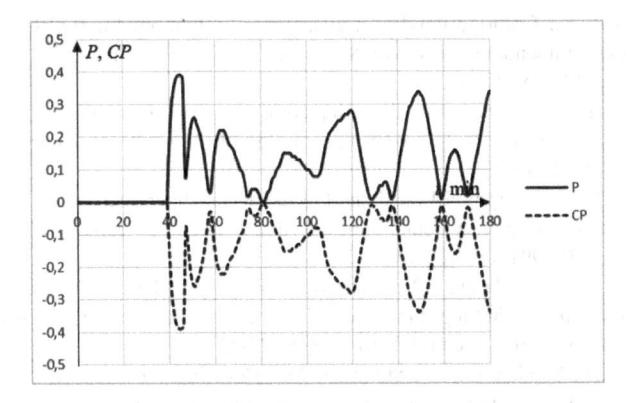

Fig. 15. The power function of the heating element P and the power compensation function of the heating element CP.

8 Conclusion

The developed model for controlling the electrical heating of the reactor allows one to simulate a multidimensional object, taking into account the influence of control loops on each other. The model can be used for reactors of various applications; it can be used to simulate a large number of complex objects.

References

1. Muravyova, E., Sagdatullin, A., Emekeev, A.: Intellectual control of oil and gas transportation system by multidimensional fuzzy controllers with precise terms. Appl. Mech. Mater. **756**, 633 (2015)
2. Muravyova, E., Bondarev, A., Kadyrov, R., et al.: The analysis of opportunities of construction and use of avionic systems based on cots-modules. ARPN J. Eng. Appl. Sci. **11**, 01 (2016)
3. Muravyova, E.A., Solovev, K.A., Soloveva, O.I., et al.: Simulation of multidimensional non-linear processes based on the second order fuzzy controller. Key Eng. Mater. **685**, 816–822 (2016)
4. Rahman, P.A., Muraveva, E.A., Sharipov, M.I.: Reliability model of fault-tolerant dual-disk redundant array. Key Eng. Mater. **685**, 805–810 (2015)
5. Muravyova, E.A.: Automated control of industrial process plants on the basis of multidimensional logic controllers: based on heat treatment processes. Dissertation, Ufa State Aviation Technical University, Ufa (2013)
6. Muravyova, E.A.: Two fuzzy controller synthesis methods with the double base of rules: reference points and training using. In: Industrial Engineering, Applications and Manufacturing: International Conference on Industrial Engineering, Applications and Manufacturing, 16–19 May. Key Eng. Mater. **685**, 186–190 (2017)
7. Muravyova, E.A.: Simulation of salt production process. In: IOP Conference Series: Earth and environmental Science, vol. 87, p. 052018. IOP Publishing (2017). https://doi.org/10.1088/1755-1315/87/5/052018

8. Muravyova, E.A.: Chlorine condenser-evaporator simulation. In: IOP Conference Series: Earth and environmental Science, vol. 87, p. 082032. IOP Publishing (2017). https://doi.org/10.1088/1755-1315/87/8/082032

9. Muravyova, E.A.: Fuzzy controller adaptation. In: IOP Conference Series: Earth and environmental Science, vol. 87, p. 082033. IOP Publishing (2017). https://doi.org/10.1088/1755-1315/87/8/082033

10. Muravyova, E.A., Sharipov, M.I.: Method of fuzzy controller adaptation. In: Proceedings of the International Conference Actual Issues of Mechanical Engineering, vol. 17, p. 82 (2017). https://doi.org/10.2991/aime-18.2018.42

11. Muravyova, E.A., Bondarev, A.V., Kadyrov, R.R., et al.: The questions of circuitry design when forming the switching functions of the control system of the matrix frequency converter. Indian J. Sci. Technol. 8(S10), 1–8 (2015)

12. Muravyova, E., Sagdatullin, A., Sharipov, M.: Modelling of fuzzy control modes for the automated pumping station of the oil and gas transportation system. In: OP Conference Series: Materials Science and Engineering, vol. 132, no. 1, p. 012028. IOP Publishing (2016)

13. Muravyova, E.A., Sagdatullin, A.M., Emekeev, A.A.: System-integrative approach to automation of the oil and gas fields design and development control. Oil Ind. 3, 92–95 (2015)

14. Muravyova, E.A.: Fuzzification concept using the any-time algorithm on the basis of precise term sets. Industrial engineering. In: Applications and Manufacturing: International Conference on Industrial Engineering, Applications and Manufacturing, 16–19 May 2017. Key Eng. Mater. 685, 129–133

15. Muravyova, E.A., Shulaeva, E.A., Charikov, P.N., et al.: Optimization of the structure of the control system using the fuzzy controller. In: 9th International Conference on Theory and Application of Soft Computing, Computing with Words and Perception, Budapest, 22–23 August 2017. Procedia Comput. Sci. 120, 487–494 (2017)

Optimization of the Process of Acoustic and Magnetic Geothermal Water Treatment Through Simulation in CoDeSys

A. V. Korzhakov, V. E. Korzhakov$^{(\boxtimes)}$, and S. A. Korzhakova

Adyghe State University, 208 Pervomayskaya Street,
Maikop 385000, Russian Federation
korve@yandex.ru

Abstract. It is proposed using the CoDeSys environment to control the process of non-chemical geothermal water treatment in real time. Non-chemical treatment of geothermal water eliminates scaling in the hydroponic greenhouse geothermal heating system and reduces the order requirements for resources. The entire intelligence system is concentrated in PLC and HMI performs the role of the thin client display. Production tests confirmed the effectiveness of acoustic and magnetic devices providing reagent-free treatment of geothermal water. The efficiency of the device was tested during three seasons of its operation. Routine maintenance showed that the amount of deposits on the pipes is absent. The method of non-chemical treatment of geothermal water makes it possible to achieve optimal temperature conditions in greenhouses by regulating the scaling on the supply pipes and heat exchanger of the geothermal heating system with low energy costs, which is largely confirmed by the introduction of the authors of the automatic control system in hydroponic greenhouses of the heating system of JSC "Raduga" (The Republic of Adygheya).

Keywords: Acoustic and magnetic device · Geothermal water · Heating system · Nonchemical treatment · Controller · CoDeSys

1 Introduction

Simulation modeling is a powerful tool for the study of complex dynamic systems. Computer simulation makes it possible to carry out computational experiments with the designed systems, to analyze, study and debug existing systems in areas where full-scale experiments are inexpedient due to danger or high cost. At the same time, this method of research is interesting for a wide range of users due to its proximity to physical modeling [1].

At the stage of synthesis of control system model we need to solve the problem of transfer of the designed control program from the simulation environment to the controller and provide the possibility of using this model for testing the target equipment. CoDeSys environment allows to solve these problems. It is equipped with a built-in visualization and operator control system. CoDeSys has the emulation mode to perform a program for PLC on the computer, where CoDeSys is running. In this mode all online functions are available. It allows to check the logical correctness of programs

© Springer Nature Switzerland AG 2020
A. A. Radionov and A. S. Karandaev (Eds.): RusAutoCon 2019, LNEE 641, pp. 1019–1033, 2020.
https://doi.org/10.1007/978-3-030-39225-3_108

without using a controller. CoDeSys provides real-time control and is less resource-intensive. All the intelligence of the system is concentrated in the PLC and the HMI acts as a thin client of display.

Research work on automation of nonchemical methods of geothermal water treatment has been carried out to increase the profitability of the JSC Raduga. The results of research work were used to control in real time the process of geothermal water reagent-free treatment by acoustic-magnetic devices. A significant reduction in scaling was achieved by acoustic-magnetic treatment of geothermal water by software connection of the required number of acoustic-magnetic devices, depending on the readings of the scale sensor. The highest efficiency of water treatment is achieved by combining ultrasound with a magnetic rotating field. The non-chemical treatment method is based on an acoustic-magnetic device that has received several patents [2–5].

2   Problem Statement

All The nonchemical treatment of geothermal water for the heating system is carried out by ultrasonic and magnetic devices. But each of these methods has its drawbacks.

So long-term operation of ultrasound has adverse effects on human health. Electromagnetic devices consume quite a lot of electricity, which is not economic [6–10].

It is proposed to use nonchemical treatment of geothermal water to eliminate scale formation in the hydroponic greenhouses geothermal heating system. To control the process of geothermal water nonchemical treatment the authors propose to use the automatic control system. To optimize the process of control and reduce operating costs, it is proposed to carry out real-time simulation of the control system.

3   Theoretical Part

Equations Let us consider the synthesis of the control system, where the control program is performed on the industrial controller OVEN PLC150 with the expansion unit of the inputs and outputs. The control object is a mathematical model of the object implemented in the CoDeSys environment. Simulation of the system takes place in real time. The main complexity of the synthesis of such a real-time control system is the mathematical model of the object and controller's interface. The controller interacts with the control object by TCP/IP network. Frame Oriented is used as communication Protocol. Values are selected from the list of "ASCII" and "RTU-mode", the default value is ASCII.

It is necessary to define a local Gateway Server and further to establish CoDeSys connection with the controller. The communication channel between the Gateway Server and the device is determined by using the Scan network command. This command finds the devices available to this server in the local network. We activate our communication channel with the PLC. Therefore, multiple channels of communication have been set with different devices in the same project. The I/O Server is organized by OPC DA 2.0 technology at the OVEN controllers. In this article we consider the OPC server for the controller (Fig. 1).

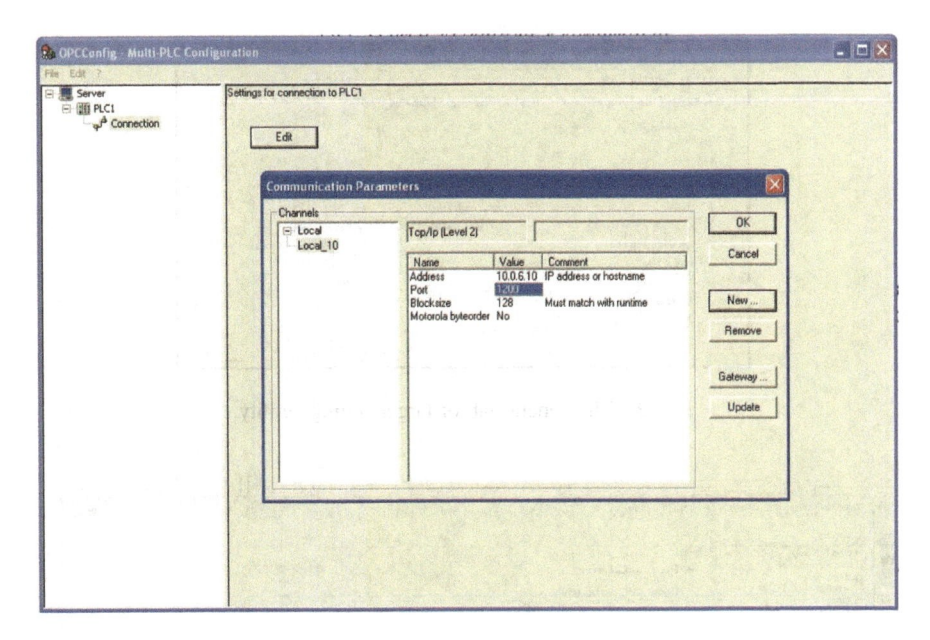

Fig. 1. The PLC connection.

OPC DA 2.0 Protocol works with TCP/IP connection. The controller and the personal computer with the installed CoDeSys Package are connected by Ethernet in the same network. CoDeSys simulates the operation of the reagent-free treatment geothermal water system by acoustic magnetic devices. The entrance of the system is geothermal water with a lot of salts, and the exit from the system is the antiscale effect. This effect is expressed by the mass fraction of scale at the thermal equipment walls.

Let's consider the connection scheme of CoDeSys controllers to the computer. When the project has been loaded into the CoDeSys environment, it is necessary to check the connection of CoDeSys controllers. Command Logout is called if the controller and the computer are connected. Target Settings is selected at the Resources tab of the CoDeSys Utility Objects Organizer [11].

Download Symbol File option is set and the selection is confirmed by pressing OK of the option Target Settings at opened screen form (Fig. 2) of the General tab.

Option Point Project is selected at the CoDeSys Main menu, and command Options is selected at the Context menu.

The option Symbol Configuration is selected at the opened screen form Options (Fig. 3) of the Category list and the flag is set at the parameter field of the Dump Symbol Entries field then the Configure Symbol File button is clicked.

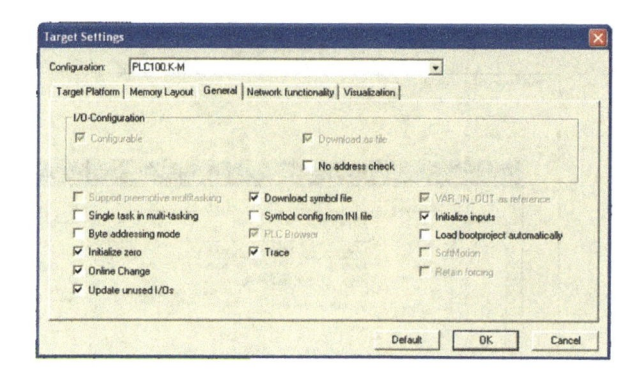

Fig. 2. The general tab of target settings utility.

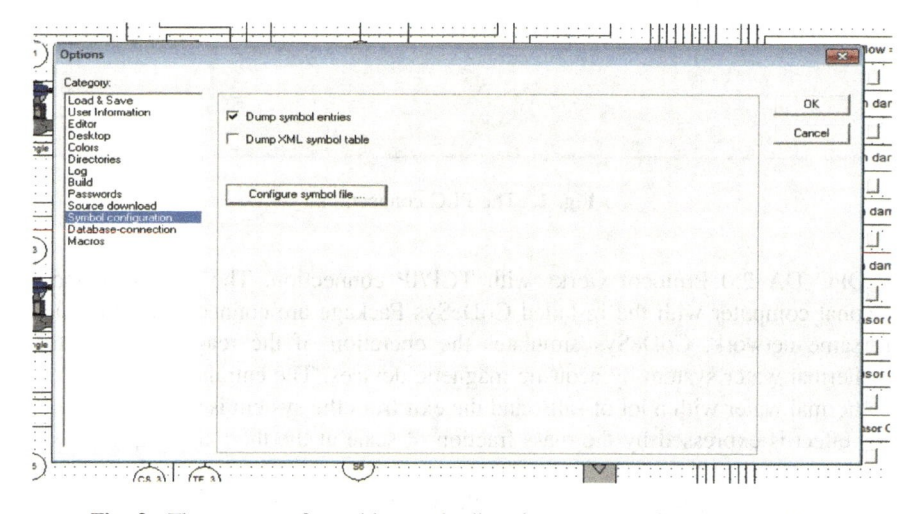

Fig. 3. The process of transition to the list of parameters of project variables.

Project objects from which we want to export variables is selected at the list of the project variables (Fig. 4), and flags must be set for them at the option fields. The flag is set from the option Export Variables of Object for ensure export of variables to the namespace of OPC_Server.

If you need to change the values of variables, you must set the flag at the Write Access option field. The project is saved (Fig. 5). The command Project Rebuild all is selected on the menu and project is recompiled.

The command Login is called and the project is loaded to the controller. OPC Configurator is run by the serial selection of the command of CoDeSys OPC_Configurator. The item Server is highlighted at the hierarchical structure (in the left field) of the opened OPC_Config screen form (Fig. 6). The value of time update is set at the parameter field Rate on the right.

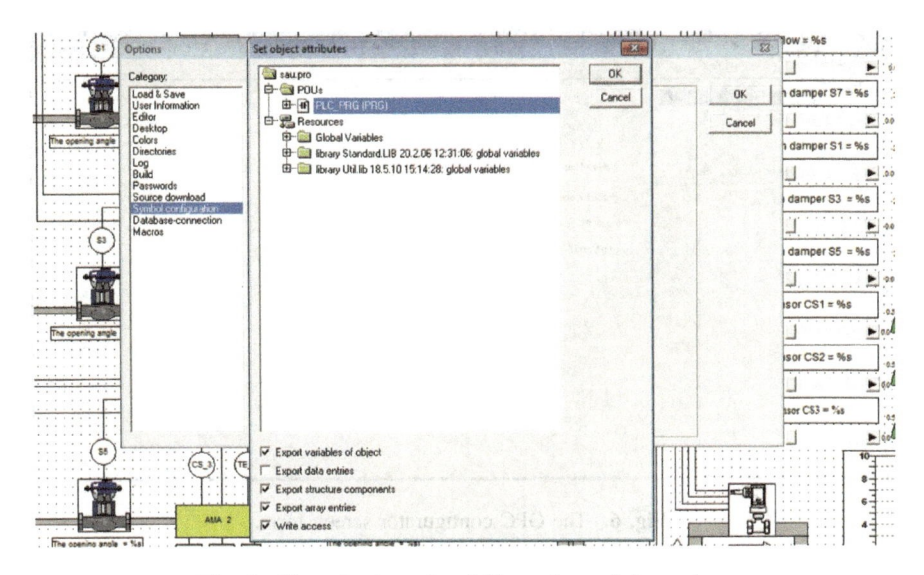

Fig. 4. The selection of variable settings of the project.

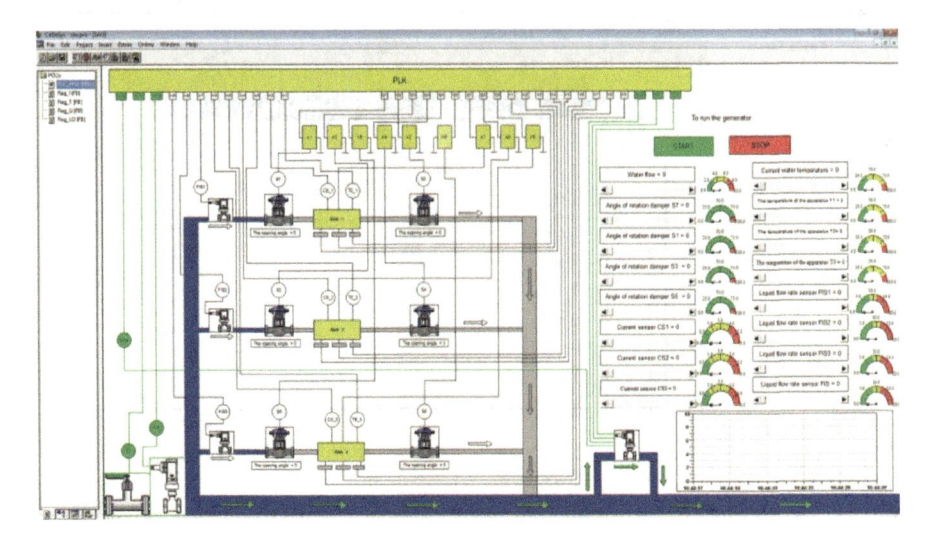

Fig. 5. The main form of the project.

The Context menu is opened by the right mouse button and the PLC option Append is selected. The connection item is selected for the PLC1 at the hierarchical structure (in the left field) of the opened screen form (Fig. 7) and the Edit button is chosen in the field of parameters, then the parameters of PLC connection Communication Parameters are set.

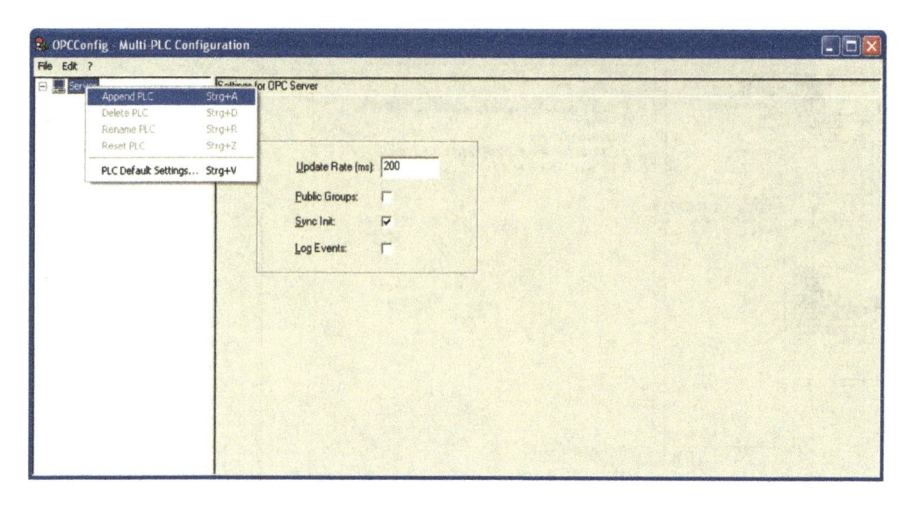

Fig. 6. The OPC configurator screen form.

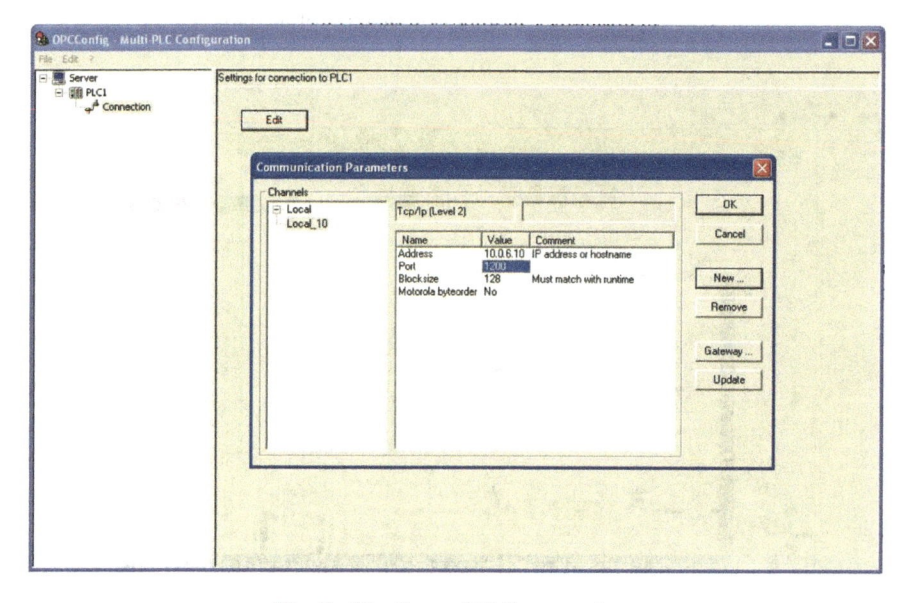

Fig. 7. The form of PLC connection.

The user confirms the choice by pressing the OK-button. So then OPC_Server has been configured and ready to work under the control of SCADA-system. The real time simulation begins with the setting of the installation parameters of the system: water velocity, water inlet temperature, initial angle of the inlet valve opening (Fig. 8).

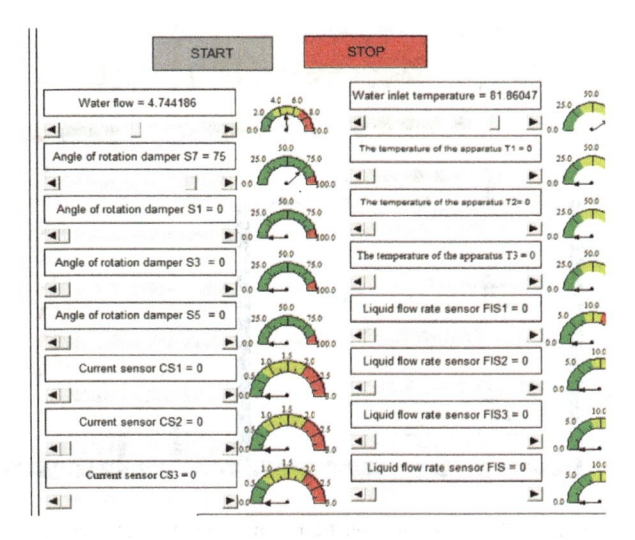

Fig. 8. The setpoint values.

After pressing the button Start, the inlet motorized valve is opened and water enters at the heating system. The motorized valve in front of acoustic-magnetic devices S1, S3, S5 are closed and the motorized valve after acoustic-magnetic devices S2, S4, S6 are fully opened. The controller polls the scale sensor installed at the heat exchanger inlet and opens the motorized valve S1, S3, S5 depending on the readings. Three motorized valves are opened if scaling process is strong. The damper closes consistently if scaling decreases. Connection of acoustic-magnetic devices occurs after stabilization of water flows in the pipes with installed acoustic-magnetic devices. The connection of one acoustic-magnetic device is demonstrated at Fig. 9. The scale sensor readings are minimal at this time.

PLC control program is based on the PD-regulator. To stabilize the water flow in the pipes with installed acoustic-magnetic devices PLC control program was developed in the CoDeSys environment.

The implementation of the PD-regulator is taken from the standard library of the CoDeSys Development environment. The advantage of the implemented PD-regulator is the ability of automatic adjustment of the control loop.

The method of periodic oscillations and the method of response to the step effect are used as methods of automatic adjustment of the controller. These methods are performed by different algorithms: Nichols-Ziegler's algorithm, Chin-Chrones-Reswick's algorithm. The developed control program is demonstrated at Fig. 10.

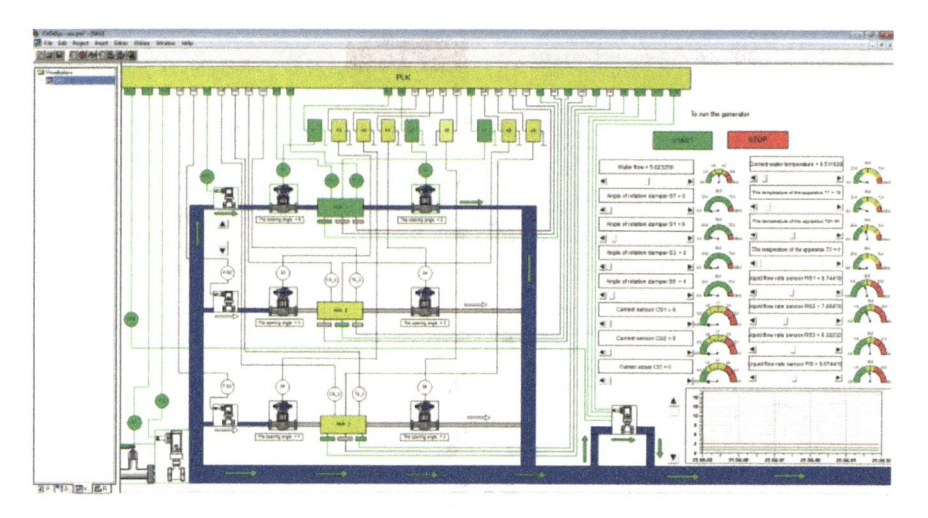

Fig. 9. The form of acoustic-magnetic device connection.

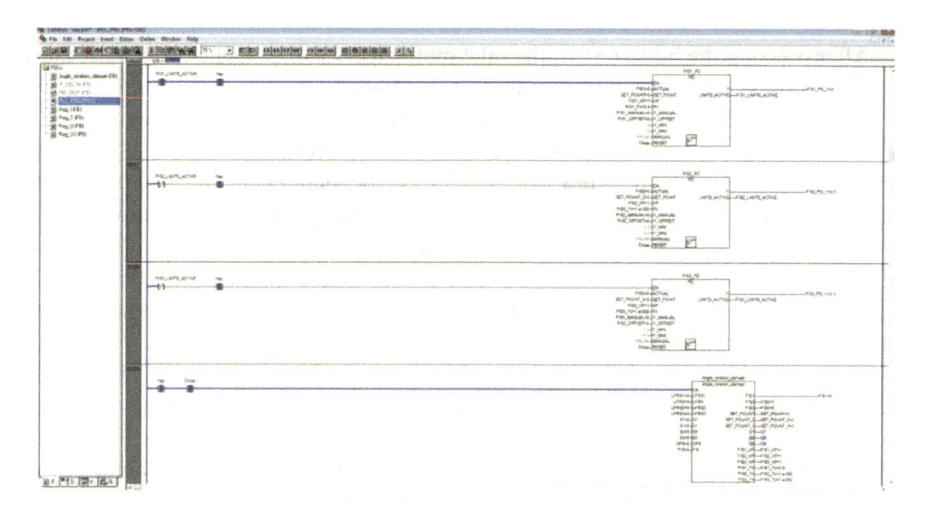

Fig. 10. The form of program code of PD-regulators.

Control program simulation was executed with a quantization step of 0.12 s. A results of the control system (PD-regulator) simulation are demonstrated at Fig. 11.

This approach of control systems modeling makes it possible to debug the developed control program at the target platform with the control object. It eliminates the possibility of equipment failure during system commissioning. The disadvantages of real time modeling by OPC-Protocol are the transport delay (that occurs when a large load on the computer network) and large CPU load during the simulation.

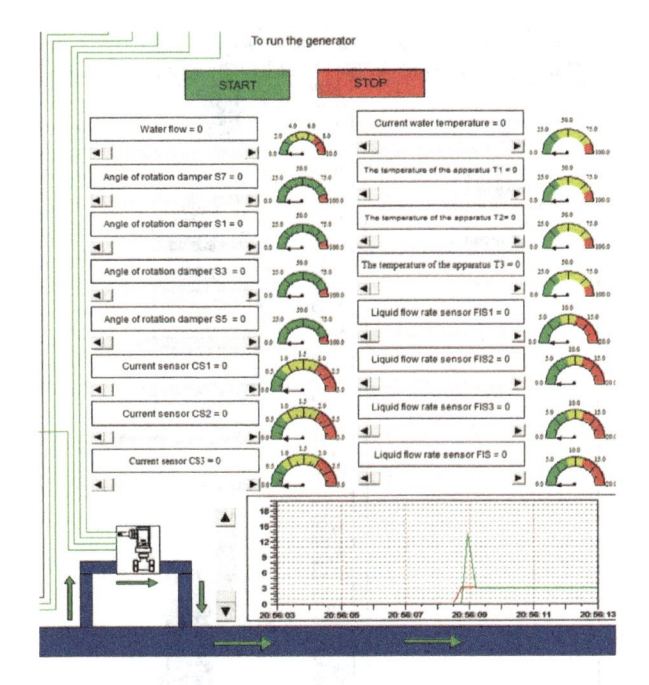

Fig. 11. Results of PD-regulator simulation.

There are all variables values of online mode in real time at screen. There are actual variables values at the announcement and code editors sections, at the watch and receipt manager and at the visualization screens. The values of function block instance variables are displayed as a hierarchical tree. For pointers, only their own values are displayed at the code section (Fig. 12).

The program provides emergency shutdowns of the acoustic-magnetic water treatment system. Current indicators are removed from the current sensors installed on all devices.

Thus, when the set current value is exceeded, the power supply is automatically removed from all devices and the dampers in front of the acoustic-magnetic devices are closed.

Also, there is a shutdown of acoustic-magnetic devices in excess of the temperature sensors installed on each acoustic-magnetic device. Figure 13 demonstrates the emergency shutdown of the system.

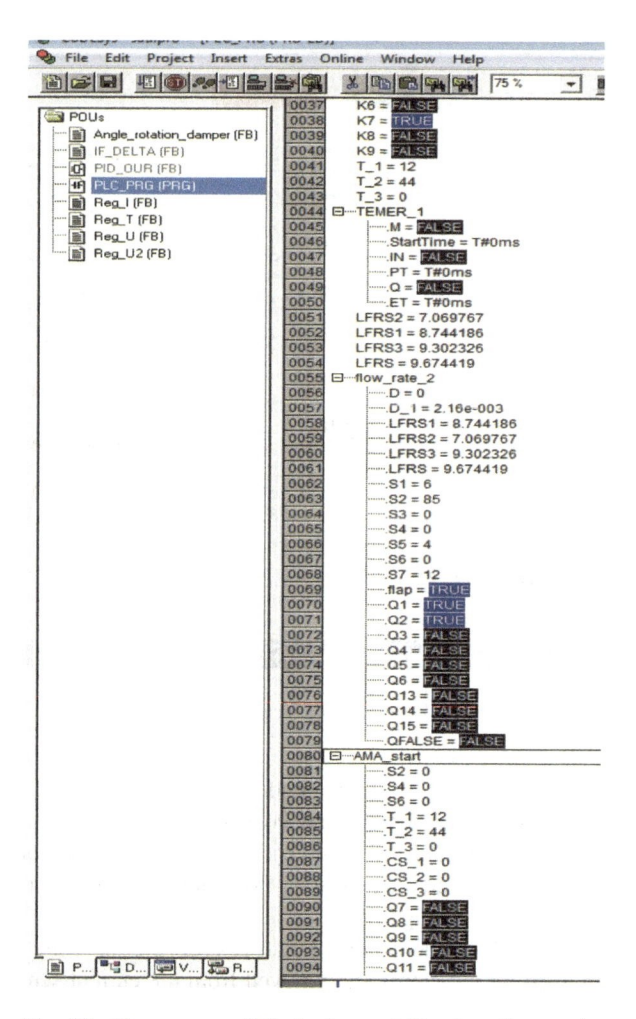

Fig. 12. The process of displaying variables in online mode.

The presented approach allows simultaneous generating of visualization, compiling the program and loading it into the controller. The user can view and correct the received code if necessary. This approach allows to move from routine programming to modular design of applied projects. It allows the users who are well aware of the device of machines, the technology of appropriate production, but are not good at programming, to become full participants of the design of applied projects [12, 13].

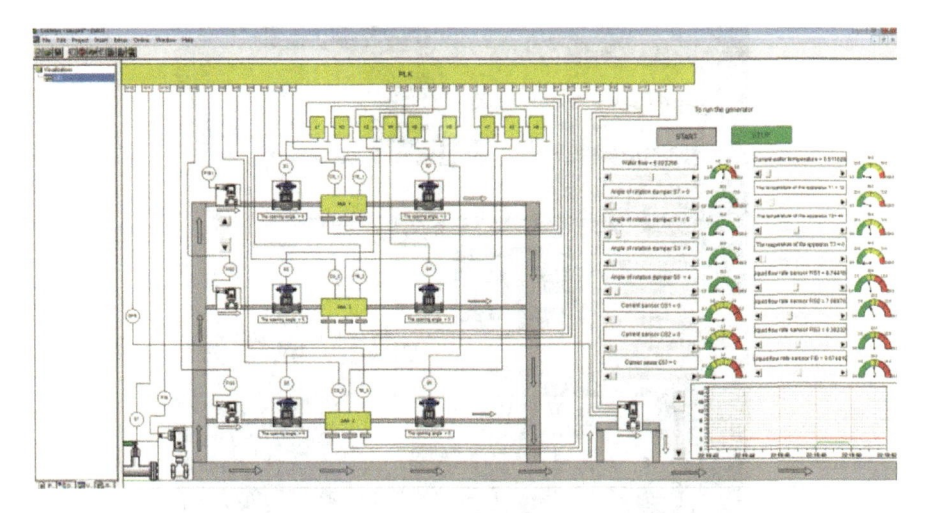

Fig. 13. Demonstration of emergency system shutdown.

4 Practical Significance

The automatic control system for geothermal water acoustic magnetic treatment in the greenhouses heating system is implemented in hydroponic greenhouses of JSC "Raduga" (The Republic of Adygheya). The appearance of the cabinet with automatic control system elements and devices is shown in Fig. 14 [14].

The acoustic magnetic device is installed in the greenhouse heating system (Fig. 15). This pipe supplies water from a geothermal source.

Production tests confirmed the efficiency of the device. Figure 16 shows a section of the pipe that was rejected in the result of routine maintenance work performed after the heating Season [14].

The efficiency of the device was tested during the year of its operation. As a result of routine maintenance it was found that the amount of deposits on the pipes began to tend to zero. Figure 17 shows a damper that has been used throughout the heating season [15].

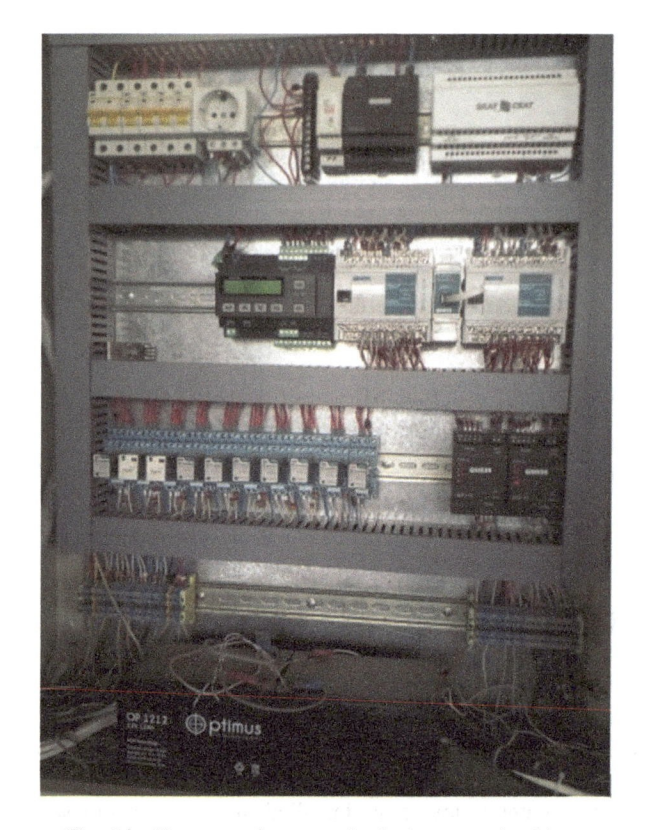

Fig. 14. The acoustic magnetic device control cabinet.

Fig. 15. The acoustic magnetic device is installed in the pipe of the greenhouse heating system.

Fig. 16. The pipe section with the salt deposits before the installation of the device.

Fig. 17. Damper in the disassembled form after the heating season with the device.

5   Conclusion

The experimental studies of metal equipment protection from corrosion and scaling have been carried out at the geothermal heat supply system of JSC "Raduga" in 2016-2018. These scientific developments have demonstrated considerable anticorrosive and antiscale effect of acoustic and magnetic water treatment in the greenhouse heating system.

Acoustic magnetic technology is a reagent-free method of scale and sediment preventing. It has intensified water treatment processes by water physical and chemical properties changing.

The process of acoustic magnetic treatment geothermal heat supply system automation has made it possible to adjust the parameters of the acoustic magnetic device in real time. Also, this automation has allowed modes of operation of acoustic magnetic devices adjustment. The use of acoustic magnetic system automation has reduced scaling by 30%.

References

1. Persin, S., Tovornik, B., Muskinja, N.: OPC-driven data exchange between MATLAB and PLC-controlled system. Int. J. Eng. Educ. **19**(4), 586–592 (2003)
2. Korzhakov, A.V., Korzhakov, V.E., Kirillov, N.P.: Device for reagentless water treatment. Author's Certificates Inventions **1724594**, 7 (1992)
3. Korzhakov, V.E., Kramarenko, B.D., Pleshakov, V.V., et al.: Device for the non-reagent treatment of liquids. Author's Certificate Invention **1514726**, 15 (1989)
4. Korzhakov, A.V., Korzhakov, V.E., Oskin, S.V.: A device for protection against the formation of deposits on the surfaces of pipelines of the heat supply systems. RU Patent 2635591, 14 November 2017
5. Korzhakov, A.V., Korzhakov, V.E., Oskin, S.V.: A device for protection against the formation of deposits on the surfaces of pipelines of the heat supply systems.RU Patent 2641137, 16 January 2018
6. Laotischen, N.P.: Magnetic water treatment, the prospects for its application in thermal power plants. Water treatment, water treatment and chemical control at steam power plants, vol. **2**, pp. 117–124 (1966)
7. Klassen, V.I.: Problems of theory and practice of magnetic treatment of water and water mixtures. Science, Moscow (1971)
8. Klassen, V.I.: Magnicifent water systems. Chemistry, Moscow (1982)
9. Minenko, V.I.: Magnetic water treatment at chemical water treatment. KN publishing house, Kharkov (1962)
10. Tebenikhin, E.F.: Reagentless methods of water treatment in power plants. Energoizdat, Moscow (1985)
11. Online help. OPC UA Server (2019). https://www.codesys.com. Accessed 29 Apr 2019
12. Petrov, I.V.: Codesys is everyday PLC programmer's tool. Autom. Ind. **8**, 3–8 (2012)
13. Kountouris, A.A., Wolinski, C.: Hierarchical conditional dependency graphs as a unifying design representation in the CODESIS high-level synthesis system. In: Proceedings 13th International Symposium on System Synthesis, pp. 66–72 (2000)

14. Korzhakov, A.V., Korzhakov, V.E., Korzhakova, S.A.: Automatization of geothermal water acoustic and magnetic treatment process in hydroponic greenhouse heating system. In: International Russian Automation Conference, p. 5619 (2018)
15. Korzhakova, Oskin S.: Identification of most effective form of pulse voltage supply of electric windings of acoustic magnetic device processing liquid in water pipes. Latvia University of Agriculture, Jelgava (2018)

Estimating the Cost of Implementing Virtual Desktops as a Stage of Project Management in the Field of Cloud Technologies

K. Makoviy and Yu. Khitskova[(✉)]

Voronezh State University,
1 Universitetskaya Square, Voronezh 394018, Russian Federation
prosvetovau@list.ru

Abstract. The article considers the problem of estimate the project in the field of cloud technologies: value determination virtual desktops in the activity of the educational organization. The introduction of desk-top virtualization leads to the appearance of many effects in the functioning of the enterprise: financial, technical, social, organizational. But the introduction of virtualization is associated with significant upfront investments. This often stops organizational leaders from implementing cloud projects. Therefore, it is important to correctly evaluate investments in such IT initiatives. The grouping of costs depends on the goals of the organization. The separation of costs into fixed and variable helps to calculate the break-even point. Grouping, which involves the division into capital and operational allows you to understand the amount of one-time investments in the project. Grouping costs according to the product development and sales allows you to calculate the planned taxes. In IT projects under different conditions, various grouping methods and costing methods are relevant. At the design stage, preparation for implementation, most are interested in capital and operating expenses. Which in turn are also divided into groups. The capital costs of implementing cloud technologies can be divided into: costs of equipment, infrastructure, installation. Operating costs: energy costs, operating costs of the virtualization system.

Keywords: Estimate the cost · Virtualization of desktops · Structure of costs · Cloud projects

1 Introduction

Project management in varying degrees, use many organizations. Different methods of project management are implemented by executive bodies of state power and state institutions at various levels of the economy. Most institutions of higher education are also budget organizations that use project management. Project management in the field of information technologies, which contribute to the formation of digital space in Russian society, is especially relevant. Information systems are moving away from computing in a client-server model to various cloud-based solutions.

Managing project costs is extremely important at all stages of project implementation, since it is largely from an adequate and accurate assessment of the costs of a project that the decision of the project initiator about implementation is largely

A. A. Radionov and A. S. Karandaev (Eds.): RusAutoCon 2019, LNEE 641, pp. 1034–1043, 2020.
https://doi.org/10.1007/978-3-030-39225-3_109

dependent. It is the financial means that ensure the implementation of the project, the specifics of which may differ within the framework of the approved budget [1–5]. Managing project costs includes phases that differ from different authors, but mainly include the following steps:

1. Planning of human, material and production resources that are necessary to carry out the work on the project;
2. Estimation of the cost of resources required for the project;
3. Project budgeting;
4. Management of changes in the project budget items.

The types of project valuation associated with its features that underlie each individual project and therefore the complexity of developing a unified system of valuation of various projects [6–9]. We have identified the main types of valuation, which can be used both sequentially and individually, independently of other types.

1. A rough estimate of the project cost based on a combination of cost and income assessment methods;
2. Evaluation of the project concept using the comparative method;
3. Preliminary assessment of the project using the cost method;
4. The final assessment, which is desirable to carry out using various methods.

All this brings the project participants closer to understanding the situation with its value.

The management of projects related to cloud technologies involves certain steps, one of which is the calculation of the cost of the project.

For the management of the organization at the first stage of the implementation of large-scale projects in the field of information technologies, financial costs and the resulting consequences as well as the economic efficiency of the project as a whole are important.

There are different methods for calculating costs, both Russian and foreign, and their classification. Often the final financial result of a project depends on the classification of costs.

Among the methods there are cost methods, the JIT calculation system, which provides for minimizing the volume of inventories [10, 11], accounting for costs and results for life cycle costing (LCC) [11–13], strategic cost analysis [14], etc. It seems to be interesting to use both cost accounting for the product lifecycle stages and strategic cost analysis (SCA) for virtual desktop implementation project. Strategic analysis considers activities within the project as a consumption value chain. Each link of the chain is considered both from the point of its need in the production process and from the position of resources consumed by it. Further, according to the method of strategic analysis, it is possible to identify the main control factor that influences the cost of the virtualization project more than others and consider various options minimizing costs with this factor. The total cost of ownership and implementation of the virtualization project depends on many factors that are often not taken into account because their influence to the cost of implementing cloud technologies is insignificant. Analysis of various methods [2, 5, 14] of the effectiveness evaluation of the cloud technologies implementation has shown that at the present time, in the digital economy reality, the problem of determining TCO for the IT-projects becomes particularly urgent. This is especially true at the initial stage of the project when a cost approach can be very useful.

2 Cost Approach in Determining the Expenses of Implementation of the Desktop Virtualization Project

2.1 Types of Costs

To use the cost approach in determining the cost of implementation, it is necessary to know the types of applications that are planned to be serviced within virtual desk-tops and their performance characteristics. In the future their predicted annual growth is important, they also have great importance for measuring and determining how much and what type of hardware resources need to be purchased at various stages of project implementation and their final cost.

At the first stage of implementation, the cost of equipment plays an important role in total costs, but then its impact on the total cost of project implementation is reduced, and marginal costs are reduced as production increases. Variable costs associated with wages and taxes on it, as well as consumables for maintenance of the system, grow with the growth of production volume [11, 14].

The costs incurred by the organization and used to calculate the full cost of the virtualization project can be structured as follows.

2.2 Capital Costs

Capital costs cover the initial investment in the creation and preparation for the work and the immediate launch of the virtualization project for workplaces and can be divided into three main parts: the cost of IT equipment, the cost of the infrastructure and the cost of installing both. Cost of IT equipment. Cost - IT equipment - is the sum of all costs associated with the acquisition of equipment [15]. This number can be calculated by multiplying a certain amount of each type of equipment by their prices.

$$C_{it} = \sum_{k=0}^{n} V_k P_k, \tag{1}$$

n- number of component types (processors, disks, servers), V_k – number of equipment (number of units); P_k – price of equipment; k – type of equipment.

The cost of upgrading the PC and repairing existing equipment constitute a significant part of the costs of higher education institutions [13, 16, 17]. Annually in high school it is necessary to update a significant number of computers.

Virtualization he cost of upgrading the PC and repairing existing equipment constitute a significant part of the costs of higher education institutions. Annually in high school it is necessary to update a significant number of computers. Virtualization makes it possible to use outdated computer models. Accordingly, the life of the computer will increase. It is usually assumed that it will double. In preparation for the implementation of the desktop virtualization project, educational organization has collected information about computers located in different audiences of the united University. By several criteria: the amount of RAM, processor, motherboard allocated 25 audiences, with 400 computers, which need to be replaced in the coming year. The cost of renewal of this number of computers is shown in Table 1. Accordingly, the cost of the equipment necessary for the virtualization of 400 jobs is calculated. The cost of equipment is calculated in Table 1.

Table 1. Costs for updating computers.

Initial hardware costs in case of refusal to implement desktop virtualization	Quantity, pcs	Costs, €	Total price, €
Thin client	400	375	150000
Windows license	400	128	51250
Total, initial equipment costs			201250

If the desktop virtualization project is implemented, the initial hardware costs will include the new server hardware necessary for the system to function. These costs are presented in Table 2.

Table 2. Initial hardware costs in the case of a desktop virtualization project.

Initial hardware costs in case of virtualization implementation	Quantity, pcs	Costs, €	Total price, €
Thin client	0	0	0
Windows license	400	128	51250
Lenovo storage V370V2	1	37 500	37500
Server system Flagman TX217/5-008SH	1	62500	62500
Licenses for virtual workplace (Rosa virtualization)	400	8438	33750
Total, initial equipment costs			185000

Cost of the infrastructure. The next important cost element is the cost of the infrastructure. The cost of infrastructure (the cost of installing equipment for cooling and the cost of equipment to provide power to the system). The infrastructure also includes uninterruptible power supplies, network filters. Infrastructural elements, cooling elements of equipment and providing it with electricity, are one of the main components of the data processing center. The power of these elements necessary for the functioning of the system is estimated in advance, but usually changes in the first years of the system [18, 19]. It is difficult enough to estimate in advance the workload on server systems and data storage. And, accordingly, the necessary capacity of cooling and other equipment. The costs associated with the infrastructure, that is, with the purchase and installation of refrigeration and power plants, are one-time constant costs, which can be represented as the (2).

$$K\&C_{In} = K_{EE} \times K_{IT} \times K_R, \qquad (2)$$

where K_{EE} – energy efficiency, varies from 1 to 2; K_{IT} – total energy consumption of IT equipment; K_{EE} – measures how efficiently energy is used, considering the total energy, including the power used for IT components, cooling, lighting and other overheads compared to the power consumed by IT loads. One of the most important cost elements is the cost of equipment installation.

Cost of installation. Acquired IT equipment should be installed in the appropriate place within the educational institution, with proper connection, both to network devices and to power distribution units. The cost of installation depends on the number of workers required for installation, their hourly wage rate, as well as the time to in-stall each equipment. This cost can be represented as the (3).

$$C_{Ins.it} = Sal_t \sum_{k-0}^{n} V_k T_k, \qquad (3)$$

where Sal_t – hourly wage of installers; n – the number of types od infrastructure elements (processors, disks, servers); V_k – the number of equipment each type; T_k – time of installation of equipment; k – types of equipment.

The costs of installing the equipment necessary for the implementation of the virtualization of workplaces are presented in Table 3.

Table 3. Costs for installing hardware and software required for virtual desktops.

Equipment for installation	Number of employees	Month salary 1 employee, € per person	Taxes on salary fund, € per month, (30% on salary)	Installation time, months	Cost of work, €
Lenovo storage V370V2	3	375	112,5	0,5	731,25
Server system Flagman TX217/5-008SH	3	375	112,5	0,5	731,25
Windows license	2	375	112,5	0,1	97,50
Licenses for virtual workplace (Rosa virtualization)	2	375	112,5	1	975,00
Setting up virtual workstations	3	375	112,50	2	2 925,00
Testing the system	3	375	112,50	2	2 925,00
Total installation costs					8385,00

One of the main elements of capital expenditure is the cost of the premises. The cost of the premises. In some cases, to implement projects related to cloud technologies, it is necessary to purchase an additional building where the data center will be located or the organization should make some initial investment to restructure the building that is planned to be rented in the long term. The building must be suitable for the specific operation of the data center. In this case, any related investments are considered as part of the capital costs and should be added to the estimated cost of the factors mentioned above. However, in cases where the building is leased, the cost of the necessary space can be considered zero in the initial year, and the cost of rent can be added to the operating costs.

In our case, the cost of the premises for a higher educational institution is provided by the state, so the costs for it were not calculated.

2.3 Operating Costs

Relate to the costs that arise during the operation of the desktop virtualization project for a specified period of time.

Energy costs. An important element of costs is the cost of energy, the price of which is quite high. Paying electricity is one of the main problems of an organization that uses desktop virtualization. It can be calculated by summing up the cost of electricity, which is necessary for the operation of all IT equipment during the life cycle of the project. In addition, an estimate of the energy consumption of additional equipment (energy consumed by lighting and refrigeration units) should be included in calculating the cost of energy using the coefficient.

$$E_n = H_{year} \times K_{EE} \sum_{l=0}^{L} \mathrm{Pr}_i \sum_{k=0}^{n} N_{ik} P_{ik}, \tag{4}$$

where H_{year} – the number of working hours in a year; Pr_i – energy price per kW per hour; i – year; n – number of component types; N_{ik} – quantity of each type of component; P_{ik} – power consumption in kW per year. Energy costs are not considered by us in the framework of this article.

The costs of project support. Constant monitoring and maintenance is necessary to maintain the equipment and information infrastructure in the required operating condition. It includes monitoring and testing of equipment, software updates (including renewal of licenses, where necessary), as well as the updating of many auxiliary components such as network elements, cooling appliances, etc. The cost of maintaining the IT infrastructure consists of labor costs, as well as the cost of the auxiliary components. [20] However, this approach makes it difficult to assess these costs, it is possible to consider the relationship between the cost of service and capital costs. In this paper, we have taken into account only the labor costs of hired workers.

Moreover, it is important to compare the costs of the information infrastructure before the implementation of the virtualization project and after the implementation of this project.

The main current costs for the information infrastructure of the educational organization before the introduction of the virtualization project are presented in Table 4.

Table 4. Current labor costs before the introduction of desktop virtualization for six months.

Current labor costs before the implementation of VDI	Number of employees	Salary 1 employee, € in months	Expenses per month, €	January	February	March	April	May	June
Salary of employees, who support the process of functioning system, €/month	5	375	1875	1250	1250	1250	1250	1250	1250
Taxes on salary fund, € per month, (30% on salary)	5	112,5	562,5	375	375	375	375	375	375
Total costs per month			2438	1625	1625	1625	1625	1625	1625
Total cumulatively				1625	3250	4875	6500	8125	9750

The costs presented in Table 4 can be compared with the costs after the introduction of desktop virtualization, presented in Table 5. Savings on wages are 1137,5 € per month. For half a year, the savings will be 6825 €, for 3 years of the project implementation, the savings will be 40950 €.

Table 5. Current labor costs after the introduction of the desktop virtualization system.

Current labor costs after implementation of VDI	Number of employees	Salary 1 employee, € in months	Expenses per month, €	January	February	March	April	May	June
Salary of employees, who support the process of functioning system, €/month	1	375	375	375	375	375	375	375	375
Taxes on salary fund, € per month, (30% on salary)	1	113	113	113	113	113	113	113	113
Total costs per month				488	488	488	488	488	488
Total cumulatively				0	975	1 463	1 950	2 438	2 925

The cost of renting a room that is necessary directly for the equipment. When implementing cloud technologies, there are two options for ensuring the availability of the required specialized area, i.e. buy/build a building or rent it. In the latter case, the cost of renting a room is an annual rent paid by the project owner to accommodate its equipment. It also includes the rental of the space needed to accommodate the infrastructure. Therefore, we first determine the required area, estimating the total number of racks in which the servers are located, which are necessary for servicing certain workloads, also determine the space required for the placement of cooling and power objects. Then this number is multiplied by the average rent for the year.

In our case, the cost of the premises for a higher educational institution is provided by the state, so the costs for it were not calculated.

Cost management of malfunctions in work. The costs of troubleshooting, such as replacing defective components or repairing them, where possible, are also part of the operating costs. However, estimating the cost of malfunction management is a very difficult task and deserves a separate study. The costs of troubleshooting, such as replacing defective components or repairing them, where possible, are also part of the operating costs.

However, estimating the cost of malfunction management is a very difficult task and deserves a separate study. This type of costs has not been analyzed by us at the present time.

3   Cost Estimate Results

As a result of the implementation of the virtualization project, the savings on the content of the information infrastructure of the educational institution can be structured according to the cost groups and presented in Table 6.

Table 6. Savings in the implementation of the desktop virtualization project in an educational institution.

Savings on fixed costs	€
Annual saving on operating costs	16 250
Saving on operating costs for three years	13 650
Total, savings for 3 years of project implementation	40 950

Table 6 shows that the savings in the implementation of the desktop virtualization project is significant for the educational organization. When we take into account that we did not take into account some types of costs, such as costs for managing failures, some infrastructure costs, energy costs, with direct implementation, the savings will be even more significant. Additional positive characteristics of the virtualization project of workstations are effects not considered by us in this article: social, technological, investment, etc. The article analyzes the main costs necessary for the implementation of the desktop virtualization project in the educational institution, structured, structured, capital and current expenses are identified, taking into account the specifics of this industry. It was revealed that the structuring and understanding of all expenses incurred by the organization in the process of deployment and operation of virtual workstations are the most important components of the evaluation of the cost-effectiveness of virtual desktops.

Currently, the ratio between the power of information technology and related costs is already large, in the future it will be difficult to provide the capacity required for the introduction of cloud technologies, using the current capabilities and strategies of organizations.

References

1. Suetin, S.N., Titov, S.A.: Projects and project management in the modern economy. Econ. Entrep. **5–7**, 469–499 (2014)
2. YoO, B.: Evaluation of the effectiveness of managing regional innovations. Manag. Econ. Syst.: Electron. Sci. **22** (2010)
3. Mahmood, S., Niazi, M., Hussain, A.: Identifying the challenges for managing component-based development in global software development: preliminary results. In: Science and Information Conference, pp. 933–938 (2015). https://doi.org/10.1109/sai.2015.7237254
4. Park, J.G., Lee, J.: Knowledge sharing in information systems development projects: explicating the role of dependence and trust. Int. J. Proj. Manag. **32**(1), 153–165 (2014). https://doi.org/10.1016/j.ijproman.2013.02.004
5. Niazi, M., Mahmood, S., Alshayeb, M., et al.: Toward successful project management in global software development. Int. J. Proj. Manag. **34**(8), 1553–1567 (2016). https://doi.org/10.1016/j.ijproman.2016.08.008
6. Govdya, V.V., Degaltseva, Z.V.: Innovative methods of cost management in the accounting and analytical cluster of agrarian formations. News Nizhnevolzhsky Agro-Univ. Complex Sci. High. Prof. Educ. **1**(37), 1–6 (2015)
7. Astarkina, N.R.: Effective ways to optimize costs in an enterprise (Practical aspects). Bus. Strat. **2**, 6–9 (2017)
8. Ivashkovskaya, I.V., Kukina, E.B., Penkina, I.V.: Economic value added. Concepts Approaches Tools Corp. Financ. **2**(14), 103–108 (2010)
9. Volkov, D.L.: Value management: indicators and assessment models. Russ. Manag. J. **3**(4), 3–42 (2005)
10. Osipov, V.S.: Economic-theoretical approaches to the definition of the chain of value and value. Econ. Sci. **97**, 55–58 (2012)
11. Kuvshinov, M., Kireeva, N.: Analysis of the correspondence of methods of managing costs to actual management tasks. Manag. Account. **22**(316), 14–24 (2014)
12. Yoo, S., Kim, S., Kim, T., et al.: Economic analysis of cloud-based desktop virtualization implementation at a hospital. BMC Med. Inform. Decis. Mak. **12**(119) (2012). https://doi.org/10.1186/1472-6947-12-119
13. Chang, B.R., Tsai, H.F., Chen, C.M.: Empirical analysis of server consolidation and desktop virtualization in cloud computing. Math. Probl. Eng. **2013** (2013). https://doi.org/10.1155/2013/947234
14. Perry, R., Waldman, B.: Measuring the business value of VMware horizon view (2013). http://www.vmware.com/files/ru/pdf/view/IDC-Quantifying-Business-Value-VMware-View-WP.pdf. Accessed 23 Aug 2017
15. Mahloo, M., Soares, J.M., Roozbeh, A.: Techno-economic framework for cloud infrastructure: a cost study of resource disaggregation. In: Federated Conference on Computer Science and Information Systems, pp. 733–742 (2017). https://doi.org/10.15439/2017f111
16. Khan, M.J., Mahmood, S.: Assessing the determinants of adopting component-based development in a global context: a client-vendor analysis. IEEE Access **6**, 79060–79073 (2018). https://doi.org/10.1109/ACCESS.2018.2878798

17. Ghobadi, S.: What drives knowledge sharing in software development teams: a literature review and classification framework. Inf. Manag. **52**(1), 82–97 (2015). https://doi.org/10.1016/j.im.2014.10.008
18. Prikladnicki, R., Audy, J.L.N.: Managing global software engineering: a comparative analysis of offshore outsourcing and the internal offshoring of software development. Inf. Syst. Manag. **29**(3), 216–232 (2012). https://doi.org/10.1080/10580530.2012.687313
19. Niazi, M., Ikram, N., Bano, M., et al.: Establishing trust in offshore software outsourcing relationships: an exploratory study using a systematic literature review. IET Softw. **7**(5), 283–293 (2013). https://doi.org/10.1049/iet-sen.2012.0136
20. Misra, S.C., Mondal, A.: Identification of a company's suitability for the adoption of cloud computing and modelling its corresponding return on investment. Math. Comput. Model. **53** (3–4), 504–521 (2011). https://doi.org/10.1016/j.mcm.2010.03.037

Automation of the Opal Colloidal Films Obtaining Processes

E. V. Panfilova[(⊠)] and V. A. Dyubanov

Bauman Moscow State Technical University, 5 Vtoraya Baumanskaya Street,
Moscow 105005, Russian Federation
panfilova.e.v@bmstu.ru

Abstract. This paper describes an equipment for the automated deposition of
photonic crystal opal colloidal films. Due to its unique properties opal films and
opal based structures have attracted great attention in the wide range of applications such as photonics, electronics and optoelectronics, laser techniques,
plasmonics, quantum computing, medicine, civil engineering and solar energy
devices. We present the equipment that combines two methods of opal films
formation - a vertical deposition and an electrophoresis. The equipment consists
of five main modules: a power supply unit, a mode setting module and a process
time reference, a display and indication module, an electrophoresis module and
vertical deposition module. Since deposition process is well controlled the
device may be used for optimizing process conditions and obtaining necessary
films' structures properties the results of an atomic force microscopy investigation of deposited colloidal films show that for silica particles success has
already been achieved by using automated device. The atomic force microscopy
images indicated that the structure of colloidal films is highly periodic.

Keywords: Opal film · Photonic crystal · Self-assembly · Automation ·
Electrophoresis · Vertical deposition

1 Introduction

Opal films consist of microspheres of silica $SiO2$ or polystyrene with similar diameter
in range from 100 to 1000 nm [1, 2]. Because of high periodic regular structure, that is
called super lattice, this films can block the light spreading in wavelength range, which
is forbidden for this structure. This wavelength range is named photonic band gap
(PBG) [3]. Therefore, the whole structure is called photonic crystal (PhC) [4].

Physical properties of PhC-based structures are used wherever light must be applied
[5–9]. Opal films and opal based structures became widespread in optics, electronics,
lasers, medicine, biology and so on. They are applied in waveguides [10] for light
transmission from source to recipient without alteration. In [11] these films were used
for luminescence microscopy of biological objects. Opal films are used in the visible or
near-infrared spectral range optical devices, including planarized opal-based
microphotonic crystal chips, switches, mirrors, filters, light-emitting diodes (LEDs),
lasers or superprisms, solar cells [12–17]. Monolayer opal film may be used to form

© Springer Nature Switzerland AG 2020
A. A. Radionov and A. S. Karandaev (Eds.): RusAutoCon 2019, LNEE 641, pp. 1044–1052, 2020.
https://doi.org/10.1007/978-3-030-39225-3_110

high-periodic layer of catalyst inside pores of structure for follow growing single walled carbon nanotubes [18].

Opal structures are obtained by deposition of colloidal microsphere particles. The process of deposition is simple and cheap because of self-assembly phenomenon [19–21]. Colloidal particles approach to the surface of the substrate, find vacant position in already formed film and then deposit in this seat. This process lead to formation high periodic uniform film (Fig. 1).

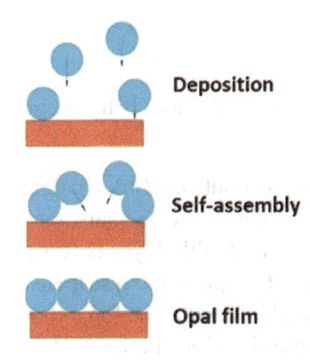

Deposition

Self-assembly

Opal film

Fig. 1. Self-assembly phenomenon.

The process of opal films deposition requires special laboratory equipment that is sometimes impossible to find. Therefore, this article is focused on the development of automated equipment for colloidal films deposition.

2 Methods of Opal Thin Films Obtaining

Opal films are mostly fabricated by four self-assembly methods: gravitational sedimentation, moving meniscus method, centrifugation, electrophoresis and vertical deposition method [22, 23]. The first one proceeds for a long time and is not suitable for mass fabrication of opal films. The last two have gained great interest because of its simplicity and ability to vary the process conditions.

2.1 Electrophoresis Method

This method requires electrochemical bath, power source, substrates, substrate holder. The holder connects substrate to the power source electrically and fixes it mechanically. Process starts after dipping of substrates into colloidal solution in electrochemical bath (Fig. 2).

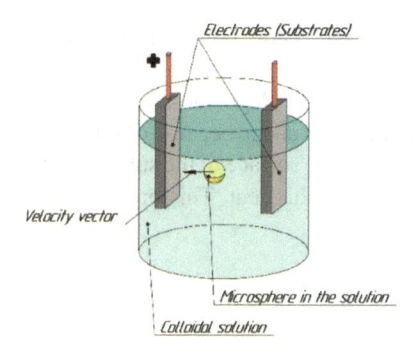

Fig. 2. The schematic diagram of the electrophoresis method.

The film formation velocity is controlled by changing potential difference between electrodes. The colloidal particles move to the substrate because they have an external double electric layer [24, 25]. The double layer is formed from solvent molecules and impurities. When an electric field is applied to a suspension of charged particles, a force appears on both parts of the double layer. This force moves the particles with respect to the liquid with a velocity proportional to the applied field. [26].

Therefore, the velocity of the electrochemical deposition process is determined by the applied potential. Because of it electrophoresis method may be used for obtaining monolayer opal film. The periodicity of the structure is determined by the ability of the particles to coagulate (adhere to each other), which is determined by zeta potential or surface external charge of the microspheres.

Since the surface charge is proportional to the particles velocity deposited films can be destructed and cracked. Therefore, the problem of process conditions optimizing is a complex task aimed to obtaining qualitative PhC colloidal films and minimizing the associated cracking.

2.2 Vertical Deposition Method

In this process, the substrate is placed vertically in a particles suspension and is gradually exposed by evaporation or other slow continuous phase removal or by substrate lifting. For this process, a bath for colloidal solution, a pump for pumping the solution, and a substrate holder for mechanical fixation of the substrates are required (Fig. 3). This method is based on capillary effect. The substrate is placed vertically in a particles suspension and is gradually exposed by evaporation or other slow continuous phase removal at a rate in range 0.1 to 10 mm/min [2, 23, 27]. While pumping the capillary forces attract the nearest to the substrate microspheres to its surface.

The particles deformation and the film's areal density is inversely proportional to the velocity of the solution removal. The velocity is to vary from 0.1 mm/min to 10 mm/min. It was demonstrated in [23] that for fixed conditions, when the concentration of the suspension is sufficiently low, a monolayer film [28] will always be deposited. It makes possible to obtain sufficiently large-area colloidal opal films by

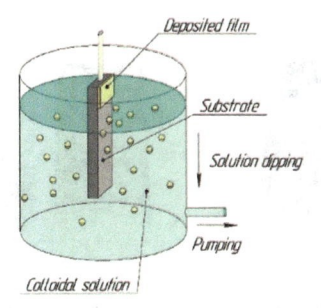

Fig. 3. The schematic diagram of the vertical deposition method.

vertical deposition method and to parameterize the deposition conditions for optimal film growth. This method is applicable for obtaining monolayer of colloidal particles.

2.3   Combination of Methods

Both methods have gained great interest for laboratory investigations and for the industrial production of PhC because for optimal conditions high quality of opal films has been observed. Each method has its advantages and disadvantages. For the further development of equipment for deposition of opal films an option that allows combining both methods were chosen. Thus, it is possible to eliminate some of the shortcomings of each of the methods, such as coagulation of colloidal particles during their deposition on a substrate. The device is useful for optimizing the film deposition conditions and obtaining desired film properties.

The developed equipment should be able to carry out both processes simultaneously and separately. It becomes a universal mechanism for growing opal structures.

3   Equipment Description

3.1   Design of the Equipment

The control unit of the equipment consists of five main modules: a power supply unit, which is used to supply power to all other modules, a modes setting module and a process time reference, a display and indication module, an electrophoresis module and vertical deposition module. General views of the equipment are presented on Fig. 4(a).

The equipment provides air cooling of the parts that are typified by the highest heat generation, such as the driver of stepper motor and the stepper motor.

Air intake is carried out through a 50 mm fan, then the air flow cool the necessary parts and is blown out through the power supply outside of the shell. The front panel of the shell is not used for fastening elements and wires, as it is used for visual inspection of the interior of the equipment and maintenance. For greater reliability and durability the equipment shell itself is made of 6 mm thick polycarbonate sheet, painted in gray blue.

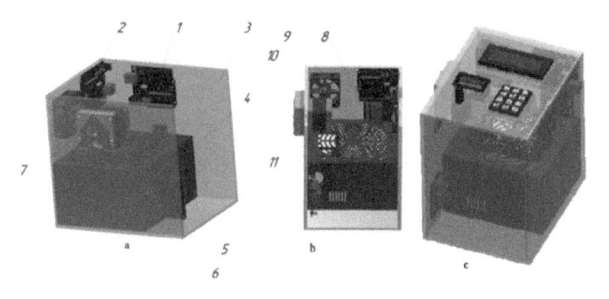

Fig. 4. (a) Isometric view: 1,2 – Arduino Uno, 3 – plastic shell, 4 – power source, 5 – voltage reducing circuit, 6 – electromagnetic relay, 7 –peristaltic pump head; (b) Front view: 8 – electromagnetic relay, 9 – cooling fan, 10 – stepper motor drive, 11 – stepper motor; (c) Overview of the device.

For convenience all the equipment control elements are placed on the front panel (Fig. 5). Therefore, the management of setting up the conditions of processes, as well as tracking the current stage of the process, is carried out without any particular difficulty. The display also shows tips that helps in managing of the equipment, so it would be easy to start out the film deposition.

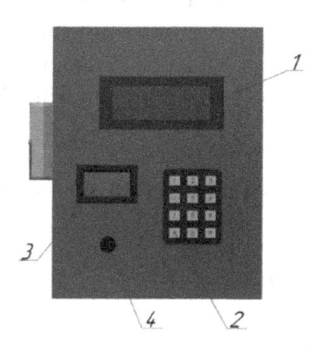

Fig. 5. Front panel of the device: 1 – liquid crystal display, 2 - membrane keyboard, 3 – digital voltmeter, 4 – precision potentiometer.

3.2   Principle of Operating of the Equipment

The equipment is implemented on two Arduino microcontrollers, one of which (first one) is responsible for setting the operation modes and time of the process, and the other (second one) controls the stepper motor.

The operating principle of the equipment is follows. Power is supplied from the power source to the Arduino board and activate it. The Arduino board is responsible for setting the modes and counting the time. The digital display connected to the first microcontroller is initialized to display the equipment interface. Data input is made through the matrix keyboard, and the "#" button plays the role of the button accepting the entered data, and the "*" button is responsible for resetting. The process time in the range from 1 to 999 min in 1 s increments is set by means of the first menu. After accepting the process time the process modes can be tuned.

Electrochemical deposition mode (on/off) is choosing and the Boolean variable stores the accepted value. In the case of the applying of electrochemical method, the display also displays a prompt that sends the operator to adjust the potential difference applied to the substrates. The potential difference adjustment is carried out using a precision multi-turn potentiometer and a voltage reducing circuit. The circuit acts as a voltage divider. When the potentiometer knob is turned the voltage is divided into parts, thus it is possible to adjust the output voltage in the range from 0.1 to 10 V. Current value is displayed on a digital voltmeter with 0.01 V resolution.

Then the switching on vertical deposition is requested. If this mode is activated the electromagnetic relay is triggered and the second microcontroller is initialized. The velocity of the substrate exposing by the solution pumping out is set by means of the second menu in the range from 0.1 to 10 mm/min. After the entered value is accepted, it is recalculated due to a predetermined formula into the motor rotational speed. Then byte data is transmitted through the RX/TX port from the first to the second micro-controller. It saves data entered and waits for the process start signal.

After all entered data are confirmed the process starts. First microcontroller sends signals to the relays, they are triggered and the process starts. One relay closes the circuit to supply the electrical potential difference to the substrates and start the elec-trophoresis process. The second relay is responsible for power transmission to the stepper motor. If the vertical deposition mode is turned on, the first microcontroller sends a signal to the second one to send a signals to the stepper motor driver and to start pumping.

Pumping is carried out using a peristaltic pump, which is connected to a stepper motor. The diameter of the tube of the peristaltic pump is selected in the way that ensures the micro-flow of the fluid to achieve the lowest rate of the solution pumping out and film deposition. For more smooth pumping a microstep mode was designed and 1/16 order step was chosen.

After a specified time the first microcontroller interrupts all signals applied to the relays, they closes the circuits and turn off the power from the rest of the system. At the same time a sound pre-programmed signal is transmitted to the built-in speaker and the process completion is indicated. The equipment display shows a message indicating that the process is complete too. After pressing any button the controller restarts, resets all entered data and after that it is possible to carry out the process again.

4   Results and Discussion

The designed automated equipment was implemented and adjusted. To test the suit-ability and operability of the device an experiment was carried out. A series of colloidal film samples was fabricated by electrophoresis process.

The automated device has been shown to produce qualitative silica opal films. We used copper polished substrate with 25 nm roughness.

The colloidal solution was synthesized by hydrolyze tetraethyl-orthosilicate in ethanol, and ammonia was used as catalyst. The silica SiO_2 particles were 250 nm diameter and they have relative size variations of less than 10%. The best results were obtained in the following conditions. Applied to the substrates potential was 3 V and a distance between the substrates was 20 mm. The experiment was carried out for 15 min, then the substrates were removed from the solution and dried for 10 h.

The structure of deposited films was investigated by atomic force microscopy (AFM) techniques. It was found that opal structure has a good periodicity, uniformity and high packing density. It indicates the layer-by-layer growth of the film. Figure 6 shows the AFM peak histogram that is peculiar to such films.

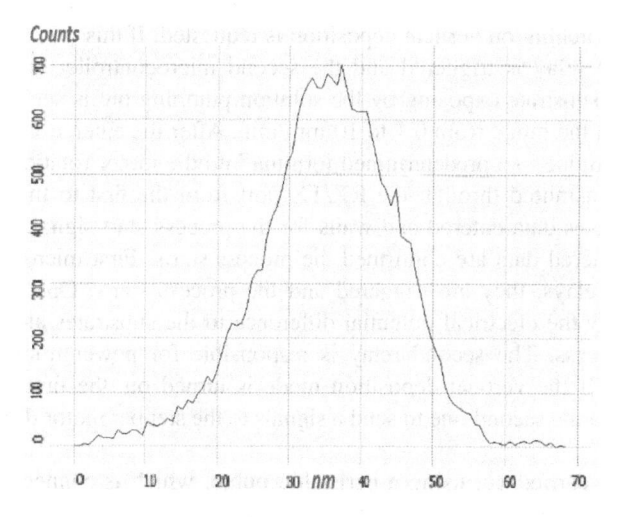

Fig. 6. AFM peak histogram of the silica opal film surface.

By using this equipment, it is possible to research the phenomenon of self-assembly. By this device one can control the process modes and optimize film deposition conditions and PhC properties.

5   Conclusion

Presented equipment provides convenient and fast deposition of highly ordered colloidal opal films. The combination of electrophoresis with vertical deposition with gravity is useful for the formation not only PhC films but PhC-based composites. Since deposition process is well controlled the device may be used for optimizing process conditions and obtaining necessary structures properties.

Now the device is successfully used in students scientific and research projects in the Chair of Electronic Technologies in Mechanical Engineering of Bauman Moscow State Technical University The results acquired with presented equipment for the experimental study of self-assembly phenomenon show that the idea of combining of two methods is very utility for research work and educational process.

References

1. Busch, K., Lolkes, S., Wehrspohn, R.B., et al.: Photonic Crystals: Advances in Design, Fabrication, and Characterization. Wiley, New York (2006)
2. Kuleshova, V.L., Panfilova, E.V., Prohorov, E.P.: Automated device for vertical deposition of colloidal opal films. In: International Russian Automation Conference, pp. 1–5 (2018). https://doi.org/10.1109/rusautocon.2018.8501782
3. Miguez, H., Blanco, A., Lopez, C., et al.: Face centered cubic photonic bandgap materials based on opal-semiconductor composites. J. Light. Technol. **17**(11), 1975–1981 (1999). https://doi.org/10.1109/50.802983
4. Armstrong, E., Dwyer, C.: Artificial opal photonic crystals and inverse opal structures – fundamentals and applications from optics to energy storage. J. Mater. Chem. **3**, 6109–6143 (2015). https://doi.org/10.1039/C5TC01083G
5. Katyba, G.M., Zaytsev, K.I., Chernomyrdin, N.V., et al.: Sapphire photonic crystal waveguides for terahertz sensing in aggressive environments. Adv. Opt. Mater. **6**(22), 1800573 (2018). https://doi.org/10.1002/adom.201800573
6. Gorelik, V.S., Yashin, M.M., Bi, D., et al.: Transmission spectra and optical properties of a mesoporous photonic crystal based on anodic aluminum oxide. Opt. Spectrosc. **124**(2), 167–173 (2018). https://doi.org/10.1134/S0030400X18020078
7. Bakhia, T., Baranchikov, A.E., Gorelik, V.S., et al.: Local optical spectroscopy of opaline photonic crystal films. Crystallogr. Rep. **62**(5), 783–786 (2017). https://doi.org/10.1134/S1063774517050029
8. Baburin, A.S., Ivanov, A.I., Trofimov, I.V., et al.: Highly directional plasmonic nanolaser based on high-performance noble metal film photonic crystal. In: Nanophot VII, vol. 10672, p. 106724D (2018). https://doi.org/10.1117/12.2307572
9. Lu, L., Joannopoulos, J.D., Soljacic, M.: Topological photonics. Nat. Photonics **8**(11), 821 (2014). https://doi.org/10.1038/nphoton.2014.248
10. Boyko, V., Dovbeshko, G., Fesenko, O., et al.: New optical properties of synthetic opals infiltrated by DNA. Mol. Cryst. Liq. Cryst. **535**(1), 30–41 (2011). https://doi.org/10.1080/15421406.2011.537888
11. Yang, S.M., Miguez, H., Ozin, G.A.: Opal circuits of light–planarized microphotonic crystal chips. Adv. Funct. Mater. **12**(6–7), 425–431 (2002). https://doi.org/10.1002/1616-3028(20020618)12:6/7%3c425:AID-ADFM425%3e3.0.CO;2-U
12. Xia, Z., Le, S., Jing, L., et al.: Tunable magneto-optical Kerr effect in Fe films underneath a two-dimensional array of polystyrene spheres covered by Au nanocaps. Chin. Phys. Lett. **30**(3), 037801 (2013)
13. Geng, C., Wei, T., Wang, X., et al.: Enhancement of light output power from LEDs based on monolayer colloidal crystal. Small **10**(9), 1668–1686 (2014). https://doi.org/10.1002/smll.201303599
14. Nelson, E.C., Dias, N.L., Bassett, K.P., et al.: Epitaxial growth of three-dimensionally architectured optoelectronic devices. Nat. Mater. **10**(9), 676 (2011). https://doi.org/10.1038/nmat3071
15. Arsenault, A.C., Puzzo, D.P., Manners, I., et al.: Photonic-crystal full-colour displays. Nat. Photonics **1**(8), 468 (2007)
16. Xia, Y., Gates, B., Yin, Y., et al.: Monodispersed colloidal spheres: old materials with new applications. Adv. Mater. **12**(10), 693–713 (2000). https://doi.org/10.1002/(SICI)1521-4095(200005)12:10%3c693:AID-ADMA693%3e3.0.CO;2-J
17. Wehrspohn, R.B., Upping, J.: 3D photonic crystals for photon management in solar cells. J. Opt. **14**(2), 024003 (2012)

18. Syritskii, A.B., Panfilova, E.V.: Investigation of opal nanostructures using scanning probe microscopy. IOP Conf. Ser.: Mater. Sci. Eng. **443**(1), 012035 (2018). https://doi.org/10.1088/1757-899X/443/1/012035

19. Freymann, G., Kitaev, V., Lotsch, B.V., et al.: Bottom-up assembly of photonic crystals. Chem. Soc. Rev. **42**(7), 2528–2554 (2013)

20. Dommelen, R., Fanzio, P., Sasso, L.: Surface self-assembly of colloidal crystals for micro- and nano-patterning. Adv. Colloid Interface Sci. **251**, 97–114 (2018). https://doi.org/10.1016/j.cis.2017.10.007

21. Li, Z., Wang, J., Song, Y.: Self-assembly of latex particles for colloidal crystals. Particuology **9**(6), 559–565 (2011)

22. Liu, G., Zhou, L., Fan, Q., et al.: The vertical deposition self-assembly process and the formation mechanism of poly (styrene-co-methacrylic acid) photonic crystals on polyester fabrics. J. Mater. Sci. **51**(6), 2859–2868 (2016)

23. Diao, J.J., Hutchison, J.B., Luo, G., et al.: Theoretical analysis of vertical colloidal deposition. J. Chem. Phys. **122**(18), 184710 (2005). https://doi.org/10.1063/1.1896352

24. Dyubanov, V., Ezenkova, D., Mozer, K.: Development of the electrochemical method of production of metamaterials. In: Technology & systems, pp. 277–282 (2018)

25. Wilms, T.T.: Direct electrohydrodynamic simulation of electrophoretic particle mobility. Eindhoven University of Technology (2012)

26. Medrano, M., Perez, A.T., Lobry, L., et al.: Electrophoretic mobility of silica particles in a mixture of toluene and ethanol at different particle concentrations. Langmuir **25**(20), 12034–12039 (2009). https://doi.org/10.1021/la900686a

27. Meijer, J.M., Hagemans, F., Rossi, L., et al.: Self-assembly of colloidal cubes via vertical deposition. Langmuir **28**(20), 7631–7638 (2012). https://doi.org/10.1021/la3007052

28. Lotito, V., Zambelli, T.: Approaches to self-assembly of colloidal monolayers: a guide for nanotechnologists. Adv. Colloid Interface Sci. **246**, 217–274 (2017). https://doi.org/10.1016/j.cis.2017.04.003

An Intelligent Automated Control System of Micro Arc Oxidation Process

P. Golubkov, E. Pecherskaya$^{(\boxtimes)}$, and T. Zinchenko

Penza State University, 40 Krasnaya street, Penza 440026, Russian Federation

Abstract. The intelligent automated control system of management of technological processes of micro arc oxidation is developed enabling to implement controlled synthesis of oxide coatings with required properties. The structure of the intelligent system of micro arc oxidation process consists of hardware, software and information support. The controlled synthesis of oxide coatings is reached by using techniques developed by the authors. The system realizes an intelligent choice of optimum technological mode based on present information of theoretical and empirical laws of process of micro-arc oxidation contained in the knowledge bank, as well as the intelligent algorithm of identification of the alloy composition on the angular factor of forming curve. Presence "parameter of technological process - property of coating" in the system of feedback in combination with the ability to adjust the technological current over a wide range, promotes maintenance of optimum technological mode throughout all the time of processing of the item. The device can be used in industries, where valve metals and alloys, (aluminum, titanium, etc.) as well as in scientific researches are used.

Keywords: The micro arc oxidation · The intelligent automated control system · Controlled synthesis · Measuring transducers · Intelligent application

1 Introduction

One of the most promising ways of increase of the resistance to the wear and corrosion of the details of machines and devices from valve metals (aluminum, titanium, magnesium etc.) and alloys is a micro arc oxidation (MAO).

It is a plasma-chemical process of hardening of the surface of preparation at the expense of transformation of outside layer of metal to high-temperature crystal aluminum oxide - corundum. Coatings produced by given way, possess high values of wearing property, corrosion stability, dielectric properties (electric strength, resistivity constant) thermal insulating properties and biocompatibility [1, 2] that opens ample opportunities to their application in engineering and transport, aviation (including drones), space-rocket industry, instrument engineering, medicine, as well as production of articles of household and special assignment [3–5].

The main problem of the MAO technology, decision of which plenty of scientific works [6–15] is devoted, is low controllability connected with insufficient level of scrutiny of process of coatings drawing and with big number of diverse factors [16, 17], in aggregate affecting physical processes proceeding in the system that makes difficult

A. A. Radionov and A. S. Karandaev (Eds.): RusAutoCon 2019, LNEE 641, pp. 1053–1061, 2020.
https://doi.org/10.1007/978-3-030-39225-3_111

to obtain oxide coatings with required properties and results in a factory defects. Under such conditions finding of optimum technological mode, in which coatings of the best quality are obtained, becomes a complicated problem, therefore problems with automation of considering technology arise. Besides being used processing equipment [18–29] does not allow to produce the control of properties of formed coatings in a real time that limits opportunities of supervision for changing of these properties and revealing of features of the MAO process due to different influencing factors.

To address the issues noted authors offer the intelligent automated control system of technological process of micro arc oxidation realizing controlled synthesis of oxide coatings with given properties. It assumes presence of feedback, in which control of processing equipment (the current source) will be produced with the aid of intelligent algorithms on the basis of measurement of the MAO process parameters and properties of MAO-coatings under conditions of real time.

2   The Structure of the Intelligent Control System of Micro Arc Oxidation Process

The intelligent control system of MAO process (Fig. 1) consists of hardware, software and information support. Hardware represents the installation of the MAO and computer.

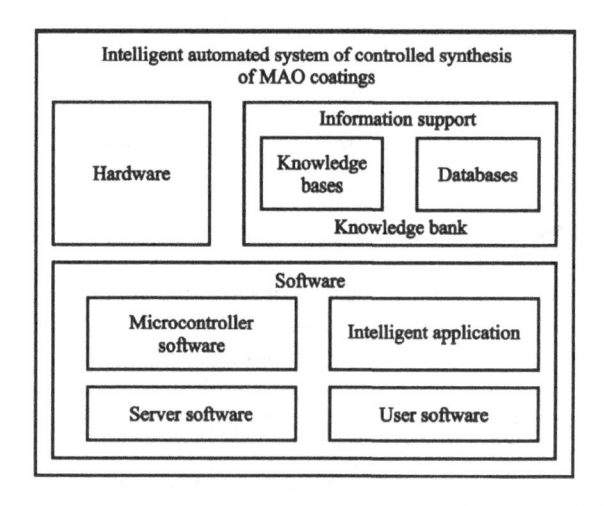

Fig. 1. The structure of the intelligent control system of micro arc oxidation process.

The installation of the MAO (Fig. 2) consists of the technological current source, measuring circuit board, microprocessor module, galvanic cell with protective fencing, the electrolyte cooling and mixing system, the power supply unit of low-voltage circuits and it is connected to the computer.

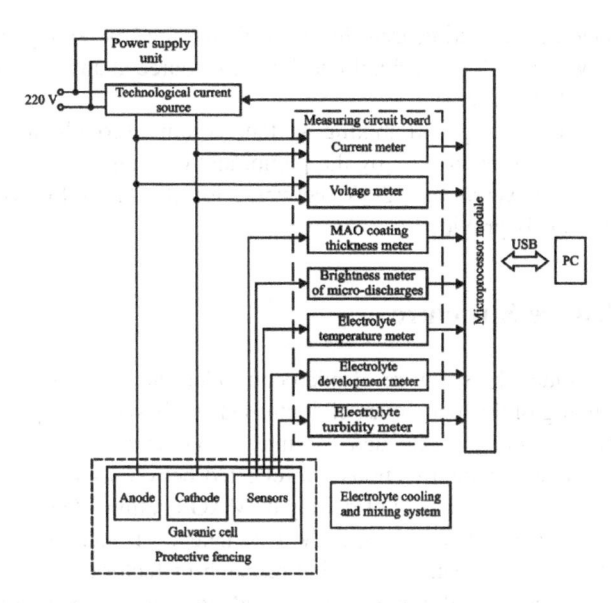

Fig. 2. The structure of hardware.

The technological current source represents a high-voltage (600 V) pulsing voltage converter. It is intended for formation the pulses of technological current coming to sample in galvanic cell, under action of which oxide coating is formed, and supplies regulation of technological current over a wide range.

Measuring circuit board unites the set of measuring transducers enabling to measure parameters of technological process (current and voltage in galvanic cell, micro-discharges brightness), properties of formed coatings (thickness) and external conditions (temperature, turbidity and electrolyte development). Signals from measuring transducers which past computer processing, are used for construction of characteristic curves of MAO process: forming curve, dynamic current-voltage characteristics, time dependences of influencing factors and properties of MAO-coatings.

Measuring transducers as well serve by the element of feedback "technological parameter of MAO process - property of MAO-coating - quality parameter", by means of which it is exercised control for process of coating formation.

Microprocessor module consists of microcontroller, digital synthesizer of signals, the converter of USB-UART interfaces and it is connected to the computer through galvanically isolated USB-port. Microprocessor module forms controlling signals for the technological current source, measuring circuit board and outlines performing service functions, and transmits information from measuring transducers to computer for further processing.

Galvanic cell represents tank from corrosion-proof steel (galvanic bath), in which there are two electrodes - the anode (processed part) and the cathode which bath itself can serve. The bath is filled with electrolyte solution usually containing sodium or potassium hydroxide and liquid glass (Na_2SiO_3). Galvanic cell is equipped by the temperature sensors, turbidity, development of electrolyte, brightness of microdischarges connected with the appropriate measuring transducers. Galvanic cell is equipped by protective fencing which switches off the technological current source at any infringements of her tightness during MAO process.

The electrolyte cooling and mixing system is flow-through, at the same time electrolyte mixing is implemented by the pump, and cooling – by the fan.

The device is powered by 220 V network, for powering low-voltage outlines separate supply unit is provided.

3   The Software Structure

The software includes the software of microcontroller and server software, intelligent application realizing of the techniques of controlled synthesis developed by the authors and user software which is intended for interaction with user. It has interface as oriented graph, on which interrelations are reflected between factors of technological process, properties and parameters of quality of MAO-coatings. On graph the operator can choose studied interrelation, specify parameters of technological mode and required specifications of coating. If during research or factories unaccounted interrelation is found out, additional graph arches are added. The microcontroller software is intended to control the power part of the installation (the technological current source), processing of signals from measuring transducers and transfer of received information on PC, as well as for performance of service functions, (service of protective fencing, error messages input, indication and so forth). Server software is responsible for set-up of the system and control of microcontroller. The data about optimum operating mode for obtain particular properties of determined sample are formed by intelligent application on the basis of algorithms of controlled synthesis and are loaded to microcontroller. Optimum technological mode is supported at the expense of feedback, at which by the ground for change of work of the equipment by intelligent application information of properties of formed coating (for example, of thickness) produced serves at processing of signals from measuring transducers.

Information support (Fig. 3) represents the knowledge bank containing knowledge bases and databases. It is conventional the contents of the knowledge banks and can be segregated into four subsystems.

The subsystem of MAO-coatings contains information of properties and parameters of quality of MAO-coatings, as well as of influencing factors of MAO process. In the subsystem of theoretical researches knowledge of physical and chemical laws loaded additionally, and they are concentrated at research of MAO process, as well as of mathematical expressions existing at present, describing interrelations between technological parameters of MAO process, properties and parameters of quality of MAO-coatings. To the subsystem of experimental researches information of methods and the measuring aids of parameters of technological process of the MAO and properties of MAO-coatings, their metrological characteristics, as well as about used technological modes is included. Reference subsystem contains information of the mechanism of MAO process and MAO-coatings application. All knowledge bases and databases

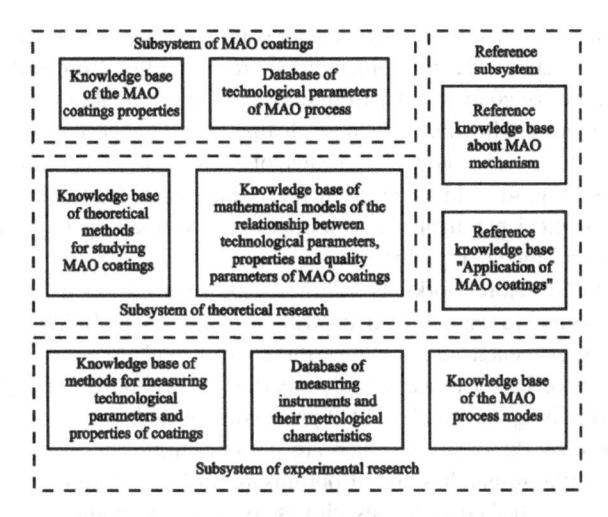

Fig. 3. The structure of the knowledge bank.

presented in the knowledge bank, has opportunity of addition that allows to add new one and to specify existing mathematical models and technological modes of MAO process, to correct methods of measurement, in such a manner, improving work of all system.

4 The Algorithm of Software Work

The scheme of information exchanging between the components of software is presented on the Fig. 4. Analog signals from measuring transducers come to microcontroller, transform into digital form through ADC. Then software of microcontroller sends these data through USB-interface to computer, where server software produces recalculation of these data to customary units of measurements, and results of measurements are displayed as schedule on the screen of the computer in a real of the time through user software.

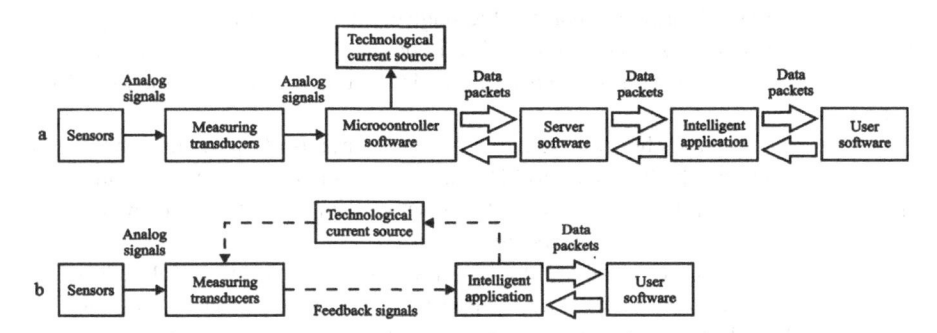

Fig. 4. (a) exchange of information between the components of software without feedback; (b) exchange of information between the components of software with feedback.

Simultaneously with that intelligent application analyses these data, and if measured values of technological parameters of process differ from optimum, and the values of properties of coating differ from present mathematical models designed for to the base, intelligent application rigs these distinctions, change the operating mode of the technological current source by means of formation of controlling signals for microcontroller (Fig. 4a). In such a manner intelligent feedback is implemented between technological parameters and properties of formed coating supporting optimum technological mode (Fig. 4b).

The software structure works as follows. Two software operating mode are supported: experimental researches and controlled synthesis of MAO-coatings. Due to conditions of experimental researches verification of mathematical models of interrelations of factors of MAO process and properties of produced coatings is produced. During the work in given mode MAO-processing is led pursuant to technological parameters given by the operator which are not optimum. The given mode is mainly applied for scientific researches and at optimisation of the technology on production.

The mode of controlled synthesis allows to define optimum parameters of technological mode using present information of interrelations of factors of MAO process and properties of produced coatings contained in the knowledge bank on the basis of intelligent algorithms.

At the same time the operator should enter required properties of MAO-coating, specifications of the detail and limiting conditions specifying range of the values of used technological parameters and determined on the basis of design-technological and technical and economical requirements.

The result of work of the program is a few sets of the values of technological parameters, appropriate to optimum modes of processing, in which MAO-coating with required properties turns out.

The algorithm of the alloy composition identification based on time dependences of the voltage drop on the sample (forming curve) has a different angular factor in the initial stage of process of the MAO for different alloys (Fig. 5) is realized in intelligent application. Angular factor for studied alloy can be defined on expression:

$$k = \frac{\partial U}{\partial t},\qquad(1)$$

where U is measured forming voltage at the stage of anodizing (area I on the Fig. 5), t – time of MAO processes. The studied alloy is defined by the way of comparison of received angular factor with the values of angular factors contained in the knowledge bank further.

Further the operator chooses one of the received variants of optimum technological mode, pursuant to which further MAO process is taking place. At the same time intelligent application forms and transmits the set of commands to the microcontroller which controls all devices in the system.

Before the beginning of MAO process a series of checks of initial state of the system is executed: test for checking composition of electrolyte and for tightness of galvanic cell. If electrolyte is strongly depleted with ions, it is inexpediently to conduct MAO process owing to possible article defect, and at disconnected terminal switch of

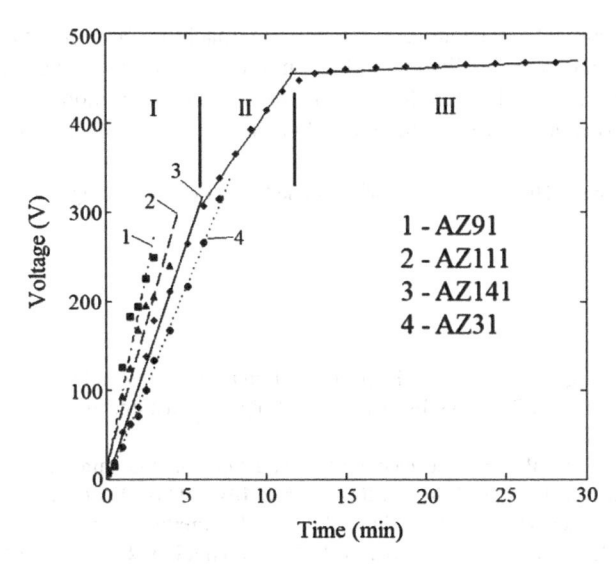

Fig. 5. Forming curves of the MAO process for different magnesium alloys: 1 - AZ91, 2 - AZ111, 3 - AZ141, 4 - AZ31. Stages of process of the MAO: I is stage of anodizing; II is sparking; III – micro arc discharges.

protective fencing of galvanic cell the technological current source is switched off automatically to avoid defeating by electric current.

After successful completion of all checks the microcontroller sets up the technological current source and begins to execute control commands received from intelligent application, realizing optimum technological mode of MAO process. At this time the microcontroller can execute a series of independent tasks connected with measurements of technological parameters and properties of formed coatings. Optimum values of technological parameters are supported by constant throughout all MAO-processing by means of intelligent feedback described earlier. MAO-processing ends either at achievement of required MAO-coating thickness, or at transition of process to stage of arc discharges that is registered according the change of their brightness.

After the completion of MAO process all used data (sample parameters, used alloy, the modes of processing etc.) are preserved in the knowledge bank for further research and application in production of coatings.

5 Conclusion

Offered intelligent automated control system of technological process of micro arc oxidation allows to increase quality of produced coatings due to algorithm of intelligent choice of optimum technological mode enabling to obtain oxide layers on the surface of detail with required properties. Feedback "technological parameter - property of MAO-coating" promotes maintenance of optimum technological mode throughout all processing of the detail.

Offered technical decisions can be used in machine building, instrument engineering, textile industry, medicine, production of drones and oil and gas equipment, the articles of special and household assignment and in other industries, where valve metals and alloys, as well as in the area of scientific researches are used.

Acknowledgments. The reported study was funded by RFBR according to the research project № 19-08-00425.

References

1. Yao, Z., Shen, Q., Niu, A., et al.: Preparation of high emissivity and low absorbance thermal control coatings on Ti alloys by plasma electrolytic oxidation. Surf. Coat. Technol. **242**, 146–151 (2014)
2. Chung, C.J., et al.: Plasma electrolytic oxidation of titanium and improvement in osseointegration. J. Biomed. Mater. Res., Part B **101**, 1023–1030 (2013)
3. Liu, Y., Liskiewicz, T., Yerokhin, A., et al.: Fretting wear behavior of duplex PEO/chameleon coating on Al alloy. Surf. Coat. Technol. **352**, 238–246 (2018)
4. Gnedenkov, S.V., Sinebryukhov, S.L., Mashtalyar, D.V., et al.: Composite fluoropolymer coatings on the MA8 magnesium alloy surface. Corros. Sci. **111**, 175–185 (2016)
5. Chien, C., Hung, Y., Hong, T., et al.: Preparation and characterization of porous bioceramic layers on pure titanium surfaces obtained by micro-arc oxidation process. Appl. Phys. A **123**, 204 (2017)
6. Krishtal, M.M., Ivashin, P.V., Polunin, A.V., et al.: The effect of dispersity of silicon dioxide nanoparticles added to electrolyte on the composition and properties of oxide layers formed by plasma electrolytic oxidation on magnesium 9995A. Mater. Let. **241**, 119–122 (2019)
7. Wei, F., Zhang, W., Zhang, T.: Effect of variations of Al content on microstructure and corrosion resistance of PEO coatings on Mg-Al alloys. J. Alloy. Compd. **690**, 195–205 (2017)
8. Mohedano, M., Mingo, B., Arrabal, R., et al.: Role of particle type and concentration on characteristics of PEO coatings on AM50 magnesium alloy. Surf. Coat. Technol. **334**, 328–335 (2018)
9. Darband, B., Aliofkhazraei, M., Hamghalam, P., et al.: Plasma electrolytic oxidation of magnesium and its alloys: Mechanism, properties and applications. J. Magnes. Alloy. **5**, 74–132 (2017)
10. Friedemann, A.E.R., Thiel, K., Hablinger, U., et al.: Investigations into the structure of PEO-layers for understanding of layer formation. Appl. Surf. Sci. **443**, 467–474 (2018)
11. Nabavi, H.F., Aliofkhazraei, M., Rouhaghdam, A.S.: Morphology and corrosion resistance of hybrid plasma electrolytic oxidation on CP-Ti. Surf. Coat. Technol. **322**, 59–69 (2017)
12. Cheng, Y., et al.: The effects of anion deposition and negative pulse on the behaviours of plasma electrolytic oxidation. A systematic study of the PEO of a Zirlo alloy in aluminate electrolytes. Electrochim. Acta **225**, 47–68 (2017)
13. Sowa, M., Worek, J., Dercz, G., et al.: Surface characterization and corrosion behavior of niobium treated in a Ca- and P-containing solution under sparking conditions. Electrochim. Acta **198**, 91–103 (2016)
14. Xia, Q., et al.: Effects of electric parameters on structure and thermal control property of PEO ceramic coatings on Ti alloys. Surf. Coat. Technol. **307**, 1284–1290 (2016)

15. Cheng, Y., Cao, J., Mao, M., et al.: Key factors determining the development of two morphologies of plasma electrolytic coatings on an Al–Cu–Li alloy in aluminate electrolytes. Surf. Coat. Technol. **291**, 239–249 (2016)
16. Golubkov, P.E., Pecherskaya, E.A., Shepeleva, Y.V., et al.: Methods of applying the reliability theory for the analysis of micro-arc oxidation process. J. Phys.: Conf. Series **1124**, 081014 (2018)
17. Golubkov, P.E., Pecherskaya, E.A., Karpanin, O.V., et al.: Automation of the micro-arc oxidation process. J. Phys.: Conf. Series **917**, 092021 (2017)
18. Mashtalyar, D.V., Gnedenkov, S.V., Sinebryukhov, S.L., et al.: Composite coatings formed using plasma electrolytic oxidation and fluoroparaffin materials. J. Alloy. Compd. **767**, 353–360 (2018)
19. Lin, C.S., Fan, Z.H., Chen, P.C., et al.: The study of remote monitoring and real-time signal processing of the pulse generator for thin film coating. J. Mater. Sci.: Mater. Electron. **28**, 3234–3242 (2017)
20. Egorkin, V.S., Vyaliy, I.E., Sinebryukhov, S.L., et al.: Composition, morphology and tribological properties of PEO-coatings formed on an aluminum alloy D16 at different duty cycles of the polarizing signal. Compos. Multipurp. Coat. **42**, 12–16 (2017)
21. Bolshenko, A.V., Pavlenko, A.V., Puzin, V.S., et al.: Power supplies for microarc oxidation devices. Life Sci. J. **11**, 263–268 (2014)
22. Bolshenko, A.V., Pavlenko, A.V., Grinchenkov, V.P., et al.: Current controllers for devices of microplasm oxidation. Russ. Electr. Eng. **83**, 260–265 (2012)
23. Borikov, V.N., Baranov, P.F., Bezshlyakh, A.D.: Virtual measurement system of electric parameters of microplasma processes. In: Proceedings of the SIBCON Conference, pp. 275–279 (2009)
24. Borikov, V.N.: Measurement system for coating quality control during high-current process in electrolyte solution. In: Proceedings of the ISMQC Conference, pp. 287–291 (2007)
25. Vagaska, A., Gombar, M.: Comparison of usage of different neural structures to predict AAO layer thickness. Tech. Bull. **24**, 333–339 (2017)
26. Lomakin, V.V., Zaitseva, T.V., Putivzeva, N.P., et al.: Implementation of the decision-making support in the management of microarc oxidation process on the basis of artificial neural networks. Sci. Bull. **244**, 124–133 (2016). Series of Economics and Informatics
27. Vagaska, A., Michal, P., Gombar, M., et al.: Simulation of technological process by usage neural networks and factorial design of experiments. MM Sci. J. **3**, 999–1003 (2016)
28. Borikov, V.: Neural method alloys identification by the microplasma oxidation process in the electrolyte solutions. Materialwiss. Werkstofftech. **37**, 915–918 (2006)
29. Golubkov, P., Pecherskaya, E., Karpanin, O., et al.: Intelligent automated system of controlled synthesis of MAO-coatings. In: Proceedings of the 24th Conference of FRUCT Association, pp. 96–103 (2019)

Improving the Parametric Reliability of Automatic Control Systems in Electric Drives

A. V. Saushev$^{(\boxtimes)}$, P. V. Adamovich, and O. V. Shergina

Makarov State University of Maritime and Inland Shipping,
5/7 Dvinskaya Street, St. Petersburg 198515, Russian Federation
saushev@bk.ru

Abstract. The paper dwells upon the methodology of improving the parametric reliability of automated electric drives. One of the basic ways to address this issue consists in optimal parametric synthesis. Apparently, the most important indicator of parametric reliability of electric drives is the margin of operability. Its value must be defined in the space of internal system parameters. Analysis shows that parametric synthesis of electric drives is a multi-attribute optimization problem. Margin of operability is proposed as the optimality criterion. An heuristic optimization method is considered applicably to automatic controls in electric drives; the method is supposed to synthesize the exclusive most optimal solution. The paper presents an electric drive state control algorithm. The proposed methods are based on information on the electric drive operability domain. The proposed methodological approach and algorithms can provide state control and margin-of-operability optimized parametric synthesis at any stage of the drive or control system life cycle.

Keywords: Parametric reliability · Margin of operability · Parametric synthesis · State control · Electric drive · Control systems

1 Introduction

The operability of various technical systems is a long-studied problem. With respect to automated electric drives (AED), it has become relevant after semiconductor chips found wide-scale use in control systems. Application of electronics greatly improves the functional characteristics of control systems. At the same time, it complicates the AED design and leads to higher failure rate with a greater proportion of progressive failures. The parametric instability of AED is the primary cause of lower operational reliability. It is apparently imperative to consider the parametric reliability in the development and production of automatic control systems (ACS) for AED or their elements [1–3]. Doing so reveals that statistical data on patterns in the parameters of AED elements is either too limited or unknown. This attributed to the emergence of functional-parametric research area in the reliability theory, which implies that the operating reliability of a system could be controlled, and the technical condition of such system could be found even if there is only scarce data or not data at all [4].

© Springer Nature Switzerland AG 2020
A. A. Radionov and A. S. Karandaev (Eds.): RusAutoCon 2019, LNEE 641, pp. 1062–1070, 2020.
https://doi.org/10.1007/978-3-030-39225-3_112

2 Statement of Problem

One of the primary ways to improve the parametric reliability of an AED consists in optimal parametric synthesis, which boils down to solving two basic problems: find the rated values of the internal system parameters; and find their permissible range. The internal parameters are such AED parameters that describe the system state and properties. When designing a system, these parameters define the vector X of controllable (variable) parameters. These parameters are often referred to as primary parameters as changing them will affect the state of any technical system. The mathematical functional AED model is an algorithm for computing the vector of output parameters Y given the vector of internal parameters X and the vector of external parameters V.

The external parameters describe the properties and influence the functioning of the environment external to the AED. For most AED, this influence is quite significant and diverse [5].

The output parameters are the AED properties of consumer's interest. These are functional parameters, i.e. functional dependencies of phase variables and the boundary values of external parameters, within which an AED remains operational.

At the stage of parametric synthesis, the output parameters are the intended use, parametric reliability, and cost-effectiveness [6].

Papers [6–9] show that for most electrotechnical systems, including electric drives, it is the margin of operability that could be used as the parametric reliability given limited statistical data on the distribution of internal AED parameters. This value is determined by the domain, in which an AED remains operational, or the operability domain. The operability domain $G = P \cap M$ specifies a set of permissible values for internal parameters, at which the AED output parameters are compliant with all applicable requirements; the domain is defined by the operability conditions [10]:

$$Y_{j\min} \leq Y_j = F_j(X) \leq Y_{j\max}, j = \overline{1, m};$$
$$X_{i\min} \leq X_i \leq X_{i\max}, i = \overline{1, n},$$

(1)

where $Y_{j\max}(X_{i\max})$, $Y_{j\min}(X_{i\min})$ are, respectively, the permissible minimum and maximum of the jth output parameter Y_j (ith internal parameters X_i); F is the operator that links output and internal parameters; D and P are the domains of tolerance defined by inequalities 1 and 2, respectively (1). The domain D in the space of internal parameters corresponds to the tolerance domain M.

The primary challenges of AED operation are to predict the electric drive state at a given time as well as at the coming time.

Paper [3] states that the design and operation of electrotechnical systems are two problems that must be considered from the common standpoint of parametric and structural control over the state of such systems. For parametric synthesis and state diagnosis, the margin of operability is the most important indicator.

This paper considers the margin of operability applicable to AED ACS as a basic indicator of parametric reliability; it also solves the problem of improving the parametric reliability of AED by optimizing the ACS parametric synthesis. The solution uses data on the AED operability domain.

3 Theory

3.1 Reliability Indicators

For AED and their ACS, the most important reliability-defining properties are fail-safety, durability, maintainability, and persistence. With respect to parametric synthesis, the core properties are fail-safety and durability [5].

Fail-safety is the ability of a system to run continuously in a fail-safe condition while staying fully functional over a certain period of time in a certain application. Durability is the AED ability to function as required in a certain application under certain maintenance and repairs until reaching the limit state.

The key indicators of fail-safety and durability that can be used to generate the objective function are the probability of fail-safe operation (i.e. the chance that the facility will not fail over a certain operating time) and mean service life, i.e. its mathematical expectation. This the literature often uses the concept of residual resource, it is the mean (mathematical expectation of) residual service life that can be used as the indicator.

As noted above, optimal parametric synthesis of an AED is about optimizing the internal parameters of the electric drive. Only the fail-safety is directly controllable; it correlates with the durability indicators to a certain degree. Indeed, adjusting the AED internal parameters will only maximize the time an electric drive stays operational; it cannot maximize the total operating time from the operation onset (or the onset of condition monitoring) to limit state. Thus, the probability of fail-safe operation is the most important indicator of AED reliability for parametric synthesis.

This indicator can be written as the probability of satisfying the conditions of operability

$$
\begin{aligned}
P_{par}(T) = P_T(X_{nom}, I) = P\{Y_j(X(t) \in [Y_{jmin}, Y_{jmax}]\}, \\
j = \overline{1, m}, \forall t \in [0, T],
\end{aligned}
\tag{2}
$$

or as the probability that the vector of the primary AED parameters $\mathbf{X}$ will be in the operability domain G, which is the set of parametric values, at which an AED meets all the requirements [10]:

$$
P_{par}(T) = P_T(\mathbf{X}_{nom}, \mathbf{I}) = P\{(\mathbf{X}(t) \in G, \forall t \in [0, T]\},
\tag{3}
$$

$X_{nom} = [X_{1nom}, \ldots, X_{inom}, \ldots, X_{nnom}]^T, I = [I_1 \ldots, I_i, \ldots I_n]^T$ are the vectors of rated parameter values and relative tolerances (scattering fields), respectively, pertaining to the accuracy classes; $X(t) = [X_1(t) \ldots, X_i(t), \ldots X_n(t)]^T$ is a random vector process of changes in the parameters of AED elements over time $[0, T]$.

Using the probability of fail-safe operation as a quality indicator when stating the problem of AED parametric synthesis requires statistical data on the distribution of primary parameters. Analysis shows such data is not available for most AED.

Margin of operability is another common indicator of mass-manufacturability and reliability with respect to progressive failure. Multiple papers [8, 11, 12] define the margin of operability as the degree to which the operating conditions match the operability conditions; these papers evaluate the indicator by each of the output parameters Y_j, $\overline{1,m}$. The margin of operability $\lambda_j^c(\mathbf{X})$ is usually calculated by the formula:

$$\lambda_j^c(X) = (Y_{j\lim} - Y_{jnom}(X))/\delta_j - 1, j = \overline{1,m}, \tag{4}$$

where $Y_{j\lim}$ is the limit (maximum Y_{jmax} or minimum Y_{jmin}) permissible value of the j-th output parameter Y_j; $Y_{jnom}(\mathbf{X})$ is the rated value of the parameter Y_j; $\delta_j = Y_{jnom}(\mathbf{X}) - Y_{pj}(\mathbf{X})$ is the normalization parameter that characterizes the scattering of the parameter Y_j via the quantile $Y_{pj}(\mathbf{X})$ of its distribution at $P \approx 0$.

This definition of the margin of operability does not convey the essence of this concept and is not applicable to the problem under consideration [6]. The margin of operability shall be evaluated at the material and structural level, as changes in Y_j are merely a consequence of the altered primary AED parameters, whereas the dependency $Y_j(\mathbf{X})$ is usually nonlinear. Such a slight change in the primary parameters due to natural aging and wear may significantly alter the AED output parameters and render it in operational.

Define the margin of operability as the degree of approximation of the vector $\mathbf{X}_t$ (actual system state) to the maximum permissible value $X_{\lim}$. The set of the maximum permissible values of $X_{\lim}$ is determined by the boundaries of the AED operability domain. The degree of approximation of the vector X_t is set as the distance from the vector end to the nearest boundary point. Let $X_d = [X_{d1}, X_{d2}, \ldots, X_{dh}]$ be a boundary point. The minimum distance of the primary-parameter vector $X_t = [X_{1t}, X_{2t}, \ldots, X_{nt}]$ from the vector $\mathbf{X}_d$ by all boundary-point values is what constitutes the margin of operability ρ.

$$\rho = \min_{[X_d]} \sqrt{\sum_{i=1}^{n} (X_i - X_{di})^2}. \tag{5}$$

If the rated internal parameters $X_{nom} = [X_{1nom}, X_{2nom}, \ldots, X_{nnom}]$ are known, one can find the rated margin of operability ρ_{nom}, which is better expressed in relative units $\lambda(\mathbf{X})$:

$$\rho_{nom} = \min_{[X_d]} \sqrt{\sum_{i=1}^{n} (X_{inom} - X_{di})^2}, \ \lambda(X) = \rho/\rho_{nom}. \tag{6}$$

Unlike $\lambda_j^c(X)$, the margin of operability $\lambda(X)$ considers both external and internal AED operating conditions.

Literature review indicates that high reliability is of paramount importance for most AED. Thus, the probability of fail-safe operation or the margin of operability are the most recommended objective functions for parametric optimization. Direct use of the most important parametric reliability indicator (fail-safe operation probability) as the objective functions for internal parametric optimization may be inefficient due to high computational intensity and low sensitivity unless in the vicinity of the operability boundaries.

Besides, if there is now statistics on the AED parameter distribution, this probability cannot be used as an objective function at all. The margin of operability $\lambda(X)$ and the minimum margin of operability $\lambda_i(X)$ i.e. min $\lambda_i(X)$ do not have this disadvantage. Besides, unlike other criteria, these can produce any Pareto-optimal solution [13, 14]. Paper [6] proves that in the operability domain, the probability $P_i(X)$ of ensuring the i-th tolerance condition is a monotonically increasing function of the margin of safety $\lambda_i(X)$ and $\max_{X \in G} P_i(X) = P_i(\lambda_i(X_0))$, where $\lambda_i(X) = \max_{X \in G} \lambda_i(X)$.

3.2 Quality Indicators and Optimality Criteria

Parametric synthesis of AED control systems usually boils down to configuring the controllers. Such configurations are often based on standard settings adjusted for technical and symmetric optima [15]. This is not a universally good solution. The approach only considers two dynamic indicators: the transient time and the maximum overshoot. Optimal synthesis of the gain coefficients and controller time constants uses the main indicator [15].

Increasingly common are state-observer controllers (SOC). These feature a fairly flexible controller structure; besides, a controller of this type requires only a single sensor of the controlled object's output coordinate. Parametric optimization of such systems often uses model synthesis based on standard distributions of characteristic polynomial roots. Such calculations ignore the parametric reliability and possible limitations on coordinates inherent in AED.

Papers [2, 16] note that using parametric optimization allows the engineer to comply with all the requirements to AED ACS. SOC synthesis requires not indirect optimality criteria; rather, it uses AED control quality requirements, including performance, accuracy, energy efficiency, and parametric reliability [1].

Thus, optimal parametric synthesis of AED and control systems is a multi-attribute (vector) optimization problem. The methodology of solving multi-attribute optimization problems of electrotechnical systems remains relevant.

3.3 Objective Function

Paper [16] proposes an extended quality criterion that includes normalized performance $\bar{Y}_1$, accuracy $\bar{Y}_2$, parametric roughness $\bar{Y}_3$ and control-related energy costs $\bar{Y}_4$:

$$\bar{Y}_m = 1 - \left(\prod_{i=1}^{m} \bar{Y}_i \right)^{1/m} \tag{7}$$

where $m = 4$ is the number of control quality indicators included in the criterion. The performance indicator $\bar{Y}_1$ here is the transient response rise time t_{H}. The accuracy indicator $\bar{Y}_2$ is the mean absolute value of the relative deviation in output coordinates at t_{H} to $t_{\text{П}}$:

$$\bar{\sigma}_y = \frac{1}{N - T} \sum_{j=T}^{N} \left| \frac{Y_j - Y_3}{Y_3} \right| \cdot 100\%, \tag{8}$$

where $T = t_{\text{н}}/T_0$ и $N = t_{\text{п}}/T_0$ is the relative rise/transient time, T_0 is the quantization period.

The proposed robustness $\bar{Y}_3$ is the mean absolute value of the deviation of transient response as caused by varying the controlled parameters against their design values over the time of the transient.

$$\bar{\sigma}_R = \frac{1}{N} \sum_{j=1}^{N} \left| \frac{Y_j - Y_j^*}{Y_j} \right| \cdot 100\%, \tag{9}$$

where Y_j, Y_j^* are the output coordinates at design/altered parameters.

For control energy costs $\bar{Y}_4$, the researchers propose using the peak motor current, which is limited to the overloading capacity of the power system.

The most important drawback of the criterion (7) is its multiplicative form, which is subjective and limited [14].

Literature review indicates that high reliability is of paramount importance for most AED. For parametrically unstable AED, the parametric reliability requirements are of essence. With regard to the problem at hand, this means that optimization must consist in maximizing the AED margin of operability. Paper [17] explains and analyzes the selected objective function with respect to the problems of synthesizing electrotechnical systems.

4   Implementation

4.1   Parametric Synthesis by the Margin of Operability

With respect to electrotechnical systems, there has been developed an optimal parametric synthesis method that uses data on the operability domain boundaries [6]. The method is guaranteed to produce the exclusive correct solution for the most general case. This idea behind this method, referred to as the decremental domain method, is to sequentially decrement the operability domain until the optimal solution is found. Tests and full-scale experiments prove it efficient as a solution for parametric synthesis of AED ACS. Consider the essence of the method.

First, use the known methods [10] to find the boundary of G, which consists of a finite number of hypersurfaces Φ_j, each of which can to a certain degree of error by described by equations $\Phi_j(X) = 0$, where $\Phi_j(X) = Y_{j\max} - F_j(X)$ or $\Phi_j(X) = F_j(X) - Y_{j\min}$ are the constraint functions in the system of inequalities (1). A convenient way to do so is to use experiment planning methods [18]. In the internal-parameter space R^n, introduce the metric l, which is a function of coordinates of two any points in this space, e.g. A and B. If, for instance, Point A is a boundary point of the domain G, while B is within this domain, and its coordinates describe the AED state at a time, this metric will define the margin of operability λ of the electric drive and will be used as the criterion to decrement the initial domain $G^{(0)}$ so as to locate the point of optimum. The domain $G^{(0)}$ is the decremented by the criterion $l = \lambda$. To that end, the criterion alters Δl i.e. $l^{(1)} = l^{(0)} + \Delta l$; then describe the boundaries of the domain $G^{(1)}$ analytically and define the necessary set of boundary points. Then generate equations $\varphi_j^{(1)}$ to describe the domain $G^{(1)}$ [6].

Use logical R-functions [10, 19] and the obtained equations $\varphi_j^{(1)}$ to write a single equation that has a pre-specified procedural error in the analysis of the tolerance domain $G^{(1)}$ at search step 1:

$$\begin{cases} G^{(1)} = 0,5\left(G_{2(m+n)-1}^{(1)} + \varphi_{2(m+n)}^{(1)} - \left|G_{2(m+n)-1}^{(1)} - \varphi_{2(m+n)}^{(1)}\right|\right); \\ G_j^{(1)} = 0,5\left(G_{j-1}^{(1)} + \varphi_j^{(1)} - \left|G_{j-1}^{(1)} - \varphi_j^{(1)}\right|\right); G_1^{(1)} = \varphi_1^{(1)} \end{cases} \quad (10)$$

Similarly find the tolerance domain $G^{(2)} \in G^{(1)}$ and repeated decrementing the original domain $G^{(0)}$ in cycles until optimal solution is found at $N = 1$.

When to stop searching shall be determined based on such internal parameter values, at which the domain $G^{(\mu)}$ degenerates to a point at the given error. The values of X_{opt} at this point determine the optimal (maximum) value of l, which corresponds to the condition $N = 1$.

4.2   AED State Assessment Methods

Assessing the technical condition of AED or their ACS consists in finding whether the vector $Y(t)$ is part of D or $X(t)$ is part of G, as well as finding the margin of operability at $Y(t) \subset D$. What makes the control problem difficult to solve is the need to plot the boundary of the operability domain, which may have a very complex configuration; another complication is the need to compute l_t and l_0. Besides, finding the vector $X(t) = \{X_1(t), X_2(t), \ldots, X_{nom}(t)\}$ requires monitoring all the n parameters of X, which might be problematic if the space R^n is a high-dimension space. To simplify the recognition problem, the operability domain is usually approximated by the largest-volume inscribed hyper parallelepiped. Thus, the permissible range of primary parameters is set for each parameter independently. Analysis shows [10] that such approach results in excess error, which grows nonlinearly as a function of the number of controlled parameters. Besides, solutions produced by such methods may be ambivalent for non-simply connected domains of operability. This makes imperative developing a simple yet reliable AED control method.

To solve the problem, link the primary parameters X to the controlled (measurable) parameters Z. The proposed approach is essentially as follows. It is known that any dynamic system with a given error can be approximated by a second-order system and identified using the transient response $h(t)$ or the frequency transfer function $W(j\omega)$. In this case, X and Z are related as $Z = \psi(X)$, and the domain G in the space R^h of the parameters Z corresponds to the tolerance domain F [6].

During parametric synthesis [6], the domain G breaks down into multiple subdomains G_i, each of which defines the margin of operability λ_i. Based on the projection $\Phi_Z : G \rightarrow F$, the domains F_i correspond to the domains G_i in the parameter space Z.

Obtain the equation for the domain F. To that end, for each hypersurface f_j of the boundary of G, generate a set N_q of boundary points equal to the number of significant coefficients in the desired equations. Use these points to obtain equations for the hypersurfaces comprising the domain F. Consider that

$$X_i^{(1)} = X_i^{(0)} + \left(\partial f_j^{(0)}(X) \Big/ \partial X_i \right) \Delta l_1 \Big/ grad \partial f_j^{(0)}(X). \tag{11}$$

Similarly find the boundaries of the subdomains F_i. Each subdomain $G_i(F_i)$ corresponds to a specific AED margin of operability. For instance, for subdomain $G_1(F_1) : \lambda_1 \in [0; l_1/l_0]$, for $G_2(F_2) : \lambda_2 \in [l_1/l_0; l_2/l_0]$.

Assessing the AED condition boils down to recognizing the parameters Z in the space R^c, where Z are the parameters that show whether the current AED state vector is part of this or that subdomain F_β, $\beta = \overline{1, S}$, is the number of subdomains, each of which had its margin of operability identified. The recognition problems are solve as follows. If substituting the monitored values Z_q, $q = \overline{1, k}$ in the latest equation produces $F < 0$, the AED is fully functional; then check whether the inequalities $F_\beta < 0$, $\beta = 1, 2, \ldots, (S+1)$ hold. If $F_\beta \leq 0$ and $F_{\beta+1} > 0$, then the vector of the current AED state lies within F_β, and the margin of operability equals λ_β. If $F > 0$, the device is non-operational. This requires parametric adjustment of the configurable AED parameters [6].

5   Conclusion

Modern electric-drive ACS are complex-structured solutions that keep the AED dynamic quality indicators and regulated properties within the specified limits. However, improving the reliability of such systems with respect to progressive failure is not a well-researched topic. To indicate the parametric reliability of AED ACS, this paper proposes the margin of operability, which describes the probability of fail-safe operation in the absence of statistical data on patterns in the primary electric drive parameters. Parametric synthesis of AED ACS for the margin of operability will improve the parametric reliability of electric drives. The paper proposes a method that guarantees an optimal solution. The considered approach to AED state assessment is reliable and convenient.

References

1. Anisimov, A.A., Tararikin, S.V., Apollonsky, V.V.: Parametrical optimization of controllers and state observers in electromechanical systems. Bull. Ivanovo State Power Eng. Univ. **2**, 21–26 (2016)
2. Selivanov, V.A.: The criteria for the optimization and the necessity for the construction of electric drive parametric systems. Bull. Belarusian-Russ. Univ. **1**(30), 120–124 (2011)

3. Saushev, A.V.: Electrotechnical system state control methods. SPGUVK, St. Petersburg (2004)
4. Abramov, O.V.: Functional-parametrical direction of risks theory: possibilities and prospects. Bull. Far-East. Branch Russ. Acad. Sci. 4(188), 96–101 (2016)
5. Saushev, A.V., YeV, B., Demidova, G.L.: Reliability indicators at parametrical synthesis of the automated electric drives. Bull. Admiral Makarov State Univ. Marit. Inland Shipp. 10(3–49), 597–607 (2018)
6. Saushev, A.V.: Parametrical synthesis of electrotechnical devices and systems. Admiral Makarov State University of Maritime and Inland Shipping, St. Petersburg (2013)
7. Saushev, A.V.: Methods and algorithms of parametrical synthesis of technical systems on the basis of areas of working capacity. Inf. Technol. 12, 24–29 (2012)
8. Abramov, O.V., YaV, K., Nazarov, D.A.: Optimal parametric synthesis with respect to working capacity criterion. Control. Sci. 6, 64–69 (2007)
9. Abramov, O.V.: Parametric synthesis of stochastic systems with respect to reliability. Science, Moscow (1992)
10. Saushev, A.V.: Operability domains in electrotechnical systems Polytechnika, St. Petersburg (2013)
11. Antashev, G.S.: Methods for parametric synthesis of complex technological systems. Science, Moscow (1989)
12. Leonov, D.V., Fin, V.A.: Experimental estimation of margin of operability in wireless electronics. Fundam. Probl. Radioengineering Device Constr. 13(4), 71–73 (2013)
13. Kuznetsova, O.A.: Multi-attribute optimization of electromechanical systems with asynchronous motors. Drive Technol. 6, 20–26 (2010)
14. Saushev, A.V., Bova, Ye.V., Belousov, I.V.: Electrotechnical design. Admiral Makarov State University of Maritime and Inland Shipping, St. Petersburg (2015)
15. Stashinov, Y.: Absolute-value optimal configuration of DC motor controls. Electrotechnics 1, 2–7 (2016)
16. Anisimov, A.A., Tararikin, S.V.: The forming of optimization criterion in the problems of parametric synthesis of the state regulators in electromechanical systems. Mechatron. Autom. Manag. 10, 36–41 (2009)
17. Saushev, A.V., Shoshmin, V.A.: Objective function in the synthesis of onshore and ship-based electromechanical systems. Inland Shipp. Transp. 5, 79–81 (2010)
18. Saushev, A.V.: Experiment Planning in Electrical Engineering. SPGUVK, Saint Petersburg (2012)
19. Rvachev, V.L.: Geometric applications of logic algebra. Technics, Kyiv (1967)

Transportation Management Systems for Airport Ground Handling

A. Dorofeev[1,2] and O. Nastasyak[3(✉)]

[1] Higher School of Economics, 20 Myasnitskaya street,
Moscow 100100, Russian Federation
[2] Financial University Under the Government of the Russian Federation,
49 Leningradskiy avenue, Moscow 125993, Russian Federation
[3] Gazpromavia Aviation Company Ltd, 71/32 Novocheremushkinskaya street,
Moscow 117420, Russian Federation
ipatyevaolga@yandex.ru

Abstract. Specifics of airport ground vehicle management in Russia lie in the fact that their operation is strictly regulated by the national legislation. In all these cases, we may distinguish between the conditions of the transportation process organization and the requirements of normative legal documents. This stipulates the use of special standards and rules for fuel consumption calculation and a certain document flow. The systems of higher functionality help solve the problems of planning and accounting of the transportation operation, technical maintenance and repair, consumption of spare parts, fuel and tires. A highly important TMS class embraces the software products intended for planning, organizing and accounting the work of motor transportation businesses. This paper considers TMS "Autobase" of our development which facilitates accounting of ramp vehicle operation and maintenance. Depending on the conditions of their transport activity, their architecture may be formed on a modularity basis, increasing the functional features to the required limits and creating complex TMS system. The TMS "Autobase" implements the tasks of keeping records of repairs and reserve vehicles, cost analysis; it presents more than 300 reports, OLAP applications.

Keywords: Fleet management · Airport ground handling · Fuel consumption · Airport support vehicle

1 Introduction

The modern enterprise control systems are known to have originated from the MRP concept of the 1960s. The main purpose of that concept was to ensure material demand planning, based on Bill of Materials (BOM) and Master Production Schedule. As time passed, the MRP concept transforming consistently to solve the ever-broadening range of tasks, developed into the corresponding MRP II/ERP/ERP II stages, and ended up as solutions of such companies as SAP, Oracle, etc. for enterprise control in many industries. One of the solutions currently included in ERP/ERP II concept is Transportation Management System (TMS) designed to control and plan transportation activities [1, 2]. Nowadays, the optimization of logistics and supply chains for

A. A. Radionov and A. S. Karandaev (Eds.): RusAutoCon 2019, LNEE 641, pp. 1071–1078, 2020.
https://doi.org/10.1007/978-3-030-39225-3_113

industrial, commercial, and construction companies enables cost saving, improved production process reliability and sustainability, ensuring timely delivery of goods. Accordingly, the vast majority of TMS is focused on solving cargo transportation issues (recording applications for transportation, application assignment by vehicles, vehicles and cargoes monitoring) related mainly to cargo transportation. Part of the issues related directly to vehicles operation is addressed by Fleet Management Systems (FMS), where all actions and events occurring with the vehicle are recorded, i.e. runs, standing time, fueling, tire wear, repairs, maintenance [3], etc. In [4] it is noted that the most urgent tasks for cargo carriers these days are: optimization of routes and schedules, optimization of the fleet composition and size.

However, some road transport enterprises have solved specific tasks not related to cargo transportation. Such enterprises include airports with fairly large fleets of special ramp vehicles designed for aircrafts ground servicing. These vehicles provide aircraft towing and refueling, power supply for airborne systems, compressed air supply, drinking water servicing, waste removing from the toilet, passenger and baggage transportation, catering [5], etc. For these services, aircraft owners make payment included in airport charge. According to the rules and regulations, each type of aircraft is serviced by a certain set of special vehicles in a regulated sequence: catering trucks, cleaning trucks, Air Start Unit (Engine start) mobile, toilet servicing truck, potable water truck, towtractor, cargo loaders, bus [6] etc. Analysis of the references shows that there is a number of issues associated with optimal control of ground handling of aircrafts, in particular, ramp vehicles [7, 8].

It is well-known that ramp vehicles service aircrafts landing at the airport and taking off according to the schedule. Their ramp-time is extremely limited. Therefore, some articles study the possibility of ramp vehicle operation optimization based on their most effective movement at the aircrafts parking, taking into account strict security measures and the rules of being close to the aircraft, with the location of vehicles at the parking strictly prescribed. Upon that, we should also note such important factors as speed limit at parking, as well as short turning radii. Therefore, the article analyzes optimal paths of service vehicle motion. Hoy following studies deal with ramp vehicles real-time dispatching using GPS and wireless data. In this concept, the central dispatcher board, seeing the whole situation at the airport, sends commands to special vehicles in order to organize their operation [9–11]. However, in view of frequent delays of arrivals and departures, the ramp vehicles fleet centralized control is becoming quite difficult, as the resource of free vehicles is limited, and in view of timetable irregularities, delays may occur when using the vehicle. In [12] this regard, considers an approach that solves the issue of planning on the basis of current state of aircrafts, certainly, considering the central timetable. Solution of a similar issue for de-icing vehicles is suggested in [13]. Due to optimal routing, it significantly reduced the aircrafts waiting time for this procedure before departure [14].

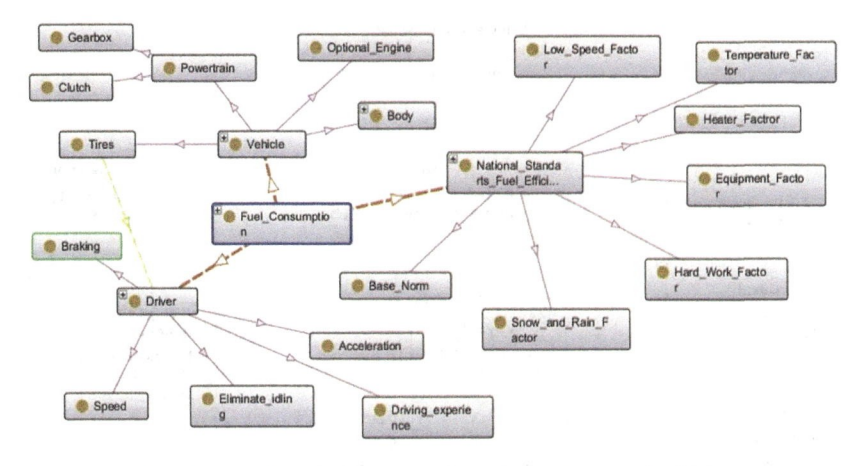

Fig. 1. Ontology fuel consumption of airport ground vehicle.

2   Fuel Consumption of Airport Ground Handling Vehicles

Currently, there is a number of solutions for the airports fleet activities control, such as INFORM GroundStar, Zebra Technologies Zebra Enterprise Solutions, including modules for planning, monitoring, and accounting. However, these solutions are not used in Russian airports due to the peculiarities of Russian legislation in the field of operation of both public transport vehicles and special-purpose vehicles. For example, there is the special legal act in Russia regulating the accounting of vehicle fuel consumption. There is also the regulatory legal act "MANUAL on organization of work and maintenance of Russian Federation airports special vehicles" (ROROS-95) regulating the document flow associated with the operation of ramp vehicles. Upon that, it should be noted that the average share of special vehicles in the fleets of Russian airports is slightly more than 50%. The share of passenger cars is 18%. The share of trucks is 12%. The share of tractors is 11%. And the share of buses is 7%. Moreover, the share of special vehicles maintains at various airports, regardless of the size of airports or climatic conditions. It is also worth noting that the ratio of vehicles of Russian and foreign production is presented differently in different airports of Russia. In this regard, the actual issue is the implementation of airport fleet control IT solution, taking into account Russian legislation and allowing keeping a record of fleet vehicles operation by different type, age, country of origin.

We have developed an airport fleet information control system, which includes the following main tasks:

- Aircraft ground handling requests management
- Fuel consumption accounting
- Repairs and maintenance accounting
- Analysis of vehicle operation costs

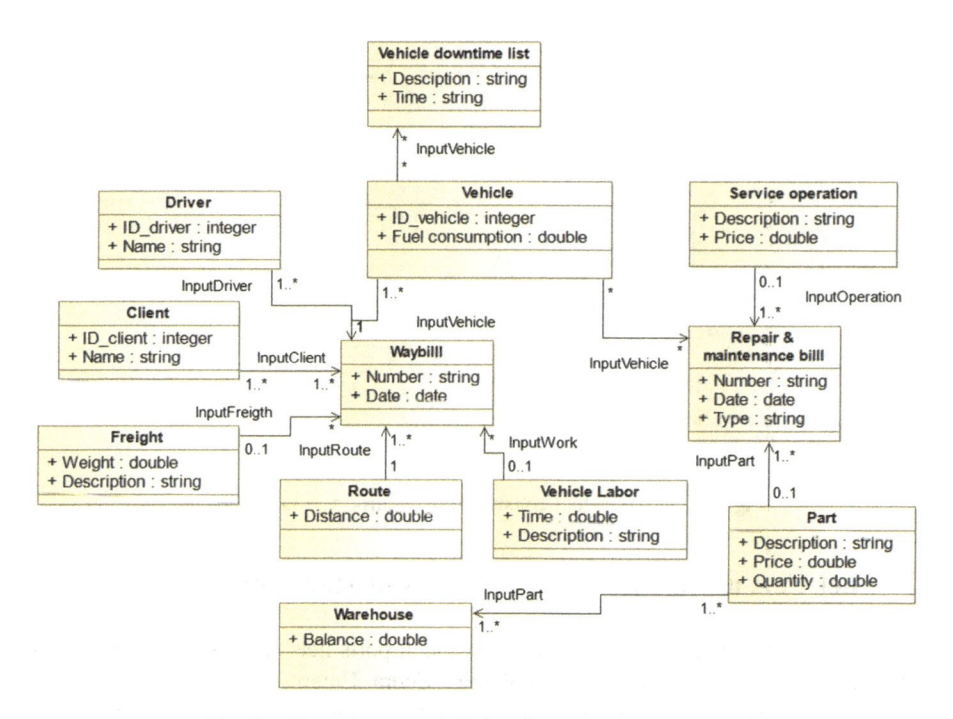

Fig. 2. Class diagram TMS for airport ground handling.

Special ramp vehicles fuel consumption accounting is quite different from trucks and passenger cars fuel consumption accounting. A number of ramp vehicles is manufactured on the basis of trucks chassis. For example, SCHMIDT AS 990 truck-mounted airport sweepers is manufactured on the basis of Mercedes-Benz Actros [15] truck-mounted airport sweeper and has two engines – main and optional - to drive various equipment. This vehicle can carry out the cleaning of flight strips using snow plough or front-mounted sweepers. It is also possible to clean waste with circular brushes and a fan, to collect water from the flight strip or wash the strip with a jet of water, as well as other operations. It is known that mileage fuel consumption of the main engine depends on weather and climatic conditions, skill and experience of the driver, vehicle age and other factors. Besides, fuel consumption of the main engine also depends on endurance hours. And the fuel consumption of the optional engine depends on endurance hours in accordance with technological operations performed by this vehicle during the flight strip cleaning. Thus, the total fuel consumption Qs of a special vehicle per shift will be

$$Q_s = Q_{bm} + Q_{bw} + Q_{aw} \tag{1}$$

where Q_{bm} is mileage fuel consumption of the main engine; Q_{bw} is fuel consumption of the main engine on endurance hours; Q_{aw} is fuel consumption of the optional engine on endurance hours depending on the type of operations.

In Russia, as well as in some other countries [16] of the world standards of fuel consumption (rate) for trucks, passenger cars, buses and some types of special machines have been developed and are currently in force. According to these standards, mileage fuel consumption is determined as follows

$$Q_n = 0,01 \times (Hsan \times S + Hw \times W) \times (1 + 0,01 \times D), \tag{2}$$

wherein Q_n is the standard fuel consumption (litres); S is the mileage of motor transport vehicle or caravan (km); $Hsan$ is the fuel allowance for the mileage of a motor transport vehicle or caravan (kerb weight without cargo).

$$Hsan = Hs + Hg \times Gr \, \text{liters}/100\text{km}, \tag{3}$$

wherein Hs is the basic fuel allowance for the mileage of motor transport vehicle (tractive vehicle) kerb weight without cargo, 1/100 km ($Hsan = Hs$, 1/100 km, for a motor transport vehicle (tractor); Hg is the consumption of fuels per additional mass of the trailer or semi-trailer, litres per 100,000 km; Gr is the dead weight of the trailer or semi-trailer, t; Hw is the fuel allowance for transportation work, 1/100 tkm; W is the volumetric capability of transportation work, thousands of km: W = Ggr × Sgr (where Ggr is the weight cargo, t; Sgr is the mileage with cargo, km.

3   TMS for Airport Ground Handling Vehicles

Upon that, the state fiscally obliges companies operating the vehicles to strictly follow these tax rates, regardless of what data on consumption are shown by the monitoring system sensors. We have previously developed software allowing transport companies' managers to calculate fuel consumption for trucks depending on the different operating conditions and tonnage of the cargo transported.

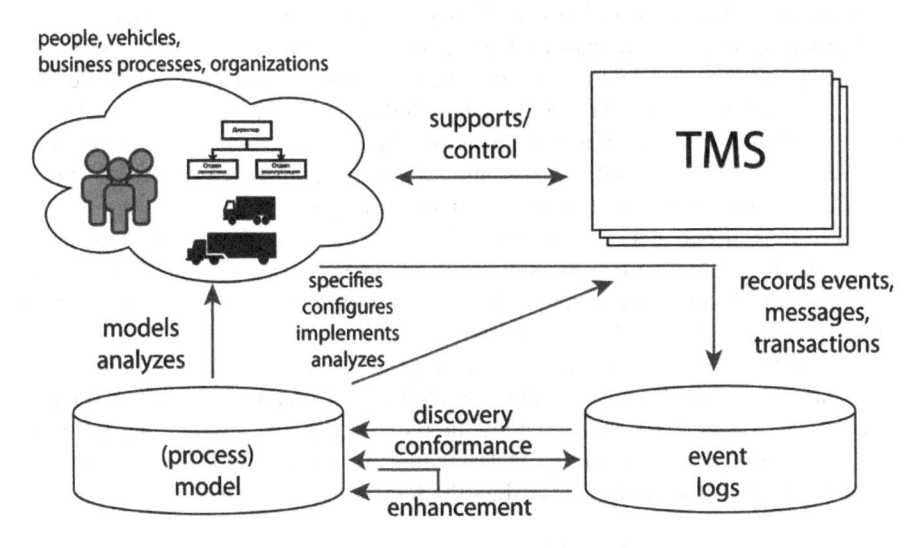

Fig. 3. Concept process mining [19].

In the implemented version of our TMS "Autobase" for airports, depending on the types of operations, the ability to consider the optional engine fuel consumption was added to the calculation of main engine fuel consumption. An ontology (Fig. 1) was developed to model this software solution. Due to this ontology, it was possible to formalize the accounting of various factors impact on fuel consumption. It should be noted that, although fuel costs reach 50% of the ramp ground machinery total operating costs, a record of the cost of spare parts, repairs, and maintenance should also be kept. Thus, the aim of planning the ramp vehicles activities, their downtime in repairs and maintenance should be considered. Details of airfield vehicles faults, repairs and maintenance are entered into TMS manually by managers. Based on these data, certificates and supporting documents, indicating the serviceable technical condition of vehicles, are generated. Maintenance planning is carried out on the basis of accumulated data on vehicle mileage and equipment operating time. Developed TMS architecture model is shown in Fig. 2. An important feature of the implemented maintenance planning subsystem is a capability of accounting of hours worked for optional engine. Run of each tire installed on the vehicle is also accounted. Strict control and accounting measures allow preventing possible accidents which may influence the airport operation.

Based on data presented in TMS, the information is generated, allowing for operational, tactical and strategic decisions on the ramp vehicles fleet management [17]. The developed analytical reports and dashboards represent the key performance indicators of the ramp ground support, form forecasts for the replacement of consumables, components and assemblies of vehicles and equipment [18]. For example, the technical readiness coefficient gives an idea of reliability, demonstrating how efficiently the machinery can perform.

This indicator depends on various factors which are also analyzed. Thus, applications for spare parts are accounted in the database, where number of days for application fulfillment is monitored. Accordingly, a conclusion is made on influence of spare parts supply terms on ramp vehicles downtime. The second factor with significant influence on the technical readiness coefficient is number of days for repair.

The information system analyses the best place for repair – external car service or own garage. Obviously, appropriate equipment and trained mechanical engineers are required for repairs in the own garage. Their work is accounted on the basis of repair sheets. As the result, the cost of services in external car services and own repair area is compared in the ratio of vehicles downtime. Aggregate costs for the ramp vehicles operation influence the airport tariffs for aircraft handling.

All actions of operators during work with the information system are recorded in the log file. TMS "Autobase" is built on client-server architecture. Accordingly, the log file contains SQL transactions represented by sequence of SQL requests, which may be used to analyze activities of operators and reveal their behavior.

This approach, which is currently gaining momentum among researchers and practitioners, is called "process mining" [19]. Specially developed application for TMS "Autobase" reviews and analyzes the log file providing statistics of all transactions by each user for the management (Fig. 3). Using this statistics, actions of unconscientious operators were revealed aimed at stealing the fuel and consumables.

4   Conclusion

The developed TMS "Autobase" has been successfully operated through many years in different organizations of Domodedovo, Sheremetyevo airports, in the airports of Arkhangelsk, Nadym, Zhukovsky. A single information space covering managers who form applications and route sheets, managers and technicians for vehicles repairs and maintenance, warehouse managers, allowed to quickly interact with all stakeholders and reduce the overall labor load of the current operating activities. Analysts and managers, up to top managers, could receive real-time data on the work carried out and costs to effectively form the tariff policy of the airport. An important feature of TMS implementation is compliance with regulatory legal requirements of public services of the Russian Federation for the ramp ground machinery operation. Software interface supports a single operational accounting of buses, passenger cars and trucks, as well as special vehicles and equipment.

References

1. Daithankarand, J., Pandit, T.: Transportation Management with SAP TM 9: A Hands-on Guide to Configuring, Implementing, and Optimizing SAP TM. Apress, New York (2014)
2. Wycoff, D.: Implementing a SAP Transportation Management System Solution: A Case Study. Regis University, Denver (2009)
3. Vivaldini, M., Pires, S., Souza, B.F.: Improving logistics services through the technology used in fleet management. J. Inf. Syst. Technol. Manage. 3, 541–562 (2012)
4. Bielli, M., Bielli, A., Riccardo, R.: Trends in models and algorithms for fleet management. Procedia Soc. Behav. Sci. 20, 4–18 (2011)
5. Kazda, T., Caves, B.: Airport Design and Operation. Emerald, Bingley (2015)
6. Stimac, I., Vince, D., Jaksic, B: Model of environment-friendly aircraft handling–case study: Zagreb airport. In: 16th International Conference on Transport Science, Portoroz, 27 May 2013 (2013)
7. Bevilacqua, M., Ciarapica, F.E., et al.: The impact of business growth in the operation activities: a case study of aircraft ground handling operations. Prod. Plan. Control 26(7), 564–587 (2015)
8. Tabares, D.A., Mora-Camino, F.A.C.: Aircraft ground handling: analysis for automation. In: 17th AIAA Aviation Technology, Integration, and Operations Conference, Denver, June 2017
9. Wang, L., Wei, J., Liu, X., et al.: Design of command and dispatch system of airport support vehicle. In: 2010 International Conference on Computing, Control and Industrial Engineering, Wuhan, 5–6 June 2010 (2010)
10. Deng, L., Gao, D.: Design of flight service vehicle dispatching management system. In: International Conference on Communication Software and Networks, Macau, 27–28 February 2009 (2009)
11. Shi, S., Qu, S., He, L.: Design of vehicle dispatching and monitoring system in airdrome. In: International Conference on Measuring Technology and Mechatronics Automation, Changsha City, 13–14 March 2010 (2010)
12. Trabelsi, F., Mora-Camino, F., Astorga, S.P.: A decentralized approach for ground handling fleet management at airports. In: International Conference on Advanced Logistics and Transport, Sousse, 29–31 May 2013 (2013)

13. Norin, A., Yuan, D., Granberg, T.A., et al.: Scheduling de-icing vehicles within airport logistics: a heuristic algorithm and performance evaluation. J. Oper. Res. Soc. **63**(8), 1116–1125 (2012)
14. Fitouri-Trabelsi, S., Nunes-Cosenza, C.A., Moudani, W.E., et al.: Managing uncertainty at airports ground handling. In: Airports in Urban Networks, Service Technique de l'Aviation civile, Paris, April 2014
15. Getting more out of your truck. For operators and the environment. http://tools.mercedes-benz.co.uk/current/trucks/brochures/environment-technology/fuel-efficiency.pdf. Accessed 12 Sept 2019
16. Jin, Y.-F., Wang, Z., Gong, H.-M., et al.: Review and evaluation of China's standards and regulations on the fuel consumption of motor vehicles. Mitig. Adapt. Strat. Glob. Change **20** (5), 735–753 (2015)
17. Dorofeev, A.: Development of internet-based applications for fleet management and logistics. In: 15th IEEE International Conference on Business Informatics, Vienna, 15–18 July 2013 (2013)
18. Dorofeev, A.: The application of decision-making support systems for fleet management in small and medium sized transport enterprises in Russia. In: 1st Virtual Conference on Intelligent Transportation Systems. EDIS - Publishing Institution of the University of Zilina, 26–30 Aug 2013 (2013)
19. van der Aalst, W.: Process Mining: Data Science in Action. Springer, Heidelberg (2016)

Automated Text Classification System Based on Statistical Unified Model

S. Skorynin[(✉)] and A. Surkova

Nizhny Novgorod State Technical University, 24 Minina street,
Nizhny Novgorod 603950, Russian Federation
skorynins@gmail.com

Abstract. The paper is devoted to the automated text classification system based on a unified model. The main text mining statistical and linguistic approaches are considered. The architecture of the developed system is given. The automated system is analyzed in details. The text data unified model consists of statistical elementary models: substrings, cumulative and the finite difference characteristics. The structure and features of each model are considered. The unified model flexibility is achieved by assigning weights for each elementary model. According to this logic, the system can be modified for different text types. The automated system for literary text classification has been tested. The classification quality was evaluated by precision, recall and f-measure. The model was trained and evaluated on the 6 classes. The total texts number in training set and test set is 600 and 60 respectively. The automated text classification system shows good results, low scores for some texts are explained. The advantages and limitations of the proposed system are shown. In addition, there is research area on the linguistic models inclusion in order to improve the classification quality of the proposed automated system.

Keywords: Text mining · Classification · Text data · Unified model ·
Automated systems · Statistical model · Linguistic model · Symbolic diversity

1 Introduction

The growth of scientific and technological progress has led to a multiple increase of created and used information, including text data. The amount of information is increasing at an astonishing rate. People are unable to solve the issues associated with this growth. Text remains one of the main types of information in most electronic repositories. As a result, it became necessary to process large amounts of information in an automated mode. Humanity needs intelligent automated electronic systems that can cope with classification, clustering, identification. Text classification is the texts distribution task to categories from a predefined set [1–3]. The quality of the automated system depends on the classification model. During model development, you should determine significant text elements, language rules, features and requirements which should be taken into account. Traditional statistical and linguistic texts models are designed for certain areas, types and genres of text data (literary texts, scientific articles, patent texts, etc.). Thus, there is no single model that would be suitable for different texts.

A. A. Radionov and A. S. Karandaev (Eds.): RusAutoCon 2019, LNEE 641, pp. 1079–1087, 2020.
https://doi.org/10.1007/978-3-030-39225-3_114

The paper proposes to consider and analyze the unified model in the automated text data classification system. This system can be used for wide classification areas. Combining models is based on bagging [3–5].

In [6] the text data analysis methodology and modeling were proposed, which includes:

- A text data models set that characterize different text parameters and reflect the different texts types characteristics
- Methods for solving classification, clustering and identification issues with the modifications possibility
- The method of rational choice from the models and methods set
- The modification procedure of models and methods depending on the specific tasks and additional conditions (specified accuracy, execution time, etc.)

The proposed methodology has shown good results for solving many practical issues: for the text streams categorization in real time [7]. The use of recurrent neural networks with unsaturated logarithmic activation function for the automatic abstraction texts tasks. However, the classification task for different texts types, requires detailed consideration.

2 Text Data Modeling Approaches

For building a unified model, it is proposed to use several elementary models or their modifications. Bagging multiple text data models is a multidimensional representation or a summary model. Bagging is a classification technology where all elementary classifiers are trained and work in parallel (independently of each other). The idea is that classifiers do not correct each other's mistakes, but compensate them when voting. Base classifiers must be independent, they can be classifiers based on different methods groups or trained on independent data sets. In the second case, you can use the same method. Bagging is an ensemble method. Ensemble methods use multiple models to obtain a better predictive performance than could be obtained from any of the constituent models. In other words, an ensemble is a technique for combining many weak learners in an attempt to produce a strong learner. Evaluating the prediction of an ensemble typically requires more computation than evaluating the prediction of a single model, so ensembles may be thought of as a way to compensate for poor learning algorithms by performing a lot of extra computation. Empirically, ensembles tend to yield better results when there is a significant diversity among the models. Many ensemble methods, therefore, seek to promote diversity among the models they combine. Using a variety of strong learning algorithms, however, has been shown to be more effective than using techniques that attempt to dumb-down the models in order to promote diversity. As a result, each feature of the model reflects a specific text characteristic. The summary model is a more approximate text representation, compared with each of the models separately. There are two main approaches to the text data analysis: statistical and linguistic models.

The first approach includes linguistic analysis of text semantics, grammar and syntax. Additional resources can be used to extract information about the meaning, language words relation with the help of ontologies and thesauruses (WordNet). Data modeling consists in the texts analysis as a sequence of sentences with their syntactic analysis [8, 9]. Linguistic models help to identify the content structure of a coherent text. Based on this knowledge it is possible to extract additional information about the text that is not explicitly available. We can extract this information because text contains many related words, and has an internal hierarchical structure [10–12].

The second approach is based on probabilistic (statistical) methods of natural language processing. The model is based on the text statistical characteristics, which include mathematical expectation, word frequency, probability distribution, information characteristics, etc. [6]. In the context of classification tasks for each of the classes is assigned a certain characteristics value. Each feature is assigned a weighting factor that characterizes the feature significance for current class. Statistical models ignore semantic context, grammatical and syntactic constructions, and only take into account the individual elements. Models that use probabilistic prediction of the neighboring words appearance, symbolic time series coding, models based on mutual information, N-grams [6, 13, 14] are widely used nowadays.

3 Text Data Unified Model

Consider a unified text data model (it is based on a statistical approach Methods such as kNN, SVM, decision trees and others can be used as a classifier. Decision - making is based on fuzzy logic and combining the classification results (bagging) [15, 16].

The unified model uses the concept of hidden parameters. The concept of hidden parameters [6] is based on the assumption that the author uses some a priori structural elements or structural invariants (lexical, grammatical, thematic). The author selecting specific words, phrases to describe his thoughts. Structural invariants are features inherent in all objects that have the same properties. The concept of hidden parameters allows revealing the text structures patterns and creating a text-generalizing model as a multidimensional object.

The proposed unified model is based on structural invariants and allows: (a) to improve the classification quality, (b) to adjust the model depending on the task. The texts unified model considered in this paper consists of the following components:

(1) A statistical model using symbolic time series coding of word combinatorics methods. The model calculates the shift entropy estimation as a sliding window length function. In the text, a character sequence of fixed length over the alphabet is distinguished. It is obvious that total count all such structural elements is calculated as "(1)":

$$M = k^m \tag{1}$$

where k - is the power of the alphabet. For a fixed value m, an arbitrary numbering of structural elements is entered. To calculate the estimates the number counters of the character sequences c_i are introduced. The initial values are supposed to be zero [17–19].

Thus, the window (length = n) is moved along the character sequence. For each structural element in window positions, the corresponding character sequence counter c_i is incremented. As a result, the words entropy estimate is calculated as "(2)":

$$C(m) = - \sum_{i=1}^{M} \left(\frac{c_i}{n-m+1} \right) \log_M \left(\frac{c_i}{n-m+1} \right) \tag{2}$$

This function is the basis for calculating peak and cumulative symbolic diversity unified characteristics. Significantly, the first proposed unified characteristic reflects the structural elements diversity boundary over a fixed alphabet. The second characteristic—the entropy function average value (shifts up to a certain minimum threshold). In this case, the text is interpreted as a continuous characters sequence without dividing into separate elements. On the basis of C(m), the concept of finite difference is calculated as "(3)":

$$\mu_k(T) = \frac{1}{\Delta C(m)} \tag{3}$$

where denominator is the function argument. The text cumulative symbolic characteristic is calculated as "(4)":

$$\mu_s(T) = \frac{1}{\tilde{m}} \sum_{i=1}^{\tilde{m}} C(i) \tag{4}$$

(2) Statistical model based on substrings.

The substring model is based on N-grams. Let a elements sequence $S = \{s_1, s_1, \ldots, s|s|\}$ be given. An n-gram is any subsequence $S = \{s_1, s_1, \ldots, s|s|\}$ of $|S|$ consecutive elements. In general N-grams elements can be both letters and words [14].

However, researchers in the field were limited with a fixed value N. The substring model removes this restriction (since it does not allow to reflect all the structural text connections). Model is based on the elements subsequence, where the N value is not fixed. For the current configuration, the minimum substring length is 2 and the maximum is 5. The text is analyzed at the letter level. The text will be interpreted as a continuous characters sequence without dividing it into separate words. Thus, substrings are built for the whole document (as a single sequence of characters).

For fuzzy text classification by categories we choose k-nearest neighbor (kNN) algorithm. In pattern recognition, the k-nearest neighbors algorithm (k-NN) is a non-parametric method used for classification. The input consists of the k closest training examples in the feature space. The output is a class membership. An object is classified by a plurality vote of its neighbors, with the object being assigned to the class most common among its k nearest neighbors (k is a positive integer, typically small).

If $k = 1$, then the object is simply assigned to the class of that single nearest neighbor. The neighbors are taken from a set of objects for which the class is known. This can be thought of as the training set for the algorithm, though no explicit training step is required. kNN method was modified considering fuzzy logic principles. kNN method used distance function between objects. Distance function usually using Euclidean distance or other measures. So, any object generates a numbered sample. The result is a k neighbors who vote for categories. To make a decision, we introduce the necessary barrier. The barrier is selected depending on the classification tasks. In other words, we need to find minimum number of neighbors, voted for a determined category. If the vote is sufficient, classifier makes a decision, otherwise the document does not belong to any of the categories [15].

The final classification model is based on the bagging of three models combination. The "decision-making" block makes a classification decision taking into account hidden parameters and weight coefficients. Texts classification is based on the fuzzy nearest neighbor method. The structure of the proposed automated system is shown in "Fig. 1". The input can be either a training set or text for classification. The first type is used for the training model; the second type is used for classification purposes. The source text passes through the preprocessing block where the stop word is removed and

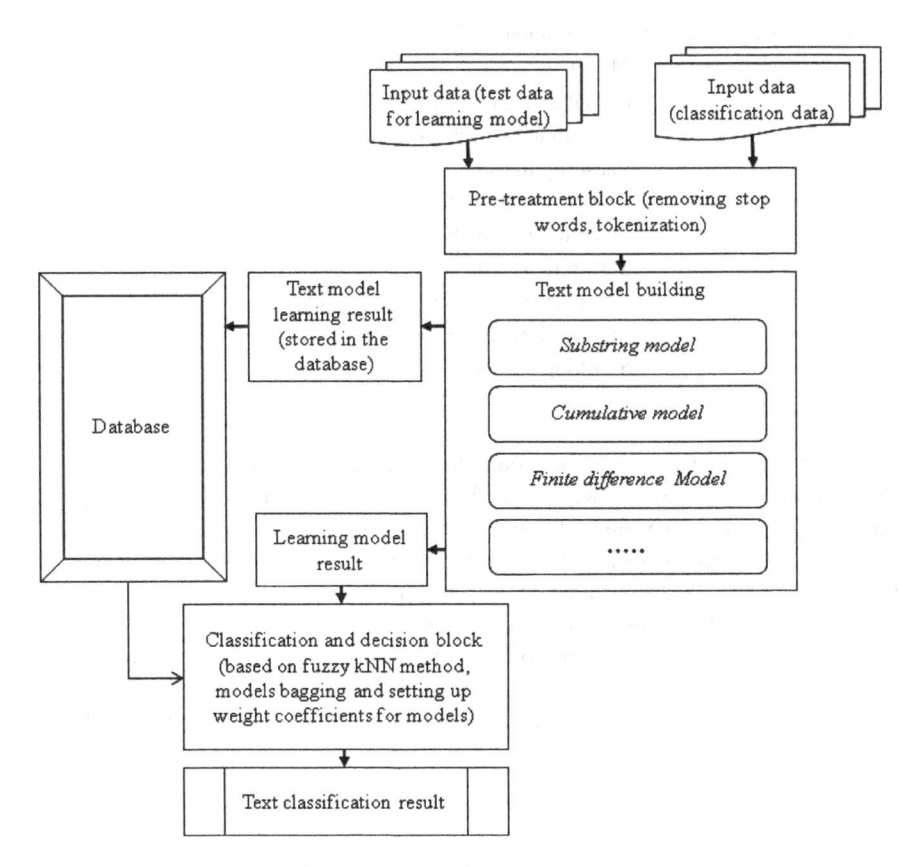

Fig. 1. Automated system structure.

the porter-stemming algorithm is used After that, the system creates 3 separate models (substring, cumulative and finite difference model). If the system is used in training mode, the models will be saved in the database (the system creates necessary etalons).

If the system is in classification mode, the resulting models are transferred to the classification (decision) block. In the classification block, the system obtains the required etalons from the database and model parameters depending on the text type. Based on the received data the system makes classification decision using fuzzy nearest neighbor algorithm. The classification result is displayed in the user interface.

4 Unified Model Results

To test the proposed unified model, the classification of literary works (6 classes in total) were chosen. The total number of texts in training set is 600. The total number of texts in test set is 60. The weights influence on the final classification quality is presented in Table 1.

Table 1. Weight coefficients influence on the final classification quality.

Configuration number	Classification quality		
	Model	Weighting factor	F-measure
1	Cumulative characteristic	0,5	0,8
	Finite difference characteristic	0,5	
	Substring model	1	
2	Cumulative characteristic	1	0,71
	Finite difference characteristic	1	
	Substring model	1	
3	Cumulative characteristic	0,5	0,67
	Finite difference characteristic	1	
	Substring model	0,5	

Based on Table 1, it can be noted that the highest classification quality was obtained with the following weight coefficients configuration: substrings – 1, cumulative characteristic – 0.5, finite difference characteristic – 0.5. This is because entropy characteristics (at the word level) give a worse result than substrings for literary texts. Therefore, for the substring model should be put higher weight. Look at the configuration number 3. It shows results deterioration compared with substring model. For further improving the classification quality, we should add new models. However, the classification quality should be comparable (or higher) with substrings. The averaged results (for the first weight configuration) for different models are presented in Table 2.

Table 2. Classification quality for different models.

Model	Classification quality		
	Accuracy	Completeness	F-measure
Cumulative characteristic	0,65	0,6	0,62
Finite difference characteristic	0,6	0,55	0,57
Substring model	0,74	0,75	0,74
Unified model	0,79	0,82	0,8

The classification quality of the unified model is increased by f-measure (on average 5–7%). The accuracy value indicates that the final unified model increases the number of correctly classified documents. The completeness value indicates that the number of classification errors has decreased. It should also be noted that:

- Increased resistance to errors in texts classification
- Ability to customize the unified model by weight coefficients (depending on the task requirements)

A unified multi-dimensional model is extensible. This means that new models can be added to the unified model. However, you should take into account the limitations:

- The scope for each model should be defined (for which tasks and texts types it is suitable)
- The experiment showed that if one model shows good classification quality, and the other does not, the final result may be unsatisfactory. In this case, you should override the weight function in the configuration block or even exclude the model from the ensemble
- Therefore, when you add new model, you should carefully test its behavior in different scenarios. Based on the results, it is necessary set correct weighting factor in the decision-making block

5 Conclusion

The paper considers a text data unified model as part of an automated system. Specific modifications of the unified model are achieved by different weights for elementary models. The classification quality of the proposed system is shown. It is promising test the classification system on the other different types and genres (for example patent texts, source code of programs). In future to increase the classification quality of the proposed automated system, it is necessary to include linguistic models.

References

1. Sebastiani, F.: Machine learning in automated text categorization. ACM Comput. Surv. **34**(1), 1–47 (2002)
2. Reinsel, D., Gantz, J., Rydning, J.: Data age 2025. The evolution of data to life-critical. Don't focus on Big Data; Focus on the data that's big. IDC White Paper, Sponsored by Seagate (2017). https://www.seagate.com/files/www-content/our-story/trends/files/idc-seagate-dataage-whitepaper.pdf. Accessed 11 Sept 2019
3. Manning, Ch.D., Raghavan, P., Schutze, H.: Introduction to Information Retrieval. Cambridge University Press, Cambridge (2008)
4. Berry, M.W., Kogan, J.: Text Mining. Applications and Theory. Wiley, New York (2010)
5. Feldman, R., Sanger, J.: The Text Mining Handbook. Advanced Approaches in Analyzing Unstructured Data. Cambridge University Press, Cambridge (2007)
6. Lomakina, L.S., Surkova, A.S., Zhevnerchuk, D.V.: Text structures synthesis on the basis of their system-forming characteristics. IV International research conference Information technologies in science, management, social sphere and medicine. Adv. Comput. Sci. Res. **72**, 108–113 (2017)
7. Lomakina, L.S., Subbotin, A.N., Surkova, A.S.: Naive Bayes modification for data streams classification. In: Proceedings of the 13th International MEDCOAST Congress on Coastal and Marine Sciences, Engineering, Management and Conservation, vol. 2, pp. 805–814 (2017)
8. Loukachevitch, N.V., Dobrov, B.V.: RuThes linguistic ontology vs. Russian WordNets. In: Proceedings of Global WordNet Conference, Tartu, pp. 154–162 (2014)
9. Dobrov, B., Loukachevitch, N., Nevzorova, O., Fedunov, B.: Methods of automated design of application ontology. J. Comput. Syst. Sci. Int. **43**(2), 213–222 (2004)
10. Ageev, M., Dobrov, B., Loukachevitch, N.: Sociopolitical thesaurus in concept-based information retrieval: ad-hoc and domain specific tasks. In: Peters, C., Quochi, V. (eds.) Cross-Language Evaluation Forum. Results of the Cross-Language System Evaluation Campaign, pp. 141–150. Springer, Heidelberg (2006)
11. Nokel, M., Loukachevitch, N.: An experimental study of term extraction for real information-retrieval thesauri. In: Proceedings of 10th International Conference on Terminology and Artificial Intelligence, pp. 69–76 (2013)
12. Mikolov, T., Yih, W., Zweig, G.: Linguistic regularities in continuous space word representations. In: Proceedings of Conference of the North American Chapter of the Association for Computational Linguistics: Human Language Technologies, pp. 746–751 (2013)
13. Lomakina, L.S., Sementsov, M.S., Surkova, A.S.: Authorship attribution of GIS software source codes. In: Proceedings of the 13th International MEDCOAST Congress on Coastal and Marine Sciences, Engineering, Management and Conservation, vol. 2, pp. 1225–1234 (2017)
14. Zecevic, A.: N-gram based text classification according to authorship. In: Proceedings of the Student Research Workshop Associated with RANLP, Hissar. Association for Computational Linguistics, pp. 145–149 (2011)
15. Surkova, A.S., Skorynin, S.S., Chernobaev, I.D.: Comparative analysis fuzzy algorithms of text classification. In: Kravet, O. (ed.) Modern Informatization Problems in Simulation and Social Technologies. Proceedings of the XX-th International Open Science Conference, Yelm, January 2016, pp. 213–218. Science Book Publishing House (2016)
16. Salton, G., Buckley, C.: Term-weighting approaches in automatic text retrieval. Inf. Process. Manag. **24**(5), 513–523 (1998)

17. Smetanin, Yu.G., Ulyanov, M.V.: Entropy characteristics of diversity in the symbols presentation of time series. Mod. Inf. Technol. IT Educ. **10**, 426–436 (2014)
18. Smetanin, Yu.G., Ulyanov, M.V.: On the number of possible reconstructions of words using subwords with Windows of different shift. Inf. Technol. **24**(4), 233–238 (2018). https://doi.org/10.17587/it.24.233-238
19. Uljanov, M., Smetanin, Y.: On a characteristic functional for words over a finite alphabet. Inf. Technol. **23**(5), 333–341 (2017). (In Russian). https://elibrary.ru/item.asp?id=29213868. Accessed 24 May 2018

Fundamentals for the Automation of Information Processing in the Identification of Chemical-Technological Systems

I. V. Germashev[1(✉)], E. V. Derbisher[2], and T. P. Mashihina[1]

[1] Volgograd State University, 100 Universitetskiy avenue,
Volgograd 400062, Russian Federation
germashev@volsu.ru
[2] Volgograd State Technical University, 28 Lenina avenue,
Volgograd 400005, Russian Federation

Abstract. The theoretical and methodological aspects of the formalization of an applied task of studying chemical-technological systems in an automated mode under uncertainty to create promising scientific and technical developments are considered. Information processing is carried out by using fuzzy mathematics. Fuzzy numbers are proposed to formalize the source data. This approach has led to dimensionless relative values, which, allow us to build aggregate estimates of a set of criteria, under conditions of multicriteriality. This was the basis for the algorithmization of the decision support procedure when choosing the optimal composition of the polymer composition. As an example, the proposed task of selecting and identifying functional additives to create the optimal polymer composite material in the conditions of technological and informational constraints. Based on the classification of additives, a training set was created, which allowed to generate data for building a decisive rule for identifying the level of activity of the additive.

Keywords: Identification · Chemical-technological systems · Fuzzy sets · Automation · Information processing · Pattern recognition · Polymer · Composite material

1 Introduction

Recall that the pattern recognition in automated mode when solving theoretical and applied computer science problems, such as expertise, identification, classification, design, prediction, and others, has become more and more popular in recent years. The theoretical platform for solving these problems is its mathematical formulation, the formation of a database of signs and a comparison of signs of new classes of technical objects in the multidimensional space of these signs by one or another optimal decision rules with known ones. Thoroughly verified a priori information is used for this purpose. If it is not sufficiently resorted to self-learning in the process of building a decision rule [1, 2].

© Springer Nature Switzerland AG 2020
A. A. Radionov and A. S. Karandaev (Eds.): RusAutoCon 2019, LNEE 641, pp. 1088–1097, 2020.
https://doi.org/10.1007/978-3-030-39225-3_115

Today it is impossible to do without computer preparation and analysis of information to solve complex specific and multifactor tasks, in the analysis of chemical-technological systems, in particular. Here, the original models themselves require considerable effort to compile and harmonize them, since in most cases they are significantly different from the known models used by mathematics and computer science, and even the formalization of the problem itself requires careful profiling. In this, vectors for considering chemical-technological systems can be extremely diverse: searching for new objects, assessing consumer qualities, designing and designing chemical-technological systems with theoretical (for example, analyzing and designing chemical structures with "given" properties [3, 4]), technological (for example, forming polymer compositions of the type: polymer matrix – filler – functional additives [5, 6]), technological (for example, the choice of extrusion mode, physical and chemical modification [7, 8], etc.), expert (for example, environmental score chemical-technological systems [9, 10]), and others.

Uncertainty inevitably arises, in the analysis of complex chemical-technological systems, which characterizes the factors of unpredictability of the behavior of such systems. In this case, the use of fuzzy logic allows you to take into account these factors and provide more reliable control. For example, it is proposed to use fuzzy logic to control the temperature of the polycondensation reactor, in [11]. This made it possible to take into account temperature measurement errors caused by noise and interference, more precisely adjust the operation mode to the technological process, which ultimately increased the efficiency of the reactor operation.

There are quite a lot of such examples when fuzzy mathematics allows to take into account the emerging uncertainties in the behavior of the system, not only in management, but also in other aspects of the functioning of systems. So, studies of the degradation of polymers are given, in [12]. The proposed model was described by non-linear regression, which was also used for extrapolation due to the use of fuzzy numbers.

Generally speaking, the use of fuzzy mathematics for the study of complex systems is quite general, which allows applying this approach not only to solve quite different problems in chemical technology, but also to transfer it to other subject areas [13–17]. This allows us to conclude about the possibility of developing formal models in terms of fuzzy mathematics for solving a certain class of problems abstracting from a specific subject area. And then, we'll get a specific solution in a given subject area interpreting the formal results in appropriate terms.

In the present work, such a study is proposed, although for illustration it is aimed at solving problems in chemical technology.

Here, there can be three approaches: deterministic, self-learning and probabilistic, and the task is to objectively, when searching for new objects, evaluating consumer qualities, designing and designing chemical-technological systems, that is, in some sense the best way, carry out identification, that is, to build a rule high, which according to the vector of signs would indicate where the chemical-technological system belongs. We note that to create new technical solutions and actively develop this area, from our point of view, the most urgent area is the further rapid development of automating the processing of information on complex chemical structures and substances, since they lay the foundation for the creation of future-oriented technologies.

2   Methodological Bases of Analysis and Preparation of Information

There are obstacles associated with the phenomenon of information uncertainty in the analysis of the chemical-technological systems. It is possible to designate the following: subjectivism, calculated errors, direct errors, outdated data, biased techniques, inaccurately chosen boundary conditions, etc. All this in the methodical plan can be considered as fuzzy sets and to apply the means of fuzzy mathematics even in such a highly specific area as chemical structures. Moreover, there are already works when fuzzy sets are used in chemistry: as a tool for modeling and automation [18–20]. Highlight the factors uncertainty related to chemical-technological systems:

- Ambiguity of the applicability of the concept of quality: a set of criteria may vary greatly in composition, content, boundaries, technology, consumer functions, etc.
- Heterogeneity and type of information: for example, incomparability of criteria among themselves, data representation in the form of numbers, sets, verbal descriptions
- Multi-criteria selection of boundary conditions
- Incompleteness of data: for example, their absence, inaccuracy, infidelity

You can specify other features. In general, we can say that we have seen that, using fuzzy sets, as a form of representing the arising uncertainties in a single information space, it is possible to adapt known methods of solving the problems endowed with them. On the other hand, when attempting to use direct classical mathematical methods for studying chemical-technological systems and the chemical structures themselves, difficulties arise, for example, with the use and creation of various often disparate models of the representation of chemical and mathematical objects. Methods that take into account the specifics of the tasks are designed to alleviate these difficulties and create conditions for the application of standard mathematical methods. Here the use of fuzzy sets gives as a tool for the formalization of chemical information in mathematical terms.

In connection with the above, the main task in the study of chemical-technological systems is the formalization of chemical information in mathematical terms to automate their identification. The study of chemical-technological systems under the conditions of the above limitations can be contained in the following fairly standard actions:

- Formulation of a chemical problem
- Selection of boundary conditions
- Preparation of the information array
- Formalization of the initial information in terms of fuzzy sets
- Formulation of a profiled mathematical problem
- Development or adaptation of mathematical methods to the profiled problem
- Selection of decision rules and methods of their application
- Creating a set of solutions
- Interpretation of results
- Decision-making

3   Theoretical Bases of Formation of Fuzzy Sets

Consider the method of creating fuzzy sets in the processing of information about the chemical-technological systems in the most general form.

Let $S = \{s_i | i = 1, \ldots, n\}$ – is a set consisting of n chemical compounds. For s_i, m characteristics $Q_{ij}, j = 1, \ldots, m$, are defined.

Depending on the conditions of use, the set of characteristics may expand and contract. For each characteristic Q_{ij} we construct a fuzzy set, $\hat{Q}_{ij}$, $i = 1, \ldots, n$, $j = 1, \ldots, m$. To do this, we define the variables x_j with the range of values x_j of G_j. Next, we select such a value q_{ij} of the variable x_j, which to the greatest extent satisfies the characteristic Q_{ij}, and a radius $\delta_{ij} > 0$, within which the values of x_j satisfy the criterion Q_{ij}. Thus, we obtain the set $X_{ij} = [q_{ij} - \delta_{ij}; q_{ij} + \delta_{ij}]$ of x_j values satisfying the characteristic Q_{ij}.

Next, for the criterion Q_{ij}, we choose the membership function μ_{ij}. Based on the construction of X_{ij}, we find that in q_{ij} the function has a maximum point; within X_{ij}, the function takes values not less than 0.5, and outside X_{ij} – less:

$$\mu_{ij} : G_j \to [0; 1], \ \mu_{ij}(q_{ij}) = 1,$$
$$\mu_{ij}(x_j) \geq 0.5 \Leftrightarrow x_j \in X_{ij}.$$

The choice of the membership function is very ambiguous and largely depends on the preferences of the researcher.

For example, in [21] the class of membership functions with a number of properties is revealed. For example, any of these functions is used to calculate v (the level of compliance of the chemical-technological system with the requirements is described in more detail below) when solving the ranking problem will arrange the objects in the same order in the series of increasing the value of v, i.e. the choice of the membership function affects the value of v only within the limits that do not swap the chemical-technological systems in the specified series, which is sufficient for relative evaluation and reasonable conclusion.

Thus, the final choice of the membership function of this type is reduced to ensuring the non-overestimation and non-underestimation of the absolute estimates, on the basis of which the rank is assigned.

So, in [21] it is proposed to use the following functions that satisfy the above requirements

$$\mu(x) = \exp\left(-\frac{\ln 2}{\delta^2}(x - q)^2\right),$$

$$\mu(x) = 1 - \text{th}\left(\frac{\ln 3}{2\delta^2}(x - q)^2\right),$$

$$\mu(x) = \delta^2 \Big/ \left((x - q)^2 + \delta^2\right).$$

In our case, as the membership function, you can, take the following, for example:

$$\mu_{ij}(x_j) = \exp\left(-\frac{\ln 2}{\delta_{ij}^2}(x_j - q_{ij})^2\right), \ i = 1, \ldots, n, \ j = 1, \ldots, m.$$

From where, we get fuzzy sets

$$\hat{Q}_{ij} = \{(x_j, \mu_{ij}(x_j))\}, i = 1, \ldots, n, \ j = 1, \ldots, m.$$

As a result, for each chemical compound, a set of fuzzy sets are constructed, describing the properties of chemical systems in terms of fuzzy sets.

4 Methodical Bases of Solving the Problem

Taking into account the above, and as a generalization of the material, we present a routing map for preparing and processing fuzzy information for automating the process of polymer composite materials identification.

1. Statement of the problem.
2. Determination of parameter space.
3. Determination of the values of the parameters Q_{ij} for each substance.
4. The choice of the "ideal" substance no. 0.
5. Determination of the values of the parameters Q_{0j} for substance no. 0.
6. Identification of the average value of q_{ij} and the allowable deviation from it δ_{ij} of the parameter no. j of the substance no. i.
7. The choice of membership function.
8. Definition of a fuzzy set for each value of Q_{ij}.
9. Calculation of the index of equality v_{ij} of fuzzy sets $\hat{Q}_{ij}$ and $\hat{Q}_{0j}$ according to the formula

$$v_{ij} = \max_{G_j} \ \min(\mu_{ij}(x_j), \ \mu_{0j}(x_j)).$$

10. Interpretation of v_{ij} values.

It should be noted that the number v_{ij} shows how the connection no. i corresponds to the connection no. 0 in the parameter no. j. As can be seen, many studies in the field of chemistry and chemical technology fall under these schemes and, therefore, these studies are quite general in nature, but will be considered on the example of active additives for polymer composite materials.

Thus, in the course of analyzing information on functional additives in the composition of polymer composite materials, the following steps were implemented:

1. Formalization of the initial information in terms of fuzzy sets;
2. Automated examination of the properties of substances and on this basis the identification of the activity of additives;

3. Automated prediction of the properties of substances and on this basis the construction of chemical structures with "given" properties (virtual additives) to polymer compositions and their sorting.

As you can see, the key step here is the formalization of chemical information.

5   Computational Experiment

Polymer composite materials are a good example for using the above method, since in addition to the polymer matrix and filler, they can contain up to thirty classical and completely new functional additives: stabilizers, plasticizers, antioxidants, amplifiers, biocides, dyes, flame retardants, antislips, etc. From the foregoing it is clear that the task of designing such complex compositions of polymer composite materials can be solved as fuzzy.

A small example, without computational details, demonstrates the use of the route outlined above for the classification of flame retardants in the composition of the polymer composite materials (polymer matrix – polyurethane) is given in Table 1.

Table 1. Example of classification antipyrene of polyurethane.

No. Additives (i)	Equality index of fuzzy sets (v_i)	Subgroup additives
1	0.52	Little active
2	0.53	Little active
3	0.66	Moderately active
4	0.70	Highly active

Here, the classification refers to the assignment of the object being diagnosed (flame retardant) to one of the sets (including fuzzy sets), which has a certain property or characteristic. The procedure can be carried out according to the type of technological and consumer function (stabilizer, flame retardant, antifreeze, anti-slip, etc.), according to the degree of the ingredient's impact on the final product, by cost (cost), by mass fraction in polymer composite materials, and environmental parameters [22] and others. For example, it is possible to distinguish subgroups of highly active, moderately active, few active, indifferent substances, also as already discussed above.

If you leave these four subgroups of additives, then for the technologist they can be identified as follows:

1. highly active – additives that produce a pronounced effect;
2. moderately active м additives that produce effects at high concentrations;
3. little active – the effect is minimal;
4. indifferent (such as fillers) – no effect.

For classification according to the four gradations, we adopted the following indicators v:

1. highly active – 0.70–1.00;
2. moderately active – 0.60–0.69;
3. little active – 0.40–0.59;
4. indifferent – 0.00–0.39.

Thus, the use of fuzzy sets allows the technologist to make decisions for designing polymer composite materials of the optimal composition, based on quantitative data.

You can create a training set to solve the problem of identifying the activity of a chemical compound according to the following scheme, based on the above classification.

The chemical structure of the compound is represented as descriptors. Physico-chemical properties, structural fragments, or other description methods can serve as descriptors. We used structural fragments. For each descriptor, the frequency of its occurrence in all classes of functional additives was counted and a statistical model of the ingredient was formed in the form of a catalog of descriptors.

In this directory, depending on the structure, all descriptors are divided into p types (for example, chemical elements, types of chemical bonds, etc.). Then, based on the analysis of the corresponding descriptors, the compound belonged to one or another class of additives.

In general terms, we describe the identification algorithm described in more detail in [3]. We introduce the notation n_i – the number of descriptors of the type i, $i = \overline{1, p}$. First, descriptor statistics are read from the directory.

Then the vectors are calculated $\bar{p}_{ij}$, each of the 4 coordinates of which shows an estimate of the probability of meeting the j-th descriptor of the i-th type in the corresponding class of additives (highly active, moderately active, little active, inactive).

For each of the p types of descriptors, it is separately decided to assign the compound to the class $k = 0, 1, 2, 3$:

$$p_{ik} = \sum_{j=1}^{n_i} p_{ijk} \bigg/ \sum_{k=0}^{3} \sum_{j=1}^{n_i} p_{ijk}, \ k = \overline{0, 3}, \ i = \overline{1, p}.$$

$r_i = \arg\max_{k=\overline{0,3}} p_{ik}$, – indicates to which class the connection type i descriptors.

At the last step, voting on p types of descriptors is performed:

$$p_k = \begin{cases} \sum_{r_i=k} p_{ir_i} \bigg/ \sum_{i=1}^{p} p_{ir_i}, & \text{if } \exists i = \overline{1, p} : \ r_i = k, \\ 0, & \text{if } \forall i = \overline{1, p} : \ r_i \neq k, \end{cases}$$

$r = \arg\max_{k=\overline{0,3}} p_k$, – connection belongs to class r.

In this method, uncertainty is processed by statistical methods, and fuzzy mathematics is involved in the formation of the training sample, on the basis of which the descriptor catalogs are built. As a result, we obtain not only the class of the additive to which the compound probably belongs, but also an estimate of the probability of such belonging. More precisely, this is the weighted average of the probabilities of the occurrence of compound descriptors among the descriptors of compounds of this class.

6 Conclusion

The method described above provides the basis for the automated design of virtual chemical compounds with "given" (environmental, technological, consumer, etc.) properties according to the following scheme.

Based on the statistical image, the most frequently encountered structural descriptors are selected, with the help of which the structure of the virtual chemical compound is created, and the result is checked by using the forecast module.

Thus, the developed methods of the theory of fuzzy sets allow us to process physicochemical information and make informed decisions under uncertainty. And on this basis to create automated expert systems and CAD of chemical structures.

The development of software and the information system itself for the optimal management of the composition of the polymer composition using the methodological foundations of fuzzy mathematics implies, in our opinion, the following actions:

1. Analysis of the subject area and identification of parameters and conditions corresponding to optimal composition of the polymer composite materials.
2. Construction of a fuzzy mathematical model for computational experiment.
3. Development of a polymer composite materials analysis method according to a mathematical model.
4. Analysis of the mathematical model, solving problems and creating a set of solutions.
5. Interpretation of analysis results and model refinement.
6. Testing the methodology on model (virtual or real) examples.
7. Creating a profiled information system.

This information system has a set of opportunities to solve problems arising in the course of the engineer-technologist. Here are some of them:

- Analysis and interpretation of initial data
- Systematization of data
- Evaluation of alternatives in the choice of polymeric compositions formulations
- Examination of active additives
- Management of the polymer composition
- Search for new additives for polymer compositions

On this basis, we are developing a decision-making method for use in designing new technologies and optimizing polymer composite materials formulations, as well as analyzing other multicomponent systems, including chemical-technological systems. Fuzzy models of source data, problems of mathematical programming, algorithms that implement these methods, which are the basis of computer modules and actually make up this information system for analyzing objects of chemical technology for decision-making. In world practice, work in this direction is also being carried out [2]. Avoid hyphenation at the end of a line. Symbols denoting vectors and matrices should be indicated in bold type. Scalar variable names should normally be expressed using italics. Weights and measures should be expressed in SI units. All non-standard abbreviations or symbols must be defined when first mentioned, or a glossary provided.

References

1. Germashev, I.V., Derbisher, V.E., Losev, A.G.: Analysis and Identification of the Properties of Complex Systems in the Natural Sciences. Volgograd State University Publishing House, Volgograd (2018)
2. Emami, M.R.S.: Fuzzy logic applications in chemical processes. J. Math. Comput. Sci. 1(4), 339–348 (2010). https://doi.org/10.22436/jmcs.001.04.11
3. Germashev, I.V., Derbisher, V.E., Vasilyev, P.M.: Prediction of the activity of low-molecular organics in polymer compounds using probabilistic methods. Theor. Found. Chem. Eng. 32(5), 514–517 (1998)
4. Khidhir, B.A., Al-Oqaiel, W., Kareem, P.M.: Prediction models by response surface methodology for turning operation. Am. J. Model. Optim. 3(1), 1–6 (2015). https://doi.org/10.12691/ajmo-3-1-1
5. Germashev, I.V., Derbisher, V.E., Tsapleva, M.N., et al.: Sorting of additives to polyethylene based on the non-distinct multitudes. Russ. Polym. News 6, 2 (2001)
6. Germashev, I.V., Derbisher, V.E., Tsapleva, M.N., et al.: Computer aided design of chemical compounds with controlled properties. Theor. Found. Chem. Eng. 38(1), 86–91 (2004)
7. Chandra, P.H., Kalavathy, S.M.S.T., Jayaseeli, A.M.I., et al.: Mechanism of fuzzy ARMS on chemical reaction. In: Snasel, V., Abraham, A., Kromer, P., et al. (eds.) Innovations in Bio-Inspired Computing and Applications. Advances in Intelligent Systems and Computing, vol. 424, pp. 43–53. Springer, Heidelberg (2016). https://doi.org/10.1007/978-3-319-28031-8_4
8. Germashev, I.V., Derbisher, V.E., Zotov, Y.L., et al.: Computer design of active additives for PVC. Int. Polym. Sci. Technol. 29(4), 78–81 (2002). https://doi.org/10.1177/0307174x02-02900413
9. Germashev, I.V., Derbisher, V.E., Orlova, S.A.: Evaluation of the activity of flame retardants in elastomeric compositions using fuzzy sets. Kauchuk i Rezina 6, 15–17 (2001)
10. Egorov, A.F., Savitskaya, T.V.: Methods and models for the risk analysis and security management of chemical plants. Theor. Found. Chem. Eng. 44(3), 326–338 (2010)
11. Reza, Z.A., Mehdi, R.: Fuzzy optimization approach for the synthesis of polyesters and their nanocomposites in in-situ polycondensation reactors. Ind. Eng. Chem. Res. 56(39), 11245–11256 (2017). https://doi.org/10.1021/acs.iecr.7b02307
12. Gonzalez-Gonzalez, D.S., Praga-Alejo, R.J., Cantu-Sifuentes, M.: A non-linear fuzzy degradation model for estimating reliability of a polymeric coating. Appl. Math. Model. 40 (2), 1387–1401 (2016). https://doi.org/10.1016/j.apm.2015.06.033
13. Wei, G.: Some induced geometric aggregation operators with intuitionistic fuzzy information and their application to group decision making. Appl. Soft Comput. J. 10(2), 423–431 (2010). https://doi.org/10.1016/j.asoc.2009.08.009
14. Elbarkouky, M., Fayek, A.R.: Building a roles and responsibilities structure for project owners in a managing contractor model using the concept of fuzzy consensus. In: Annual Conference - Canadian Society for Civil Engineering, vol. 2, pp. 1014–1024 (2009)

15. Jiang, Z., Feng, X., Shi, J.: An extended fuzzy AHP based partner selection and evaluation for aeronautical subcontract production. In: 2009 9th International Conference on Hybrid Intelligent Systems, vol. 1, pp. 367–372 (2009). https://doi.org/10.1109/his.2009.78
16. Liao, Y.: Evaluation method for the location selection of railway passenger station based on triangular fuzzy number. China Railw. Sci. **30**(6), 119–125 (2009)
17. Chen, J., Gao, X., Ding, L.: A comprehensive threat assessment method for group aircrafts cooperative air defensive combat under command of AWACS. J. Northwest. Polytech. Univ. **27**(5), 624–629 (2009)
18. Utkin, V.S.: Expert assessment of the quality of materials using fuzzy sets. Build. Mater. **6**, 34–35 (2001)
19. Bakhitova, R.H., Spivak, S.I.: Fuzzy interval estimates in the kinetics of chemical reactions. Proc. Univ. Chem. Chem. Tehnol. **42**(3), 92–96 (1999)
20. Kuznetsova, I.M., Kharlampidi, H.E., Ivanov, V.G., et al.: General Chemical Technology. Basic Concepts of Designing Chemical and Technological Systems. Lan, St. Petersburg (2014)
21. Germashev, I.V., Derbisher, V.E.: Properties of unimodal membership functions in operations with fuzzy sets. Russ. Math. **51**(3), 72–75 (2007). https://doi.org/10.3103/S1066369X07030115
22. Derbisher, E.V., Pogorelov, P.I., Germashev, I.V., et al.: Apriori ranking of factors when calculating the index of environmental hazard of substances using fuzzy sets. Chem. Ind. Today **8**, 48–56 (2006)

R&D in Collection and Representation of Non-structured Open-Source Data for Use in Decision-Making Systems

A. I. Martyshkin$^{(\boxtimes)}$, I. I. Salnikov, and E. A. Artyushina

Penza State Technological University, 1/11 Baydukova avenue,
Penza 440039, Russian Federation
alexey314@yandex.ru

Abstract. The paper covers the preparation and actual analysis of unstructured data sourced from social media (SM) by means of Big Data technology, which shall mean any dataset too large and complex to be processed by conventional methods. The researchers have analyzed highly functional data collection systems and services, including SM. Those include rather large systems as well as projects that legally collect information. Experiments have produced Python code to optimally collect open-access data from VK.com for specific applications. The paper presents the web interface of this software. It describes the intended use and functionality of the system, with extra details on the interface tabs. Note that the collected data can characterize any registered SM user to a certain extent. A person's interests can be suggested from their posts and subscriptions. The study is aimed to design tools to collect and represent unstructured open-source data so as to construct a person's social profile (SP). The conclusion is that the collected data can characterize SM users to this or that extent.

Keywords: Data visualization · High-level structures · Unstructured data · Software system · Social profile · Social media · Big Data

1 Introduction

Big Data is an umbrella term defined as any data set too large and complex for conventional processing [1, 2].

Data science covers the use of big data mining methods. Machine learning, data science, and big data are related in the same way as, for instance, crude oil and oil refineries.

Big data is based on the rule of 3 Vs: variety, velocity, and volume, see Fig. 1: volume is the amount of data contained in a dataset; variety is the types of collected and stored data; velocity is the rate, at which new data is generated.

Over time, there emerged new V-rules: veracity, viability, value, variability, and visualization.

In its turn, data science is a data-based research field that uses theoretical, mathematical, computational, and other applied methods for data analysis and evaluation. Data science is directly linked to big data and ability to use them, as well as to the

© Springer Nature Switzerland AG 2020
A. A. Radionov and A. S. Karandaev (Eds.): RusAutoCon 2019, LNEE 641, pp. 1098–1112, 2020.
https://doi.org/10.1007/978-3-030-39225-3_116

knowledge of machine learning, computation, and algorithms. This knowledge is what distinguishes a data scientist from an ordinary statistics specialist [2].

What makes this research relevant is the universal and common use of IT in the modern society's life. This can have both positive and negative effects, as aside from new potential threats, this gives rise to new opportunities in various social aspects [3, 4].

The «digital footprint» human makes when using e-communication can affect all humanities: social studies, demographics, socio-economic geography, and even history. Considerable data-collection capacities are wielded by mobile carriers: in many countries, there are many more cell contracts than people [5] whereas data collected from the numerous base stations can quite precisely locate a user. Of course, such data is protected by law; however, illegal use has taken place quite a few times already [6]. There are other potentially useful sources, e.g. social media (SM). In the context of this research, SM data shall mean user pages made openly accessible upon the owners' consent.

SM date back to 1978 when Bulletin Board System, or BBS, was created. The system was PC-hosted; users had to dial through the mainframe modem to communicate via landlines. BBS was the first system to enable user-to-user communication via the Internet. Social e-media that appeared in the early 2000s can be described as platforms, online services, or websites that help build, represent, or arrange social relationships in the Internet [7, 8].

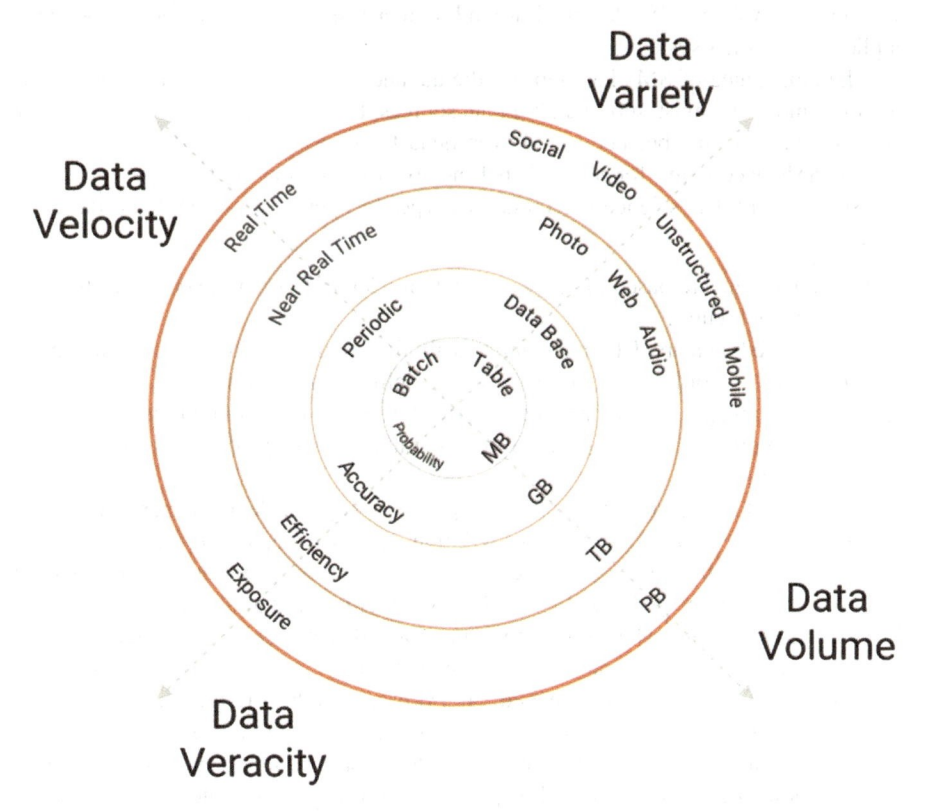

Fig. 1. Basic Big Data rules.

Technology development altered the Internet and the SMs. Classmaster, SixDe-grees, Friendster, and MySpace were made for making friends yet all of them became something else. The today's SMs are extremely diverse in terms of their target audience and purpose. For instance, LinkedIn is designed for making business contacts and has over 30 million users, while Instagram was made purely for sharing pictures. Beside the basic details provided when signing up, user-medium interaction continually adds more data on the user and their interests. This shall be referred to as a person's social profile (SP) [9].

2   Statement of Problem

Analysis of openly accessible, or open-access data published in the Internet, in particular by means of social media, could be of use in handling various practical problems. No such solution would be possible without special data mining methods and algorithms [10–12]. The research object is a piece of state-of-the-art software for collection and representation of data and SP.

The research objective is to create tools for collecting unstructured open-source data for human social profiling in decision-making systems. Thus, the research team seeks to design software for collecting and making human SP. This paper also reviews the software systems (SS) designed around similar tasks and focuses on the potential application of such systems.

The emergence of SMs has boosted the advancement of big data, as social media are now among the most important big data sources. Industrial experts believe that 90% of the world's data has been collected over the last two years, with unstructured sources such as SMs accounting for 80% of such newly collected data [13–16].

Big data and data science use numerous types of open-access data. Consider them in detail.

1. Structured data are based on a specific data model and stored in specially allocated fields in each entry;
2. Unstructured data are data that cannot easily fit a model due to variations in context and nature. An ordinary email address exemplifies unstructured data;
3. Natural language data represents a special type of unstructured data, which is difficult to process, as it requires both substantial knowledge of linguistics and data science skills;
4. Machine data are computer-, software-, process-, app-, or device-generated without human intervention. According to Wikibon, the total population of nodes in the network of Web-connected physical devices generating new data will exceed the human population by 2020 [17]. Such network is referred to as Internet of Things;
5. Graph data is a somewhat confusing term, as any data can be represented as a graph. Graph here means the graph is of mathematical origin. Consider a graph representing the friends and subscribers of an SM user, let it be VK.com. In this case, the nodes are the user page and their friends' pages, while the edges are the page-to-page relations. The same pages can be used to construct a few more graphs based on interests in specific fields. Graph superimposition might reveal new information of interest for the researcher;

6. Video, audio, and graphical data are complicated for data scientists;
7. Streaming data can be of any of the above types and have one important feature: the information is received by the system when a certain even occurs therein. Analysis is mainly complicated by the need to tailor data science methods to the stream and its properties;

Let us now review big data processing SS. Consider the basic tasks of the modern software in direct contact with all the diverse SM-published data.

1. Monitoring and analysis.
2. Forecasting and management.

SS can be classified in terms of the end user [10].

1. By the analyzed SM levels. SS can handle analysis, monitoring, forecasting, and management;
2. By SM models. SS might use different models. Those include the network structure model or the data propagation model;
3. By data analysis methods. SS use statistical and graph methods as well as semantic text analysis;
4. By scope of analysis. SS mainly focus on analyzing SMs, communities, users, information messages, or comments;
5. By analysis conditions. A system may not provide analytical data for this or that reason; it can analyze data retrospectively or in real time;
6. By data collection conditions. See analysis conditions. However, data can be collected from the entire population or on a subject basis;
7. By scope of sources. Software can use a single source or more. Those can be SM pages, blogs, forums, file hostings.
8. By volume of processed data. SS can be designed around big data or modeled data.

Large-scale systems include:

1. Search engines.
   a. search.twitter.com;
   b. blogs.yandex.ru;
2. Alert engines.
   a. google.ru/alerts;
   b. twilert.com.
3. Aggregators.
   a. trends.google.ru/trends;
   b. wordstat.yandex.ru.
4. Data collectors.
5. Data collectors and aggregators.
   a. FeedsApi – feedsapi.com;
   b. Quadrigram – quadrigram.com.

These systems have a number of advantages: they are free and easy to use. However, their limited functionality prevents professional use.

Over time, there emerged new V-rules: veracity, viability, value, variability, and visualization.

3   Commercial SM Monitoring and Analysis Systems

Such systems are designed to help companies address internal and external development challenges such as employee analysis, transforming business processes, enhancement of attitudes in the collective, market studies, finding new partners and others.

To be competitive, the today's systems must feature some core functionality, namely.

1. Monitoring the brand mentions.
2. Determination of market risks and opportunities.
3. Web analytics.
4. SM support.
5. SM forecasting and management.

There exist numerous multifunctional data collection systems and services, some of which collect SM data. Those include rather big systems such as Social Studio, AlterianSM2, BrandSpotter, and Hadoop. These products are designed for use in large-scale commercial projects. There are other projects that help legally collect data: Webhose.io, Scrapinghub, ParseHub, VisualScraper, etc. [1]. Unfortunately, many of such systems are either paid, have excessive or lackluster functionality.

Paragraphs below review several software products for commercial use.

Social Studio. This software tracks trademark mentions in social media and online; it uses morphological features to evaluate the connotations of such mentions.

IQBuzz. This system provides round-the-clock SM monitoring for real-time output. Collective access is possible; analyzed data can be made publicly accessible. The system can determine the tone of a message, draw conclusions, and compile social and demographic data on the message author from their SM page.

Brand Analytics. This system offers broad functionality and can collected data by keywords, geotags, and authors. Collection of meta data enables data quantification, tone analysis, search for duplicates, and language determination; it takes into account the author's gender and location. Another advantage is that this software is developed in Russia and fully supports the Russian language.

All of these systems are great for SM analysis. They provide in-depth monitoring of brand mentions, consider the connotations, and support user feedback by integrating the company's accounts. They have a common drawback though: such systems are mainly intended for serious business analytics, management, and long-term forecasting.

A small project could make better use of something made specifically for it. Python is a programming language suitable for data collection and analysis [3]. Python code is arranged in functions and classes that can be merged into modules, which themselves can be merged into packages. A data collection problem can be analyzed by parcing a VK.com group, i.e. by retrieving the members' IDs for further use. A parcer is a program or a script that uses a pre-configured algorithm to collect the required data from a website. Parcing is the process of algorithmic data search, analysis, and systematization that produces output in a format suitable for storage and use: Excel, csv, or txt. To collect data, this research uses a Python parcer that interacts with the VK.com API to collect user data.

Any data science project undergoes multiple stages or life cycles from a statement of problem to a visualized solution. See Fig. 2 for a flowchart.

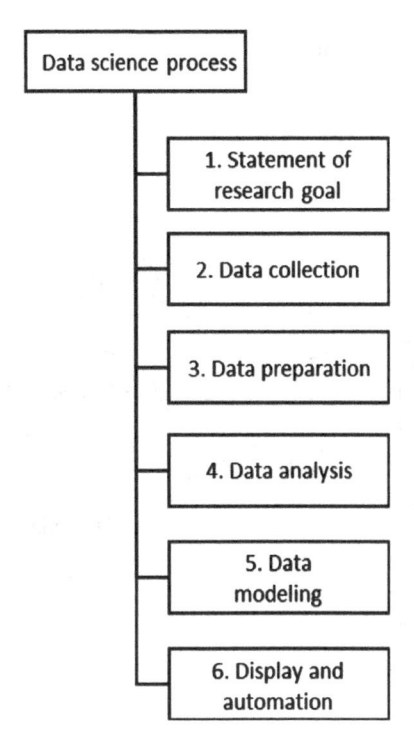

Fig. 2. Data science stages.

4   Research Materials and Results

The subject area of this R&D effort is the problem of hiring a person. Hire is a multistep process intended to find and engage a candidate that has skills the company needs.

Consider handing this by means of data science. Let data be sourced primarily from SMs. A person's user profile can quite precisely characterize them.

If a person is interested in IT or programming, such interests will likely be reflected by their SM page, if they have any. SP data can be limited to a minimum set:

1. Surname.
2. Name.
3. Gender.

4. Date of birth.
5. Profile URL.
6. Interests.
7. Groups and communities.

Consider the architecture of the proposed solution, see Fig. 3. As can be seen in the flowchart below, this SS is an applied modular program.

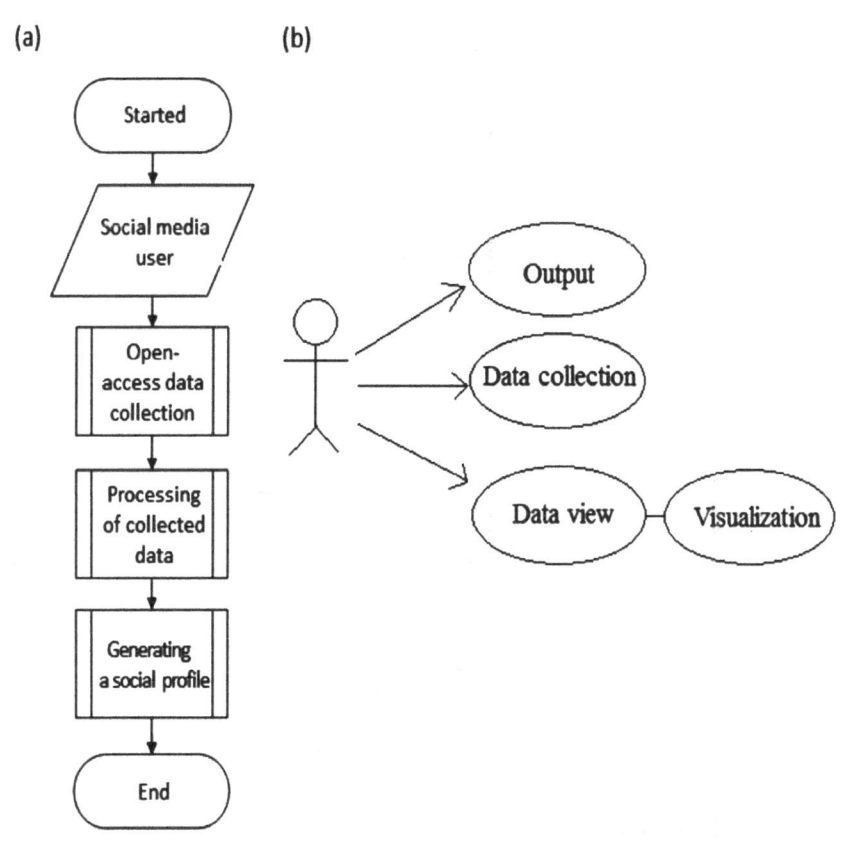

Fig. 3. Solution architecture (a) algorithm flowchart; (b) options for use.

The client is a user that uses the program GUI, see Fig. 3a.

The application has several functions: it collects data from groups and communities and writes it as unstructured .csv data; it also collects data on the group members to also write a .csv. Since it is unstructured data that this program will use, conventional databases are unnecessary, and .csv will do.

Collected data is visualized and the user's SP is compiled on a web page opened in the browser. Consider the basic functionality available to the user via the system GUI, see Fig. 3b.

Everything is quite simple now; to access the app, one must have a VK.com account and a developer status. Then create an app (see Figs. 4 and 5) to freely access the VK IP.

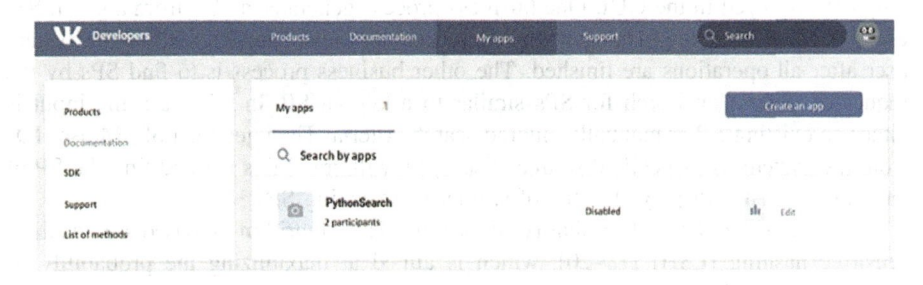

Fig. 4. Application section.

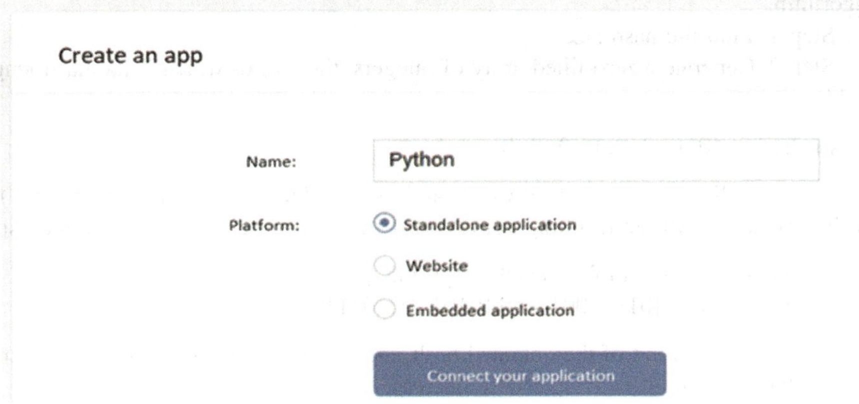

Create an app

Name: **Python**

Platform: ⦿ **Standalone application**

            ○ **Website**

            ○ **Embedded application**

            Connect your application

Fig. 5. Creating an app.

This will generate the token you need. A token is an alphanumeric combination, e.g. 3dsf84ds5g48gd45sddsdf48fg4h4s84a548sdf41b.

A registered user has access to all the functions of the system; basic functions include:

1. Logoff;
2. Data collection;
3. SP generation;
4. SP search.

Now that we have covered the GUI functionality, consider the business logic of the most important business processes implemented in this SS. The basic business process is to generate the SP of a job candidate, see Fig. 3a. As can be seen in the flowchart, the algorithm receives an SM group/community ID or name, then sequentially calls the following operations: collect open-access data from the group, process the collected data, collect data on the members, write data in a table, generate the SP. After the SP is fully generated, it is checked for presence in the collected data. If the profile is found, it is complemented with additional data; otherwise, a new profile is created.

The two other business processes are based on the process above; unlike the latter, they are displayed in the GUI. One business process generates an SP from a given SM user ID. Its only distinguishing feature is to display the sought-after SP to the system user after all operations are finished. The other business process is to find SPs by the requested criteria or search for SPs similar to a known SP. In this case, the input is either an existing SP or manually entered search criteria. Then get a list of SM user IDs from the system user-specified source. The cycle will crawl this list and find the SP of each user. Then it displays the list of similar or matching SPs.

Now consider the SM similarity algorithm. The algorithm is based on locality-sensitive hashing (LSH) [18–20], which is aimed at maximizing the probability of collisions of similar function arguments.

We use SimHash. This is a hashing algorithm that converts a text into a list of values that ultimately represents the text signature. Consider an example of running the algorithm.

Step 1. Find the hash size.

Step 2. Generate a zero-filled array of integers, the size of which is the hash length in bits.

var hash = [0, 0, 0, 0, 0, 0, 0, 0];

Step 3. Split the document by words and use any known algorithm (md5, sha1) to hash each word. All the resulting values must be equal in length to a predefined size.

var word = ["algorithm", "search", "profile"];
var wordHash = [01111001, 00110101, 00101110];

Step 4. For each bit of the obtained hash, increase the corresponding array element by 1 if the original bit is 1; otherwise decrease.

01111001
00110101
00101110
hash = [−3, −1, 3, 1, 1, 1, −1, 1];

Step 5. Use the resultant array to generate a hash as follows: if an array element is > 0, the corresponding bit in the hash is 1; otherwise, 0.

hash = [0, 0, 1, 1, 1, 1, 0, 1];

Such hashes are used to fine inter-text similarity. To find the similarity of two signatures, find the number of dissimilar positions and calculate the ratio of matching positions y to the length.

00111101 XOR 00111000 = 00000101, similarity = 6/8 = 0.75

In this case, the signature similarity is 75%.

The following languages, technologies, and frameworks are proposed for implementing this application: Python 3.6, HTML, CSS, JavaScript, PyQT, VK library, and VkAPI.

Let us review this software as well as the procedure of generating a user's social profile. In particular, consider how an SP could be interpreted in this process. The first thing to review is the main page, see Fig. 6.

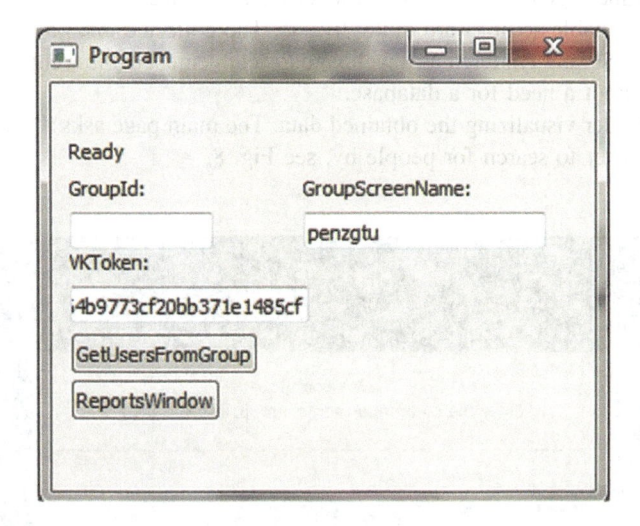

Fig. 6. Main page.

In this window, the user shall enter their VK.com API token and the group/community to collect information from. To start collecting the data and generate a user's SP, the system user must enter the group ID or name and then click GetUsersFromGroup [21]. The program will send a get query to the VK.com servers via the provided API; the server provides a JSON response, see Fig. 7.

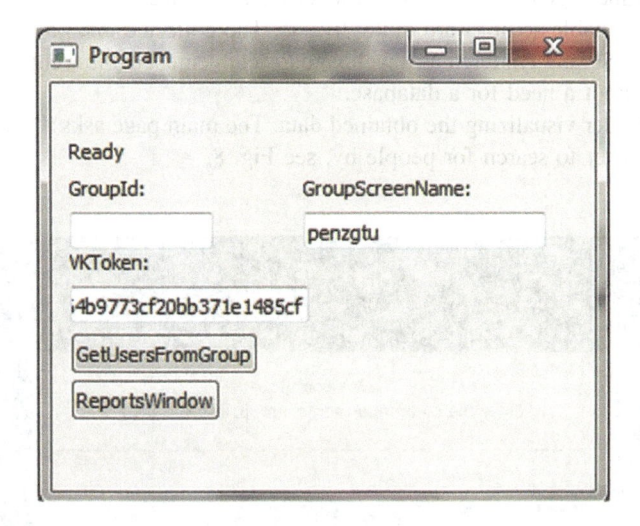

Fig. 7. JSON response from the server.

The collection process is not displayed; however, the results are written as unstructured csv data, thus converting JSON into csv. The collected data is written in two files: group.csv and users.csv; the former stores group IDs, descriptions, names, screen names, number of members, etc., the latter contains data on the group users: IDs, names, surnames, gender, dates of birth, etc. Data collection can be configured by scripts; the available settings are quite diverse. The collected data, also referred to as the SM user's SP, are rendered as a webpage for convenience; the webpage is linked to a csv file without a need for a database.

Now consider visualizing the obtained data. The main page asks the system user to enter a parameter to search for people by, see Fig. 8.

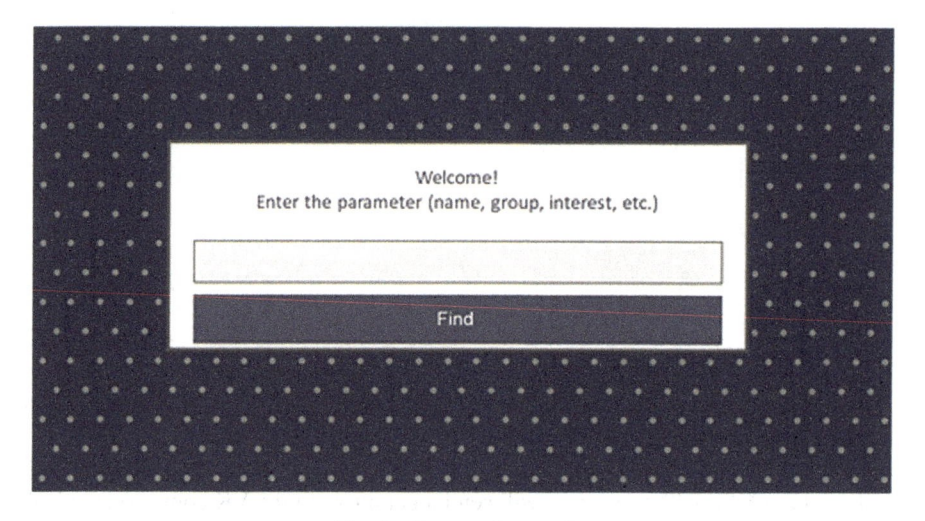

Fig. 8. Parametric search.

Clicking "Find" without any parameters will open the general table that contains all the collected data, see Fig. 9.

User ID *	Name -	Surname .	Country .	City *	Gender *	Date of birth .
5595	Marina	Gmyrina	Russia	Penza	Female	October 6th
50243	Lyudmila	Plakhina	Russia	Penza	Female	November 7th
53197	Lyubov	Krysina	Russia	Penza	Female	June 29th
57296	Andrey	Andriyanov	Russia	Moscow	Male	April 15th
134121	Pavel	Makarov	Russia	Penza	Male	Sep 20, 1989
231404	Maria	Sasikova	Russia	Moscow	Female	March 15th
317947	Maria	Antimonova	Russia	Penza	Female	August 16th
4791527	Anna	Evensen	Russia	Penza	Female	Feb 21, 1992
733919	Ilya	Kirillov	Russia	Penza	Male	October 12th

Fig. 9. Unstructured general data table.

Then the user lands on the social profile page. For instance, the query is Ruslan Nazirov, see Figs. 10 and 11.

Fig. 10. SP page.

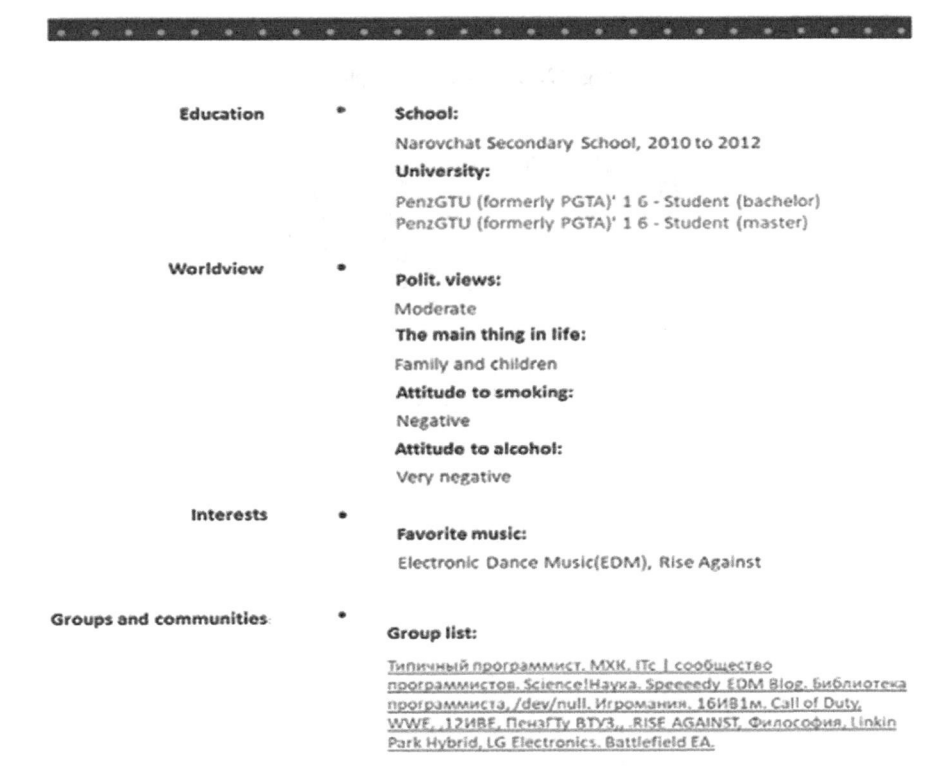

Fig. 11. Profile details.

Besides, the program can build connectivity graphs, i.e. retrieve the found user's friends and render their relations if there are any. The graph can grow in depth to see friends' friends, etc.; an example graph is shown in Fig. 12.

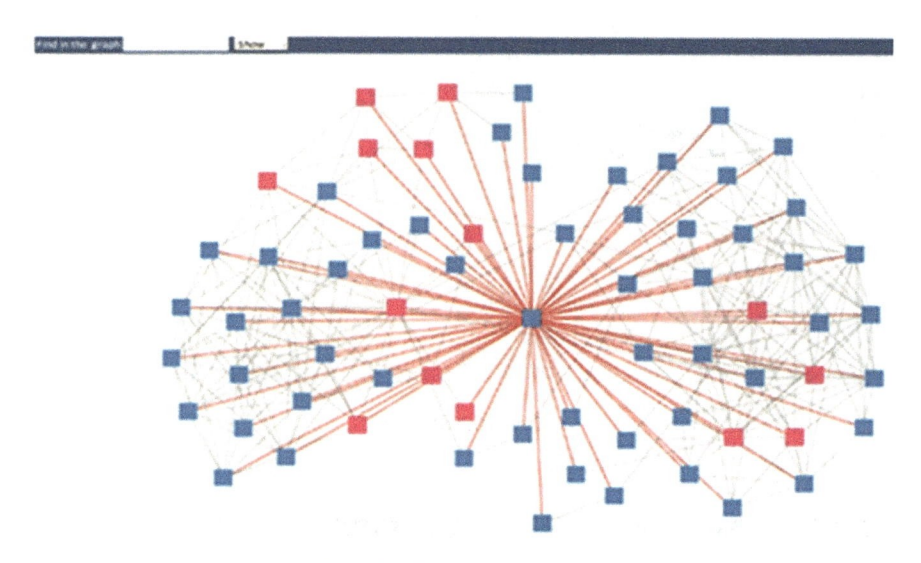

Fig. 12. Ego network graph.

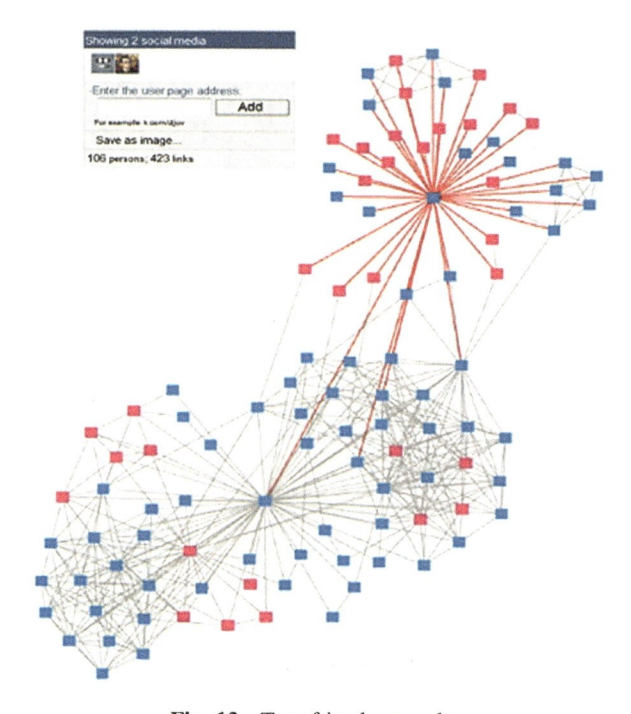

Fig. 13. Two friend networks.

The graph clearly shows how the user's friends are related; analysis reveals two outstanding groups. This is a page of our department's master student; one group is his university friends; the other group is his friends from school and hometown. The graph can be analyzed in-depth to see how Ruslan's friends relate to his friends' friends, see Fig. 13.

This graph can be used to test the six degrees of separation as well as the known proverb: a man is known by the company he keeps.

5 Conclusion

The conclusion is that the collected data can characterize SM users to this or that extent. Analysis of the above data reveals the user resides in Penza. Reviewing his communities reveals that he is interested in IT, in particular in programming.

The research has produced tools to collect and represent unstructured open-source data so as to construct a person's SP. The ultimate problem it tackles is employee hiring, or headhunting for a specific job.

The authors have analyzed the subject area, reviewed and classified the existing SS aimed at information retrieval from social media. Emphasis is made on data science processes and methods; basic DS stages and steps are considered in detail. The paper presents problems and popular solutions pertaining to each step.

It should be noted that the volume of open-access data generated by Internet users is growing ever faster. This contributes to the advancement of big data and data science. This is what necessitates the development of such SS.

Acknowledgements. Research was supported by the RFBR Grant for Best Projects of Basic Research, Grant No. 19-07-00516 A.

References

1. Schmi, A.V.: New Approaches to Big Data: Winning Management Strategies in Business Intelligence. PALMIRA, Moscow (2016)
2. Cielen, D., Meysman, A.: Introducing Data Science: Big Data, Machine Learning, and More, Using Python Tools. Piter, St. Petersburg (2017)
3. Korshunov, A., Beloborodov, I., Buzun, N., et al.: Social network analysis: methods and applications. In: Proceedings of ISP RAS, vol. 26, pp. 439–456 (2014)
4. Churakov, A.N.: Analysis of social networks. Sociol. Stud. **1**, 109–121 (2001)
5. Gubanov, D.A., Novikov, D.A., Chkhartyshvili, A.G.: Social Media: Models of Information Influence, Management, and Confrontation. Fizmatlit, Moscow (2010)
6. OSINT (2019). https://ru.wikipedia.org/wiki/OSINT. Accessed 12 Aug 2017
7. Federal Law152-FZ: On personal data. Russian codes, vol. 31, p. 3451 (2006)
8. Davern, M.: Social networks and economic sociology: a proposed research agenda for a more complete social science. Am. J. Econ. Sociol. **56**(3), 287–302 (1997)
9. Martyshkin, A.I., Salnikov, I.I., Pashchenko, D.V., et al.: Development of software for analysis of publication of users in social networks. XXI Century: Resumes of the Past and Challenges of the Present Plus, vol. 7, no. 4, pp. 35–44 (2018)

10. Timonin, A.Y., Bozhday, A.S.: Use of Big Data to construct person's social profile from open-access data. Bull. Penza State Univ. **2**(10), 140–144 (2015)
11. Bazenkov, N.I., Gubanov, D.A.: Information systems for social networks analysis: a survey. Large Scale Syst. Control. **41**, 357–394 (2013)
12. Batura, T.V.: Methods of social networks analysis. Bull. NSU Ser. Inf. Technol. **10**(4), 13–28 (2012)
13. Aggarwal, C.C.: Social Network Data Analytics. Springer, Cham (2011)
14. Content analytics for structured and unstructured data analysis: IBM Big Data competence center, Moscow (2014)
15. Bershadskaya, E.G., Nazirov, R.R.: Collection and representation of unstructured open-source data. In: Proceedings of the 16th All-Russian R&D Conference on State-of-the-Art Spacetime Signal Processing Methods and Means, pp. 64–68 (2018)
16. Martyshkin, A.I., Bershadskaya, E.G.: Software development for collection and representation of unstructured open-source data. Actual IT Technol. **28**(28), 6–17 (2018)
17. Martyshkin, A.I., Bershadskaya, E.G.: Software system for collecting non-structured information to construct the social profile of human. XXI Century: Resumes of the Past and Challenges of the Present Plus, vol. 8, no. 1, pp. 27–33 (2019)
18. Vidiasova, L., Novikov, D., Bershadskaya, E.: Do social networks help to organize a community around E-participation portals in Russia? In: ACM International Conference. Proceeding Series Challenges in Eurasia. Proceedings of the International Conference on Electronic Governance and Open Society: Challenges in Eurasia, pp. 62–69 (2017)
19. Martyshkin, A.I.: Designing the stages of collection, analysis and representation of data from the communities in social networks. XXI Century: Resumes of the Past and Challenges of the Present Plus, vol. 8, no. 45, pp. 53–57 (2019)
20. Martyshkin, A.I., Salnikov, I., Paschenko, D.V.: The stages of collection and presentation of Big Data to build a social profile of the person. In: Proceedings of the TSU, Technical Sciences, vol. 9, pp. 617–628 (2018)
21. Bershadskaya, E.G., Vidyasova, L.A.: Automated opinion mining tool as a text processing tool. In: Proceedings of the 16th All-Russian RD Conference on State-of-the-Art Spacetime Signal Processing Methods and Means, pp. 46–50 (2016)

Modeling of Interaction of the Mechatronic Unit Segments in the Adaptable Part of an Aircraft Wing

N. Sharonov[1,2], A. Makarov[1(✉)], A. Ivchenko[1], and A. Gorelova[1]

[1] Volgograd State Technical University,
28 Lenina avenue, Volgograd 400005, Russian Federation
`amm34@mail.ru`
[2] Innopolis University,
1 Universitetskaya Street, Innopolis 420500, Russian Federation

Abstract. The work represents the results of modeling and studying the interaction of the mechatronic unit segments in an aircraft wing when an external impact is applied. The basis of the model of an adaptable wing main frame structure proposed is a mechatronic unit, which is designed to ensure the preset angle between the elements in the adaptable wing structure; a coordinated control of the model; and a possibility to vary the airfoil geometry with due regard to the aerodynamic and technical requirements. When modeling, displacements, stresses, and deformations were determined, which arisen in various operating modes with external loads. As a result of the research, computer models of a fragment of the mechatronic unit have been obtained, stresses and deformations under various loads have been determined, and recommendations for enhancing the efficiency of the structure under study have been developed. The dynamics of the damages indicates that the weakest element of the structure is the link in the place of bending. Also, stress risers were detected, which became the round parts of the segments. The results of the research can be used for designing and manufacturing configurable aircraft wing structures.

Keywords: Modeling · Mechatronic unit · Adaptable wing · Configurable structure

1 Introduction

Adaptable wing technologies come into use to improve the aerodynamic properties of aircrafts by means of varying the airfoil geometry, depending on a flight mode [1]. The basis for the adaptable wing main frame design proposed is a mechatronic unit, which is designated to ensure the preset angle between the elements in the adaptable wing structure, with their coordinated control allowing to vary the airfoil geometry.

The adaptable controlled wing is a wing of an airplane type vehicle with the profile, which takes the form close to the optimum in each preset flight mode. Modern designs of a wing like this allow to smoothly float the leading-edge and trailing-edge assemblies and change the wing curvature, depending on the height, speed, and overload.

© Springer Nature Switzerland AG 2020
A. A. Radionov and A. S. Karandaev (Eds.): RusAutoCon 2019, LNEE 641, pp. 1113–1123, 2020.
https://doi.org/10.1007/978-3-030-39225-3_117

Ideally rigid, non-deforming wing structures exclude any adaptation thereof to varying conditions. Devices, such as flaps, slats, ailerons, and shields lead to limiting changes in the common form with minor advantages in comparison with those non-deforming. The optimum advantages can be obtained from a wing design, which is transformable and adaptable in its nature [2].

One can find several concepts of an adaptable wing of various complexity in the literature, depending on specific applications and goals. Adjustable flaps and slats, which are designated to optimize the takeoff and landing parameters, are most wide-spread [3, 4]. Wings provided with a variable swirl angle [5] or a general possibility to vary the curvature profile [6, 7] are used, which allows to control the aerodynamic flow and reduce the resistance. Also, wing segments with the variable chord [8] and wings with a variable aspect ratio [9] are used, which allows to control an airplane shape during a flight.

This work considers modeling and studying the interaction of a mechatronic unit segments.

2   The Materials and Research Methods

The basis of the model of an adaptable wing main frame structure proposed is a mechatronic unit, which is designed to ensure the preset angle between the elements in the adaptable wing structure; a coordinated control of the model.

A possibility to vary the airfoil geometry with due regard to the aerodynamic and technical requirements. Various options of the structural implementation are considered and proposed in [10–18].

The mechatronic unit under study represents successive combined double-action joints [19]. To study its operation modes, a 3D-assembly model of the product was carried out in the format, which was necessary for computer modeling.

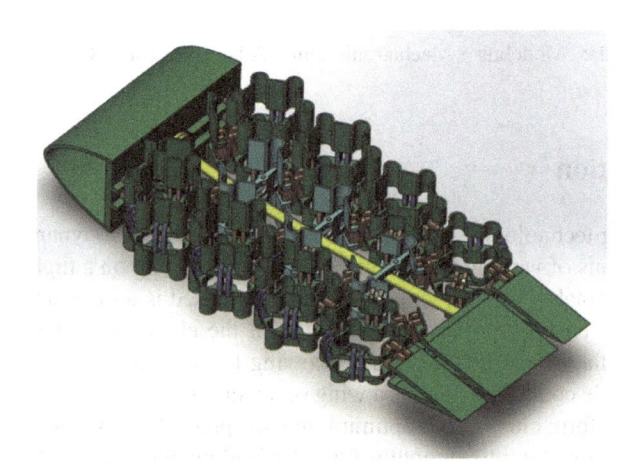

Fig. 1. 3D solid assembly made in Solidworks.

SolidWorks program complex was used for the modeling, which enables developing products with any complexity grade and designation and allows to carry out the analysis of the motion of and interaction between solid products [20].

3D solid elements were created and the assembly of the product model was performed on the basis of the mechatronic unit prototype design developed [18] in "TMPK-VOLGOGRAD" Ltd. by engineer Ivchenko A.V. (Fig. 1).

The following modeling problems were solved for the model developed:

- To determine the stresses, displacements, and deformations, which arise in the links between the mechatronic unit segments under various loads
- To determine the maximum damages (weak elements) and the service life of the structure elements, depending on the stress applied.

To solve these problems, a simplified model was used (Fig. 2), which consists of two segments (1) and two links (2).

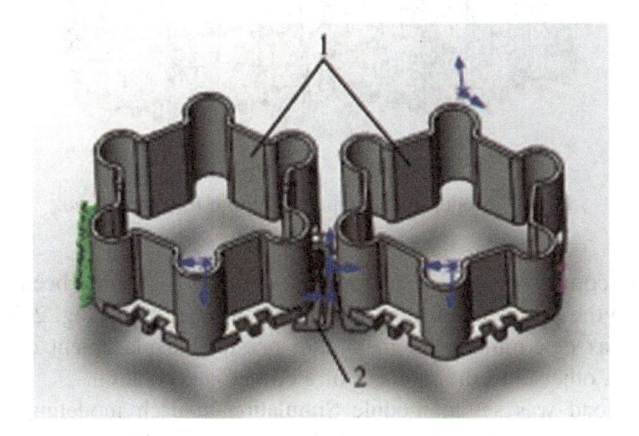

Fig. 2. Model of two segments with links.

To model the loads, module Simulation is used in SolidWorks program complex where the interaction between the modules and the links was preset as that of solid bodies, with the "anchorage" type fixation and a face for the application of the distributed force. The parameters of the material and computational mesh are set out in Table 1 and in Fig. 3.

Table 1. The parameters of the material and computational.

Material name	ABS
Model type	Linear elastic isotropic
Ultimate tensile stress	$3e + 007$ N/m^2
Elastic modulus	$2e + 009$ N/m^2
Poisson ratio	0.394
Mass density	1020 kg/m^3

(continued)

Table 1. (*continued*)

Material name	ABS
Shear modulus	$3.189e + 008$ N/m^2
Total of network nodes	16281
Total of network elements	7860
Maximum aspect ratio	16.393
% of elements with aspect ratio <3	34.1
% of elements with aspect ratio >10	1,34

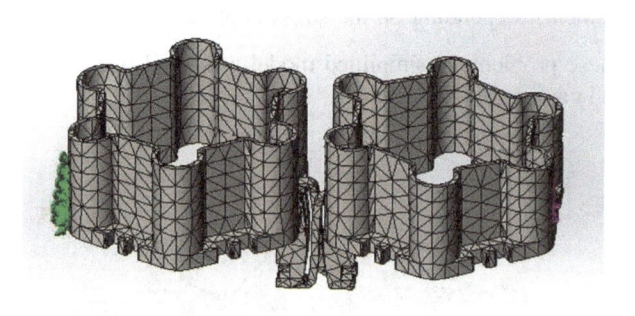

Fig. 3. Model with computational mesh being placed.

A series of computer experiments for force $F = 1$; 1,5; 2 N has been carried out for the model with the parameters from Table 1, which is set out in Fig. 2. The aim of the experiments was determining the maximum possible displacement of the segments relative to each other until the moment of a critical deformation.

A certain load was set in module Simulation at each modeling stage and the following parameters were recorded with the use of tool "Profiles":

- The distribution of deformation D in the model with the respective force
- The distribution of stresses G in the model with the respective force
- The distribution of displacements S in the model with the respective force.

At the next modeling stage, it was determined how the structure destruction took place: from what elements the process started and which the service life was when the load was symmetrical. To this end, research "fatigue" was created in module Simulation, with the symmetrical load for the forces: $F = 50$, 100, 150, 200, 250, 300, 500 N.

3   The Result and the Discussion Thereof

The summary data of the modeling results for any parameters provided are set out in Table 2.

Table 2. Modeling results.

No. p/p	Force F, N	Maximum displacement S, mm	Maximum stress G, N/m²	Maximum stress G, N/m²	Maximum stress G, N/m²
1	1	22.845	44761108	0.015	47.936
2	1.5	31.689	63911508	0.021	50.106
3	2	38.858	80275136	0.027	62.818

Displacement profile S in the model with the force of 2 N is set out in Figs. 4 and 5 represents the dependence of maximum displacement S of the model on force F.

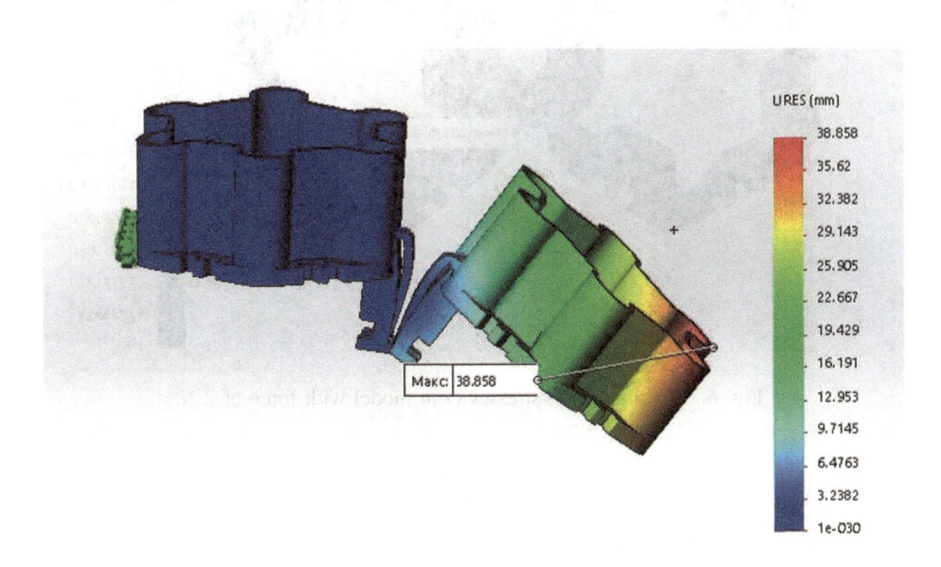

Fig. 4. Displacement profile S in model with force of 2 N.

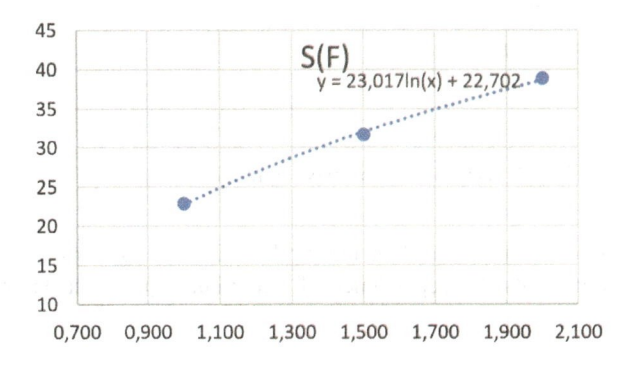

Fig. 5. Dependence of maximum displacement S of model on force F.

Based on the research findings, the critical slope angle of the segments relative to each other has been determined for the model, which is 62.818 grade under the load of 2 N. With the further increase in the load on the face, the proportional limit of the link takes place.

Distribution of stresses G in the model with the force of 2 N is set out in Figs. 6 and 7 represents the dependence of maximum stress G on force F.

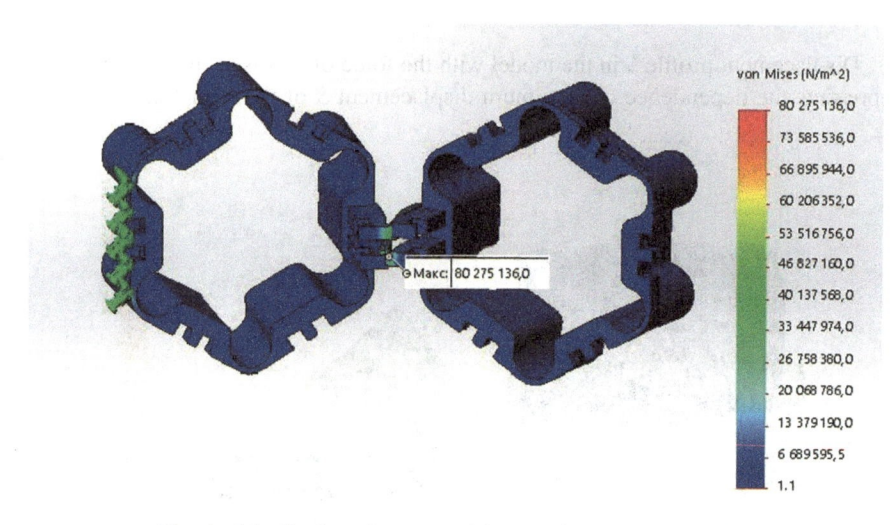

Fig. 6. Distribution of stresses G in model with force of 2 N.

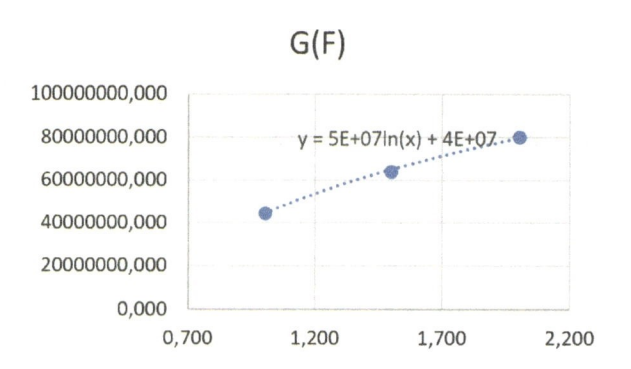

Fig. 7. Dependence of maximum stress G on force F.

Distribution of deformation D in the model with the force of 2 N is set out in Figs. 8 and 9 represents the dependence of maximum equivalent deformation D on force F.

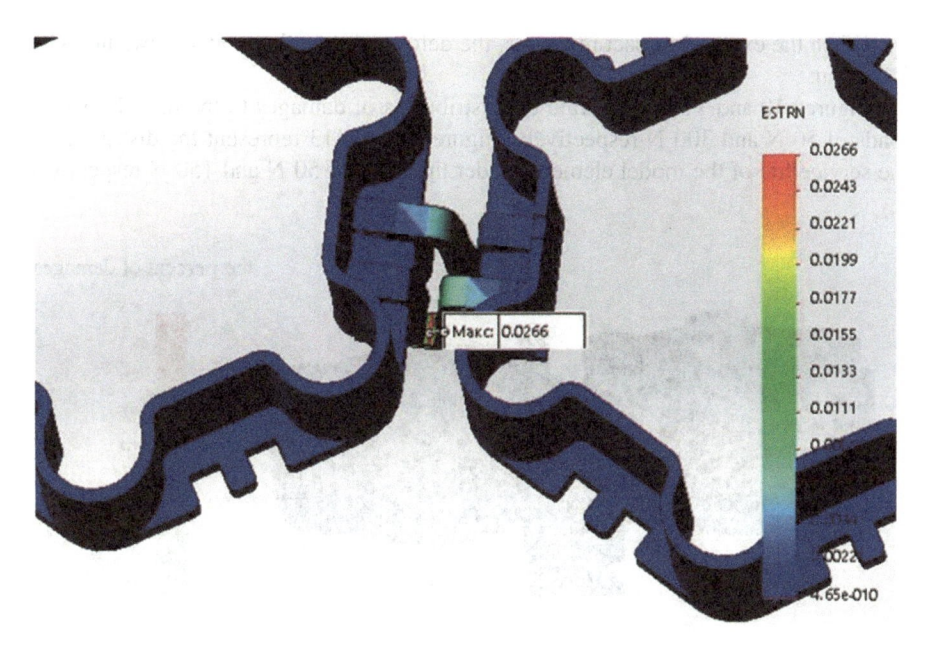

Fig. 8. Distribution of deformation D with force of 2 N.

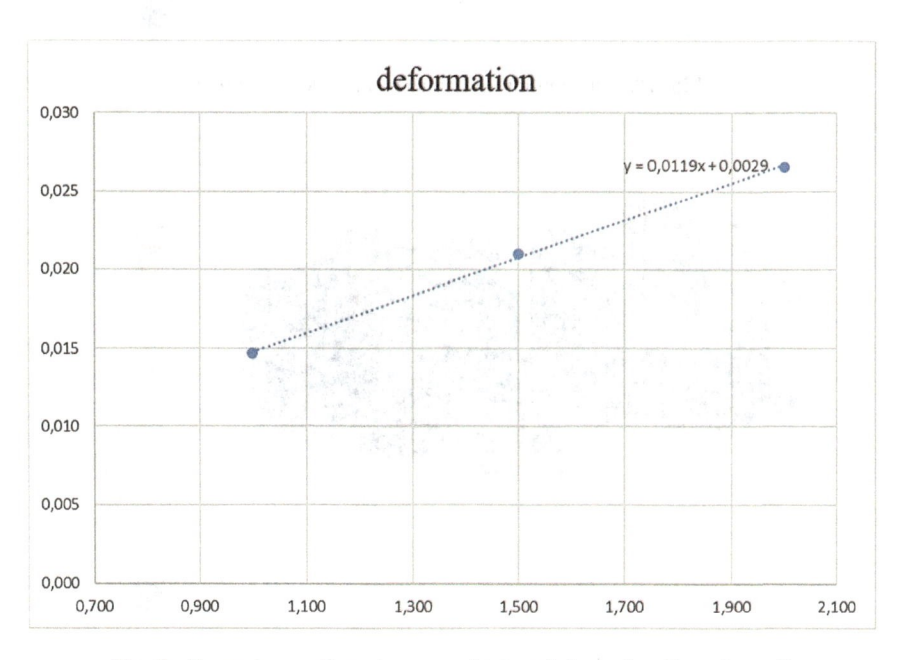

Fig. 9. Dependence of maximum equivalent deformation D on force F.

When the external impact increases, the deformation in the model grows up, which is logical.

Figures 10 and 11 demonstrate the distribution of damages to the model under the loads of 50 N and 300 N respectively. Figures 12 and 13 represent the distribution of the service life of the model elements under the loads of 50 N and 150 N respectively.

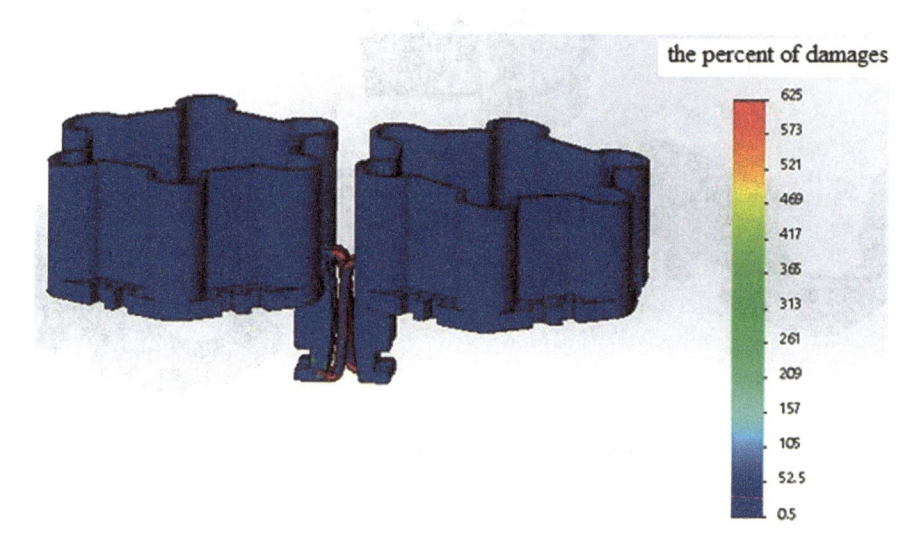

Fig. 10. Distribution of damages under load of 50 N.

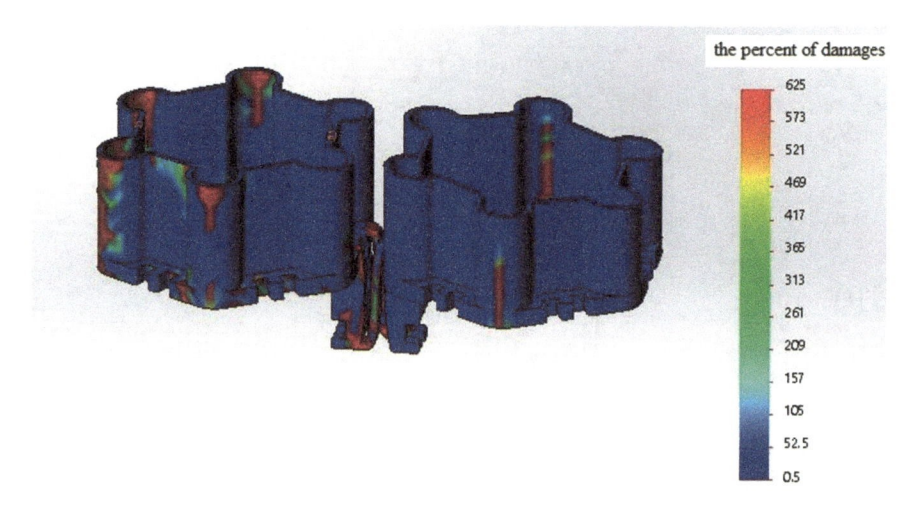

Fig. 11. Distribution of damages under load of 300 N.

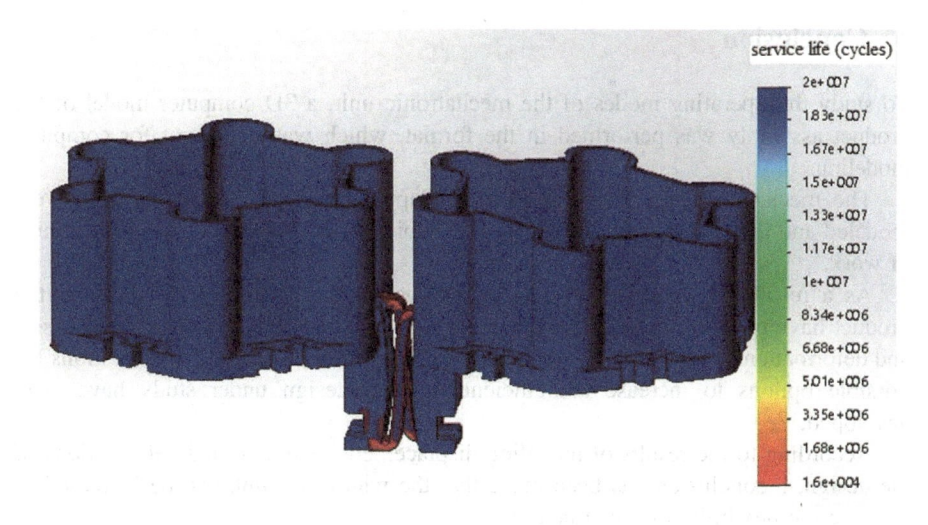

Fig. 12. Distribution of service life of model elements under load of 50 N.

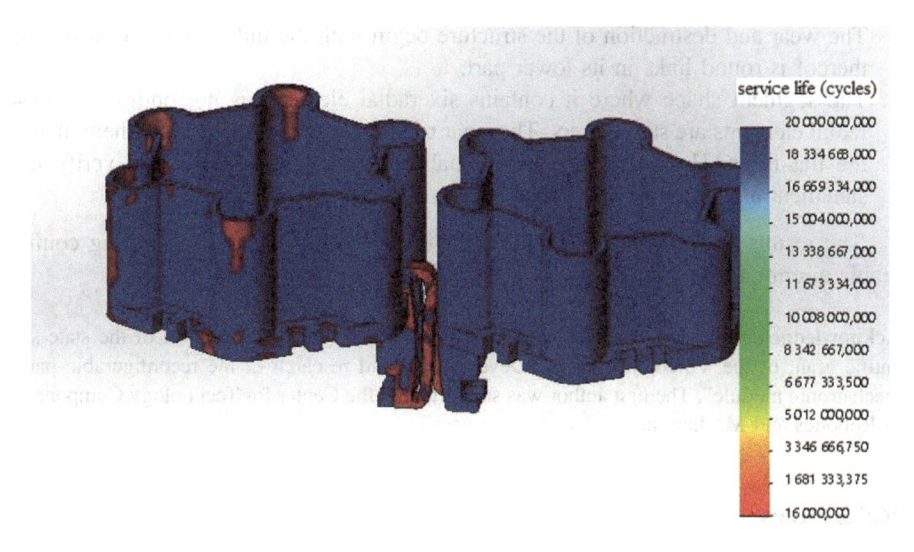

Fig. 13. Distribution of service life of model elements under load of 150 N.

The dynamics of the damages indicates that the weakest element of the structure is the link in the place of bending. This is it where the destruction of the structure starts from and it, respectively, has the minimum service life of 16000 cycles.

Also, stress risers were detected (Figs. 11 and 13), which became the round parts of the segments. This result allowed to determine that the destruction of the structure does not come equally, therefore, there are requirements for searching for a more optimum shape of the segments.

4   Conclusion

To study the operating modes of the mechatronic unit, a 3D computer model of the product assembly was performed in the format, which was necessary for computer modeling.

The mechatronic unit segments of the adaptable part of an aircraft wing were modeled and the statistics and interaction dynamics thereof were studied in the course of work.

As a result of the research, computer models of the unit and assembly of the product have been obtained in the static and dynamic modes; displacements, stresses, and deformations under various loads have been determined, and recommendations for possible options to increase the efficiency of the design under study have been developed.

According to the results of modeling displacements, stresses, and deformations of the design, a conclusion has been made that the width of a link can be increased to enhance the flexibility of the structure.

As a result of modeling the fatigue limit, the following conclusions have been made:

1. The wear and destruction of the structure begin with the links. The "weakest" part thereof is round links in its lower part.
2. The segment shape where it contains six radial elements is not optimum. These radial elements are stress risers. The wear of the structure goes faster in them than in the flat faces. However, some additional researches are necessary to verify this statement and ascertain the shape of the segments.

The results of the research can be used for designing and manufacturing configurable aircraft wing structures.

Acknowledgements. The research was carried out with the financial support of the state scientific grant of the Volgograd region "Development and research of the reconfigurable panel mechatronic module". The first author was supported by the Center for Technology Components in Robotics and Mechatronics.

References

1. Ivchenko, A.V., Sharonov, N.D.: The mechatronic unit and frame adaptable wing. Prog. Veh. Syst. 176–177 (2018)
2. Barbarino, S., Bilgeno, O., Ajaj, R.M., et al.: A review of morphing aircraft. J. Intell. Mater. Syst. Struct. **22**, 823–877 (2011)
3. Monner, H.P., Bein, T., Hanselka, H., et al.: Design aspect of the adaptive wing – the elastic trailing edge and the local spoiler bump. In: Proceeding of Royal Aeronautical Society Symposium on Multidisciplinary Design and Optimization, pp. 15.1–15.9. Royal Society Publishing (1998)
4. Pecora, R., Barbarino, S., Concilio, A., et al.: Design and functional test of a morphing high-lift device for a regional aircraft. J. Intell. Mater. Syst. Struct. **22**(10), 1005–1023 (2011)

5. Pecora, R., Amoroso, F., Lecce, L.: Effectiveness of wing twist morphing in roll control. J. Aircr. **49**(6), 1666–1674 (2012)
6. Barbarino, S., Pecora, R., Lecce, L., et al.: Airfoil structural morphing based on SMA actuator series: numerical and experimental results. J. Intell. Mater. Syst. Struct. **22**, 987–1004 (2011)
7. Stanewsky, E.: Adaptive wing and flow control technology. Prog. Aerosp. Sci. **37**(7), 583–667 (2001)
8. Perkins, D.A., Reed, J.L., Havens, E.: Morphing wing structures for loitering air vehicles. In: Proceedings of the 45th AIAA Conference on Structures, Structural Dynamics and Materials, AIAA Paper, p. 1888 (2004)
9. Blondeau, J., Pines, D.: Pneumatic morphing aspect ratio wing. In: Proceedings of the 45th AIAA Conference on Structures, Structural Dynamics and Materials, AIAA Paper, p. 1808 (2004)
10. Ivchenko, A.V.: Flexagon grids: dynamic forms of the future. Inventor **9**(201), 37–39
11. Ivchenko, A.V.: Flexagon grids: Combinatorics of reversive tiling patterns based on the synthesis of tritetraflexagons. Math. Des. Tech. Aesthet. **4**, 6–21
12. Ivchenko, A.V.: Flexagon grids: dynamic forms of the future. Stanochny Park **2**, 37–39 (2017)
13. Ivchenko, A.V.: Flexagon grids and a method to create robotic spatial forms on their basis. Inventions **3**, 29–36 (2017)
14. Ivchenko, A.V., Ziatdinov, R.: Flexagon grid: a novel method of creating dynamic spatial forms based on the synthesis of tritetraflexagons. Eng. Sci. Tech. Int. J. (2017). https://doi.org/10.1016/j.jestch.2017.12.001
15. Ivchenko, A.V.: Mechatronic unit and adaptable wing main frame. Inventions **7**, 34–38 (2018)
16. Howell, L.L., Nelson, T.: US patent 20160177605, 23 June 2016
17. FlexSys morphing wing (2018). https://www.liveleak.com/view?t=fb6_1478428583. Accessed 23 Sept 2018
18. Ivchenko, A.V.: Configuring honeycomb filler for an adaptable wing based on a network combination of joints with hidden topological surfaces. Inventions **8–9**(224–225), 41–44 (2018)
19. Ivchenko, A.V., Tulaev, A.I.: Honeycomb panel and filler design. Utility model RU patent 173383, 24 August 2017 (2017)
20. Alyamovsky, A.A.: Engineering Calculations and SolidWorks Simulation. DMK Press, Moscow (2015)

Modelling Steel Casting on a Continuous Unit

A. Galkin[✉], P. Saraev, and D. Tyrin

Lipetsk State Technical University,
30 Moskovskaya Street, Lipetsk 398055, Russian Federation
avgalkin82@mail.ru

Abstract. The paper considers the process of steel casting on a curved type unit of continuous steel casting. A mathematical model of temperature distribution in a bar during casting is proposed. The model consists of the equation of thermal conductivity and boundary conditions depending on the location of the secondary cooling nozzles, the type of nozzles, the flow rate and the temperature of the supplied coolant. The solution of the model equations is carried out by the finite element method implemented in the developed software. The software also provides bar cooling process visualization. Various alternatives of the heat transfer coefficient between the metal surface and the coolant are investigated. In addition, the stress-strain state of the bar in the continuous-casting unit is investigated. The developed software can be used by technologists to select bar cooling modes in the continuous-casting unit, as it allows to predict the appearance of defects. Defects predicting is based on the comparing of the bar crust temperature in a curved region of the continuous-casting unit and the temperature of embrittlement for specified steel grades.

Keywords: Steel casting · Mathematical modeling · Thermal regime · Stress-strain state

1 Introduction

Continuous casting of steel is an essential attribute of modern metallurgical production. Unlike casting in molds, this approach allows to reduce energy consumption, the time of bar solidification, the consumption of metal which has to be cut on head and bottom parts. The nature of cooling inside the casting unit has a direct impact on the output and quality. The solidification process consists of a period of solid crust formation in the crystallizer and further cooling in the injectors zone [1].

The purpose of the study is to determine technological parameters of continuous-casting of steel, in which there is the highest output with the lowest defects probability.

Cutting device typically is burning unit, separated from a bar the slab of the required length. The scheme of such unit is represented on Fig. 1.

The study of heat exchange processes in a bar during steel casting is carried out by solving partial differential equations (heat conduction equations) with given heat exchange conditions at the boundaries.

They form a mathematical model of the process, solution of which is carried out by the finite element method, being absolutely stable in the context of the problem under consideration [2–9].

A. A. Radionov and A. S. Karandaev (Eds.): RusAutoCon 2019, LNEE 641, pp. 1124–1137, 2020.
https://doi.org/10.1007/978-3-030-39225-3_118

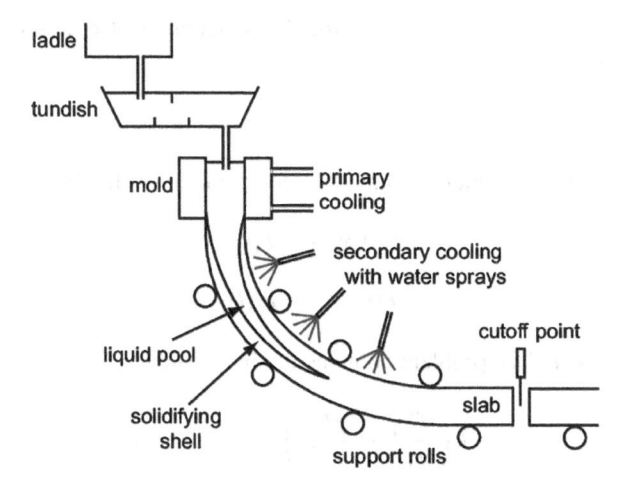

Fig. 1. Scheme of continuous-casting unit of steel.

In addition, the stress-strain state of the bar in the continuous-casting unit is considered. The casting process is the initial formation of a solid crust on the surface of the metal melt after it enters the crystallizer and further solidification by cooling the nozzles during movement along the guide roller system. In this regard, there is a ferrostatic pressure, which has a significant impact on possible defects.

2   Modeling the Thermal Regime of Casting Process

2.1   Temperature Distribution Model

The temperature field of the bar section is constant at a fixed time. At the initial moment, the molten metal enters the crystallizer. Heat removal is carried out through the surface of the bar, and its intensity reflects the efficiency of the cooling system. Thus, the equation describing the bar cooling process in a continuous casting of steel could be described by the partial differential equation of parabolic type:

$$\frac{\partial T}{\partial t} = a^2 \left(\frac{\partial^2 T}{\partial x^2} + \frac{\partial^2 T}{\partial x^2} \right) + f(x, y, t) \tag{1}$$

or

$$c(x)\rho(x)\frac{\partial T}{\partial t} = \frac{\partial T}{\partial x}k(x)\frac{\partial T}{\partial x} + \frac{\partial T}{\partial y}k(y)\frac{\partial T}{\partial y} + f(x, y, t) \tag{2}$$

where x, y are spatial coordinates; T is the vector of the temperature distribution (unknown); t is time; k is the coefficient of thermal conductivity (W/m · °C), c is the heat capacity coefficient (J/°C); ρ is the superficial density (kg/m^3).

The initial conditions are set only for the temperature itself, since the equation contains only the first derivative:

$$T(x, y, t_0) = \phi(x, y). \tag{3}$$

The first boundary problem contains the temperature of the diffusing substance:

$$T(0, y, t) = T_1(t), \tag{4}$$

$$T(l, y, t) = T_2(t). \tag{5}$$

The second boundary problem contains the heat flow:

$$-k\left(\frac{\partial T}{\partial x} l_x + \frac{\partial T}{\partial y} l_y\right) = \alpha(T_f - T), \tag{6}$$

where l_x, l_y is the length of the boundary section (m).

The thermal conductivity coefficient (cf. Fig. 2) and the heat capacity coefficient (cf. Fig. 3) change when the temperature changes. In order to increase the reliability of the model, the corresponding dependences were constructed.

Calculation of the heat transfer coefficient of the bar surface with sprayed water is based on the temperature difference between them [10]:

$$\alpha = \frac{2}{\sqrt{\pi}} \int_0^{V_s} e^{-t^2} \left(245V_s\left(1 - \frac{V_s \Delta T}{58223}\right) + 4.3\Delta T^2\left(1 - th\left(\frac{\Delta T}{115}\right)\right)\right) \tag{7}$$

or [11]

$$\alpha = 190 + tgh\left(\frac{V_s}{8}\right)\left(140V_s\left(1 - \frac{V_s \Delta T}{72000}\right) + 3.26\Delta T^2\left(1 - th\left(\frac{\Delta T}{128}\right)\right)\right), \tag{8}$$

where $V_s = V/S$ is the surface water consumption (m/s); V is the static injector performance (m^3/s); S is the area of irrigated surface (m^2); ΔT is the temperature difference (°C); α is the heat transfer coefficient (W/(m^2 · °C)).

The values of the heat transfer coefficient applying in calculations models (7) and (8) differ at low and high temperatures. In this case, the surface water flow rate is a derivative of injectors characteristics (such as static performance and opening angle).

An example of the injectors location in the cooling system is shown on Fig. 4.

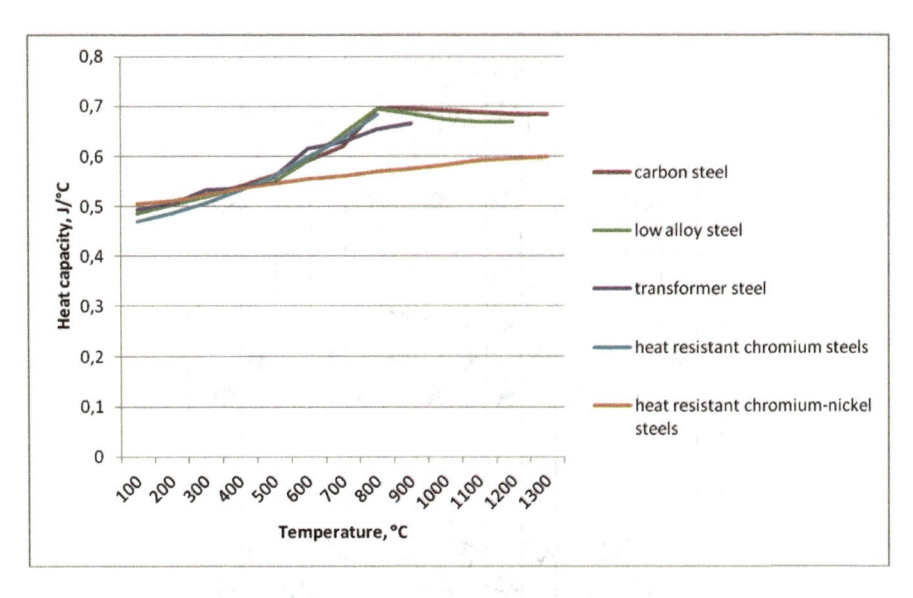

Fig. 2. Dependence of heat capacity on temperature.

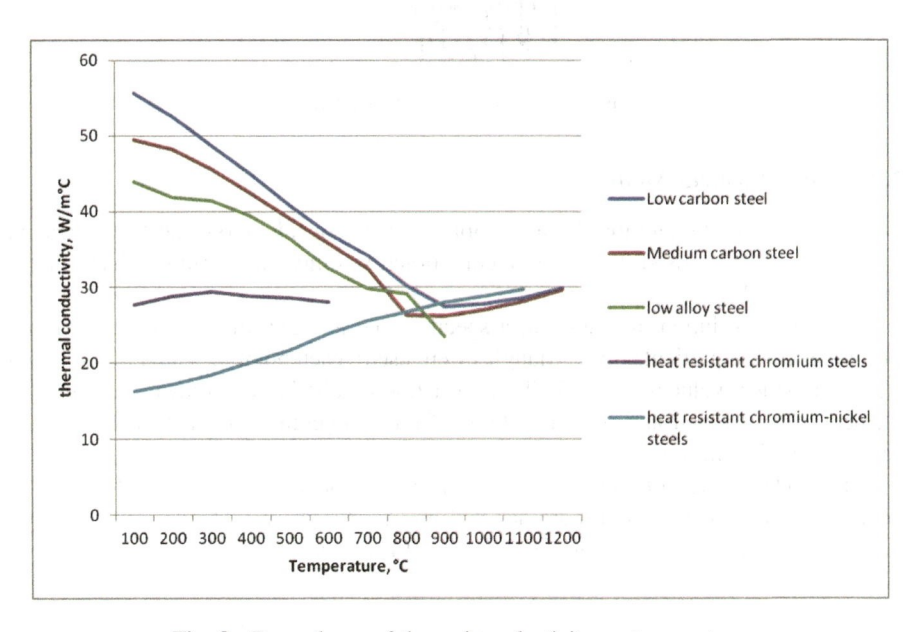

Fig. 3. Dependence of thermal conductivity on temperature.

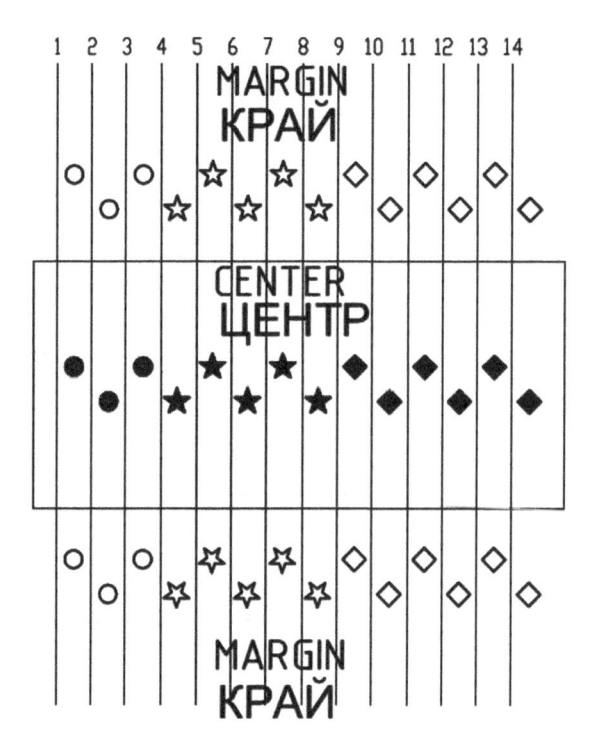

Fig. 4. Scheme of injectors position.

2.2 Finite Element Method

To obtain a discrete solution for the problem under consideration, the finite element method (FEM) was used due to its unconditional stability and arbitrary shape of the treated area [12–18].

During the solution, a section at a specific time is fixed, the area is divided into primitives (in the particular case, triangles were used), each point of which is associated with a numerical value (cf. Fig. 5). This formation is called a two-dimensional simplex element. Then formed and solved systems of linear algebraic equations with the subsequent iteration in time.

The nodal values of the scalar value T are denoted by T_i, T_j and T_k, and the coordinate pairs of three nodes are denoted by (X_i, Y_i), (X_j, Y_j), (X_k, Y_k).

Thus, the initial equation in the context of FEM has the following discrete form:

$$[C]\left\{\frac{\partial T}{\partial t}\right\} + [K]\{T\} - \{Q\} = 0, \tag{9}$$

where $[C]$, $[K]$, $\{Q\}$ are heat capacity matrix, heat conductivity matrix and heat load vector respectively.

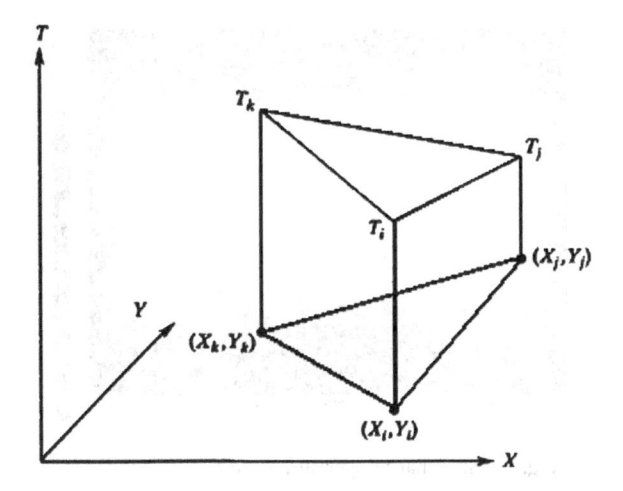

Fig. 5. Two-dimension simplex-element.

2.3   Modelling the Thermal Regime of the Casting Process

The concept of the study is to discretize the one-dimensional motion time and the two-dimensional space of the bar section. The values of the temperature field within the continuous casting of steel unit change sequentially as the bar section moves from a conditionally uniform temperature to the melt of the metal in the center and a solid crust at the border.

The developed software takes into account the interaction of the bar with such structural elements as the crystallizer and rollers, and also takes into account the intensity of heat exchange with water sprayed by secondary cooling injectors.

Due to the heterogeneous contact between the bar and water sprayed by injectors, it is also necessary to take into account the differentiation of heat exchange depending on the removal of the point for which the temperature is calculated tilting from the injector to the bar surface.

The purpose of the developed software is to test new modes of the equipment operation without production process itself. This approach allows to minimize the probability of defects during the casting by analyzing the changes in the parameters of the casting unit, which leads to an increase in production capacity with constant quality.

The picture of the temperature field on the exit of the unit is shown in the "result" tab (cf. Fig. 6).

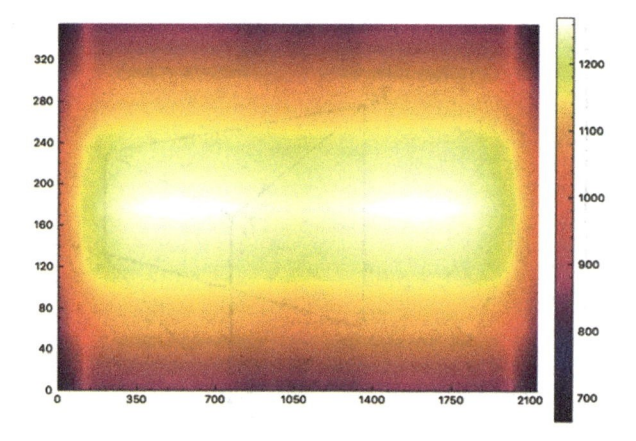

Fig. 6. Steel bar temperature on the exit of the unit.

The embrittlement tab has been added to estimate the temperature change along the longest side of the bar most prone to cracks (cf. Fig. 7).

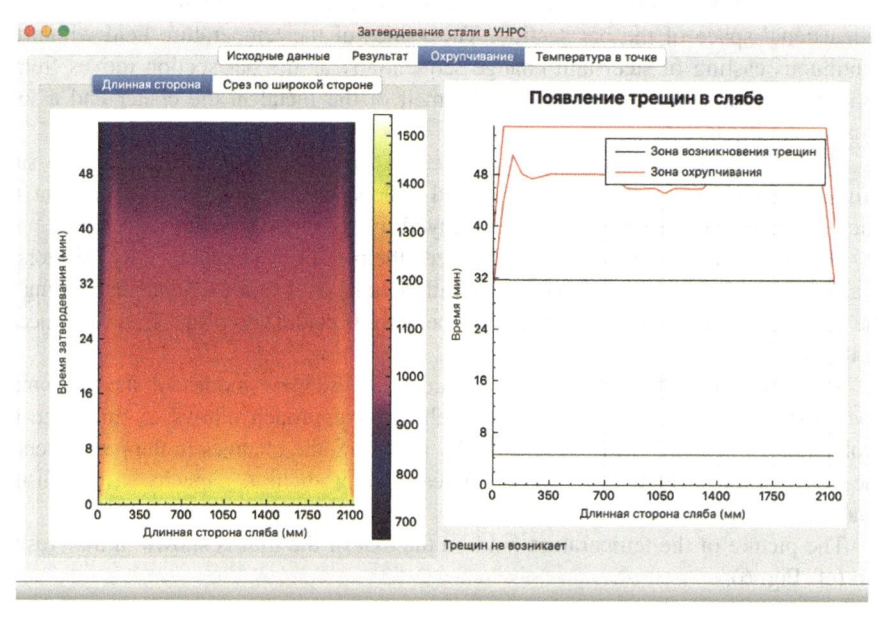

Fig. 7. Temperature change on the long side of the slab.

The intersection of the crack and embrittlement occurrence areas (temperature range at which deformation of the bar entail the occurrence of structural defects) will indicate the possibility of cracks occurrence within the selected mode.

It is also added the opportunity to plot temperature changes at specific points (cf. Fig. 8).

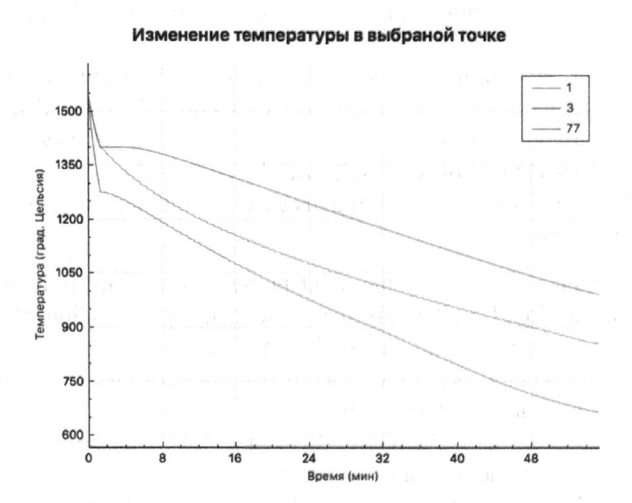

Fig. 8. Point temperature graphs.

2.4 Analyzing Results of the Modelling

For the steel grade 08YU at an initial temperature of 1500 °C and casting time 55.4 s, the calculated temperature values on the third of the long side is presented in Table 1 [19].

Comparing simulation results it is possible to note, that the heat transfer coefficient applying model (7) at temperatures above 1200 °C is slower than the real heat, and at temperatures below 1000 °C is faster. When using the heat transfer coefficient applying the model (8) at temperatures above 1000 °C, the heat exchange goes faster than the real one, and at temperatures below 1000 °C is close to the real one.

Table 1. Table types steel.

Output	Temperature, °C		
	14th minute	18.4th minute	35.6th minute
Real value	1250	1130	970
Obtained by the model (7)	1420	1200	910
Obtained by the model (8)	1140	1010	960

3   Modelling of the Stress-Strain State

Considering the bar temperature field to be known, it is possible to construct the mathematical model containing equations connecting stresses and strains at any point of the bar.

Generally, for modelling process it is necessary to know the bar geometric dimensions, the temperature distribution over its section, the dependence of the steel mechanical properties on temperature, parameters of the process, as well as the criteria using which the correctness of the results will be evaluated.

Studing bar behavior in the continuous-casting unit, the following assumptions were taken into account [20]:

- The stress-strain state of only the solid phase of the metal is modeled, and the presence of the liquid melt is taken into account by the action of the boundary conditions on the crust as the ferrostatic pressure
- The metal is considered as a homogeneous isotropic medium within the solid phase
- Deforming rollers are absolutely rigid
- Residual stresses in the solid phase after compression in the transition from one deforming section to another are not taken into account
- The inner surface temperature of the solid phase is equal to the solidus temperature
- The temperature distribution over the thickness of the solidified melt is linear

Modeling of the steel slab was carried out in three-dimensional space using ABAQUS software complex. The model consists of non-deformable rollers and a deformable bar that models only the hardened part of the steel bar. The rollers are modeled as analytical non-deformable cylindrical surfaces. A fragment of the steel bar in the pull-correct rollers is shown on Fig. 9.

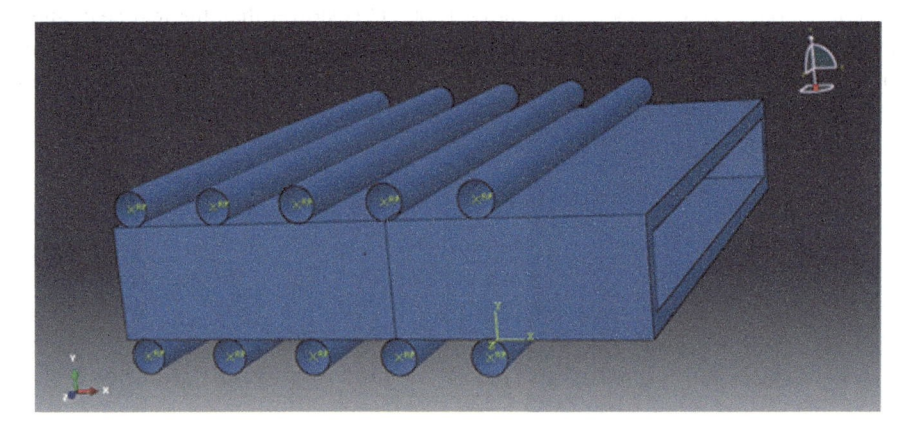

Fig. 9. Steel bar in pulling-correct rollers.

Continuous-cast steel bar according to its internal structure can be divided into two parts:

- Area containing a solid (the crust) and the liquid phase
- Area containing a solid structure over the entire section

Thus, in the first part of the steel bar, it is hollow, i.e. only the crust is considered, and instead of the liquid phase, a pressure corresponding to the ferrostatic pressure is applied to the inner surface of the crust. In the second part the steel bar has a solid structure.

The problem of deformable solid body mechanics is formulated as follows: there are external influences on the body, fixed or moving in space, as a function of coordinates and time. It is necessary to find some coordinate and time function systems describing the state of this body.

From a mathematical point of view, the calculation of the structure is reduced to solving boundary problems for systems of equations including the relations of stresses and strains theory (equilibrium equations, compatibility equations, relations between displacements and deformations), as well as defining equations, i.e. the relationship between stresses and strains.

The initial point is to determine the most accurate model of the object material behavior. Taking into account the high temperatures of the steel bar crust and the level of loads from ferrostatic pressure, it should be assumed the possibility of elastic deformation.

For the elastic model, the stress field has to satisfy the differential equilibrium equations

$$\nabla_i \sigma_{ij} = 0 \tag{10}$$

and relations connecting the components of the strain tensor ε_{ij} and the displacement vector u_i are

$$\varepsilon_{ij} = \frac{1}{2}\left(\frac{\partial u_i}{\partial x_j} + \frac{\partial u_j}{\partial x_i}\right). \tag{11}$$

Stresses and strain components are related to each other by Hooke's law

$$\sigma_{ij} = G\varepsilon_{ij}. \tag{12}$$

To obtain the only solution of the problem, it is necessary to add its boundary conditions:

- Internal boundary (internal pressure is ferrostatic)

$$p = \rho g l, \tag{13}$$

- Outer border (wide side)

$$u = \hat{u}, \tag{14}$$

and initial conditions $u = u_0$, where u is the displacement vector at the entrance of rollers, u_0 is the vector of prescribed displacements of surface points, p is the ferrostatic pressure, ρ is the metal density, g is the acceleration of gravity, l is the height of the melting column.

On Fig. 10 there are shown the equivalent von Mises stress arising in the process of steel casting in a continuous unit. The von Mises stress is a combination of six stress components occurring in a three-dimensional solid.

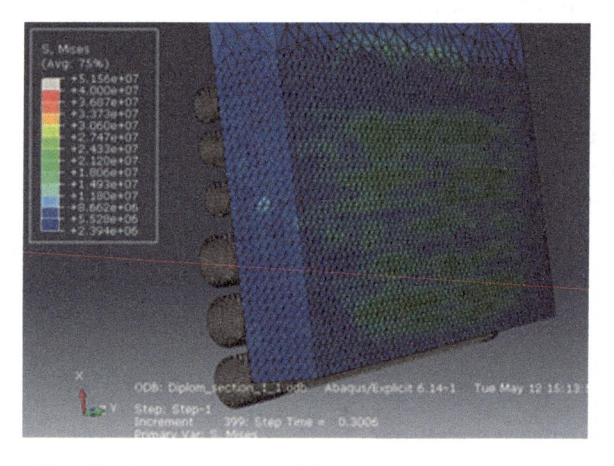

Fig. 10. The stress state of the steel bar during casting.

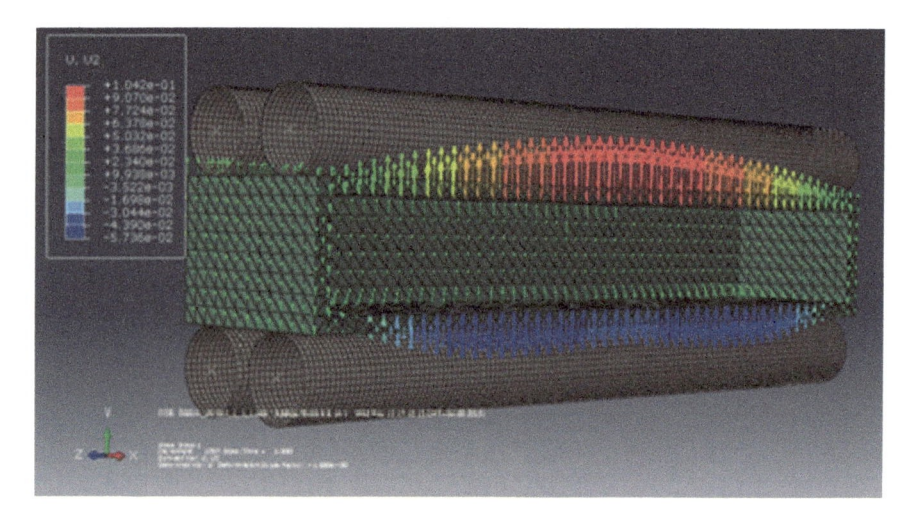

Fig. 11. Deformation in the direction of ε_{yy}.

Figure 11 shows the deformation occurring in the direction ε_{yy} in the process of steel casting in a continuous unit. The maximum value ε_{yy} achieved in the middle of the wide edge, which contribute to the occurrence of transverse cracks. It is possible also to notice that the deformation at the point of contact with the roller is less than the deformation in the inter-roller space.

4 Conclusion

The aim of the study was to explore the features of the process of steel hardening and the stress-strain state of the steel bar within the continuous-casting unit. The model describing the formation of steel bar in the continuous unit with curvilinear vertical section was constructed. It has a researching interest, because in the process of solidification the bar changes its shape, which can lead to the cracks formation.

The process of steel hardening is mathematically described by a partial differential equation of parabolic type with boundary conditions of the 3rd type, which describe convective heat transfer. To solve this problem, the finite element method, leading the original equation to the System of linear equations solved by the Gauss method was used.

There was developed the software implementation of the steel hardening model in the continuous unit. This program allows to obtain data on the temperature field of the steel bar at the exit of the continuous unit, based on the initial information on steel, initial temperatures and splitting intervals. All the information obtained is presented numerically and graphically. All results correspond to real data.

To approximate the model closer to the reality, the possibility of taking into account defects in the cooling system, in particular, the presence of defected injectors is used. In this case, it was proved that the efficiency of heat transfer is reduced compared to the ideal case, and the cooling of steel occurs unequally.

Also, the study deals with the problem of cracks formation on the steel bar. It is associated with a decrease of steel plasticity in a certain temperature range. If a change in the bar shape begins with weak plasticity, the probability of cracks is very high, i.e. the bar will be defective. If this problem occurs, it is necessary to change the cooling system and check new bar for cracks.

The stress-strain state of the steel bar was described mathematically using differential equations and boundary conditions. To solve this problem there was used the finite element method, which is widely applied in the mechanics of the deformable body. There were also derived matrix stiffness, stress and strain.

Studying steel bar behavior in the casting unit, some assumptions were made, that helps to implement modeling in the software package ABAQUS, while taking into account all the features of the steel bar.

Build in ABAQUS models together simulate the process of continuous casting of steel in curved casting unit. The generalized analysis of the obtained results allows to assert that on the basis of the developed mathematical model, with the known mechanical and plastic properties of the spilled steel grade, as well as the conditions of the process, it is possible to make a comparative analysis of the continuously cast steel bar VAT as a whole, or its individual sections with the steady-state casting process, in order to assess the degree of probability of defects.

Acknowledgments. This work is partially supported by Russian Foundation for Basic Research (RFBR) and Lipetsk regional administration. Grand 19-48-480009 r_a.

References

1. Reardon, A.: Metallurgy for the Non-Metallurgist. ASM International (2011)
2. Chown, L.H.: The influence of continuous casting parameters on hot tencile in low Carbon, Niobium and Boron steels. Dissertation, University of the Witwatersrand, Johannesburg (2008)
3. Crowser, D.N.: The effects of microalloying elements on cracking during continuous casting. In: Proceedings of the International Symposium on Vanadium Application Technology, pp. 99–131 (2001)
4. Kim, S.K., Kim, N.J., Kim, J.S.: Effect of Boron on the hot ductility of Nb-containing steel. Metall. Mater. Trans. **33**, 701–704 (2002)
5. Bimacombe, J.K., Sorimache, K.: Crack formation in the continuous casting of steel. MTB **8–2**, 489–505 (1977)
6. Grosseiber, S., et al.: Influence of strain on hot ductility of a V-Microalloyed steel slab. Steel Res. Int. **83–5**, 445–455 (2012)
7. Minz, B.: The influence of composition on the hot ductility of steels and to problem of transverse cracking. ISIJ Int. **39–9**, 833–855 (1999)
8. Presoly, P., Pierer, R., Bernhard, C.: Identification of defect prone Peritectic steel grades by analyzing high-temperature phase transformation. Metall. Mater. Trans. **44–12**, 5377–5388 (2013)
9. Chen, L., et al.: Effects of second particle dispersion on kinetics of isothermal peritectic transformation in Fe-C alloy. ISIJ Int. **52–3**, 434–440 (2012)
10. Viscorova, R., Scholz, R., Spitzer, K.-H., et al.: Spray water cooling heat transfer under oxide scale formation condition. In: WIT Transactions on Engineering Sciences. WIT Press, Southampton (2006)
11. Bellet, M., Salazar-Bbetancourt, L., Jaouen, O., et al.: Modeling of water spray cooling. Impact on thermomechanics of solid shell and automatic monitoring to keep metallurgical length constant. In: European Continuous Casting Conference. Austrian Society for Metallurgy and Materials, pp. 1202–1210 (2014)
12. Zienkiewicz, O.C., Morgan, K.: Finite Elements and Approximation. Dover Publications, New York (2006)
13. Bathe, K.J., Wilson, E.L.: Numerical Methods in Finite Element Analysis. Prentice-Hall, New York (2012)
14. Silvester, P.P., Ferrari, R.L.: Finite Elements for Electrical Engineers. Cambridge University Press, Cambridge (1983)
15. Suli, E.: Finite Element Methods for Partial Differential Equations. University of Oxford Press, Oxford (1988)
16. Reddy, J.N.: An Introduction to the Finite Element Method. McGraw-Hill Education, New York (2006)
17. Smith, I.M., Griffiths, D.V., Margetts, L.: Programming the Finite Element Method. Wiley, New York (2014)

18. Kiritsis, D., Eemmanouilidis, Ch., Koronios, A., et al.: Engineering Asset Management. Proceedings of the 4th World Congress on Engineering Asset Management, pp. 591–592 (2009)
19. Saraev, P.V., Galkin, A.V., Tyrin, D.Yu.: Constructing and researching mathematical model of bar cooling in a continuous casting unit. In: Proceedings of XIII International Conference. Institute of Control Science RAS, Stary Oskol (2018)
20. Galkin, A.V., Filippov, D.A., Pimenov, V.A.: Mathematical modeling of steel bar stress-strain state in the continuous-casting unit. In: Proceedings of the XII Russian Workshop for Young Researchers. Institute of Control Science RAS, Volgograd (2015)

Deployment of Intelligent Tools into the Distributed Translation System of Models

M. Polenov$^{(\boxtimes)}$, V. Guzik, and A. Kurmaleev

Southern Federal University, 44, Nekrasovsky lane,
Taganrog 347922, Russian Federation
mypolenov@sfedu.ru

Abstract. The paper discusses the organization of intellectual support for converting models in the Distributed translation system of models used for robotic systems simulation. Such conversion allows researchers to reuse of existing models for required simulation packages. To automate the process of model's conversion it was proposed to develop an instrumental translation tool. Such tools were created and called Multitranslator. It allows automating the multilanguage translation of models' source program code into target language in required format. Next, the Distributed translation system was created on the basis of Multitranslator. This system proved to be quite effective, but in some cases, there were situations of uncertainty of decisions during the translating process, which required the involvement of researcher. To resolve such problem, it was proposed to use an expert system that relies CLIPS tools and so it was synthesized. It led into improvement of multilanguage model's translation by reducing the time costs of conversion procedure for modeling tools. The structure of Distributed translation system of models with CLIPS application programming interface was reviewed and also the code samples of interaction with interface in third-party applications. Translating of node models demonstrated expediency to use suggested approach for mobile robotic platform simulation.

Keywords: System models · Translation of models · Multitranslator · Distributed translation system · Expert system · CLIPS · Mobile robotic platform

1 Introduction

Nowadays various simulation packages are already in wide use in complex systems research and development. It led researchers and engineers into reuse and rewrite of models for required simulation environments. To automate the process of converting models it was proposed to create automated translation tools.

It ended up in creation of instrumental environment for multilanguage translation of models that was named Multitranslator (MT) [1]. It was based on the development of the translating modules as a set of rules containing description of languages.

MT allowed automating the translation of source program codes for not just modeling languages but also various programming languages.

© Springer Nature Switzerland AG 2020
A. A. Radionov and A. S. Karandaev (Eds.): RusAutoCon 2019, LNEE 641, pp. 1138–1146, 2020.
https://doi.org/10.1007/978-3-030-39225-3_119

As a result of practical use of the Multitranslator [2] it was decided to improve its functionality:

- Add the possibility to use the MT with simplified interface for the end user, which allows only the translation of models without installing the main application
- Reduce the time shutters caused by the additional load on the workstation that processed the translation
- Provide remote storage for models and translating modules to have backup and share remote models files for each researcher.

And so, the Distributed Translation System of Models with Storage of Models [3] was developed. It is based on a distributed architecture [4], on model reuse approach [5] and implemented as a client-server application.

To tell in short about the organization of multilanguage models translation using the Multitranslator, we have to consider translating modules principles:

- The developed translating module is one of the main components of the MT
- Each translating module contains two sets of instructions, one for input language and another one for output language
- One part of translating module is set for parsing the input model's language description
- Another is used for generating the output language structures in the required format for target package.

But even though, sometimes the translation of models is failed to process in cases of insufficient input data or with the uncertainty of decisions occurred during translation [2]. It can happen when different results are possible when parsing the code.

To resolve such problem, it was decided to use an expert system [6, 7]. It has obviously reduced the time costs of translation in certain cases even more that was proved by experiments.

In the researches, we came to the simulation of mobile robotic platforms topic. What brought us to converting models from MATLAB [8] into models for the Robot operating system (ROS) [9].

Therefore, application of Distributed translation system of models becomes relevant again.

Effective solution with usage of intellectual support for multilanguage translation might solve such modeling problems. But the current system needs additional development which is discussed further.

2 Models Translation for Mobile Robotic Platforms

Multilanguage translation allows to reuse the models even of MATLAB and Octave (format is well compatible with the MATLAB syntax) [10] environments.

In order to implement this approach, it was proposed to use the developed translating module for converting MATLAB models into C++ code, and an additional knowledge base for the expert system (ES) [7], focusing on the features of the description of models in ROS [11].

When implementing the proposed solutions, it became obvious that for further development it is necessary to create a software subsystem of interaction between the user, the expert system implemented on CLIPS (C Language Integrated Production System) [12], and the Multitranslator. A structure of the Distributed Translation System of Models was suggested as a result. It is shown in Fig. 1.

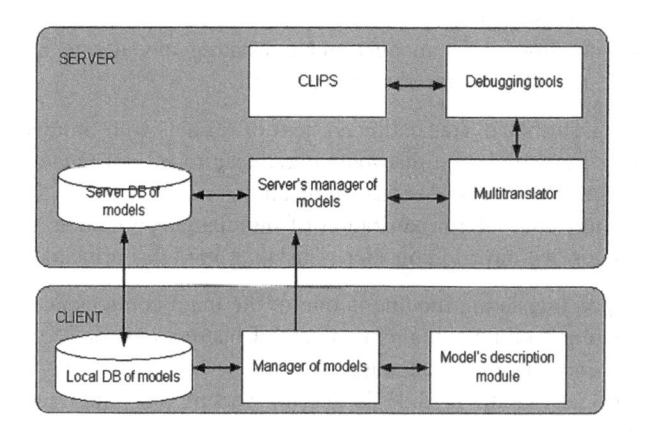

Fig. 1. Basic structure of Distributed Translation System of models.

All this require solving the following tasks:

- Organization of two-way data exchange between the MT and the expert system
- Simplify the process of debugging the system by the developer, since there is no need rerun the expert system output process to obtain debug information
- Creating a dialogue system with the user
- Development of an intelligent editor for the user, which will allow editing and adding rules for the expert system if the user has the necessary skills

Since the CLIPS environment works in dialogue with the user, therefore, it is necessary to use the CLIPS API to achieve the objectives.

3   Usage of CLIPS Interface

Let us consider the features of interaction organization of Distributed Translation System of Models with CLIPS via API. It is shown in Fig. 2.

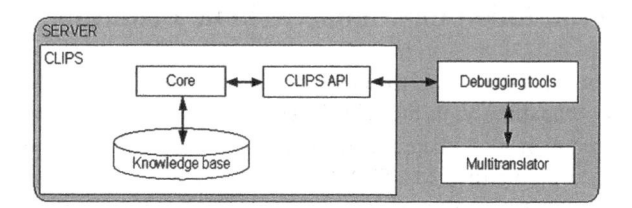

Fig. 2. Structure of interaction of Distributed Translation System of models with CLIPS.

In figure above Multitranslator sends the untranslated data to our debugging tool that transfers it further to the CLIPS via API. Only then, data is being processed by expert system and so the corrected translated data is put into memory of CLIPS and then is read by debugging tool and sent to the MT to finish off the translation process.

Since the CLIPS API library is written in C++, first of all, to initialize CLIPS in C#, within the framework of the created Distributed Translation System of Models and User Interface complex, it is necessary to configure marshaling (the process of transforming the representation of an object format suitable for storage or transfer).

For libraries (* .dll) that are not written according to the .NET standards, this is possible using a special method, this method is contained in .NET in the "DllImportAttribute" class, which reflects the transfer of the dynamic library as a static entry point:

```
[System.AttributeUsage(System.AttributeTargets.Method, Inherited = false)]
[System.Runtime.InteropServices.ComVisible(true)]
public sealed class DllImportAttribute : Attribute
```

This class is applied to connect methods for C# and C++ languages. Despite this, Visual Basic uses this class for the "Declare" method. However, for connecting methods that contains fields:

- "BestFitMapping"
- "CallingConvention"
- "ExactSpelling"
- "PreserveSig"
- "SetLastError"
- "ThrowOnUnmappableChar"

This class will be applied directly to represent Visual Basic methods.

It is also worth clarifying that the "DllImportAttribute" class does not support marshaling of universal types.

At its core, the described class detects and calls the exported function and, if necessary, marshals the arguments of this function across the interaction boundaries.

To use the required method of this class, it is necessary to:

- Define methods in the library. It requires passing the name of the called method and the name of the library containing the method
- Create a class that will contain library marshaling functions. To do this, you can use the existing class as well as create classes for each marshaled method or create one class where the methods will be placed
- Prototyping in managed code. In the case of C#, you must use the "DllImportAttribute" to define the class and its methods. You also need to add the static and extern modifiers
- Call the DLL function. Now you can call class methods in the same way that a non-marshaled method is called. However, it is worth remembering that passing structures and implementing callback methods are exceptional cases

After describing the method used, let's consider an example of a code fragment for the environment creation method for the CLIPS API:

```
[DllImport("CLIPSDynamic64.dll", EntryPoint = "__CreateEnvironment")]
public static extern IntPtr CreateEnv();
```

In this example, the method takes as an argument the path to the required library (in our case it is added to the project and is in the same folder with it, so the full path is not required) and the exact name of the desired method in the library itself. After that, you must specify the method for import and the arguments to be passed.

In this case, the "__CreateEnvironment" method is declared as a void * method in C ++, therefore, the "IntPtr" prefix is added to correctly interpret typing, since this type is a special Environment type in the CLIPS API library. In this case, nothing is passed as arguments.

Thus, it is necessary to apply marshaling for each method that is planned to be used in the framework acting with the CLIPS API, not forgetting to correctly interpret these methods and their arguments. Let us give one more example of starting the expert system operation:

```
[DllImport("CLIPSDynamic64.dll", EntryPoint = "__EnvRun")]
public static extern void EnvRun(IntPtr Env, long runLimit);
```

In this example, the environment created by the method above and the number of rules to execute are passed as a parameter. To execute all the rules, you must pass as an argument "−1".

After implementing an environment for interacting with CLIPS, it is necessary to transfer to the creation of the ES itself. To do this, use the following method:

```
[DllImport("CLIPSDynamic64.dll", EntryPoint = "__EnvBuild")]
public static extern int EnvBuild(IntPtr Env, string constrStr);
```

An example of its use:
CLIPS.EnvBuild(Env, CLIPS.test);
Here CLIPS is the conditional name of the class to be accessed, and test is the specified expert system for checking the operation of the interaction and building the environment and ES.

After building, you should start cleaning the environment from unnecessary components, namely the "__EnvReset" method:

```
[DllImport("CLIPSDynamic64.dll", EntryPoint = "__EnvReset")]
public static extern void EnvReset(IntPtr Env);
```

And it's executing:
CLIPS.EnvReset(Env);
Only now, it is possible to launch the expert system itself for testing. The declaration of the corresponding method was indicated above, and below is an example of its use:

CLIPS.EnvRun(Env, −1);

When used in C++, the method automatically interprets production rules and outputs the result to the console, in this case the result will be displayed in the appropriate user interface window along with other debugging information.

At the end of the work, it is necessary to launch the CLIPS environment destructor and use the following method:

```
[DllImport("CLIPSDynamic64.dll", EntryPoint = "__DestroyEnvironment")]
public static extern void DestroyEnvironment(IntPtr Env);
```

And it is called when the work of ES is completed:

CLIPS.DestroyEnvironment(Env);

To organize data exchange between CLIPS and external applications, it is necessary to use global variables. They are CLIPS objects that store data outside the defining constructs of the CLIPS language (any constructions with "def").

In CLIPS, global variables can be defined using the "defglobal" operator, for example:

(defglobal ?*name* = expression)

The value of a global variable can be changed anywhere in the code using the "bind" operator. It is used similarly to the "defglobal" construction.

Changing global variables is allowed from external applications using the following CLIPS API functions:

```
GetDefglobalValue("name",&pointer)
SetDefglobalValue("name",&pointer)
```

The CLIPS API also allows to edit kernel memory, depending on the user's responses, using the "FunctionCall" operand:

```
FunctionCall("function", "arguments", &pointer)
```

It calls a function named "function", passing in the parameter names "arguments". All functions of the expert system are based on the CLIPS language. In this case, the CLIPS API loads the expert system and all internal memory during initialization.

Therefore, data exchange between CLIPS and external applications is achieved by reading and writing global variables by calling the CLIPS API core functions.

4   Debug Simplification

The expert system allows improving the system for the developer interface. It was decided to use as a basis the element of the multi-document interface (MDI), which allows to create child objects within a single window and is used in many modern development environments. Because of the algorithmization of the interaction process, two main paths of the debug loop were made to the interface:

- By pressing the button in the drop-down menu and debugging information is displayed in the child interface window. By pressing a key, the data is saved and processed by the expert system. At the end of processing, the results are recorded in a file opened in another child window for viewing. This method eliminates the possibility of data transmission to the Multitranslator, since this data is needed to debug the expert system itself
- When the Distributed Translation System of Models transfers data from the MT to the expert system, the data is displayed in a separate window for debugging. After that, you need to press a separate key in the drop-down menu, which will cause the data transfer to the expert system for analysis and will display a window with the results automatically. In this case, after the completion of the debugging process, it is possible by pressing a single key to see how the results of the expert system will affect the translation process of the source model code

At the same time, you can close the internal windows, if necessary, and you can open them again, which allows the implementation of the MDI class.

All this allows to develop an expert system for translating models with greater accuracy and much faster, as well as debugging the work of the whole complex using a systematic method.

5   Testing Sample

We've made a set of experiments with model's translation by Distributed Translation System of Models. The one was to review in this paper. We choose the lidar component model as a source one. Such model code example for Octave is:

```
sub = rossubscriber('/scan');
scan = rosmessage('sensor_msgs/LaserScan')

count = scan.ScanTime / scan.TimeIncrement;
disp("New laser scan msg!");

% convert
for i = 1:count
    degree = rad2deg(scan.AngleMin + scan.AngleIncrement * i);
    disp("[%f, %f]", degree, scan->ranges[i]);
end
```

After parsing this model the success message is printed and preset output file is filled with following C++ code of Node in ROS format:

```
#include "ros/ros.h"
#include "sensor_msgs/LaserScan.h"
#define RAD2DEG(x) ((x)*180./M_PI)

// Callback function
void scanCallback(const sensor_msgs::LaserScan::ConstPtr & scan)
{
    int count = scan->scan_time / scan->time_increment;
    ROS_INFO("New laser scan msg!");
    % convert
      for(int i = 0; i < count; i++) {
          float degree = RAD2DEG(scan->angle_min + scan->angle_increment * i);
          ROS_INFO("[%f, %f]", degree, scan->ranges[i]);
      }
}

// Main function to set up ROS node
int main(int argc, char **argv)
{
    // Set up ROS
    ros::init(argc, argv, "rplidar_node");
    ros::NodeHandle n;

    // Create a subscriber
    ros::Subscriber sub = n.subscribe< sensor_msgs:: LaserScan>("/scan", 1000, scanCallback);

    ros::spin();
    return 0;
}
```

From the example of model conversion above, the test was successful, and generated model for ROS are working correctly. Translating of node models demonstrate expediency to use suggested approach for mobile robotic platform simulation.

6 Conclusion

In the result of this research, the proposed approach of intellectual support for multi-language translation of models for the development of mobile platforms models, based on the use of the Distributed Translation System of Models and, translating modules of the MT and the expert system, was implemented. As part of this work, a new knowledge base was developed for the expert system implemented on CLIPS. The interaction of data exchange between the Multitranslator and the CLIPS core was organized, which allowed to configure the user.

The proposed approach uses the Multitranslator and the expert system within the Distributed Translation System of Models as a tool of converting models, significantly expanding its functionality and making this approach more versatile and in demand when modeling complex systems.

Acknowledgments. The reported study was funded by the Ministry of Science and Higher Education of the Russian Federation (Project part of State task No. 2.3928.2017/4.6).

References

1. Chernukhin, Yu.V., Guzik, V.Ph., Polenov, M.Yu.: Multilanguage translation for virtual modeling environments. Publishing House of Southern Scientific Center of Russian Academy of Sciences, Rostov-on-Don (2009)
2. Chernukhin, Yu., Guzik, V., Polenov, M.: Multilanguage translation usage in toolkit of modeling systems. WIT Trans. Inf. Commun. Technol. **58**, 397–404 (2014). https://doi.org/10.2495/icte130491
3. Polenov, M., Guzik, V., Gushanskiy, S., et al.: Development of the translation tools for distributed storage of models. In: Proceedings of International Conference on Application of Information and Communication Technologies, pp. 30–34. IEEE Press (2015)
4. Tanenbaum, E., Van Sten, M.: Distributed Systems: Principles and Paradigms. Prentice-Hall, Upper Saddle River (2006)
5. Robinson, S., Nance, R.E., Paul, R.J., et al.: Simulation model reuse: definitions, benefits and obstacles. Simul. Model. Pract. Theory **12**, 479–494 (2004). https://doi.org/10.1016/j.simpat.2003.11.006
6. Polenov, M., Guzik, V., Gushansky, S., et al.: Intellectualization of the models translation tools for distributed storage of models. In: Proceedings of 16th International Multidisciplinary Scientific GeoConference, Albena, vol. 1, pp. 255–262 (2016)
7. Polenov, M., Gushanskiy, S., Kurmaleev, A.: Synthesis of expert system for the distributed storage of models. In: Software Engineering Trends and Techniques in Intelligent Systems. Advances in Intelligent Systems and Computing, vol. 575, pp. 220–228 (2017)
8. Matlab (2019). https://www.mathworks.com/products/matlab.html. Accessed 08 Apr 2019
9. Robotic operation system (2019). http://www.ros.org/. Accessed 08 Apr 2019
10. Octave (2019). https://www.gnu.org/software/octave. Accessed 15 Apr 2019
11. ROS: Nodes explanation (2018). http://wiki.ros.org/Nodes. Accessed 09 Apr 2019
12. CLIPS (2019). http://www.clipsrules.net. Accessed 03 May 2019

Enterprise Information Security Assessment Using Balanced Scorecard

R. Fatkieva[1(✉)] and A. Krupina[2]

[1] Russian Academy of Sciences Saint Petersburg Institute of Informatics and Automation, 39 14th Line of Vasilievsky Island, St. Petersburg, Russian Federation
rikki2@yandex.ru
[2] Saint Petersburg Electrotechnical University LETI, 5 Professora Popova Street, St. Petersburg 197022, Russian Federation

Abstract. The paper deals with an algorithm of an enterprise balanced scorecard development and implementation. The balanced score card provides comprehensive assessment of all aspects of an enterprise thus resulting in its control as a whole. The approach has several advantages: it gives a complete vision of the processes at the enterprise administration disposal; it helps to avoid critical situations and security breaches, in particular, unauthorized access; it facilitates interaction at all organizational levels and ensures understanding of strategic goals by all participants in the production process. Using enterprise production process as an example the description was given to strategic goals, objectives, critical indicators and strategy map of their interrelations in order to reduce the amount of damage from unauthorized access and other security violations as well as to enhance the production process. To minimize the impact of intrusions on the enterprise productivity the set of countermeasures was elaborated. Indicators changes due to the above measures and expressed in the form of their additive convolution were evaluated and compared under different states of the system. Measures efficiency with regards to information security is assessed according to the values obtained. The measures make also possible monitoring downtime and working time at the enterprise, reducing the risk of intentional equipment damage including that due to unauthorized access, and achieving strategic goals.

Keywords: Balanced scorecard · Security · Strategy map · Additive convolution

1 Introduction

Post-industrial society is characterized by the involvement of information technologies in all spheres of human activities thus resulting in the increase of information flows both in the internal media and in the external environment of companies, enterprises, etc. Growing number of enterprises and their scaling rates predetermines the need for information security assessment and control. One of the principles of security assurance consists in the use of strategic planning methods for risks determining which is reasonable to start with an analysis of the economic potential of production growth

A. A. Radionov and A. S. Karandaev (Eds.): RusAutoCon 2019, LNEE 641, pp. 1147–1157, 2020.
https://doi.org/10.1007/978-3-030-39225-3_120

through the use of own resources. Certain difficulties are encountered while determining the risks associated with the failure in the supply chain or in the production process. There is a need to develop requirements to the technology process, security criteria, identification and assessment of the risks of their violation. Static methods described in [1–3] proved to be insufficient for the assessment of security risks. Dynamic character of risk assessment is to be taken into account in order to assure security performance without any unacceptable data damage. One of the methods providing such an assessment is an approach to security control by means of balanced scorecard [4]. The methods are based on the analysis of the concept developed by David Norton and Robert Kaplan and considered in [5–12]. However, the above methods lack the means of assessing the dynamics of indicators changes and those of modeling information security breaches. Therefore, the creation of a methodological framework for security control based on a balanced scorecard taking into account indicators and production dynamics is rather a relevant task. Such a method for security monitoring and control of information systems of enterprises is presented below.

2 Principles of Creating a System for Monitoring and Managing Information Security

The integration of various models and methods in assessing enterprise performance enables a creation of information security control system. To build such a system it is necessary to solve the following particular problems:

1. Development of a structural scheme of the enterprise functioning;
2. Creation of a generalized description of various classes of models and methods in order to establish their relationships and correspondences using various metrics;
3. Description, classification and selection of a system of indicators that provides monitoring the states of objects and processes;
4. Development of combined methods for the evaluation of quality indicators of models being specified in both numerical and non-numerical (nominal, ordinal) forms;
5. Formalization of methodological principles for solution of problems of multi-criteria evaluation for the purposes of security assurance.

Solution of the above tasks makes it possible to create a system of information security management that includes subsystems for monitoring and forecasting enterprise functioning and reflects the dynamics of the processes when those or other indicators change their values.

In these conditions the choice of a method for indicators selection and evaluation is of paramount importance as it would enable monitoring dynamics of the system changes, analyzing its current state and creating a set of control actions aimed at the correction of deviation from the specified value. One of the key tasks consists in formalized construction of a vector of indicators reflecting the current state of the object being analyzed. The task of indicators selection is complicated in particular by the following circumstances: the number of levels required to be involved, the number of indicators themselves (projections that provide the lower threshold of controllability), the ability to measure indicators and the frequencies of those measurements (since there is a different time period for evaluation of one or the other indicator) and a need to take into account indicators' sustainable cause-effect relationship.

3 Problem Statement

Let us consider formulation of a problem by considering a practical example of the functioning of a machine-building enterprise. It is required to develop an approach to the real- time assessment of the dynamics of the indicators changes under the influence of destructive behavior during the production cycle. To build a vector of indicators we apply the methods for creating a balanced scorecard (BSC) consisting of the following steps:

S1. Formation of a list of target strategic goals W aimed at the production process enhancement by solving particular optimization tasks. The latter are traditionally reduced to the search and selection of development trajectory at a time interval meeting either the requirement of reaching minimum of integral losses or that of reaching maximum net profit in the production of products.

Formation of a list of target strategic goals enables to determine the balanced scorecard (BSC) elaboration depth level as well as to form the methods for the BSC creation at the enterprise. At this point it is necessary to foresee the following stages: the choice of a development strategy (goals, objectives); identification of levels involved in strategic planning; development of generalized performance indicators in the form of additive or multiplicative convolution (to be defined below) for BSC implementation. Ignoring this stage can lead to a failure of the adjustment of the chosen strategy under indicators dynamic changes. If strategic goals are set in quantitative terms, then the dynamics of the goal achievement is also could be assessed numerically. Selection of generalized performance indicators is based on the following criteria: measurability and ease of calculation; controllability (attainability is possible to be influenced); consistency with a common goal (the presence of cause-effect relationships and the organizational structure of the enterprise).

S2. Decomposition of strategic goal into sub-goals in order to reach generalized performance indicators values. Decomposition implementation results in a more detailed consideration of the process of a goal achievement while using the mathematical apparatus of information theory and system analysis describing possible operating conditions of an enterprise from the standpoint of information security assurance. If the object in question is poorly structured or does not have a hierarchical structure, then this step may be skipped. Let us consider the stages of determining the levels involved in strategic planning:

1. Creation of a structural scheme and of a functional one of the production process being necessary to determine the number of levels in the set of levels L during production. The use of the structural scheme gives possibility for a more detailed goal-setting as well as for identification of the goals that were not been considered previously.

2. Decomposition of the structural scheme and identification of the objective functions of the structural units at all the levels $W^L = \{W^l\}$ where $W^l = \left(W_1^l, W_2^l, \ldots W_N^l\right)$ – objective functions of the level 1, N – is the number of objective functions thus making possible to form a relevant structure of indicators set for the corresponding function of production. In order to simulate the production process and to decompose the cycle into subtasks it is possible to use an apparatus of oriented graphs and Petri nets reflecting the dynamics of functioning.

S3. Definition of arguments set D for the assessment of the degree of their influence on the goal achievement included in the function $W_i^l = \left(D_{i1}^l, D_{i2}^l, \ldots D_{iK}^l\right)$, where $i = 1 \ldots N, K$ is a number of arguments.

S4. Creation of a strategy map of the relationship of functions $W^l = \left(W_1^l, W_2^l, \ldots W_N^l\right)$ which shows the degree of influence of arguments D_i^l on the final result. The strategy map gives possibility not only the to trace the trajectory required to achieve the goal but also to develop a set of measures to adjust the strategy in case of deviation from the trajectory and to assess the periods needed for the indicators evaluation.

S5. Definition of a set of indicators to assess the achievement of a goal $X_i^l = (X_{i1}^l, X_{i2}^l, \ldots X_{iP}^l)$, where $i = 1 \ldots N, X_{i(1 \ldots P)}^l = E\left[W\left(D_{i(1 \ldots K)}^l\right)\right]$ is the mean value of the objective function argument.

S6. Monitoring and evaluation of indicators $X_i^l(t)$ in a real time mode by direct measurement depends on the applied methods of measurement, collection and evaluation of parameters. In this case the control task both for an enterprise as a whole and for each l-th level consists in the simultaneous meeting three conditions [13, 14]:

1. Measurement and evaluation of indicators for each of the levels at a given moment of time;
2. Prediction of some indicators values at a fixed moment of time t from a given interval;
3. Formation of control actions with the adjustment of measures on achievement of strategic objective in case of deviations of the observed parameter from the specified values. There is a relation between process states at a time interval $(t, t + \Delta t)$: $\tilde{X}_i(t) = f(X(t, t + \Delta t), Y(t, t + \Delta t), M(t, t + \Delta t))$, where $X(t, t + \Delta t)$ is the state of the observed object at time interval $X(t, t + \Delta t)$; $Y(t, t + \Delta t)$ are control actions to ensure production security; $M(t, t + \Delta t)$ are measures for information security assurance.

In some cases, in order to form a function $\tilde{X}_i(t) = f(X(t, t + \Delta t), Y(t, t + \Delta t), M(t, t + \Delta t))$ it is required to perform a convolution of indicators to provide a rapid response to a particular incident. The most frequently used one is an additive convolution with constant weight coefficients for particular indicators which makes it possible to evaluate a generalized performance indicator (GPI):

$$GPI = \sum_{i=1}^{P} \gamma_i \tilde{X}_i \tag{1}$$

where γ_i – normalized weight coefficients, determined in the related literature or by the expert estimates, $\tilde{X}_i$ – normalized indicators. Indicators normalization is reasonable to carry out with dependence on the following features [15]:

(a) if the increase in X_i enhances the performance of the analyzed object then the normalization is calculated according to the formula $\tilde{X}_i = \frac{X_i - X_{i\min}}{X_{i\max} - X_{i\min}}$;

(b) if on the contrary the decrease in X_i improves the performance of the analyzed object then according to the formula $\tilde{X}_i = \frac{X_{i\max} - X_i}{X_{i\max} - X_{i\min}}$.

S7. Security management made by means of forecasting mechanisms and of the event probability evaluation. Assessment of damage from identified risk events and the formation of measures to eliminate detected violations.

4 The Results of Information Security Breaches Simulation Using the Balanced Scorecard

Let us proceed with the same example of machine-building enterprise the state space of which can be described as a graph network process represented in Fig. 1, where 1 – receiving the order; 2 – development of technical specifications (TS); 3 – negotiating the price; 4 – coordination of working drawings; 5 – identification of details and components; 6 – employment regulation; 7 – inventory of existing documentation in accordance with the specifications; 8 – development of equipment and special tools; 9 – purchase of materials for special tools; 10 – procurement of tools for production; 11 – transfer of material for equipment and special tools to production; 12 – manufacture of equipment and special tools; 13 – control of equipment and special tools by the Technical Control Department; 14 – transfer of equipment and special tools to the warehouse; 15 – transfer of material of detail manufacturing into production; 16 – manufacture of details; 17 – delivery of details in the Technical Control Department; 18 – transfer of details to the warehouse; 19 – assembly of the entire product; 20 –product testing; 21 – product packing; 22 – delivery to the customer.

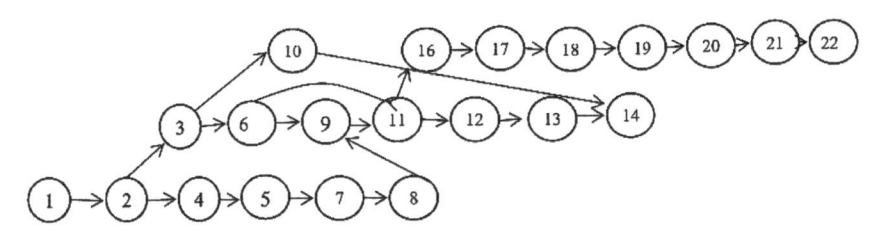

Fig. 1. Network graph of the enterprise functioning.

The task of modeling consists in determination of the space of indicators thus making it possible to assess the dynamics under the influence of destructive behavior during the production cycle. One of the processes disrupting the work of the enterprise are attacks that are implemented through the various networks and communication means so it is necessary to consider the manufacturing process in the general case as varying with time due to changes in the design of the components due to the threats, to the changes in the labor regulation, etc. For this purpose, it is expedient to represent the scheme of the enterprise functioning in the normal mode both in the analytical and algorithmic forms:

$$F = \text{Alg} < W(t), Y(t), X(t), R(t) > ,$$

where $W(t)$ – is the enterprise development strategy in the form of goal vector $W^L = \{W^l = (W_1^l, W_2^l, \ldots W_N^l)\}$ (see Problem Statement); $Y(t)$ – control of state space; $X(t)$ – performance indicators; $R(t)$ – resources of the enterprise; and then to make an assessment of the impact of destructive effects on the implementation of objective functions W.

S1. Formation of a list of strategic goals aimed at the production process enhancement by virtue of losses minimization. The above strategy is the most relevant choice as the manufacturing equipment of the enterprise is rather expensive and so in order to increase the profits in the future it is necessary to renovate and repair it timely as well as to prevent its wear-out including that due to the third-party destructive activities thus minimizing its downtime. It is necessary as well to take into account the losses associated with the information leakage. In this case the objective can be formulated numerically: the reduction of enterprise losses (which can be affected) by 20%.

S2. Decomposition of the strategic objective into the form of a set of sub-goals being necessary to achieve the objective (a fragment of which is presented in the Table 1).

Table 1. Definition of indicators of a sub-goal "reducing damage from defective parts".

Indicator	Calculation formula and arguments
The average number of the eliminated security threats during planning period (a year)	$X_1 = \dfrac{\sum_1^{12} F_1}{12}$, where F_1 is the number of the eliminated security threats for each of 12 months
The ratio of the number of defective parts to the total number of parts produced for the planning period	$X_2 = \dfrac{F_2}{F_3}$, where F_2 is the number of defective parts, F_3 is total number of parts

(*continued*)

Table 1. (*continued*)

Indicator	Calculation formula and arguments
Changes in enterprise productivity due to the changes in the equipment performance under unauthorized access	$X_3 = T_{ef} \cdot a \cdot H(t + \Delta t) - T_{ef}^* \cdot a^* \cdot H^*(t)$, where T_{ef}, T_{ef}^* - effective fund of equipment operation with and without violations correspondingly; a, a^* are the numbers of devices of the same type, machines, units, with and without violations; $H(t + \Delta t)$, $H^*(t)$ are hourly rates of performance of a piece of equipment according to the manufacturer's passport with and without violations
Change in equipment performance time under unauthorized access	$X_4 = T_{cal} - T_{UA} - T_{tech}$, where T_{cal} is equipment production resource, T_{UA} is downtime caused by unauthorized access, T_{tech} is downtime for technological reasons
The ratio of the number of eliminated equipment failures due to unauthorized access to the total number of failures	$X_5 = \frac{F_4}{F_5}$, where F_4 is the number of eliminated failures, F_5 is total number of equipment failures
The ratio of the number of processes with security responsible persons to the total number of production processes	$X_6 = \frac{F_6}{F_7}$, where F_6 is the number of processes with assigned to them security responsible, F_7 is total number of part manufacturing processes

S3–5. Defining for each level both the set of functions $W_{1...N}^l$ being necessary to achieve the goal and their arguments enabling to form a set of indicators that determine the fulfilment of the task. At this stage one of the difficulties consists in the determination of the target values of the indicators of the realistically achievable level. At the same time the development of indicators should be carried out both for the long-term and for the short-term planning: target values for a long-term period are determined for deferred indicators.

The selection of indicators and the definition of their target values as well as the presence of logical links between them make it possible not only to evaluate the production process and to monitor its dynamics but also to form a strategy map (Fig. 2).

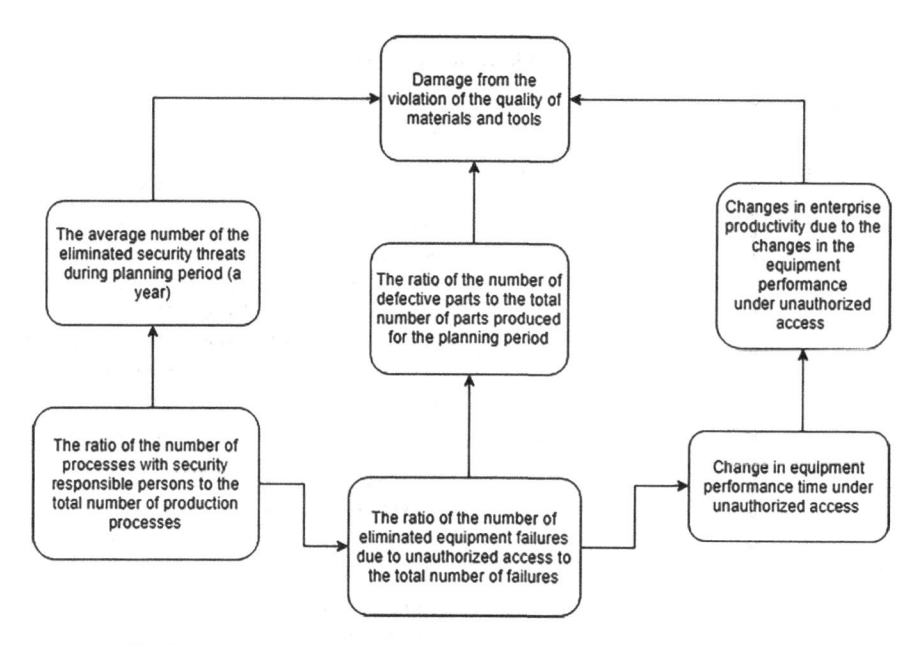

Fig. 2. Building a strategy map of the process "details manufacturing".

Building a strategy map facilitates the assessment of the impact of each indicator on the fulfillment of the objective function, indicators convolution with corresponding weight coefficients, damage assessment and development of a set of measures capable to reduce the damage. As an example, let us consider the dynamics of the change of the indicators presented in Fig. 2.

The least impact on the capability of damage minimization has the indicator expressed by the ratio of the number of processes with the assigned persons responsible for them to the total number of the parts manufacturing processes (Fig. 3, curve 1). The most reduction of damage amount can be achieved by undertaking measures to reduce the number of defective parts (Fig. 3, curve 6).

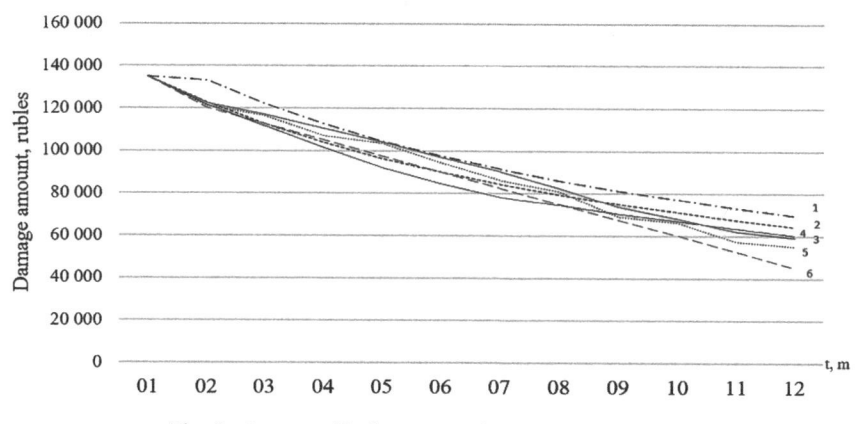

Fig. 3. Impact of indicators on the amount of damage.

In general, the analysis of the behavior of the curves in Fig. 3 demonstrates the impact of indicators on the change of damage amount to the enterprise as a whole but it cannot always reflect a complete picture of the dynamics at all levels. In this case it is necessary to estimate the performance with regard to multiplicative or additive convolution (formula 1). Graph 1 in Fig. 4 reflects the dynamics of changes of the generalized measure of damage recovery. The change in the influence of the ratio of the number of defective parts during the planning period to the total number of parts produced during that period (Fig. 4, graph 2) has the strongest impact on the additive convolution while the indicator of changes in enterprise productivity under unauthorized access has the weakest impact (Fig. 4, graph 4). Thus the simulation results being presented in Fig. 3 and taking into account the totality the whole package of the measures undertaken do not always accurately reflect the actual damage assessment and require consideration of all cause-effect relationships in the analyzed processes.

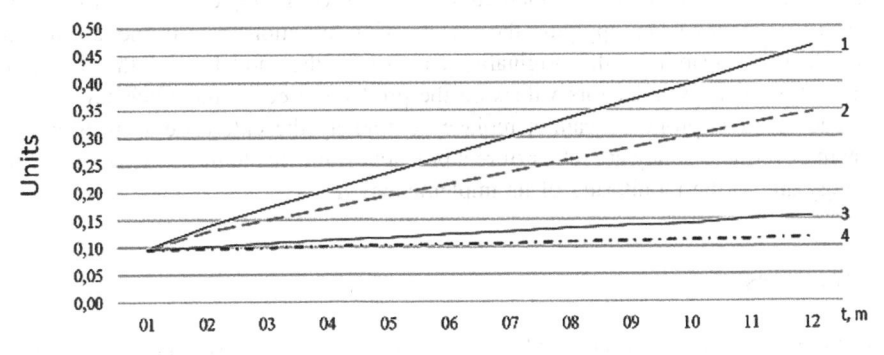

Fig. 4. Dynamics of change of additive indicators convolution.

To reduce the impact of unauthorized access on enterprise performance let us consider a set of measures presented in the Table 2.

Table 2. Security package.

Security package	Percentage of implementation of the security package	Percentage of change in the generalized performance indicator (GPI) of the enterprise
Increasing the number of closed-circuit television installations at the enterprise	20%	40%
Restrictions of employees access to strategic equipment	20%	20%
Control of equipment operation personnel	20%	10%
Strengthening security at the enterprise	20%	30%

The dynamics of changes in the indicators presented in Table 2 showed that the most efficient are measures on video surveillance and security staff strengthening. The measures undertaken provide monitoring both working time and downtime at the enterprise and reducing the risk of deliberate equipment breakdown including that due to unauthorized access thus resulting in the achievement of the previously set goal.

5   Conclusion

The paper discusses the algorithm of development and implementation of a balanced scorecard and of creation of a strategy map of its indicators relationship for the purpose of enterprise information security assurance. Enterprise operation efficiency is assessed by comparing the conditions of information security for various system states. It is noted that the use of the balanced scorecard lets filling the gap between the development of a strategy and its implementation by virtue of the rapid response to security indicators changes when applying the system of information security monitoring and control. The research results originality consists in the modeling of the impact of information security indicators values on the production economic efficiency.

The methods proposed cannot replace completely the enterprise security control system nevertheless they are rather of real practical value due to the visualization of the strategy and to the monitoring of its implementation.

References

1. Yusupova, N.I., Smetanina, O.N., Agadullina, A.I., et al.: Modeling issues in the organization of information and intellectual support for management decisions in complex systems. Fundam. Res. **2**, 107–113 (2017)
2. Zyryanova, T.Yu.: Comparative analysis of methods for risk assessment and risk forecasting for information systems. Bull. Ural Fed. Dist. Secur. Sphere Inf. **1**(23), 28–35 (2017)
3. Fatkieva, R.R.: Modeling automated technological processes under conditions of information threats. Sci. Bull. Novosib. State Tech. Univ. **1**(70), 167–176 (2018). https://doi.org/10.17212/1814-1196-2018-1-167-176
4. Kaplan, R.S., Norton, D.P.: Balanced Scorecard. Olimp-biznes, Moscow (2003)
5. Minaev, V.A., Vayts, E.V., Grachyova, Yu.V.: Balanced scorecard while ensuring information security. In: Fundamental System Security Issues. Materials III Seminar-School of Young Scientists, pp. 173–177. Elec State University, Elec (2016)
6. Gryshko, V., Zos-Kior, M., Zerniuk, M.O.: Integrating the BSC and KPI systems for improving the efficiency of logistic strategy implementation in construction companies. Int. J. Eng. Technol. **7**(3), 131–134 (2018)
7. Harihayati, T., Lubis, R., Atin, S.: The company's performance assessment using balanced scorecard. In: IOP Conference Series: Materials Science and Engineering, vol. 407, no. 1, p. 012067 (2018). https://doi.org/10.1088/1757-899X/407/1/012067
8. Khanova, A.A., Urasaliev, N.S., Usmanova, Z.A.: The method of situational control of complex systems based on the balanced scorecard. Sci. Bull. NSTU **60**(3), 69–82 (2015)
9. Rahimi, H., Bahmaei, J., Shojaei, P.: Developing a strategy map to improve public hospitals performance with balanced scorecard and dematel approach. Med. J. **19**(7) (2018). https://doi.org/10.5812/semj.64056

10. Leksono, E.B., Suparno, S., Vanany, I.: Development of performance indicators relationships on sustainable healthcare supply chain performance measurement using balanced scorecard and DEMATEL. Int. J. Adv. Sci. Eng. Inf. Technol. **8**(1), 115–122 (2018). https://doi.org/10.18517/ijaseit.8.1.3852
11. Yoshikuni, A.C., Albertin, A.L.: IT-enabled dynamic capability on performance: An empirical study of BSC model. RAE Revista de Administracao de Empresas **57**(3), 215–231 (2017). https://doi.org/10.1590/S0034-759020170303
12. Perez, C., Montequin, V.R., Fernandez, F.O.: Integration of balanced scorecard, strategy map, and fuzzy analytic hierarchy process for a sustainability business framework: a case study of a Spanish software factory in the financial sector. Sustainability **9**(4) (2017). https://doi.org/10.3390/su9040527
13. Lakshmanan, S., Christopher, S.E., Kinslin, D.: Analyzing the business performance of an ERP systems in automotive ancillary industries-balanced scorecard perspective. Int. J. Eng. Technol. **7**(36), 599–603 (2018)
14. Vorobeichikov, S.E., Konev, V.V.: On sequential confidence estimation of parameters of stochastic dynamical systems with conditionally Gaussian noises. Autom. Remote Control **78**(10), 1803–1818 (2017)

Bottom Induction Stirrer for Induction Crucible Furnace with Graphite Crucible

K. Bolotin$^{(\boxtimes)}$ and D. Brazhnik

Ural Federal University, 19 Mira Street,
Yekaterinburg 620002, Russian Federation
bolotinke@gmail.com

Abstract. Paper is devoted to study of the possibility of using a bottom induction stirrer with a rotating electromagnetic field for a force on melt of precious metals in a graphite crucible of an induction crucible furnace. The option of liquid metal stirring after melting is completed to achieve temperature uniformity was investigated. Numerical simulation of three depended problems was carried out: electromagnetic, hydrodynamic, and heat transfer in liquids. The optimization of supply current frequency magnitude was carried out by the Nelder-Mead method; the difference between the maximum and minimum temperatures in the melt volume four seconds after the start of stirring was chosen as the objective function, its minimization was achieved. Range of optimal frequencies of the supply current from 250 to 260 Hz was determined. The most effective frequency is f = 256.2 Hz, with it the temperature difference was $\Delta T = 8.6$ °C. At the next stage, the process of mixing the metal during its melting will be considered.

Keywords: Magnetohydrodynamic · Computational fluid dynamics · Finite element method · Induction · Electromagnetic · Induction crucible furnaces · Stirring · Rotating magnetic field · Nelder-Mead method · Frequency · Graphite crucible

1 Introduction

Induction crucible furnaces (ICF) are widely used in metallurgical industry in production of precious metals, including alloys of gold and platinum [1–3]. Depending on the electrical conductivity of crucible material, distinguish ICF with direct and indirect heating. In the first case, additional force from the electromagnetic (EM) field affects on the melt, which allows to speed up the smelting process and to achieve uniformity of chemical and temperature properties. Crucibles in such ICF are made of refractory concrete, which collapses and is wetted by the aggressive and heavy melt of precious metals.

In ICF with indirect heating, graphite crucibles are used, being a conductor and having an electrical conductivity equal to $\sigma = 1.25 \cdot 10^5 - 3 \cdot 10^5$ S/m. This crucible is resistant to wetting by metal, which allows to increase its service life and ensure the absence of impurities in the melt [4]. Graphite is a reducing agent, this helps prevent oxidative reactions in the liquid metal during its preparation. But from another side

A. A. Radionov and A. S. Karandaev (Eds.): RusAutoCon 2019, LNEE 641, pp. 1158–1166, 2020.
https://doi.org/10.1007/978-3-030-39225-3_121

Being a conductor, the crucible almost completely shields the melt from the external electromagnetic field induced by the ICF [5, 6]. In view of this, one of the main drawbacks of a graphite crucible is small degree of force applied by the EM field to melt, which does not allow the metal to be stirred during smelting. As a result, a significant temperature difference in the near-wall and central regions of melt, which can lead to boiling and large losses due to this. In addition, the process time of melting increases due to the fact that heat transfer and convection in liquid metal are of natural nature [7].

Based on the foregoing, solving the problem of increasing the force effect of the electromagnetic field on the liquid metal, in order to intensify the stirring process, will reduce the time and energy costs of preparing the melt. Currently, stirring is carried out using an inert gas supply, this allows for equalization of temperature and chemical composition in the volume of the melt. At the same time, this method cannot be applied at the initial stage of melting, which leads to additional losses in case of overheating and does not completely solve the previously listed problems.

The use of an induction magnetohydrodynamic (MHD) stirrer operating at a low frequency will make it possible to influence the metal from the initial moment of melting, allowing it to distribute the temperature evenly, thereby reducing the overall process time [8–12]. In addition, the absence of external exposure will allow for a pure product.

MHD stirrers can be divided according to the type of placement on the near-wall and end edges. The application of the first type is limited due to the heating inductor encircling the crucible along its entire height. End edge inductor can be placed above or under the crucible. The placement of the stirrer above the crucible was considered in papers [13, 14]. The main disadvantage of this method is the increased temperature effect on the stirrer inductor. As a result, the option of placing the MHD stirrer under the bottom remains.

A laboratory bottom three-phase MHD stirrer with a rotating magnetic field, previously designed by a scientific group, was considered as a test facility [15, 16]. The problem including electromagnetic, hydrodynamic and heat and mass transfer processes occurring when the liquid metal is stirred in a graphite crucible after melting was considered. Since this work only one possibility of using this stirrer in conjunction with a graphite crucible was investigated.

All numerical simulations were performed in the Comsol Multiphysics software package. introduce the paper. The paragraphs continue from here and are only separated by headings, subheadings, images and formulae. The first paragraph after a heading is not indented (Bodytext style).

2   Numerical Model

Figure 1 present three-dimensional schematic model of bottom MHD-stirrer, which phases were connected according to the scheme AAZZBBXXCCYY [17–20].

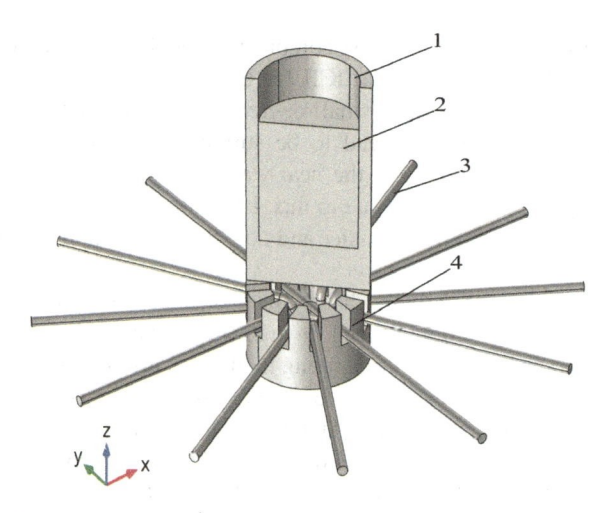

Fig. 1. Bottom MHD-stirrer with rotating field (air hidden): 1 – graphite crucible (partially hidden), 2 – molten metal (partially hidden), 3 – coils, 4 – magnetic core.

Table 1 contains the basic physical and geometrical parameters of the computer model.

Table 1. Physical and geometrical parameters of MHD-Stirrer.

	Magnetic core	Coils	Molten gold	Graphite crucible
μ, o.e.	$1.2 * 10^3$	1	1	1
Σ, S/m	10^6	$5.7 * 10^7$	$3.4 * 10^6$	$3 * 10^3$
Current, A	0	320	0	0
Dynamic viscosity, Pa·s	0	0	0.003	0
Diameter, mm	100	–	80	100
Height, mm	60	–	100	170

Size of working gap coincides with thickness of graphite crucible bottom and it's equal to $D = 35$ mm. A problem that includes three related parts was formulated.

During the numerical simulation of electromagnetic problem, the following assumptions were made:

- The outer boundary is a magnetic insulator
- Current waveform - pure sine wave
- Electrical and magnetic properties of all domains do not depend on temperature
- The magnetic circuit is given a solid geometric shape
- The field of forces induced in the metal is quasistationary
- The magnetic field of a moving melt affects the field of the inductor

The Reynolds number for this model is equal to Re = $2 \cdot 10^4 – 3 \cdot 10^4$, indicating the turbulent character of the motion, so hydrodynamic problem was solved with the use of turbulence k-omega SST model.

Were also designed two types of boundary conditions:

- A walls with metal no slip to the side wall and bottom
- A walls with metal slip to the top

For the hydrodynamic part, the following assumptions were made:

- The physical properties of the areas do not depend on temperature changes
- Gravity affects mass transfer processes
- The formation of a free surface is not taken into account
- The amount of substance in the calculated area does not change

The heat flux outlet through the upper boundary of the melt, imitating free convective air flow, was chosen as the boundary conditions. And through the side and bottom incoming, imitating the flow of heat from the heating inductor.

The initial condition was the temperature distribution after the completion of the smelting process, its appearance is shown in Fig. 2.

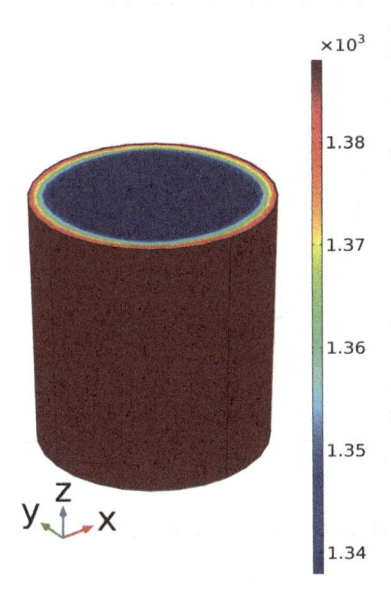

Fig. 2. The initial temperature distribution in the volume of the melt.

The following assumptions were made:

- The physical properties of the domains do not depend on temperature
- Thermal convection, which is caused by temperature differences, is not taken into account

For the formulated numerical model, the "Frequency-transient" mode was used. It allows solving time-dependent hydrodynamic and heat-mass transfer problems with updating the electromagnetic problem in the "Frequency domain".

In this regard, one computational mesh was built for all three parts, including 119,000 elements, including 23,000 elements in the metal, condensed to its lower part for a more accurate calculation of the main influence of electrodynamic forces.

Optimization was carried out on basis of Nelder-Mead method. Control variable was selected value of the magnitude of the supply current frequency from 5 Hz to 500 Hz.

The choice of the integral electromotive force in the volume of liquid metal does not allow to reliably estimate the stirring efficiency for different values of frequency [21]. So as the objective function, the time of the technological process was chosen to equalize the temperature in the volume of the melt.

The initial temperature gradient equal $\Delta T = 50$ K, 4 s after the stirring is turned on, its value drops below 10 K, the intensity of the temperature change decreases and ΔT for a long time reaches a value close to zero, a typical graph of temperature versus time with stirring is presented in Fig. 3.

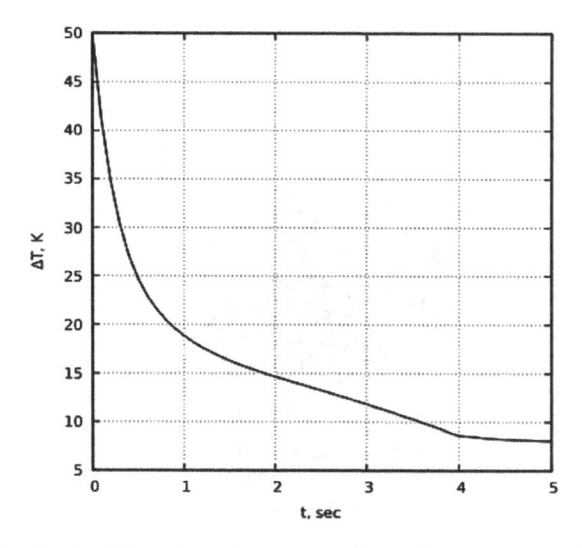

Fig. 3. Graph of dependence between gradient of temperature and time.

In this regard, it was proposed to use the value of the temperature gradient on 4 s after the start of stirring as a objective function, thus taking into account the influence of the electromagnetic field generated by the movement of the melt on the electromagnetic field of the MHD-stirrer. The objective function as follow:

$$S = T_{\max}(4) - T_{\min}(4)$$

3   Results

Figure 4 shows a graph of objective function versus supply frequency value, as you can see, the optimum frequency lies in the range from 250 to 260 Hz, the smallest value of the temperature gradient at 4 s is fixed at the frequency $f = 256.2$ Hz. In the future, all results are given for this value.

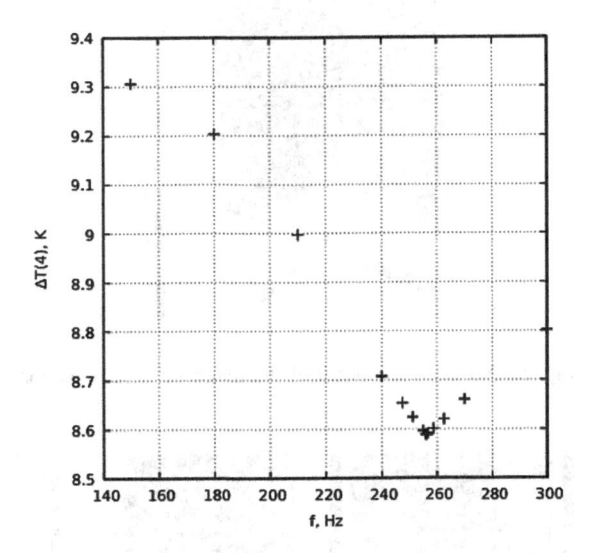

Fig. 4. Graph of dependence between objective function and supply frequency.

The distribution of the normal component of the magnetic field induction in the molten metal and the crucible is shown in Fig. 5. As you can see, its greatest value is in a graphite crucible, which corresponds to theory. When the frequency decreases, the depth of penetration of the EM field into the metal increases, but the degree of its impact decreases, which is why it was necessary to optimize the frequency.

Figure 6 shows the change in the distribution of the temperature field from 1 to 4 s with a step of 1 s in the XZ-plane of the liquid metal.

As you can see, the stirring of metal begins in the lower part of the domain, the stirring area gradually rises along the Z-axis, this is consistent with the ideas about the processes occurring during the operation of bottom MHD-stirrer with a rotating magnetic field.

The dependence of the maximum value of velocity as a function of time is shown in Fig. 7. In this work, velocity was not used as an objective function, since its maximum value does not always guarantee the most effective stirring. At the same time, it can be seen that the speed increases in steps, possibly due to the large slip between the layers of molten metal with high density.

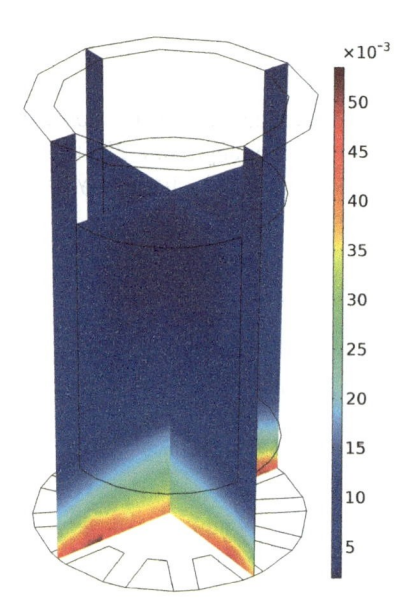

Fig. 5. Normal component of the magnetic field induction.

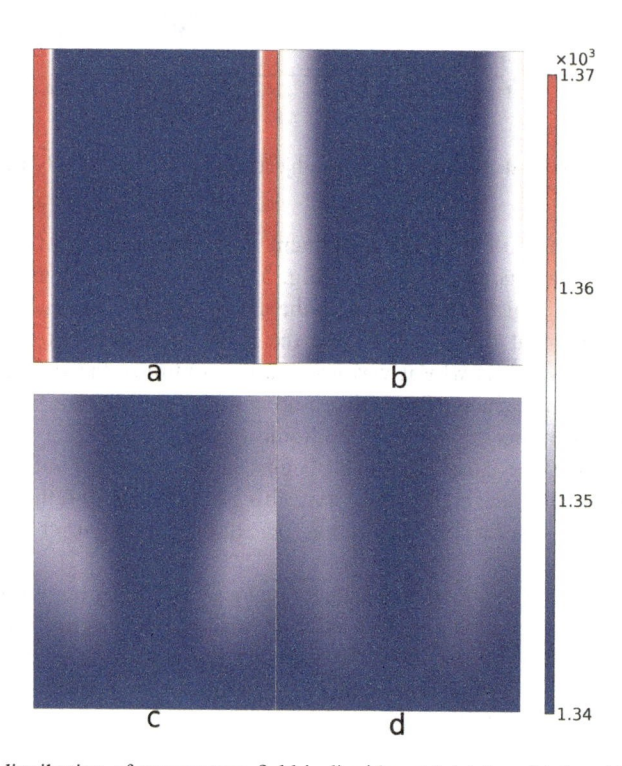

Fig. 6. The distribution of temperature field in liquid metal (a) 1 s; (b) 2 s; (c) 3 s; (d) 4 s.

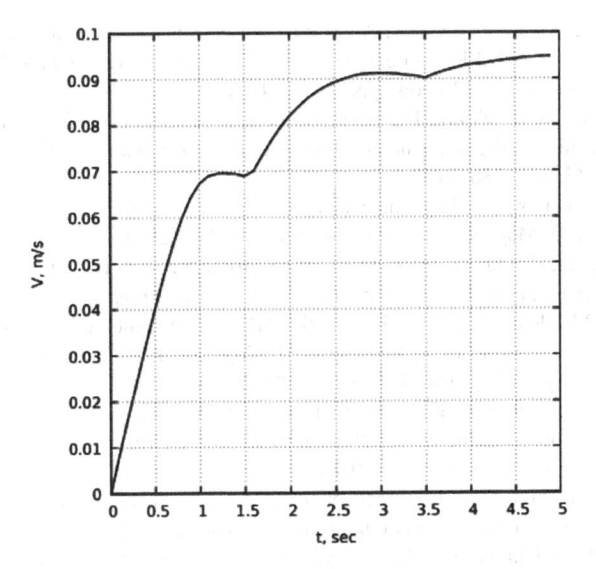

Fig. 7. Graph shown dependence of melt velocity on time.

4   Conclusions

It has been demonstrated that a bottom MHD-stirrer with a rotating magnetic field can be used to stirring molten metal in a graphite crucible to equalize its temperature throughout the volume.

Optimization was carried out, during which the range of optimal frequencies of the supply current from 250 to 260 Hz was determined. The most effective frequency is $f = 256.2$ Hz. At the next stage, the process of mixing the metal during its melting will be considered.

Acknowledgments. The work was supported by Act 211 Government of the Russian Federation, contract № 02.A03.21.0006.

References

1. Morgan, M.: Future looks hot for SA platinum manufacturing industry. J. S. Afr. Inst. Min. Metall. **105**, 28 (2005)
2. Craw, D., Henry, W.: A laboratory scale chill-casting vacuum induction furnace. J. Sci. Instrum. **33**, 22–23 (1956)
3. Henderson, G.: Continuous casting of precious metals. Wire Ind. **69**(70), 609 (2002)
4. Heumannskaemper, D., Pavoni, M., Vohra, A., et al.: Coating on performance. In: Proceedings of the Materials Science and Technology Conference and Exhibition, vol. 1, pp. 163–170 (2015)
5. Hamri, B.: The effect of electromagnetic stirring during solidification on the copper structure. Int. J. Appl. Eng. Res. **11**, 4667–4675 (2016)

6. Ivanov, A., Bukanin, V., Zenkov, A.: Simulation of induction heating processes in ICF. In: Proceedings of the IEEE Conference of Russian Young Researchers in Electrical and Electronic Engineering, Moscow, pp. 981–984 (2019)
7. Lu, G.Q., Ma, W.H., Wang, H., et al.: Numerical simulation on the flow behavior of the melts of metallurgical-grade silicon refining in a vacuum induction furnace. Trans. Mater. Heat Treat. **34**, 181–186 (2013)
8. Barglik, J., Smagor, A.: Mathematical modelling of induction stirring of liquid metal in crucible furnace. Magnetohydrodynamics **53**, 699–706 (2017)
9. Yang, G., Zhao, E., Qin, L., et al.: Effect of electromagnetic stirring on melt flow velocity of laser melt pool and solidification structure. Infrared Laser Eng. **46**, 0906006 (2017)
10. Slazhniev, M., Kim, K., Sim, H., ct al.: MHD-equipment and technologies of semi-continuous billet casting of high-strength Al-alloys. In: IOP Conference Series: Materials Science and Engineering, vol. 424, p. 012079 (2018)
11. Yoshikawa, N., Watanabe, K., Igarashi, T., et al.: Fundamental studies on induction heating and stirring of non-metallic molten fluid. In: IOP Conference Series: Materials Science and Engineering, vol. 424, p. 012080 (2018)
12. Liu, Z., Liu, X., Zhu, T., et al.: Effects of electromagnetic stirring with low current frequency on re distribution in semisolid aluminum alloy. Acta Metall. Sinica **51**, 272–280 (2015)
13. Berezina, A., Monastyrska, T., Davydenko, O., et al.: Effect of melt treatment in magnetohydrodynamic facility and severe plastic deformation on structure and properties of silumins. Metall. Phys. Newer Technol. **31**, 1417–1426 (2009)
14. Dubodelov, V., Pogorsky, V., Goryuk, M.: Magnetodynamic mixer-batcher for overheating and pouring of cast iron. Key Eng. Mater. **457**, 481–486 (2011)
15. Idyatulin, A., Saparulov, S., Saparulov, F., et al.: Simulation of flank induction rotator of liquid metal. Russ. Electr. Eng. **80**, 392–397 (2009)
16. Idyatulin, A., Saparulov, S., Saparulov, F., et al.: Simulation of the multifunctional melting unit. In: Proceedings of IFOST 3rd International Forum on Strategic Technologies, Moscow, pp. 425–428 (2008)
17. Bolotin, K., Smolyanov, I., Shvydkiy, E.: Numerical simulation of the electromagnetic stirrer adapted by using magnetodielectric composite. Magnetohydrodynamics **53**, 723–730 (2017)
18. Bolotin, K., SmolyanovI, T.F., et al.: Numerical study of the possibility of using cermet inserts in electromagnetic stirring application. Acta Technica CSAV **63**, 709–720 (2018)
19. Bolotin, K., Frizen, V., Shvydkiy, E.: Numerical and experimental simulation of a bottom electromagnetic stirrer with a rotating field. In: Proceedings of 18th International Conference on Computational Problems of Electrical Engineering, CPEE, Czech (2017). 8093072
20. Bolotin, K., Shvydkiy, E., Sokolov, I.: Experimental investigation of the bottom MHD stirrer with the working gap compensated by magnetodielectric composite. Magnetohydrodynamics **55**, 23–30 (2019)
21. Bolotin, K., Sokolov, I., Bychkov, S.: Influence of optimality criterion choice on the electromagnetic stirrer optimization results. In: Proceedings of the IEEE Conference of Russian Young Researchers in Electrical and Electronic Engineering, ElConRus, St. Petersburg, pp. 939–941 (2019)

Approximation Method for Probability Density Function of Non-Gaussian Random Processes in Telecommunication and Data-Measuring Systems

A. B. Semenov[1($\boxtimes$)] and V. M. Artushenko[2]

[1] Moscow State University of Civil Engineering, 26 Yaroslavskoe Highway, Moscow 129337, Russian Federation
semenovab@mgsu.ru
[2] University of Technology, 42 Gagarina Street, Korolev 141074, Russian Federation

Abstract. The article considers the method of constructing two-dimensional probability density function of non-Gaussian random processes using one-dimensional probability density function. The method is based on the use of a specially constructed transition probability density function. Kullback-Leibler probability estimation is applied to analyze how close the approximating function approaches the initial function. The property of monotonic approximation of the approximating function to the initial one with the growth of the correlation coefficient in a wide range of variation of the dispersion values is substantiated. A special procedure for constructing the transition function is proposed, which is based on the Gaussian curve and gives a minimum approximation error in a wide range of parameters. Examples of application of the proposed method to the particular case of Laplace distribution, which is often found in real telecommunication and measurement systems, are given. It is shown that in a wide range of variations of the correlation coefficient, the value of the divergence of the initial and approximating functions is quite small. This property allows the proposed method to be widely used in the study of non-Gaussian processes in various broadband radio and cable systems of local and public communication.

Keywords: Approximation · Non-Gaussian random processes · One-dimensional probability density function · Kullback-Leibler probability estimation

1 Introduction

Modern stage of development of data transmission and processing technology is characterized by a rapid growth of applications in combination with the process of joining individual solutions into complex computing and information systems. One of the required steps in practice to build the hardware base for the implementation of these solutions consists in designing optimized systems and devices for measuring and processing incoming data signals at the input of network interfaces and other electronic

A. A. Radionov and A. S. Karandaev (Eds.): RusAutoCon 2019, LNEE 641, pp. 1167–1174, 2020.
https://doi.org/10.1007/978-3-030-39225-3_122

devices. The process of designing and further optimization of such equipment involves analysis of the probability density function (PDF) of both these signals and additive noise which cannot be avoided at the receiver input and which jeopardizes valuable data transferred via the communication channel.

The problem of analysis and further optimization, associated with it, raises not only when creating new equipment but also adapting the existing hardware base to new applications. This approach allows for less time to be spent on development. Preservation of familiar interface for interaction with the conventional devices for technicians and engineers also becomes important. As an example, so called "Long Ethernet" devices can be mentioned. They are gaining popularity these days due to the wide implementation in the engineering practice of such complex systems as Smart City and Smart House [1–8]. These devices serve those parts of the system which tend towards the centralized architecture.

Analysis of receivers of such systems is usually based on conventional statistical approaches of radio engineering. As it was shown in many theoretical and experimental studies, except for small number of model cases, PDFs of real signals and interference signals are quite often have clearly defined non-Gaussian nature, which makes it rather difficult to obtain accurate results [9–11]. As a result, in synthesizing of optimized data-measuring devices and systems based on them, one of major problems is approximation of probability density functions of non-Gaussian processes.

An additional requirement to the approximating function consists in that it must be simple and at the same time provide the most adequate and complete description of probability density functions of the signal and interference.

2 Initial Conditions

The best accuracy can be assured by description of the PDF in the form of a multi-dimensional multivariable function. A weak spot of such description is a great complexity of calculations, even in two-dimensional theoretical analysis. Awareness of this fact resulted in a number of methods allowing for the elimination of this drawback. The most popular among them is based on the representation of PDF as a series which is built using specially selected orthogonal basis functions.

Another way to solve this problem – approximation of PDF applied for non-Gaussian random processes.

Hereafter we consider and analyze a simple and easy-to-use method of mathematical description of probability density function based on the latter approach. The method consists in rejection of two-dimensional representation of PDF of non-Gaussian random processes and describing it as a combination of one-dimensional probability density function and a specially constructed transient PDF. As it was proposed, real probability density function of non-Gaussian process $\{n_h\}$ is approximated by a known one-dimensional PDF $W(n_{h-1})$ and a specially constructed transient probability density function $W^A(n_h|n_{h-1})$. In this case, n_h, n_{h-1} designate values of the non-Gaussian random process $\{n_h\}$ at discretization step h and $h-1$ [9].

An advantage of using one-dimensional description of the random process consists in simplicity of further theoretical analysis.

3 Basic Functions

As a basis, we use known one-dimensional PDF $W(n_{h-1})$, which is influenced by a specially constructed transient function $W^A(n_h|n_{h-1})$, in accordance with the proposed method. In this approach, approximation of the non-Gaussian process based on the probability density function will have the following form:

$$W^A(n_h, n_{h-1}) = W(n_h) \cdot W^A(n_h|n_{h-1}) \tag{1}$$

As a transient PDF in (1), it is appropriate to use the following function:

$$W^A(n_h|n_{h-1}) = \left(2G^2\right)^{-0.5} \cdot \exp\left\{-[n_h - M(n_{h-1})]^2/2G^2\right\} \tag{2}$$

In (2), parameter G^2 characterizes the intensity of the random process $\{n_h\}$. $M(n_{h-1})$ designates a special function:

$$M(n_{h-1}) = n_{h-1} - \frac{d}{dn_{h-1}}0.5G^2\ln W(n_{h-1})$$

If the process $\{n_h\}$ is described by Gaussian distribution $W(n_{h-1}) = N(0,\ \sigma^2)$ (2) can be directly transformed into known relation:

$$W^A(n_h|n_{h-1}) = \left(2G^2\right)^{-0.5} \cdot exp\{-(n_h - rn_{h-1})^2/2G^2\} \tag{3}$$

where σ^2 – dispersion of the random process $\{n_h\}$; r in (3) designates correlation coefficient of process $\{n_h\}$.

If the intensity of the random process $G^2 = \sigma^2(1 - r^2)$, then we obtain the expression for the Gaussian probability density function.

4 Approximation Accuracy Measure

The studied objects relate to probability distribution. Considering this fact, in order to make numerical estimation of approximation accuracy we use so called an informational criterion commonly applied in this field:

$$\min I_k\left(W, W^A\right); n_h, n_{h-1}\hat{I}\Pi \tag{4}$$

where I_k designates Kullback-Leibler information.

The latter actually characterizes mean information contained in the domain Π of variation of components n_h and n_{h-1} of the randomly correlated process when distinguishing the hypotheses:

$$H_0 : W(n_h|n_{h-1})$$

and

$$H_1 : W^A(n_h|n_{h-1})$$

Actually, (4) represents approximation accuracy measure of the initial and approximating functions and can be considered as a measure of their similarity.

5 Kullback-Leibnic Estimation

Estimation based on Kullback–Leibler information can be made in two ways:

$$I_{12.K}(W, W^A) = \iint_{-\infty}^{\infty} W(n_h, n_{h-1}) \cdot \ln \frac{W(n_h, n_{h-1})}{W^A(n_h, n_{h-1})} dn_h dn_{h-1}; \tag{5}$$

$$I_{21.K}(W, W^A) = \iint_{-\infty}^{\infty} W^A(n_h, n_{h-1}) \cdot \ln \frac{W^A(n_h, n_{h-1})}{W(n_h, n_{h-1})} dn_h dn_{h-1}; \tag{6}$$

Hereafter, when condition (4) is additionally satisfied, we will use (5), (6) to check the description of PDF of non-Gaussian processes with (1) and (2).

As testing PDFs, we consider the following:

$$W(n_h) = \frac{\nu}{\Gamma(\nu^{-1})\sigma} \left[\frac{\Gamma(3/\nu)}{\Gamma(\nu^{-1})}\right] \cdot exp\left\{-\left[\frac{\Gamma(3/\nu)}{\Gamma(\nu^{-1})}\right]^{\nu/2} \left[\frac{|n_h|^{\nu}}{\sigma^{\nu}}\right]\right\}, \quad \nu \geq 0.5 \tag{7}$$

$$W(nh|n_{(h-1)}) = \frac{\nu}{2\Gamma(\nu^{-1})\sigma} \left[\frac{\Gamma(3/\nu)}{(1-r^2)\Gamma(\nu^{-1})}\right]^{0.5} \cdot exp\left\{-\left[\frac{\Gamma(3/\nu)}{(1-r^2)\Gamma(\nu^{-1})}\right]^{\nu/2} \left[\frac{|n_h - rn_{h-1}|^{\nu}}{\sigma^{\nu}}\right]\right\} \tag{8}$$

where $\Gamma(\cdot)$ – gamma-function; ν – distribution parameter.

(7) and (8) are convenient for further analysis due to the fact that they can be transformed into known distributions in particular cases. For example, based on (7), when $\nu = 2$, we obtain the well-known function of the Gaussian distribution, while (8), when $\nu = 1$, is transformed into the Laplace distribution.

As a transient approximating PDF we use the following:

$$W^A(n_h|n_{h-1}) = (2\pi G^2)^{-0.5} \cdot exp\left\{-[n_h - M(n_{h-1})]^2/G^2\right\}$$

In this case, it is assumed that

$$M(n_{h-1}) = n_{h-1} - 0.5G^2 Z_A(n_{h-1}) \tag{9}$$

and

$$Z_A(n_{h-1}) = -\frac{d}{dn_{h-1}} \ln W(n_{h-1}) = \frac{v}{2^{0.5v}\sigma^v} |n_{h-1}|^{v-1} \operatorname{sgn}(n_{h-1})$$

where sgn – a mathematical sign, which designates a piecewise constant function of actual argument.

6 Modeling Results

The following shows real two-dimensional PDFs (Fig. 1) and isolines (Fig. 2) which characterize their correlation properties.

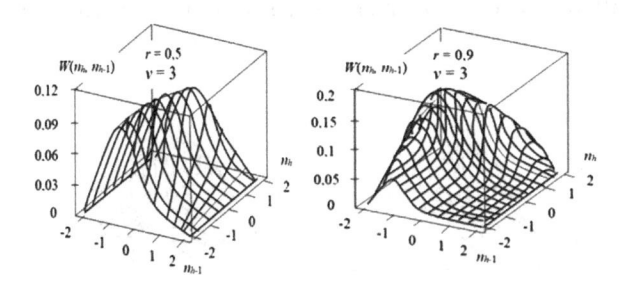

Fig. 1. Real two-dimensional PDFs for $r = 0.5$ (on the left) and $r = 0.9$ (on the right).

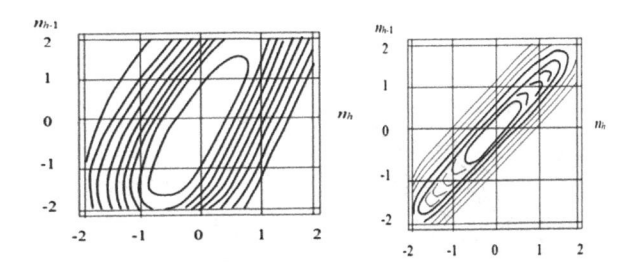

Fig. 2. Isolines for $r = 0.5$ (on the left) and $r = 0.9$ (on the right).

Let us consider an example of constructing two-dimensional PDFs of non-Gaussian process, when only real one-dimensional PDF $W(n_{h-1})$ in the form of (7) is known. According to (2), we can determine transient PDF. Following (9), we write function $M(n_{h-1})$:

$$M(n_{h-1}) = n_{h-1} - 0.5G^2 \frac{d}{dn_{h-1}} \left\{ -\left[\frac{\Gamma(3/v)}{\Gamma(v^{-1})}\right]^{v/2} \left[\frac{|n_{h-1}|^v}{\sigma^v}\right] \right\}$$
$$= n_{h-1} - 0.5G^2 \frac{v}{2^{0.5v}\sigma^v} |n_{h-1}|^{v-1} \operatorname{sgn}(n_{h-1})$$

Next calculations are easy to make using equivalent correlation coefficient r_e It is determined as:

$$r_e = 1 - G^2/\sigma^2$$

When r_e is determined in this way, the transient PDF will take the following form:

$$W^A(n_h|n_{h-1}) = [2\sigma^2(1 - r_e)]^{-0.5} \times \exp\{-\frac{n_h - n_{h-1} + 0.5\sigma^2(1 - r_e)z_e \cdot n_{h-1}}{\sigma^2(1 - r_e)}\} \quad (10)$$

where $z_e(n_{h-1}) = \frac{v}{2^{0.5v}\sigma^v}|n_{h-1}|^{v-1}\text{sgn}(n_{h-1})$.

The results of modeling two-dimensional non-Gaussian PDFs in accordance with (1), (7) and (10) for different correlation coefficients r_e and v are shown in Figs. 3 and 4.

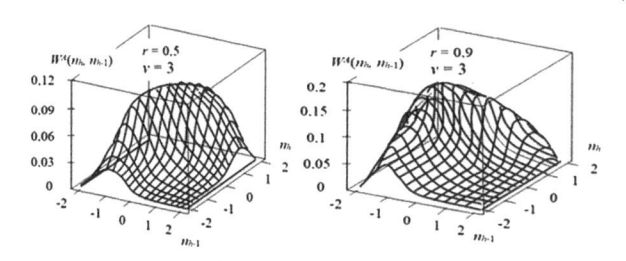

Fig. 3. The constructed two-dimensional PDF for $r = 0.5$ (on the left) and $r = 0.9$ (on the right).

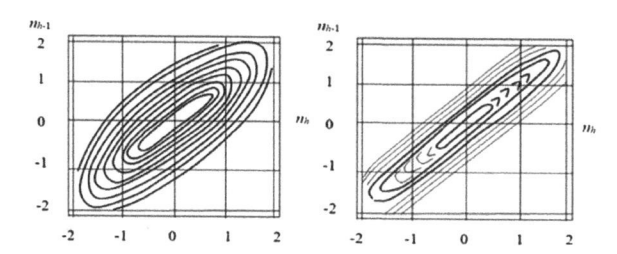

Fig. 4. Isolines of the constructed two-dimensional PDF for $r = 0.5$ (on the left) and and $r = 0.9$ (on the right).

Figures 3 and 4 show real and constructed two-dimensional PDFs. From the results obtained, it immediately follows that correlation coefficient r_e significantly influences the approximation accuracy: as the correlation coefficient grows, monotone growth of the degree of proximity of the approximating function $W^A(n_h, n_{h-1})$ to the real function $W(n_h, n_{h-1})$ is observed.

7 Degree of Proximity for PDF

Let us make qualitative estimation of the degree of proximity between real and approximating PDFs. In order to do this, we use a procedure for determination of the similarity measure for probability distribution functions (4), (5) and (6). For the sake of simplicity we select a particular case and consider the Laplace density. In order to obtain this probability density function, it is enough to assume $v = 1$ in (8).

Figures 5 and 6 shows the results of these calculations in the form of dependencies $I_{12.K}$ and $I_{21.K}$, and iso-level lines of the shown surfaces.

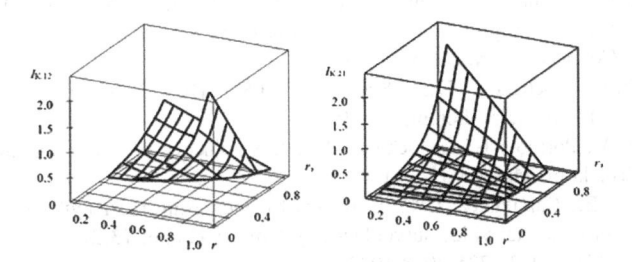

Fig. 5. $I_{12.K}$ (on the left) and $I_{21.K}$ (on the right) as a function of r and r_e.

As we see from Figs. 3, 4 and 5, the most informative dependency is $I_{21.K}$. It shows that, as correlation coefficients r and r_e grow, the degree of proximity of the PDF grows.

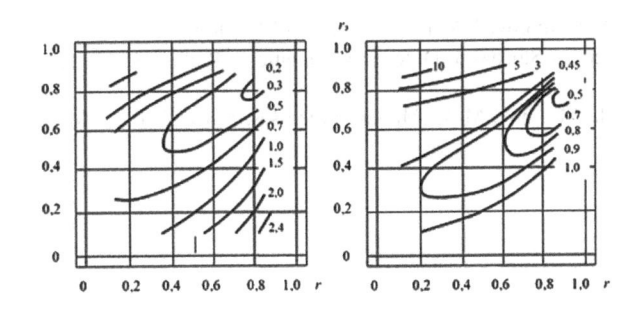

Fig. 6. Iso-level lines for $I_{12.K}$ (on the left) and $I_{21.K}$ (on the right).

8 Conclusion

1. Two-dimensional probability density function of the Laplace distribution, which suits well for describing non-Gaussian random processes, can be approximated using a combination of two functions: one-dimensional probability density function and a specially selected transient function.

2. The degree of divergence between the actual and approximating function uniformly decreases as the correlation coefficient grows.
3. It is appropriate to construct the transient function based on the Gaussian curve, and a special procedure is required for its construction.

References

1. Volkov, A.: Cybernetics of building systems. Cyber physical construction systems. Ind. Civ. Eng. **9**, 4–7 (2017)
2. Semenov, A.B.: Structured Cabling Systems for Data Centers. DMK Press, Moscow (2001)
3. Kuleshov, V.N.: Theory of Telecommunication Cables. State Publishing on Questions of Radio and Connections, Moscow (1950)
4. Iossel, Yu.J., Kochanov, E.S., Strunskiy, M.S.: The Calculation of the Electrical Capacitance. Energoizdat, Leningrad (1981)
5. Andrev, V.A., Portnov, E.L., Kochanovskiy, L.N.: Theory of Telecommunication Cables. Hotline - Telekom, Moscow (2011)
6. Semenov, A.B., Kandzyuba, E.V.: Prospects for increasing the length of the symmetrical cable channel of digital video surveillance systems. T-Comm **13**(2), 25–30 (2017). https://doi.org/10.24411/2072-8735-2018-10232
7. Semenov, A.B., Bylina, M.S., Kandzuoba, E.V.: Structure of symmetrical "long" ethernet cable. T-Comm **2**, 25–30 (2019)
8. Kandziouba, E.V., Portnov, E.L., Semenov, A.B.: A quick method of calculation the Shannon performance of twisted pair channel on "long" ethernet. In: Systems of Signals Generating and Processing in the Field of on Board Communications, p. 1 (2018)
9. Artushenko, V.M., Samarov, K.L.: Construction of two-dimensional correlated models of additive and multiplicative non-Gaussian noise. Electrotech. Inf. Complexes Syst. **9**(4), 83–93 (2013)
10. Klovsky, D.D., Kontorovich V.Ya., Shirokov, S.M.: Models of Continuous Communication Channels Based on Stochastic Differential Equations. Radio and Communication, Moscow (1984)
11. Tikhonov, V.I., Mironov, M.A.: Markov Processes. Soviet Radio, Moscow (1977)

System to Capture Movements of Buyers and to Determine Quality of Store Employees

A. D. Ulyev, V. L. Rozaliev$^{(\boxtimes)}$, Yu. A. Orlova, and A. V. Alekseev

Volgograd State Technical University, 28 Lenina Avenue,
Volgograd 400005, Russian Federation
vladimir.rozaliev@gmail.com

Abstract. The article presents the method of recognition of sales consultants on the basis of a neural network to determine the posture of a person, as well as methods of monitoring the behavior of the sales consultant and analysis of its interaction with the buyer by means of video. The article also discusses the methods of establishing the missing key points on the basis of the physiological structure of the person and the implementation of inter-chamber tracking using key points and segments. A brief overview of similar systems capable of tracking customers or store employees based on video or other related technologies is made. The methods of establishing the identity of the employees of the store and locking the customers to save purchase history. The article presents a description of the proposed methods of operation of the cascade of neural networks to solve the problem, shows the results obtained, as well as the ways of its further improvement.

Keywords: Neural network · Artificial intelligence · Recognition of human pose · Behavior monitoring

1 Introduction

The modern era is characterized by a transition from the economy of producers to the economy of consumers. In the conditions of toughening competition in the sphere of trade and rendering services, client-oriented services acquire special importance.

The main problem of introducing such services is the human factor, control of which is problematic due to the lack of ready-made software products.

Ensuring the proper quality of service delivery becomes the main objective of the market strategy for business development.

To improve the quality of service, it is proposed to develop and implement a software product to monitor the activities of consultant salesmen through the analysis of their work with the use of equipment for video fixing.

The work of the software product is based on the algorithm Pose Estimator [1], which allows you to determine the position of a person, clarifying algorithms, auxiliary neural networks that help to identify the seller-consultant, and also determine the quality of services provided to them [2].

A. A. Radionov and A. S. Karandaev (Eds.): RusAutoCon 2019, LNEE 641, pp. 1175–1184, 2020.
https://doi.org/10.1007/978-3-030-39225-3_123

2 The Proposed Methodology

To solve this problem, we suggest using a cascade of two neural networks, as well as a number of methods and algorithms:

- Fast convolutional neural network FastPoseEstimator, trained on mobilenet architecture
- Algorithm of stabilization of "key" points, allowing determine those points of the body, which could not recognize the neural network
- Neural network for determine the behavior of the sales assistant and store employee
- Algorithms for tracking people in the frame
- Library "Face Recognition" to identify buyers and sellers

3 The First Stage, the Use of the Neural Network "Fast Pose Estimator"

The main task of the neural network is to establish a person's pose through a non-parametric representation called the Part Affinity Fields (PAFs) by developers, to further determine the location of the seller's uniform of the consultant (branded T-shirt, cap, etc.).

The main advantage of the neural network is the high speed of the work, 1 frame in 1–2 s on GPU and 5–8 s on CPU. The main drawback is the decrease in the quality of work when compared with the classic version of Pose Estimator.

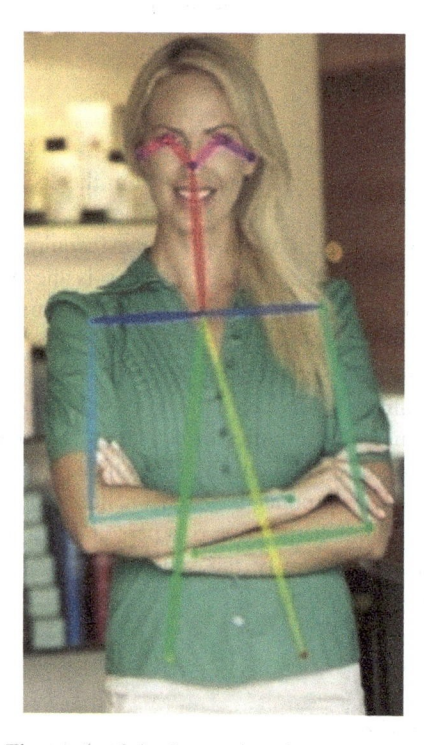

Fig. 1. The result of the fast pose estimator neural network.

Input data for the algorithm "Pose Estimator" is a graphic image of the sales consultant, on the output - an image with the selected parts of the human body. The result of this network can be seen in Fig. 1.

4 The Second Phase, a Stabilization Algorithm "Key" Points of a Person Body

The skeleton is built from the starting point located under the throat, then the eyes, shoulders and pelvis are recognized. From the shoulders completed the hands, from the points of the pelvis of the foot, from the eyes ears. Thus, the key point for the construction is the point under the neck, then we call it the initial one.

It is necessary to determine the dominant color of the t-shirt seller. To improve the quality of work, the following algorithm is proposed for finding the missing "key" points, based on information about the structure of the human structure. Consider one of the options for finding the missing "key" point of the body.

If only one shoulder of a person is recognized, the second shoulder can be completed by depositing an equal segment from the projection of the starting point on the normal of the point of the found shoulder to the side of the not found shoulder equal to the distance from the projection of the point to the normal of the point of the found shoulder to the point of the found shoulder.

Thus, using the information about the structure of the human structure it is possible to determine the approximate location of the missing points of the shoulders, pelvis points. If there is only a starting point, it is proposed to determine some area of a fixed size on the human's body by depositing the area below the key point by N pixels. An example of the algorithm is shown in Fig. 2.

Fig. 2. Example of detection of the second shoulder.

5   The Third Stage, the Determination of the Dominant Color in the Uniform Section

The main task of this stage is to establish a dominant colour in the area of the uniform of a person to determine it in the group of sales consultants. Within the frame-work of this algorithm, an image from the "Pose Estimator" with the tops of the human body parts is input. On the basis of which there is a selection of the necessary clothing of a person. To implement the definition of dominant colour in an established area, there are several methods: determining the ratio of a pixel to a given set of colours and clustering by the k-means method [3–5].

In the first method, the image is converted to HSV colour space, after which all pixels of the image are analyzed and based on the Hue, Saturation, Value data, the colour is set.

The idea of the k-means method [6] is to minimize the total quadratic deviation of the cluster points from the centers. At the first stage, you select points (three dimensional RGB space) and determine whether each point belongs to this or that center. Then at each stage, the centers are redefined until a single center is found.

6   The Fourth Stage, Training the Appearance of Sales Consultants

To determine only the colour of a person's clothing is not enough to classify him as a sales consultant group.

It is necessary to take into account the conditions of the difference in the illumination of the room at different times of the day, as well as the likelihood that there may be clarified and dark areas in the room. Thus, the recognized colour of the shape can vary.

To solve this problem, it is necessary to teach the system all possible colors that can be "read" from the clothing of the seller-consultant.

The operator of the software product at the beginning of work with the program should start the training mode, with by means of which prevailing color uniforms randomly moving around the store sales assistant.

On the basis of the prevailing colors of the clothes is formed an "average dominant color", which is the central point in the formation of the color range for referring a person to a sales consultant [7].

7   Determine the "Field of View"

After the classification of people in the frame to groups of buyers and sellers, the program automatically controls the quality of services rendered by sellers-consultants. To assess the quality of the seller's communication with the client, it is proposed to use a set of algorithms and a cascade of neural networks.

The main task of the algorithms is to determine the location of the seller next to the buyer, as well as control over the personalization of the employee's appeal to the buyer.

In order to exercise control over the personalization of the employee's appeal to the buyer, it is proposed to determine the "field of view" of people in the frame.

The seller must always interact with the buyer, be in the "review" area of the buyer and talk about the benefits of the goods. Thus, to accomplish the task, it is necessary to build a "field of view" based on the location of the eyes and ears obtained from Pose Estimator, after which the sector of intersection of these areas and the "angle of interaction" between the buyer and the seller should be determined.

If the areas do not intersect or the "interaction angle" is less than the angle set by the operator of the software product, it is considered that the seller is near the buyer, but does not interact with it. Examples of definition algorithms for the "field of view" are presented in Fig. 3.

To determine the interest and emotionality of the store employee when interacting with the buyer, it is proposed to use neural networks that define the above parameters by the store employee behavior: hand speeds, moving around the store to accompany the buyer, movements to show goods, etc.

In addition, it is planned to use a separate neural network, which, by facial expression and behavior, will determine customer satisfaction with the services provided by the seller's. The conclusion about the work sellers of the store is made on the basis of all the factors that determine the interaction with the buyer, after which at the end of the month an estimate is calculated for each store employee.

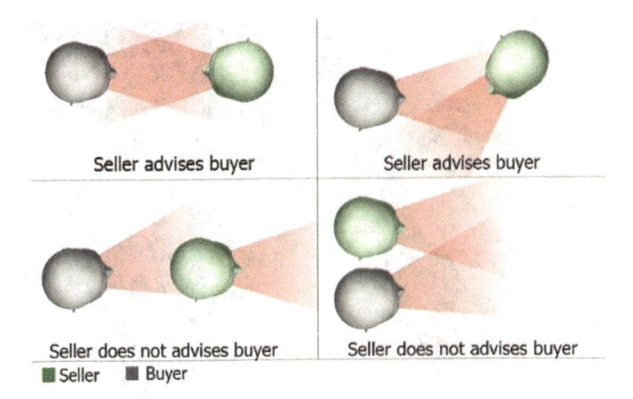

Fig. 3. An example of the definition of "field of view".

8   Inter-chamber Tracking

To control the movement of employees and customers in the premises, it is proposed to use a number of algorithms and methods for inter-camera tracking. The specified module will allow to "lead" a person "between frames", as well as to recognize a person if he has left the control zone of one camera and entered the control zone of another camera.

Tracking is based on the data returned from the PoseEstimator algorithm in the first stage and the data returned from the stabilization algorithms of the "key" points returned in the second stage. Thus, we do not use third-party neural networks, do not develop new ones, but use the data that was obtained earlier. This helps us to significantly accelerate the speed of the program.

It is proposed to divide the body area obtained from the "key" points stabilization algorithm into five equal parts, set the dominant color for each part and form the color range for each part (these algorithms are described in stages 3 and 4, they are used again). In addition, PoseEstimator returns the coordinates of points and other parts of the body. But since the legs and arms do not have a solid contour and the area cannot be formed on the basis of two points (left upper and lower right corners of the area), it is proposed to form a small area around a point of N pixels. We used this approach in the stabilization algorithm of "key" points at the second stage.

For the formed areas, the dominant color is also set and the color range is calculated. After that, an array with the found color ranges is assigned a key that identifies the person in the frame.

At the stage of tracking the area are the above-mentioned algorithm, one now checks the occurrence of the dominant color of the area in the previously found color range. In the case of such a border, the color range is refined to improve the quality of work. If there is more than 50% in the color area, the found person is recognized by the color and the key generated at the first recognition stage is assigned to it. You can see an example of tracking in Fig. 4.

Fig. 4. An example of the tracking algorithms.

9   Identification of Employees and Customers

The task of identifying sellers and buyers is quite difficult in its decision due to the low quality of images from video cameras. We offer a number of modules and hardware to solve the problem.

It is proposed at the entrance to the store to install a camera with a resolution of at least 720px, with which customers will be identified by face using the python library "Face Recognition". This library is easy to connect and use, it includes modules for searching a face in a photo and recognizing a face from an existing base. Thus, after the first call, the client will get into the database and will be identified with each new visit to the store. And the control of its movement will be provided by the inter-chamber tracking module.

For sellers, it is proposed to equip a "starting" point in the hall (for example, when leaving the service room), where the 720px camera will also be located, which allows identifying an employee. Further movement of the employee will be monitored by the inter-camera tracking module.

10 Overview of Analogues

It should be noted that the finished software products that allow to solve the problem discussed in this article are not present. Similar software products perform only part of tasks.

The simplest example of intelligent video surveillance is motion detection. One detector can replace several video surveillance operators. And in the 2000s, the first video analytics systems began to appear, capable of recognizing objects and events in the frame.

Most of the solutions work with face recognition technologies. Solutions in this area include Apple, Facebook, Google, Intel, Microsoft and other technology giants. Surveillance systems with automatic passenger identification are installed in 22 US airports. In Australia, they are developing a biometric system of face recognition and fingerprinting within a program designed to automate passport and customs control.

An interesting project of NTechLab company showed a system capable of real-time recognition of sex, age and emotions using the image from a video camera. The system is able to evaluate the audience's reaction in real time, so you can identify the emotions that visitors experience during presentations or broadcasts of advertising messages. All NTechLab projects are built on self-learning neural networks. In our system, we do not yet use data on a person's face. We plan to process this information at the next stages of the project development.

In other systems, the object tracking function is used - tracking. The operation of the tracking modules is related to the operation of the motion detector.

To construct the trajectories of the movement, a sequential analysis of each frame is carried out, on which moving objects are present. In the general case, several moving objects can be present in one frame, so the program needs not only to construct trajectories, but also to distinguish objects and their movements. The simplest implementation of tracking considers two frames and builds trajectories along them. First, the movements on the current and previous frame are marked, then, by analyzing the speed, the direction of movement of objects, and also their sizes, the probabilities of the transition of objects from one point of the trajectory of the previous frame to another point of the current are calculated. The most probable movements are assigned to each object and added to the trajectory. Objects in the frame can move in different ways:

their trajectories may intersect, they can disappear and arise again. To improve the accuracy of tracking, some manufacturers use the technology of sequence analysis and continuous post-processing of the results obtained. The program builds graphs - it analyzes the transitions of objects from one state to another. In order to understand which, object the movement corresponds to, the speeds and directions of motion, position, color characteristics are also analyzed. As a result, a set of the most probable displacements of the object is formed, forming a trajectory. We have planned to use this approach in our system.

Another analogue of our system - GPS-trackers. These systems work based on the definition of geolocation. To implement this solution, each employee must be equipped with a separate GPS tracker, the data from which will be sent to the server at some interval. However, this solution has a number of drawbacks:

- The solution is not cost-effective, since it is necessary to purchase GPS trackers for all personnel
- We can't exclude the situation in which the seller can give his GPS-tracker to a partner to deceive the system
- Such a solution is not universal. When identifying sales consultants through the camera, it is possible to expand the functionality, determine the level and time of interaction of the seller with the buyer, and much more

Also, analogs include systems for counting the number of visitors on a video stream. These systems also have a number of shortcomings, the main one of which is the impossibility of identifying sales consultants and the quality of their services. An example of the work of such products is shown at Fig. 5.

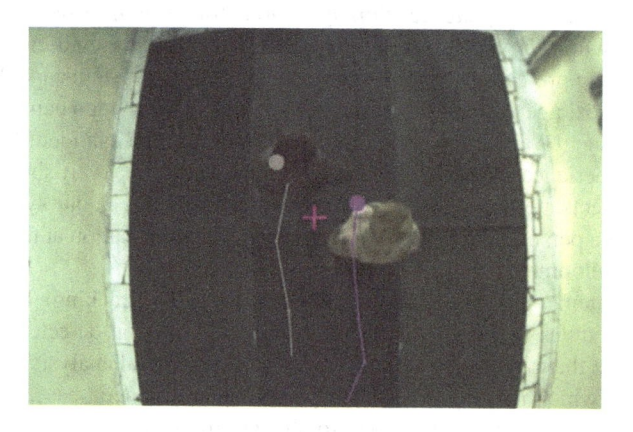

Fig. 5. Example of a program for counting the number of visitors.

11   Conclusion

In order to improve the technological process of detecting the seller's consultant, it is possible to develop additional functionality.

To more accurately determine the seller's consultant, it is possible to analyze several elements of the uniform at once (for example, a yellow T-shirt and black pants).

In addition, it is possible to search for the company logo on the uniform, the location of which will allow us to identify with confidence the person as the seller-consultant.

Another factor that allows to detect the seller, can serve as a definition of behavior, characteristic for the seller-consultant. To solve this problem, you will need to create another neural network.

It is also possible to identify additional factors that determine the quality of the seller's work in the store, such as the presentation goods to customers who have not stopped at the rack with the goods, but passing by.

In addition, it is planned to implement the ability to install "dead zones" in the program. This function allows you to set the "non-human" type for those objects that the Fast Pose Estimator neural network defines as people.

Modules for customer identification and tracking allow you to delve into the marketing segment. We expect to identify the necessary goods and goods that are popular based on past visits to the store by the buyer. For example, when buying a short when visiting the store in the new, we will recommend the employee to advertise a T-shirt to the shorts already purchased.

Thus, the developed software will make it possible to qualitatively improve the work of the sales assistant and, as a result, will lead to an improvement in the customer focus of the business. Figure 6 shows an example of the program.

Fig. 6. An example of the program when recognizing an employee of the store.

This work is a continua of the work, where the features and possibilities of determining the post-sense of its semantic distinctive feature were considered [8–10].

Acknowledgment. This work was partially supported by RFBR and administration of Volgograd region (grants 17-07-01601, 18-07-00220, 19-47-343001, 19-47-340003, 19-47-340009, 19-07-00020).

References

1. Cao, Z., Simon, T., Wei, S.E. et al.: Real-time multi-person 2D pose estimation using part affinity fields (2018). http://arxiv.org/abs/1611.08050. Accessed 25 July 2019
2. Ulyanova, O.: Psychological features of sales consultants network marketing. SSTU, Samara (2019)
3. Cao, Z.: Real-time multi-person 2D pose estimation using part affinity fields (2018). http://www.ri.cmu.edu/wpcontent/uploads/2017/04/thesis.pdf. Accessed 28 July 2019
4. Iqbal, U., Gall, J.: Multi-person pose estimation with local joint-to-person associations (2018). https://arxiv.org/pdf/1608.08526.pdf. Accessed 25 July 2019
5. Insafutdinov, E., Pishchulin, L., Andres, B., et al.: A deeper, stronger, and faster multi-person pose estimation model (2018). https://arxiv.org/pdf/1605.03170.pdf. Accessed 23 July 2019
6. Osipova, Y., Lavrov, D.: Application of cluster analysis by k-means method for classification of scientific texts (2018). https://cyberleninka.ru/article/n/primenenie-klasternogo-analiza-metodom-k-srednih-dlya-klassifikatsii-tekstov-nauchnoy-napravlennosti. Accessed 29 July 2019
7. Khorunsjiy, M.: Method for quantifying color differences in the perception of digital images. Science, St. Petersburg (2008)
8. Rozaliev, V., Orlova, Y.: Recognition of gesture and poses for the definition of human emotions. In: 11th International Conference of Pattern Recognition and Image Analysis: New Information Technologies Conference proceedings, Samara, 23–28 September 2013, vol. 2, pp. 713–716 (2013)
9. Bobkov, A., Rozaliev, V.: Fuzzification of data describing the movement of a person. open semantic technologies for the design of intelligent systems. In: Materials International Scientific Technology Conference, Minsk, 10–12 February 2011, pp. 483–486 (2011)
10. Black, M., Jacobs, D.: End-to-end recovery of human shape and pose (2018). https://www.researchgate.net/publication/321902575_Endtoend_Recovery_of_Human_Shape_and_Pose?discoverMore=1. Accessed 25 July 2019

Automating the Detection of Sarcastic Statements in Natural Language Text

A. V. Dolbin, V. L. Rozaliev$^{(\boxtimes)}$, Yu. A. Orlova, and A. D. Ulyev

Volgograd State Technical University,
28 Lenina Avenue, Volgograd 400005, Russian Federation
vladimir.rozaliev@gmail.com

Abstract. This article is devoted to the sarcasm recognition in the text written in a natural language. The main goal is to increase the accuracy of sentiment analysis. The sentiment level determination of a text that describes the appearance of a person was chosen as a domain area for the experiment. At first, references to the personality and elements that describes appearance from text are detected using the method of latent semantic analysis. The next step is to evaluate the attitude to a person in text using pre-labeled sentiment dictionary. At this stage, the method of recognizing sarcastic sentences that contains a description of the appearance is used. The sentiment level should be re-evaluated in the person information model. The results of the experiment showed that the recognition of sarcasm based on the morphological features of words and the frequency characteristics of the sentences does not effectively increase the accuracy of sentiment level determination.

Keywords: Sentiment analysis · NER · Text mining · Machine learning

1 Introduction

Sentiment analysis of the text belongs to the category of information retrieval tasks. The importance of an effective solution to this problem grows over time, since the amount of information that needs to be processed by semantic analysis systems is continuously increasing. At the moment there are quite effective methods for sentiment analysis of the text, but there are a number of directions, the solution of which will make it possible to achieve greater accuracy of correct recognition. One such direction is the recognition of sarcasm. Sarcasm can be classified as an implicit approach to the expression of opposing emotions. However, even a person cannot always determine reliably whether this phrase is a sarcasm.

The application of sentiment analysis focused on different domains, from management to computer science, social sciences and business due to its importance to society as whole and different tasks [1].

Sentiment analysis is a complex process that involves 5 various steps to analyze data:

- Data collection. The first step of analysis consists of collecting data from a user created content
- Text preparation. It is used to clear the extracted data before analysis

A. A. Radionov and A. S. Karandaev (Eds.): RusAutoCon 2019, LNEE 641, pp. 1185–1194, 2020.
https://doi.org/10.1007/978-3-030-39225-3_124

- Sentiment determination. In this step sentences from reviews and opinions are studied
- Sentiment classification
- Results presentation: the main goal of the analysis is to transform unstructured text into meaningful information [2].

There are five specific problems within the field of sentiment analysis:

- Document-level sentiment analysis
- Sentence-level sentiment analysis
- Aspect-based sentiment analysis
- Comparative sentiment analysis
- Sentiment lexicon acquisition [3].

This topic is quite popular in computer science. Such conclusion can be made on the basis of a large number of articles devoted to the study of the problems of sentiment analysis of a text written in natural language. And Twitter is the real quintessence for conducting experiments related to semantic analysis.

Tweets (other types of posts with a limited size can also be considered) differ from reviews in the first place because of their purpose: while reviews represent the authors generalized thoughts, tweets are more random and limited to 140 characters of text. Thus, solution to use distant supervision, in which training data consists of tweets with emoticons, was proposed in [4]. Authors in [5] report on the construction of a new Twitter NEL dataset that remedies some inconsistencies in prior data for linking recognized named entity. And the authors in [6] successfully compiled a training collection for analyzing sentiments using the valence vocabulary for social and political texts in Russian. According to their study, 93% of texts were classified correctly with an error of ± 1 class, which closely matches the result of the initial use of SentiStrength to tweets in English. There are a large variety of different papers devoted to deep learning models, such as [7] and [8].

The task of automating the definition of sarcasm itself is of little practical value. Typically, you need a limited application area to apply the sentiment analysis. And most often the development is carried out in the following areas:

- Sentiment analysis of users reviews
- Analysis of comments posted on social media resources [9].

The problem of recognizing sarcastic sentences in the text in natural language was considered in the context of searching for elements of a person's appearance and determining the sentiment class. This named entity was chosen not by chance, since it is a quite complex task to recognize it with high accuracy due to a large number of approaches to co-referencing through pronouns in the third person.

Supervised machine learning models learn to make predictions by training on labeled examples and their expected outputs. They can be used to replace human curated rules. Hidden Markov Models (HMM), Support Vector Machines (SVM), Conditional Random Fields (CRF), and decision trees can be used as basis for machine learning systems for named entity recognition [10].

The aim of this work was to examine modern methods for determining the author's relationship to the described person by performing the sentiment analysis. The most obvious area of application of the development, considered in this article, is the analysis of comments on photos in social networks. Using the methods of machine learning, it is possible to construct a model that is able to recognize a positive or negative attitude to the appearance of the person depicted in the photograph. The main contribution of the authors of the article is the adaptation of existing methods of assessing the sentiment in the field of recognition of a person's appearance in the text in natural language [11].

2   Information Model

First of all, it was required to develop an information model of a person's appearance. This model must meet the following requirements:

- Extensibility
- Visibility
- Completeness of the description.

A frame presentation of knowledge is perfectly suitable for this description. Figure 1 shows the final model of a person's appearance using the frame representation language notation. There are the main components on which it is possible to compose a complete description of a person's appearance in the frame slots. Slots for the model were compiled by the authors of the article. In this figure, "M" is the set of valid values for the description elements of the appearance for each slot. The specialty of the FRL notation is that it is permissible to join special procedures-demons to it. The only procedure is the determination of sentiment level with the subsequent resolution of sarcasm. It is worth noting that each non-empty slot must correspond to the sentences from which the facts were extracted for the frame. This is required for further recognition of the presence of sarcasm in the text.

A person is one the most difficult named entity to recognize in the text. It's not so difficult to determine person if text contains the name or surname of the entity. One common approach is to use contextual rules. Every rule represents a standard regular expression, which is constructed from the training sample as follows:

- Any mention of a person should be replaced with a special word {PERSON}
- If the word in a training sample is an element of a human appearance, it should be set in its initial form
- All other words should be replaced with its parts of speech
- After processing the entire training sample, similar contextual rules should be combined using special characters. The '?' symbol means that this position can be omitted, the '+' symbol means that the position can be repeated one or more times in a row and the '|' symbol represents logical "or"
- Optionally, some phrases can be listed at the end of the training sample, which indicates that the sentence excludes the possibility of containing the entity.

Thus, the outputs are kind of regular expressions, which are applied to the text to define a named entity with it.

The next step is to resolve the reference of pronouns in the third person. Reference resolution of pronouns in the third form is one of the most common, but at the same time the simplest case. This task can be considered as a problem of binary classification. Therefore, it is a good opportunity to use support vector machines (SVM).

```
(frame Human_Appearance
 (Height      ( value (M) ) ( IF_ADDED(sentiment_analysis) ) )
 (Body        ( value (M) ) ( IF_ADDED(sentiment_analysis) ) )
 (Head        ( value (M) ) ( IF_ADDED(sentiment_analysis) ) )
 (Hair        ( value (M) ) ( IF_ADDED(sentiment_analysis) ) )
 (Face        ( value (M) ) ( IF_ADDED(sentiment_analysis) ) )
 (Forehead    ( value (M) ) ( IF_ADDED(sentiment_analysis) ) )
 (Eyebrows    ( value (M) ) ( IF_ADDED(sentiment_analysis) ) )
 (Eyes        ( value (M) ) ( IF_ADDED(sentiment_analysis) ) )
 (Eyelashes   ( value (M) ) ( IF_ADDED(sentiment_analysis) ) )
 (Nose        ( value (M) ) ( IF_ADDED(sentiment_analysis) ) )
 (Lips        ( value (M) ) ( IF_ADDED(sentiment_analysis) ) )
 (Chin        ( value (M) ) ( IF_ADDED(sentiment_analysis) ) )
 (Teeth       ( value (M) ) ( IF_ADDED(sentiment_analysis) ) )
 (Neck        ( value (M) ) ( IF_ADDED(sentiment_analysis) ) )
 (Shoulders   ( value (M) ) ( IF_ADDED(sentiment_analysis) ) )
 (Chest       ( value (M) ) ( IF_ADDED(sentiment_analysis) ) )
 (Back        ( value (M) ) ( IF_ADDED(sentiment_analysis) ) )
 (Legs        ( value (M) ) ( IF_ADDED(sentiment_analysis) ) )
 (Arms        ( value (M) ) ( IF_ADDED(sentiment_analysis) ) )
 )
```

Fig. 1. Person appearance information model.

The list of parameters for support vector machines which were used for training:

- Number of sentences between antecedent and anaphora
- Whether the antecedent is in the nominative
- Position of the anaphora in the sentence
- Position of the antecedent in the sentence
- Number of nouns and pronouns, which are located in sentences with antecedent and anaphora
- Are the antecedent and anaphora case matches
- Are the antecedent and anaphora genus matches
- Are the antecedent and anaphora both in a plural or singular form.

To fill the frame, a method of latent semantic analysis was used, or abbreviated LSA, as it has proved itself in the field of machine learning. Methods that do not use a pre-tagged training sample for the learning process show a smaller effectiveness in terms of recognition. The method of latent semantic analysis can be characterized as establishing the relationship between the vectors of the features of the analyzed documents to the words that serve as the keys. Thus, to use the method of semantic analysis of text in natural language, the slots of the frame should be used as search keys.

The latent-semantic algorithm is as follows:

- Create a list of all keywords that will be searched in the text
- Create a frequency matrix A, in cells of which the count of how many times does this word occur in the text
- Apply TF-IDF method on a frequency matrix to ensure that results are relevant
- Apply a singular matrix decomposition: algorithm divides the transformed frequency matrix A into three composite matrices U, Vt and S according to (1)

$$A = U \cdot S \cdot Vt \tag{1}$$

- Matrix U contains the coordinates of keywords and Vt – coordinates of documents.

Singular value decomposition of the matrix allows you to get rid of unnecessary noise, which significantly increases the efficiency of the method. The number of rows and columns that can be discarded before the subsequent analysis can be selected experimentally. It is now possible to obtain the nearest documents, which has the same semantic meaning as specified keyword, and then fill in the frame slots.

3   Vocabulary Based Sentiment Analysis

All approaches to determination of sentiment class are divided into three main groups:

- Compilation of a sentiment vocabulary
- The use of various classifiers
- The use of compiled contextual rules.

A rule-based approach shows the most accurate results, but it requires very high costs and colossal linguistic work for compilation. The main drawback of this approach is that it is extremely difficult to compose universal rules that are suitable for all domains. To achieve the most effective evaluation of the tonality, the rules are compiled for a specific application area.

In this experiment, an approach based on the valence dictionary was applied, since it shows a fairly high percentage of correct recognition. The task is greatly simplified if there is a source for compiling a dictionary of valences belonging to the domain under study. Such dictionary was compiled on the basis of the corpus of the Russian language OpenCorpora. All the phrases that are marked as "Qual" were chosen from this dictionary. Further, only those word forms that can be used for describing a person's appearance have been filtered out. To simplify the task of sentiment analysis, it was decided that the valences would correspond to available sentiment class. Table 1 shows an example of a sentiment dictionary.

Table 1. Part of compiled sentiment dictionary.

Keyword	Valence
Friendly	2
Unfriendly	−2
Shy	0
Impartial	1
Mean	−1

For the study, the basic five sentiment classes were compiled:

- Negative
- Strongly negative
- Positive
- Strongly positive
- Neutral.

To determine the sentiment, the naive Bayes method was used. This method has proved itself in the field of machine learning. A naïve Bayesian algorithm is a classification algorithm based on the Bayes theorem with the assumption of independence of features. The classifier assumes that the presence of any feature in the class is not related to the presence of any other attribute. Let the P(d|c) be the probability of finding a document in all the documents of a given class. The basis of the naive Bayesian classifier is the corresponding theorem (2). In (2) P(c) is the probability of certain document can be found among all data set and P(d) – probability the document occurs throughout the whole corpus:

$$P(c|d) = \frac{P(d|c) \cdot P(c)}{P(d)} \tag{2}$$

Thus, the naive Bayes method is based on the problem of finding the maximum probability of a document belonging to a certain sentiment class. Thus, the sentiment level for each key element of a person's appearance can be determined by (3):

$$P_{\max} = \arg\max \left[P(c) \prod_{i=1}^{n} P(\omega_i|c) \right] \tag{3}$$

Classification using naïve Bayes is easy and fast and requires less training data. Also, it is better suited for classification based on categories (sentiment analysis with separate defined classes refers to such cases). However, if there is some value of a category characteristic in the data set that was not found in the training samples, then the model will assign a zero probability to this frame slot. Sentiment class for each key element of a person's appearance can be determined by (3) where P(w|c) is probability of occurrence of a certain term in a document.

Experimentally, it was found that the hierarchical classification gives better results than the flat one, because for each classifier, you can find a set of features that allows you to improve results. However, it requires a lot of time and effort for training and testing. Figure 2 shows the final classifier based on the naive Bayes method.

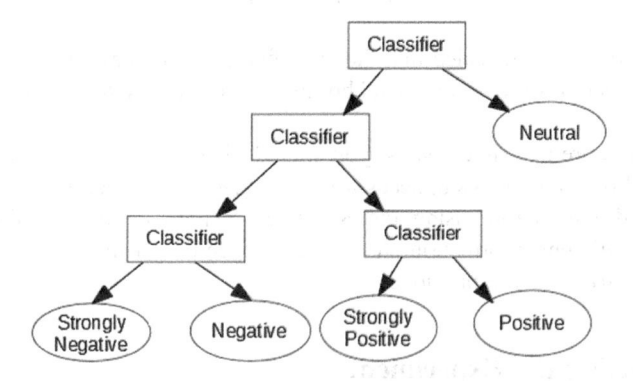

Fig. 2. The hierarchical structure of sentiment classes.

4   Approach to Sarcastic Sentences Determination

The issue of sarcasm recognition in a sentence requires the training of another classifier. To solve this issue, the method of k-nearest neighbors is used. To classify each of the test sample objects, you must perform the following steps sequentially:

- Calculate the distance to each of the training sample objects
- Choose k training sample objects, the distance to which is minimal
- Class of the object being classified is the class most often encountered among the k nearest neighbors
- The following set of parameters for the vector of singularities was compiled
- The presence of word forms, which are specific for sarcasm (such expressions include common words from the Internet slang)
- The presence of quotes in the text (if there are quotes, it is most likely that the text contains a certain degree of irony)
- High frequency of punctuation
- The presence in the text of words that are most often used in conjunction with sarcasm for a particular language, which are taken from training samples.

For this case, the weight is given as a function of the distance to the nearest neighbors. In (4) $d(x, x(i))$ is a function which determines the distance between elements in a vector space. Equation (5) finally determines whether or not the text being analyzed contains sarcasm, where Zi is a sum of weights for all of the available classes. If so, then the class of the slot must be changed to the opposite.

$$\omega(x(i)) = \omega(d(x, x_i)) \tag{4}$$

$$C = \arg \max Z_i \tag{5}$$

Empirically, it was revealed that the classifier gives the best efficiency in terms of accuracy if it analyzes the nearest-neighbor number K equal to the number of sentiment classes.

To obtain more plausible results, you should filter out the most frequent words in the model. This step removes unnecessary noise that could affect the final result of the study. In addition, before using the K nearest neighbors method, the volume of aggregated sentiment information should be considered. In this study the results for unigrams and trigrams are provided.

5 Conducting the Experiment

To decide whether a sentiment recognition effectiveness is better or worse using the method of sarcasm determination, a numerical metric is needed. For most modern algorithms based on machine learning, metrics of accuracy and completeness of search are used. The accuracy of the search determines the proportion of documents that really belong to a given sentiment class across all documents of this class. The completeness of the search determines the ratio of the found classifiers of documents belonging to this class to all documents in the sample. Since in real practice of machine learning the maximum accuracy and completeness of search are unattainable simultaneously, the analysis of results using the F-measure will be the most acceptable. The F-measure is calculated using (6).

$$F = 2 \cdot \frac{Precision \cdot Recall}{Precision + Recall} \tag{6}$$

A training set consists of 500 samples was compiled: 150 of them were marked as "containing sarcasm" and 350 were marked as "not containing sarcasm". This ratio between classes was not chosen randomly, since the likelihood of evaluating the sentiment class of the text as positive or negative is much higher than sarcastic. The experiment was conducted on a sample of 100 texts, which are supposed to be a description to the different photos with no more than 200 words in length and contains only the information about person's appearance.

Table 2. Part of compiled sentiment dictionary.

	Recall	Precision	F
Unigrams, without sarcasm	0.80	0.82	0.810
Trigrams, without sarcasm	0.85	0.84	0.844
Unigrams, with sarcasm	0.86	0.68	0.760
Trigrams, with sarcasm	0.87	0.77	0.820

As can be seen from the obtained results (Table 2), the method of sarcasm recognition in the text slightly lowers accuracy due to a relatively large number of false positives. It can be concluded that lexical features and punctuation signs are not enough to train the classifier at a sufficient level. Most often, sentences have a complex structure, which cannot be treated as a "bag of words" and requires the use of contextual syntactic rules.

6 Conclusion

Sentiment detection approaches can be classified in machine learning and hybrid lexicons. The machine learning approach is used to predict the polarity of moods based on both trained and test data sets. Although the lexicon-based approach does not require prior preparation for data collection. In this study an approach using machine learning with labeled valence dictionary was used.

As a result of the experiment, it can be concluded that the resolution of the task of recognizing sarcasm in a text containing a description of the appearance of a person cannot be effectively resolved only using the methods of machine learning with supervision. As a further study, it requires the development of contextual rules based on the syntactic structure of the text.

At this stage, the F-measure estimation showed that the method slightly reduces effectiveness due to a relatively large number of false positives. It may be worth considering a deeper approach to analyzing emotions, as suggested by the authors.

There are alternative approaches to solving the problem of determining the tonality class of the analyzed text. For example, the use of a neural network for text analysis can significantly expand the boundaries of tonality classes. This is achieved by applying a suitable output function, as a result of which the output is the probability with which a text fragment belongs to each class.

Acknowledgements. This work was partially supported by RFBR and administration of Volgograd region (grants 18-07-00220, 19-47-343001, 19-47-340003, 19-47-340009, 19-47-343002, 19-07-00020).

References

1. Dandrea, A., Ferri, F., Grifoni, P., et al.: Approaches, tools and applications for sentiment analysis implementation. Int. J. Comput. Appl. Technol. **125**, 8 (2015)
2. Medhat, W., Hassan, A., Korashy, H.: Sentiment analysis algorithms and applications: a survey. Ain Shams Eng. J. **5**, 1093–1113 (2014)
3. Feldman, R.: Techniques and applications for sentiment analysis. Commun. ACM **56**, 82–89 (2013)
4. Alec, G., Lei, H., Richa, B.: Twitter sentiment classification using distant supervision. Standford University, Stanford (2009)
5. Derczynski, L., Maynard, D., Rizzo, G., et al.: Analysis of named entity recognition and linking for tweets. Inf. Process. Manage. **51**, 18 (2014)

6. Koltsova, O., Alexeeva, S., Kolcov, S.: An opinion word lexicon and a training dataset for Russian sentiment analysis of social media. In: Computational Linguistics and Intellectual Technologies: Proceedings of the International Conference, p. 11 (2016)
7. Volkova, L., Kotov, A., Klyshinsky, E., et al.: A robot commenting texts in an emotional way. Commun. Comput. Inf. Sci. **754**, 256–266 (2017)
8. Olsher, D.: Full spectrum opinion mining: integrating domain, syntactic and lexical knowledge. In: Sentiment Elicitation from Natural Text for Information Retrieval and Extraction, pp. 693–700 (2012)
9. Ritter, A., Clark, S., Mausam, M.: Named entity recognition in tweets: an experimental study. In: Proceeding of the Conference on Empirical Methods in Natural Language Processing, pp. 1524–1534. Association for Computational Linguistics (2011)
10. Yadav, V., Bethard, S.: A survey on recent advances in named entity recognition from deep learning models. In: Proceedings of the 27th International Conference on Computational Linguistics, pp. 2145–2158 (2018)
11. Dmitriev, A., Zaboleeva-Zotova, A., Orlova, Y., et al.: Automatic identification of time and space categories in the natural language text. In: Proceedings of the IADIS International Conference, pp. 23–25 (2013)

Required Coke Quality Influence on a Coking Coal Mix Price Research. Linearization of a Coking Coal Procurement Optimization Model

A. Lipatnikov[1(✉)] and D. Shnayder[2]

[1] PJSC Magnitogorsk Iron and Steel Works,
93 Kirova Street, Magnitogorsk 455019, Russian Federation
lipatnikov.av@mmk.ru
[2] South Ural State University,
76 Lenina Avenue, Chelyabinsk 454080, Russian Federation

Abstract. The article performs the problem of simplification of a previously developed coking coal mix procurement optimization model. The model was developed at PJSC MMK (Magnitogorsk Iron and Steel Works), one of the leading Russian metallurgical plants, for improving the cooperation between technological and economical departments. It includes their background and can be used as an expert system for the decision making process. However, interconnections between coal quality indicators and coke quality indicators resulted in nonlinear mathematical program task. Therefore, the weak point of the model was the solution algorithm. Linearization method was used to fight the problem, which brought the task to linear program problem. Linearizing the problem allowed to employ the sensitivity analysis so that it became possible to answer several questions essential for the procurement department. What is the inclusion price for the supplier? How much does the coke quality improvement cost? Which suppliers are essential, which are surplus, which are lacking and so on. The simplex method used for solving the task improved the solution speed from twenty minutes to seconds and garantied the global solution for the problem.

Keywords: Linearization · Industrial optimization models · Coking coal procurement · Sensitivity analysis · Linear programming · Metallurgy · Decision support systems

1 Introduction

In a modern competitive world big industrial companies have to employ all the possibilities of automation in their areas of production to be effective. Especially it concerns those, which involve several processing divisions. Magnitogrsk Iron and Steel Works (MMK) is one of Russian leading steel making plants. It has started to implement strategic initiative "Industry 4.0" aimed at digitalization, which will integrate all the optimization projects, technological process models, databases, automatic control systems. In the context of the initiative a model of coking coal procurement optimization was developed [1].

© Springer Nature Switzerland AG 2020
A. A. Radionov and A. S. Karandaev (Eds.): RusAutoCon 2019, LNEE 641, pp. 1195–1205, 2020.
https://doi.org/10.1007/978-3-030-39225-3_125

The model is used for the decision-making process automation in the field of coking coal procurement. It involves background of several departments such as R&D center, commercial department, coking plant and economical department. The main purpose of the model is to find the cheapest coking coal blend, which will provide the required coke quality taking into account capabilities of the suppliers and the contractual limitations [2].

The model includes several nonlinear regression equations reflecting the interconnections between the coke quality indicators and the coal quality indicators. The feasible region was nonlinear and, therefore, gradient-based algorithms were used to find the solution. Those algorithms do not allow employing sensitivity analysis for the solution, and it takes significant time to find the solution. Therefore, it is essential to reduce the problem statement to make the simplex method applicable for it.

2 Linearization of a Coking Coal Procurement Optimization Model

Coke is the most essential and the most expensive fuel for the blast furnace process. It takes a significant part in iron cost structure [3].

Coke consumption for a ton of iron produced is the main efficiency indicator for the blast furnace shop. In its turn coke quality, which has a great influence on that indicator, vastly depends on coking coal blend quality, since coke is made of coal blend [4].

However, it is possible to prepare coal blend of the same quality in several different ways. Thus, the variables for the model are the suppliers' shares [5]. The coal blend price was chosen as a target function. There are also two types of constraints in the model: technological and economical.

Technological constraints are provided by the R&D center and Coke Plant. They include the required coke quality, capacities of the suppliers and the coal blend brand structure.

Economic constraints are provided by the commercial department and economical department. They include contract limitations and the amount of iron to be produced.

All in all, the model can be described as follows:

$$F = \sum_{i=1}^{n} c_i r_i \rightarrow \min \tag{1}$$

$$\begin{cases} M_{10} < M_{10}^*, \\ M_{25} < M_{25}^*, \\ Ad < Ad^*, \\ Sd < Sd^*, \\ N_i^{\min} < N_i < N_i^{\max}, \\ \sum_{i=1}^{n} f_i \leq f^*, \\ r_i > 0. \end{cases} \tag{2}$$

where r_i – the share of the i-th vendor in the overall supply, %; c_i – the one ton of coal concentrate price of the i-th vendor; N_i – tons quantity for the i-th vendor; N_i^{min} – minimum tons quantity constraint for the i-th vendor (contract limitation); N_i^{max} – maximum tons quantity constraint for the i-th vendor (contract limitation or production capacity); M_{10} – abrasion coke strength, % (star for the required quality); M_{25} – impact coke strength, % (star for the required quality); Ad – ash content, % (star for the required quality); Sd – sulfur content, % (star for the required quality); f_i – shares of fat and gas-fat coal types, % (star for the required quantity); n – number of vendors.

Coke quality indicators are calculated on the basis of coal quality indicators, which include the moisture content (Wr), the ash content (A), volatile substances output (Vd), sulfur content (S), the thickness of the plastic layer (Y), vitrinite reflectance (R_o), the sum of inert components (Sok) and the technological value coefficient (TVC).

In general, the equations between coke quality indicators and the coal concentrate quality indicators are as follows: coke quality indicator $= f$ (coal concentrate quality indicators). First equations were nonlinear in several components like Vd, Y, R and TVC. R^2 level for them was about 80%. For the simplification purposes nonlinear component was left only for the Vd regressor, what was appreciated by the technologists, since other parameters didn't have extremum in the feasible area. There was a small R^2 reduction to the level of 60–70%. The equations are as presented below:

$$M_{10} = b_0 + b_1 \cdot Wr + b_2 \cdot A - b_3 \cdot Vd + b_4 \cdot S - b_5 \cdot Y - b_6 \cdot R_0 + b_7 \cdot Sok + b_8 \cdot Vd^2 - b_9 \cdot TVC,$$

$$(3)$$

$$M_{25} = b_0 - b_1 \cdot A + b_3 \cdot Vd + b_4 \cdot Y + b_5 \cdot R_0 - b_6 \cdot Sok - b_7 \cdot Vd^2 + b_8 \cdot TVC, \quad (4)$$

$$Ad = b_0 + b_1 \cdot Wr + b_2 \cdot A - b_3 \cdot Vd - b_4 \cdot R_0 + b_5 \cdot Sok, \quad (5)$$

$$Sd = b_0 + b_1 \cdot Wr + b_2 \cdot Ad - b_3 \cdot Vd + b_4 \cdot S \quad (6)$$

The coefficients' values (the absolute values here) are not shown since they are corrected from month to month, as soon as new data appears. However, the signs reflect influence direction for the target value.

Coal quality indicators are calculated according to the formula below (7), where correcting coefficient represents the average difference between the real laboratory analysis data and the weighted mean values of the coal indicators according to the factual consumption structure. Quality indicators for each vendor are provided by the laboratory analysis.

$$K = \sum_{i=1}^{n} (k_i \cdot r_i) + k_{corr}, \quad (7)$$

where r_i – the share of the i-th vendor in the overall supply; k_i – quality indicator of the i-th vendor (Wr, A, Vd, S, Y, Ro, Sok, TVC); k_{corr} – correcting coefficient for a certain indicator; K – weighted mean value of a coal quality indicator; n – number of vendors.

Since the equations are nonlinear the gradient algorithms for the solution were used (the first equation types were more complex then presented above). The R language with «nloptr» library for nonlinear and not convex optimization were employed [6]. The Augmented Lagrangian algorithm was used to turn constraints into the target function penalties. It took about twenty minutes to get the solution.

However, the commercial department wished to know the following:

- How do the prices influence the solution?
- What is the inclusion price for the supplier, which is not in the list?
- What is the exclusion price for the supplier, which is in the list?
- Which supplier should we change the contract limitations for?

This information was essential since this model can not only be used as an inner decision support system, but can also help for holding negotiations with the suppliers, whereas market conditions, such as prices and limitations, permanently changing.

There are two types of linearization for this model.

The first one supposes that we just investigate the equations themselves to study which coal quality indicators has a good influence on the coke quality and then just set coal quality as constraints instead of coke quality. Such a problem will be linear since the formula (7) is linear.

If we look through the Eqs. (3–6) it is obvious that we need to increase such indicators as the thickness of the plastic layer (Y), vitrinite reflectance (R_o) and the technological value coefficient (TVC), and decrease such indicators as the moisture content (Wr), the ash content (A), sulfur content (S), the sum of inert components (Sok) to get a better coke quality, which is not on the contrary with the physics of the process. As for the volatile substances output (Vd), we can admit that we have the optimal value for the M_{10} and M_{25} indicators since their part are quadratic. The optimal value for the M_{10} is $-b_3/(2b_8)$ and $b_3/(-2b_7)$ for the M_{25}. It is essential, to know that, since M_{10} and M_{25} are the most meaning blast furnace process figures [7].

From this point of view, it is satisfying when the Vd is between the required Vd and the optimal form both sides (see Fig. 1). The point "A" reflects the Vd^* which is the required Vd parameter. The point "B" reflects the optimal Vd value according to the formulas mentioned above. The point C reflects the Vd'^* which provides the same coke

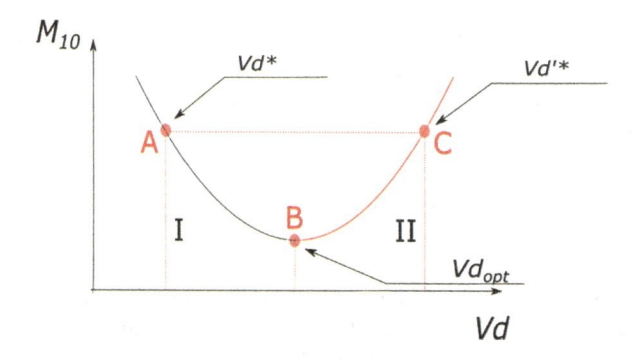

Fig. 1. The optimal part for the Vd.

quality as the Vd in the "A" point. Any point between A and C will provide a better coke quality then in the required variant. Therefore, it involves two constraints for the Vd instead of one to make the problem linear.

At first glimpse, it seems that the area I and the area II are equal for the problem solution. It is true for the quality parameters and the coke consumption in the blast furnace, but counting economic effect showed that Vd has a great influence on the result in the coke making division.

Vd reflects the loss of materials during the coke-making processing therefore has a great influence on the economic effect. That's why it is practical not to take the right part of the function (BC). Thus, the Vd constraints can be described with two inequalities. Since M_{10} is more important than M_{25}, the optimal value is the optimum for M_{10}.

All in all, the constraint part of the problem (2) can be described as follows:

$$
\begin{cases}
Wr \leq Wr^*, \\
S \leq S^*, \\
A \leq A^*, \\
Sok \leq Sok^*, \\
Y \geq Y^*, \\
R_0 \geq R_0^*, \\
TVC \geq TVC^*, \\
Vd \leq Vd_{opt}, \\
Vd \geq Vd_{opt} - \left| Vd_{opt} - Vd^* \right|, \\
N_i^{min} \leq N_i \leq N_i^{max}, \\
\sum_{i=1}^{n} f_i \leq f^*, \\
r_i \geq 0
\end{cases}
\tag{8}
$$

Such problem description allowed employing the prices factor analysis. It resulted in a table with 3 prices: current, minimum and maximum (see Table 1). It is not easy to interpret these prices, when the current price doesn't reach contractual limitations, either high or low.

The example of this analysis usage is as follows:

Table 1. The examples of factor analysis usage.

Supplier	Calculated quantity, ton	Minimum quantity, ton	Maximum quantity, ton	Supplier price	Minimum price	Maximum price
Example 1						
Supplier A	70000	70000	100000	4200	4030	–
Example 2						
Supplier B	50000	30000	50000	4700	–	4843
Example 3						
Supplier C	14843	0	25000	4550	4520	4565

When the minimum contractual limitation is equal resulting quantity (example 1) it means that the company is forced to buy more coal from this supplier than it needs at such price. The minimum value in the factor analysis will let the decision maker know what the maximum price should be for this amount of coal becomes efficient in this conditions. If the price is lower than the minimum price, than the amount of coal, required from this supplier, will arise. Minus in the maximum value means, that increasing the price will not influence the solution anyhow since the contractual limitations makes the company buy this amount at an overestimated price. Thus, it becomes possible to distinguish which contracts are efficient and which are not, and gives the prices for the procurement department to orientate when negotiating.

The second example shows the opposite situation, when the maximum contractual limitation is equal calculated quantity. This means that the quality of the coal provided by the supplier is too high for the price or the price is too low for such quality. The company needs to take as much product as possible and the quantity is only limited by the supplier capacity or strategy. In such situation the maximum value in factor analysis will let the decision maker know the real price which suits such product quality.

The solution will contain less volume of this supplier, if he rises the price more than the maximum price in the factor analysis. Minus in the minimum value means that decreasing the price will not influence the solution, since all the available volume was already taken.

The last example reflects the situation when the contract limitations don't influence the solution. This situation appears when the supplier can be easily substituted by other brands, and its quantity depends only on technological constraints. In this case both values mean the interval when the price change for the supplier will only influence the total price of a coal mix but not its structure.

It is obvious that just linearization of the model provides additional commercial information besides global solution requiring just half of a minute for calculating.

However, a great disadvantage of this system is that the solution can be even cheaper since different types of coal quality figures have different influence on the coke quality figures. It is possible to make some of indicators worse when some others get better if the resulting coke quality will be at a required level.

The second reason for changing the system was a wish of technologists to get a similar factor analysis for the coke quality indicators, to answer the question how much will it cost to increase the required coke quality? The global purpose of the model is to reduce the costs, but not making the best coke ever.

All of the above leads to a second type of linearization, when we use the same coke quality indicators as constraints, but change the nonlinear parts of the equations to linear in some current regions of the solution.

Integrally we can't change the equations, but we have some constraints for Vd, for example, that narrows feasible region therefore the values in that region are close for the linear and nonlinear equations. The example of such linearization is presented on the Fig. 2.

As we have seen earlier the target area for the *Vd* lies between the "A" and "B" points. It is possible to approximate the parabolic part of the curve with a line with some accuracy loss of course. Therefore, the task was to estimate the difference between these two functions. This difference is represented by the grey shading on the Fig. 2.

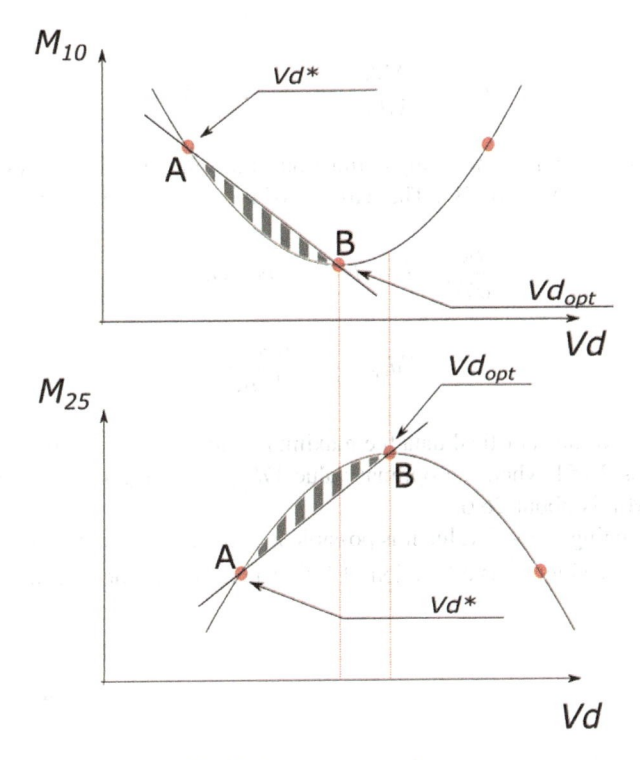

Fig. 2. Linearization scheme.

As far as the coke quality parameters are measured with the one decimal places accuracy, and the month mean is reported with two decimal places accuracy, the difference of three decimal places accuracy will suit our approach.

The reconstructed equations' parts for the *Vd* are looking as follows:

$$\frac{M_{10} - M_{10}^*}{M_{10}^{opt} - M_{10}^*} = \frac{Vd - Vd^*}{Vd^{opt} - Vd^*};$$

(9)

$$\frac{M_{25} - M_{25}^*}{M_{25}^{opt} - M_{25}^*} = \frac{Vd - Vd^*}{Vd^{opt} - Vd^*}$$

(10)

To estimate accuracy a distance function was used (we will speak about M_{10}, because the situation for the M_{25} is similar):

$$d(Vd) = b_3' \cdot Vd + b_0' - (b_0 - b_3 Vd + b_8 Vd^2); \tag{11}$$

$$b_3' = \frac{M_{10}^{opt} - M_{10}^*}{Vd^{opt} - Vd^*}; \tag{12}$$

$$b_0' = -\frac{M_{10}^{opt} - M_{10}^*}{Vd^{opt} - Vd^*} \cdot Vd^* + M_{10}^*; \tag{13}$$

It is possible to use this simple function since the line is always higher then parabola between "A" and "B". The extreme of the distance function appears when:

$$\frac{\partial d}{\partial Vd} = b_3' + b_3 - 2 \cdot b_8 \cdot Vd = 0; \tag{14}$$

$$Vd_{\max(d)} = \frac{b_3' + b_3}{2 \cdot b_8} \tag{15}$$

According to the statistical data the maximum value for the Vd^* was 24.78 and the minimum was 26.51 when the optimal value Vd_{opt} according to the equations' coefficients for M_{10} is about 26.6.

When changing the Vd^* value it is possible to estimate the maximum possible value for the difference since b_3' is a function of Vd^*. The results of the calculation are shown on the Fig. 3.

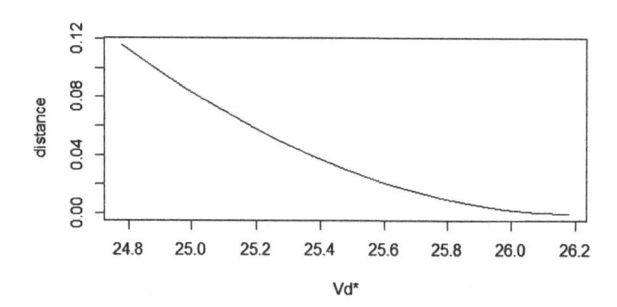

Fig. 3. The maximum difference value between the functions.

The maximum difference is not bigger than two decimal places in 93% cases. Technologists are trying not to decrease much this parameter; otherwise, they will have a negative influence on the equipment.

Therefore, it is possible to employ such linearization for the task. With this assumption, we can also employ the factor analysis for the coke quality indicators, thus telling the technologists, how expensive their desirers are.

So, it is possible to reconstruct the Eqs. 3–4 as follows:

$$M_{10} = b_0' + b_1 \cdot Wr + b_2 \cdot A - b_3' \cdot Vd + b_4 \cdot S - b_5 \cdot Y - b_6 \cdot R_0 + b_7 \cdot Sok - b_9 \\ \cdot TVC; \tag{16}$$

$$M_{25} = b_0' - b_1 \cdot A + b_3' \cdot Vd + b_4 \cdot Y + b_5 \cdot R_0 - b_6 \cdot Sok + b_8 \cdot TVC; \tag{17}$$

After getting the new equations, it is possible to set the linearized task for the problem. The definition for the constraint problem in context of the second type linearization looks like:

$$\begin{cases} M_{10} \leq M_{10}^*, \\ M_{25} \geq M_{25}^*, \\ Ad \leq Ad^*, \\ Sd \leq Sd^*, \\ Vd \leq Vd_{opt}, \\ Vd \geq Vd_{opt} - \left| Vd_{opt} - Vd^* \right|, \\ N_i^{\min} \leq N_i \leq N_i^{\max}, \\ \sum_{i1=}^{n} f_i \leq f^*, \\ r_i \geq 0 \end{cases} \tag{18}$$

The development of the system was based on the R language with the "lpSolve" package, providing the linear problems solver [8].

3 Results

The conducted research showed the two ways of a coal procurement model simplification. The first one is using the coal quality constraints instead of coke quality constraints. The second is changing the nonlinear parts of coal-coke equations to linear, adding some extra constraints. Of course, the second type of model linearization is more preferable, as it gives better solutions, though it has some shortcomings.

First, this involves some extra constraints, which narrow the feasible region. Also, in this very problem, it appeared to be organic as these constraints are required for the economic effect increase.

The second is employing the linear solver does not allow finding the solution, when the feasible region is empty, just telling there is no solution. The usage of gradient methods with turning constraints into penalty functions allowed the model to do that.

The last thing is that the model has a small loss of accuracy, which is not meaning according to the precision the coke quality indicators are measured with, that was proofed in the article.

However, model linearization has many advantages comparing to the gradient methods.

First of all, the linear problem definition always returns in the global optimum, so that the solution is the best for such constraints. Simplex-method is easier interpreted than the gradient-based methods. When the feasible region is empty, it is easy to understand which constraints are too strict.

The second is that linear solution allowed employing both types of factor analysis: prices and technology. It can provide extra information for commercial department and R&D department, which can broad the problem understanding much further than the inner business-process.

The last advantage is speed. Linearized tasked is solved much faster, thus it saves the user's time. There is no need to count gradients or Jacobians many times so that the decision support system can be provided on a local machine.

The model now is accepted by the R&D department, economical department, commercial department and coke plant on PJSC MMK. It is implemented as an automated advisor system.

4   Conclusion and Outlook

The next step for the model development is its realization as an automated managing system. It involves integration into the corporate information system and further integration into the blast furnace managing system and other "Industry 4.0" projects [9–13]. The system should automate the business process on all its stages: coal blend quality definition, coke quality definition, price and economical effect calculation, optimization of the blend structure.

Employing this system will provide all of these functions for all the departments involved, so that each department will be able to control the whole process from their point of view.

The further developing of the mathematical part of the model lies in the area of the store-managing model. It would be the add-on for the model, which provides the information for the commercial department, which structure should the reserves have to provide the best quality at a lowest price in a long-term period.

References

1. Lipatnikov, A.V., Shmelyova, A.E., Stepanov, E.N., et al.: Mathematical modeling and optimization of raw coal consumption in PJSC «MMK». Bull. Nosov Magnitogorsk State Tech. Univ. 16(4), 30–38 (2018). https://doi.org/10.18503/1995-2732-2018-16-3-30-38
2. Lipatnikov, A.V., Stepanova, A.E., Stepanov, E.N., et al.: Optimization model of raw coal supply and consumption in PJSC "MMK". Scientific and technological progress in ferrous metallurgy. In: Materials of the III International Scientific Conference, pp. 240–247 (2017)
3. Diez, I.M., Alvarez, R., Barriocanal, C.: Coal for metallurgical coke production: predictions of coke quality and future requirements for coke making. Int. J. Coal Geol. 50, 389–412 (2002)
4. Stankevich, A.S., Zolotukhin, Y.A.: Determining the technological value of coal on the basis of coke-quality predictions. Coke Chem. 58(7), 233–244 (2015)

5. Stepanov, E.N., Melnikov, I.I., Gridasov, V.P.: Research of physical, chemical and strength properties changes of skip and tuyere coke. Steel **4**, 2–4 (2009)
6. NLopt package (2018). ab-initio.mit.edu/wiki/index.php/NLopt_Algorithms. Accessed 25 Oct 2017
7. Urin, N.I., Morozov, O.S., Lihacheva, O.L., et al.: Influence of coke quality on blast furnace operation. Steel **11**, 9–11 (2011)
8. Crawley, M.J.: The R Book. Wiley, New York (2013)
9. Lipatnikov, A.V., Astratova, E.V., Shnayder, D.A.: Optimization model of iron ore supply and consumption in PJSC "MMK" Scientific and technological progress in ferrous metallurgy. In: Materials of the III International Scientific Conference, pp. 15–22 (2017)
10. Shnayder, D.A., Kazarinov, L.S., Barbasova, T.A.: Data mining and model-predictive approach for blast furnace thermal control. In: Intelligent Systems Conference, IntelliSys, pp. 653–660 (2017)
11. Trofimova, V.Sh., Ivanova, T.A.: The use of decision trees for optimization of iron ore production process. Actual problems of modern science, technology and education. In: Abstracts of the 77th International Scientific and Technical Conference, p. 156 (2019)
12. Kazarinov, L.S., Barbasova, T.A.: Elliptic component analysis. In: Proceedings of 2nd International Conference on Industrial Engineering, Applications and Manufacturing (2016). https://doi.org/10.1109/icieam.2016.7910936
13. Kazarinov, L.S., Barbasova, T.A.: Identification method of blast-furnace process parameters. In: V International Conference for Young Scientists. High Technology: Research and Applications (2016). https://doi.org/10.4028/www.scientific.net/KEM.685.137. Key engineering materials 685:137–141